Exploring Earth and Life through Time

Exploring Earth and Life through Time

Steven M. Stanley
The Johns Hopkins University

W. H. Freeman and Company
New York

Cover image: This fossilized tree trunk has weathered out of the Triassic Chinle Formation in the Petrified Forest National Park in Arizona. *(Photograph by Harry Foster. Reproduced by permission of the Canadian Museum of Nature, Ottawa, Canada.)*

Library of Congress Cataloging-in-Publication Data

Stanley, Steven M.
Exploring earth and life through time / Steven M. Stanley.
p. cm.
Includes bibliographical references and index.
ISBN 0-7167-2339-5
1. Historical geology. I. Title.
QE28.3.S74 1992 92-14871
551.7—dc20 CIP

Printed in the United States of America

1 2 3 4 5 6 7 8 9 0 RRD 9 9 8 7 6 5 4 3 2

CONTENTS

PREFACE

We humans are inherently curious about our roots — not only about our ancestry during the past few decades or centuries but also about our animal origins and the history of the world our four-legged ancestors inhabited. Now that we are damaging the global environment in ways that may harm our own species along with many others, our curiosity heightens. This book, unlike other textbooks that introduce historical geology, shows the student how Earth's ecosystem has changed through time and how events of the past provide a perspective for dealing with changes that are taking place now or may take place in the future.

In fact, the geologic record uniquely documents not only the history of Earth's habitats far back in time but also the events of the past few million years that have shaped the modern ecosystem. What lessons can we learn from "natural" changes of the past? One issue is how we exercise our unrivaled powers to alter environments on Earth. Another is how we cope with the changes that we do produce. Thanks to exciting advances in the study of ancient oceans, climates, and land areas, the geologic record now reveals that many of the kinds of change that may loom ahead have occurred before. In addition, the fossil record reveals many ways in which life has responded. Ultimately, extinction has been the helpless response of nearly all species to certain environmental changes that have been beyond their control: Only a tiny fraction of the species that were alive halfway through the Age of Mammals survive today. Farther back in time, mass extinctions have occasionally decimated life on Earth. Thus, the geologic record reveals that environments and life are transitory. As the French adage has it, the more things change, the more they stay the same. Nonetheless, we cannot justify the crisis that we may soon inflict on rainforests and grasslands and coral reefs by arguing that it will simply mimic "natural" mass extinctions of the past. Events of the distant past do not provide an ethic for our behavior, but they do offer us the opportunity to predict some of the consequences of our potential actions so that we can make choices and adjustments.

Study of Earth's history has other practical benefits as well. Geologists have been able to locate important natural resources such as coal and petroleum by understanding the record of rocks and fossils. They have also come to understand the origins of valuable but nonrenewable sources of fresh water far below Earth's surface. By learning how events of the past have created and concentrated resources, a student of Earth history comes to appreciate the limited supply and inevitable depletion of materials that he or she might otherwise take for granted.

ORGANIZATION AND CONTENT

My aim in writing this concise version of *Earth and Life through Time* has been to make the history of our planet and its inhabitants accessible to a wider range of students. To accomplish this goal, I have eliminated specialized topics, especially ones focusing on regional geologic events outside North America. I have also abbreviated the discussion of sedimentary environments, confining it to a single chapter, and the review of Precambrian geology, covering it in two chapters rather than three.

Exploring Earth and Life through Time is not simply a reduced version of its parent, however. New boxes in Chapters 2 to 16 show students how Earth's history sheds light on contemporary issues. Nine of the boxes discuss modern environmental issues that relate to subjects in the chapters where the boxes appear; the issues range from loss of wetlands to future extinctions to global warming. Three of the remaining boxes shed light on the process of biological evolution, and the other two focus on natural resources.

A Major Events diagram on the third page of each of the final eight chapters helps students comprehend the global events that took place during each major interval of Proterozoic and Phanerozoic time. Each diagram provides a graphic summary of events that occurred during an interval of time, and it can serve as a handy guide to the rest of the chapter.

Scientific terms are printed in boldface where they first appear and are also defined in the Glossary or, if they are names of major groups of organisms, in Appendix III.

Numerous innovative features of *Earth and Life through Time* that have been well received are retained in this book. Chapter 2 reviews the structure of modern ecosystems. Without such an introduction, no student who is not already well versed in ecological principles could understand environments and life of the past. Chapter 3 explains how we recognize ancient environments. Lacking this kind of review, no student could appreciate how we reconstruct the ancient world. Similarly, Chapters 6 and 7 explain plate tectonics and mountain building in ways that teach students how lateral and vertical movements of the lithosphere have caused Earth's geography to evolve over millions of years. As in the parent volume, these and other early chapters provide the raw material for understanding the history of Earth and its biota. Appendix I and Appendix II provide additional background on minerals, rocks, and fossils for the student who has not previously studied physical geology.

Chapters 8 through 16, like comparable chapters of *Earth and Life through Time*, integrate the history of life with the physical history of Earth, including plate-tectonic events. During the past decade, paleogeography, paleoceanography, and paleoclimatology have emerged as new disciplines that are revolutionizing the Earth sciences. Without labeling these disciplines, I have attempted to accord them their newly earned status.

SUPPLEMENTS

The following supplemental materials for *Exploring Earth and Life through Time* are available to adopters:

The **Instructor's Manual,** prepared by Robert D. Merrill of California State University, Fresno, contains chapter outlines, summaries, and objectives; teaching tips; answers to end-of-chapter text exercises; additional questions and answers; and additional resource references, including audio visual aids, for each chapter.

A set of **120 slides** contains a selection of color line diagrams from the text.

For more information and to request copies of these supplements, please contact

Professor Services Department
W. H. Freeman and Company
4419 West 1980 South
Salt Lake City, UT 84104

Telephone: 801-973-4660
FAX: 801-977-9712

ACKNOWLEDGMENTS

Authors do not produce books without a great deal of help, and I have been most fortunate to have had the assistance of an exceptionally able publishing team. As always, my friend Jerry Lyons served not only as an acquisitions editor but also as a fount of wisdom in the conception of this book. Kay Ueno proved to be highly skilled and very understanding as a development editor, Barbara Salazar was as good a copy editor as I have ever encountered, and Georgia Lee Hadler, whose appointment as project editor was wonderful news, lived up to her previous stellar performances. Susan Stetzer, as production coordinator, dealt ably with the difficult problems of length and scheduling. Finally, the outstanding work of people who contributed to the design and art program will be apparent to anyone who glances through the finished product. These include Alice Fernandes-Brown, designer; Bill Page, illustration coordinator; John Hatzakis, production artist; and Travis Amos, photo researcher.

A large group of colleagues offered advice and reviewed text as this volume was assembled. I thank them for their hard work and valuable suggestions, but accept full responsibility for any errors that may remain. These colleagues are:

Barbara Brande, University of Montevallo
William Cornell, University of Texas, El Paso
Louis Dellwig, University of Kansas
Jay M. Gregg, University of Missouri, Rolla
Bryce Hand, Syracuse University
James Jones, University of Texas, San Antonio
Karl Koenig, Texas A&M University
David Liddell, Utah State University
Robert Merrill, California State University, Fresno
Cathryn R. Newton, Syracuse University
Anne Noland, University of Louisville
Lisa Pratt, University of Indiana, Bloomington
Randall Spencer, Old Dominion University

Steven M. Stanley

TO THE STUDENT

I use my own understanding of the past to gain insight into the present and the future. Nearly all students will live much longer than I, so they will have greater opportunities to employ lessons from the past. I want to address a few comments to you, the student, to help you gain useful insights as you use this book to explore Earth and life through time.

First, I urge you not to treat this book as a compendium of facts. It is the major events and concepts and themes that matter most, and I exhort you to focus on them. Here are some of the kinds of important questions that you should ask yourself in organizing the material that the book presents: When did major ice ages begin in the course of Earth's history, and how were continents arranged when they began? When and where have mountains risen up in North America and what caused them to form? What evolutionary breakthroughs have led to dramatic expansions of various groups of plants and animals, and how have these expansions changed the biosphere? What have been the patterns and probable causes of the great extinctions of the past? Forty million years ago, southern England was tropical and Alaska was subtropical; how did the world's climate come to be as it is today?

The Major Events figure near the beginning of each of the final eight chapters will help you to identify and review major events and themes, and so will the exercises at the ends of chapters. If you can complete these exercises, you will have mastered the broad outlines of the subject.

As you first venture into the subject of Earth's history, the vastness of geologic time may seem daunting, but this reaction is misleading. It is actually quite easy to come to grips with the geologic time scale. There is no need to connect intervals that span millions of years to the much shorter intervals by which we measure our everyday lives: seconds, minutes, days, weeks, months, and years. In fact, we seldom try to relate seconds or minutes to years, but instead maintain separate time scales in our heads. You must simply make this kind of leap to establish the scale of geologic time in your thinking. Except for the most recent interval of geologic time, the basic unit is a million years, and it is often convenient to round geologic dates or intervals off to units of tens or hundreds of millions of years. At the start of your study of Earth's history, you should learn the elements of the geologic time scale—the eons, the Phanerozoic periods, and the Cenozoic epochs. By this strategy you will establish a framework for positioning major events so that a general picture of Earth's history will emerge.

Many forms of life have come and gone during our planet's history. The basic types are easier to keep track of if you have a sense of how they lived. I have therefore tried to convey useful information about the mode of life of each major group of animals or plants, either in the text or in a figure caption, when I first introduce it. Appendix III also allows you to see where a particular group fits into the general classification of life. Boldfaced names of groups of animals or plants that do not appear in this appendix are defined in the Glossary.

I wish you a pleasant voyage through the history of our planet and its inhabitants. This is as close as you will get to entering a time machine!

Steven M. Stanley

Exploring Earth and Life through Time

CHAPTER 1

Introduction to Earth, Life, and Geologic Time

Few people recognize, as they travel down a highway or hike along a mountain trail, that the rocks they see around them have rich and varied histories. Unless they are geologists, they have probably not been trained to identify a particular cliff as rock formed on a tidal flat that once fringed a primordial sea, to read in a hillside's ancient rocks the history of a primitive forest buried by a fiery volcanic eruption, or to decipher clues in lowland rocks telling of a lofty mountain chain that once stood where the land is now flat. Geologists can do these things because they have at their service a wide variety of information gathered during the two centuries that the modern science of geology has existed. The goal of this book is to introduce enough of these geologic facts and principles to give students an understanding of the general history of our planet and its life. The chapters that follow describe how the physical world assumed its present form and where the inhabitants of the modern world came from. They also reveal the procedures through which geologists have assembled this information. Students of Earth history inevitably discover that the perspective this knowledge provides changes their perception of themselves and of the land and life around them.

Knowledge of Earth's history can also be of great practical value. Geologists have learned to locate petroleum reservoirs, for example, by ascertaining where the porous rocks of these reservoirs tend to form in relation to other bodies of rock. Geologists have also helped discover deposits of coal and ore and other natural resources buried within Earth.

An understanding of Earth history helps us to address problems caused by changes that are now taking place in the world, or that will be occurring soon. The rock record also sheds light on ways in which a broad array of factors cause environmental change and on the rates at which various kinds of change occur. The shifting of coastlines as sea level rises or falls represents one example. The geologic record of the past few thousand years documents a general rise in sea level as huge glaciers have melted and released water into the ocean. The geologic record near the edge of the

A geyser in the Black Rock Desert of Nevada: The green cyanobacteria that thrive in the warm water that issues from the geyser resemble ones that lived more than 2 billion years ago and are preserved in Precambrian rocks. *(Stephen Trimble.)*

sea reveals how coastal marshes have shifted as sea level has changed. These marshes are very important to humankind; they cleanse marginal marine waters and sustain valuable forms of animal life. Study of the geologic history of coastal marshes will help us to predict their fate as sea level continues to change in the decades and centuries to come.

The geologic record of the history of life also provides a unique perspective on the numerous extinctions of animals and plants that are now resulting from human activities. Humans are causing extinction by destroying forests and other habitats, but our collective behavior also affects life profoundly in less direct ways. Very soon human activities will cause average temperatures at Earth's surface to rise in many areas of the world. The geologic record of ancient life reveals how climatic change has affected life in the past—how some species have survived by migrating to favorable environments, for example, and how others that failed to migrate successfully have died out. To the surprise of many biologists, geologic evidence has revealed that many of the natural assemblages of species that populate the world today are not ancient associations of interdependent species. Instead, they are associations that have developed very recently (on a geologic scale of time) as climates have changed in ways that have caused species to shift their distributions. Today spruce trees grow naturally in the United States only in the cold climates of eastern New England. Pollen preserved in clay shows that about 20,000 years ago spruce trees grew as far south as southern Georgia. At that time, areas close to Washington, D.C., were covered with frozen tundra, closely resembling the habitat that now occupies broad areas of northern Canada. Early humans—members of our own species—lived through similar changes that were occurring simultaneously in Europe.

As we come to understand the speed and profundity of natural environmental change and the transience of assemblages of species, we begin to appreciate the fragility of the world we live in. More generally, having studied the past, we can make more intelligent choices as we contemplate the future of our changing planet.

Before we launch into our detailed examination of the history of Earth and its life, however, an introduction to some of the basic facts and unifying concepts of geology is in order. The first seven chapters lay this groundwork.

THE PRINCIPLE OF UNIFORMITARIANISM

Fundamental to the modern science of geology is the principle of **uniformitarianism**—the belief that there are inviolable laws of nature that have not changed in the course of time. Of course, uniformitarianism applies not only to geology but to all scientific disciplines—physicists, for example, invoke the principle of uniformitarianism when they assume that the results of an experiment will be applicable to events that take place a day, a year, or a century after the experiment is conducted—but geologists hold this principle in particularly great esteem because, as we shall see, it was the widespread adoption of uniformitarianism during the first half of the nineteenth century that signaled the beginning of the modern science of geology.

Actualism: The Present as the Key to the Past

The principle of uniformitarianism governs geologists' interpretations of even the most ancient rocks on Earth. It is in the present, however, that many geologic processes are discovered and analyzed, and the application of these analyses to ancient rocks in accordance with the principle of uniformitarianism is sometimes called **actualism.** When we see ripples on the surface of an ancient rock composed of hardened sand (sandstone), for example, we assume that these ripples formed in the same way that similar ripples develop today—under the influence of certain kinds of water movement or wind. Similarly, when we encounter ancient rocks that closely resemble those forming today from volcanic eruptions of molten rock in Hawaii, we assume that the ancient rocks are also of volcanic origin.

Actualism is commonly expressed by the phrase "The present is the key to the past." This idea is only partly true, however. Although it is universally agreed that natural laws have not varied in the course of geologic time, not all past events have been duplicated within the time span of human history. Many researchers believe, for example, that the impact of very large meteorites may explain certain past events, such as the extinction of the dinosaurs 65 million years ago. They can calculate that the impact of a huge meteorite — one something like 10 kilometers (6 miles) in diameter — if it were to land in the ocean, would produce a huge wave that would crash over coastlines thousands of kilometers from the impact site. Nonetheless, because we have never observed the arrival of such a large meteorite, we do not know exactly what else would happen. It has been suggested that the fine dust injected into the upper atmosphere might block the sun's rays from Earth's surface for many days. As we will learn in Chapter 14, there is some evidence to support this contention, but because we cannot observe the consequences of such an event today, the idea is difficult to verify. In other words, in this case actualism does not apply.

Similarly, geologists have found that certain types of rocks cannot be observed in the process of forming today. In such cases, geologists usually assume that

1 The rocks in question formed under conditions that no longer exist;

2 The conditions responsible for the formation of these rocks still exist, but at such great depths beneath Earth's surface that we cannot observe them; or

3 The conditions exist today but produce the rocks only over a long interval of geologic time.

Many iron ore deposits more than 2 billion years old, for example, are of types that cannot be found in the process of forming today. It is believed that when the iron ore formed, chemical conditions on Earth differed from those of the present world and, furthermore, that the rocks underwent slow alteration after they were formed. The existence of these iron ore deposits does not necessarily negate the principle of uniformitarianism inasmuch as there is no evidence that natural laws were broken. It does, however, present geologists with a problem they cannot solve by applying the principle of actualism, because a human lifetime is only a small fraction of the time needed to study the rocks' development.

In an attempt to address some of these problems, geologists have learned to form certain kinds of rocks in the laboratory by duplicating the conditions that prevail at great depths within Earth. They expose simple chemical components to temperatures and pressures many times greater than those at Earth's surface. Such experiments indicate the range of conditions under which a particular type of rock could have formed in nature. In conducting these experiments, geologists are, in a sense, expanding the domain of actualism by using as a model not only what is happening in nature today, but also what happens under artificial conditions and may have happened under natural conditions long ago.

The Uniformitarian View of Rocks

Until the early nineteenth century, many natural scientists subscribed to a concept known as **catastrophism.** According to this idea, floods caused by supernatural forces formed most of the rocks visible at Earth's surface. Late in the eighteenth century, Abraham Gottlob Werner, an influential German professor of mineralogy, claimed that most rocks formed as a result of the precipitation of minerals from a vast sea that periodically flooded and retreated from the surface of Earth. These ideas were largely speculative.

Not long after Werner published his ideas, James Hutton, a Scottish farmer, established the foundations of uniformitarianism by writing about the origins of rocks in Scotland. Hutton concluded that rocks formed as a result of a variety of processes currently operating at or near the surface of Earth — processes such as volcanic activity and the accumulation of grains of sand and clay under the influence of gravity. It was only after extensive debate that Hutton's interpretation of the origins of rocks was generally accepted by the scientific community. Once established, however, uniformitarianism soon dominated the science of geology, gaining almost total acceptance after Charles Lyell, an Englishman, popularized it in the 1830s in a three-volume book titled *Principles of Geology.* Let us briefly examine the uniformitarian view of how rocks form.

Rocks consist of interlocking or bonded grains that are typically composed of single minerals. A **mineral** is a naturally occurring inorganic solid element or compound with a particular chemical composition or range of compositions and a characteristic internal structure. Quartz, which forms most grains of sand, is probably the most familiar and widely recognized mineral; other minerals constitute the materials we call limestone, clay, and asbestos. Most rocks consist of two or more minerals. Rocky surfaces that stand exposed and are readily accessible for study are generally designated as **outcrops** or **exposures.** Scientists also have access to rocks that are not visible in outcrops. Well drilling and mining, for example, allow geologists to sample rocks that lie buried beneath Earth's surface.

Basic Kinds of Rocks On the basis of modes of origin, many of which can be seen operating today, early uniformitarian geologists, led by Hutton and Lyell, came to recognize three basic types of rocks: igneous, sedimentary, and metamorphic. **Igneous rocks,** which form by the cooling of molten material to the point at which it hardens, or freezes (much as ice forms when water freezes), are composed of interlocking grains, each consisting of a particular mineral. The igneous rock most familiar to the nongeologist is granite. Molten material, or **magma,** that turns into igneous rock comes from great depths within Earth, where temperatures are very high. This material may reach Earth's surface through cracks and fissures in the crust and then cool to form **extrusive, or volcanic, igneous rock** (Figure 1-1), or it may cool and harden within Earth to form **intrusive igneous rock** (Figure 1-2). Igneous rock that solidifies deep within Earth is sometimes uplifted by subsequent movements of earth and eventually exposed at Earth's surface by **erosion,** which is a group of processes that loosen rock and move pieces of it downhill.

Sedimentary rocks form from **sediments,** which are materials deposited at Earth's surface by water, ice, or air. Most sediments are accumulations of distinct mineral grains. Some of these grains are products of weathering (i.e., decay and breakup) of older rocks, while others result from the chemical precipitation of minerals from water. Grains of sediment seldom become bonded to form hard rock until long after they have accumulated. The two important agents of this rock-forming process, which is known as **lithification,** are **compaction** of sediment under the influence of gravity and **cementation** of grains by the pre-

FIGURE 1-1 Mount St. Helens erupting in 1980. The cone of the volcano is itself formed of volcanic igneous rock extruded from the volcano. *(U.S. Geological Survey.)*

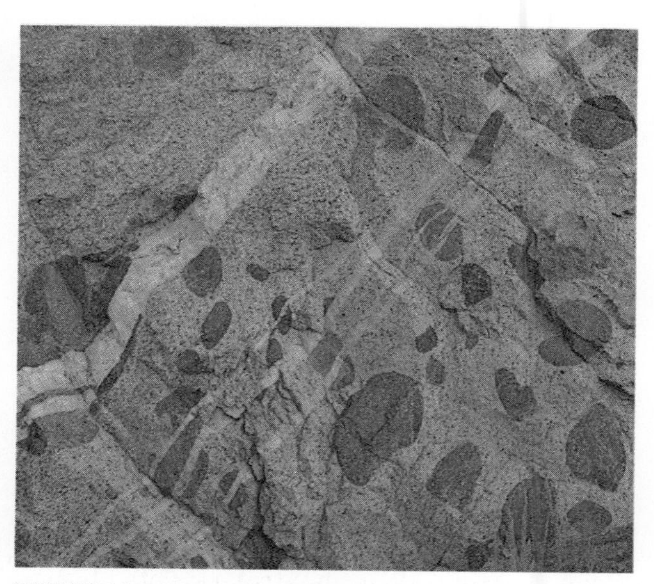

FIGURE 1-2 Intrusive igneous rock. The dark bodies are pieces of the surrounding rock that the magma that formed the igneous rock incorporated before it solidified. The light-colored diagonal bands on the left are veins that formed when a second body of magma intruded the main body of igneous rock. *(Martin G. Miller.)*

FIGURE 1-3 **Horizontal bedding in sedimentary rocks bordering the Grand Canyon.** *(Peter Kresan.)*

cipitation of mineral cement from solutions that flow between the grains. Lithification is a form of **diagenesis,** which is the full set of processes, including dissolution, that alter sediments at low temperatures after they have been buried. Cementation that transforms sand into sandstone is an example of diagenesis.

Most igneous rocks consist of silicate minerals (see Appendix I) and so do most sedimentary particles, or **clasts,** derived from them. Sedimentary rocks formed primarily of silicate minerals are thus known as **siliciclastic** rocks, and these are the most abundant sedimentary rocks of Earth's crust.

Sediments usually accumulate during discrete episodes, each of which forms a tabular unit known as a **stratum** (plural, **strata**). Strata tend to remain distinct from one another even after lithification because the grains of adjacent beds usually differ in size or composition. Because of their differences, the contacting surfaces of the strata usually adhere to each other only weakly, and sedimentary rocks often flake or fracture along these surfaces. As a result, sedimentary rocks exposed at Earth's surface often can be seen to have a steplike configuration when they are viewed from the side (Figure 1-3). **Stratification** is the word used to describe the arrangement of sedimentary rocks in discrete layers. **Bedding** is stratification in which layers exceed 1 centimeter (~0.4 inch) in thickness, and **lamination** is stratification on a finer scale.

Metamorphic rocks form by the alteration, or **metamorphism,** of rocks within Earth under conditions of high temperature and pressure. By definition, metamorphism alters rocks without turning them to liquid. If the temperature becomes high enough to melt a rock and the molten rock later cools to form a new solid rock, this new rock is by definition igneous rather than metamorphic. Metamorphism produces minerals and textures that differ from those of the original rock and that are characteristically arrayed in parallel wavy layers (Figure 1-4). The two groups of rocks that

FIGURE 1-4 **Metamorphic rock.** This is a coarsely crystalline kind of rock known as gneiss. *(Peter Kresan.)*

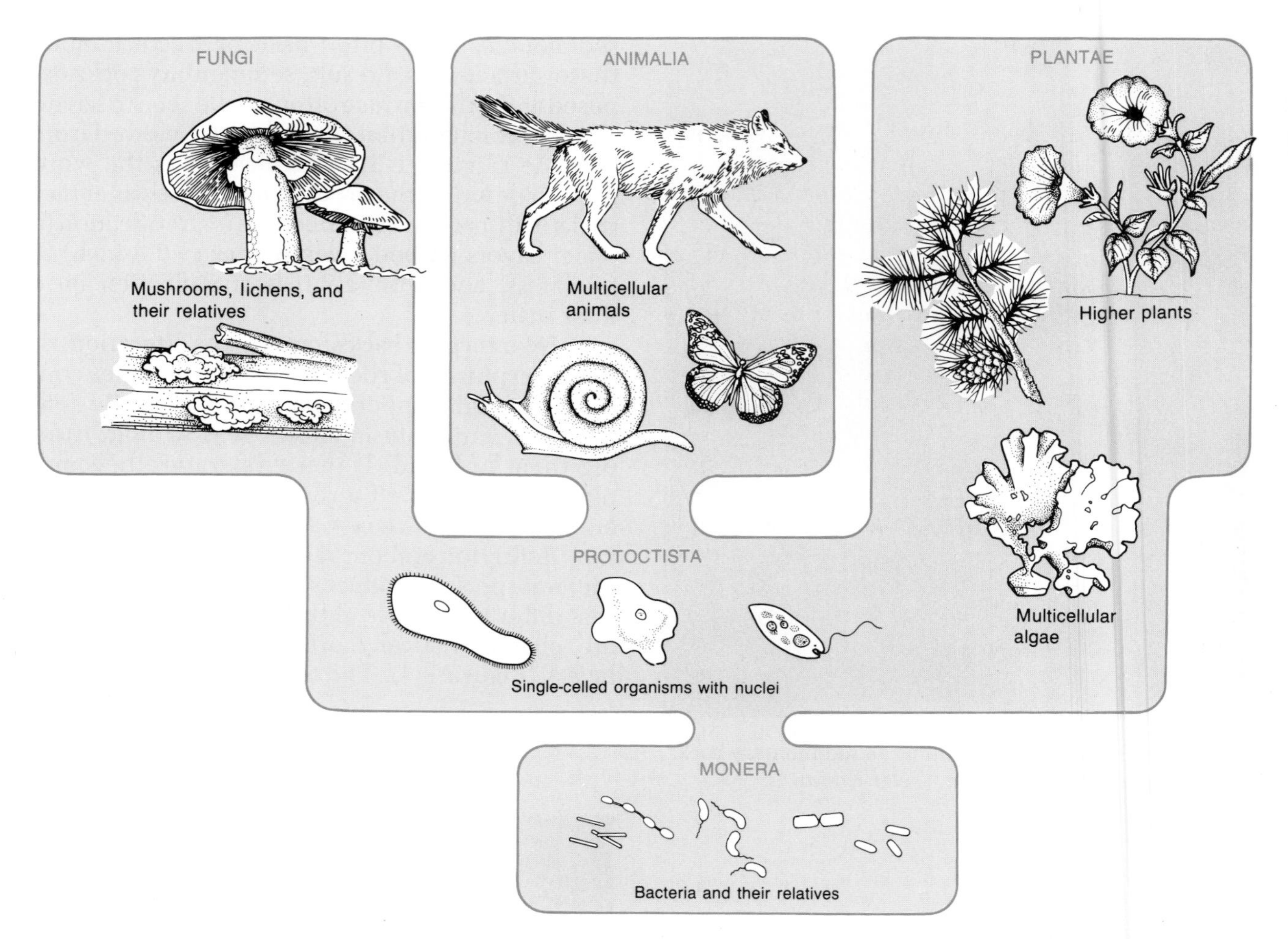

FIGURE 1-5 The five kingdoms of living things. The Monera include simple forms whose cells lack the internal organization represented by subcellular bodies such as nuclei and chromosomes. Some experts divide the Monera into two kingdoms and hence recognize six kingdoms altogether. The Protoctista include single-celled organisms that possess nuclei and chromosomes; animal-like protoctists eat other organisms, while plant-like protists manufacture their own food. Red, green, and brown algae are multicellular, but their cells are not differentiated into tissues of discrete cell types; some experts classify these algae as protoctists, while others classify them as members of the Plantae. Plantae are multicellular organisms that manufacture their own food, and animals are multicellular organisms that ingest food and digest it within their bodies. Fungi, which include mushrooms, lichens, and molds, absorb food from their environment.

form at high temperatures—igneous and metamorphic rocks—are sometimes referred to as **crystalline rocks.**

Classification of Bodies of Rock Geologists also classify rocks into units called **formations.** Each formation consists of a body of rocks of a particular type that formed in a particular way—for example, a body of granite, of sandstone, or of alternating layers of sandstone and shale. A formation is formally named, usually for a geographic feature such as a town or river where it is well exposed. Smaller rock units called **members** are recognized within some formations. Similarly, some formations are united to form larger units termed **groups,** and some groups, in turn, are combined into **supergroups.**

More about the nature and origin of minerals and the three basic types of rocks can be found in Appendix I.

LIFE ON EARTH

Organisms that have inhabited Earth in the course of geologic time have left a partial record in rock of their presence and their activities. This record reveals that life has changed dramatically since it first arose on Earth and that its transformation has been intimately associated with changes in physical conditions on Earth — in climates or in the positions of continents, for example.

It is not easy to provide a precise definition of *life*, but two attributes that are generally regarded as essential to life are the capacity for self-replication and the capacity for self-regulation. Viruses are simple entities that can replicate themselves (or reproduce), but they do not regulate themselves — that is to say, they do not employ raw materials from the environment to sustain orderly, internal chemical reactions. Thus viruses are not considered to be living things. On Earth today, all entities that are self-replicating and self-regulating are also cellular; that is, they consist of one or more discrete units called cells. A living cell is a module that includes a number of distinct features, including apparatuses that facilitate certain chemical reactions. The chemical "blueprint" for a cell's operation is coded into the chemical structure of the gene. An essential feature of this blueprint is the cell's built-in ability to duplicate itself so that a replica can be passed on to another cell or to an entirely new organism.

Taxonomic Groups

Until well into the nineteenth century, scientists divided all living things into two categories: the animal kingdom and the plant kingdom. As various forms of life came to be better understood, however, these distinctions became increasingly difficult to maintain. Today five kingdoms are recognized — the Monera, Protoctista, Fungi, Animalia, and Plantae (Figure 1-5).

A more detailed classification of many forms of life can be found in Appendix III. As this appendix indicates, each of the five kingdoms is divided into numerous subordinate groups. These kingdoms and their subordinate groups are known as **taxa,** or **taxonomic groups,** and the study of the composition and relationships of these groups is known as **taxonomy.** Taxa within kingdoms range from the broad category known as the phylum (plural, phyla) to the narrowest category, the **species** (Table 1-1), which consists of a group of individuals that interbreed or have the potential to interbreed in nature and that do not breed with other interbreeding groups. The basic categories of higher taxa — the kingdom, phylum, class, order, family, and genus (plural, genera) — are sometimes supplemented by categories such as the subfamily and the superfamily. Names of genera are printed in italics, as are species designations. Actually, the name of the species consists of two words, the first of which is the name of the genus to which the species belongs.

TABLE 1-1 Major taxonomic categories within a kingdom, as illustrated by the classification of humans

Kingdom: Animalia
  Phylum: Chordata
    Class: Mammalia
      Order: Primates
        Family: Hominidae
          Genus: *Homo*
            Species: *Homo sapiens*

NOTE: Between these categories, intermediate ones (e.g., superorders, suborders, superfamilies, and subfamilies) are sometimes recognized.

Figure 1-6 further illustrates how humans are classified within the order Primates of the class Mammalia. In general, the narrower the taxonomic category, the greater the biological similarity of its members. Humans and gorillas, for example, have enough in common to be assigned to the same superfamily, but monkeys differ from these groups in several ways and are assigned to other superfamilies. All of these superfamilies are nonetheless similar enough to be united in a single suborder. Often one or a small number of biological features serve to distinguish one higher taxon from other closely related taxa of the same rank. Dinosaurs, for example, are divided into two orders on the basis of pelvic structure (Figure 1-7).

PRIMATES

SUBORDER ANTHROPOIDEA

SUPERFAMILY HOMINOIDEA

FAMILY Hylobatidae

Hylobates (gibbon)

FAMILY Hominidae

Homo (human)

FAMILY Pongidae

Gorilla (gorilla)

Pan (chimp)

Pongo (orangutan)

SUPERFAMILY CERCOPITHECOIDEA

Old World monkeys

SUPERFAMILY CEBOIDEA

New World monkeys

FIGURE 1-6 The taxonomic position of the human genus *Homo* within the order Primates and the family Hominidae. There are four other genera in the superfamily Hominoidea: three ape genera of the family Pongidae and one genus of the family Hylobatidae.

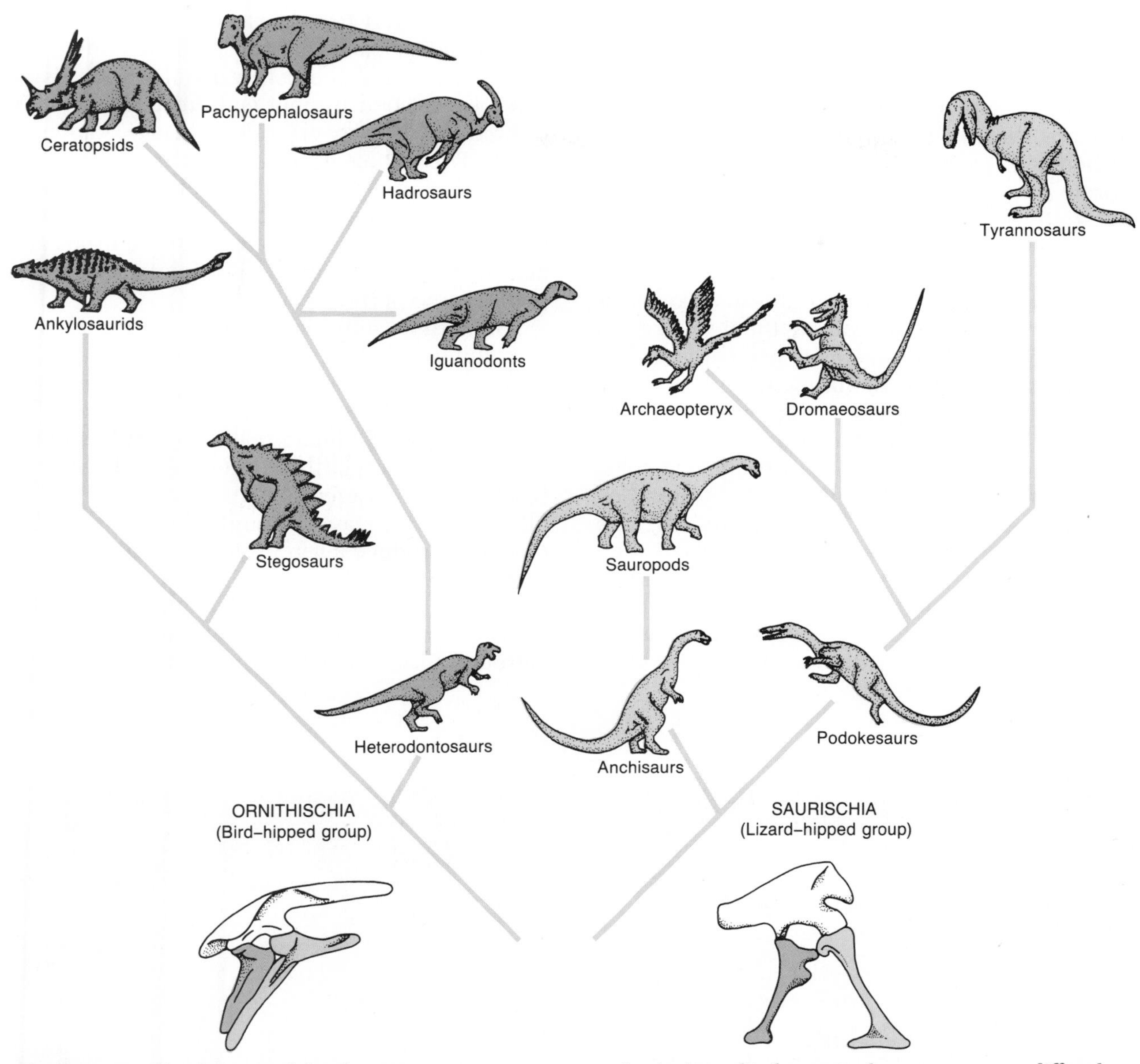

FIGURE 1-7 The division of the dinosaurs into two orders: the ornithischian, or "bird-hipped," dinosaurs and the saurischian, or "lizard-hipped," dinosaurs. Of the three large pelvic bones (shown beneath the two orders), the pubic bones in these two groups differ the most both in shape and in position. *(After E. H. Colbert, Evolution of the Vertebrates, John Wiley & Sons, Inc., New York, 1980.)*

Fossils

Most of our knowledge about the life of past intervals of geologic time is derived from fossils. The term **fossil** is usually restricted to tangible remains or signs of ancient organisms that died thousands or millions of years ago. **Fossilization** is the process of becoming a fossil. Because few fossils can survive the high temperatures at which igneous and metamorphic rocks form, almost all fossils are found in sediments and sedimentary rocks. Of the five kingdoms, only the fungi are poorly represented in the fossil record. Fossils of the other four kingdoms depict many aspects of the history of life on Earth — a history that, as we shall see in Chapter 5, encompasses the evolution of living things.

The most readily preserved features of animals are the structures that are informally described as "hard parts" — teeth and bones of vertebrate animals and comparable solid structures of invertebrate animals. Many groups of invertebrates lack skeletons and therefore have poor fossil records or none at all. Some invertebrate animals, on the other hand, have internal skeletons embedded in soft tissue; among them are the sea lilies (Figure 1-8). Others have protective external skeletons; among them are the bivalve and gastropod mollusks, whose tissues are housed inside skeletons popularly known as seashells. Hard parts are often preserved with only a modest amount of diagenetic alteration, but at times they are completely replaced by minerals that are unrelated to the original skeletal material.

Although terrestrial plants do not generally have mineralized skeletal structures per se, their cells have rigid walls of cellulose. As a consequence, woody tissue and even many leaves are much more likely to be preserved than the flesh of animals. After plants are buried in sediment, the spaces left inside the cell walls of woody tissue may be replaced by inorganic materials — most commonly by finely crystalline quartz, known as chert. This filling process, which produces the fossils that are often called petrified wood, is known as **permineralization.** Permineralization is not restricted to woody plant tissue. Porous animal skeletons, such as the bones of vertebrates, also become permineralized.

Sometimes solutions percolating through rock or sediment dissolve fossil skeletons, leaving a space within the rock that is a three-dimensional negative imprint of the organic structure. This type of imprint, which is called a **mold,** is sometimes filled secondarily with minerals. If this process has not occurred in nature, the mold can be filled with wax, clay, or liquid rubber in the laboratory to produce a replica of the original object (Figure 1-9).

Fossils called **impressions** might be viewed as squashed molds. Impressions usually preserve in flattened form the outlines and some of the sur-

FIGURE 1-8 Fossil crinoids (sea lilies) from Cretaceous strata in Kansas. The globular bodies, to which arms are attached, are about the size of golf balls. *(Chip Clark.)*

FIGURE 1-9 Molds of Devonian starfishes. The molds occupy the underside of a slab of sedimentary rock. The sediment that formed these molds settled on top of the starfishes and filled the depressions that they left in the underlying sediment when they decayed. The arms of these fossils are about the length of a human thumb. *(Chip Clark.)*

FIGURE 1-10 A fossil impression, with traces of carbon produced by the carbonization of original tissues, of a Cenozoic-age butterfly. *(Leo F. Hickey, Smithsonian Institution.)*

face features of soft or semihard organisms such as insects or leaves (Figure 1-10). A residue of carbon remains on the surface of some impressions after other compounds have been lost by the escape of liquids and gases. This process of carbon concentration is known as **carbonization.**

Tracks, trails, burrows, and other marks left by animal activity are known as **trace fossils** (Figure 1-11). Because they are the direct results of activity, trace fossils can reveal a great deal about the behavior of extinct animals—although the animal that made a particular kind of trace cannot always be identified with certainty.

Fleshy parts of animals, or "soft parts," are occasionally found in the fossil record, but only in sediments that date back a few millions or tens of millions of years. Chemical residues of tissues and cells can be identified in much older rocks. The deposit most famous for preservation of soft parts is the Eocene Geiseltal of Germany, which is more than 40 million years old. In the nearly impermeable Geiseltal sediments, which are rich in oily plant debris, the skin and blood vessels of long-extinct frogs can still be studied, fossil leaves are still green, and insects retain their iridescent color (Figure 1-12). Protection from oxygen is the

FIGURE 1-11 Dinosaur tracks in Connecticut. A dinosaur formed these tracks by walking across wet mud that hardened into rock. *(Dinosaur State Park, Rocky Hill, Connecticut.)*

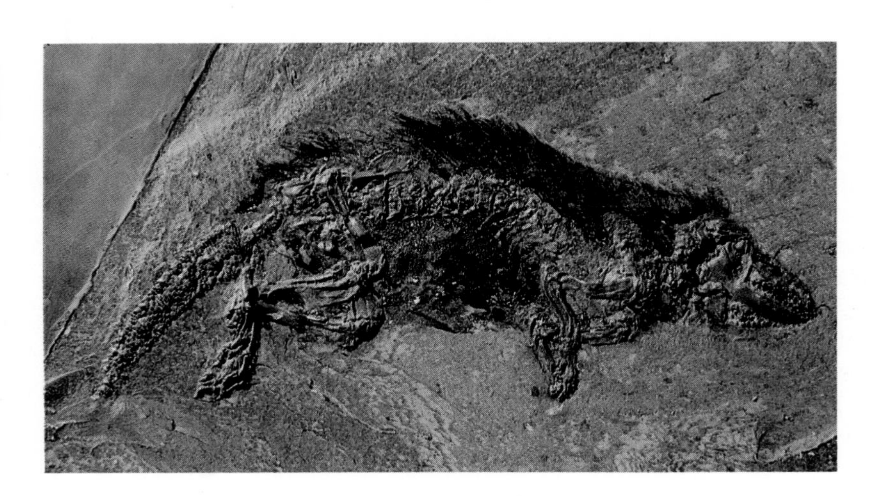

FIGURE 1-12 Remarkable preservation of an Eocene mammal. This creature, which resembled a hedgehog, became buried in the bottom of a lake, where its flesh and fur were partially preserved because so little oxygen was present that scavenging animals and bacteria of the kind that cause decay were absent. This fossil is about 40 centimeters (16 inches) long. *(Nature Museum, Senkenberg, Germany.)*

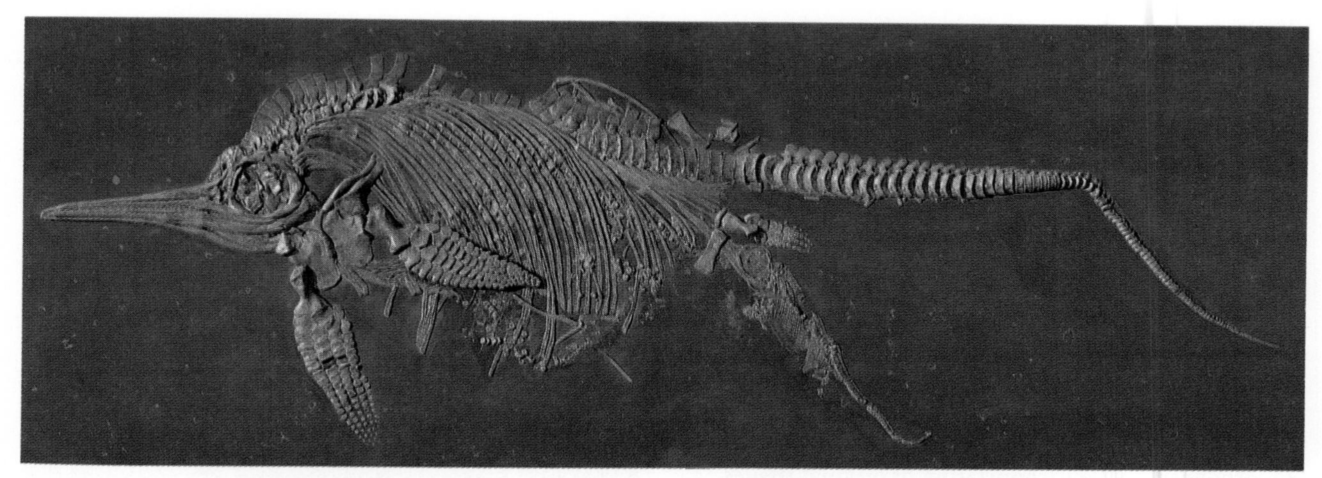

FIGURE 1-13 An ichthyosaur that died in the act of giving birth. The infant's head apparently stuck in the mother's birth canal and both animals died. These swimming reptiles were preserved in a Jurassic deposit in Germany. The mother was about 2 meters (6 feet) long. *(Staatliches Museum für Naturkunde.)*

secret for survival of soft tissue: It is most likely to be preserved when organisms are buried in fine-grained, relatively impermeable sediment, especially if oily water-repellent organic matter is also present.

Some unusual fossils reveal the outlines of soft parts even though those parts are not themselves preserved (Figure 1-13). Most of these fossils were preserved under special conditions such as the absence of oxygen, rapid burial in fine-grained sediment, or protection in a nodular structure called a **concretion,** formed by diagenetic alteration of sediment. Although the origin of concretions is not fully understood, those that form around fossils seem to have resulted from localized diagenesis caused in some way by the partial decay of the fossilized organism.

Less spectacular fossils—especially hard parts and molds—are much more common in sedimentary rocks than most people realize. They are especially abundant in sedimentary rocks that were formed in the ocean, where animals with skeletons abound. To appreciate how common fossils are, one need only recognize that many limestones consist largely of skeletal fragments of marine organisms.

Organisms often lose their identity in contributing to **fossil fuels,** which are condensed and altered forms of organic matter that can be burned to supply energy for human use. **Coal** is a form of altered organic matter that serves as a fossil fuel. Coal forms when concentrated plant debris is buried and subjected to moderately high temperatures and pressures that drive out many compounds in liquid and gaseous states. Most low-grade coals are full of carbonized fossil plants that can still be identified as particular genera or even species. **Peat** is sediment composed primarily of plant debris that has not yet been altered to form coal.

Although fossils occur with great frequency in many sedimentary rocks, it is important to recognize that most species of animals and plants have never been discovered in the fossil record. Rare species and those that lack skeletons are especially unlikely to be found in fossilized form. Even most species with skeletons have left no permanent fossil record. A variety of processes destroy skeletons. Animals that scavenge carcasses, for example, may splinter bones in the process. Also, many bones, teeth, and shells are abraded beyond recognition when they are transported by moving water before they finally become buried. Because marine organisms live in a vast basin in which sediments accumulate, these forms of life are more frequently preserved than terrestrial plants and animals. Even after burial, many fossils fail to survive diagenesis, metamorphism, and erosion of the sedimentary rocks in which they are embedded. Finally, many fossil species remain entombed in rocks that have never been exposed at Earth's surface or sampled by drilling operations.

In Chapter 2 we will learn some of the ways in which fossils reveal information about ancient environments; in Chapter 4 we will consider how fossils are employed to date rocks; and in Chapter 5 we will examine how fossils document the evolution of life.

MOVEMENTS OF EARTH

Rocks are not motionless. They not only move but also break, and they can even change shape without breaking. Rocks that are buried deep within Earth under great pressure can be folded or squeezed into new shapes like bread dough. Most deformed rocks that we now see at Earth's surface acquired their present shapes when they were buried far below the surface (Figure 1-14).

Under other conditions, rocks remain brittle and break instead of folding or deforming into new shapes. A surface along which rocks of any kind have broken and moved past each other is called a **fault.** Appendix II reviews the nature of faults, folds, and other structures resulting from the deformation of rocks and illustrates how these and other features of rocks enable us to reconstruct the geologic events within Earth's interior.

Most of our knowledge about the structure of Earth's interior derives from the study of oscillatory movements called **seismic waves,** which travel through Earth as a consequence of natural or artificial disturbances. An earthquake is a natural seismic disturbance that results from the sudden movement of one portion of Earth against another along a fault. Artificial explosions at Earth's surface also produce seismic waves.

An earthquake always begins at a **focus,** a place within Earth where rocks move against other rocks along a fault and produce seismic waves. Earthquake foci lie within Earth's mantle and crust, far from its center—but the waves emitted from foci often pass great distances through Earth to emerge at the surface, where they can be detected with instruments called seismographs. Geophysicists can then evaluate the activity of Earth's internal structure by recording the times at which the waves of an earthquake arrive at many locations.

The denser the material, the more rapidly seismic waves travel through it. Study of the rates at which waves travel in various directions from a focus reveals that the materials that form the central part of Earth are much more dense than those near the surface. The density gradient from the surface to the center of Earth is not gradual, however; instead, the planet is divided into several discrete concentric layers (Figure 1-15). At

FIGURE 1-14 Folded sedimentary rocks in the Rocky Mountains of Canada. *(Geological Survey of Canada.)*

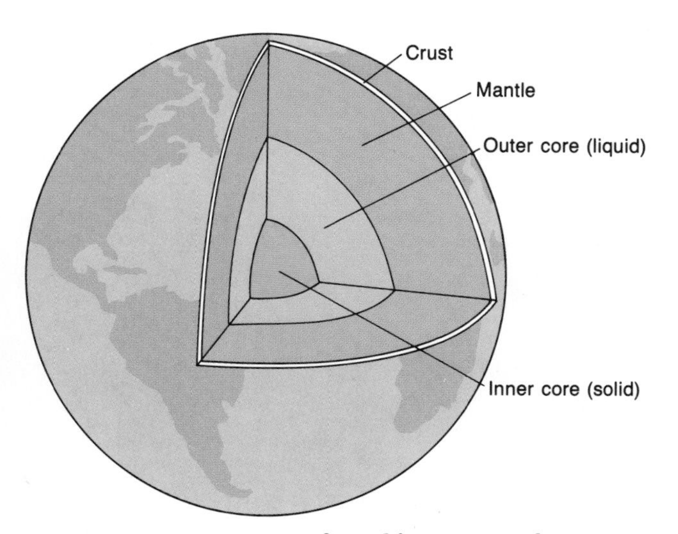

FIGURE 1-15 Zonation of Earth's interior. The crust, which includes continents at the surface of Earth, rests on the mantle. The mantle in turn rests on the core. The outer core is liquid, but the inner core is solid. *(After W. J. Kauffman, Planets and Moons, W. H. Freeman and Company, New York, 1979.)*

Earth's center is the **core,** whose solid, spherical inner portion and liquid outer portion are thought to consist primarily of iron. Forming a thick envelope around the outer core is the **mantle,** a complex body of less dense rocky material. Finally, capping the mantle is the **crust,** which consists of still less dense rocky material. As we shall see in Chapter 8, the density gradient from the core to the crust developed early in Earth's history, when molten materials of low density rose to float on materials of higher density. The passage of seismic waves from the rocks of the crust to the denser rocks of the mantle is signaled by an abrupt increase in velocity known as the **Mohorovičić discontinuity,** or **Moho** for short (Figure 1-16). Because continental crust is much thicker than the crust beneath the oceans, the Moho dips downward beneath the continents.

The rocks that form oceanic crust are the type known as **mafic**—a label whose first three letters indicate that these dark rocks are rich in magnesium (Mg) and iron (Fe). Mafic rocks are much less common in continental crust than are the lighter-colored, less dense rocks known as **felsic**—an adjective derived from the first three letters of *feldspar,* the name of the most common mineral of continental crust. In comparison with mafic rocks, felsic rocks are rich in silicon and aluminum and poor in the heavier elements magnesium and iron. Rocks of the mantle are even richer in magnesium and iron than the oceanic crust—hence their great density—and they are known as **ultramafic** rocks.

Continental crust not only stands above oceanic crust but also extends farther down into the mantle (Figure 1-16). The continental crust extends even farther down beneath a mountain range than it does elsewhere. **Isostatic adjustment,** or the upward or downward movement that keeps crust in gravitational equilibrium as it floats on the mantle, is responsible for this phenomenon. In effect, the root beneath a mountain acts to balance the mountain (Figure 1-17).

Although the crust and the upper mantle differ in density, they are firmly attached to each other, forming a rigid layer known as the **lithosphere.** Below the lithosphere is the **asthenosphere,** which is also known as the "low-velocity zone" of the mantle because seismic waves slow down as they pass through it. This property tells us that the asthenosphere is composed of partially molten rock—slushlike material consisting of solid particles with liquid occupying the spaces in between. Although the asthenosphere represents no more than 6 percent of the thickness of the mantle, the mobility of this layer allows the overlying lithosphere to move. The lithosphere does not move as a unit, however; instead it is divided into **plates** that move in relation to one another. Some plates carry continents with them as they move, while others carry only oceanic crust.

Plates move over the surface of Earth about as rapidly as your fingernails grow. Slow as this rate may seem, the progress of plates over millions of years has been considerable. Many have moved as far as 500 or even 1000 kilometers (~300 to 600

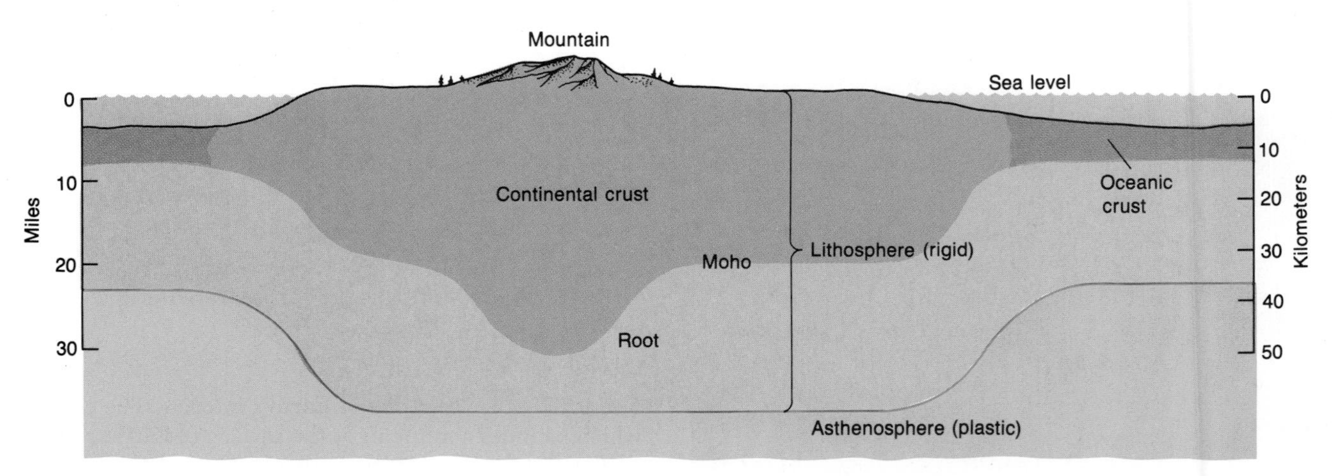

FIGURE 1-16 The structure of the upper part of Earth. Continental crust is much thicker than oceanic crust, and it is especially thick where there are mountains, beneath which lie deep roots. The Moho separates the crust from the mantle. The crust and the upper mantle together form the rigid lithosphere.

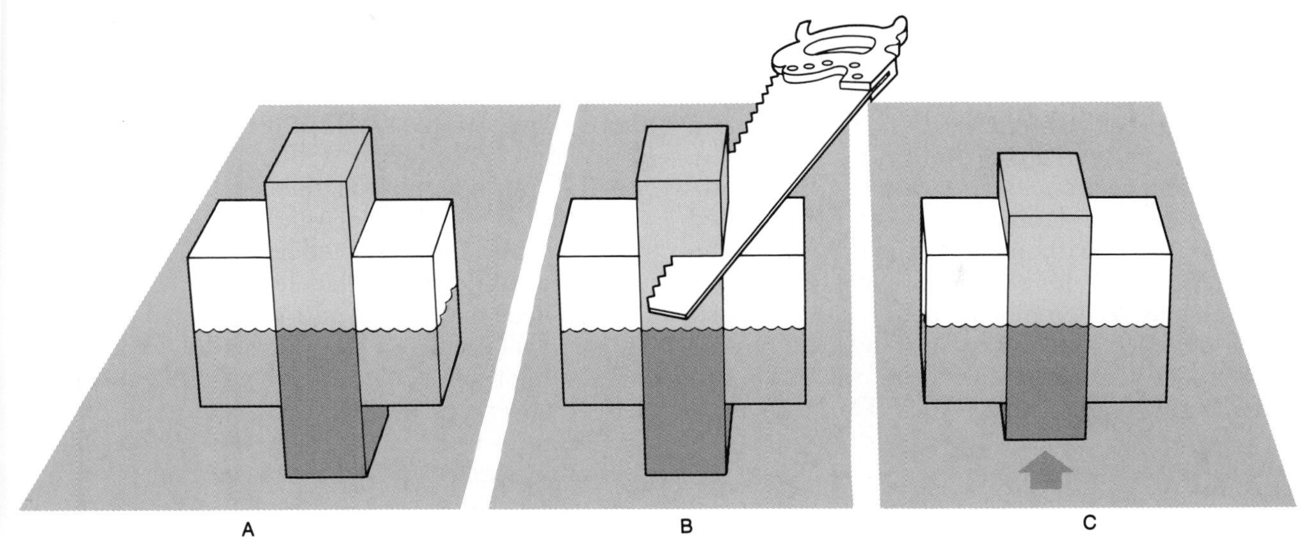

FIGURE 1-17 The principle of isostasy, illustrated by blocks of wood in water. *A.* Three blocks of wood float adjacent to one another. Because the wood is half as dense as the water, each block lies half above the surface of the water and half below. This means that the weight of a block of wood is equivalent to the weight of the volume of water that it displaces. When the top of the center block is cut off *(B)*, the weight of this block no longer balances the weight of the displaced water. As a result, the block bobs upward until it is balanced once again, lying half above and half below the water *(C)*. The central block is like the part of a continent where a large mountain is present—the low-density elevated crust must be balanced by a root.

miles) in 10 million years. Since the early 1960s it has been recognized that many earthquakes can be attributed to the motions of plates. Movement of the edge of one plate over the edge of another is, in fact, a major cause of mountain building. We will learn more about plate movements in Chapter 6 and about mountain building in Chapter 7.

FUNDAMENTAL PRINCIPLES OF GEOLOGY

In keeping with the principle of uniformitarianism, processes operating on and within Earth follow physical and chemical laws. Several principles derived from physical laws are fundamental to our method of learning about Earth's history—a method that relies heavily on information gained from studying the relative positions of bodies of rock.

In the seventeenth century, Nicolaus Steno, a Danish physician living in Florence, Italy, formulated three axioms for interpreting stratified rocks. The first is the **principle of superposition,** which states that in an undisturbed sequence of strata, the oldest strata lie at the bottom and successively higher strata are progressively younger. In other words, in an uninterrupted sequence of strata, each bed is younger than the one below and older than the one above. This, of course, is a simple consequence of the law of gravity, as is Steno's second principle, the **principle of original horizontality,** which states that all strata are horizontal when they form. As it turns out, this principle requires some modification. We now recognize that some sediments, such as those of a sand dune, accumulate on sloping surfaces, forming strata that lie parallel to the surface on which they were deposited. Sediments seldom accumulate at an angle greater than 45° to the horizontal, however, so a reasonable restatement of Steno's second principle would be that almost all strata are initially more nearly horizontal than vertical. Steno's third principle is the **principle of original lateral continuity,** which he invoked to explain the occurrence on opposite sides of a valley (or some other intervening feature of the landscape) of similar rocks that seemed once to have been connected. Steno was, in effect, pointing out that strata originally are unbroken flat expanses, "pinching out" laterally to a thickness of zero or abutting against the walls of the natural basin in which they formed. The original continuity of a stratum can later be broken either by erosion, as in the development of a river valley, or by faulting.

Complex sequences of events can be "read" from geometric relationships more complicated than those dealt with in Steno's principles. **Intrusive relationships,** for example, reveal the relative ages of rocks. When magma penetrates preexisting solid rock, it eats its way into the solid rock or pushes part of the rock aside before cooling to form a body of solid igneous rock. Thus the invading igneous rock is always younger than the rock that it intrudes (Figure 1-2).

Faults, of course, are also younger than the rocks that they transect, and when one fault is offset by another, the fault that is offset is older (Figure 1-18). This is the **principle of crosscutting relationships.**

Finally, when fragments of one body of rock are found within a second body of rock, the second is always younger than the first. The second body of rock may be either a sedimentary rock in which some particles have come from another body of rock or an igneous body that incorporated pieces of surrounding rock before cooling (Figure 1-2). This relationship, which is known as the **principle of components,** can be stated more succinctly: A body of rock is younger than another body of rock from which any of its components are derived.

These simple principles have enabled geologists to establish the relative ages of most bodies of rock that lie adjacent to one another at the surface of Earth. When the rocks being compared lie at great distances from one another, however, other methods must be employed. Fossils provide one valuable means of establishing the relative ages of rocks that lie far apart. William "Strata" Smith, a British surveyor, noted late in the eighteenth century that fossils are not randomly distributed in rocks. When Smith studied large areas of England and Wales, he found that fossils in sedimentary rocks occurred in a particular vertical order ("vertical" in terms of the succession of one layer above another). To the surprise of less experienced observers, Smith could predict the vertical ordering of fossils in areas he had never visited. We now recognize that this ordering, known as **fossil succession,** reflects the sequence of organic evolution—the natural appearance and disappearance of species through time. Although Smith, like nearly all of his contemporaries, did not believe in evolution, he was able to use his knowledge of fossil succession to determine where isolated outcrops of sedimentary rocks fitted into the general sequence of strata in England and Wales.

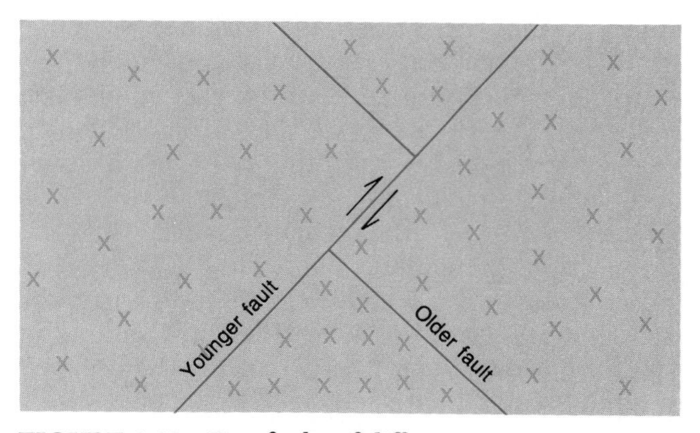

FIGURE 1-18 Two faults of different ages. In accordance with the principle of crosscutting relationships, the fault that is offset by the other is the older of the two.

GEOLOGIC TIME

Although the principles outlined in the preceding section allow us to establish the relative ages of many bodies of rock, they do not permit us to determine the actual ages of rocks measured in thousands or millions of years. As we will see in Chapter 3, some sedimentary beds are produced annually, like the rings in a tree trunk. Unless the latest of a continuous sequence of annual beds is currently forming, however, it is impossible to count backward to determine precisely how many years ago an older bed formed. In other words, if a sequence of this type formed long ago, we cannot tell the actual ages of its beds. Fortunately, "geologic clocks" in the form of minerals that undergo **radioactive decay** provide us with a means of approximating the actual ages of ancient rocks. Naturally occurring radioactive materials decay into other materials at known rates. By measuring the amount of radioactivity in a radioactive material that has been decaying since it became part of a rock, we can estimate the age of the rock. We will learn more about this technique in Chapter 4.

During the last century, long before the discovery of radioactivity, it became apparent that very old sedimentary rocks contain no identifiable fossils. Beginning with these rocks and examining progressively younger rocks in any region, early geologists discovered that fossils became abundant at a certain level. This level became the boundary at which all of geologic time was divided into two major intervals. The oldest rocks with conspicuous fossils were designated as Cambrian in age, and still older rocks became known as Precambrian rocks. Today the Precambrian designation is still used informally, but the Precambrian interval is formally divided into the Archean Eon and the Proterozoic Eon, with the boundary between these two eons placed at 2.5 billion years ago. Subsequent geologic time, from Cambrian on, constitutes the Phanerozoic Eon, meaning the "interval of well-displayed life." An **eon** is the largest formal unit of geologic time.

Phanerozoic time is divided into three primary intervals, or **eras,** which the history of life on Earth serves to define. The earliest is the "interval of old life," or the Paleozoic Era. This is followed by the "interval of middle life," or the Mesozoic Era, which is commonly called the Age of Dinosaurs, and by the "interval of modern life," or the Cenozoic Era, which is informally designated as the Age of Mammals. Figure 1-19 depicts these eras and the intervals within them, known as the geologic **periods.**

Figure 1-19 also indicates when each period began and ended, as determined by radioactive materials in rocks whose ages approximate period boundaries. Note that the Phanerozoic interval began about 570 million years ago. A human lifetime is so short in comparison with this figure that geologic time seems too vast for us to comprehend; experience does not permit us to extrapolate from the time scale we are familiar with, measured in seconds, minutes, hours, days, and years, to a scale suitable for geologic time. Geologists therefore use a separate scale when they think about geologic time—one in which the units are millions of years. If the Phanerozoic interval of time were compressed into a year, we would find animals with backbones crawling up onto the land for the first time in mid-April, dinosaurs inheriting Earth in early July but then suddenly dying out in late October, and humans appearing on Earth within 2 hours of midnight on New Year's Eve.

Even during the nineteenth century, before geologists had any means of measuring the actual

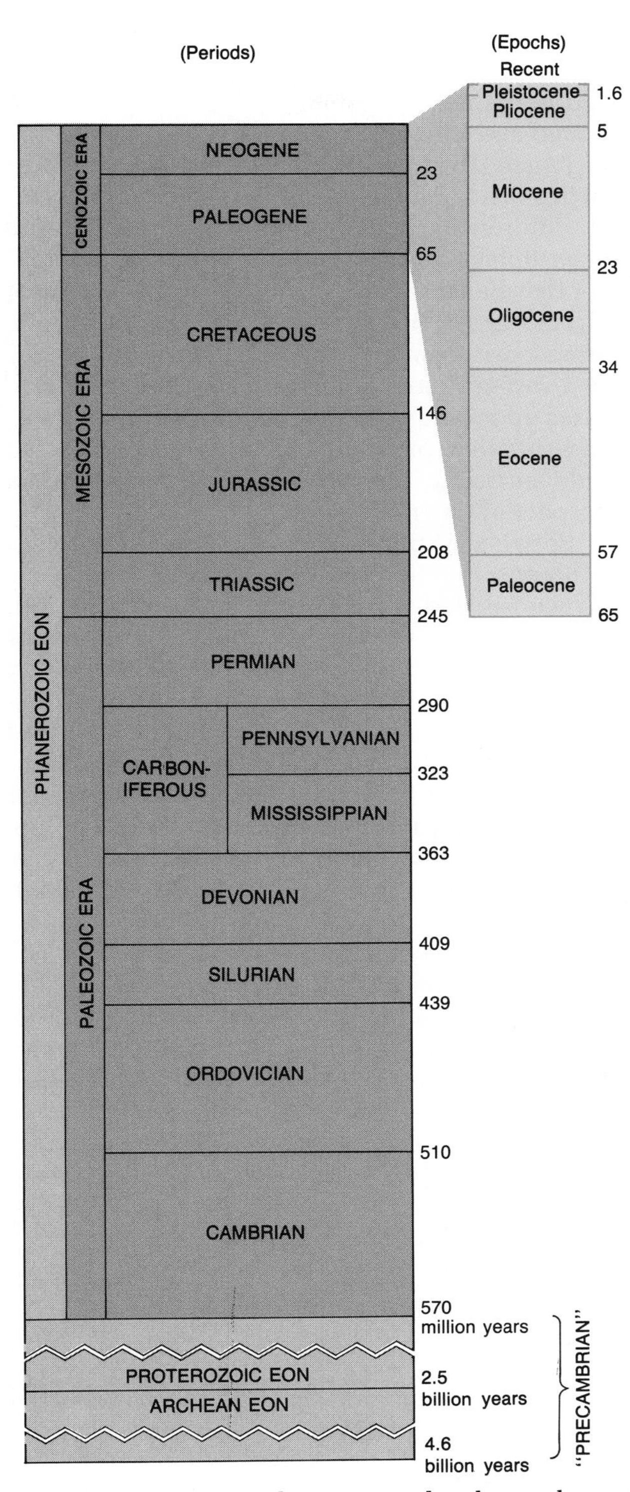

FIGURE 1-19 The geologic time scale. The numbers on the right represent the ages of the boundaries between periods and epochs in millions of years. The Recent Epoch (the past 10,000 years or so) is also known as the Holocene.

ages of rocks, they were aware that Earth was hundreds of millions of years old. By noting how slowly geologic processes had operated during the years since geologic features had first been recorded, they knew that most rocks could not be formed and then destroyed quickly.

James Hutton, in presenting his view of Earth's history, recognized that one of the greatest differences between uniformitarianism and catastrophism lay in the uniformitarian assumption that Earth was very old—that the sequence of rocks at Earth's surface could not have been formed by a few brief upheavals or floods; they must instead have been formed slowly by processes operating over vast stretches of time. Hutton wrote that, in viewing Earth in these terms, he could envision "no vestige of a beginning, no prospect of an end."

Between 1865 and the beginning of the twentieth century, the physicist Lord Kelvin and his followers issued a challenge to uniformitarianism, arguing that Earth was only about 20 million, or perhaps 30 or 40 million, years old. Lord Kelvin and his followers based their assertion on calculations of the rate at which Earth could be expected to cool after its formation at an initially high temperature. Their assumption was that the planet had formed in a molten state and had been cooling rapidly ever since. The fact that Earth's interior was still very hot (temperatures rose as a person descended through a mine shaft) seemed to indicate that Earth could not be very old.

It was not until the discovery of naturally occurring radioactive material that these physicists' calculations were disproved. It is now known that radioactive decay releases heat, which has reduced the rate at which Earth has cooled since its origin. Thus the interior of the planet has remained quite hot over more than 4 billion years. The energy of this heat moves the huge plates of rock, some of which carry continents, over Earth's surface.

Dating of rocks by means of radioactive materials reveals that some rocks on Earth are more than 3.8 billion years old. Many major geologic events span millions of years, but in the context of geologic time, these events are of relatively brief duration. We now know, for example, that the Himalayas, the tallest mountains on Earth, formed within the past 15 million years or so, but this period of time represents less than one-third of 1 percent of Earth's history. Destructive processes have also yielded enormous changes within a tiny fraction of Earth's lifetime. Mountains that were the precursors of the Rockies in western North America were leveled just a few million years after they formed, and most of the Grand Canyon of Arizona was cut by erosion within just the past 2 or 3 million years. We will examine these events in greater detail in later chapters.

Partly because episodic movements of earth elevate basins in which deposition occurs, rocks are not deposited continuously anywhere; one or more breaks interrupt any local record of deposition. An **unconformity** is a surface between a group of sedimentary strata and the rocks beneath them; it represents an interval of time during which erosion occurred rather than deposition. Because rock deformation is episodic, only rocks of a certain age will have been affected by a particular deformational episode. The surface between the deformed and undeformed rocks represents one kind of unconformity: When a group of rocks has been tilted and eroded and younger rocks have been deposited on top of them, the eroded surface is termed an **angular unconformity** (Figures 1-20 and 1-21*A*). Other unconformities are less dramatic. Sometimes the beds below an eroded surface are undisturbed, and only the irregular surface between groups of beds reveals a past episode of erosion. This kind of unconformity is called a **disconformity** (Figure 1-21*B*). An unconformity in which bedded rocks rest on an eroded surface of crystalline rocks (Figure 1-21*C*) is sometimes called a **nonconformity.**

GROWTH OF THE GEOLOGIC TIME SCALE

In the nineteenth century, when the geologic periods were first distinguished as unique intervals of time, geologists did not know even approximately how long ago each period had begun or ended. Each period was defined simply as the undetermined interval of time represented by a body of rock called a geologic **system.** The Cambrian Period, for example, was simply an interval of time that corresponded to those rocks that were designated as the Cambrian System. Geologists at that time did not have the means to study the entire sequence of rocks on Earth, from the most

FIGURE 1-20 An angular unconformity in the Grand Canyon. Horizontal Cambrian strata rest on tilted strata of Precambrian age. *(Peter Kresan.)*

ancient to the most modern; they could only study whatever promising rock sequences were accessible to them. Thus the Cretaceous System was formally designated in France in 1822, whereas the much older Cambrian and Silurian systems did not gain formal recognition until 1835. System after system was added up and down the sequence until, finally, all the Phanerozoic rocks of Europe were included.

Although the order in which the geologic systems were designated was haphazard, the total body of rock assigned to each system was not chosen arbitrarily. Two criteria were most important in these decisions. One was the occurrence of unique groups of fossils. Most systems contain many fossil taxa that differ considerably from the taxa found below and above them. Major extinctions have caused the most striking contrasts between systems. A system representing an interval that followed a great extinction lacks many fossil groups that are well represented in the preceding system, and the younger system contains many new taxonomic groups that evolved to replace those that died out.

Another trait that led early workers to recognize some bodies of rock as systems was the nature of the rocks themselves. Most of the distinctive lithological features of systems relate to the history of life. The Cretaceous System, for example, was designated to include the thickest deposits of chalk in the world. Chalk is soft, fine-grained limestone. The abundance of chalk in the Cretaceous System reflects the fact that during the

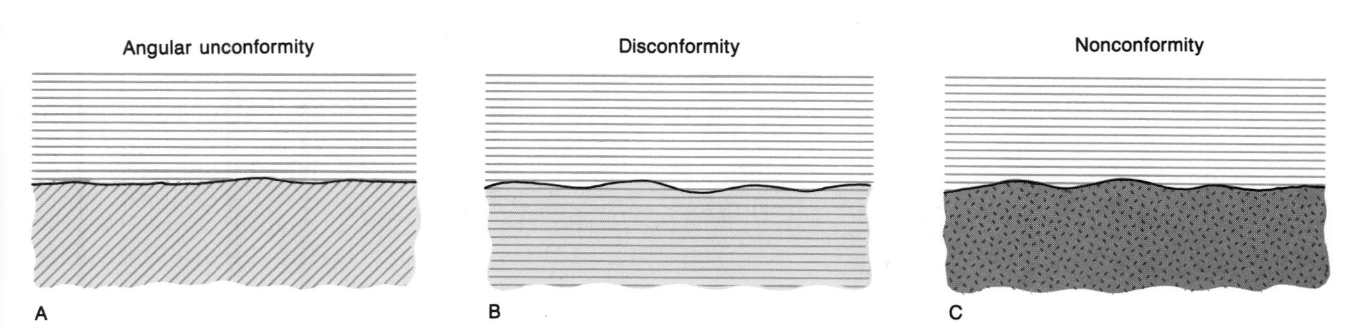

FIGURE 1-21 Diagrams of an angular unconformity, a disconformity, and a nonconformity. An angular unconformity *(A)* separates tilted beds below from flat-lying beds above. A disconformity *(B)* separates flat-lying beds below from other flat-lying beds above, but the upper beds rest upon an erosion surface that developed after the lower beds were deposited. A nonconformity *(C)* separates flat-lying beds from crystalline (igneous or metamorphic) rocks.

Cretaceous Period there was a great proliferation of the kinds of organisms that produce the particles of calcium carbonate that form chalk: small, single-celled organisms that still float in the sea today, but in reduced abundance.

The study of early Paleozoic rocks illustrates how new systems were designated to fill gaps between others that had been recognized earlier. In 1835 Adam Sedgwick and Roderick Murchison presented a joint paper in which they established the Cambrian and Silurian systems, primarily on the basis of geologic studies in Wales. The term *Cambrian* was derived from the Roman name for Wales (Cambria), while *Silurian* was named for the Silures, an ancient tribe that had once inhabited Wales. Murchison defined the Silurian System primarily on the basis of its fossils, and Sedgwick noted that the Cambrian, which is the oldest Phanerozoic system, was less fossiliferous; he recognized, as we do today, that during Cambrian time animal life on Earth was in a primitive stage of development.

In 1879 Charles Lapworth, a Scottish schoolmaster, showed that in many areas of the world the lower part of the Paleozoic interval, which includes the Cambrian and Silurian systems, actually displayed a succession of three distinctive groups of fossils. He suggested that the label *Cambrian System* be retained for the rocks containing the oldest group of fossils and that the label *Silurian System* be applied to the rocks containing the youngest group. For the intervening rocks, with their own distinctive fossils, Lapworth proposed the name *Ordovician,* for the Ordovices, an ancient Welsh tribe that was the last in Britain to submit to Roman domination. Lapworth's threefold division of the lower portion of the Paleozoic interval is still accepted, although today we have a more refined view of the formal boundaries between the three systems that he recognized.

It seems remarkable today that all of the geologic systems of the Phanerozoic Eon were first designated during a brief interval of the nineteenth century in one small region of the world: Great Britain and nearby areas of western Europe. It was a lucky circumstance that modern geology came into being in this particular region. Few other geographic areas of comparable size display such large volumes of sediment and rock representing all of the Phanerozoic systems.

CHAPTER SUMMARY

1 The principle of uniformitarianism, which is fundamental to natural science, asserts that the laws of nature do not vary in the course of time.

2 Actualism is the uniformitarian procedure whereby events and processes of the geologic past are interpreted in light of events and processes observed in the modern world or recreated in the laboratory.

3 A mineral is an inorganic element or compound that is characterized not only by its chemical composition but also by its internal structure. Rocks are aggregates of mineral grains.

4 Igneous rocks form by the cooling of liquid rock; sedimentary rocks are layered rocks that accumulate under the influence of gravity; metamorphic rocks form by the alteration of preexisting rocks at high temperatures and pressures.

5 Living things are grouped into species, which encompass individuals that can or do interbreed; species are grouped into categories called higher taxa.

6 Fossils are remains or tangible evidence of ancient life found within rocks.

7 Rocks break under stress and move along faults; deep within Earth they also bend and flow.

8 Earthquakes and artificial explosions create seismic waves that reveal much about the structure of Earth's interior.

9 Earth's interior is divided into concentric layers. A central core of high density is surrounded by a less dense mantle, which is blanketed by a still less dense crust.

10 The parts of Earth's crust that form continents are thicker and less dense than the parts that lie beneath the oceans.

11 The crust and upper mantle constitute the rigid lithosphere, which is divided into discrete plates that move laterally over a partially molten zone of the mantle.

12 The relative ages of rocks that come into contact with one another can often be determined by the principles of superposition, original horizontality, lateral continuity, intrusive relationships, crosscutting relationships, and components.

13 Changes in life on Earth during the course of geologic time are reflected in the rock record. The fossil succession in the rock record reveals the relative ages of rocks in different regions.

14 The decay of naturally occurring radioactive materials reveals the actual ages (in years) of some rocks.

15 The scale that is employed to divide the rock record into units representing discrete intervals of geologic time was developed in Europe during the nineteenth century.

EXERCISES

1 Give general examples of the use of actualism to interpret ancient rocks.

2 In which of the three basic kinds of rocks do nearly all fossils occur? Why?

3 Name and describe as many modes of fossil preservation as you can.

4 What evidence is there that Earth's outer core is liquid?

5 What is isostasy and how does it explain why mountains have roots?

6 What kinds of features distinguish one system of geologic time from another?

7 Name three different kinds of unconformities. How does each type form?

8 Describe three kinds of relationships between two bodies of rock that indicate which of the two bodies is the younger one.

ADDITIONAL READING

Berry, W. B. N., *Growth of a Prehistoric Time Scale,* Blackwell Scientific Publications, Palo Alto, Calif., 1987.

Decker, R., and B. Decker, *Volcanoes,* W. H. Freeman and Company, New York, 1989.

Faul, H., and C. Faul, *It Began with a Stone,* John Wiley & Sons, Inc., New York, 1983.

Hallam, A., *Great Geological Controversies,* Oxford University Press, Oxford, England, 1989.

Press, F., and R. Siever, *Earth,* W. H. Freeman and Company, New York, 1986.

van Andel, T. H., *New Views on an Old Planet,* Cambridge University Press, Cambridge, England, 1985.

PART

I

THE ENVIRONMENTAL SETTING

Environments at Earth's surface have changed continuously in the course of geologic time. Sedimentary rocks and the fossils they contain reflect the nature of these environments and thus provide clues that allow geologists to unravel the history of environmental change. These clues make it possible to reconstruct both local structures and broad geographic features of the past.

The most fundamental division of environments on Earth — one that is reflected in the distribution of both biological and sedimentary features — is the division between land and sea. Among the nonmarine environments that can be recognized through the study of rocks and fossils are deserts, rivers, lakes, forests, and grasslands; the marine environments include river deltas, lagoons, sandy beaches, reefs built by plants and animals, and deep seafloors. Within both the terrestrial realm and the marine realm, climatic conditions exert strong control over the geographic distribution of plants and animals. Other physical factors, as well as interactions between organisms, determine the distribution of life on a smaller scale.

This alluvial fan in Pakistan is formed of sediment eroded from the Himalaya Mountains and deposited at the mouth of a valley. *(Art Wolfe.)*

CHAPTER

2

Environments and Life

Organisms are able to live only in environments where they can find food, tolerate physical and chemical conditions, and avoid natural enemies. These were requirements for life in the past just as they are requirements for life in the modern world. Climate is the environmental factor that, on a global scale, exerts the most profound control over the distribution of species, influencing conditions not only on land but also in bodies of water. Temperature and other climatic conditions strongly affect the distribution of plant species, and plant species in turn influence the distribution of many animal species. In this chapter, therefore, we will pay close attention to the mechanisms that create the prevailing weather patterns on Earth today. Our major focus, however, will be on the relationships between living things and the environments that they inhabit.

If the planet had undergone little change in the course of geologic time, we could directly apply what we know about life and environments today to ancient fossils and rocks. But Earth has changed dramatically over the course of its history: The planet's materials — including living matter — have changed continually both in composition and in location. Life, of course, has undergone vast evolutionary changes, and many conspicuous physical features of the modern world — including the polar ice caps of Greenland and Antarctica and the ice-cold body of water that now forms the deep ocean — were not present 100 million years ago. Although this situation does not violate the principle of uniformitarianism (p. 2), inasmuch as natural laws are not broken, it does require us to take changing environmental conditions into account when we interpret the rock record. What we learn in this chapter will serve as a starting point for our exploration of environments and life through the eons, beginning with the planet's origins and moving forward until we reach the present.

A savannah is a grassy plain that supports few trees and is generally populated by numerous species of large animals. This savannah, with zebras in the foreground and wildebeest in the distance, is in East Africa. *(Frans Lanting/Minden Pictures.)*

LAND AND SEA

One means of gaining an understanding of environments on Earth is to examine the configuration of the planet's surface. You will recall that Earth's crust is divided into the thin, dense oceanic crust and the thicker, less dense continental crust — a distinction that accounts for Earth's external shape, with continental surfaces standing above the seafloor. A **hypsometric curve** illustrates what proportions of Earth's surface lie at various altitudes above and depths below sea level (Figure 2-1).

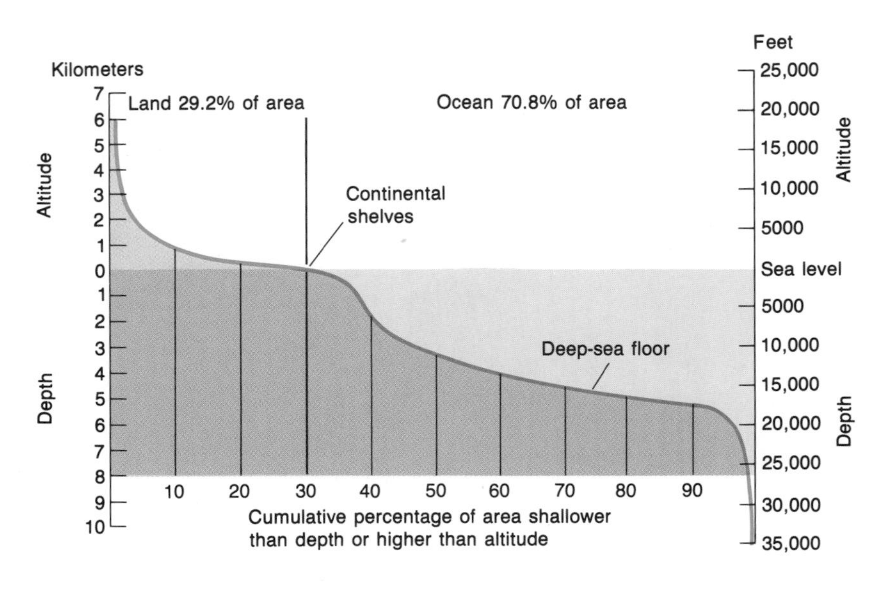

FIGURE 2-1 The hypsometric curve, which represents the surface of Earth. The curve shows the relative amounts of land and seafloor that lie at various distances above and below sea level. The plot is cumulative, depicting the total percentage of land that lies above each depth or altitude. About 70 percent of Earth's surface lies below sea level, and most of this area forms the deep seafloor. Continental shelves are borders of continents flooded by shallow seas. The left side of the diagram shows that mountains account for relatively little of Earth's surface.

Those environments on or close to Earth's surface that are inhabited by life are called **habitats.** Nearly all habitats can be classified as terrestrial or aquatic. Aquatic habitats are further divided into freshwater habitats (e.g., lakes, rivers, and streams) and marine habitats (e.g., those within oceans and seas). Birds, bats, and insects use the atmosphere above the ground as a part-time habitat, but virtually all of these creatures also conduct some activities, such as feeding, sleeping, and reproduction, at the surface. Later we will examine the nature of the various habitats on Earth together with the forms of life that occupy them. First, however, we will examine some of the principles that govern the distribution of species within habitats in general.

PRINCIPLES OF ECOLOGY

Ecology is the study of the factors that govern the distribution and abundance of organisms in natural environments. Some of these factors are conditions of the physical environment, and others are modes of interaction between species.

A Species' Position in Its Environment

The way a species relates to its environment defines its **ecologic niche.** The niche requirements of a species include particular nutrients or food resources and particular physical and chemical conditions. Some species have much broader niches than others. Before human interference, for example, the species that includes grizzlies and brown bears ranged over most of Europe, Asia, and western North America, eating everything from deer and rodents to fishes, insects, and berries. The sloth bear, in contrast, has a narrow niche. It is restricted to Southeast Asia, feeding mainly on insects, for which its peglike teeth are specialized, and on fruits. The ecologic niches of many other closely related species present similar contrasts.

We speak of the way a species lives within its niche as a **life habit.** A species' life habit is its mode of life—the way it obtains nutrients or food, the way it reproduces, and the way it stations itself within the environment or moves about.

Every species is restricted in its natural occurrence by certain environmental conditions. Among the most important of these **limiting factors** are physical and chemical conditions. Most ferns, for example, live only under moist conditions, whereas cactuses require dry habitats. The salt content of water is an important limiting factor for species that live in the ocean. Few starfishes and sea urchins, for example, can live in lagoons or bays where normal ocean water is diluted by fresh water from rivers.

Almost every species shares part of its environment with other species. Thus, for many species, **competition** with other species—or the process in which two or more species vie for an

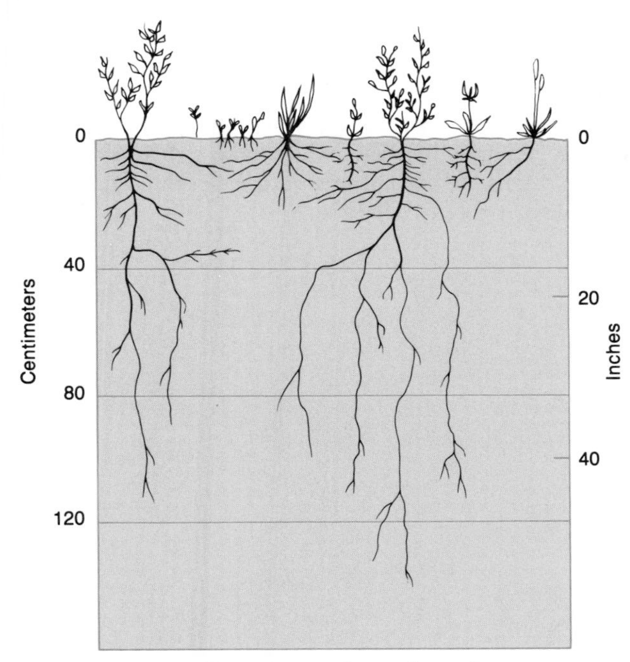

FIGURE 2-2 Differences in the niches of coexisting species of plants. The roots of different species occupy different depth zones of the soil and thus avoid competing for water and nutrients. *(After H. Walter, Vegetation of the Earth in Relation to Climate and Eco-Physiological Conditions, Springer-Verlag, Stuttgart, 1973.)*

environmental resource that is in limited supply—is a limiting factor as well. Among the resources for which species commonly compete are food and living space. Often two species that live in similar ways cannot coexist in an environment because one species competes more effectively, thereby excluding the other. Plants that grow in soil, for example, often compete for water and nutrients in that soil; as a result, plant species living close together often have roots that penetrate the soil to different depths (Figure 2-2).

Predation, or the eating of one species by another, is another limiting factor. An especially effective predator can prevent another species from occupying a habitat altogether.

Communities of Organisms

Populations of several species living together in a habitat form an **ecologic community.** In most ecologic communities, some species feed on others. The foundation of such systems consists of organisms called **producers,** which are plants or plantlike organisms that manufacture their own food from raw materials in the environment. In contrast, animals and animal-like organisms, known as **consumers,** feed on other organisms. Consumers that feed on producers are known as **herbivores,** and consumers that feed on other consumers are known as **carnivores.** Terrestrial herbivores include such diverse groups as rabbits, cows, pigeons, garden slugs, and leaf-chewing insects. Terrestrial carnivores include weasels, foxes, lions, and ladybugs.

The organisms of a community and the physical environment they occupy constitute an **ecosystem.** Ecosystems come in all sizes, and some encompass many communities. Earth and all the forms of life that inhabit it represent an ecosystem, but so does a tiny droplet of water that is inhabited by only a few microscopic organisms. Obviously, then, large ecosystems can be divided into many smaller ecosystems, and the size of the ecosystem that is treated in a particular ecologic study depends on the type of research that is being conducted. The animals of an ecosystem are collectively referred to as a **fauna** and the plants as a **flora.** A flora and a fauna living together constitute a **biota.**

One of the most important attributes of an ecosystem is the flow of energy and materials through it. When herbivores eat plants, they incorporate into their own tissue part of the food that these plants have synthesized. Carnivores assimilate the tissue of herbivores in much the same way. In most ecosystems, carnivores that eat herbivores are eaten in turn by other carnivores; in fact, several levels of carnivores are often present in an ecosystem. An entire sequence of this kind, from producer to **top carnivore,** constitutes a **food chain.** Because most carnivores feed on animals smaller than themselves, the body sizes of carnivores often increase toward the top of a food chain (Figure 2-3).

Simple food chains—sequences in which a single species occupies each level—are uncommon in nature. Most ecosystems are characterized by **food webs,** in which several species occupy each level. Most species below the top carnivore level serve as food for more than one consumer species. Similarly, most consumer species feed on more than one kind of prey.

Parasites and scavengers add further complexity to ecosystems. **Parasites** are organisms that

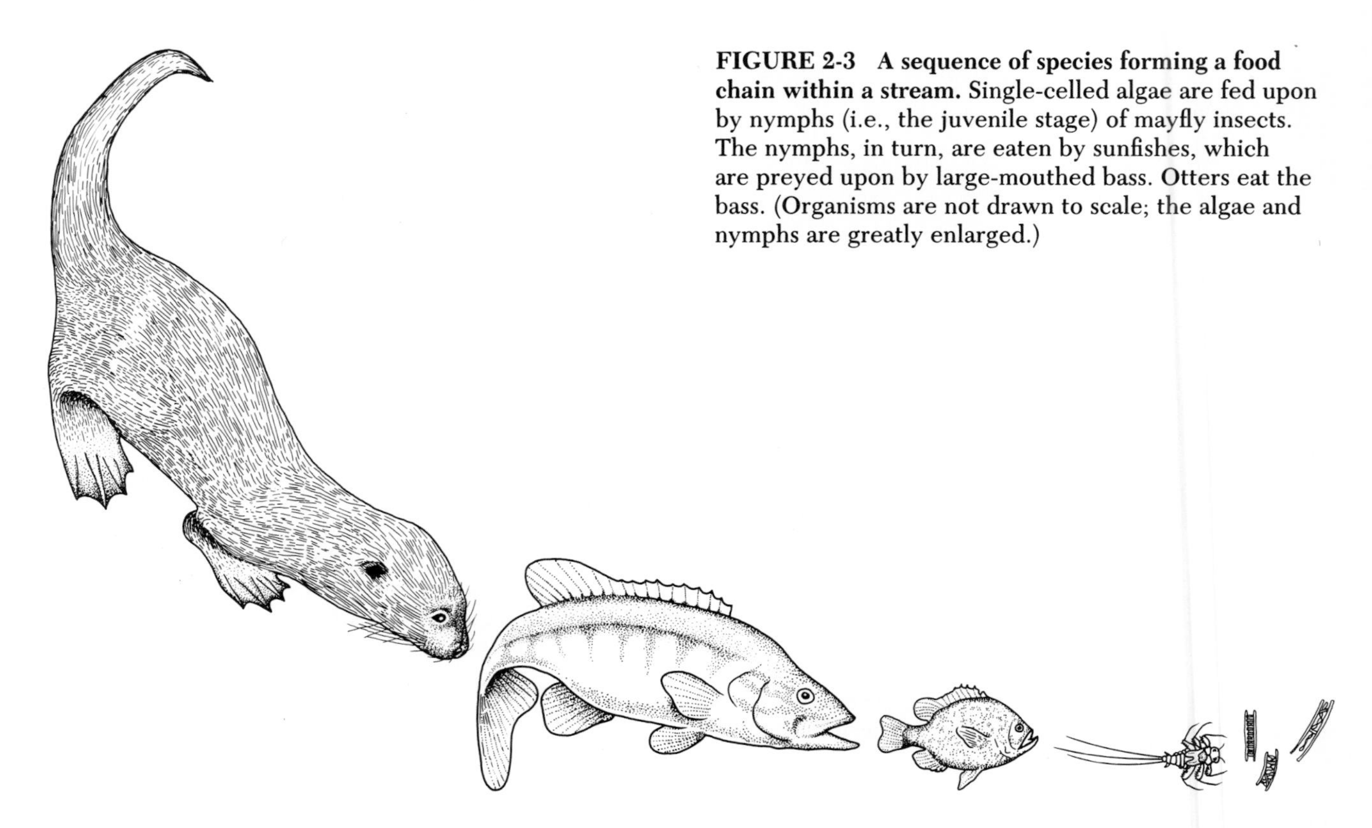

FIGURE 2-3 A sequence of species forming a food chain within a stream. Single-celled algae are fed upon by nymphs (i.e., the juvenile stage) of mayfly insects. The nymphs, in turn, are eaten by sunfishes, which are preyed upon by large-mouthed bass. Otters eat the bass. (Organisms are not drawn to scale; the algae and nymphs are greatly enlarged.)

derive nutrition from others without killing their victims, and **scavengers** feed on organisms that are already dead. A flea that feeds on the blood of a dog, for example, is a parasite, as is a tapeworm that lives within a human. A vulture, in contrast, is a scavenger, as is a maggot, which feeds on dead flesh.

Although material flows from one level of a food web to the next, it does not stop at the highest level. In fact, materials are cycled through the ecosystem continuously, with single-celled bacteria completing the cycle (Figure 2-4). Some of these bacteria decompose dead animals and plants of all types into simple chemical compounds, while others transform decomposed material, liberating nutrients to be reused by plants.

The term **diversity** is used to describe the number of species that live together within a community. Diversity can be measured in several ways, some of which take into account the relative abundance of species, but the simplest measure of diversity is nothing more than a count of the number of species present.

Diversity is normally low in habitats that present physical difficulties for life. Because plants require water to make food, for example, deserts contain fewer species of plants than do moist tropical forests. Only a few types of plants, such as cactuses, can sustain themselves in desert environments.

Predation is another factor that influences the diversity of a community. When a predator disappears from a region, a species that has served as its prey may become much more abundant. A predator can even eliminate a species from a community. Conversely, by eliminating a species' competitors, a predator can allow the species to survive in a community.

Certain types of physical disturbances can also prevent strongly competitive species from excluding less competitive species. Storm waves, for example, may tear animals and plants from rocky shores, leaving bare surfaces for the invasion of species that are weak competitors. Species that specialize in invading newly vacated habitats—land cleared by fire, say, or new shore areas formed along rivers that change course at flood stage—are aptly called **opportunistic species.** Populations of opportunistic species seldom survive for long in the face of better competitors, however. Because opportunistic species tend to be good invaders, while some of their populations are disappearing from one area, others are becoming established elsewhere. The plants that we

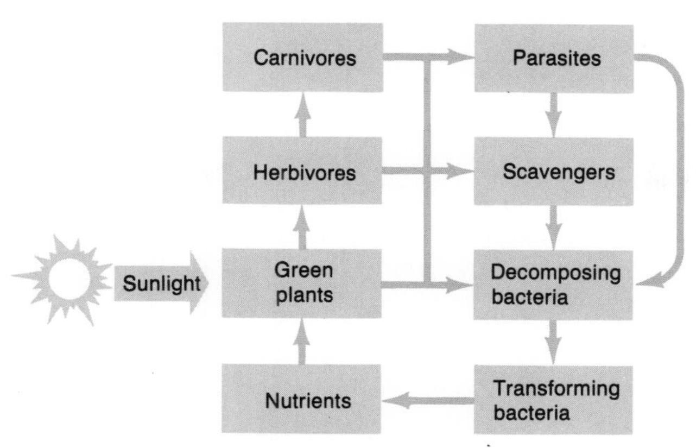

FIGURE 2-4 The cycle of materials through an ecosystem.

call weeds are opportunistic species par excellence. In gardens and lawns, many types of weeds come and go in the course of a few seasons.

Biogeography

The distribution and abundance of organisms on a broad geographic scale are studied within the field known as **biogeography.** The limiting factor that plays the largest role here is temperature. Many species are restricted to polar regions, others to tropical regions near the equator, and still others to temperate regions in between.

Among most large, widespread taxonomic groups of animals and plants, more species are adapted to the tropics than to cold climates. In addition, tropical communities are, on the average, more diverse than communities situated at high latitudes. In general, species increase in number toward the equator, where many more kinds of animals and plants can live than the number that can survive the harsher conditions at higher latitudes.

Temperature, however, is not the only control of biogeographic patterns of occurrence, as evidenced by the fact that most species are not found in all the habitats that meet their particular ecologic requirements. Dispersal of most species is also restricted by barriers, the most obvious of which are land barriers for aquatic forms of life and water barriers for terrestrial forms of life. Of course, these barriers change with time, and the geographic distributions of species shift accordingly. *Mammuthus*, the genus that includes the extinct members of the elephant family known as mammoths, evolved in Africa about 5 million years ago, during the early part of the Pliocene Epoch. Blocked by northern oceans, mammoths were unable to migrate to North America until the Pleistocene Epoch. During the Pleistocene Epoch, however, large volumes of water were locked up on the land as glaciers (or ice sheets), and sea level fell throughout the world. Consequently, a land bridge emerged between Siberia and Alaska, allowing mammoths to invade the Americas, where they survived until several thousand years ago (Figure 2-5).

The survival of mammoths in the Americas represents an interesting biogeographic phenomenon—the development of a **relict distribution,** or the presence of a taxonomic group in one or two locations after it has died out elsewhere. By late in the Pleistocene Epoch, mammoths had disappeared from most of the area they had previously occupied and remained in only a relatively small area of North America.

THE ATMOSPHERE

The **atmosphere,** or the envelope of gases that surrounds Earth, serves life primarily as a reservoir of chemical compounds that are used within living systems. The atmosphere has no outer boundaries; it merges gradually into interplanetary space. The dense part of the atmosphere, however, forms a thin envelope around Earth, so that more than 97 percent of the mass of the atmosphere lies within 30 kilometers (~19 miles) of Earth's surface. (To place this figure in perspective, note that it resembles the average thickness of Earth's continental crust.)

Chemical Composition of the Atmosphere

Nitrogen is the most abundant chemical component of the atmosphere, making up about 78 percent of the total volume of atmospheric gas. It is an important component of proteins, which stimulate chemical reactions and serve as important building blocks in all living things. Second in abundance is oxygen, which forms about 21 percent of the volume of the atmosphere. Both

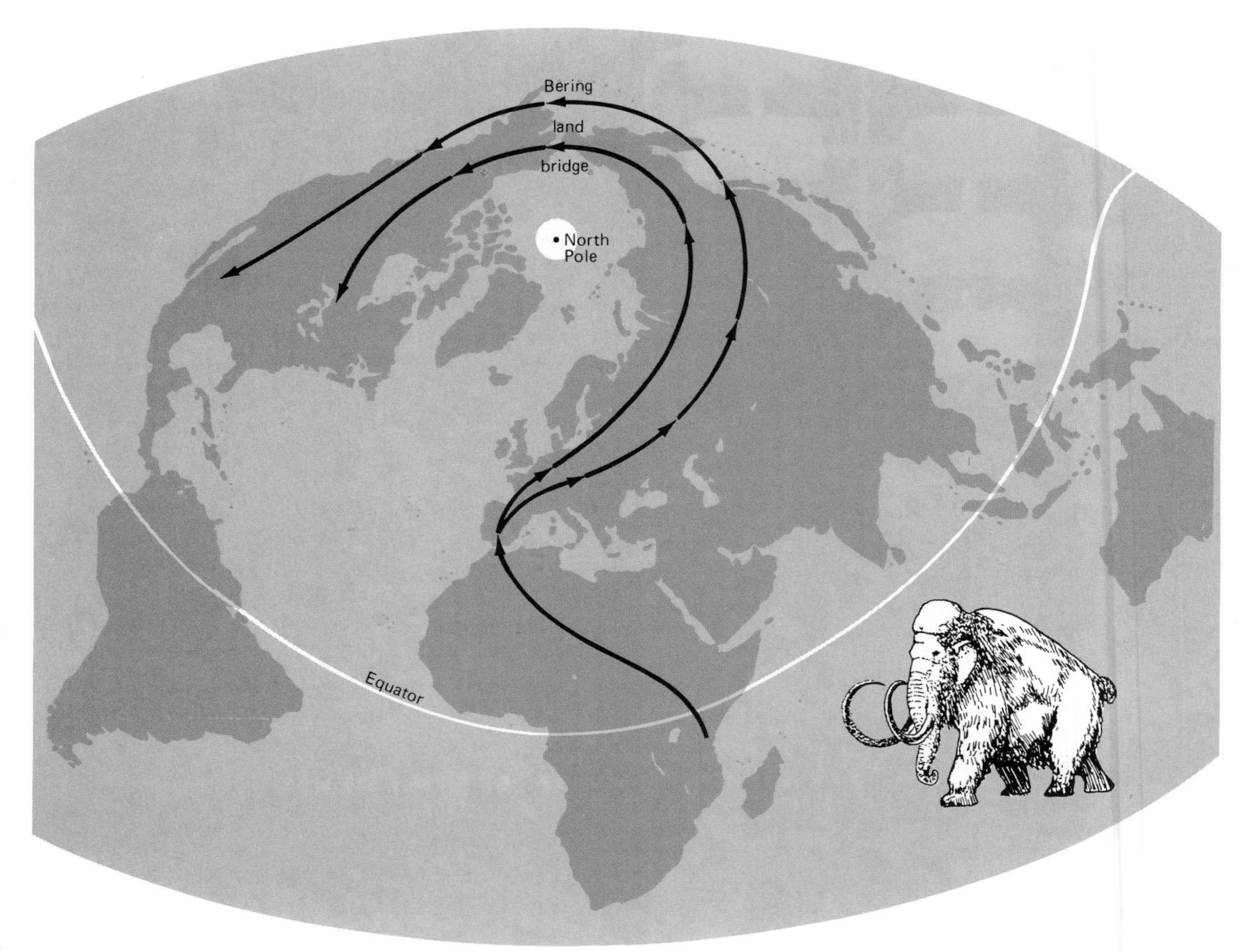

FIGURE 2-5 Changes in the geographic distribution of mammoths. Mammoths evolved in Africa during the Pliocene Epoch and spread to Eurasia. Later, during the Pleistocene interval, when sea level was lowered, they crossed the Bering land bridge, which emerged between Eurasia and North America. *(After V. J. Maglio, Amer. Philos. Soc. Trans. 63:1–149, 1986.)*

nitrogen and oxygen exist in the atmosphere as molecules that consist of two atoms (N_2 and O_2); this is why we speak of them as compounds rather than as elements.

Atmospheric nitrogen and oxygen are maintained at consistent levels by being cycled through living organisms continuously and returned to the air. Figure 2-6 illustrates the **global oxygen cycle.** Most oxygen enters the atmosphere from plants, which produce it through **photosynthesis**—the process by which water and carbon dioxide are converted to sugar and oxygen. The green compound chlorophyll acts as a catalyst in this process, while the energy that is necessary for the conversion is derived from the sun. A smaller amount of oxygen comes from the upper atmosphere, where sunlight breaks down water vapor (H_2O) into oxygen and hydrogen.

Carbon dioxide (CO_2), from which plants produce oxygen, is contributed to the atmosphere by animal respiration and by the burning of fossil fuels. It forms only about $\frac{3}{100}$ of 1 percent of the atmospheric volume, but in recent history its abundance has been much more variable than that

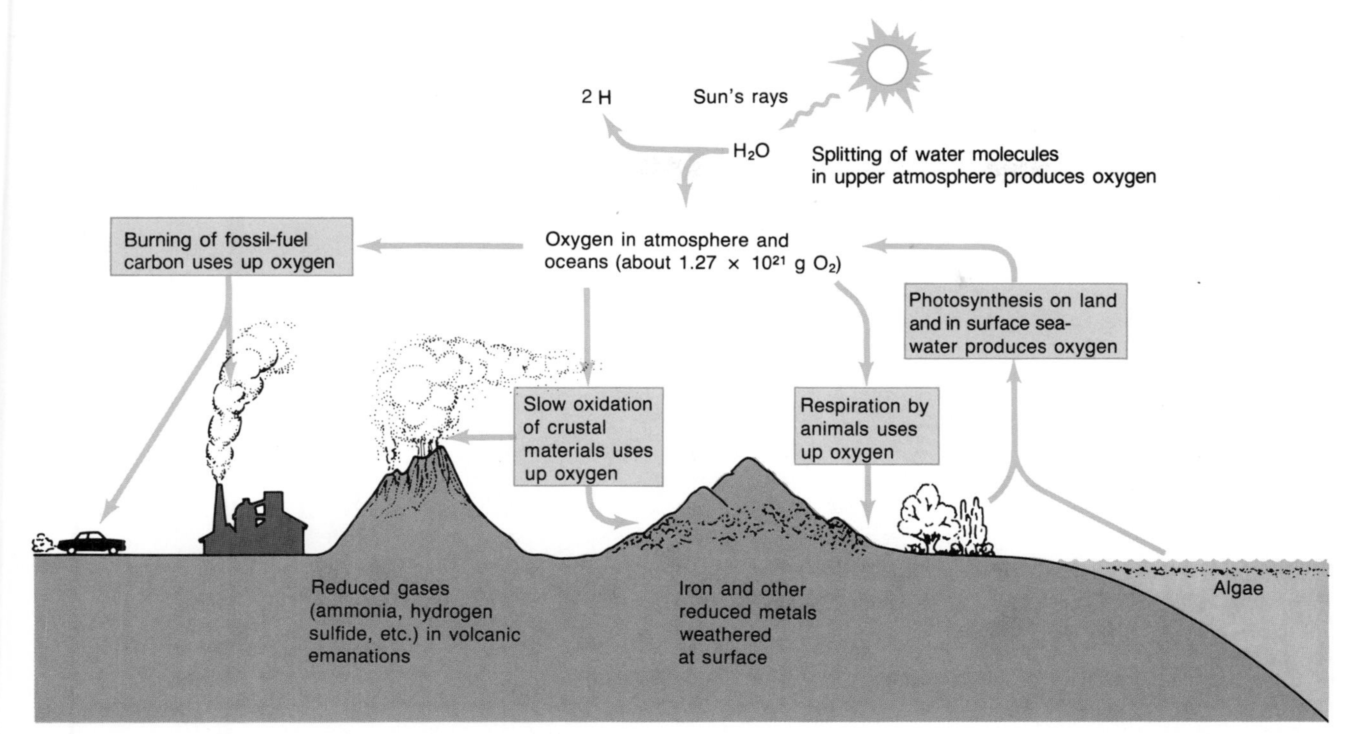

FIGURE 2-6 The cycle of oxygen through the atmosphere. The most rapid input of oxygen is from the photosynthesis of producers on the land and in the sea. Sunlight decomposes water in the upper atmosphere to produce oxygen at a lower rate. *(After F. Press and R. Siever, Earth, W. H. Freeman and Company, New York, 1986.)*

of nitrogen and oxygen. Since 1900, the burning of coal, oil, and wood has increased the volume of carbon dioxide in the atmosphere by about 10 percent. If this buildup continues, it will soon cause Earth's atmosphere to warm up through the so-called **greenhouse effect.** This term refers to the manner in which a greenhouse traps heat. Solar radiation of short wavelength passes through the glass of a greenhouse and warms the soil within it. The soil then radiates heat of long wavelength, which is partly trapped by the glass. This is the same mechanism by which the interior of an automobile heats up on a sunny day. Carbon dioxide in the atmosphere acts in the same manner as the glass of a greenhouse, allowing solar radiation to pass to Earth's surface and then preventing much of the resulting heat from escaping from the lower atmosphere. If carbon dioxide continues to accumulate in the atmosphere as a result of the burning of fossil fuels, global temperatures will eventually increase, and local climates will then change in ways that will greatly affect the patterns of life on Earth.

Temperatures and Circulation in the Atmosphere

Both the atmosphere and the ocean consist of fluid that is in constant motion in relation to the earth beneath them, and most of the energy that produces this motion comes ultimately from the sun. Thus solar radiation is responsible for much of what takes place in the atmosphere and in the ocean.

When solar energy reaches Earth, a good deal of it is absorbed and turned into heat energy. The amount of solar radiation that is absorbed at Earth's surface varies from place to place according to the percentage of solar radiation reflected from the surface, a factor known as the **albedo.** Where sunlight strikes the surface at a low angle, so that a large percentage is reflected, it generates relatively little heat. Sunlight also generates less heat when it strikes ice than when it strikes water, soil, or vegetation, because ice reflects more radiation. The albedo ranges from 6 to 10 percent for the ocean; from 5 to 30 percent for forests, grassy

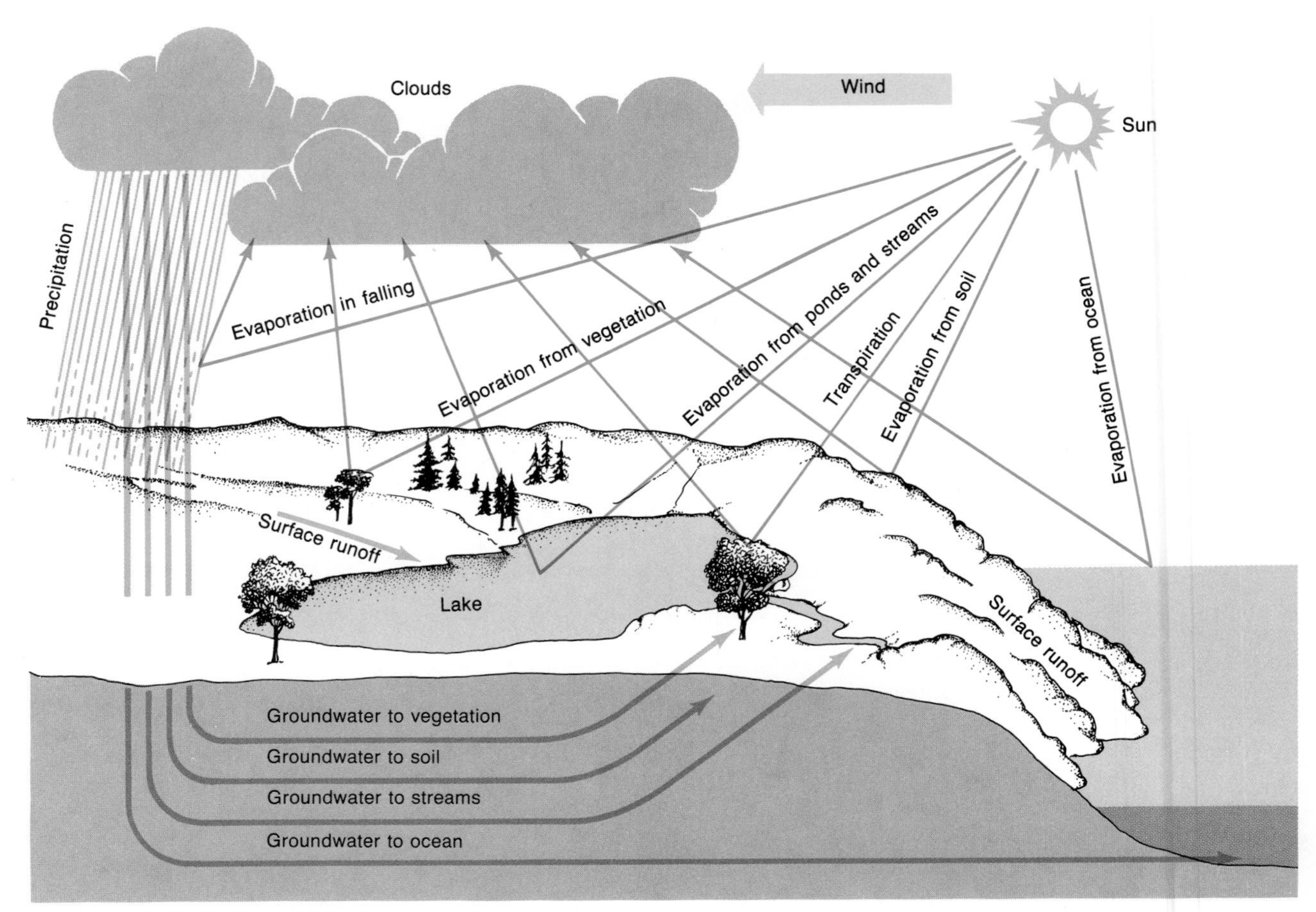

FIGURE 2-7 The global hydrological cycle.

surfaces, or bare soil; and from 45 to 95 percent for ice or snow.

Heat from the sun causes water at Earth's surface to evaporate and to enter the atmosphere as water vapor (Figure 2-7). Under the cool conditions that characterize high altitudes, water vapor condenses to form clouds. Precipitation from clouds carries water back to Earth, where it evaporates immediately or after moving through the ground or flowing through streams and rivers. Some water reaches the atmosphere again by **transpiration,** or emission from plants that have collected the water through their roots. Evaporation from these sites of accumulation completes the **global hydrological cycle,** or the cycle of water movement.

The movement of water and water vapor through the hydrological cycle involves only a few of the many types of fluid movements driven by solar energy near Earth's surface. Oceans and lakes are warmed by the absorption of solar radiation, and the atmosphere is warmed primarily by heat that rises from the land. This warming produces fluid movement.

Much of the transfer of heat from place to place in the ocean or atmosphere occurs by **convection,** which results from the fact that a liquid or gas is less dense when it is warm than when it is cool. Thus, if a kettle of water is heated from below, the warmed water near the bottom rises through the cooler water above; in other words, the water in the kettle "turns over," and continues to do so as cooler water sinks and is warmed. Similarly, warm water in a glass with floating ice cools at the top and undergoes convective turnover (Figure 2-8).

Convection operating in conjunction with other forces produces major movements within Earth's atmosphere. Because sunlight strikes Earth's surface at a low angle near the poles, the

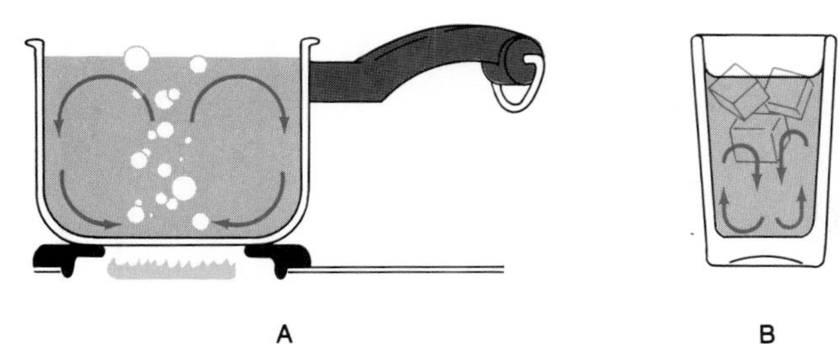

FIGURE 2-8 Convection in liquid. *A.* Water in a pan heated from below turns over as the warm water rises from the bottom and the cooler water at the top descends. *B.* Cooling of water by ice at the top of a glass results in a similar convective motion.

poles are much cooler than the equatorial region, where the sun's rays impinge on the surface more directly. The atmosphere also becomes cooler with altitude above Earth's surface, from which most atmospheric heat is derived.

To understand the actual pattern of atmospheric motion, let us consider what might happen if Earth did not rotate but were heated evenly on all sides by a sun that rotated in the plane of Earth's equator. This imaginary system would produce the pattern of atmospheric convection shown in Figure 2-9: Warm air rises near the equator while cool air sinks near the poles. The real system, in which Earth rotates, produces a much more complicated pattern of atmospheric circulation (Figure 2-10). To understand this pattern, it is necessary to take into account the **Coriolis effect,** which results from Earth's rotation. As Earth rotates, air currents are deflected clockwise in the Northern Hemisphere and counterclockwise in the Southern Hemisphere.

The presence of the Coriolis effect prevents Earth's atmosphere from circulating in the simple pattern shown in Figure 2-9. To understand why, consider what happens near the equator. Although air in this region rises from the warm surface of Earth and spreads in a general way to the north and south, the air that spreads poleward from the equator at high elevations does not move *directly* to the north and south because the Coriolis effect deflects it toward the east. As a result, the air piles up north and south of the equator more rapidly than it can escape toward the poles, producing belts of high atmospheric pressure between about 20 and 30° north and south of the equator. This high pressure bears down on Earth's

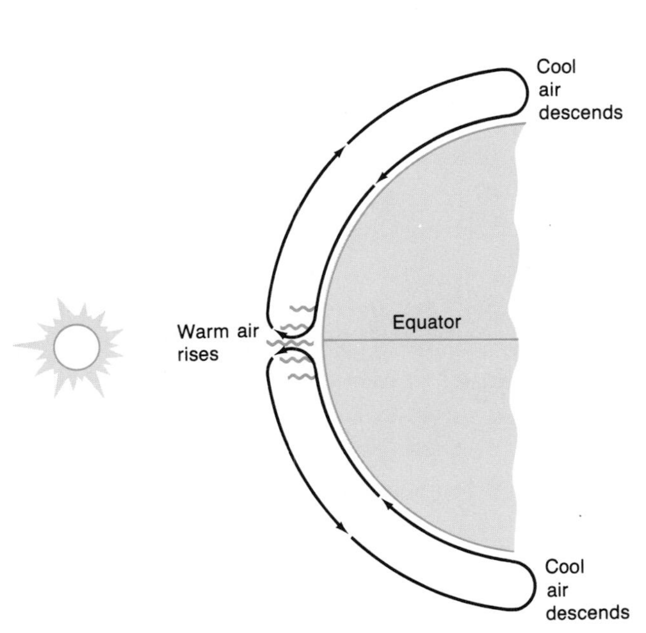

FIGURE 2-9 Circulation of atmosphere around an imaginary Earth that receives equal amounts of sunlight on all sides from a source rotating in the plane of the equator. One convection cell occupies the Northern Hemisphere and another the Southern Hemisphere.

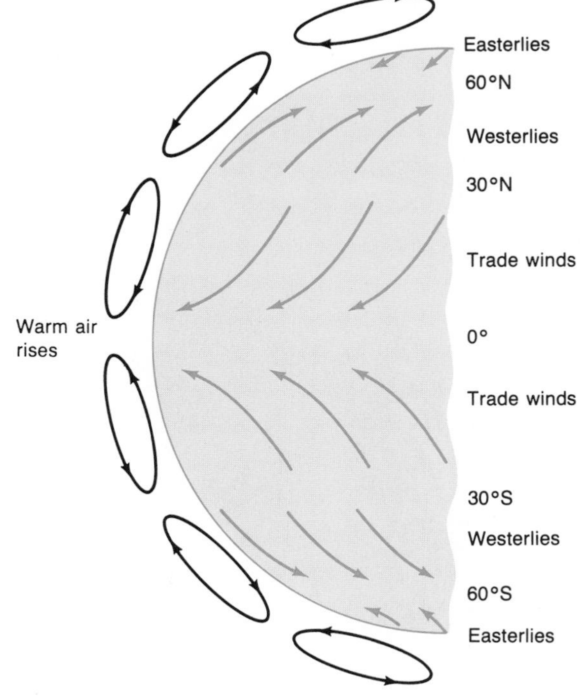

FIGURE 2-10 The major gyres of Earth's atmosphere. The lower segments of these gyres represent the major wind systems, labeled at the right.

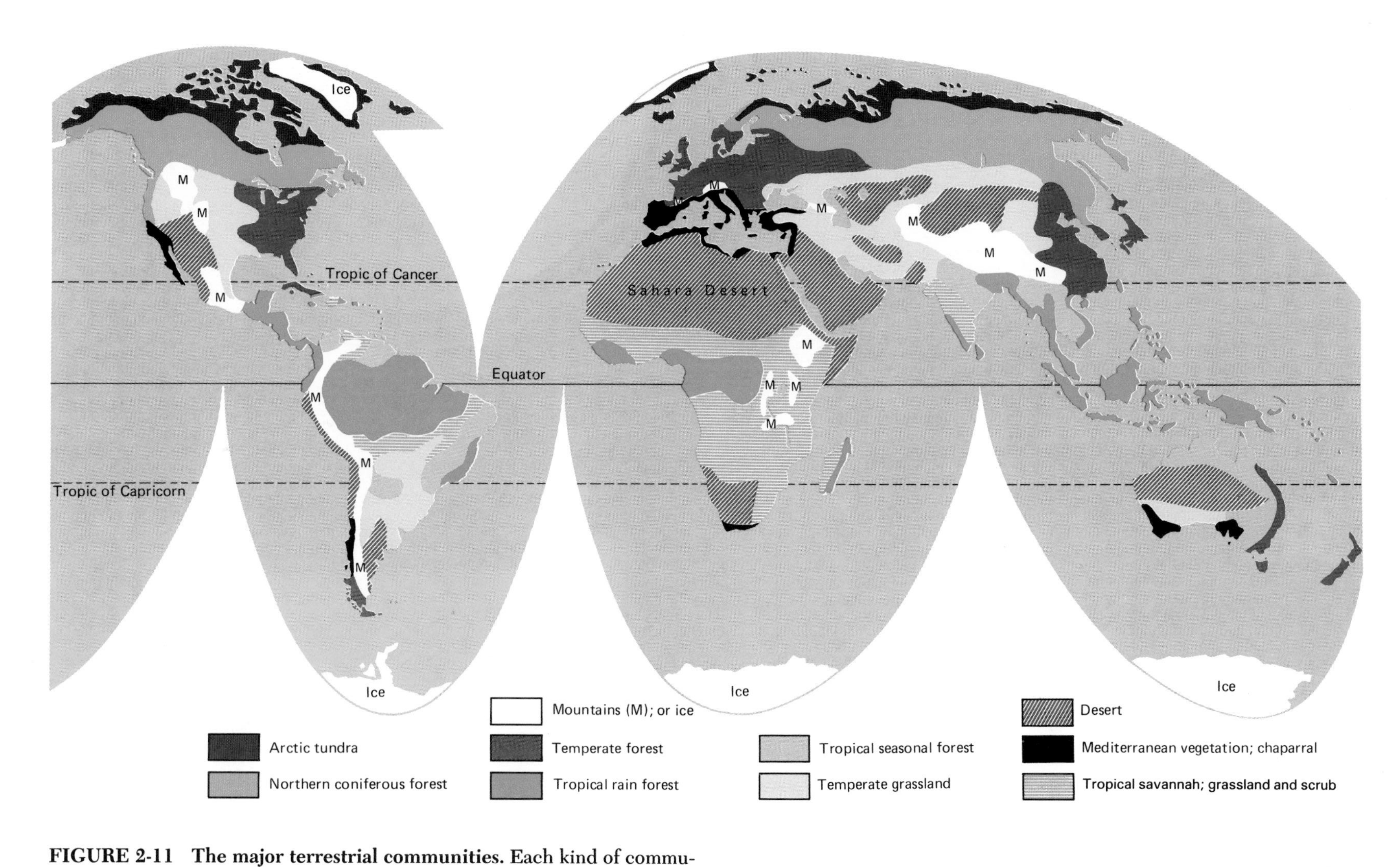

FIGURE 2-11 The major terrestrial communities. Each kind of community is characterized by a particular association of plants adapted to particular climatic conditions. *(After C. B. Cox and P. D. Moore, Biogeography: An Ecological and Evolutionary Approach, John Wiley & Sons, Inc., New York, 1980.)*

surface, pushing winds at the surface toward the north and south. These winds are also deflected by the Coriolis effect. Thus the surface winds that flow toward the equator, known as **trade winds** (Figure 2-10), move diagonally westward. Note that the trade winds replace the warm air that rises at the equator, completing a cycle, or **gyre,** of airflow on either side of the equator.

Another gyre is positioned poleward of the trade winds. Here **westerlies** flow toward the northeast. Like the trades, these winds originate from the high-pressure system where air piles up between 20 and 30° from the equator. Still farther from the equator in each hemisphere is yet another gyre, which originates where cool air descends near the pole to produce **easterlies.** The section that follows explains how major air movements in the atmosphere influence the distribution of climates and organisms.

THE TERRESTRIAL REALM

At the present time, the continents of the world stand at relatively high elevations above sea level (Figure 2-1). Thus the expanses of land are broader today than they were during most of Phanerozoic time. Another unusual feature of the modern world is a steep temperature gradient between each pole and the equator. Whereas tropical conditions prevail near the equator, the average summer temperature near the north and south poles is well below freezing. Since climatic conditions have a profound effect on the distribution of organisms on the land, climates will be our first consideration in discussing the distribution of life on the broad continental surfaces of the modern world.

Climates and Vegetation

It is remarkable how closely the distribution of terrestrial vegetation corresponds to the geographic pattern of climates. This correspondence, coupled with the fact that plants are the dominant producers of the food web and thus strongly affect the distribution and abundance of animals, makes climate an especially significant factor in terrestrial ecology. Plants, in fact, serve not only as sources of food but also as habitats for many animals; numerous insects, for example, spend their entire lives on certain types of trees, and insects account for most species of organisms living in the world today.

Terrestrial climates near the equator are not only very warm but often very moist as well. When air in this region rises after being heated at Earth's surface, it cools (Figure 2-10); because cool air cannot hold as much water vapor as warm air, it loses moisture in the form of rain. In South America and Africa, the only large continents with equatorial regions (Figure 2-11), the warm, moist conditions that characterize **tropical rain forests** allow so many kinds of plants to thrive (Figure 2-12) that they form what are informally called jungles. These plants provide food and shelter for a wide variety of animals.

Tropical climates—those in which the average air temperature ranges from 18 to 20°C (~64 to 68°F) or higher—are usually found at latitudes within 30° of the equator. You will recall that between 20 and 30° north and south of the equator, the air that rises from the equator piles up and, after cooling, descends to form trade winds (Figure 2-10). The air of these winds is dry, having dropped much of its moisture at high altitudes as rain. As a result, the trade winds drop little rain; instead, they pick up moisture from the surface of Earth, leaving **deserts** on broad continental areas that lie about 30° north and south of the equator. The Sahara is the largest of these deserts, but

FIGURE 2-12 A tropical rainforest in Peru. *(Andre Bärtschi.)*

FIGURE 2-13 The Sonora Desert of Arizona. *(Peter Kresan.)*

broad deserts also occupy southern Africa, central Australia, and southwestern North America (Figure 2-13).

Most deserts receive less than 25 centimeters (~10 inches) of rain per year. Because only a few types of plants can live under such conditions, the desert environment is characterized by sand and bare rock rather than by dense vegetation (Figure 2-13). Some desert plants, such as cactuses, have the capacity to store water, and nearly all have small leaves so that a minimum of water is lost by evaporation. Few species of animals can survive in the desert environment, and a large percentage of those that do are nocturnal; many are small rodents that find refuge from the hot sun by remaining in burrows during the day.

Deserts also form at latitudes where moisture is more plentiful, but only where regional conditions minimize the moisture in the air. The Sierra Nevada near the west coast of the United States, for example, leaves what is known as a **rain shadow** to their east. Winds rising from the west over the Sierra Nevada cool and lose their moisture rapidly, so even the highest elevations of the mountain range receive less rain than the lower western slopes. Even less rain is shed on the eastern side, where the air heats up as it descends. There, in the rain shadow, lies the Great Basin, a desert that includes most of the state of Nevada. Deserts also exist at high latitudes in central Asia, but only in regions so far from oceans that they receive little moisture (Figure 2-11).

Savannahs and **grasslands** form in areas where rainfall is sufficient for grasses to thrive, but not sufficient to allow the growth of forests. Many savannahs and grasslands are, in fact, positioned between dry deserts and wet woodlands. The Great Plains of the United States is one such region; others are found in Africa, where the savannahs are noted for their populations of large animals

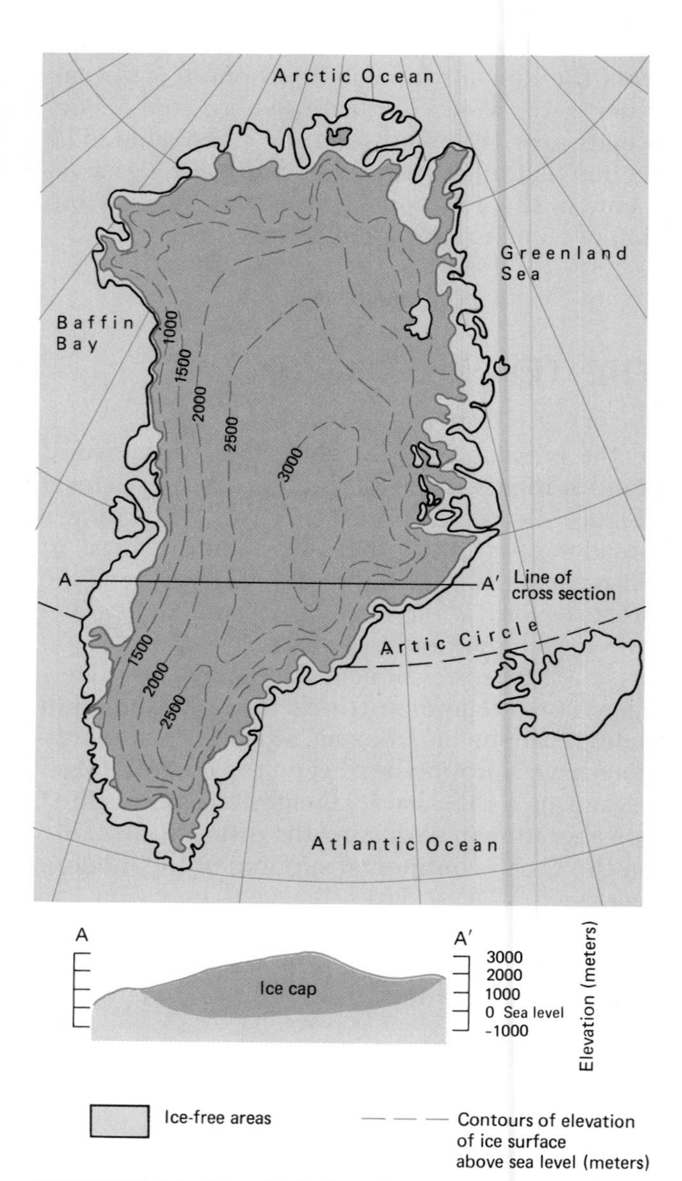

FIGURE 2-14 The thick glacial ice cap of Greenland, which depresses the continental crust. *(After F. Press and R. Siever, Earth, W. H. Freeman and Company, New York, 1986.)*

(p. 24). Most of the herbivores found in savannahs and grasslands — including bisons, antelopes, zebras, and wildebeests — are relatively large animals that graze on grasses and have enough stamina to flee from carnivores such as jackals, lions, and cheetahs. The majority of carnivores in these regions are also large animals that have the ability to capture the large grazers. Many savannahs and grasslands also support scattered trees, and these habitats intergrade with open woodlands.

In sharp contrast to the rich biotas of tropical rain forests, warm savannahs, and woodlands are the meager biotas located near the north and south poles. Today large ice caps cover Greenland and Antarctica, the two large continents of polar regions. Such ice caps have been absent from Earth in past times, when no large continents have occupied polar regions or when the polar regions have been warmer. The ice caps of Greenland and Antarctica are now so heavy that they actually depress the continental crust (Figure 2-14). These ice caps are continental glaciers.

Glaciers are among the most impressive physical structures on Earth. They are not simply masses of ice; they are masses of ice in motion (Figure 2-15). Glaciers form from snow that accumulates until it is so thick that the pressure of its weight recrystallizes the individual flakes into a solid mass. Glaciers slide slowly downhill, but they also spread over horizontal surfaces because of internal deformation. This movement resembles the flowing of any solid material as it nears its melting point. Glaciers form not only at high latitudes but also at high elevations near the equator, where the atmosphere and surface of Earth are cool. They occupy mountain valleys, through which they flow downhill. When they encounter warmer temperatures closer to sea level, they usually melt, though near the poles they may reach the sea before they can do so. Large chunks of ice then break from their terminal portions and float off in the form of icebergs.

Nearly all of Antarctica is covered by ice, but no other large continent in the Southern Hemisphere has large areas that experience very cold conditions year round. In the Northern Hemisphere and in many tropical areas above the tree line, however, there is a type of subarctic ecosystem known as **tundra.** Tundra exists in areas where a layer of soil beneath the surface remains frozen even though air temperatures rise above freezing during the summer. Under these conditions, water is never available in abundance. The dominant plants in tundras are not grasses and tall trees but rather plants that need little moisture, such as mosses, sedges, lichens (associations of algae and fungi), and low-growing trees and shrubs. A broad belt of tundra stretches across the northern margins of North America and Eurasia, supporting a low diversity of animal life (Figure 2-16). Rodents and snowshoe hares are present in tundras today,

FIGURE 2-15 The Hubbard glacier of Alaska. A large piece of the glacier is plunging into the water. Pieces that broke off earlier float in the foreground as icebergs. *(Tom Bean.)*

FIGURE 2-16 A caribou grazing in North American tundra. *(Michael Francis/Wildlife Collection.)*

and the dominant herbivores of larger size are the caribou, reindeer, and musk ox. Foxes and wolves are the primary hunters.

South of the tundra in the Northern Hemisphere, in areas where moisture is sufficient, forests rather than deserts or grasslands are found. The cold regions adjacent to tundra are cloaked in **evergreen coniferous forests,** dominated by such trees as spruce, pine, and fir. These trees are successful in areas with short summers because of their ability to conduct photosynthesis (i.e., to make food) year round. The diversity of animals in cold evergreen forests is not so great. To the south, in slightly warmer climates with longer summers, **temperate forests** replace evergreen forests. Some evergreen trees can be found in such forests, but deciduous trees such as maples, oaks, and beeches are usually present in greater abundance. Ground animals are more diverse in temperate forests than in cold evergreen forests, and birds are especially well represented.

Mediterranean climates, which are characterized by dry summers and wet winters, often prevail along coasts that lie about 40° from the equator. During the summer, the land in these areas is warmer than the ocean, so that moist air coming off the ocean is warmed over the land and retains its moisture. In the winter the land is cooler than the ocean, so that moist sea air cools over the land and drops its moisture. This type of climate characterizes much of California and southeastern Australia as well as the Mediterranean region. Mediterranean climates support chaparral vegetation, which consists primarily of shrubby plants with waxy leaves that retain moisture during summer droughts. Such climates have attracted large human populations, which have altered them greatly by decimating the native biotas.

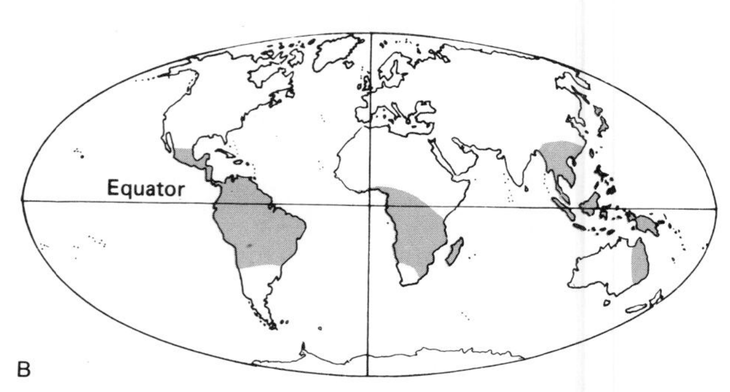

FIGURE 2-17 The cycad, a plant that was especially common during the Mesozoic Era. *A.* A living cycad. Today few cycads are found outside the tropics, and it appears that ancient cycads were also restricted to warm climates. *B.* The distribution of cycads today. *(A. Gerald Cubitt. B. After C. B. Cox and P. D. Moore, Biogeography: An Ecological and Evolutionary Approach, John Wiley & Sons, Inc., New York, 1980.)*

Fossil Plants as Indicators of Climate

Because plants are so sensitive to environmental conditions, the fossils of many plants can be used to interpret climatic conditions of the past. The **cycads,** for example, are an ancient group of plants that are now found growing only in tropical and subtropical settings (Figure 2-17). Because this distribution seems to reflect a fundamental physiological limitation of the group, it is assumed that fossil cycads also lived in warm climates.

Flowering plants, which include not only plants with conspicuous flowers but also hardwood trees (e.g., maples, beeches, and oaks) as well as grasses and their relatives, are valuable indicators of climates of the past 80 or 90 million years, the interval during which they have been abundant on Earth. The thick, waxy leaves found on some hardwood plants, for example, help these plants retain moisture and thus serve to indicate that fossil hardwood plants lived in warm climates. Leaf margins provide an even more useful means of assessing temperatures of the past. Leaves with smooth rather than jagged or toothed margins are especially common in the tropics. In fact, if a large flora occupies a region that is characterized by

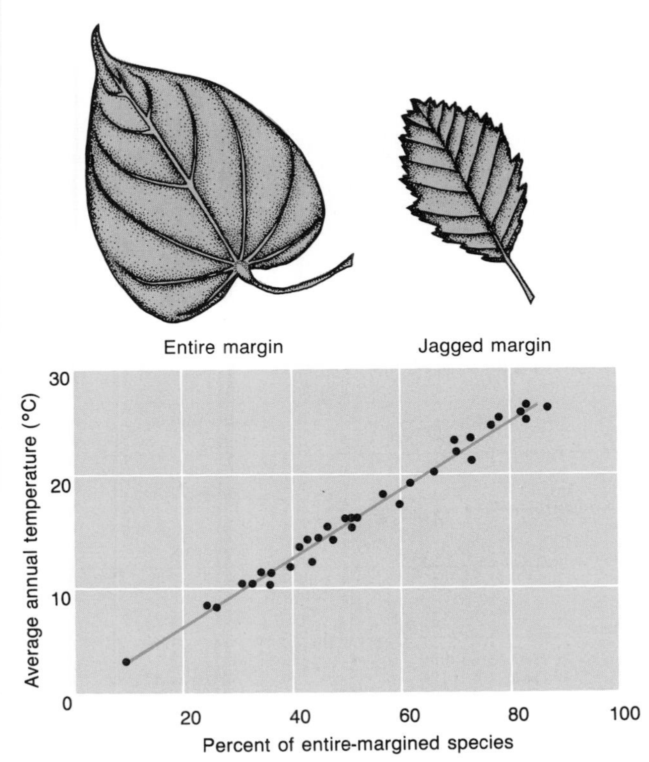

FIGURE 2-18 Relation between climate and the shapes of leaves of flowering plants. In the modern world the average annual temperature of a region is closely correlated with the percentage of plant species with entire (or smooth) margins. *(After J. A. Wolfe, American Scientist, 66:994–1003, 1978.)*

moderate or abundant precipitation, the percentage of plant species with smooth margins provides a remarkably good measure of the average annual temperature of that region (Figure 2-18). As we will see in Chapters 14, 15, and 16, this relationship has revealed a considerable amount of information about temperatures of the Late Cretaceous and Cenozoic intervals of geologic time.

Inasmuch as certain groups of animals are also restricted to warm regions, some animal fossils also serve as indicators of warm climates. Reptiles, which do not maintain constant warm body temperatures, are among the animals that cannot live in very cold climates.

THE MARINE REALM

The ocean floor is a vast basin in which most sediments have accumulated over the course of Earth's history. For this reason, and because so many types of organisms live in the ocean, the seafloor is also where most species of the fossil record have been preserved.

We have seen how the geographic distribution of terrestrial species reflects broad patterns of air movement in the atmosphere. In a similar way, the distribution of marine species reflects large-scale movements of water in the ocean.

Water Movements

The major ocean currents at Earth's surface owe their existence primarily to large-scale winds. The trade winds blow toward the equator from the northeast and southeast (Figure 2-10), pushing equatorial water westward to form the north and south **equatorial currents** (Figure 2-19). These currents pile water up on the western sides of the major ocean basins, where some of the water flows backward under the influence of gravity as **equatorial countercurrents.**

Because equatorial currents are also affected by the Coriolis effect, they are deflected toward the poles as they approach the western boundaries of ocean basins. (Recall that the Coriolis effect bends a current in the Northern Hemisphere toward the right, or clockwise, and bends a current in the Southern Hemisphere toward the left, or counterclockwise.) At the same time, the trade winds drive water along the eastern margins of ocean basins toward the equator. The result of these movements for each major ocean is a full clockwise gyre north of the equator and a counterclockwise gyre south of the equator. The Indian Ocean, most of which lies south of the equator, has only a counterclockwise gyre. The Gulf Stream, a famous segment of another gyre, carries warm water from low latitudes across the North Atlantic to warm the shores of Great Britain, where, in the southwest, palm trees survive more than 50° north of the equator. An eastern segment of the Pacific gyre has the opposite effect, bringing cool water to the coast of California.

In the Southern Hemisphere, the southern segments of the three gyres are known as the **westwind drifts.** Strengthened by westerly winds (Figure 2-10), these drifts join to form the Antarctic **circumpolar current** (Figure 2-20). The

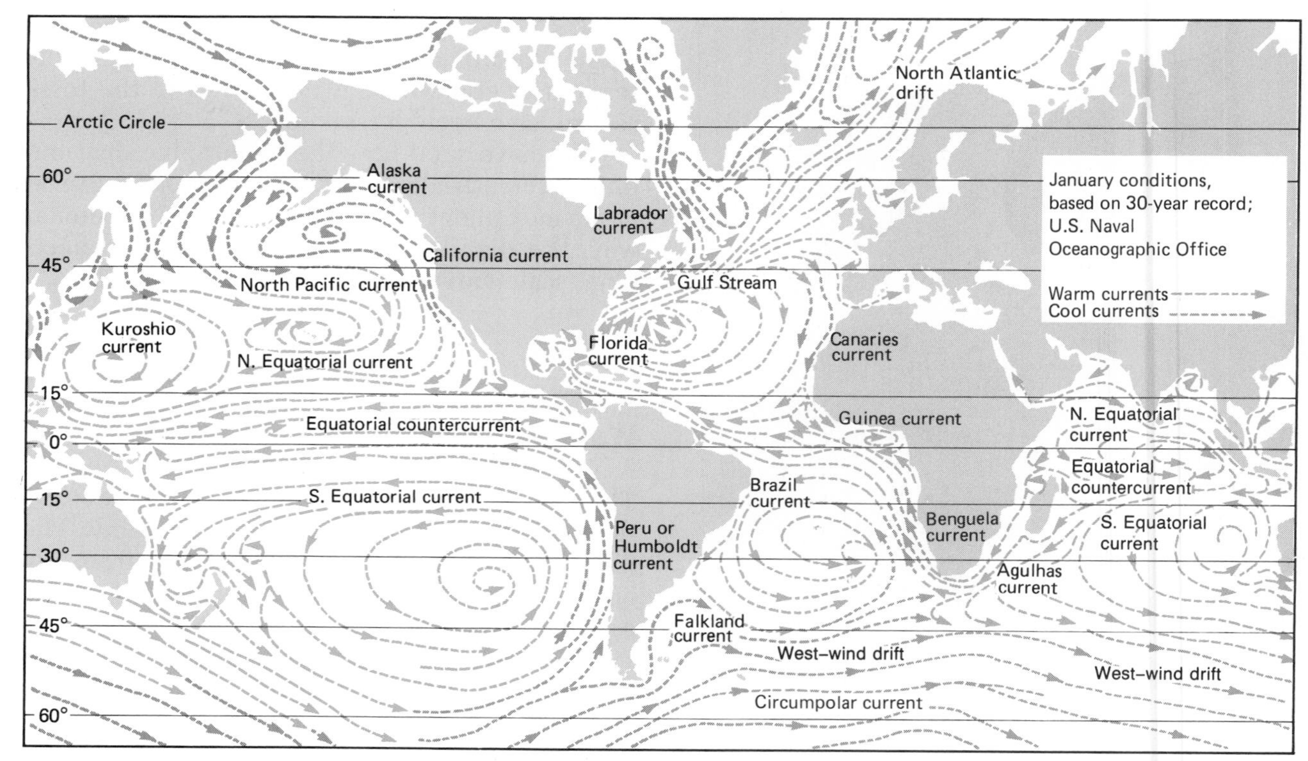

FIGURE 2-19 Major surface currents of the ocean. Note that large gyres north of the equator move clockwise, while those south of the equator move counterclockwise. *(After P. R. Ehrlich, A. H. Ehrlich, and J. P. Holdren, Ecoscience: Population, Resources, and Environment, W. H. Freeman and Company, New York, 1977.)*

landmasses of North America and Eurasia inhibit the development of a comparable circumpolar current in the Northern Hemisphere.

Near the poles, water that is dense because it is frigid and more saline than normal sinks to great depths (Figure 2-21). When this water reaches the deep seafloor, it spreads toward the equator. Antarctic water that descends in this manner is slightly colder than the Arctic water from the north, so it hugs the bottom of the sea and flows well into the Northern Hemisphere. Above this water, which remains at near-freezing temperatures, slightly warmer Arctic water flows southward. These currents supply the deep sea with oxygen from Earth's atmosphere near the poles. This oxygen permits a wide variety of animals to live in the bottom water despite the freezing temperatures.

Waves are yet another important form of water movement. **Surface waves** result from the circular movement of water particles under the influence of the wind. Because this movement decreases with depth, wave motion has no effect

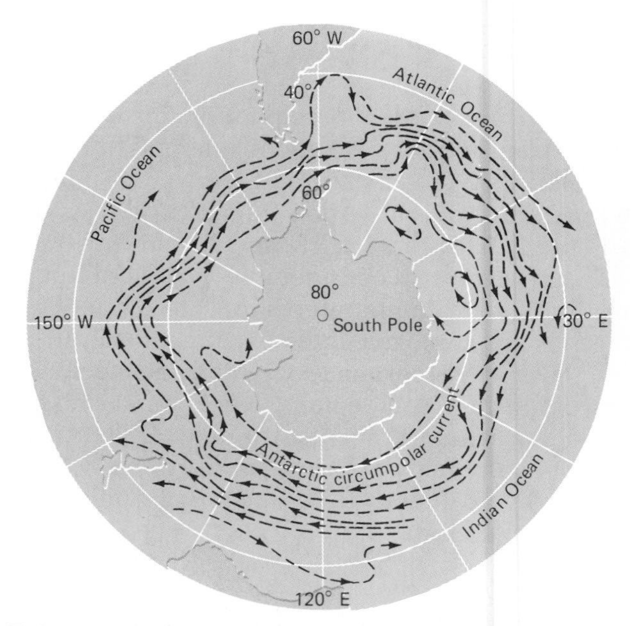

FIGURE 2-20 The eastward-flowing circumpolar current around Antarctica. Note in Figure 2-19 how this current is formed by the large counterclockwise gyres of southern oceans. *(After A. N. Strahler, The Earth Sciences, Harper & Row, New York, 1971.)*

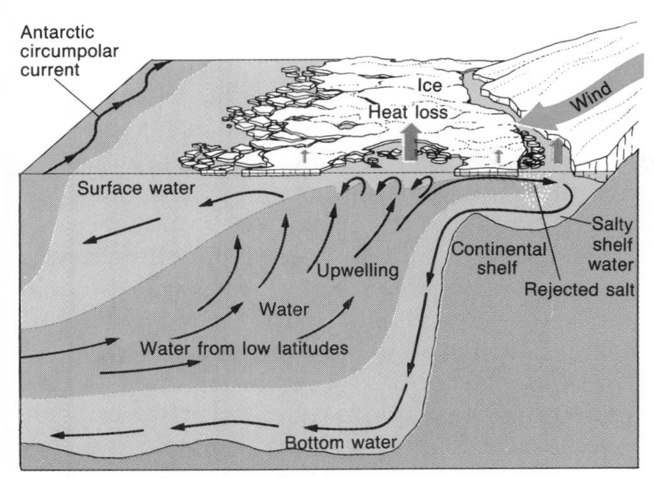

FIGURE 2-21 Formation of cold bottom water around Antarctica. Water flowing below the surface of the ocean from low latitudes rises up against the continental margin all around Antarctica. Here it becomes more dense, for two reasons. First, it loses heat to the cold atmosphere. Second, freezing of some of the water to form sea ice leaves excess salt behind, so that the remaining water is slightly hypersaline. The cold, salty, dense water formed in this way sinks to the deep seafloor, where it spreads throughout the world. *(After A. L. Gordon and J. C. Comiso, Scientific American, June 1988.)*

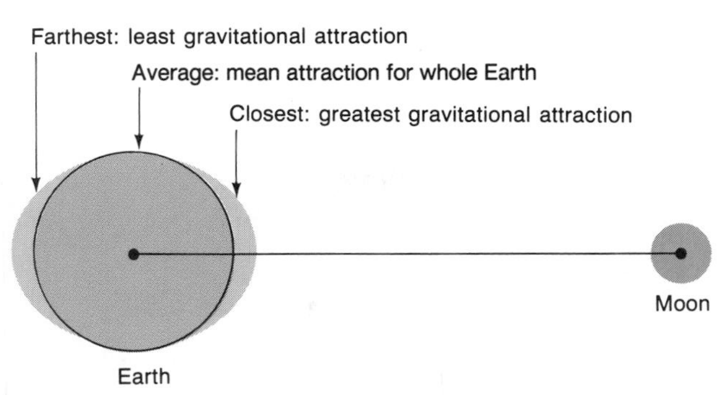

FIGURE 2-23 The origin of tides in the ocean. The moon exerts a gravitational attraction on the oceans that is strongest closest to the moon, causing the ocean to bulge out. The weakest attraction is on the side farthest from the moon, and thus the ocean bulges out there as well. The solid Earth rotates beneath these tidal bulges, causing them to move in relation to it. *(After F. Press and R. Siever, Earth, W. H. Freeman and Company, New York, 1986.)*

below a certain water depth, often several meters below the surface, depending on the size of the wave. Waves that are far from shore form swells that lack sharp crests. Very close to shore, the seafloor so greatly impedes the forward movement of waves that they steepen and break (Figure 2-22).

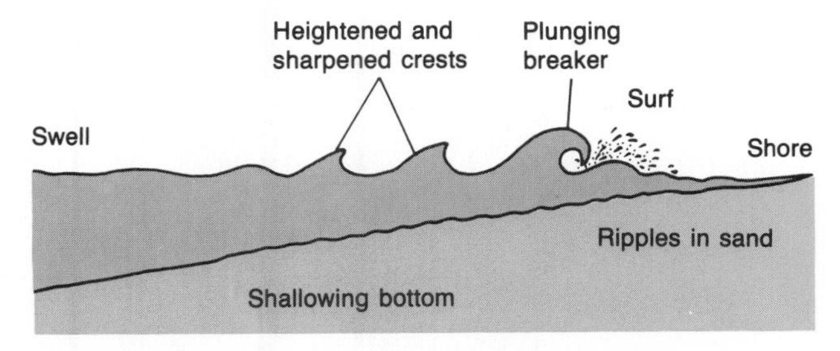

FIGURE 2-22 Waves breaking as they approach a beach. Offshore the waves take the form of swells; when the motion of a wave below the sea surface is obstructed by the seafloor, the upper part of the wave spills over toward the beach. *(After F. Press and R. Siever, Earth, W. H. Freeman and Company, New York, 1986.)*

Tides, which also cause major movements of water in the oceans, result from the rotation of the solid Earth beneath bulges of water that are produced primarily by the gravitational attraction of the moon (Figure 2-23). As tides approach a coast, they often generate strong currents. A tide flows toward a coast and then ebbs again within a few hours as Earth rotates, moving the coast away from the tidal bulge. As tides ebb and flow at the edge of the sea, they move the shoreline back and forth across an **intertidal zone.** Landward of this zone is the **supratidal zone,** a belt that is dry except when flooded by storms or strong onshore winds that coincide with high tide.

The Depth of the Sea

The depth of the sea varies from the thickness of a film at the shoreline to more than 10 kilometers (~6 miles) in the deep sea. Large areas of seafloor lie between 3 and 6 kilometers (~2 to 4 miles) below sea level (Figure 2-1). These areas constitute the **abyssal plain.**

More details of the configuration of the seafloor are shown in Figure 2-24. A **continental shelf** is nothing more than the submarine extension of a continental landmass. The **shelf break** marks the edge of the shelf; seaward of it, the

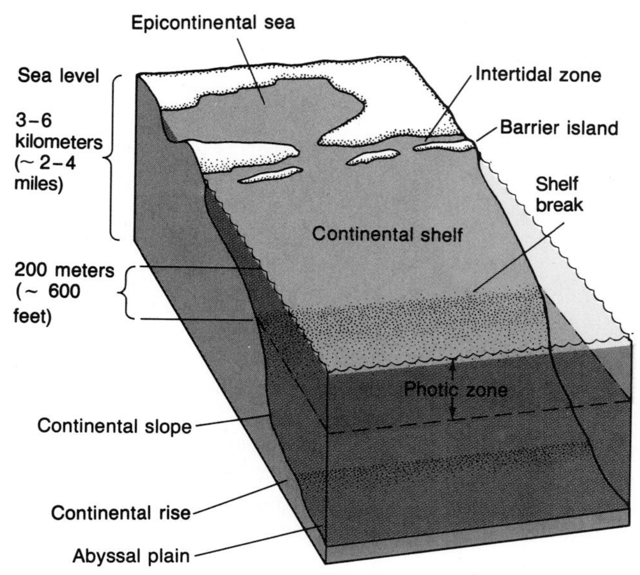

FIGURE 2-24 The aquatic environments at the edge of a continent. When the water is clear, the continental shelf—the submerged margin of a continent—usually lies within the photic zone, where enough sunlight penetrates to permit photosynthesis.

continent pinches out along the **continental slope.** Near the base of the continental slope, continental crust gives way to oceanic crust. Just seaward of this juncture is the **continental rise,** consisting of sediment that has been transported down the continental slope. Beyond the continental rise lies the abyssal plain, which is the surface of a layer of sediment resting on oceanic crust. When we speak of the **deep seafloor,** we are usually referring to the region below the shelf break; the area above the deep seafloor is often referred to as the **oceanic realm.**

Along the margin of the sea, **barrier islands** of sand heaped up by waves and wind often parallel the shoreline. In the protection of these elongate islands are relatively quiet lagoons or bays. **Marshes,** which are formed by low-growing plants that inhabit the intertidal zone, fringe the margins of these ponded bodies of water (Figure 2-25). Here plant remains accumulate as peat, which, if buried under the proper conditions of temperature and pressure, can turn to coal. In places the sea spreads farther inland over a continent, forming a broad, semi-isolated **epicontinental sea.** At the present time the seas happen to stand lower in relation to continental surfaces than they have done at most times during the past 600 million years. For this reason, continental seas are not well developed; Hudson Bay in eastern Canada is perhaps the best modern example. As we journey through Phanerozoic time in later chapters, we will examine many ancient epicontinental seas and the life they harbored.

How is life of the ocean related to the water's depth? Depth itself is probably a limiting factor for only a few species, but some significant limiting factors such as light (in the case of plants) and temperature are often closely related to depth. The upper layer of the ocean, where enough light penetrates the water to permit plants to conduct photosynthesis, is known as the **photic zone.** Although the base of this zone varies from place to place depending on the clarity of the water, it usually lies between 100 and 200 meters (~ 300 to 600 feet) below sea level. It happens that 200 meters is also the approximate depth of the shelf break in most areas.

For life of the seafloor, the most profound environmental change associated with depth takes place along the margin of the ocean. In contrast to life of the adjacent **subtidal** seafloor, which is never exposed to the air, the biota of the intertidal zone must endure large, often rapid fluctuations in environmental conditions. At some latitudes in this zone, hot dry conditions prevail at low tide during the summer, yet winter chills drop temperatures below freezing. Furthermore, in the **surf zone,** where waves break along a beach, the constant movement of the sand permits only a few species to survive. Species that live in this zone are

FIGURE 2-25 An intertidal marsh along the coast of Virginia. *(Bates Littlehales.)*

exceptionally mobile and can quickly reestablish themselves in the sand if they are dislodged by a wave.

Conditions are unusual in the deep sea, too, but here the environment is more stable. As we have seen, the frigid water that descends over the abyssal plain from polar regions approaches the freezing point. In fact, all of the oceans' waters are cool at depths below about 500 meters (~1600 feet). The environment formed by these waters, known as the **psychrosphere,** is inhabited by unique groups of species that are adapted to cold conditions.

Marine Life Habits and Food Webs

Most of the photosynthesis that takes place in the ocean is conducted by single-celled floating algae. For this reason, these algae are widely considered to be the most important producers at the base of the food web. Organisms that float in water are known as **plankton,** and plantlike organisms that belong to this group constitute the **phytoplankton.** The most important groups of phytoplankton are **dinoflagellates, diatoms,** and (in warm regions) **calcareous nannoplankton** (*nanno* = very small). The general characteristics of these groups of algae are shown in Figure 2-26. All three groups have extensive fossil records—the diatoms and calcareous nannoplankton because they have hard skeletons and the dinoflagellates because they have durable cell walls.

Feeding on the phytoplankton are floating animals known as the **zooplankton,** among which are small shrimplike crustaceans and other animals that spend their full lives afloat. Also included in this group are the floating larvae of some invertebrate species that spend their adult lives on the seafloor. These larvae allow the seafloor species to disperse over large areas of the ocean; after spending time in the plankton, they settle to the sea bottom and develop into adult animals. Some members of the zooplankton are carnivores that feed on other zooplankton.

Although many types of plankton have the capacity to swim, planktonic species move through the ocean primarily by drifting passively along, going with the flow. Animals that move through the water primarily by swimming are termed **nekton.** The most important of these swimmers are fishes. Both the plankton and the nekton include

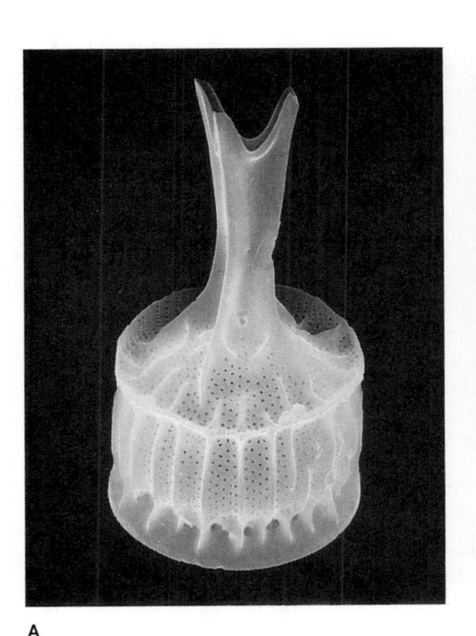

A

B

C

FIGURE 2-26 Representatives of the major types of single-celled algae in modern oceans: a dinoflagellate *(A)*, a diatom *(B)*, and a cell of calcareous nannoplankton *(C)*. Their diameters are about 10, 10, and 8 μm, respectively. *(A. Michael Hoban, California Academy of Sciences. B. C. L. Stein. C. Mitch Covington, Florida Geological Survey.)*

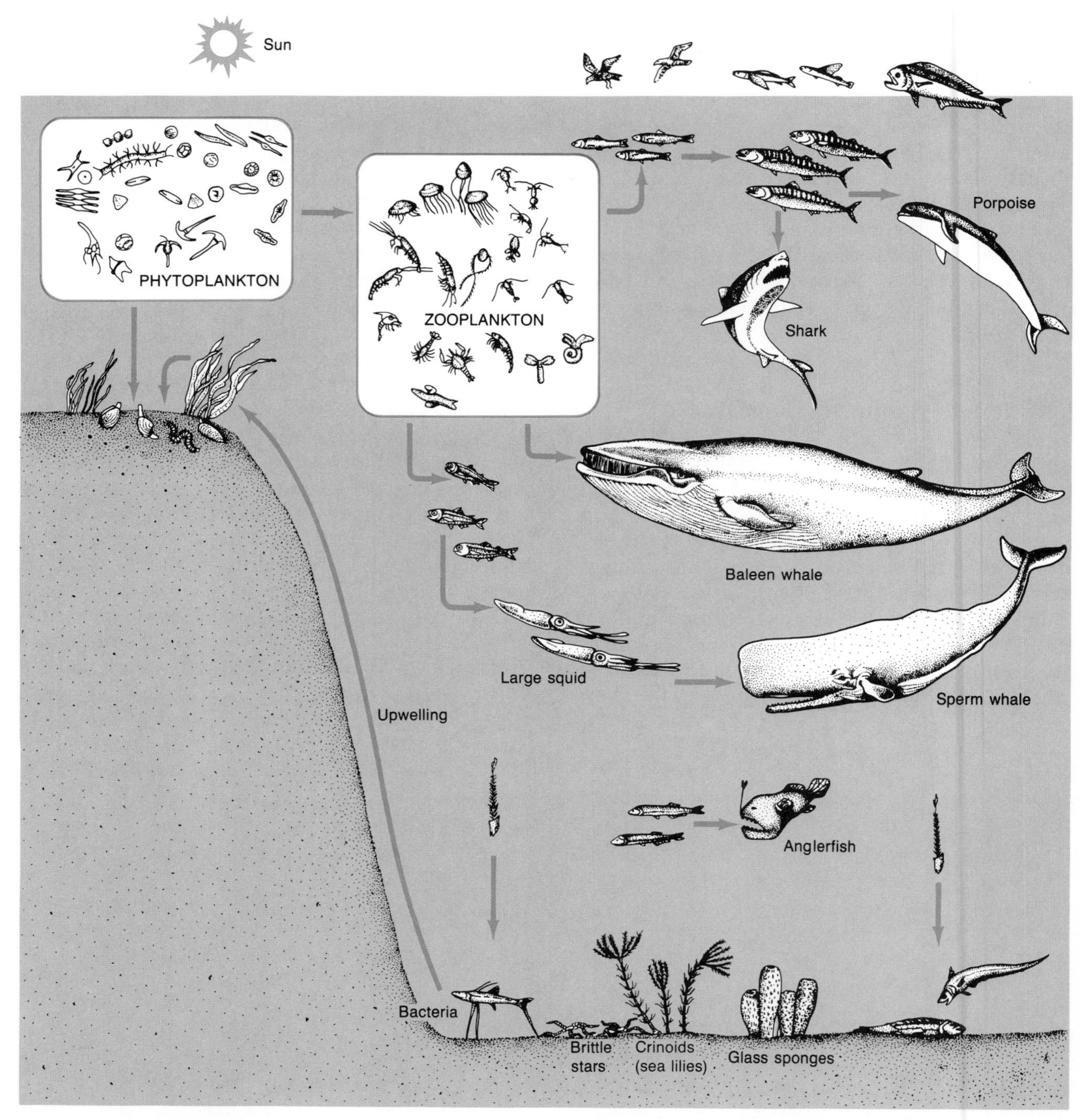

FIGURE 2-27 The food web in the ocean. (The various forms of life are not drawn to scale.) Phytoplankton occupy the photic zone of the ocean, and thus most zooplankton, which feed on phytoplankton, also live here. On continental shelves, especially near the shore, bottom-dwelling plants also contribute food to the marine ecosystem. Most species of large carnivores are fishes. Whales are warm-blooded mammals that include carnivorous porpoises and sperm whales, which feed on large animals, as well as baleen whales, which strain tiny zooplankton from the water. As the amount of plant material diminishes with depth, the abundance of animal life diminishes as well. A few herbivores that feed on plankton suspended in the water, such as sponges and crinoids (sea lilies), live on the deep seafloor, but most herbivores there extract their food from sediment. Bacteria in the deep sea turn dead organic matter into nutrients that upwelling currents carry to the surface for use by phytoplankton and other photosynthetic life.

not only herbivores, which feed on phytoplankton, but also carnivores, which feed on other animals. Planktonic and nektonic organisms together constitute **pelagic life,** or oceanic life that exists above the seafloor.

Both immobile and mobile organisms also populate the seafloor, and these organisms are known as **benthos,** or **benthic** (or **benthonic**) **life.** The seafloor itself is often referred to as the **substratum.** Some substrata are formed of rock, but they are more likely to be composed of soft substances such as loose sediment. Some benthic organisms live on top of the substratum, while others live within it.

Just as producers and consumers float in the water, both nutritional groups are found on the seafloor. Benthic producers include certain kinds of single-celled algae as well as multicellular plants. Because they require light, these photosynthetic forms must live on or close to the surface of the substratum.

Some benthic herbivores graze on plantlike forms, especially algae growing on hard surfaces. Some strain phytoplankton and plant debris from the water. Others consume sediment and digest organic matter mixed in with the mineral grains.

Bottom-dwelling carnivores of modern seas include crabs and starfishes as well as several kinds of snails and worms. In addition, many fishes swim close to the sea bottom and feed on bottom-dwelling animals.

Figure 2-27 depicts the basic features of the marine food web. The phytoplankton occupy the photic zone above both the continental shelves and the deep sea. Joining them as producers on continental shelves, where the photic zone reaches the seafloor, are bottom-dwelling plants. The high concentration of phytoplankton in the photic zone causes zooplankton to be concentrated there as well. Some herbivorous zooplankton can also be found at greater depths, where they feed on the algal cells and plant debris that rain slowly down from the photic zone.

Different kinds of swimmers occupy different depth zones in the ocean. Some, such as herring, feed on zooplankton. So do the great baleen whales, which strain zooplankton through a sieve-like bony structure. Other fishes, including nearly all sharks, are carnivorous, as are many kinds of whales. Carnivores are found at all depths of the ocean, although some species are restricted to narrow depth zones.

A wide variety of benthos are found on the shallow seafloors of the photic zone. Here most of the food at the base of the food web comes from phytoplankton, although some also derives from benthic plants. In the deep sea, however, where suspended food is scarce, only a few kinds of benthos strain food from the water: Most herbivores extract food from sediment, and so do many types of carnivorous fishes. In fact, numerous species live along the cold, dark abyssal plain. Organic debris arrives here from shallow waters at a very slow rate, however, so the density of animals is low. Thus a survey of 1000 square meters of deep seafloor might uncover dozens of species, each represented by a small number of individuals.

Bacteria live throughout the ocean but are most abundant in the deep sea, where organic debris accumulates. Some bacteria decompose this debris, while others transform some of the products of decay into simple nutrient compounds of nitrogen and phosphorus. Phytoplankton use these compounds to make food, thereby cycling the materials back through the ecosystem. A crucial step in this recycling is the physical process known as **upwelling,** or the movement of cold water upward from the deep sea to the photic zone. Upwelling tends to occur along the margins of continents, where the large oceanic gyres drag water away from the land (Figure 2-19); this water is replaced by water that wells up from the deep sea. Upwelling often brings nutrients to the photic zone in large quantities, producing an unusually rich growth of phytoplankton. The phytoplankton support large populations of zooplankton, which in turn support large populations of fishes.

Marine Temperature and Biogeography

In the marine realm, as in the terrestrial realm, temperature plays a major role in the geographic distribution of species. A pattern can be seen in the distribution of planktonic life. The calcareous nannoplankton, for example, live primarily in warm waters; individual species are found in narrow latitudinal belts where water temperatures remain within certain limits (Figure 2-28). Most species of planktonic diatoms, in contrast, live in cool waters at high latitudes.

The geographic distributions of species of the seafloor are also limited by temperature, and

Box 2-1
The Fragile Reef

(Norbert Wu.)

Living coral reefs, with their colorful assemblages of animals and plants, are among the most beautiful biotic communities on Earth. Snorkelers flock to shallow tropical seas to marvel at these magnificent underwater kingdoms. Of all living communities, however, reefs are among the most vulnerable to environmental change, and today many coral reefs are under severe environmental stress because humans are altering their habitat. The fossil record alerts us to the alarming possibility that many modern reef ecosystems may collapse in the coming decades. During the past 600 million years a succession of communities have built reefs resembling those of the modern world. The reign of each group of reef builders ended abruptly, with reefs nearly disappearing from the world's oceans. After each crisis, reefs recovered their luxuriance only after millions of years of evolution.

The reef ecosystem is fragile because reef corals have very specific ecologic requirements. They can live only in waters of normal marine salinity; they cannot tolerate brackish or hypersaline conditions. Reef corals also require warm water, apparently because warmth facilitates the corals' secretion of their calcium carbonate skeletons. Few reefs grow more than 30° north or south of the equator. Yet temperatures only slightly higher than those of the warmest tropical seas in the world today are lethal to most coral species. Reefs can flourish only where water movements are strong enough to provide a rich supply of the zooplankton upon which corals feed. The water must also be clear, because sediment in suspension hinders corals' feeding and screens out sunlight needed by their symbiotic algae. Finally, reefs suffer if the water is overly rich in nutrients, especially phosphates, because such water favors the growth of mats of fleshy algae. These algae, which are quite unlike the tiny symbiotic algae in the corals' tissues, can smother corals and other members of the reef community.

Another factor that contributes to the instability of a coral reef community is the interdependence of its species; the disappearance of one kind of organism can devastate many others. In 1983, the sudden death of sea urchins wreaked havoc on reefs throughout the Caribbean. For unknown reasons, about 98 percent of the population of the sea urchin *Diadema* vanished. While some humans might otherwise have applauded the decline of this creature, whose poisonous, needlelike spines cause frequent injuries to swimmers, the ecologic result was disastrous. Sea urchins graze on soft algae and prevent them from forming the mats that smother corals. Following the decline of *Diadema,* mats of algae quickly spread over reefs, killing many corals and other reef builders. It was several years before populations of *Diadema* expanded to their former densities, and the damaged reefs are still in the process of recovering.

Modern coral reefs have held their own against natural disasters over vast stretches of time, but today human disturbances threaten to inflict more lasting damage on many reefs, or even total destruction. Nearly 2 million people a year visit the Pennekamp Coral Reef State Park in Florida, which includes a substantial fraction of the few well-developed shallow-water coral reefs in North American waters. About half of these enthusiastic visitors plunge into the water, and many bump into corals or trample them. Boats also scrape the reefs. Coral colonies that have taken decades or even centuries to grow are being destroyed, and many areas of the reef are already dead. Reef dwellers are also being poisoned by sewage, toxic wastes, and petroleum compounds discharged by boaters and by the burgeoning population of the Florida Keys. Phosphates from fertilizers and human waste do less direct damage. They promote the growth of smothering algal mats, as does the heavy fishing in surrounding waters, because many kinds of fish graze on algae.

Human activities that are accentuating the greenhouse effect (p. 31) may soon prove damaging to coral reefs in many areas of the world. Foremost among these activities is the burning of organic compounds, which releases carbon dioxide into the atmosphere. Pulses of regional warming unrelated to human activities have killed some reef-building corals of the Pacific Ocean in recent years. These events have sounded an important warning. If, during the next few decades, human-induced greenhouse warming greatly elevates temperatures of tropical seas throughout the world, coral reefs may suffer widespread damage.

The fossil record, then, shows us that organic reefs have always tended to be ecologically fragile structures, and the ecology of modern reefs indicates that they are not exceptions to this rule.

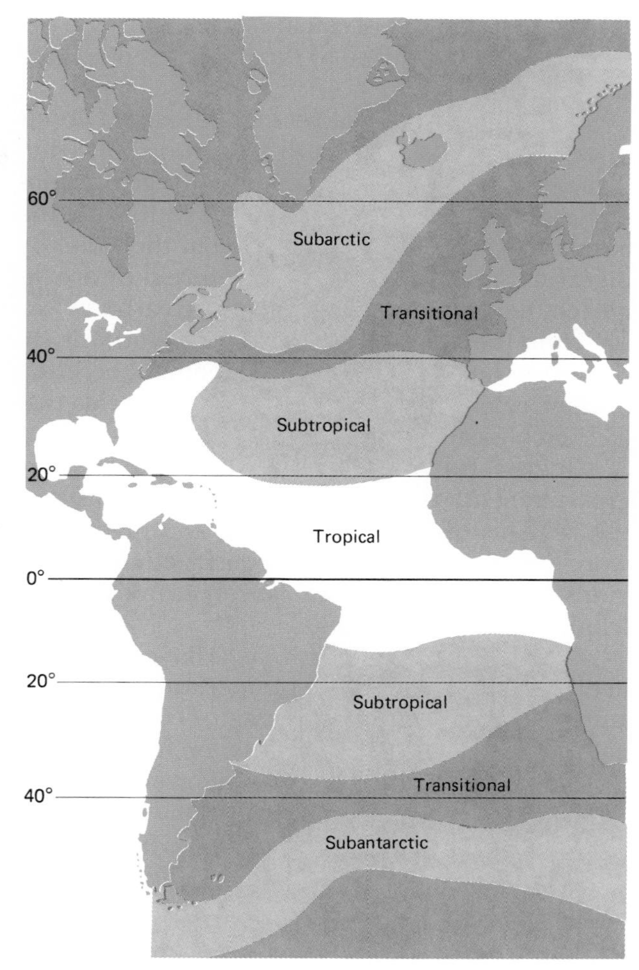

FIGURE 2-28 Latitudinal zones of calcareous nannoplankton in the modern ocean. Although these members of the phytoplankton are found in cold waters, fewer species live there than in warm-water zones. *(After A. McIntyre and A. W. H. Bé, Deep Sea Research, 14:561–597, 1967.)*

some large groups of animals and plants are restricted to certain latitudes. The reef-building **hexacorals** represent one of the most important groups of this type, forming massive reefs that are restricted largely to the tropics (Box 2-1).

Hexacorals are saclike animals with tentacles. Most of them derive nutrition from smaller animals that they capture from the water by means of stinging cells on their tentacles.

Reef-building corals are colonial animals that grow as clusters of connected individuals, or polyps. A colony forms from a single original polyp that develops from a larva. The original polyp gives rise to a colony by budding off additional polyps which, in turn, bud off others. Each polyp secretes a cup of calcium carbonate, and the adjacent cups are fused to form a large composite skeleton.

Some other types of reef dwellers join corals in contributing skeletons to the solid reef structure. Aiding corals in forming their large colonies are single-celled algae that live and multiply in the coral tissues. The corals supplement their diet by digesting some of these algae. Before they are digested, however, algae cells remove carbon dioxide from the corals for use in photosynthesis. In this way, the algae facilitate the corals' secretion of their calcium carbonate skeleton. Species that lack such algae cannot form reefs.

Reefs form their own fossil records as they grow. Beneath their living surface they consist of the remains of the dead animals and plants that were responsible for their construction. Other kinds of limestones are also largely tropical in distribution (see Appendix I).

Oxygen isotopes of fossil shells have been used in efforts to determine the temperatures at which animals lived millions of years ago. Oxygen occurs naturally in two isotopic forms, or forms of the element that have different atomic weights. Oxygen 18 has two more neutrons than the more common form, oxygen 16, and is thus the heavier **isotope.** The two isotopes have the same chemical properties, but marine organisms that secrete shells incorporate the isotopes in slightly different proportions, depending on the temperature of the environment. As temperature decreases, the percentage of oxygen 18 increases. A difficulty encountered in attempts to analyze ancient ocean temperatures by means of oxygen isotopes is that some ancient shells have suffered chemical alteration after burial. As a result, temperature estimates based on oxygen isotopes are sometimes inaccurate.

Salinity as a Limiting Factor

The saltiness of natural water is called **salinity.** Oceanic seawater contains about 35 parts of salt per thousand parts of water, or is said to have a salinity of 35 parts per thousand (3.5 percent). Salinities of 30 to 40 parts per thousand are regarded as within the normal range for seawater; water of lower salinity is called **brackish,** while water of higher salinity is termed **hypersaline.**

Brackish and hypersaline conditions are most commonly found in bays and lagoons along the margins of the ocean. Brackish conditions result from an influx of water from rivers into bays or lagoons that are partially isolated from the open ocean. Hypersaline conditions also develop in bays and lagoons, but only in those whose waters evaporate rapidly—usually in hot, arid climates.

The salinity of brackish and hypersaline waters typically changes frequently in response to changes in rainfall and in evaporation rates. The salt content of the tissues of most animals is similar to that of normal seawater. Marine animals therefore tend to find it difficult to move into a habitat where the salinity is abnormal or fluctuating. It is hardly surprising that most bays and lagoons contain fewer species of animals than normal marine habitats. Many marine animals migrate into marshes to breed, however, because the scarcity of predators here and the abundance of organic matter from decaying vegetation improve the chances that offspring will survive.

Freshwater Environments

The difficulty of living in water of low salinity is a major reason the faunas of rivers and lakes are not very diverse. Rivers and lakes are freshwater habitats, or habitats whose salinities remain below 5 parts per thousand (0.5 percent). Most freshwater animals must have ways of excreting excess water that enters their body tissues from the environment. Phytoplankton and zooplankton similar to those of the ocean also inhabit lakes, but in reduced variety. Because streams and rivers are constantly in motion, they do not sustain a planktonic community as complex as that of the ocean. Most producers occupying rivers live on the bottom, and a large proportion of consumers are immature growth stages of terrestrial insects (Figure 2-3). Both lakes and rivers differ from the ocean also in their small variety of larger animal species. Unfortunately, relatively few kinds of these animals are readily preserved as fossils; fishes and shelled mollusks are the primary exceptions.

As we shall see, most major groups of aquatic life evolved in the ocean, and only later did some invade less hospitable freshwater environments.

CHAPTER SUMMARY

1 The way a species interacts with its environment defines its ecologic niche.

2 The distribution and abundance of any species are governed by a number of limiting factors: the availability of food, the nature of physical and chemical conditions in the environment, and the presence of other species that are potential predators or competitors.

3 Communities are groups of coexisting species that form food webs. Plants, which are at the base of most food webs, are fed upon by herbivores, which are fed upon in turn by carnivores. Communities and the environments they occupy constitute ecosystems.

4 The diversity of a community is the variety of species it encompasses.

5 Bacteria decompose dead organisms and transform the products of decay into compounds that serve as plant nutrients. Thus materials are cycled through the ecosystem continuously.

6 Physical barriers to dispersal and changes in environmental temperature are the most important factors that limit geographic distributions of species; many more species exist in warm climates than in cold climates.

7 Continental glaciers (ice caps) exist near Earth's poles on Greenland and Iceland.

8 Carbon dioxide in the atmosphere produces a greenhouse effect, trapping heat and warming Earth.

9 Tropical rain forests develop near the equator, where warm air rises and loses its moisture. The dried air descends north and south of the equator and circles back toward the equator as trade winds. In many areas that lie between 20 and 30° from the equator, trade winds produce deserts and dry grasslands.

10 Because land plants are highly sensitive to climatic conditions, fossil land plants are useful indicators of ancient climates.

11 In large oceans, prevailing winds and the Coriolis effect create huge gyres of water movement.

12 The dominant food producers in the ocean are photosynthetic single-celled algae that float in shallow waters.

13 The deep sea is cold because its waters come from near the north and south poles, where frigid water sinks to great depths and flows toward the equator. Many species of animals inhabit the deep sea, but their populations are quite small because little food reaches this environment.

14 The salinity of bays and lagoons near the margin of the ocean differs from that of normal seawater and fluctuates greatly; relatively few species are able to live in these environments.

15 Freshwater environments, such as lakes and rivers, usually harbor relatively few species because life in fresh water poses physiological problems for many kinds of animals.

EXERCISES

1 Sometimes the species of a community are described as forming a food chain. Why is it usually more appropriate to speak instead of a food web?

2 Which terrestrial and marine environments characteristically contain few species? Explain why each of these environments is populated in this way.

3 How do the main kinds of producers (photosynthesizers) in the ocean differ in mode of life from those on the land?

4 What can fossil plants tell us about ancient environments?

5 What does the shape of the hypsometric curve tell us about the distribution of seafloor environments?

6 Explain the greenhouse effect.

7 How do winds affect the ocean on a large scale?

8 Rain forests are sometimes likened to coral reefs, in that both support communities of highly diverse species. Why are both communities restricted to the tropics?

ADDITIONAL READING

Cox, C. B., and P. D. Moore, *Biogeography: An Ecological and Evolutionary Approach,* John Wiley & Sons, Inc., New York, 1985.

Kormondy, E. J., *Concepts of Ecology,* Prentice-Hall, Inc., Englewood Cliffs, N.J., 1984.

Newton, C., and L. F. Laporte, *Ancient Environments,* Prentice-Hall, Inc., Englewood Cliffs, N.J., 1989.

CHAPTER 3

Sedimentary Environments

Many lessons can be learned from the depositional settings of ancient sedimentary rocks. The most general reason for studying these settings is to reconstruct geography of the past; that is, **paleogeography.** The goal is not only to learn about the distribution of land and sea at a particular time, but also to identify and reconstruct more localized environmental features, such as deserts, lakes, river valleys, lagoons, and submarine shelf breaks. In most instances, geologists can learn not only where a river valley was located, for example, but also what kind of river occupied that valley—perhaps one formed by many small, intertwining channels choked with bars of gravel and sand, or one that flowed along a single broad, winding channel. Frequently geologists can also "read" from the sedimentary record whether the terrain that once bordered an ancient river was a dry, sparsely vegetated plain or a swamp densely populated by water-loving trees and undergrowth. Thus one aspect of paleogeography is the study of ancient climates.

The identification of ancient sedimentary environments also provides geologists with a framework within which to interpret life of the past. Although we can learn some aspects of how an animal or plant species lived by studying the configuration of its fossil remains alone, a fuller understanding of that species can come only when its habitat is taken into consideration. It was once widely believed, for example, that the largest dinosaurs were too big to be fully terrestrial and therefore must have spent much of their time in water, like hippopotamuses. As will be discussed more fully in Chapter 13, the fossil record of these giant creatures contradicts this idea: Their bones are frequently found in sedimentary rocks that represent nonaquatic environments.

The study of ancient sedimentary environments is also of considerable practical value in that many sedimentary deposits occur in conjunction with important natural resources or are themselves natural resources. By understanding the environmental relationships of sedimentary rocks, economic geologists can often predict where these resources will be found beneath Earth's surface. Coal is a resource whose location can be predicted on this basis, as are petroleum and natural gas, which tend to accumulate in porous rocks such as clean sands deposited along ancient shorelines or rivers and in ancient limestone reefs constructed in shallow seas by corals or coral-like organisms.

Before geologists can interpret the origins of ancient sedimentary rocks, however, they must

In Death Valley, California, there is little vegetation in the dry climate to stabilize the surface of the land, so the water from occasional rains erodes and transports large amounts of sediment and deposits much of it as alluvial fans and sand dunes. *(Tom Bean.)*

first understand the patterns and processes by which sediment is deposited in the modern world. Unfortunately, the products of these processes are often obscured, and geologists have to excavate, tunnel, or core before they can observe them. **Coring** is a technique that can be used to examine a deposit at the center of a lake, lagoon, or deep ocean. A tube is driven into the bottom of the body of water and then withdrawn. The core of sediment thus extracted can be studied to determine the sequence of sediment deposition at the site. Coring at several locations provides a three-dimensional picture of sedimentary deposits. Similarly, geologists who wish to examine a meandering river's depositional record must either dig one or more pits in the valley floor adjacent to the channel or sample the floor by means of coring. (The reason for studying the sediments *adjacent* to a river will become clear in the course of this chapter.)

Some sedimentary features provide highly reliable information about the nature of the environments in which they formed. Many other sedimentary features, however, offer the geologist only ambiguous testimony about the environments in which the sediments were deposited. Most coal deposits, for example, represent swamps choked with vegetation—but such swamps are typical of both the banks of rivers and the shores of marine lagoons. Geologists must therefore consider the nature of the beds that lie above or below coal deposits; if sediments containing fossils of marine animals lie above or below a coal bed, it is likely that a marine lagoon, not a river, lay adjacent to the swamp or marsh in which the coal formed (Figure 2-25).

We will begin our exploration of depositional environments and their characteristic sediments with settings on land. We will then move to freshwater systems, and finally shift our focus to the marine realm.

SOIL ENVIRONMENTS

Soil can be defined as loose sediment that has accumulated in contact with the atmosphere rather than under water. Soil rests either on sediment of different characteristics or on rock. It serves as a medium for the growth of plants by supplying essential nutrients and by providing a base for the physical support of roots and underground stems. Soils form in a variety of environments throughout the world—in tropical rain forests, in arid regions, and even on mountaintops. Moreover, layers of diagenetically altered soil can often be found buried within thick sequences of ancient sediment. Recognition of these ancient soils is of great geologic significance in that they can be used to define the configurations of ancient landscapes. Certain soils also offer important evidence about the climatic conditions under which they formed.

How Soils Form

Soils develop in part by the weathering processes described in Appendix I. The upper zone of many soils, referred to as **topsoil,** consists primarily of sand and clay mixed with organic matter called **humus.** Humus, which gives topsoil its dark color, is derived from the decay of plants and in turn supplies nutrients for other plants; thus it occupies an important position in the cycling of materials through the plants of terrestrial ecosystems. Bacteria and fungi produce most of the decay. Desert soils are poor in humus because vegetation in such environments is sparse.

The type of soil that forms depends in part on climatic conditions. In warm climates that are dry part of the year, calcium carbonate accumulates in the layer of soil below the topsoil, forming what are known as **caliche** nodules, or **calcrete** (Figure 3-1). The accumulation of caliche nodules results from the evaporation of groundwater under hot, dry conditions.

In moist, tropical climates warm waters percolate through the soil, destroying humus by oxidizing it. Silicate minerals in such areas also break down and disappear quickly, leaving the soil rich in aluminum oxides as well as in iron oxides, which give it a rusty red color. Tropical soils of this type are known as **laterites.**

Ancient Soils

Ancient buried soils can be exceedingly difficult to recognize or to interpret, partly because diagenesis often alters the chemical components of soils beyond recognition. One place where ancient soil *is* likely to be found, however, is beneath

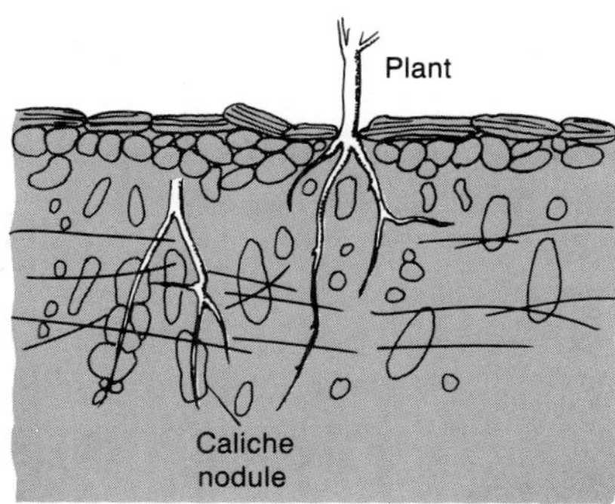

FIGURE 3-1 Nodules of caliche (calcrete) in soil. The white nodules have formed around plant roots, as shown in the drawing. *(Lawrence Hardie, Johns Hopkins University.)*

an unconformity. At such a site, soils will have formed while the rocks below the unconformity were exposed at Earth's surface — that is to say, before these rocks were buried beneath younger sediments.

Plant roots provide clues to the identity of ancient soils. Burrows made by animals such as insects and rodents are also diagnostic features. Certainly the most unusual of these excavations are the structures known as "devil's corkscrews," which are actually burrows that beavers of an extinct species dug with their teeth in the Oligocene and Miocene soils of Nebraska (Figure 3-2). Skeletons of these beavers have been found in the burrows, and scratches on the burrow walls match their front teeth! The fact that these animals lived as far as 10 meters (~ 33 feet) below ground level indicates that the level of standing water in the ancient soil stood barely above this depth; had it been higher, the beavers would have drowned.

LAKES AS DEPOSITIONAL ENVIRONMENTS

At no time in geologic history have lakes occupied more than a minute fraction of Earth's surface — but because lakes form in basins that lie at lower elevations than most soils, lake deposits are much more likely than soils to survive in the face of erosion.

Lake deposits share a number of typical characteristics by which they can be recognized. For

A

B

FIGURE 3-2 Fossil burrows made by beavers in soils more than 20 million years old in Nebraska *(A)*. Unlike modern beavers, these animals did not live near water, but instead lived in grasslands, where they formed colonies like those of prairie dogs. The burrows, which the beavers excavated with their front teeth *(B)*, extend downward as far as 10 meters (~ 33 feet) into the soil. *(A. Carnegie Museum of Natural History. B. After L. D. Martin and D. K. Bennett, Palaeogeography, Palaeoclimatology, Palaeoecology 22:173–193, 1977.)*

example, sediments around the margins of a lake tend to be coarser than those that lie toward the lake's center, partly because the current of a stream or a river slows down when it meets the waters of a lake, dropping its coarse sediment load near the shore as it does so. Furthermore, the wind-driven waves on the surface of the typical lake touch bottom only when they approach the shore, winnowing the sediment there and driving clay-sized particles into suspension. These particles later settle to the bottom toward the lake's center.

Fossils are valuable tools for distinguishing lake sediments from marine sediments. Although fish fossils are found in both kinds of deposits, the presence of exclusively marine fossils, such as corals, provides strong evidence that ancient sediments did not originate in a lake. Because burrowing animals are not as abundant in lakes as they are in many marine environments (and because waves and currents in lakes are generally weak), the fine-grained sediments that accumulate in the centers of lakes are likely to remain well layered, just as they were when they were laid down (Figure 3-3).

Another clue in the identification of lake deposits is close association with other nonmarine deposits, such as river sediments. It would be highly unusual to find lake deposits directly above or below deep-sea deposits, or even above or below sediments deposited on the continental shelf, unless the two types of deposits are separated by an unconformity.

FIGURE 3-3 Lake deposits from the Eocene Green River Formation of Wyoming. The even laminations are typical of sediments that have accumulated in lakes. *(Photograph by the author.)*

GLACIAL ENVIRONMENTS

Even more useful than soils as clues to ancient climatic conditions are certain complex suites of depositional features that are known to develop only in particular climates. The sedimentary features associated with some types of glaciers, for example, are excellent indicators of cold climates. Glaciers that form in mountain valleys seldom leave enduring geologic records because the mountains through which they move, and on which they leave their mark, stand exposed above the surrounding terrain and therefore tend to be eroded rapidly on a geologic scale of time. Continental glaciers, however, leave legible records that survive for hundreds of millions of years.

Today one continental glacier occupies most of Greenland, and an even larger one occupies nearly all of Antarctica (Figure 2-20). Each of these glaciers is thickest in the center, tapering toward the continental margin. Continental glaciers leave traces of their activity on the lowland areas over which they move, and these traces have led to the recognition of widespread intervals of continental glaciation not only in late Neogene time (i.e., during the past 3 million years or so) but also much earlier, during the late Paleozoic, Ordovician, and Precambrian intervals.

As glaciers move, they erode rock and sediment, transporting both in the direction of flow. In this process, pieces of rock become embedded in the ice at the base of a glacier, where they become tools of further erosion, commonly leaving deep scratches in the bedrock that serve to record the direction in which the glacier is moving. When such scratches are found in an area that is now free of glaciers, it is safe to assume that a glacier passed that way in the distant past (Figure 3-4).

Glaciers leave records not only of erosion but also of deposition. The mixture of boulders, pebbles, sand, and mud that is scraped up by a moving glacier is deposited partly as the glacier plows along and partly when it melts. Some of the boulders bear scratches from abrasion during their journey at the base of a glacier. This heterogeneous material is called **till.** At the farthest reach of a glacier's advance, till plowed up in front of the glacier is left standing in ridges known as **moraines.** A considerable amount of till, however, after being transported for some distance, is de-

FIGURE 3-4 A rocky surface in eastern Canada scoured by glaciers. *(Peter Kresan.)*

FIGURE 3-5 Glacial till. A mountain glacier deposited this bouldery sediment on the eastern side of the Sierra Nevada in California during the Pleistocene Epoch. *(Martin Miller.)*

posited beneath the glacier; thus till may be deposited over a broad area even as a glacier continues to melt back. In front of a moraine, sediment from a melting glacier is often deposited by streams of meltwater issuing from the retreating mass of ice. Here the sediment tends to be sorted by size into layers of gravel, cross-bedded sand, and mud, forming well-stratified glacial material known as **outwash.** Figure 3-5 shows till on the eastern side of the Sierra Nevada, which was glaciated during the recent (Pleistocene) Ice Age. Lithified till is known as **tillite.**

In many instances, the meltwaters that issue from a glacier converge to form a lake in front of the glacier. The alternating layers of coarse and fine sediment that typically accumulate in such lakes are called **varves** (Figure 3-6). Each coarse layer forms during the summer months, when streams of meltwater carry sand into the lake, whereas each fine layer is formed during the winter, when the surface of the lake is sealed by ice and all that accumulates on the lake bottom are clay and organic matter that settle slowly from suspension. Each pair of layers of varved sediment thus represents a single year's deposition, so geologists can count the number of years represented by a series of layers. In some areas, thousands of years of deposition have been tallied in this manner.

When a glacier encroaches on a lake or ocean, pieces of it break loose and float away as icebergs, some of which are immense (Figure 2-15). As chunks of glacial ice melt in a lake or an ocean, their sediment sinks to the bottom, creating a highly unusual deposit in which pebbles or even boulders rest in a matrix of finer sediments (Figure 3-7). Unlike the tightly packed, coarse material that characterizes glacial till, these **dropstones** occur either singly or scattered throughout the matrix. Very few natural mechanisms other than this so-called ice rafting bring large stones to the middle of a lake or to a seafloor far from land.

FIGURE 3-6 Varved clays in Canada. The varves are layers of clay and silt deposited in quiet lake waters. Varved clays produced during the Pleistocene Epoch at Toronto *(A)* are remarkably similar to Precambrian varves in southern Ontario that are nearly 2 billion years older *(B)*. *(A, P. F. Karrow, Ontario Department of Mines. B, F. J. Pettijohn.)*

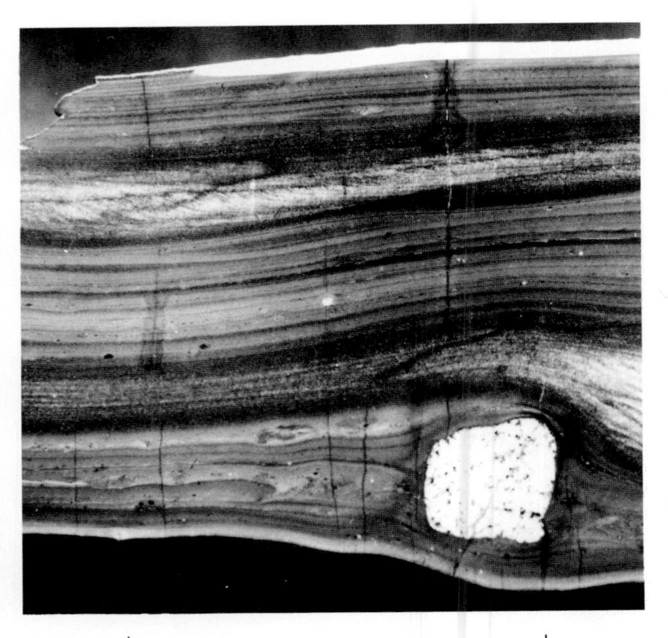

FIGURE 3-7 Fine-grained sedimentary rock of the Proterozoic Gowganda Formation, which is about 2 billion years old, in southeastern Canada. An igneous pebble apparently dropped from floating ice. The layers in the sedimentary rock appear to be varves. *(D. A. Lindsey, U.S. Geological Survey.)*

DESERTS AND ARID BASINS

Like frigid glacial environments, arid and semiarid basins are characterized by a unique suite of sedimentary deposits. Deserts contain little or no humic soil, because dry conditions support little vegetation, the source of humus. The rain that occasionally falls in deserts leads to erosion and to deposition of sediment, often carrying the chemical products of weathering to desert basins. The subsequent precipitation of evaporite minerals in these basins sets arid regions apart from those with moist climates. In the latter, permanent streams flow great distances without being absorbed into the soil or evaporating into the air; thus they are frequently capable of reaching the ocean by passing into large rivers. As a consequence, we speak of most humid regions as having **exterior drainage,** or drainage to areas beyond their borders. Runoff from arid regions, in contrast, is too sparse and intermittent to form permanent streams and rivers, and the result is **interior drainage** — a situation in which streams die out through evaporation and through seepage of water into the dry terrain.

Lakes in areas with interior drainage also tend to be temporary and are known as **playa lakes** (Figure 3-8). Temporary streams bring dissolved salts to these lakes along with suspended sediment. As the lakes shrink by evaporation, the salts precipitate as sediments called **evaporites.**

FIGURE 3-8 **A playa lake in Death Valley, California.** *(Peter Kresan.)*

Death Valley: A Modern Example

Death Valley, seen on page 50 and in Figure 3-8, typifies arid basins in several ways. Temporary streams carry sediment down valleys incised into the naked rocks of nearby highlands to form low cone-shaped structures called **alluvial fans,** which spread out onto the floor of Death Valley. These structures form where a mountain slope meets the valley floor, causing streams to slow down and drop much of their sediment. Alluvial fans consist of poorly sorted sedimentary particles that range from boulders to sand near the source area and from sand to mud on the lower, gentler slopes. Most alluvial fans include broad deposits of coarse, cross-stratified sediments laid down in a complex network of channels, or **braided streams,** that carry water and sediment during the infrequent rainy intervals. (Alluvial fans and braided-stream deposits form in moist climates, too, as we shall see.) The braided streams of these channels lead to the center of the basin, where their occasional flow of water forms temporary lakes. As the lake waters dry up, evaporite minerals accumulate. The same minerals also accumulate on those parts of the basin floor where groundwater seeps to the surface and evaporates. The evaporites of Death Valley are composed primarily of halite, gypsum, and anhydrite. Alternate wetting and drying in this basin produce large polygonal mud cracks in many areas (Figure 3-9). Around the margins of these "salt pans," calcium carbonate deposits form caliche.

FIGURE 3-9 **Large mudcracks in Death Valley, California.** *(Peter Kresan.)*

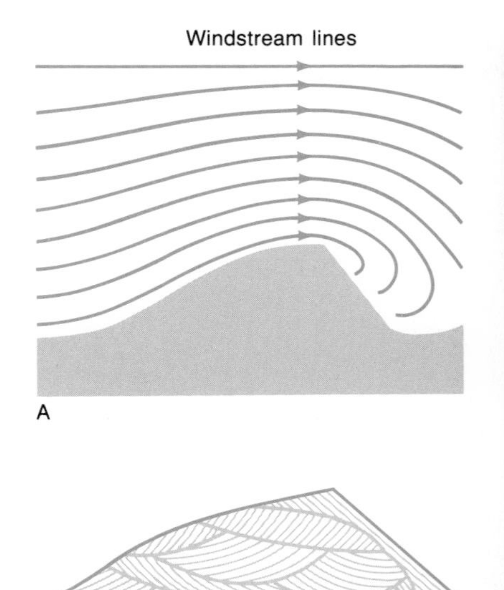

FIGURE 3-10 The internal structure of a sand dune. *A.* The windstream becomes compressed above a dune and consequently increases in velocity. The dune ceases to grow taller when its height becomes so great that it causes the windstream to move rapidly enough to transport sand. As sand passing over the dune accumulates on the steep leeward slope, the dune begins to "crawl" in that direction. *B.* This cross section of a dune shows the cross-bedding that results from shifts in the wind. As the wind direction changes, the shape of the dune is altered by removal of sand, and a new leeward slope forms. *C.* Cross-stratified dune deposits of the Jurassic Navajo Sandstone in Arizona. *(C. Tom Bean.)*

Sand Dunes

Sand dunes are hills of sand that have been piled up by the wind (p. 50). In some desert areas, including Death Valley, dunes occupy less than 1 percent of the total area; but in areas where dunes are well developed, they form magnificent landscapes. Dunes are familiar sights not only in desert terrains but also landward of the sandy beaches that border oceans and large lakes. Where the wind blows across loose sand, a dune can begin to form over any obstacle that creates a wind shadow in which sand can accumulate. Figure 3-10*A* shows how a dune tends to "crawl" downwind as sand from the upwind side moves over the top of the dune and accumulates on the downwind side. As the prevailing wind direction shifts back and forth, the direction in which the dune migrates shifts as well. This shift in direction usually leads to the truncation of preexisting deposits, which often causes a new set of beds to accumulate on a curved surface cut through older sets. Thus dunes are characterized by **trough cross-stratification.** Figure 3-10*B* and *C* show an idealized cross section through a dune and a real section through an enormous lithified dune that is more than 200 million years old.

RIVER SYSTEMS AS DEPOSITIONAL ENVIRONMENTS

In areas that receive more rainfall than those we have been considering, the precipitation is more likely to create the features of exterior drainage—small streams that meet to form larger streams, which in turn flow into still larger rivers. Ultimately the waters of most large rivers reach the sea. Figure 3-11 summarizes the sequence of aquatic environments through which water typically passes as it moves from the headwaters of river systems in hills or mountains to the deep sea, transporting and depositing sediment in the process. Each of these depositional environments is represented by many different sedimentary units in the geologic record. In the remainder of this chapter we will investigate the diagnostic features

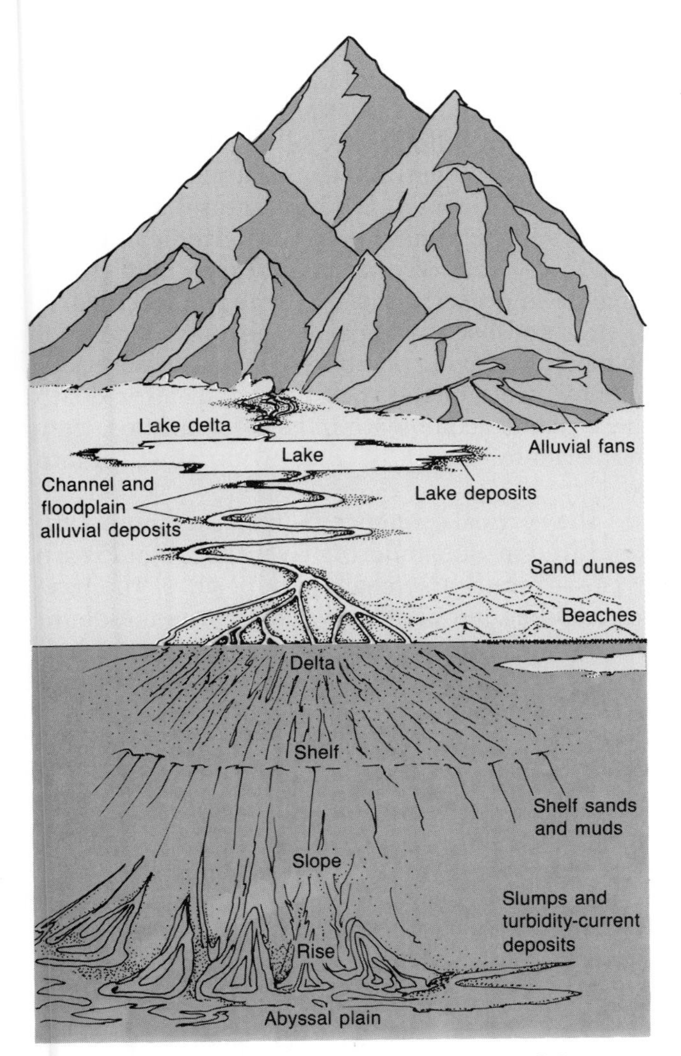

FIGURE 3-11 An idealized representation of the downhill course of sediment transported from mountains to the sea. Along the way, sediment is trapped in many different depositional environments. *(After F. Press and R. Siever, Earth, W. H. Freeman and Company, New York, 1986.)*

of ancient deposits formed in each of these environments — features that enable us to recognize each type of deposit as far back in the geologic record as a few hundred million or even 2 or 3 billion years.

Alluvial Fans in Moist Climates

Sediment in the form of alluvial fans accumulates at the feet of mountains and steep hills in both moist and arid regions. In moist climates, however, water often spreads these coarse sediments farther from their source, producing fans that slope more gently than those of dry basins.

Braided-Stream Deposits

Runoff water moves rapidly over the relatively steep, unvegetated terrain that is commonly found on the lower reaches of an alluvial fan and in the region just beyond it. As a result, this water transports large volumes of coarse sediment. In such an environment, a river does not flow as a simple winding waterway like the Mississippi or the Thames. Instead, it follows a complex network of interconnected channels, which give it the name *braided stream* (Figure 3-12). The areas between the channels are elevated "bars" of coarse sediment.

Meandering Rivers

Many rivers occupy solitary channels that wind back and forth like ribbons (Figure 3-13). Unlike braided streams, these **meandering rivers** are not choked with sediment — that is, sediment is supplied to them slowly relative to the rate at which the water flows. Rivers usually meander most

FIGURE 3-12 A braided stream in Alaska. The water and sediment emerge from the base of a melting glacier that is out of view. *(Tom Bean.)*

FIGURE 3-13 A meandering river. Sand forms point bars on the insides of the meanders. *(Peter Kresan.)*

actively in regions far from uplands and in gently sloping terrain. If there is any irregularity in this terrain, the river's path will curve, and centrifugal force will then cause the water to flow most rapidly near the outside of the bend and least rapidly near the inside. The river will then tend to cut into the outer bank of the bend, where the current is swift. On the inside of the bend, the current is so weak that sediment is deposited rather than eroded, and it accumulates there to form what is known as a **point bar.**

The sediment that forms point bars usually consists of sand. Most of this sand is cross-bedded, because large ripples along the riverbed where it accumulates migrate in the course of time. In deeper water, where the current is stronger, the sediment in the river channel is coarser (Figure 3-14). Gravel is often found along with coarse sand in the deepest part of the channel, and this gravel moves only when the river is flowing strongly.

Because mud tends to move downstream, very little of it accumulates within a meandering-river channel. When the river overflows its banks, however, it carries fine sediment laterally to the lowlands adjacent to the river channel. In these areas, known as **backswamps,** the spreading floodwaters flow slowly, allowing mud to accumulate before they recede. In keeping with the normal pattern of sediment deposition, these floodwaters become progressively slower as they flow away from the channel, and thus they tend to drop the coarser portion of their suspended sediment before they spread far from the channel. Sand is therefore dropped first, followed by silt, and together they form a gentle ridge called a **natural levee,** which borders the channel. Because natural levees and backswamps are inundated only periodically, they tend to dry out and crack. Many such mud cracks have been preserved in the stratigraphic record. Levees and backswamps also become populated by moisture-loving plants, which may leave traces of their roots in the rock record. If after death these plants accumulate in large quantities, they may even form deposits that eventually turn into coal.

The vertical sequence in which sediments are deposited in a meandering river is seen in Figure 3-14—from coarse channel deposits at the base to cross-bedded point-bar sands in the middle to muddy backswamp deposits at the top. Levee sediments sometimes lie between the point-bar and backswamp deposits. In summary, the sequence passes from coarse-grained sediments at the bottom to fine-grained sediments at the top. This

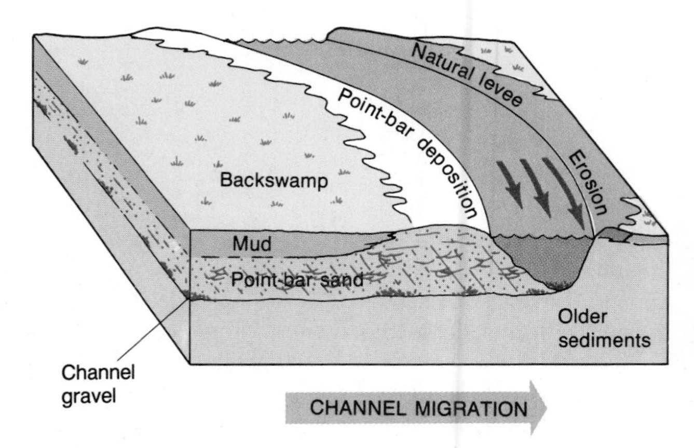

FIGURE 3-14 Deposition of sediment by a meandering river. At each bend the river migrates outward. The current flows most rapidly on the outside of a bend *(long arrow)*, so the water cuts into the outer bank. On the inner side of the bend, where the current moves slowly, sand accumulates to form a point bar. As the channel migrates outward, the point-bar sands advance over the coarser, cross-bedded sands deposited within the original channel. Pebbles often accumulate at the base of the channel. Muddy backswamp deposits, which form when the river floods its banks, migrate in their turn over the point-bar sands. As a result of this shifting of environments, coarse sediments at the base grade upward into fine sediments at the top.

coarse-to-fine sequence forms because as the channel migrates laterally, the point bar builds out over deeper gravels and the backswamp shifts over older point-bar deposits. As the channel migrates across a broad area, this sequence of deposits forms a broad composite depositional unit. The meandering-river sequence shown in Figure 3-14 illustrates **Walther's law,** which states that when depositional environments migrate laterally, sediments of one environment come to lie on top of sediments of an adjacent environment.

In a broad basin that happens to be subsiding, a river may migrate back and forth over a large area many times, piling one coarse-to-fine composite depositional unit on top of another as it goes. Each of these composite units, or **sedimentary cycles,** lies unconformably on the one beneath it, because the channel in which the basal deposits accumulate removes the uppermost sediments of the preceding cycle as it migrates. Sometimes, however, a channel cuts deep into the sediments of the preceding cycle, removing not only the uppermost deposits but some lower ones as well. Thus many of the cycles that have been preserved in the geologic record of a migrating channel are really only partial cycles. Figure 3-15 shows meandering-river cycles of the Catskill Formation in Pennsylvania.

Two final points deserve mention. First, not every river can be assigned to the braided or meandering category. Some rivers have segments or branches that are braided and others that are meandering. Second, no river deposits its entire load of sediment in the river valley. A river ultimately discharges its water and remaining sediment into a lake or the sea (Figure 3-11).

DELTAS

When a river empties into either a lake or the sea, its current dissipates, and it often drops its load of sediment in a fanlike pattern. The depositional body of sand, silt, and clay that is formed in this way is called a **delta** because of its resemblance to the Greek letter Δ. Most of the large deltas that have been well preserved in the geologic record formed in areas where sizable rivers emptied into ancient seas.

We have already seen that as moving water slows down, it drops first sand, then silt, and then

FIGURE 3-15 Cyclical meandering-river deposits of the Catskill Formation at Peters Mountain, Pennsylvania. Each cycle contains deposits that grade upward from coarse sandstone (C) to fine mudstone (F). *(F. J. Pettijohn and P. E. Potter, Atlas and Glossary of Primary Sedimentary Structures, Springer Publishing Company, Inc., New York, 1964.)*

clay. As river water mixes with standing water and begins to slow down, it, too, loses sand first. Silt, which is finer, spreads farther from the mouth of the river, and clay is carried even farther. The typical result is a delta structure that includes **delta plain, delta front,** and **prodelta** deposits (Figure 3-16).

Delta-plain beds, which consist largely of sand and silt, are nearly horizontal except where they are locally cross-bedded. Some delta-plain deposits accumulate within channels. As a river slows down on the surface of the delta, sand builds up on the bottom, causing the river channel to branch repeatedly into smaller channels that radiate out from the mainland. These **distributary channels** are floored by cross-bedded sands. Sand also spills out from the mouths of the channels, forming shoals and sheetlike sand bodies along the delta front. Between the distributaries and separated from them by levees are swamps, which are sometimes dotted by lakes. Here, as in the overbank areas of meandering rivers, muds accumulate and marsh plants often grow, contributing to future coal deposits.

Delta-front (or foreset) beds slope seaward from the delta plain, usually lying in waters that are deeper than those agitated by wind-driven surface waves. These beds consist largely of silt and clay, which can settle under these quiet conditions. Because they lie fully within the marine system, delta-front muds harbor marine faunas that often leave fossil records, but these muds usually contain fragments of waterlogged wood as well. In fact, the presence of abundant fossil wood in ancient marine muds testifies to the presence of both land and a river system near the site. In short, most ancient subtidal muds that harbor abundant fossil wood debris represent deltaic deposits.

Spreading seaward at a low angle from the lowermost delta-front deposits are the prodelta (or bottom-set) beds, which consist of clay. Even

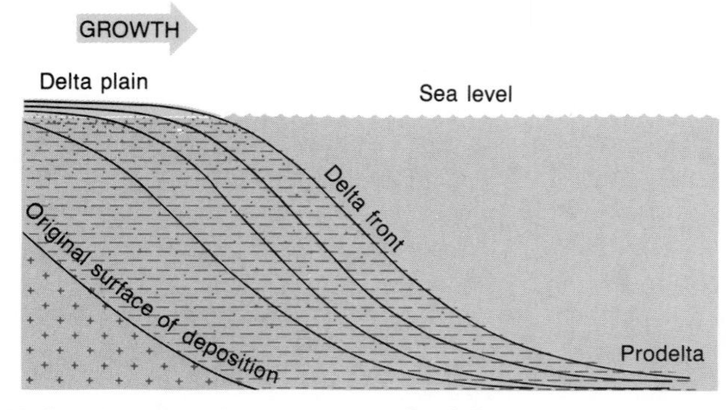

FIGURE 3-16 Cross section of a delta. As river water flows into the sea, it slows down. First sand drops from suspension in the delta plain; then silt and clay settle on the delta front (the slope is greatly exaggerated in the diagram); and finally clay settles in the prodelta. As the delta grows seaward, the sandy shallow-water sediments build out over the finer-grained sediments deposited in deeper water.

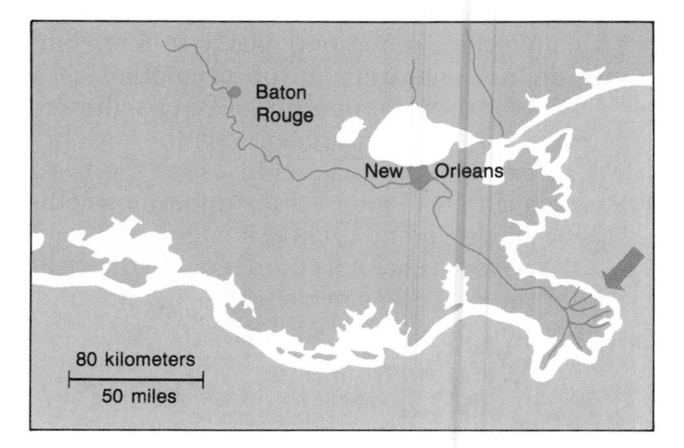

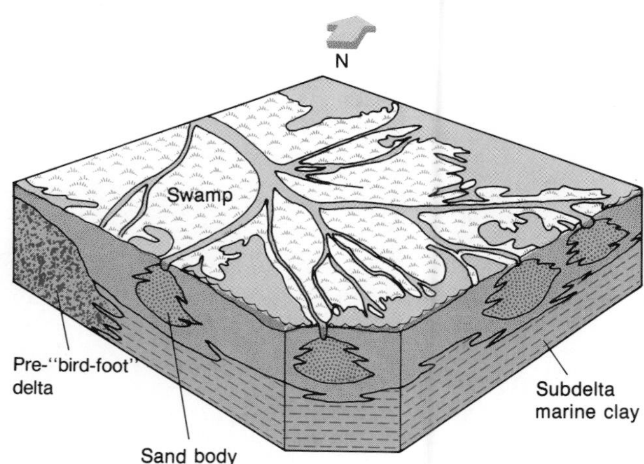

FIGURE 3-17 Structure of the "bird-foot" lobe of the modern Mississippi delta. At this lobe (identified by the arrow on the map) sediments are being deposited in the shallow water where the river meets the sea. Here, as shown in the block diagram below, the river channel divides into many distributary channels. Swamp deposits accumulate between the distributaries. Bodies of sand accumulate in front of the distributaries where the river water meets the ocean and slows down. As the bird-foot delta builds seaward, the distributaries extend over these sand bodies. *(Modified from H. N. Fisk et al., Jour. Sedim. Petrol. 24:76–99, 1954.)*

during floods, the fresh water that spreads from distributary channels slows down so abruptly that it loses its silt on the delta front. It is partly because fresh water is less dense than saline marine water that clay is carried far from the mouths of distributary channels. Because the fresh water floats on top of the more dense marine water, it does not mix in quickly; instead, it spreads seaward for some distance, carrying much of its clay.

As a delta **progrades,** or grows seaward, relatively coarse deposits on the delta plain build out over finer-grained foreset beds in accordance with Walther's law (Figure 3-14), and these delta-front beds build out over still finer-grained bottom-set beds. The result is a sequence of deposits that coarsens toward the top. As we will now see, the stratigraphic expression of this upward-coarsening sequence varies in accordance with the nature of the delta.

The famous Mississippi River delta has built out into the Gulf of Mexico in an area that is protected from strong wave action. As a result, this delta projects far out into the sea. Because construction of the delta from river-borne sediment prevailed decisively over the destructive forces of the sea, this type of delta is sometimes called a river-dominated delta.

That portion of the Mississippi delta which has grown at any given time has been much smaller than the delta as a whole. This growing portion, or **active lobe,** is the site of the functioning distributary channels (Figure 3-17). Many previously active lobes can be identified in the delta-plain portion of the Mississippi delta, and these lobes, which have been dated by the carbon 14 method (described in Chapter 4), provide a history of deltaic development during the past few thousand years (Figure 3-18). Depositional activity (or lobe growth) periodically shifted in this delta when floods caused the river to cut a new channel and to abandon the previously active channel and its distributaries.

The fate of an abandoned delta lobe is, in fact, the key to understanding the stratigraphic sequence produced by a river-dominated delta. In short, an abandoned lobe gradually sinks, for two reasons: First, the sediments of which it is formed compact under their own weight; and second, the lobe is part of the entire delta structure, which is constantly sinking as a result of the isostatic

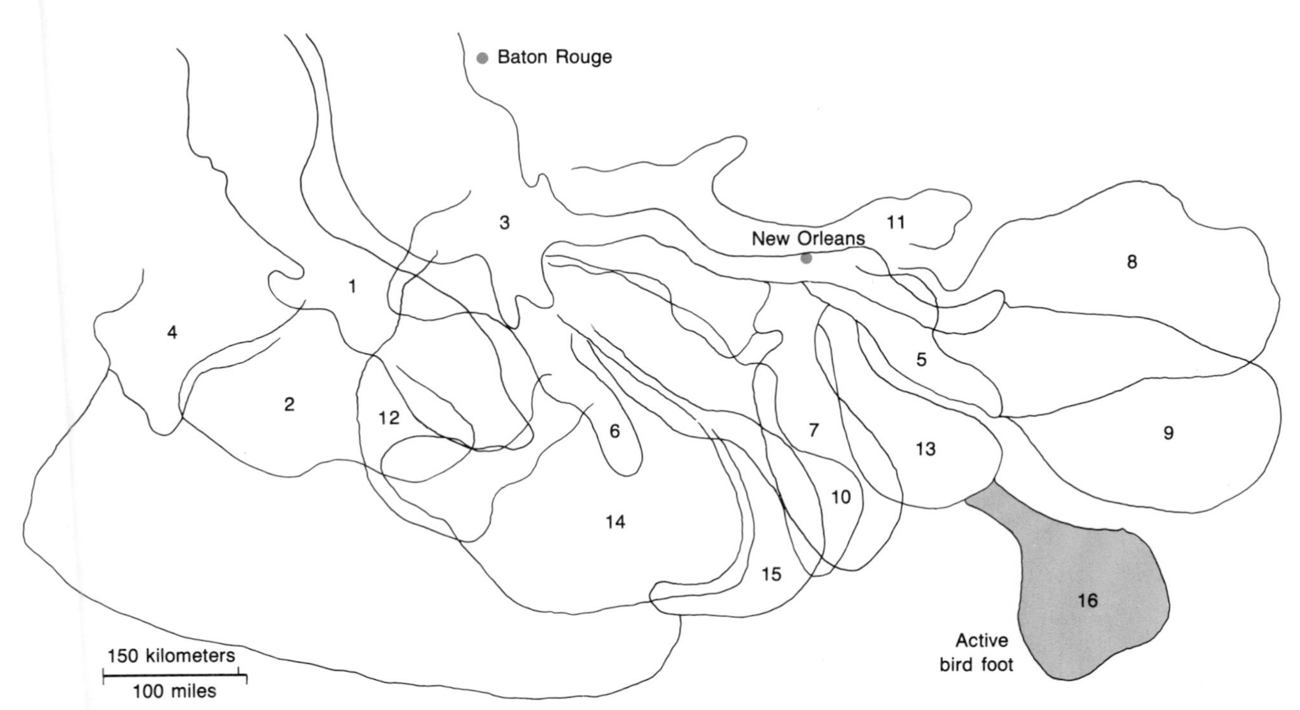

FIGURE 3-18 Lobes of the Mississippi delta. The active bird-foot lobe is numbered 16. Older lobes that are now inactive are numbered 1 through 15, with number 1 being the oldest. The older lobes are now settling, and some are already partially or entirely submerged beneath the sea. *(After T. Elliott, in H. G. Reading [ed.], Sedimentary Environments and Facies, Blackwell Scientific Publications, Ltd., Oxford, 1978.)*

response of the underlying crust to the weight of the continually growing mass of sediment. After an abandoned lobe settles, a younger lobe will eventually grow on top of it. Each lobe will then consist of the typical upward-coarsening sequence. The result is an accumulation of cycles that differ markedly from those of meandering rivers, which, it will be recalled, become finer-grained toward the top.

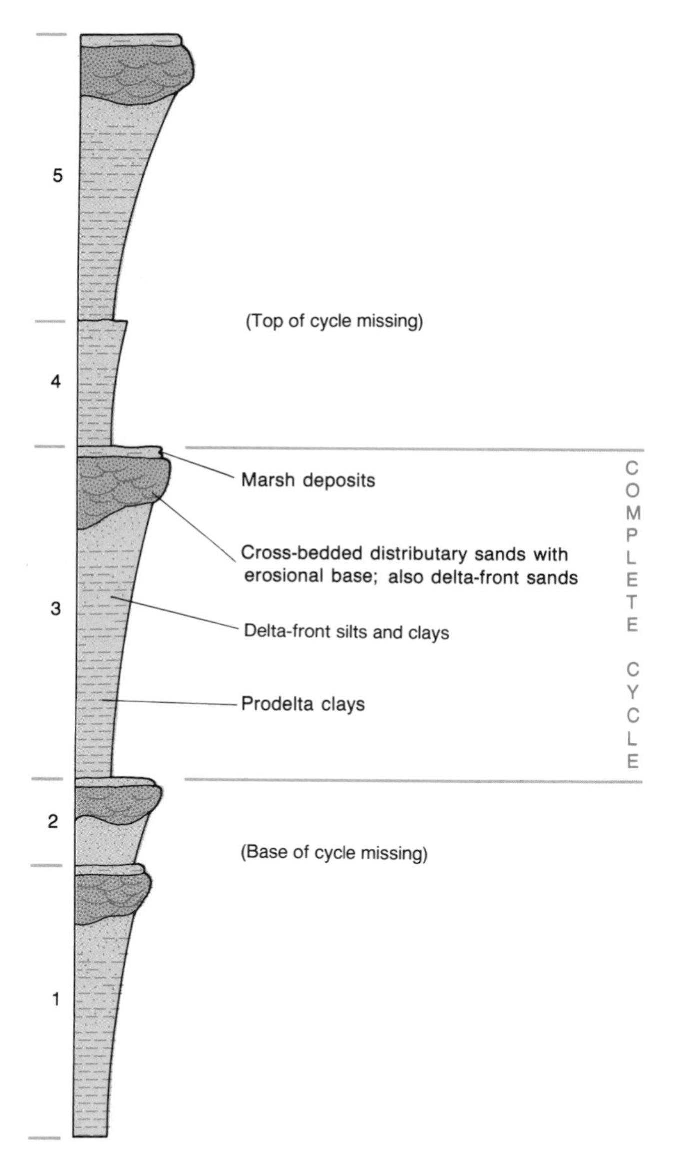

FIGURE 3-19 An idealized stratigraphic section representing five deltaic cycles. Cycle 3 is a complete cycle; within it bottom-set clays pass upward to top-set sands. Each cycle represents an accumulation of sediments resulting from the seaward growth of a deltaic lobe.

Some deltaic cycles in the rock record lack tops because sediment was eroded away before another cycle was superimposed (Figure 3-19). Because the building of a river-dominated delta is limited to the active lobe at any given time, beds representing a delta can seldom be traced very far laterally in the rock record.

Changes in the rate at which a delta sinks or the rate at which its river supplies it with sediment can alter the size of the active lobe. Today such changes are causing the Mississippi's bird-foot delta to shrink rapidly, with alarming consequences (Box 3-1).

THE BARRIER ISLAND–LAGOON COMPLEX

Because deltas form exclusively at river mouths, they occupy only a small percentage of the total shoreline of the world's oceans. Much longer stretches of shoreline are fringed by barrier islands, composed largely of clean sand that has been piled up by waves. Although some barrier islands extend laterally from wave-dominated deltas, most such islands derive their sand not from neighboring rivers but from the marine realm. They are built up as waves and shallow **longshore currents** winnow sediments and sweep sand parallel to the shoreline. In the beach zone that is washed by breaking waves, deposits tend to have nearly horizontal bedding but often dip gently seaward. Cross-bedding develops in areas where the beach surface is gently irregular and changes from time to time. Wind-blown sand often accumulates behind the beach as sand dunes, but in time these dunes are often eroded away.

Shallow **lagoons** lie behind long barrier islands such as those that border the Texas coast of the Gulf of Mexico (Figure 3-20). Because they are protected from strong waves, these lagoons trap fine-grained sediment and are usually floored by muds and muddy sands. Rivers that empty into lagoons often build small deltas.

Other depositional environments are also found along the shores of lagoons. Among them are **tidal flats** formed of sand or muddy sand whose surfaces are alternately exposed and flooded as the tide ebbs and flows. High in the intertidal

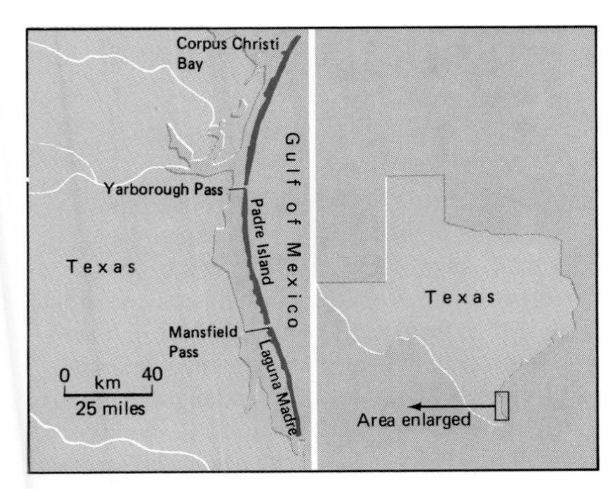

FIGURE 3-20 Barrier islands along the Texas coast. Tides along the Texas coast are weak, and there are few tidal channels or passes, so large permanent lagoons lie behind the Texas barrier islands.

zone, above barren tidal flats, marshes fringe one or both margins of many lagoons (Figure 2-25). Here plant debris accumulates rapidly and decomposes to form peat, which can become coal after a long period of burial.

Fresh water from rivers and streams tends to remain trapped in coastal lagoons for some time. Thus the waters of lagoons in moist climates are often brackish. The salinity of these waters at any given time depends on the rate of freshwater runoff from the land, which varies in the course of the year. Laguna Madre of Texas is typical of lagoons found in more arid climates, where temperatures are high (Figure 3-20). Because they receive little fresh water from rivers and suffer a high rate of evaporation, the ponded waters of this long lagoon are hypersaline (p. 48). Whether lagoons are brackish or hypersaline, however, their abnormal and fluctuating salinity excludes many forms of life that require normal marine salinity. As a result, the fossil faunas in the ancient sediments of lagoons are not very diverse. Those species that are found in lagoons, however, often occur in large numbers. Usually some of them are burrowers that disturb the muddy sediments of the lagoonal floor, leaving these sediments either mottled or homogeneous and largely devoid of bedding structures.

When a **barrier island–lagoon complex** receives sediment at a sufficiently high rate, it progrades—that is, it migrates seaward—like the active lobe of a delta. Unlike the migration of a delta, however, this progradation takes place along a broad belt of shoreline (Figure 3-21). As the shoreline of a sea migrates landward, marsh and tidal-flat deposits prograde over sediments of the lagoon and associated tidal channels. These sediments in turn build out over deposits of the barrier beaches and over the tidal deltas and marshes behind them. Thus the horizontal sequence of environments (barrier beach, marsh or tidal delta, lagoon, tidal flat, and marsh) come to be represented by a corresponding vertical sequence of sedimentary deposits, in accordance with Walther's law.

The barrier island–lagoon system illustrates particularly well how the close vertical association of sediments that represent adjacent depositional environments can be used to identify ancient depositional systems. We have seen that vertical associations of rock types are also very useful in the identification of ancient meandering-river deposits (Figure 3-15) and deltas (Figure 3-19).

ORGANIC REEFS

In tropical shallow marine settings, if siliciclastic sediments are in short supply, carbonate sedimentation usually prevails. Here coral reefs are

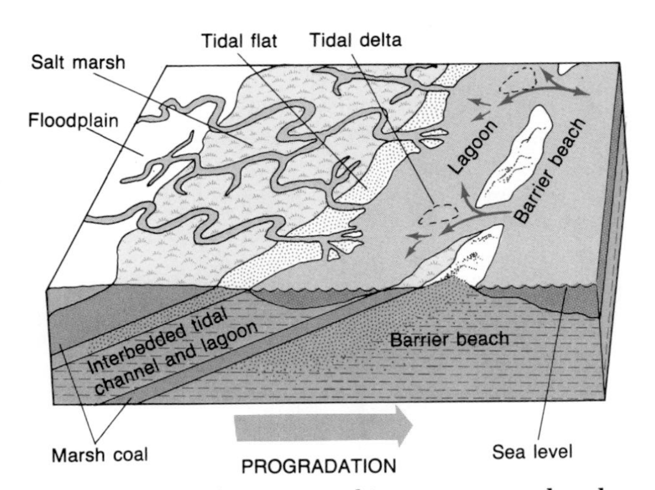

FIGURE 3-21 The stratigraphic sequence produced when a barrier island–lagoon complex progrades. Sediments of the lagoon and of the adjacent marshes and tidal flats are superimposed on beach sands of the barrier island.

Box 3-1
The Shrinking Mississippi Delta

The coastal wetlands of Louisiana provide nearly 30 percent of the United States' annual seafood harvest. Every winter these wetlands support the nation's largest population of waterfowl. Although vitally important to the entire nation, the wetlands are now disappearing at an alarming pace. Land along the Louisiana coast, which is mostly wetlands, is shrinking at a rate of about 100 square kilometers (~40 square miles) a year.

The Mississippi delta dominates the Louisiana coastline (pp. 62–64). In the past, sediment carried by the Mississippi River accumulated rapidly and built the delta out into the Gulf of Mexico. Today the river is no longer able to expand or even maintain its delta. Now the marine waters of the Gulf are drowning the delta, reducing the area available to plants and animals adapted to fresh and brackish water in marshes. This change has come about for two reasons. First, the Mississippi has been bringing less sediment to the delta every year. Second, the delta has been subsiding more rapidly than it once did.

The average volume of suspended sediment in the lower Mississippi was only half as large in 1980 as it was in 1950. Dams built on upstream tributaries of the river are largely responsible for this change. The dams have been beneficial in some ways: They have reduced flooding during times of high runoff from the land, and they have produced lakes for human recreation. On the other hand, the dams have trapped sediment that would otherwise have traveled all the way to the Mississippi delta. At the same time, artificial levees built along the lower Mississippi to reduce local flooding have kept floodwaters from carrying sediment to some areas that otherwise would be floodplains.

An increase in the rate at which the massive delta has subsided has done even more than dams and artificial levees to prevent the buildup of sediments from keeping the delta above water. In fact, the shoreline of the Gulf of Mexico, from Florida to Texas, began to sink very rapidly just before the middle of the twentieth century. The positions of ancient shorelines indicate that during the past thousand years the average rate of sinking has been only about 1.55 millimeters (~$\frac{1}{16}$ inch) a year. In recent years the rate has been nearly five times greater. Again, human activities appear to have caused the change. So much water has been pumped out of the ground around the margin of the Gulf of Mexico in recent decades that the soft sediments of the region have become compacted, and the surface of the land has been sinking much more rapidly than it did in earlier decades. As it sinks, the waters of the Gulf encroach farther and farther over the delta, drowning valuable wetlands.

Perhaps the best way to reduce the rate at which wetlands are disappearing in Louisiana would be to restrict the outflow of the river into the Gulf to a small number of distributaries, which would then deposit a large percentage of the sediment that the river carries. At least in these areas, the buildup of sediments might compensate for the sinking of the land. Wetlands would continue to shrink in other areas, but here, at least, they might be sustained. Unfortunately, even this strategy would not entirely solve the problem. The wetlands of Louisiana, which sustain such rich faunas of waterfowl and edible aquatic life, will continue to shrink for years to come.

A region of the Mississippi delta is being flooded as it subsides. The white objects are floats used to position a pipe that will carry petroleum from an offshore well. *(Donald Davis, Louisiana Geological Survey.)*

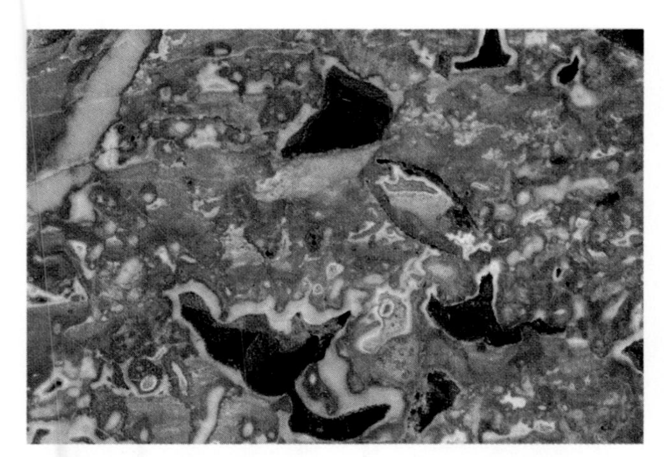

FIGURE 3-22 Polished slab of porous reef rock from the Triassic of Italy. The light-colored band in the upper left is a longitudinal section of a tubular sponge skeleton. Other visible particles are skeletal material of calcareous algae that also contributed to the reef framework. The dark areas are cavities. This portion of the slab is about 10 centimeters (~4 inches) across. *(Lyndon Yose.)*

often prominent. Modern coral reefs are rigid structures that rise above the seafloor in shallow waters of high clarity and of normal marine salinity (p. 46). As we shall see in later chapters, however, large reefs of the geologic past have been formed by organisms other than corals. Because they are produced largely by organisms that secrete calcium carbonate, organic reefs form their own distinct depositional records—as bodies of limestone.

A modern reef has several structural components that form a complex limestone deposit. The basic framework of a **reef** consists of the calcareous skeletons of organisms, primarily corals. This framework is strengthened by cementing organisms that encrust the surface of the reef. **Carbonate sediment,** which is composed of fragments of the skeletons of reef-dwelling organisms, is trapped within the porous framework, filling some voids. With their complex internal structure, reef limestones are typically either unbedded or only poorly bedded (Figure 3-22). Even with the presence of infilling debris, reef limestone is so porous that many ancient buried reefs serve as traps for petroleum, which migrates into them from sediments rich in organic matter.

Because living reefs stand above the neighboring seafloor, they alter patterns of sedimentation nearby. On the leeward side of an elongate reef there is often a relatively calm lagoon, especially if the reef has a typical **reef flat,** or horizontal upper surface, that stands very close to sea level (Figure 3-23). Below the living surface of the

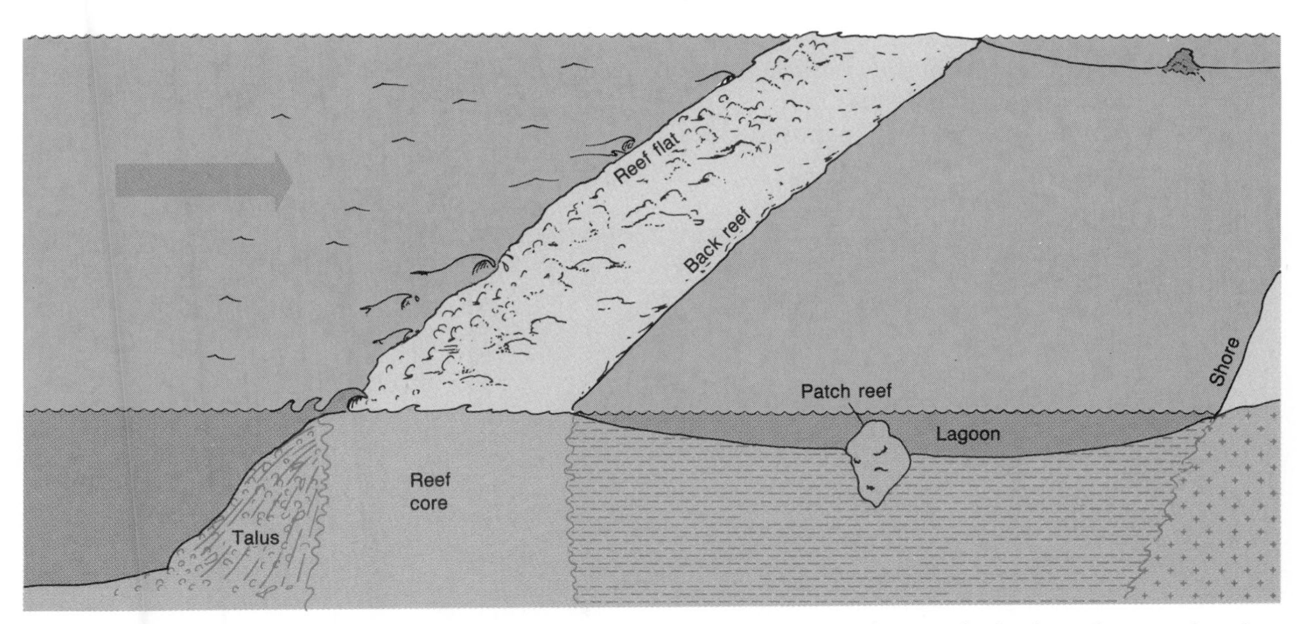

FIGURE 3-23 Diagram of a typical barrier reef. The reef grows up to sea level; thus the reef flat is exposed at low tide. Waves break across the reef flat, losing energy in the process and leaving a quiet lagoon behind. Sediment accumulates in the back-reef area and in the lagoon, and here and there patch (or pinnacle) reefs rise up from the lagoon floor.

FIGURE 3-24 Outcrop along Windjana Gorge, northwestern Australia, revealing the internal structure of a Devonian reef. The reef core consists of unbedded limestone. The talus is crudely bedded, and the beds slope away from the reef core. The back-reef facies is also crudely bedded, but the beds are approximately horizontal. *(From P. E. Playford, Geological Survey of Western Australia.)*

reef is a limestone core consisting of dead skeletal framework, dead skeletons of cementing organisms, and trapped sediment. A pile of rubble called **talus,** which has fallen from the steep, wave-ridden reef front, often extends seaward from the living surface.

Reefs build upward rapidly enough to remain near sea level even when the seafloor around them is becoming deeper. Many reefs, in fact, grow so rapidly and are so durable that they build seaward in the manner of a prograding delta (Figure 3-16). Figure 3-24 shows a spectacularly exposed cross section of a Devonian reef in Australia. Although it was built by organisms that have been extinct for hundreds of millions of years, this reef closely resembles many modern reefs in its basic structure—it displays both a seaward talus slope and a leeward reef flat.

Isolated **patch reefs,** or **pinnacle reefs,** are often found in lagoons behind elongate reefs (Figure 3-23). Elongate reefs that face the open sea and have lagoons behind them are known as **barrier reefs** (Figure 3-25). Reefs that grow right along the coastline without a lagoon behind them

FIGURE 3-25 A barrier reef. The open sea is on the right, the lagoon, on the left. *(Ralph and Daphne Keller/ NHPA.)*

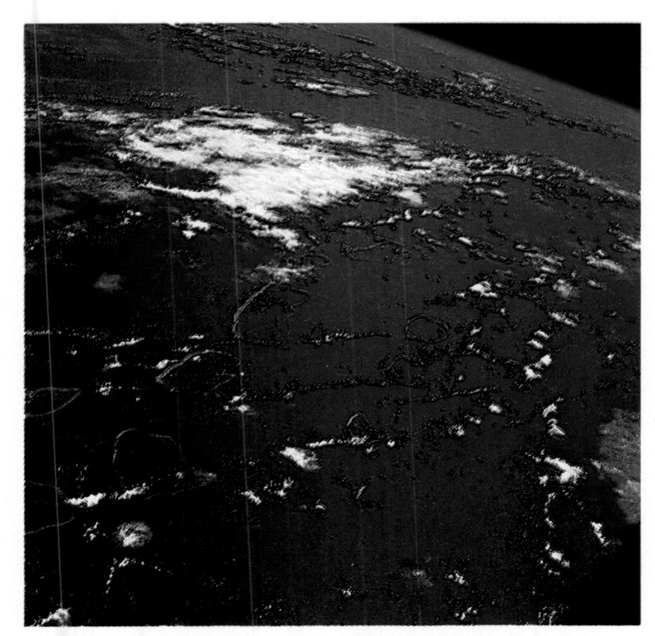

FIGURE 3-26 Coral-reef atolls of the Tuamotu Archipelago, near Tahiti in the South Pacific. The widest segments of the circular atolls face the prevailing winds from the upper left. *(NASA.)*

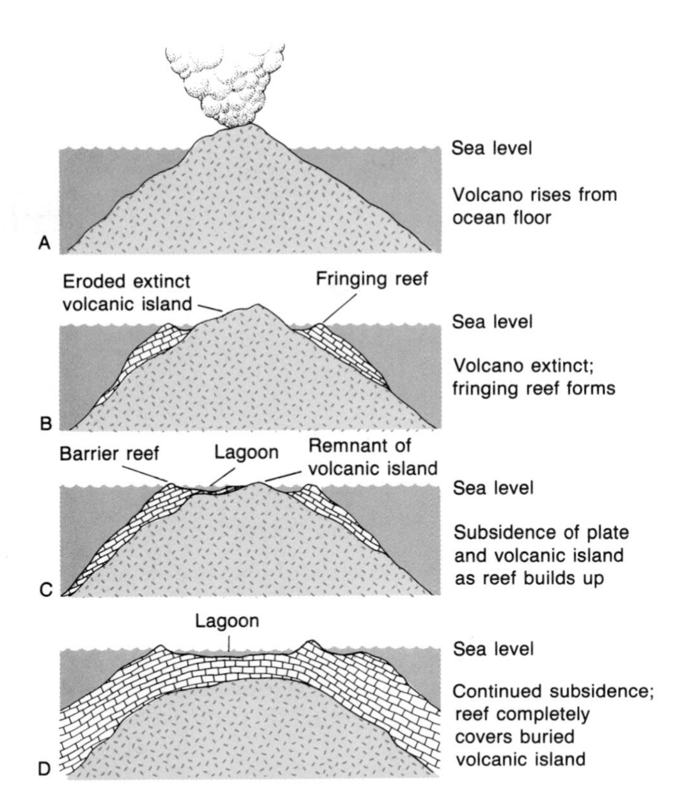

FIGURE 3-27 The development of a typical coral atoll in the Pacific as proposed by Charles Darwin. *(After F. Press and R. Siever, Earth, W. H. Freeman and Company, New York, 1982.)*

are known as **fringing reefs.** Some fringing reefs grow seaward and eventually become barrier reefs.

Perhaps the most curious reefs in the modern world are the circular or horseshoe-shaped structures known as **atolls** (Figure 3-26). Atolls frequently form on volcanic islands and thus are quite common in the tropical Pacific, which is dotted by many such islands. Charles Darwin offered an explanation for the origin of the Pacific atolls that is still accepted today (Figure 3-27). According to Darwin, each atoll was formed when a cone-shaped volcanic island was colonized by a fringing reef. The island eventually sank beneath the sea, leaving a circular reef standing alone with a lagoon in the center, where limestone now accumulates in quiet water. Often the reef does not quite form a full circle but is instead broken by a channel on the leeward side, where reef-building organisms do not thrive. Horseshoe-shaped atolls range up to about 65 kilometers (~ 40 miles) in diameter; during World War II their lagoons served as natural harbors for ships.

Ancient atolls that lie buried beneath younger sediments can be identified by the study of cores of sediment brought up from drilling operations—and because porous reef rocks often serve as traps for petroleum, drilling in the vicinity of these atolls is often a profitable venture. Figure 3-28 shows the outline of a subsurface atoll of late Paleozoic age that has yielded considerable quantities of petroleum in the state of Texas.

CARBONATE PLATFORMS

A **carbonate platform** is a structure that is formed by the accretion of carbonate sediment and that stands above the neighboring seafloor on at least one of its sides. Like organic reefs, carbonate platforms consist largely of calcium carbonate that originated in shallow tropical waters at or near the site where they accumulated. In fact, organic reefs often grow along the windward margins of carbonate platforms and form parts of them.

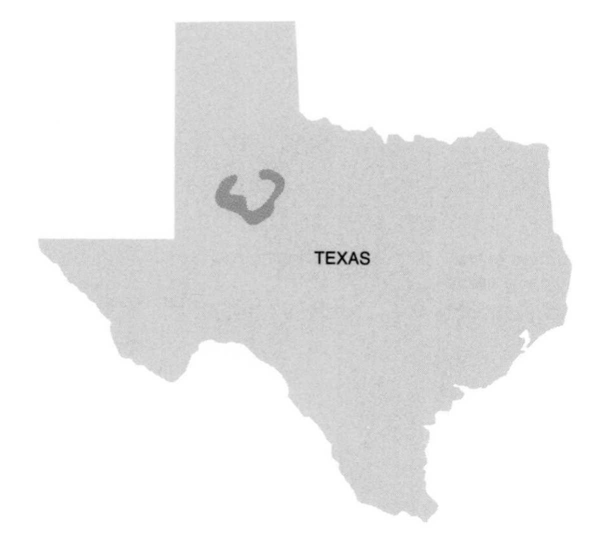

FIGURE 3-28 A late Paleozoic horseshoe-shaped atoll that lies almost a kilometer (~0.7 mile) below the surface of the land in Texas. The atoll was discovered when rocks of the region were drilled for petroleum. It would appear that the reef faced prevailing winds from the south. *(After P. T. Stafford, U.S. Geol. Surv. Prof. Paper No. 315-A, 1959.)*

In times past, carbonate platforms have stretched along most or all of the eastern margin of the United States; but because climates are cooler today than they have been during most of Earth's history, these structures are not well represented in the modern world. The tropical areas of the modern world do offer several examples, however. In the western Atlantic a large carbonate platform extends seaward from the Yucatan Peninsula of Mexico. Smaller platforms border the islands of the Antilles, and the platforms known as the Little Bahama Bank and Great Bahama Bank lie to the east and southeast of Florida (Figure 3-29).

The varied sediments that are now accumulating on the Bahama banks resemble those of many ancient carbonate platforms. Here, as on carbonate platforms generally, sediments accumulate rapidly. Since mid-Jurassic time, about 170 million years ago, some 10 kilometers (~6 miles) of shallow-water carbonates have accumulated both on the Bahama banks and in southern Florida, which was part of the same carbonate platform during Cretaceous and earlier times. The heavy buildup of carbonate sediments in this region has caused the oceanic crust to bend downward, and shallow-water Jurassic deposits now lie up to 10 kilometers below sea level. During the Cenozoic Era, when carbonate sediments ceased to be deposited in some areas, deep channels developed. The channels, known as the Straits of Florida, Tongue of the Ocean, and Exuma Sound, now separate the Florida peninsula, the Little Bahama Bank, and the Great Bahama Bank.

Bordering tidal channels in some areas of the Bahama banks are knobby intertidal structures known as **stromatolites**, which are produced by

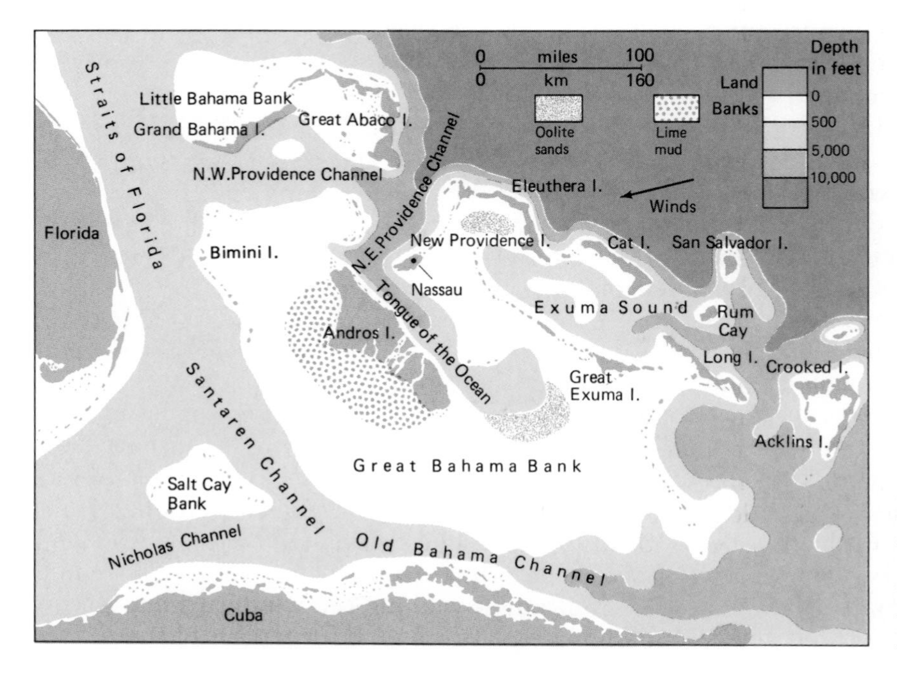

FIGURE 3-29 The Bahama banks, which are now separated from Florida by the Florida Straits. *(After N. D. Newell and J. K. Rigby, Soc. Econ. Paleont. and Mineral. Spec. Paper No. 5, 1957.)*

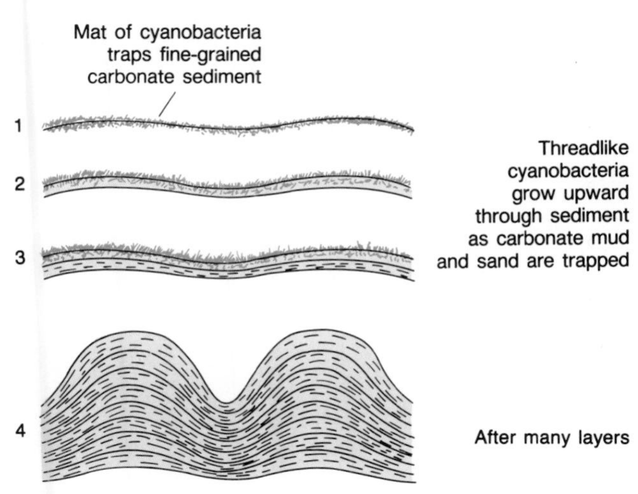

FIGURE 3-30 The growth of stromatolites. A mat of cyanobacteria traps mud, and the cyanobacteria grow through it to form another layer. The accumulation of several layers leads to the formation of a stromatolite.

FIGURE 3-31 Living stromatolites. Subtidal forms growing today in current-swept channels between the Exuma islands on the eastern Great Bahama Bank. The largest ones are about 2 meters (~6 feet) tall. *(E. A. Shinn, U.S. Geological Survey.)*

threadlike organisms. As Figure 3-30 indicates, these organisms form sticky mats by trapping carbonate mud. They then grow through the mat thus formed to produce another mat. Repetition of this process on an irregular surface forms a cluster of stromatolites. The threadlike organisms that form stromatolites are **cyanobacteria** (formerly known as blue-green algae), which, as we will see in later chapters, are among the most primitive groups of organisms in the world. The fossil record of stromatolites is unusually ancient, extending back through more than 3 billion years of geologic time.

There is a simple reason that stromatolites are found almost exclusively in supratidal and high intertidal settings. Because these environments are exposed above sea level much of the time, they become hot and dry, and relatively few marine animals can survive in them. Thus there is little to interfere with the tendency of the mats created by cyanobacteria to form layered structures. Such mats grow underwater here and there, but they are eaten by the many grazing marine animals and damaged by burrowers, so they seldom accumulate to form stromatolites or well-layered limestones. Exceptions to this rule are large column-shaped stromatolites that grow in subtidal channels in the Bahamas where tidal currents are very strong (Figure 3-31). Since few animals can survive in these current-swept areas, stromatolites flourish there. Stromatolites also flourish in Shark Bay, Western Australia, where the waters are hypersaline and animals are very rare. As we will see in later chapters, many very ancient stromatolites may have formed beneath the sea before the origin of marine animals that feed on cyanobacteria.

The backslopes of tidal channel levees in some areas of the Bahama banks dry out after occasional flooding by storm-driven seas. As a result, the surface of the sediment here, which is also bound by a cyanophyte mat, is broken by mud cracks resembling those that often form when a mud puddle on the land dries up. Mud cracks associated with ancient marine deposits usually represent intertidal or supratidal environments that were alternately wetted by the sea and dried by the sun (Figure 3-32).

FIGURE 3-32 Mud cracks in a dolomitic carbonate rock from the Carboniferous System of Ireland. *(I. M. West.)*

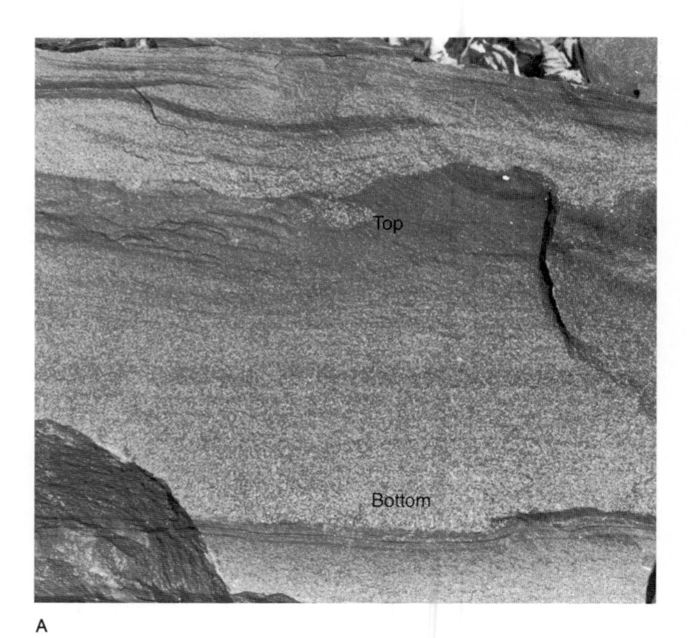

A

B

FIGURE 3-33 Ordovician turbidite beds in New York State. *A.* A turbidite bed that grades from coarse at the bottom to fine at the top; the upper surface is irregular because it was disturbed by the succeeding turbid flow and distorted by subsequent compaction. *B.* The bottom surface of a turbidite bed, showing "sole marks" produced when the sediment forming the bed filled depressions that the turbid flow depositing it scoured in the preexisting sediment surface; the current flowed toward the upper left. *(Earle F. McBride.)*

SUBMARINE SLOPES AND TURBIDITES

One of the greatest advances in the study of sedimentary rocks took place in the middle of the twentieth century with the recognition that certain sedimentary rocks have been produced by **turbidity currents.** A turbidity current is a flow of dense, sediment-charged water moving down a slope under the influence of gravity.

Turbidity currents were first noticed in clear lakes, where flows were observed to form from muddy river water that hugged the lake floor. These currents flowed for a considerable distance toward the center of the lake, slowing down and dropping their sediment only when they reached gentler slopes and spread out. In the 1930s the Dutch geologist Philip Kuenen demonstrated in the laboratory that turbidity currents can attain great speed, especially when they are heavily laden with sediment and are moving down steep slopes. Sediment suspended in a turbidity current behaves as part of the moving fluid, and its presence increases the density of this fluid by as much as a factor of two.

When a turbidity current reaches a place whose slope is more gentle than that from which it originated, it slows down and spreads out, dropping its sediment in the general sequence that we have now seen again and again: First the coarse sediment falls from suspension, and then, much later, the fine material follows. The result is a graded bed of sediments, with poorly sorted sand and granules at its base and mud at the top. Such a graded bed is known as a **turbidite** (Figure 3-33).

Large turbidity currents flow down continental slopes and deposit turbidites along continental rises and on the abyssal plain. Turbidity currents that originate near the edge of the continental shelf not only carry sediment to the deep sea but also erode both the continental slope and part of the continental rise. Such currents are also largely responsible for carving the great submarine canyons that incise many parts of the slope. The turbidity currents slowed down in front of these canyons, dropping part of their sedimentary load to form deep-sea fans that superficially resemble alluvial fans (Figure 3-34). In fact, much of the continental rise actually consists of coalescing submarine fans.

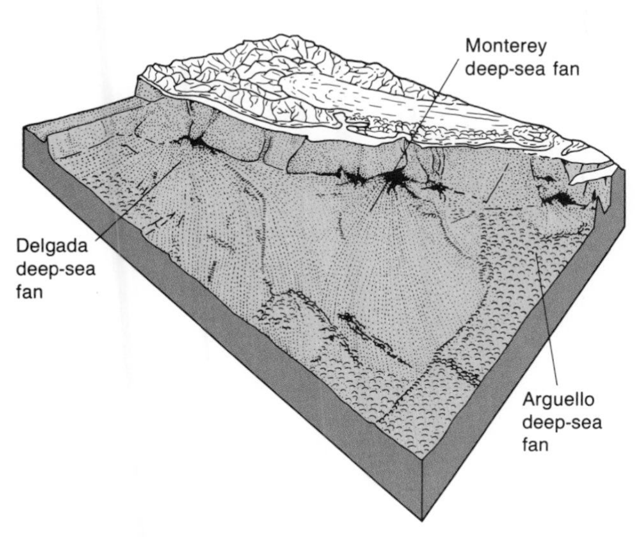

FIGURE 3-34 Deep-sea fans spreading from submarine canyons along the coast of California. The fans consist primarily of turbidite deposits. *(After H. W. Menard, Geol. Soc. Amer. Bull. 71:1271–1278, 1960.)*

FIGURE 3-35 A sequence of thick Miocene turbidite beds separated by thin shale beds along the Santerno River in northern Italy. *(From F. J. Pettijohn et al., Sand and Sandstone, Springer-Verlag, New York, 1973.)*

A closer look at turbidite deposits shows that their sands are typically graywackes (Appendix I) that are quite unlike the clean sands of meandering-river deposits. This is only one of several ways in which a cyclical sequence of turbidites differs from the meandering-river cycle, although both show a grading of sediment from coarse to fine within each complete cycle. Another difference is that a single turbidite is usually only a few centimeters and seldom as much as a meter thick (Figure 3-35), whereas most complete meandering-stream cycles measure several meters from bottom to top. In addition, turbidites lack the large-scale cross-bedding that characterizes the channels of a meandering river. In addition, the base of a turbidite is often irregular because the earliest, most rapidly moving waters of the turbid flow have scoured depressions in the sedimentary surface laid down earlier. These scours are then filled in by the first sediments that settle as the current slows. When the base of a lithified turbidite is turned over for inspection, its irregularities, or "sole marks," can reveal the direction of water flow (Figure 3-33*B*).

PELAGIC SEDIMENTS

Turbidity currents and other bottom flows carry mud to the abyssal plain of the ocean well beyond the continental rise. None of these flows, however, contributes sediment to the deep sea at a high rate; indeed, sediment in most areas of the abyssal plain accumulates at a rate of about 1 millimeter per 1000 years! In the deep sea, sparse sediments from turbidity currents are joined by clays from two other sources. One source is the weathering of rocks that have been produced by oceanic volcanoes. Such clays are less abundant in the Atlantic than in the central Pacific, where volcanoes are common. Clays also reach the deep sea by settling from the pelagic realm, which they reach after traveling through the air as windblown dust or moving seaward from the land at very low concentrations in surface waters of the ocean.

Clays that settle from the upper part of the water column are one type of **pelagic sediment.** Other types are contributed by small pelagic organisms whose skeletons become sedimentary particles when the organisms die. Whether clays predominate in any area of the deep sea depends on the extent to which they are diluted by the more rapid accumulation of biologically produced sediments. Some of the latter consist of calcium carbonate, while others consist of silica.

Calcium carbonate predominates in seafloor sediment at low latitudes (Figure 3-36), where its fine grain size has led oceanographers to refer to it

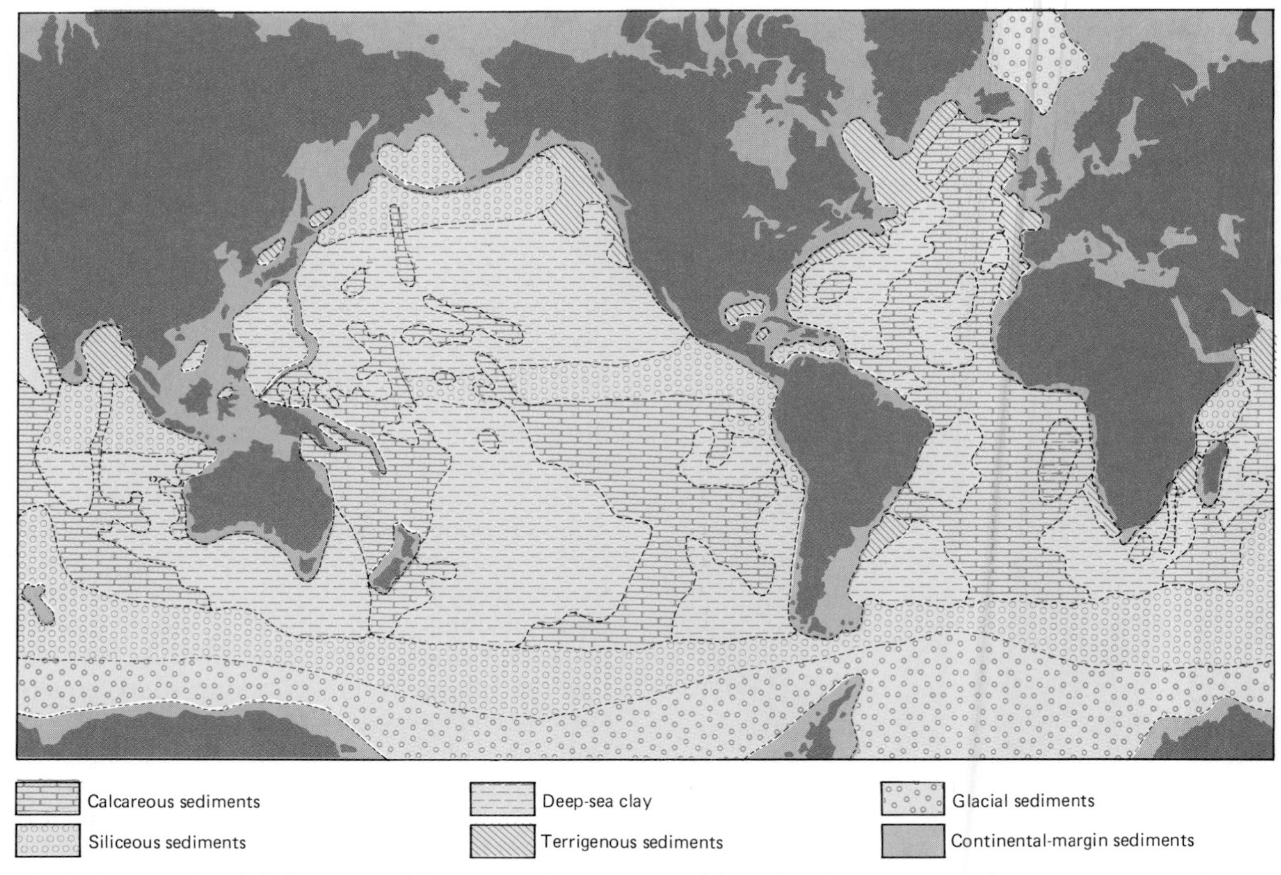

FIGURE 3-36 The global pattern of deep-sea sediments. Calcareous oozes are restricted to low latitudes. Most areas of siliceous ooze lie closer to the poles, although some occur close to the equator in areas of the Pacific and Indian oceans. *(After T. A. Davies and D. S. Gorsline, in J. P. Riley and R. Chester [eds.], Chemical Oceanography, Academic Press, Inc., London, 1976.)*

FIGURE 3-37 Siliceous ooze from the bottom of the Indian Ocean. The meshlike skeletons of radiolarians are the dominant components (× 100). *(Scripps Institution of Oceanography, University of California, San Diego.)*

as "ooze." This **calcareous ooze** consists of skeletons of single-celled planktonic organisms. Among them are the globular skeletons of planktonic foraminifera, which are amoebalike creatures. Other important constituents are the armorlike plates that surround calcareous nannoplankton, the single-celled floating algae that are major components of tropical phytoplankton (Figure 2-26; see also Figure 2-28).

Calcareous ooze is abundant only at depths of less than about 4000 meters (~13,000 feet). The reason for this limitation is that calcium carbonate tends to dissolve as it settles through the water column. As pressure increases and temperature declines with depth, the concentration of carbon dioxide increases; as a result, the water becomes undersaturated with respect to calcium carbonate and thus tends to dissolve it.

In many regions at high latitudes as well as in tropical Pacific regions characterized by strong upwelling, biologically produced **siliceous ooze** carpets the deep seafloor (Figure 3-37). This sediment consists of the skeletons of two groups of organisms that thrive where upwelling supplies nutrients in abundance: the diatoms, a highly productive phytoplankton group found in nontropical waters (Figure 2-26), and radiolarians, which, like planktonic foraminifera, are single-celled amoebalike creatures that float in the water and feed on other organisms. The skeletons of both diatoms and radiolarians consist of a soft form of silica similar to that of the semiprecious stone opal. When concentrated as siliceous sediment, these opaline skeletons tend to recrystallize and hence to lose their identity (Appendix I). In the process, the rock that they form becomes a dense, hard chert composed of finely crystalline quartz. Before recrystallization, the soft sediment is known as diatomaceous earth. The abrasive component of many scouring powders used in kitchens is diatomaceous earth.

Thick diatomaceous bodies of sediment have formed in marine areas of strong upwelling, where diatoms have been unusually abundant. Diatoms did not exist until late in Mesozoic time, however, and here we come to an important point: The composition of pelagic sediments has changed markedly in the course of geologic time as groups of sediment-contributing organisms have waxed and waned within the pelagic realm. We will examine the highlights of these changes in some of the chapters that follow.

CHAPTER SUMMARY

1 Modern environments where sediment is deposited offer valuable examples of the ways in which ancient sedimentary rocks formed. Sometimes one kind of rock alone serves to identify an ancient environment. Usually, however, suites of closely associated rock types are required for this purpose, and these are commonly organized into cycles in which one kind of sediment tends to lie above another.

2 Ancient soils are sometimes found beneath unconformities, although they may be hard to identify because of chemical alteration.

3 Lake deposits, which are much less common than marine deposits, are characterized by thin horizontal layers, by few burrows, and by an absence of marine fossils.

4 Glaciers, which plow over the surface of the land, often leave a diagnostic suite of features, including scoured and scratched rock surfaces, poorly sorted gravelly sediment, and associated lake deposits that exhibit annual layers.

5 In hot, arid basins on the land, erosion of the surrounding highlands creates gravelly alluvial fans. Braided streams flowing from the fans toward the basin center deposit cross-bedded gravels and sands. Beyond these deposits there may be shallow lakes and salt flats where evaporites accumulate. Some arid basins also contain dunes of clean, cross-bedded, wind-blown sand.

6 In moist climates, meandering rivers leave characteristic deposits in which channel sands and gravels grade upward through point-bar sands to muddy backswamp sediments.

7 Where a river meets a lake or an ocean, it drops its sedimentary load to form a delta; deltaic deposition typically produces an upward-coarsening sequence as shallow-water sands build out over deeper-water muds.

8 More widespread than deltas along the margin of the ocean are muddy lagoons bounded by barrier islands formed of clean sand.

9 Coral reefs border many tropical shorelines. A typical reef stands above the surrounding seafloor, growing close to sea level and leaving a quiet lagoon on its leeward side. Most reef limestones are supported by rigid internal organic frameworks. Coral reefs form parts of many carbonate platforms, although these platforms contain a number of other deposits as well.

10 Beyond the edge of the continental shelf, turbidity currents periodically sweep down continental slopes to the continental rise and deep seafloor, where they spread out, slow down, and deposit graded beds of sediment.

11 Still farther from continental shelves, only fine-grained sediments accumulate. Clay reaches these deep-sea areas very slowly. In some areas the deposition of clay is far surpassed by the accumulation of minute skeletons of planktonic marine life that settle to the seafloor to form calcareous or siliceous oozes.

EXERCISES

1 What kinds of nonmarine sedimentary deposits reflect arid environmental conditions?

2 What kinds of nonmarine sedimentary deposits reflect cold environmental conditions?

3 What kinds of deposits indicate the presence of rugged terrain in the vicinity of a nonmarine depositional basin?

4 In what nonmarine settings do gravelly sediments often accumulate?

5 Contrast the patterns of occurrence of sediments and sedimentary structures in the following three kinds of depositional cycles: the kind produced by meandering rivers, the kind produced by deltas, and the kind produced by turbidity currents.

6 Draw a profile of a barrier island–lagoon complex, and label the various depositional environments.

7 What features typify sediments that accumulate in the centers of lakes?

8 How do stromatolites form?

9 Describe the kind of rock found in a typical organic reef.

10 Where is a lagoon in relation to a barrier reef? Where is it in relation to an atoll?

11 Which features of carbonate rocks suggest intertidal or supratidal deposition? Which features suggest subtidal deposition?

12 What types of sediments and sedimentary structures usually reflect deposition in a deep-sea setting?

ADDITIONAL READING

Collinson, J. D., and D. B. Thompson, *Sedimentary Structures,* Unwin Hyman, London, 1989.

Davis, R. A., *Depositional Systems,* Prentice-Hall, Inc., Englewood Cliffs, N.J., 1983.

Miall, A., *Principles of Sedimentary Basin Analysis,* Springer-Verlag New York Inc., New York, 1990.

Prothero, D., *Interpreting the Stratigraphic Record,* W. H. Freeman and Company, New York, 1990.

Reading, H. G., *Sedimentary Environments and Facies,* Blackwell Publications, Palo Alto, Calif., 1986.

Reineck, H. E., and I. B. Singh, *Depositional Sedimentary Environments,* Springer-Verlag New York Inc., New York, 1986.

Selley, R. C., *Ancient Sedimentary Environments and Their Subsurface Diagnosis,* Cornell University Press, Ithaca, N.Y., 1985.

PART II

THE DIMENSION OF TIME

Rocks that represent particular intervals of the geologic time scale can often be recognized throughout the world by characteristic features. The foremost such features of Phanerozoic rocks are the fossils they contain. We can estimate the actual ages of rocks and fossils (in years) by measuring the degree to which radioactive elements within them have decayed. The kinds of organisms whose fossils are most useful for dating rocks are those that existed for short intervals of time, inhabited many different environments, and left behind fossils that are distinctive and abundant.

Of course, fossils could not be used to date rocks if life on Earth were not constantly changing. Because organic evolution is the process by which the sweeping changes in organisms have come about, the fossil record is a unique repository of information, revealing patterns and rates of both evolution and extinction.

In the Drumheller Badlands of southern Alberta, Canada, erosion has cut into horizontal strata. These deposits record the disappearance of the dinosaurs at the end of the Mesozoic Era. *(R. Hartmier.)*

CHAPTER

4

Correlation and Dating of the Rock Record

As we saw in Chapter 1, geologists of the nineteenth century divided the rock record into systems such as the Cambrian, Ordovician, and Silurian largely on the basis of the fossils found in the various strata. In fact, fossils continue to have primary importance in efforts to determine the relative ages of rock strata. In this chapter we will discuss how fossils are used to subdivide the rock record into intervals that represent still smaller units of time. We will also analyze deposits and erosional surfaces that formed over large areas during very brief intervals. These surfaces can be used, along with fossils, to determine whether certain rocks at different sites are approximately the same age. Another important approach we will consider is the use of naturally occurring radioactive materials, which, as we learned in Chapter 1, allow us to estimate the actual ages of rocks and fossils in years. In other words, this approach allows us to position strata and fossils in *absolute* time, rather than simply in *relative* time, which is just a matter of ordering them from older to younger.

A living planktonic foraminiferan. Strands of protoplasm radiate from the hollow skeleton, which is the size of a small grain of sand. Fossil skeletons of planktonic foraminifera are widely used to date sediments. *(Manfred Kage/Peter Arnold, Inc.)*

The study of the age relationships of layered rocks usually begins with an examination of a **stratigraphic section** — a local outcrop or series of adjacent outcrops that displays the rocks' vertical sequence (Figure 4-1). By the process known as **correlation,** the fossils of two spatially separate stratigraphic sections — one in England and one in France, for example — can be identified as having the same geologic age. Since William "Strata" Smith first showed that fossils could be used for correlation early in the last century (p. 16), the distribution of fossils has become much better known, and the precision of correlation has been greatly improved.

INDEX FOSSILS

Fossil species and genera that are especially well suited to correlation procedures are called **index fossils,** or **guide fossils.** Such fossils have some or all of the following desirable characteristics:

1 They are easily distinguished from other taxa.

2 They are geographically widespread and thus can be used to correlate rocks over a large area.

3 They occur in many kinds of sedimentary rocks and therefore can be found in many places.

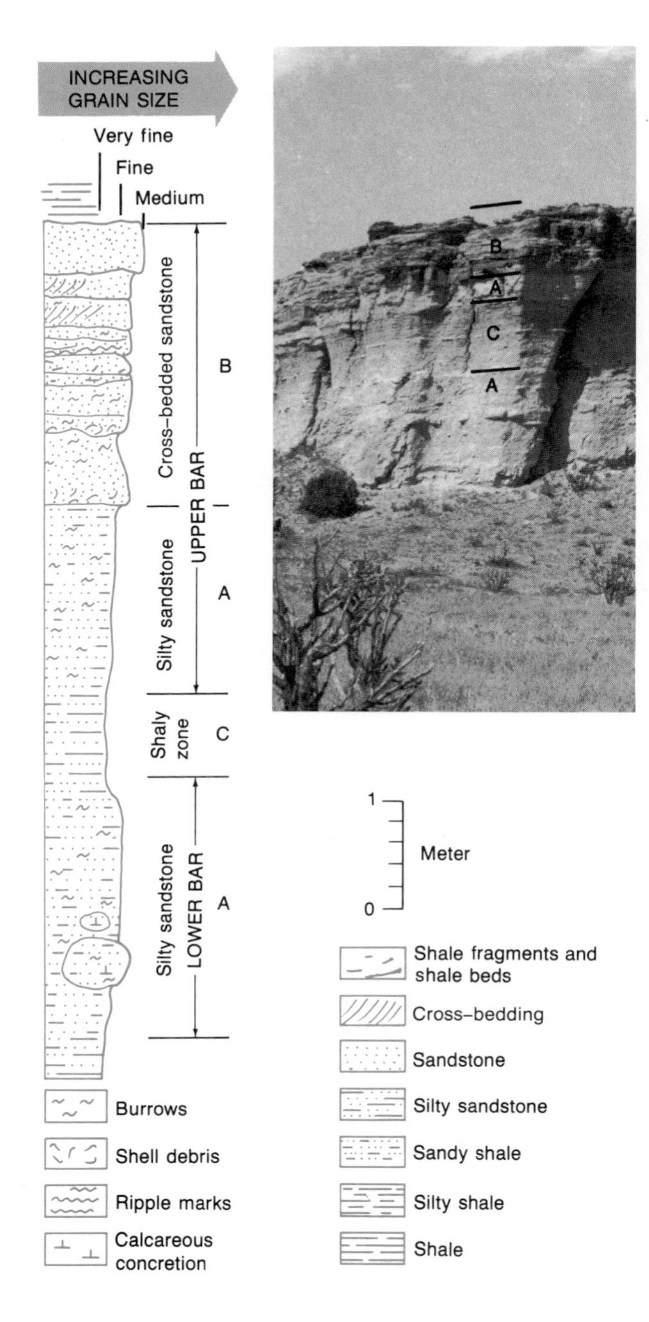

FIGURE 4-1 A stratigraphic section in the San Juan Basin of New Mexico. The sandstone exposed here (unit B) represents an offshore sandbar of Cretaceous age. On the left is a graphic representation of the section, where, in keeping with convention, sandstones are depicted in profile as projecting beyond shales. Sandstones usually stand out farther than other sedimentary rocks in actual outcrops because they are relatively resistant to erosion. *(After N. A. La Fon, Amer. Assoc. Petrol. Geol. Bull. 65:706–721, 1980.)*

4 They are restricted to narrow stratigraphic intervals, and thus allow for precise correlation.

Unfortunately, few index fossils exhibit all of these traits as strongly as we could wish. Consider the planktonic foraminiferan fossils found in late Mesozoic and Cenozoic sediments (p. 80), for instance. Like living species of planktonic foraminifera, these single-celled creatures resembled amoebas, but, unlike amoebas, they possessed skeletons, and their skeletons settled to the seafloor after the creatures died. Planktonic foraminifera meet important criteria for index fossils in two respects: First, they are easily identified under the microscope; and second, they floated across large areas of the sea and settled in a wide variety of sedimentary environments. The primary limitation of these organisms is that they lived in offshore areas and are seldom found in sediments deposited near shore. Moreover, some extinct species of planktonic foraminifera lived over much longer intervals than others. Those that survived for as long as 15 or 20 percent of the Cenozoic Era make poor index fossils. Only the earliest and latest appearances of such species provide information that is useful for correlation. Other species, however, lived for much shorter intervals so that their occurences permit far more precise correlation.

ZONES

A segment of the stratigraphic record that is characterized by particular species of index fossils may be formally recognized as a **zone.** Figure 4-2 shows zones that contain extinct fossil animals known as graptolites, many of which serve as excellent index fossils. Some zones are defined by the presence of single species; others are distinguished by the presence of two or more species. In general, the lower boundary of a zone is defined as the level at which one or more species first appears in the stratigraphic record. Correspondingly, the upper boundary is defined as the level at which the same species—or another species or group of species—disappears. Thus the lower boundary of a zone is approximately the same age everywhere, and so is the upper boundary. Zones are commonly named for particular species that

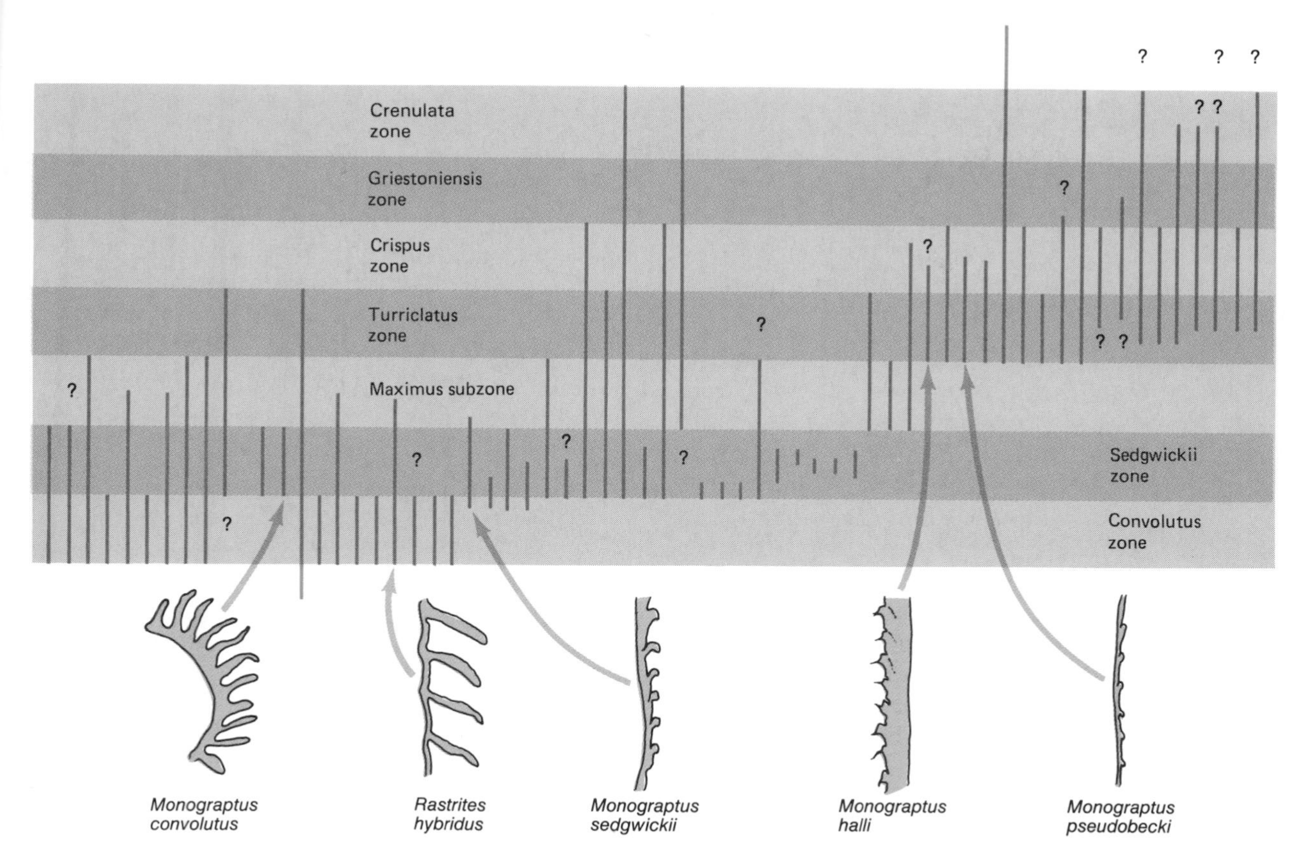

FIGURE 4-2 Zones based on the presence of fossils of planktonic graptolites, an extinct group of animals whose fossils are extremely useful for correlation. The zones represented in this bar graph are those of the lower part of the Silurian System in the British Isles. Vertical bars represent the known ranges of species, several of which are illustrated. A question mark indicates that a range may be longer than that shown here. All ranges shown are for species that first appear in the *argenteus, convolutus,* or *sedgwickii* zone or the *maximus* subzone. Illustrated fossils indicate the variety of forms among graptolite species. The stalks of these animals are about 2 or 3 millimeters ($\frac{1}{8}$ inch) wide. *(Adapted from R. B. Rickards, Geological Journal, 11:153–188, 1976.)*

characterize them, but if a zone includes the stratigraphic ranges of several species that belong to a single genus, it may instead be named for this genus. Some zones are further divided into subzones (see Figure 4-2).

Although some zones can be found in many parts of the world, others are restricted to a single continent. Moreover, no zone represents exactly the same interval of time throughout a region, because the members of no extinct species appeared or disappeared simultaneously in all of the areas they inhabited. As we will see in Chapter 5, it is not unusual for a species to originate in a small area and later greatly expand its geographic range. Often a species or genus—the mammoth, for example—persisted in a restricted region after it had died out in most of the areas it once inhabited (Figure 2-5); and many species have had complex histories of migration that have resulted largely from changing environmental conditions.

Imperfect as the process may be, the recognition of zones often results in correlation with only a small margin of error. If a species existed for only a million years, for example, its discovery in two areas cannot lead to a correlation error that exceeds a million years, even if the species appeared and disappeared at different times in different places. Any uncertainty resulting from correlation based on such a species would represent less than 2 percent of the Cenozoic Era, less than 1 percent of the Mesozoic Era, or less than 0.5 percent of the Paleozoic Era (see Figure 1-19).

RADIOACTIVITY AND ABSOLUTE AGES

In 1895 Antoine Henri Becquerel discovered that the element uranium undergoes spontaneous radioactive decay; that is, its atoms change to those of another element by releasing subatomic particles and energy. Geologists soon recognized that radioactive elements and the products of their decay could be used as geologic clocks to measure the ages of rocks. These clocks are the primary source of the absolute scale of geologic time that now allows us to calibrate the relative time scale, which is based largely on the principle of superposition (p. 15) and on fossils. Before the use of radioactive materials, only unusual circumstances, such as the presence of annual varves in lake deposits (which we saw in Figure 3-6), permitted geologists to estimate intervals of absolute time.

Atoms and Isotopes

Only a few naturally occurring chemical elements are useful for dating rocks by means of radioactive decay. A chemical element is characterized by its unique atomic number, which influences its chemical behavior. The atomic number of an element designates the number of protons—positively charged particles—in the nucleus of one of its atoms. The two elements with the lowest atomic numbers are of such low density that they are gases: Hydrogen is the element with the smallest number of protons—one—and helium is second, with two. Atomic numbers of other elements range to slightly above 100.

An atom of an element can also have one or more neutrons in its nucleus. The nucleus of a helium atom, for example, has two neutrons. A neutron has the same mass as a proton, and the total number of protons and neutrons constitutes atomic weight. Because neutrons have no charge, however, their presence does not influence the chemical behavior of elements—the way particular elements combine with others. Sometimes, however, the presence of neutrons creates instability in the nucleus, and the result is radioactive decay.

Isotopes are forms of an element that differ in the number of neutrons in their nuclei. Some isotopes are stable and others are unstable (radioactive). A radioactive isotope decays spontaneously to a lower energy level, changing to a new element in the process. The isotope that undergoes decay is known as the *parent isotope,* and the product is known as the *daughter isotope.* There are three modes of decay:

Loss of an alpha particle An alpha particle consists of two protons and two neutrons. In other words, it represents an atom of helium. Loss of an alpha particle converts the parent isotope into the element whose nucleus contains two fewer protons (Figure 4-3).

Loss of a beta particle A beta particle is an electron, which has a negative charge but no mass. Its loss turns a neutron into a proton, changing the parent isotope into the element whose nucleus contains one more proton.

Capture of a beta particle Addition of a beta particle turns a proton into a neutron, changing the parent isotope into the element whose nucleus has one less proton.

Some radioactive isotopes decay into other isotopes that are also radioactive. In fact, several steps of decay are required to yield a stable isotope from some parent isotopes.

The utility of radioactive elements for dating rocks lies in the fact that each radioactive element

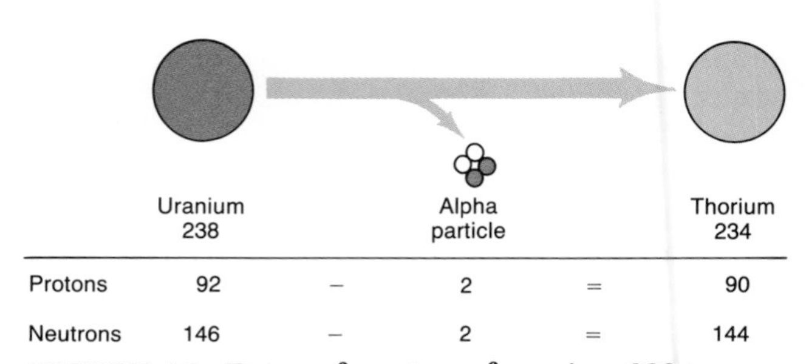

FIGURE 4-3 Decay of an atom of uranium 238 to thorium 234 by loss of an alpha particle. This loss reduces the atomic number by two and the atomic weight by four.

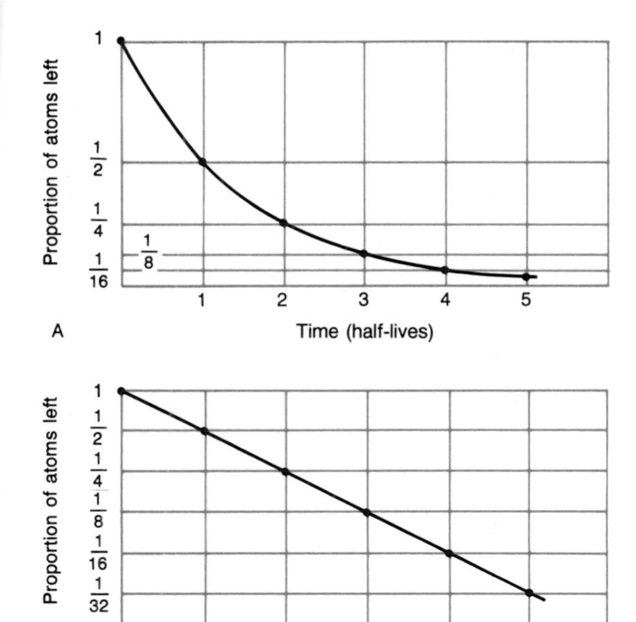

FIGURE 4-4 Arithmetic and geometric patterns formed by loss of atoms through radioactive decay. When plotted on a standard arithmetic scale *(A)*, the number of atoms can be seen to decrease more slowly with each successive interval of time. When the number of atoms is scaled as a geometric progression *(B)*, the plot forms a straight line. Half of the atoms present at the beginning of each interval (or half-life) survive to the beginning of the next interval.

decays at its own nearly constant rate. Once this rate is known, geologists can calculate the length of time over which decay in a natural system has been proceeding by measuring the amounts of both the radioactive parent isotope and the daughter isotope that remain in the rock. This procedure is known as **radiometric dating.** Radioactive isotopes decay at a constant exponential or geometric rate (not at a constant arithmetic rate), as Figure 4-4 indicates. For this reason, no matter how much of the parent element is present when it begins to decay, half of that amount will survive after a certain time. Then half of the surviving amount will remain after another interval of the same duration, and so on. This characteristic interval is known as the **half-life** of a radioactive element. Thus, in the course of four successive half-lives, the number of atoms of a radioactive element will decrease to one-half, one-fourth, one-eighth, and one-sixteenth of the original number of atoms.

Useful Isotopes

Many elements occur naturally as both radioactive and nonradioactive (or stable) isotopes. As Table 4-1 indicates, several radioactive isotopes are abundant in rocks and are therefore useful for geologic dating. Most of these isotopes occur in igneous rocks; thus, if we know the amounts of parent and daughter elements currently present in an igneous rock, we can calculate the time that has elapsed since the parent element was trapped—the date when magma cooled to form the rock. A few minerals on the seafloor incorporate radioactive elements as well, and radiometric dates obtained for these minerals represent the interval of time that has elapsed since they formed. Radiometric clocks are often reset by metamorphism, which separates radioactive isotopes from their decay products.

As Table 4-1 indicates, the half-lives of naturally occurring radioactive isotopes differ greatly, and these differences have a bearing on the ultimate use of various isotopes. More specifically, isotopes with short half-lives are useful for dating

TABLE 4-1 Properties of some radiometric isotopes that are commonly used to date rocks

Radioactive isotope	*Approximate half life (years)*	*Product of decay*
Rubidium 87	48.6 billion	Strontium 87
Thorium 232	14.0 billion	Lead 208
Potassium 40	1.3 billion	Argon 40
Uranium 238	4.5 billion	Lead 206
Uranium 235	0.7 billion	Lead 207
Carbon 14	5730	Nitrogen 14

NOTE: The number after the name of each element signifies the atomic weight of that element and serves to identify the isotope. Carbon 14, which has a very short half-life (that is, a high rate of decay), is used for dating materials younger than about 70,000 years. The other radioactive isotopes are employed for dating much older rocks.

only very young rocks, while those with long half-lives are best used to date very old rocks. These limitations are related to measurement problems associated with the various isotopes. In essence, isotopes such as carbon 14 that have short half-lives decay so quickly that their quantities in old rocks are too small to be measured. By the same token, isotopes such as rubidium 87 that have long half-lives decay so slowly that the quantities of their daughter elements in very young rocks are too small to be measured accurately. Let us examine the radiometric utility of the various decay systems listed in Table 4-1.

The rubidium-strontium system derives its value from the fact that rubidium occurs as a trace element in many igneous and metamorphic rocks and even in a few sedimentary rocks. Because of the long half-life of rubidium 87, the parent isotope, the rubidium-strontium system is generally useful only for dating rocks older than about 100 million years.

The useful radioactive isotopes uranium 235, uranium 238, and thorium 232 decay to different isotopes of lead (Table 4-1). Most useful for dating by means of uranium and thorium is the silicate mineral zircon, which is widespread in low concentrations in igneous and metamorphic rocks as well as in detrital sediments derived from them. By artificially abrading the surfaces of zircon grains to obtain unaltered material and by employing several grains to study a single rock unit, geologists are now able to date even very old (Precambrian) rocks quite precisely. Some calculated ages in the range of 2 to 3 billion years are considered to be within just a few million years of the actual ages!

The uranium-lead and thorium-lead systems have also been used to date rocks from the moon. As we shall see in Chapter 8, the ages of the oldest dated moon rocks — slightly more than 4.6 billion years — closely approximate the age of Earth and its solar system, estimated from other evidence.

Argon, the decay product of potassium 40, is an inert gas — that is to say, it is a gas that does not combine chemically with other elements. Argon does, however, become trapped in the crystal lattice of some minerals that form in igneous and metamorphic rocks. A deficiency of the potassium-argon method is that argon can leak from the lattice of a crystal, making it appear that less than the actual amount of decay has occurred. Nonetheless, as we will see in Chapter 16, this method has provided important dates for events in the evolution of humans in Africa.

In recent years, several additional radiometric techniques have been developed to determine the ages of rocks or fossils a few million years old or younger. One of these techniques is **fission-track dating,** a method for measuring the decay of uranium 238 that usually yields more accurate age estimates than conventional measurements of the daughter product (lead 206) when a rock is too young to have accumulated much lead. The key observation is that when uranium 238 decays, it emits subatomic particles that fly apart with so much energy that they penetrate the surrounding crystal lattice, producing fission tracks (Figure 4-5). These tracks can be enlarged in the laboratory by acid etching and can then be counted under a microscope. After these tracks have been counted, the remainder of the uranium can be subjected to a neutron field, which causes it to decay completely. The number of tracks thus produced can be compared to the number that formed naturally, and the resulting numerical ratio reveals the age of the mineral.

Another method of dating applies to reef-building corals (Box 2-1). It turns out that these animals incorporate a small amount of uranium in their skeletons. The fact that uranium 234 decays rapidly to thorium 230 allows accurate dating of

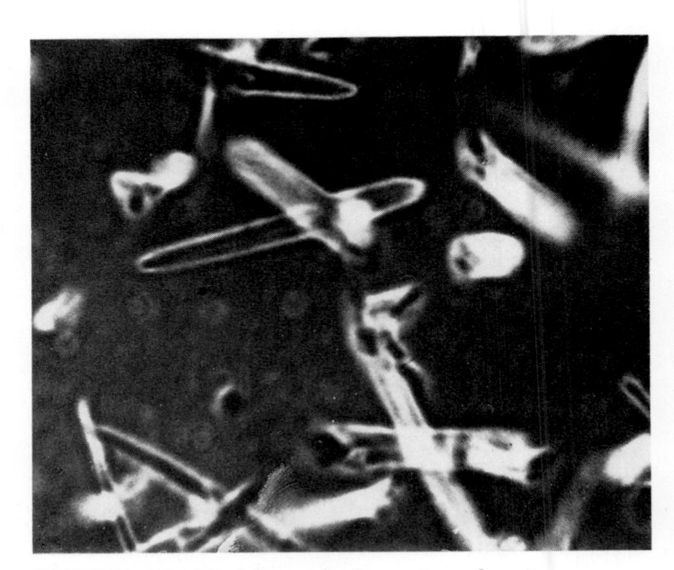

FIGURE 4-5 Fission tracks in a mineral grain enlarged by etching. The complete track oriented almost horizontally in the center is 13.5 micrometers long. *(M. K. Rosen and D. S. Miller, Rensselaer Polytechnic Institute.)*

corals that range in age from a few thousand years to about 300,000 years. Radiometric isotopes of uranium yield both lead and helium as final daughter products. Thus, by measuring the amount of helium and the amount of undecayed uranium trapped in well-preserved coral skeletons, geochemists can date corals several million years old.

Carbon 14 dating, or **radiocarbon dating,** is the best known of all radiometric techniques—but because the half-life of carbon 14 is only 5730 years, this technique can be used only on materials that are less than about 70,000 years old. Objects of biological origin, such as bones, teeth, and pieces of wood, make up the bulk of the materials dated by this method. Despite its limitations, radiocarbon dating is of great value for dating materials from the latter part of the Pleistocene Epoch—an interval so recent that most other radioactive materials found in its sediments have not decayed sufficiently for their products to be measured accurately. Fortunately, the useful range of carbon 14 extends back far enough to encompass the entire time interval during which modern humans have existed, as well as the interval during which glaciers most recently withdrew from North America and Europe at the close of the recent Ice Age (Pleistocene Epoch). Thus radiocarbon dating plays a valuable role in the study of human culture, sometimes being used to date materials that are no more than a few hundred years old.

Carbon 14 is a rare isotope of carbon that forms in the upper atmosphere, about 16 kilometers (~10 miles) above Earth's surface, as a result of the bombardment of nitrogen by cosmic rays. Both carbon 14 and the stable isotope carbon 12 are assimilated by plants, which turn them into tissue. Once a plant dies, however, carbon is no longer incorporated in its tissues, and the carbon 14 that was present when the plant died decays back into nitrogen 14. Thus the percentage of carbon 14 in the tissues of plants declines in relation to the percentage of carbon 12, and the ratio of the two can be used to determine when the tissue died. A basic assumption in radiometric dating is that the rate of carbon 14 production has been constant during the past 70,000 years and that the ratio of carbon 14 to carbon 12 in the atmosphere has also remained constant. Although no major miscalculations have resulted from this assumption, some minor errors have been noted. Wooden objects used by the ancient Egyptians at times that are well documented in historical records, for example, have yielded radiometric dates that are slightly too early. Such errors, which exceed 10 percent for material 5000 years old, apparently result from minor changes in the rate at which carbon 14 has been produced in the upper atmosphere.

FOSSILS VERSUS RADIOACTIVITY: THE ACCURACY OF CORRELATION

The fact that the half-lives of radioactive isotopes are well established should not be taken to indicate that radiometric dating permits more accurate correlation of sedimentary rocks than the use of fossils can offer. For one thing, most minerals that can be dated radiometrically are of igneous origin, and, as Figure 4-6 shows, the dating of igneous rocks often yields only a maximum or a minimum age for associated sedimentary rocks. Clasts of igneous rocks or minerals found in sedimentary rocks can also be dated radiometrically, but only estimates of the maximum ages of sedimentary rocks can be derived in this way.

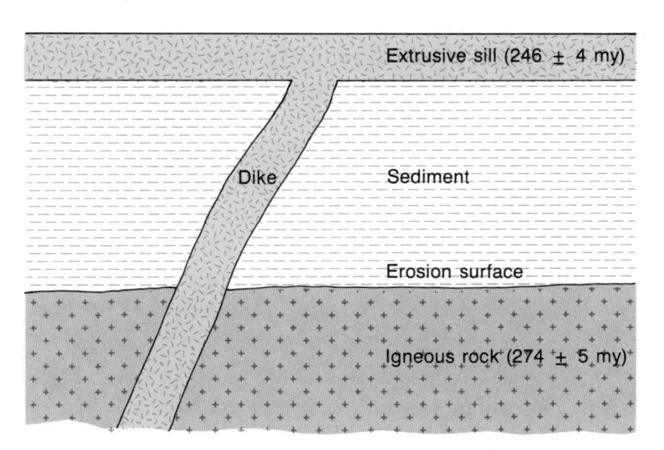

FIGURE 4-6 A cross section of a group of rocks, showing how the age of a sedimentary unit can be bracketed by the radioactive dating of associated igneous rocks. Here the sediments lie on top of a body of igneous rock dated at 274 ± 5 million years and are therefore younger than this. The sediments are also cut by a dike and covered by a sill of igneous rock dated at 246 ± 4 million years; they are therefore older than this. We can conclude that this sedimentary unit may be as old as 279 million years or as young as 242 million years.

Even more fundamental uncertainties are inherent in radiometric dating. Most published radiometric dates, for example, are followed by the symbol ± and a number representing a smaller interval of time (for example, 472 ± 7 million years). This plus-or-minus sign indicates uncertainty that is attributable to possible errors in the measurement of the quantities of parent and daughter elements.

Not indicated by these plus-or-minus figures is the potential error that results from the fact that only parent and daughter atoms actually present in a rock can be detected. In accepting a date, even with a plus-or-minus figure, we are assuming that a dated rock has remained a closed system—that it has neither lost nor received parent or daughter atoms from some other source. Unfortunately, this is not always the case. Rocks can, in fact, both gain and lose atoms, although loss is a more frequent problem than gain. This is a particular difficulty of the potassium-argon decay system, whose daughter element frequently seeps from rock, leading to underestimation of the amount of decay that has occurred and hence to an underestimation of the rock's age. Another source of error in dating important stratigraphic boundaries is the absence of appropriate radioactive isotopes in rocks positioned close to the boundaries.

These types of errors sometimes add up to sizable total errors, especially when very old rocks are being dated. This point is well illustrated by past estimates made for the beginning and end of the Silurian Period (Figure 4-7). In evaluations made between 1959 and 1968 alone, the duration of the Silurian was halved and then doubled and then halved again. It now seems likely that the Silurian began between 440 and 430 million years ago and that it ended between 410 and 400 million years ago.

For other important segments of the record, dates are quite accurate. For example, reliable dating of an igneous sill close to the boundary between the Triassic and Jurassic systems in eastern North America establishes a reliable estimate of very close to 200 million years for the age of the boundary.

Fossil graptolites (Figure 4-2) of Silurian age allow for more accurate correlation of sedimentary rocks than do radiometric dates. Many kinds of graptolites floated in the sea and thus spread quickly over large areas. In addition, graptolites evolved rapidly. These two factors together have

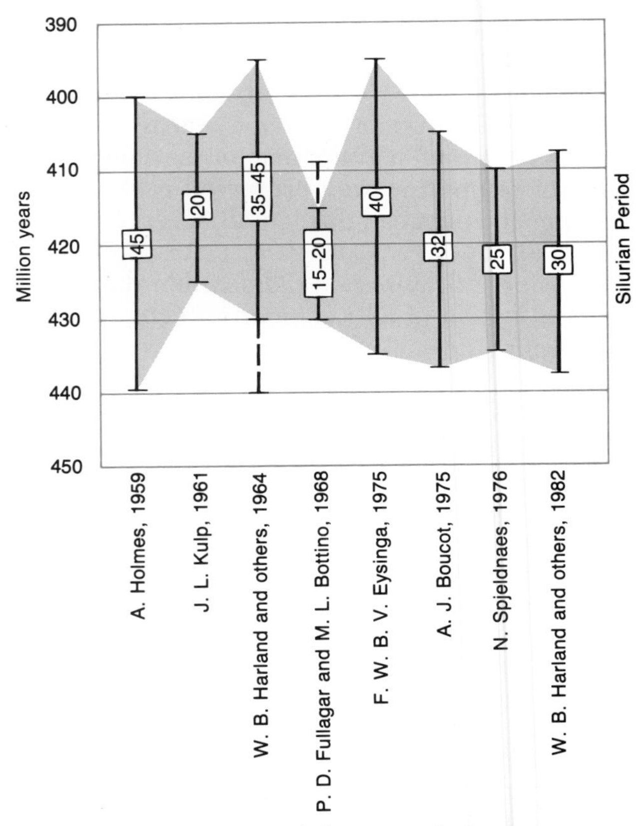

FIGURE 4-7 Estimates of the interval of time represented by the Silurian Period. The date each estimate was made and the author or authors responsible for each estimate are listed. *(Adapted from N. Spjeldnaes, Amer. Assoc. Petrol. Geol., Studies in Geol. 6:341–345, 1978.)*

made graptolites highly useful index fossils. Most individual species of graptolites existed for about a million years; thus correlations based on the occurrences of such species cannot be inaccurate by a larger interval than this. Even widely separated regions that have no fossil graptolite species in common may share certain graptolite genera, and these genera provide for correlations within the Silurian System that are more accurate than correlations based on radiometric dates. Although radiometric dates are not accurate enough to establish the exact times of appearance and disappearance of individual graptolite species, geologists can estimate the approximate lengths of time that individual species survived. It can be noted, for example, that there are about 30 successive graptolite zones in the Silurian System and that the Silurian lasted about 30 million years. Therefore, because the geologic ranges of many species coincide or nearly coincide with single zones

(Figure 4-2), it can be concluded that many graptolite species lived about a million years.

Although radiometric dating provides a special kind of geologic time scale — specifically an absolute scale, or one based on years — most geologic correlations are still based on fossils. Not only are fossils more common than radioactive elements in sedimentary rocks, but the analysis of fossils usually allows for greater accuracy. The geologic intervals at the upper and lower ends of the geologic scale of time, however, represent striking exceptions to this general rule. At the upper end, most rocks that are older than about 1.4 or 1.5 billion years contain few fossils that are well enough preserved and easily enough identified to serve as index fossils. Thus radiometric dates serve as a primary basis for correlation of these early rocks, especially from continent to continent. And at the lower end of the time scale, the interval extending from the present back to about 70,000 years ago, radiocarbon dating offers estimates of age that commonly have plus-or-minus values of only a few percent. Because relatively few species of animals or plants have appeared or disappeared during this brief interval, fossils found in the corresponding rocks have much less value in correlation.

In the following sections we will examine several other stratigraphic features that in many cases permit more precise correlation than either fossils or radioactive isotopes.

TIME-PARALLEL SURFACES IN ROCKS

Imagine that someone with supernatural powers were in an instant to spray paint over many areas of Earth — ocean floors, lake bottoms, and terrestrial lowlands. In keeping with the law of superposition, this layer would become a time-parallel surface — that is to say, it would separate deposits that had formed before it was laid down from accumulations that were laid down afterward. Similarly, if a powerful force were suddenly to lower sea level throughout the world by 200 meters, the deposition of sediments at the shoreline would shift seaward for great distances, and the first of the new shoreline deposits thus created would be of nearly the same age on all continents, so that these deposits could be accurately correlated throughout the world. These two hypothetical scenarios — the formation of a time-parallel surface and a sudden relocation of shorelines — are not far removed from actual events whose geologic records have enabled us to correlate rocks of widely separated regions. We will consider these events and others in the sections that follow.

Key Beds

A **key bed,** or **marker bed,** is a bed of sediment that resembles our hypothetical layer of paint: All parts of it are virtually time-parallel; that is, of identical age. Widespread layers of volcanic ash, for example, often function as key beds. Sometimes an ancient ash fall, which may represent either a single volcanic eruption or a series of nearly simultaneous eruptions, can be traced for thousands of square kilometers, and so provides a time-parallel surface in the stratigraphic record. Figure 4-8 shows an ash bed that formed when a

FIGURE 4-8 A thick band of volcanic ash (above the hammer head) overlain by two similiar but thinner ash beds. This conspicuous group of Cretaceous beds permits correlation over a large area of Colorado. *(Erle Kauffmann, University of Colorado.)*

volcano erupted in the region of the Appalachian Mountains more than 350 million years ago, when these mountains were just beginning to form.

Glacial tills such as those described in Chapter 3 also serve as useful marker beds; even tills formed by glaciers in different parts of the world can be useful if they represent only a brief interval of global cooling.

Certain events produce key beds that allow for correlation within particular basins of deposition. The character of deep-water evaporite deposits in a basin, for example, can reflect aquatic conditions throughout the basin. Because deep-water evaporites (Appendix I) typically form widespread horizontal beds (Figure 4-9), an individual bed that differs in chemical composition from beds above and below can often be traced over thousands of square kilometers.

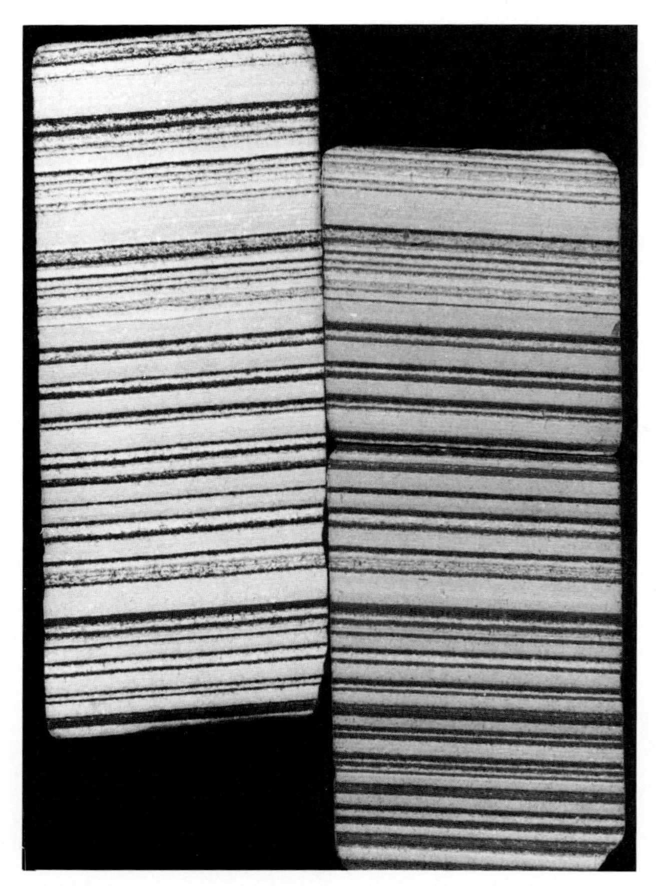

FIGURE 4-9 Two cores taken from the Castile evaporites, which were precipitated in western Texas near the end of the Permian Period. The cores are from localities 14.5 kilometers (~9 miles) apart, yet their lamination matches almost perfectly, allowing for precise correlation. The alternating dark and light bands, which range up to a few millimeters in thickness, probably represent seasonal organic-rich (winter) and organic-poor (summer) layers. If this is the case, each pair of bands represents one year, and the 200,000 or so paired bands of the Castile Formation represent about 200,000 years of deposition. *(R. Y. Anderson et al., Geol. Soc. Amer. Bull. 83:59–86, 1972.)*

Shifting of Depositional Boundaries

No particular type of environment stretches infinitely far in any direction. Instead, one environment of sediment deposition inevitably gives way to another, and when it does so, there is either an abrupt or a gradual transition from the first mode of deposition to the second. The set of characteristics of a body of rock representing a particular local environment is called a **facies,** and, accordingly, lateral changes in the characteristics of an ancient body of rock, which reflect lateral changes in the depositional environment, are known as facies changes (Figure 4-10). Thus one facies in a body of rock might consist of a reef of porous limestone containing fossil reef-building organisms in their living positions. This facies might

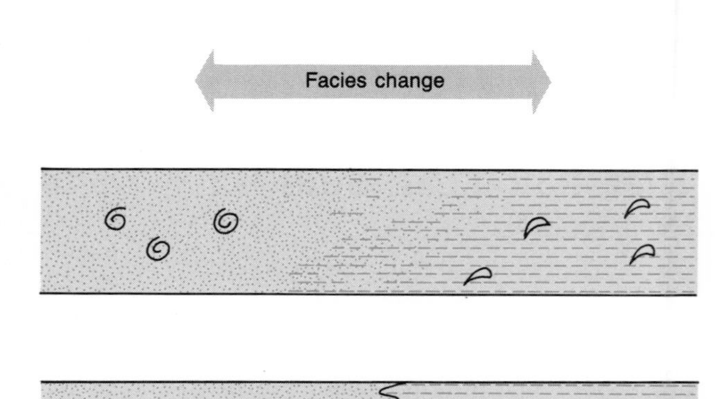

FIGURE 4-10 A facies change, from sandstone on the left to shale on the right. The facies not only consist of different kinds of rock, they also contain different kinds of fossils (here represented by symbols). The upper and lower diagrams illustrate two ways in which geologists depict facies changes in stratigraphic cross sections. The upper diagram emphasizes the gradationalness of the facies change.

pass laterally into a reef-flank facies characterized by steeply dipping talus deposits of poorly sorted rubble from the reef, as in Figures 3-23 and 3-24. This second facies might grade in turn into an interreef-basin facies composed of fine-grained clayey limestone.

We observed in Chapter 3 that when a sandy beach progrades over muddy offshore deposits, the process of progradation pushes the shoreline seaward. This migration of a shoreline is known as **regression** (Figure 4-11*A*). In other instances, the shoreline and the facies seaward of it shift inland in a process known as **transgression** (Figure 4-11*B*). In still other cases, a transgression is followed by a regression (Figure 4-11*C*), or vice versa, resulting in a more complex pattern that is very useful for correlation because it contains a time-parallel surface (or a time-parallel line if the facies are depicted in a two-dimensional cross section). This surface (or line) connects the facies boundaries where they were positioned at the time of maximum transgression. Most time-parallel surfaces of this type are useful only for correlating stratigraphic sections that represent single depositional basins, for example, sections that represent different parts of a lake or inland sea. The unconformities discussed in the next section can provide evidence for correlation over longer distances.

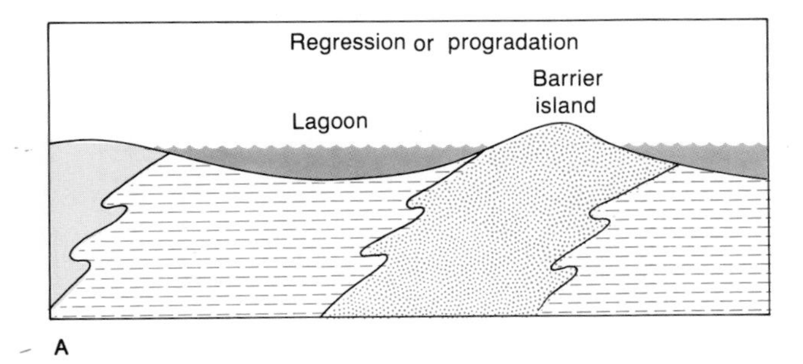

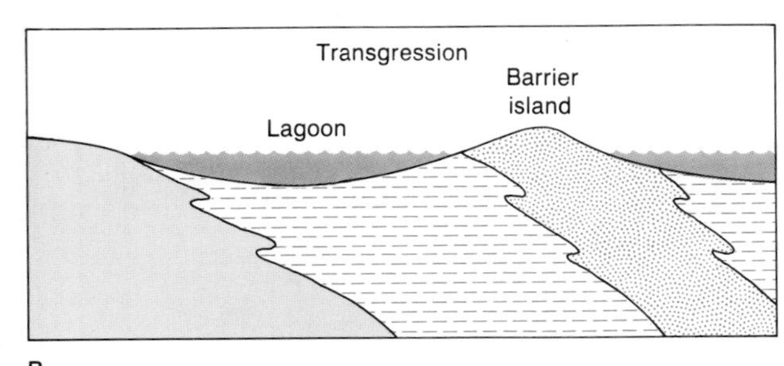

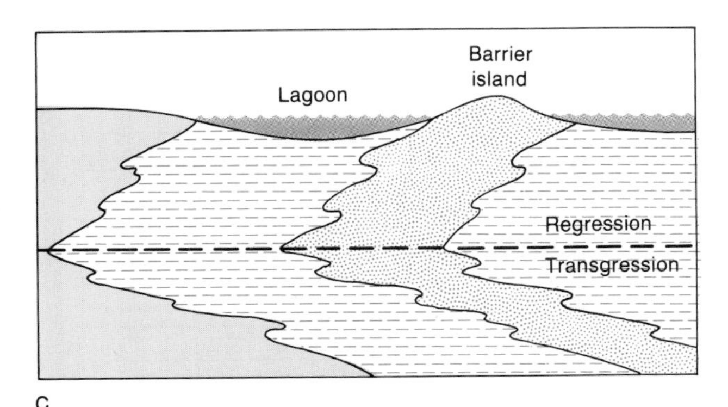

FIGURE 4-11 Correlation based on a stratigraphic pattern in which a regressive depositional sequence follows a transgressive sequence. *A.* The pattern of a regressive (progradational) sequence. *B.* The pattern of a transgressive sequence. *C.* When a regression follows a transgression, the points of maximum transgression of various facies can be connected to form a line of correlation, as indicated by the broken line in the diagram. A similar line can be constructed for a stratigraphic pattern in which a transgression follows a regression.

Unconformities, Bedding Surfaces, and Seismic Stratigraphy

Unconformities that formed during the same interval of tectonic uplift or nondeposition sometimes constitute nearly time-parallel surfaces. Such unconformities may truncate rocks of many different ages, but the sediments resting directly on top of the erosional surface are often nearly the same age in all parts of a depositional basin. The deposits resting on the surface of an unconformity will rarely be precisely the same age, however, and in some cases the ages of these deposits will vary greatly from place to place. A sea that has deserted an area, for example, may later invade it again slowly, after a period of erosion, reaching different parts of the area at different times. On the other hand, if sea level drops suddenly throughout the world and then rises again rapidly, the resulting global unconformities may represent fairly accurate time markers.

It is not known why sea level has dropped episodically in the course of Earth's history. As we will see in Chapter 16, sea level fell repeatedly by as much as 100 meters (~330 feet) within intervals of just a few thousand years during the recent Ice Age, when glaciers expanded over the land, "locking up" water and thus removing it from the global hydrological cycle. There is no evidence

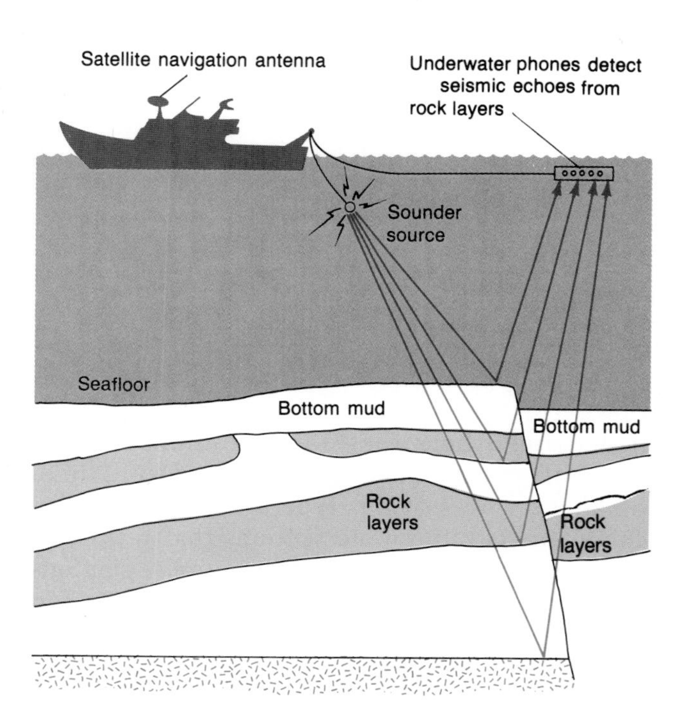

FIGURE 4-12 The use of seismic reflections to study sediments and rocks buried beneath the seafloor. Sound waves are generated by a sounder that makes a pneumatic explosion like that of a bursting balloon. The sound waves bounce off surfaces of discontinuity, which include bedding surfaces and unconformities. Underwater phones then pick up the reflections, allowing marine geologists to determine the configurations of buried features. Fossils recovered from sediment cores reveal the ages of individual sedimentary beds. *(After F. Press and R. Siever, Earth, W. H. Freeman and Company, New York, 1986.)*

that similar continental glaciation took place during some earlier intervals in which sea level fell. It has been shown, however, that swelling and subsiding of the deep seafloor has moved sea level up and down, but at a slower rate: on the order of 10 meters (~33 feet) per million years. The areas of seafloor responsible for most of this fluctuation are the midocean ridges. As we shall see in Chapter 6, these are submarine mountain chains, where new oceanic lithosphere is formed.

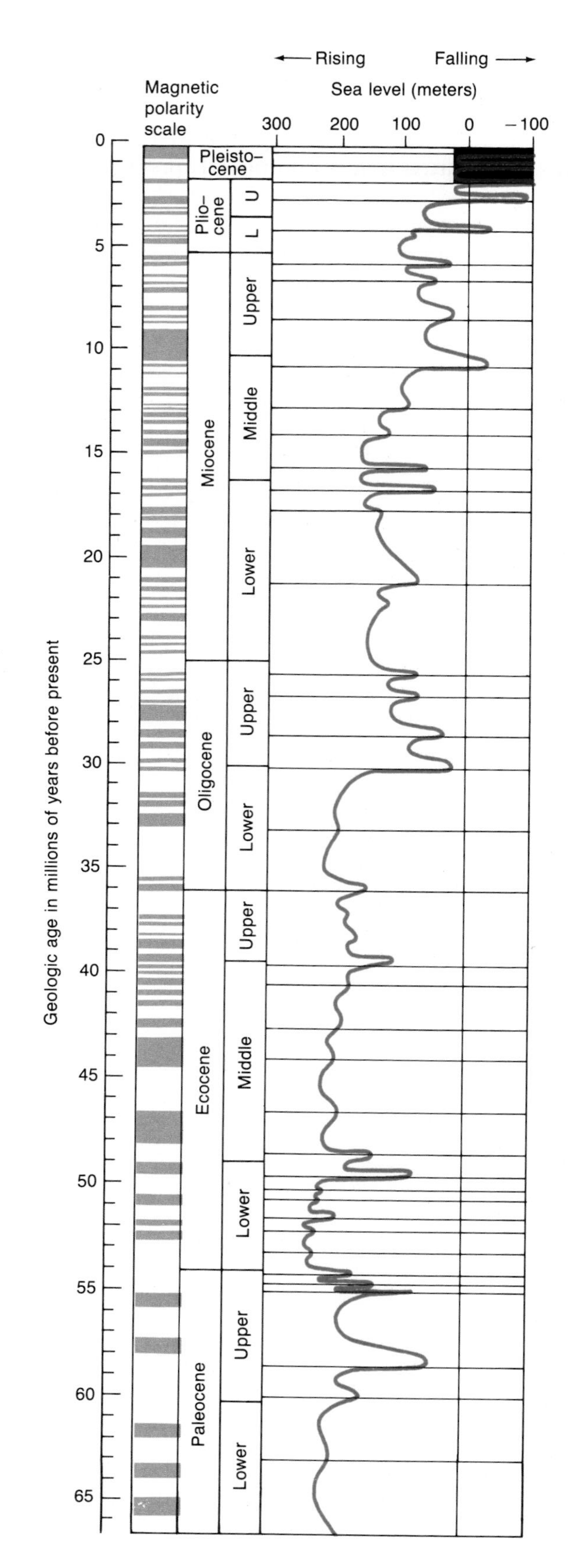

FIGURE 4-13 Estimates, based primarily on seismic stratigraphy, of relative changes in sea level during the Cenozoic Era. Horizontal segments of the sea-level curve represent sudden drops in sea level. The horizontal scale is in arbitrary units. *(After B. Haq et al., Science 235:1156–1167, 1987.)*

In general, the global unconformities that have been used most successfully as time markers are those that occur within Mesozoic and Cenozoic sediments lying along continental shelves. Most continental shelves have remained below sea level since their formation and thus have not been destroyed. Such sediments and unconformities have been studied by means of the seismic reflections that are generated when artificially produced seismic waves bounce off physical discontinuities within buried sediments (Figure 4-12). Some of these discontinuities have been found to be time-parallel bedding surfaces, and others have been identified as unconformities.

When compilations of many seismic profiles are used together with fossil evidence to date local changes in sea level, global patterns of sea-level change may be revealed. Evidence that sea level rose or fell in many widely separated areas at the same time indicates that the change occurred on a global scale. Estimates of times and amounts of change have yielded a global curve of sea-level changes during the Cenozoic Era (Figure 4-13). Less precise information for earlier intervals, based in part on rocks, fossils, and unconformities visible on the continents, has yielded a less detailed curve for the entire Phanerozoic interval.

The changes in sea level shown in Figure 4-13 are based on seismic evidence of simultaneous shifts in many parts of the world and are thus assumed to reflect global or **eustatic sea-level changes.** However, global changes in sea level are not reflected in seismic profiles where there is an offsetting tectonic change in the position of the land (Figure 4-14). If the land in one area rises in pace with a eustatic rise in sea level, for instance, the eustatic rise will not be expressed as it would be if the land were to remain stationary (Figure 4-15*A*). Moreover, if the tectonic uplift exceeds the eustatic rise in sea level, sea level in that area

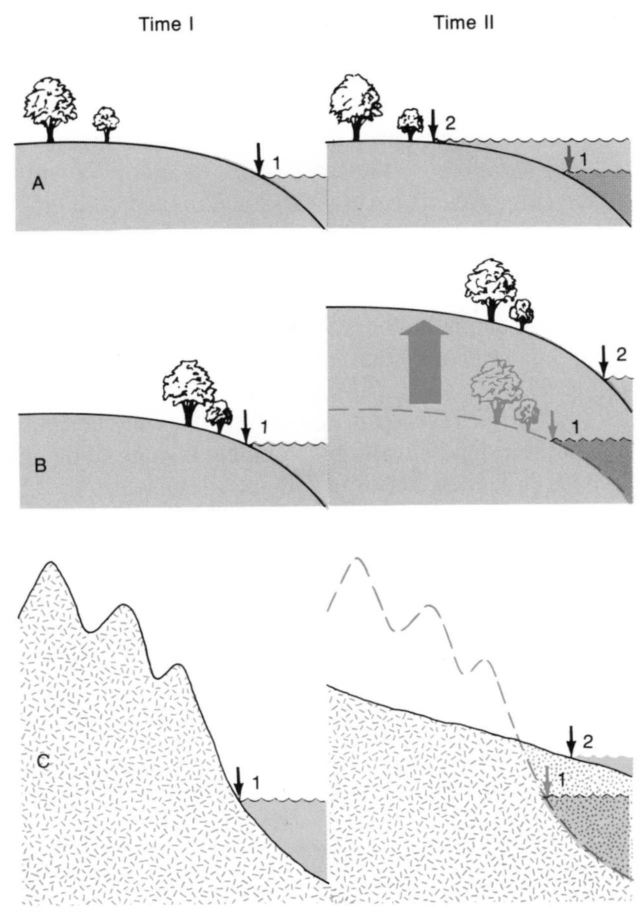

FIGURE 4-15 Cross sections of shorelines showing that a rise in sea level does not necessarily result in a transgression. In each of the three pairs of diagrams, the initial and final positions of sea level are numbered 1 and 2. *A.* The land remains unchanged, and a rise in sea level causes a transgression. *B.* A rise in sea level is accompanied by regression rather than by transgression because the land rises tectonically *(heavy arrow)* more than sea level does. *C.* A rise in sea level is accompanied by regression (progradation) rather than by transgression because sediment eroded from nearby highlands pushes the shoreline seaward.

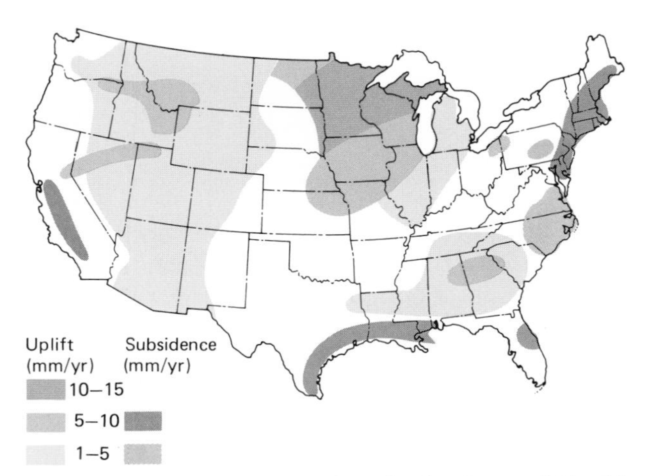

FIGURE 4-14 Rates of uplift and subsidence of Earth's crust in the United States today. *(After S. P. Hand, National Oceanic and Atmospheric Administration.)*

Box 4-1
Searching for Oil off Southern New Jersey

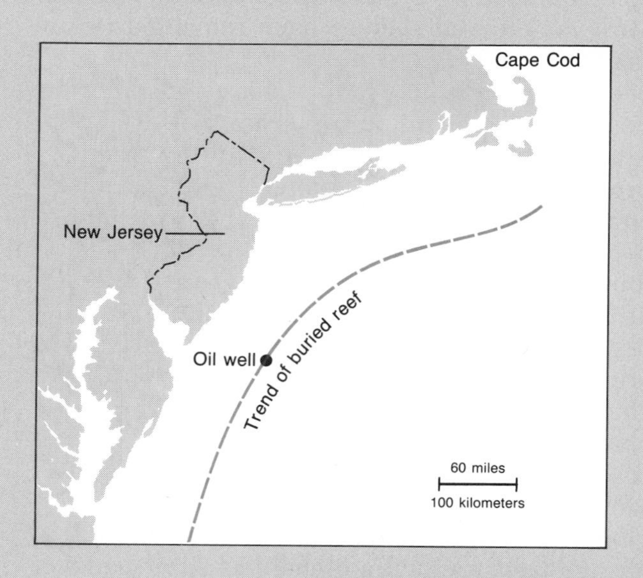

Seismic stratigraphy is a major tool in the search for petroleum and natural gas. Seismic cross sections reveal that a large ridge extends for hundreds of kilometers parallel to the Atlantic coast of the United States, deeply buried in sediment below the continental slope. The seismic sections display broad surfaces within the pile of sediments containing the ridge. These surfaces represent brief intervals when patterns of deposition changed or deposition ceased altogether. Thus the surfaces allow for correlation, and those that are deflected upward in the vicinity of the ridge indicate that it once stood above the surrounding seafloor. When the ridge was discovered, its location and configuration suggested that it might be an ancient barrier reef that long ago grew along the edge of the continental shelf. This possibility aroused the interest of the petroleum industry, because buried reefs are porous structures that frequently trap oil and gas (p. 67).

Barrier reefs grow at the surface of the ocean, but the ridge beneath the continental slope of the eastern United States lies about 4 kilometers (~2.5 miles) below present sea level—much lower than sea level has ever stood during Phanerozoic time. In fact, the entire continental margin has subsided greatly since it formed early in the Mesozoic Era. The margin formed when a huge continent broke apart, separating the landmasses that are now North and South America from Europe and Africa. This event marked the birth of the Atlantic Ocean, which has been widening ever since as the continents on either side have moved farther and farther apart. As the continents separated, the newly formed continental margins subsided, and sediments accumulated to great thicknesses along eastern North America (Chapter 6). Thus it was reasonable to suppose that a reef had once occupied the shallow waters off the newly formed North American coast and that after it died, it sank and was buried beneath younger deposits.

will actually fall in relation to the land (Figure 4-15*B*) regardless of what is happening in the rest of the world. Box 4-1 illustrates how, in a similar way, a continental margin can sink so as to lower rocks that originated in shallow water far below the position at which they formed.

In addition, vertical sea-level changes have often been mistaken for transgressions and regressions, which are lateral shifts in the position of the shoreline. Of course, a strong correlation exists both between transgressions and eustatic rises and between regressions and eustatic falls—but there is a complicating factor that weakens this correlation, making transgressions and regressions unreliable indicators of eustatic changes. As Figure 4-15*C* indicates, the rate at which sediment accumulates strongly influences whether transgression or regression will occur in a particular area. Thus regressions have often occurred locally during global intervals of rising sea

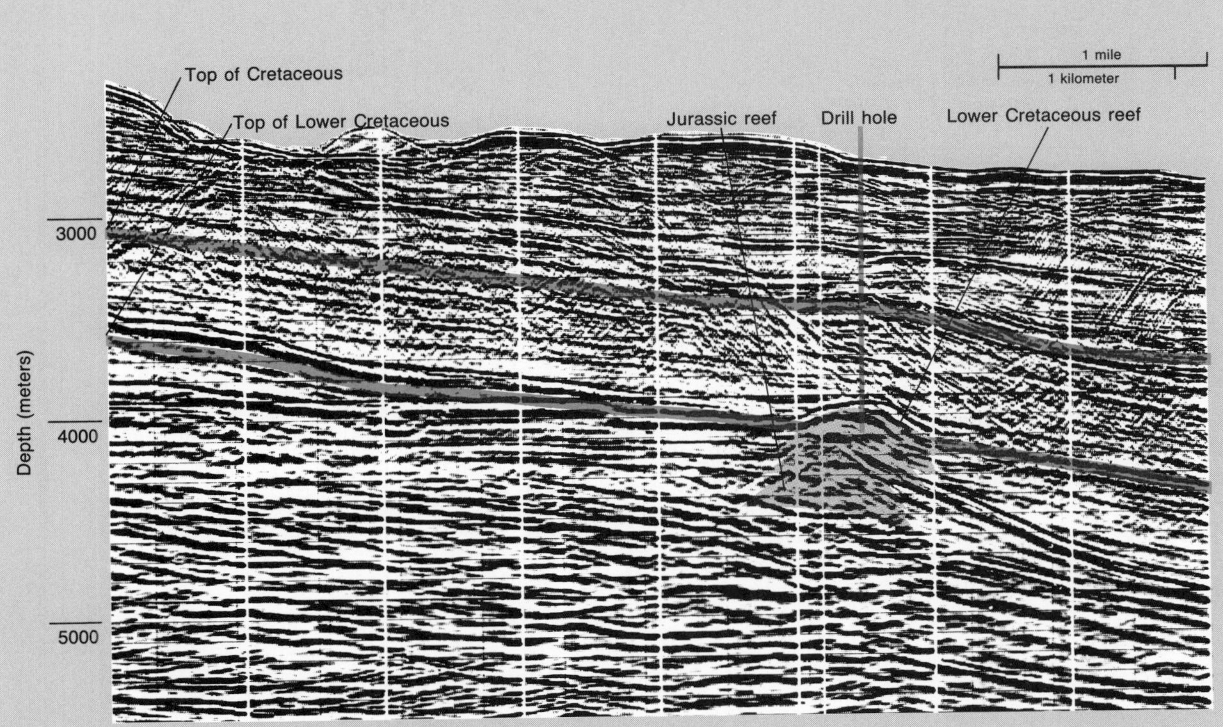

Seismic section of the continental margin off the coast of New Jersey, revealing a cross-section of an ancient reef. *(After R. N. Ehrlich et al., American Association of Petroleum Geologists, Studies in Geology, No. 27, 1988.)*

The likelihood of striking oil led three corporations to cooperate in constructing an offshore rig and drilling into the buried ridge in 1984. The drilling produced samples containing fossils that revealed the ages of the ridge and the deposits above it. For geologists the results were positive in one sense: The ridge turned out to be a barrier reef, of Late Jurassic and Early Cretaceous age. They were negative in a more important sense: The reef contained no commercially valuable oil or gas.

level simply because sediment that has been supplied from the land at a very high rate has pushed the sea back from the land.

Magnetic Stratigraphy

There is one kind of global event that provides for accurate correlation throughout the world: a reversal in the polarity of Earth's **magnetic field.** As we have seen, Earth's core consists of dense material made up of iron and other heavy substances. In the outer part of the core, this material is in a liquid state, and its motion generates a magnetic field. As a result, the planet behaves like a giant bar magnet (Figure 4-16). When iron-containing minerals form sedimentary or igneous rocks at or near Earth's surface, they often become aligned with Earth's magnetic field just as a compass needle does when it is allowed to rotate freely. As

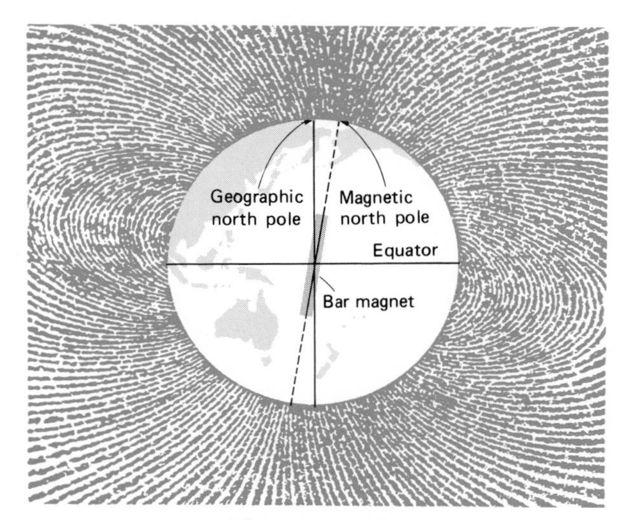

FIGURE 4-16 Earth's magnetic field. The magnetic field, represented by the lines of force surrounding the planet, resembles the one that would be produced by a bar magnet located within the planet with its long axis inclined slightly (11°) from Earth's axis of rotation. *(After F. Press and R. Siever, Earth, W. H. Freeman and Company, New York, 1986.)*

small grains of iron minerals settle from water to become parts of sedimentary rocks, they often rotate in such a way that their magnetism becomes aligned with that of the planet; and iron minerals that crystallize from lava or magma automatically become magnetized by Earth's magnetic field as they cool.

It is a startling fact that Earth's north and south magnetic poles occasionally switch positions. No one knows why. Intervals between such **magnetic reversals** vary considerably, but during the Cenozoic Era they have averaged about a half-million years. Sequences of magnetized rocks that can be dated radiometrically have revealed the history of

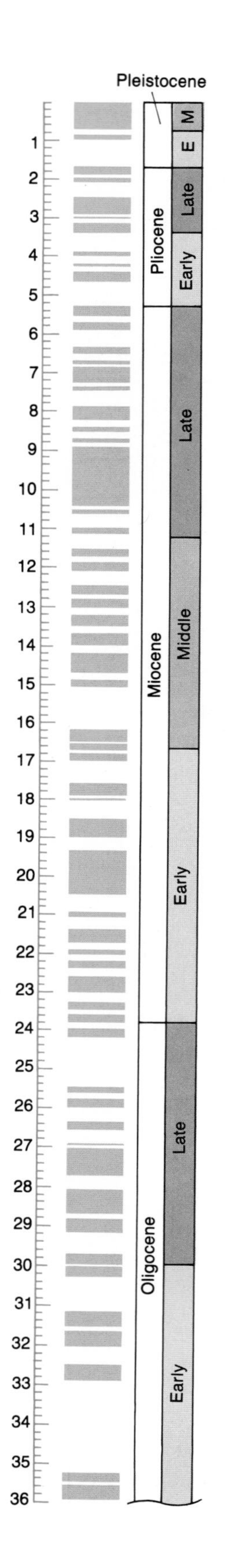

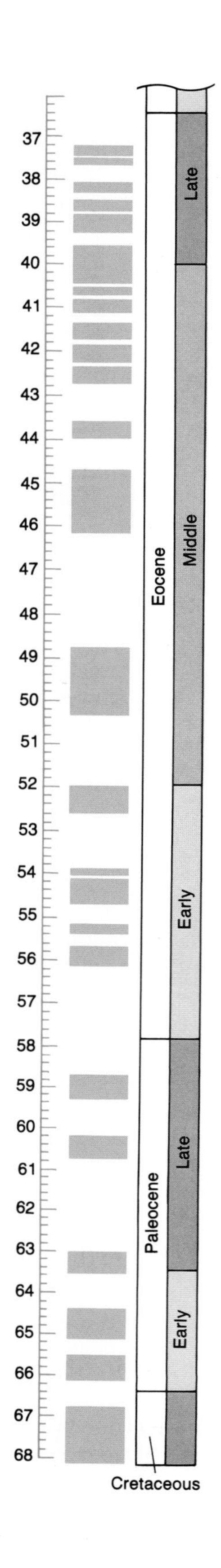

FIGURE 4-17 Magnetic polarity scale for the Cenozoic Era. Numbers represent millions of years. About half of the time the magnetic field has had normal polarity (polarity like that of the present), as indicated by the colored segments of the scale, and about half of the time the polarity has been reversed, as indicated by white space. *(After W. B. Berggren et al., Geol. Soc. Amer. Bull. 96:1407–1418, 1985.)*

magnetic reversals during the Cenozoic Era (Figure 4-17) and during much of the Mesozoic Era as well. Periods when the polarity was the same as it is today are known as *normal intervals,* and periods when the polarity was the opposite of what it is today are called *reversed intervals.* Among the rock sequences most useful for magnetic correlation are multiple lava flows, each of which was magnetized as it cooled. The piles of volcanic rocks that result often provide a detailed history of magnetic reversals.

The primary difficulty encountered in efforts to use the magnetic record for correlation lies in determining which worldwide events are reflected by a magnetic reversal in one locality. Particularly helpful in this respect is a "signature"; that is, a distinctive sequence of reversals. In rocks that are known to be of Eocene age, for example, a pattern that indicates a long normal interval flanked by two long reversed intervals can only be early Middle Eocene in age (Figure 4-17). Used in conjunction with other dating methods, magnetic stratigraphy has greatly improved geologists' ability to date Cenozoic sediments throughout the world.

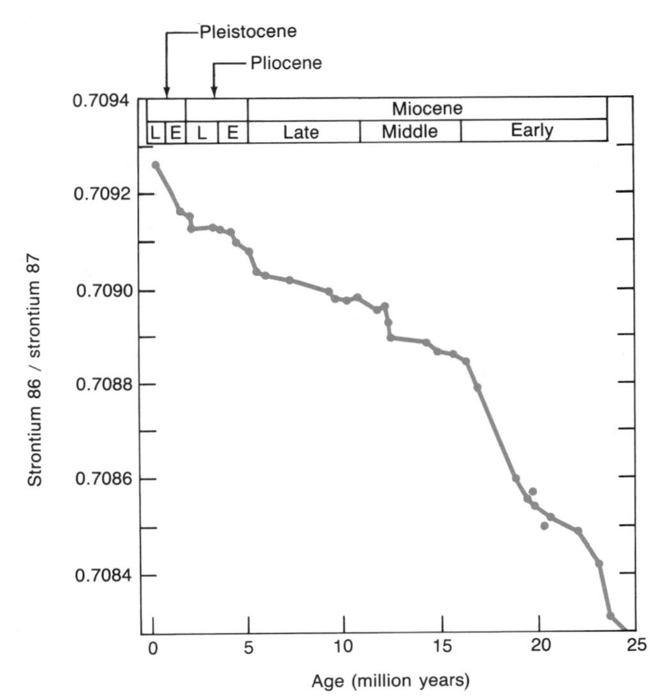

FIGURE 4-18 Changing ratio of strontium isotopes during the past 25 million years for marine fossils composed of calcium carbonate. *(After D. J. DePaolo, Geology 14:103–106, 1986.)*

Strontium Isotopes

Two stable isotopes of the element strontium that occur in all modern seas provide a special opportunity to date relatively young rocks containing fossils that have not undergone appreciable diagenetic alteration. These two isotopes, strontium 86 and strontium 87, occur in the same relative abundance in all modern seas. The ratio of the two isotopes' abundances has changed through time, however, as a result of many factors, including changes in the rates at which rocks of various isotopic composition have yielded their strontium to the ocean as a result of exposure and erosion. During the past 25 million years, the relative abundance of strontium 87 has increased slightly. Small amounts of strontium take the place of calcium in the crystal structure of the calcareous skeletons of marine organisms. This strontium has the same isotopic ratio as the seawater in which the animal lives. Thus well-dated fossils provide a record of changes in the ratio of strontium 87 to strontium 86 in seawater. Once a record of these changes has been established (Figure 4-18), geologists can date fossil skeletons by measuring their precise isotopic composition. This valuable dating method has only recently been put to use. It promises to resolve many stratigraphic problems, especially with respect to Cenozoic fossils, which often have undergone relatively little diagenetic alteration and retain their original strontium-isotope ratios.

THE UNITS OF STRATIGRAPHY

Having examined the ways in which rocks are correlated and dated, we can now appreciate how they are formally classified. As we saw in Chapter 1, geologists divide the stratigraphic record into local three-dimensional bodies of rock that are known as *formations.* Recall that formations are sometimes united into larger divisions known as

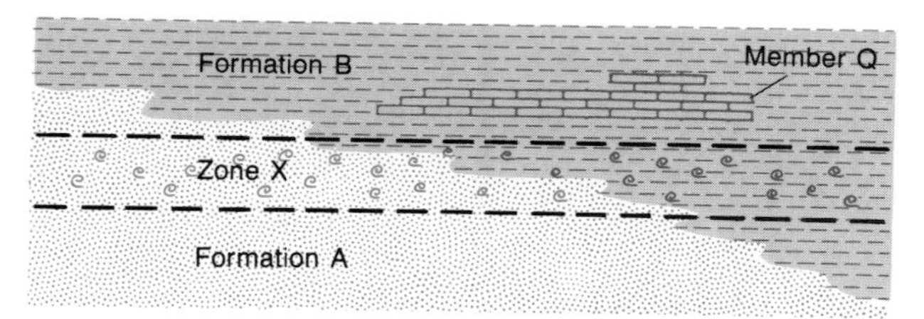

FIGURE 4-19 Cross section of two formations (A and B) separated by a surface that is not time-parallel. The boundaries of the biostratigraphic zone X are approximately time-parallel, and they pass across the boundary between the two formations at an angle. A boundary like the one between formations A and B can result from either transgression or regression (see Figure 4-11). Within formation B, which consists primarily of shale, is a member (Q), a thin layer of limestone.

groups and sometimes include smaller units called *members*. Groups, in turn, may be united into *supergroups*. All these entities are known as **lithostratigraphic,** or **rock-stratigraphic, units.** They are given formal names that are often based on local geographic features, such as rivers and towns. Formations are delineated on the basis of **lithology,** or the physical and chemical characteristics of rock. They are relatively homogeneous bodies, many of which consist of a single rock type. Others, as we have seen, consist of two or more rock types in alternating layers. Formations of this sort include the sedimentary rock cycles produced by deltas (Figure 3-19) and by meandering rivers (Figure 3-14). Some formations display modest lateral facies transitions, but a major facies change, even if gradual, usually leads geologists to divide a body of rock into two adjacent formations. A formation is assigned a **type section** at a particular locality, and it is there that its upper and lower boundaries are defined.

Lithostratigraphic units are designated without regard to time relationships—that is, their upper and lower boundaries may or may not be time-parallel. Therefore, geologists have superimposed on the lithostratigraphic classification another classification system that is based on time relationships revealed by fossils (Figure 4-19). The basic unit of this **biostratigraphic** system is the zone (Figure 4-2). As we have seen, the upper and lower boundaries of zones are not perfectly time-parallel surfaces. In most cases, however, they are nearly time-parallel, and thus the biostratigraphic classification of rocks approximates a chronological classification.

It is primarily on the basis of biostratigraphic zones that Earth's stratigraphic record is formally divided into worldwide systems that correspond to the geologic periods: the Cambrian System, the Ordovician System, the Silurian System, and so on. Systems are grouped into **erathems** and divided into **series,** and series are further divided into **stages.** These divisions of the stratigraphic record, known as **chronostratigraphic,** or **time-stratigraphic, units,** correspond to eras, epochs,

TABLE 4-2 Geologic time units and chronostratigraphic units

Time unit	*Example*	*Chronostratigraphic unit*	*Example*
Era	Paleozoic	Erathem	Paleozoic
Period	Devonian	System	Devonian
Epoch	Late Devonian	Series	Upper Devonian
Age	Famennian	Stage	Famennian

NOTE: Chronostratigraphic units are bodies of rock that represent time units bearing the same formal name. When an epoch (the time unit) is designated by the term *Early* or *Late*, the corresponding series (the chronostratigraphic unit) is identified by the adjective *Lower* or *Upper*. For example, the Lower Devonian series of rocks represents Early Devonian time.

and ages, and, like systems, are given the same formal names (Table 4-2). Important stages of the geologic column are listed in Appendix IV.

Are systems, series, and stages biostratigraphic units? It is logical to answer yes because the boundaries of these divisions are based largely on zones, which are biostratigraphic units; but because the imperfections of zonal correlations represent only small errors when they are considered as percentages of large units such as stages, series, and systems, these units have traditionally been treated as if they had time-parallel boundaries. Thus they have been designated as chronostratigraphic units.

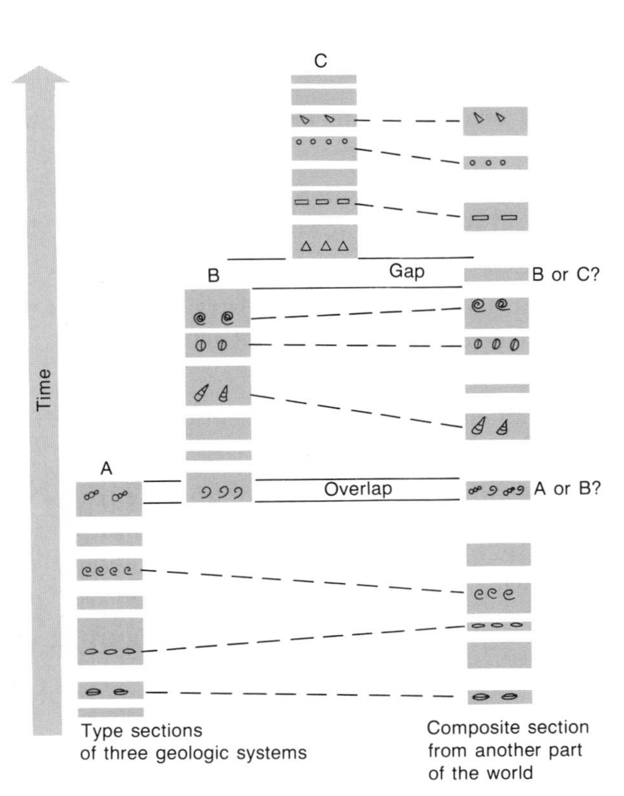

FIGURE 4-20 Problems of correlation between type sections. A, B, and C represent type sections of three geologic systems in different areas. The vertical scale is based on time rather than on sediment thickness. On the right is a composite section showing the systems in yet another, more fully exposed area. Gaps resulting from deposition or erosion are found in all these sections. Lines of correlation by means of fossil correspondence *(dashed lines)* show varying degrees of accuracy, with time lines horizontal. Systems A and B overlap in time, and there is a gap in time between systems B and C. Even if perfect correlation were possible, part of the composite section on the right could not be assigned to a system.

Traditionally, each system has been represented by a single type section, a local stratigraphic section that has served to define the system. As Figure 4-20 indicates, correlation of a type section with sections in other regions is inevitably imperfect. In fact, some systems may overlap slightly, because the geologic time scale developed haphazardly in the nineteenth century, according to the order in which early geologists happened to study rocks of various ages (p. 18). As we shall see in the introductory sections of Chapters 10 through 16, different geologic systems were established in different regions of Europe. Many were established in regions where their upper or lower boundaries—or both—are missing from outcrops because of erosion or incomplete exposure of the rock record. To reduce the resulting problems of correlation (shown in Figure 4-20), stratigraphers are now establishing a **boundary stratotype** for each boundary between two systems. This is a single stratigraphic section in which the boundary is designated (somewhat arbitrarily) and marked on the outcrop. A boundary stratotype serves as a reference point for correlation with all other regions of the world. When a particular stratotype is to be established, an effort is generally made to choose a geologic section in which a continuous sedimentary record spans the boundary and in which numerous guide fossils are present for correlation with other regions. Quite commonly this approach to international correlation, which has only recently begun, will result in the location of lower and upper boundary stratotypes for a given system in two different geographic areas. Boundary stratotypes are being designated not only for systems but also for series, stages, and even zones.

CHAPTER SUMMARY

1 In accordance with the law of superposition, stratified rocks accumulate in such a way that the

oldest beds are at the bottom and the youngest are on top.

2 Because species of plants and animals have appeared and disappeared over large areas of Earth in the course of geologic time, fossils provide a useful means of correlating widely separated bodies of sedimentary rock.

3 The most useful species for correlation are called index or guide fossils. Ideally, these species are easily identified, widely distributed, abundant in many kinds of rock, and restricted to narrow vertical stratigraphic intervals.

4 Index fossils are used singly or in groups to define biostratigraphic zones in rocks. The boundaries of most zones are nearly time-parallel.

5 The scale of geologic time based on fossils is a relative one: Bodies of rock are simply ordered from oldest to youngest. The scale based on radiometric dating, in contrast, is an absolute scale in which events are measured in years.

6 Correlations that are based on radiometric dates are often less accurate than ones based on index fossils.

7 Key beds such as ash falls, glacial tills, and evaporite beds are sedimentary layers that form almost simultaneously over large areas, and thus serve as useful time markers.

8 Some unconformities are useful for identifying times in Earth's history when sea level has suddenly dropped throughout the world. The analysis of seismic reflections often allows geologists to recognize these unconformities and their associated bedding surfaces deep within Earth.

9 Periodic reversals in the polarity of Earth's magnetic field provide excellent markers for correlation of magnetized rocks.

10 A body of rock that is characterized by a particular lithology or group of lithologies is often recognized as a formal lithostratigraphic unit (a member, formation, or group). Rock units do not necessarily have time-parallel upper or lower boundaries, so they are often transected obliquely by biostratigraphic units.

11 Chronostratigraphic units are composite bodies of rock that have been recognized on a global scale. To the extent that correlations are accurate, time-stratigraphic units (erathems, systems, series, and stages) correspond to units of geologic time (eras, periods, epochs, and ages).

EXERCISES

1 What is the difference between the relative and absolute ages of rocks?

2 For what interval of geologic time is radiocarbon dating applicable?

3 How are strontium isotopes useful in dating rocks?

4 Why does a geologic formation not necessarily have upper and lower boundaries that are time-parallel?

5 What factors prevent biostratigraphic zones from having boundaries that are perfectly time-parallel?

6 What factors lead to imperfections in radiometric dating?

7 Construct a diagram resembling Figure 4-11*C* to depict the depositional history of a barrier island–lagoon complex that has undergone a regression followed by a transgression.

8 Construct a diagram resembling Figure 4-15*B* but illustrating that a lowering of sea level need not always lead to regression.

9 How are seismic profiles used to study regional stratigraphy?

ADDITIONAL READING

Ager, D. V., *The Nature of the Stratigraphic Record,* John Wiley & Sons, New York, 1981.

Blatt, H., W. B. N. Berry, and S. Brande, *Principles of Stratigraphic Analysis,* Blackwell Scientific Publications, Oxford, 1991.

Hallam, A., *Facies Interpretation and the Stratigraphic Record,* W. H. Freeman and Company, New York, 1981.

Haq, B. U., J. Hardenbol, and P. R. Vail, "Chronology of Fluctuating Sea Levels since the Triassic," *Science* 235:1156–1167, 1987.

Matthews, R. K., *Dynamic Stratigraphy,* Prentice Hall, Englewood Cliffs, N.J., 1984.

Prothero, D., *Interpreting the Stratigraphic Record,* W. H. Freeman and Company, New York, 1990.

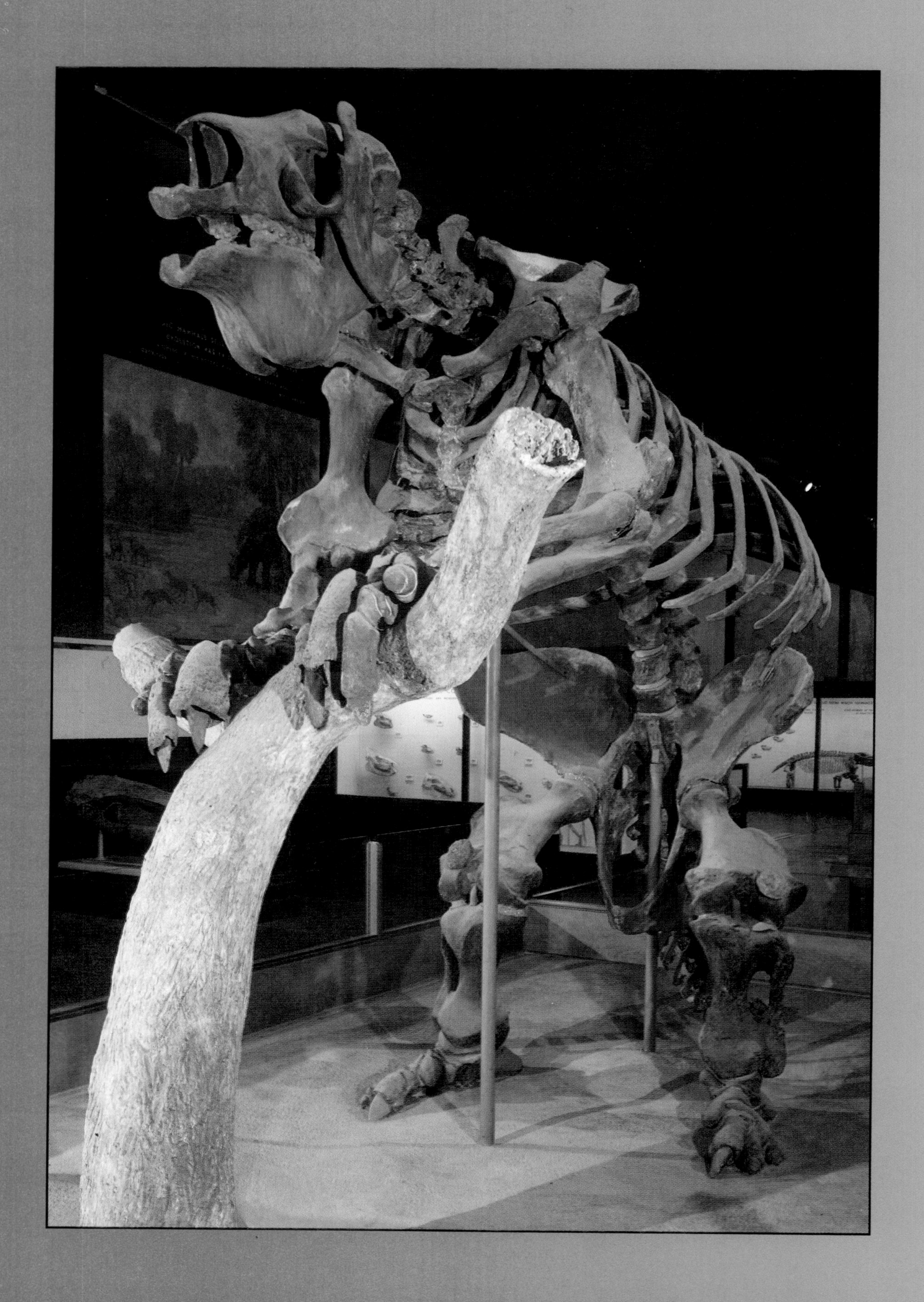

CHAPTER

5

Evolution and the Fossil Record

The central concept of modern biology is that living species have come into being as a result of the evolutionary transformation of quite different forms of life that lived long ago. Indeed, it is often maintained that very little of what is now known about life would make sense in any context other than that of organic evolution. It is important to remember, however, that while the broad definition of *evolution* is "change," *organic evolution* does not encompass every kind of biological change; instead, the term refers only to changes in **populations,** which consist of groups of individuals that live together and belong to the same species.

When we examine the broad spectrum of organisms that inhabit our planet, we cannot help but be impressed by the success with which each form functions in its own particular circumstances. Members of the cat family, for example, have sharp fangs in front for puncturing the flesh of prey and bladelike molars in the rear for slicing meat. Horses, on the other hand, are equipped with chisel-like front teeth for nipping grass and broad molars for grinding it up (Figure 5-1). Plants, too, exhibit a variety of forms and features that vary in accordance with the plants' ways of life. Most tree species that are native to tropical rain forests, for example, have leaves that are waxier than those of plants found in cool regions, and a typical tropical leaf terminates in an elongate tip called a drip point. The drip point and the waxy surfaces help the leaves shed the rainwater that falls on them daily in a rain forest; the waxy surfaces also keep them from drying out in the tropical heat. In contrast, the leaves of another rain-forest plant, the bromeliad, form a cup that acts as a private reservoir for rain. Without this feature, the bromeliad would dry up and die, because it lives high above the moist forest floor, attached to trees. Yet another rain-forest plant, the Venus's-flytrap, secretes a sweet nectar that lures insects to the midrib of its leaf (Figure 5-2). On the margins of these leaves are rows of spines that mesh when the leaves snap shut around an unsuspecting insect. The Venus's-flytrap then devours the insect in a reversal of the normal roles of plant and animal.

These specialized features, which allow animals and plants to perform one or more functions that are useful to them, are known as **adaptations.** Each individual organism possesses many adaptations that function together to equip it for its particular way of life. Before the middle of the nineteenth century, however, the nature of

Megatherium americamum was the largest species of ground sloths, all of which are extinct. This Pleistocene species, which was about 6 meters (~20 feet) long from head to tail, migrated to North America from South America before it died out. *Field Museum of Natural History, Chicago, neg. GEO8460c.)*

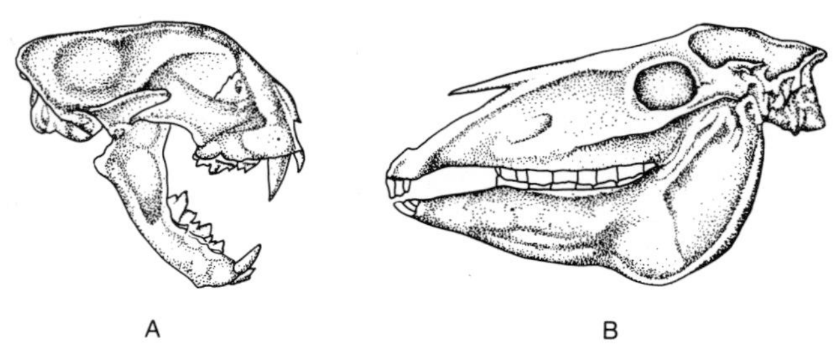

FIGURE 5-1 The teeth of a cat *(A)* and a horse *(B)*, both of Pliocene age.

adaptations was not well understood. At that time it was assumed that adaptations represented perfect mechanisms that had been specially designed to allow each species to function optimally within its own ecologic niche. Since then it has become widely acknowledged that adaptations are fraught with imperfections, many of which stem from evolutionary heritage. An animal or plant, in other words, may develop a useful new feature with which to perform a function, but the evolution of this feature will sometimes be constrained by the structure of the ancestral organism. Evolution can operate only by changing what is already present; it cannot work with the freedom of an engineer who is designing a new device from raw materials. The business of evolution, in other words, is and always has been remodeling—not new construction.

FIGURE 5-2 A Venus's-flytrap luring a bee onto a gaping pair of leaves, which will snap shut. *(Michael Viard, AUSCAPE International.)*

It becomes obvious that evolution is a remodeling process when we observe that certain organs, such as the cheek teeth of mammals and the leaves of higher plants, serve different functions in different species but nonetheless share a common "ground plan," or fundamental biological architecture. All mammals, for example, possess teeth that are rooted in bone and consist of both dentin and enamel. Similarly, certain types of cells and tissues form the leaves of nearly all flowering plants. Common ground plans suggest common origins, and this is one of the many pieces of evidence that indicate that groups of species of the modern world have a common evolutionary heritage: No matter how greatly the species of a given order or class may differ, they share certain basic features that reflect their common ancestry.

We will begin by reviewing the ideas of Charles Darwin, the man who popularized the idea of evolution. Next we will examine natural selection, which Darwin recognized to be the dominant process of evolution, and discuss current knowledge and theories of how new species come into being. Finally, we will learn more of what the fossil record has revealed about extinction and about rates, trends, and patterns of evolution.

CHARLES DARWIN'S CONTRIBUTION

Few biologists gave serious consideration to the idea of organic evolution until 1859, when Charles Darwin (Figure 5-3) published his great work, *On the Origin of Species by Natural Selection*. One of the most effective ways to appreciate the power of the basic evidence that evolution has occurred is to put yourself in Darwin's position when, in 1831, at the age of 23, he set sail as an unpaid naturalist aboard the *Beagle* on an ocean voyage that took him around the world. In the course of this trip Darwin became convinced of

FIGURE 5-3 A portrait of Charles Darwin in old age. *(Painting by John Collier, National Portrait Gallery, London.)*

the workings of evolution and also accumulated much of the evidence that subsequently enabled him to convince others of the validity of his ideas.

The Voyage of the *Beagle*

Some of the observations that Darwin made during his historic voyage were of a purely geologic nature. Darwin had been well tutored in geology by Adam Sedgwick, an expert on the early Paleozoic rocks and fossils of Britain. Moreover, just before he set sail, Darwin was given the first volume of Charles Lyell's *Principles of Geology* by one of his teachers, the botanist J. S. Henslow. Although Henslow advised Darwin to read the book but not to believe it, Darwin, like most others who studied geology at the time, became convinced that Lyell's uniformitarian approach represented a valid interpretation of Earth's history (p. 3). On the voyage of the *Beagle*, Darwin witnessed some remarkable geologic processes—including earthquakes and volcanic activity—that corroborated this idea and enabled him to appreciate more fully than many of his contemporaries how Earth could be transformed continuously by everyday processes. Darwin's conversion to uniformitarian geology paved the way for his conclusion that natural processes transform life in the same way that they transform Earth's physical features.

While Darwin's adherence to the uniformitarian view of Earth's history provided a framework for his acceptance of evolution, it was his observation of the geographic distributions of living things that ultimately led him to theorize that many different forms of life possess a common biological heritage. Darwin was surprised to find South America inhabited by animals that differed substantially from those of Europe, Asia, and Africa. The large flightless birds of South America, for example, were species of rheas (Figure 5-4), which belonged to a different family from

FIGURE 5-4 The rare flightless bird *Rhea darwinii*, named after Charles Darwin. *(Painting by John Gould from C. Darwin, Zoology of the Voyage of H.M.S. Beagle, Smith and Elder, London, 1838–1843.)*

the superficially similar birds of other continents — the ostriches of Africa and the emus of Australia. Among other unique South American creatures were the sloths and the armadillos. Not only was South America the home of living representatives of these groups, but it was there that Darwin dug up the fossil remains of extinct giant relatives of the living forms (see p. 102 and Figure 5-5). Why, he asked, were the rhea birds as well as all living and extinct members of the sloth and armadillo families found nowhere but in South America?

Darwin was also intrigued to find that species of marine life on the Atlantic side of the Isthmus of Panama differed from those on the Pacific side. In places the isthmus is only a few miles wide, and it struck Darwin as strange that the marine creatures on opposite sides of this narrow neck of land should differ from each other — unless the various species had somehow come into being where they now lived. If the species had instead been scattered over the planet by an external agent, he reasoned, many should have landed on both sides of the isthmus.

Perhaps the most striking of Darwin's observations concerned life forms on oceanic islands. Darwin noted that no small island situated more than 5000 kilometers (~ 3000 miles) from a continent or from a larger island was inhabited by frogs, toads, or land mammals unless they had been introduced by human visitors. The only mammals native to such islands were bats, which could originally have flown there. This observation led Darwin to suspect that species could originate only

A

FIGURE 5-5 Unusual South American mammals. A. A living three-toed sloth, hanging upside down in its normal mode of life. This animal is the size of a small dog. *B.* A reconstruction of two of the Pleistocene mammals of South America that Darwin unearthed as fossils. *Megatherium*, the giant ground sloth, was more than 6 meters (~ 20 feet) in length — larger than an elephant. It ranged northward into the United States. The giant armadillo is *Glyptodon*. (*A. Jerry Ellis/The Wildlife Collection. B. Courtesy of the Field Museum of Natural History, Chicago, neg. CK20T. Painting by C. R. Knight.*)

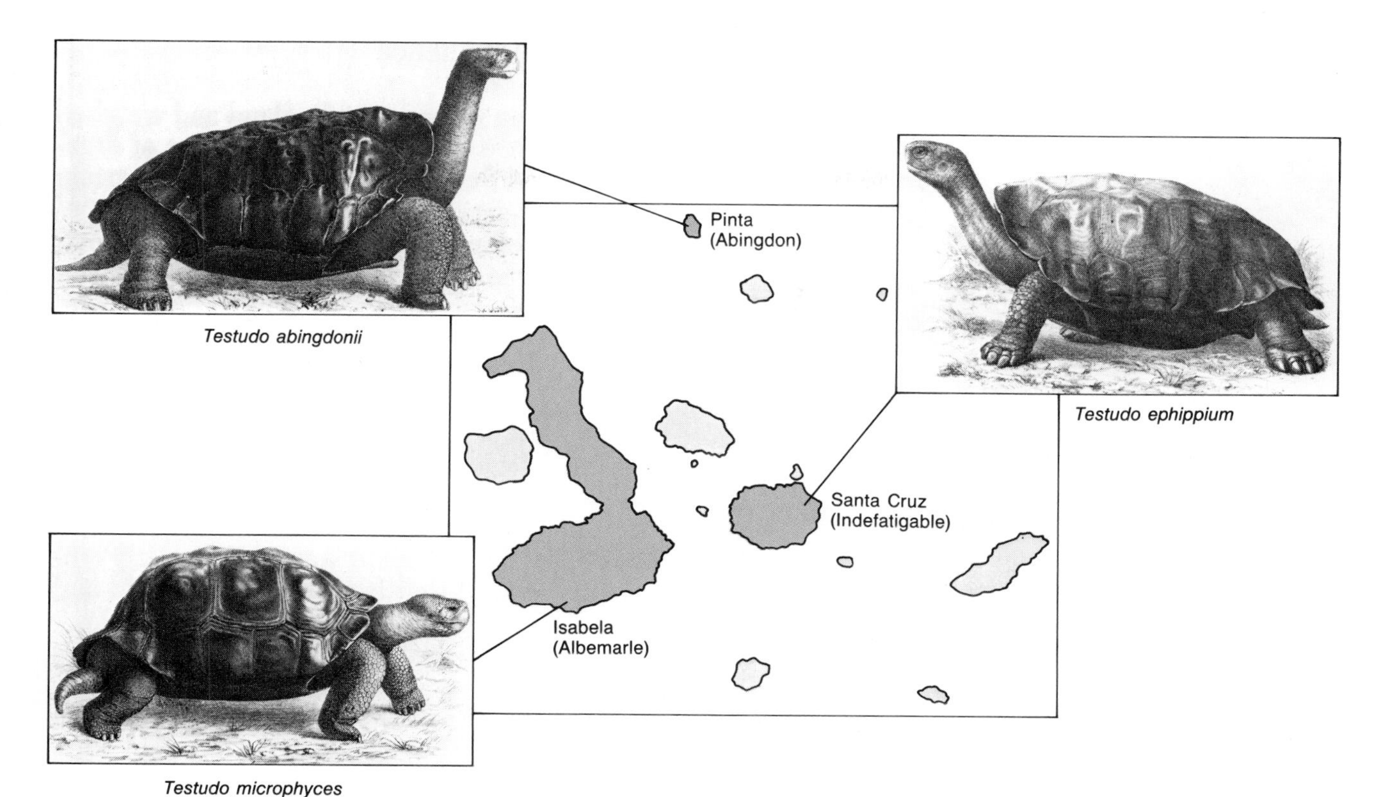

FIGURE 5-6 Three tortoises, each of which inhabits a different island of the Galápagos. *Testudo abingdonii,* which inhabits Pinta Island, has a long neck and a shell that is elevated in the neck region; these features represent adaptations for reaching tall vegetation. The shells of these animals can exceed 1 meter (~ 3 feet) in length. *(T. Dobzhansky et al., Evolution, W. H. Freeman and Company, New York, 1977.)*

from other species. Otherwise, why would isolated areas of land be left without important forms of life?

The Galápagos Islands, which lie astride the equator about 1100 kilometers (~ 700 miles) from South America, played an especially large part in the development of Darwin's new ideas. Darwin found the Galápagos to be inhabited by huge tortoises, and he thought it curious that the people who lived on the islands could look at a tortoise shell and immediately identify the island from which it had come. The fact that different races of giant tortoises occupied different islands (Figure 5-6) led Darwin to suspect that these distinctive populations of tortoises shared a common ancestry but had somehow become differentiated in form as a result of living separately in different environments. Even more striking were the various kinds of finches that Darwin found in the Galápagos; some types had slender beaks, others had somewhat sturdier ones, and still others had very heavy beaks, which served the function of breaking seeds (Figure 5-7). One kind of Galápagos finch behaved like a woodpecker but used a cactus spine as a woodpecker uses its long beak to probe for insects in wood. Furthermore, all of the finches in the Galápagos resembled a species of finch on the South American mainland (the closest large landmass) rather than the finches found in other regions of the world.

Darwin began to ponder whether a population of finches from the South American mainland might have reached the islands and become altered in some way to assume a wide variety of forms. It seemed that the finches had somehow differentiated in such a manner that they were able to pursue ways of life that on the mainland were divided among several different families of

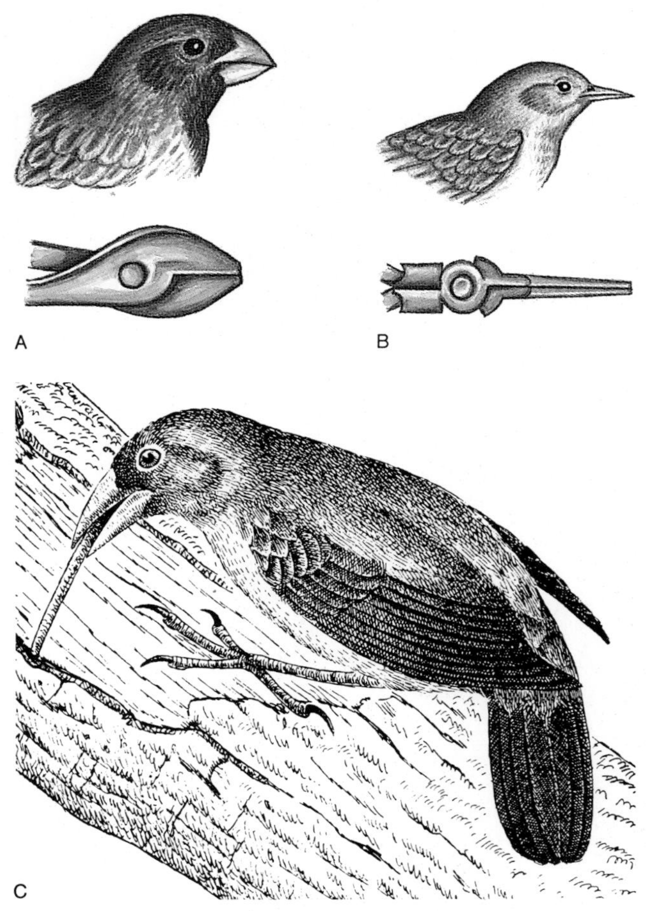

FIGURE 5-7 Three of the finch species that Darwin observed in the Galápagos Islands. *A.* The large tree finch's parrotlike beak operates like heavy pliers in crushing fruits and buds. *B.* The warbler finch's beak operates like needlenose pliers in catching insects. *C.* The woodpecker finch, which excavates tree bark with its chisel-like beak, uses a cactus needle as a tool to probe for insects. *(A and B. From P. R. Grant, "Natural Selection and Darwin's Finches," Scientific American. Copyright © 1991 by Scientific American, Inc. B. From D. Lack, "Darwin's Finches," Scientific American. Copyright © 1953 by Scientific American, Inc. All rights reserved.)*

birds. As Darwin put it in the journal in which he described his scientific work on the voyage:

> Seeing this gradation and diversity of structure in one small, intimately related group of birds, one might really fancy that from an original paucity of birds in this Archipelago, one species had been taken and modified for different ends.

Anatomical Evidence

When Darwin returned to England and weighed other evidence indicating that one type of organism evolved from another, he found that certain anatomical relationships seemed to build an especially compelling case. One such piece of evidence was the remarkable similarity of the embryos of all vertebrate animals (Figure 5-8). Darwin was intrigued by the admission of Louis Agassiz, a noted American scientist, that he could not distinguish an early embryo of a mammal from that of a bird or a reptile. This, Darwin reasoned, was exactly what could be expected if all vertebrate animals had a common ancestry: Although adult animals might become modified in shape as they adapted to different ways of life, early embryos were sheltered from the outer world and would thus undergo less change.

Equally convincing to Darwin was the evidence of **homology**—the presence, in two different groups of animals or plants, of organs that have the same ancestral origin but serve different functions. The principle of homology is illustrated by the variations in teeth and leaves discussed earlier in this chapter. Another example is the common origin of the toes of land-dwelling mammals and the wings of bats (Figure 5-9). Bats' wings are actually formed of four toes whose external appearance and bone configuration resemble those of walking mammals. If bats' wings did not share a common biological origin with the feet of walking mammals, why do the two types of organs have similar bone configurations? Such evidence of common origins abounds in both the animal world and the plant world.

The existence of **vestigial organs**—organs that serve no apparent purpose but resemble organs that do perform functions in other creatures—further supported Darwin's argument in favor of evolution. One of the most striking aspects of vestigial organs is that they are usually smaller and less complex than the functional organs they resemble. Whales, for example, have apparently useless bones that resemble the functional pelvic bones of other mammals (Figure 5-10). Why would whales possess such bones if they were not remnants of once functional organs that are now in the process of disappearing? The vestigial bones of whales reflect a biological past in which these mammals, too, had legs.

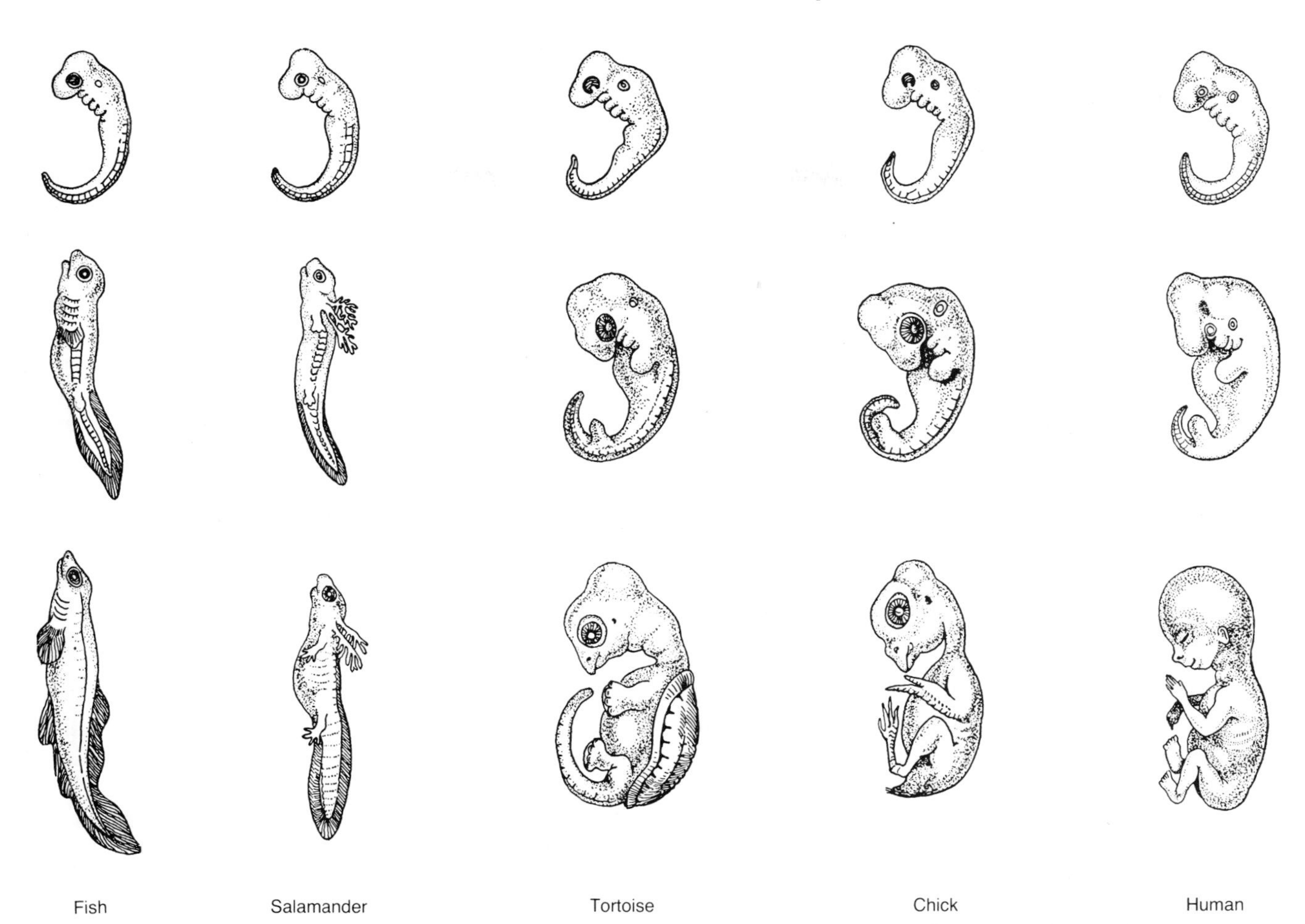

FIGURE 5-8 Early stages in the embryology of various groups of vertebrate animals. Note that in the earliest stage shown here *(top)*, the embryos of all the groups are nearly identical. *(After G. Hardin, Biology: Its Human Implications, W. H. Freeman and Company, New York, 1949.)*

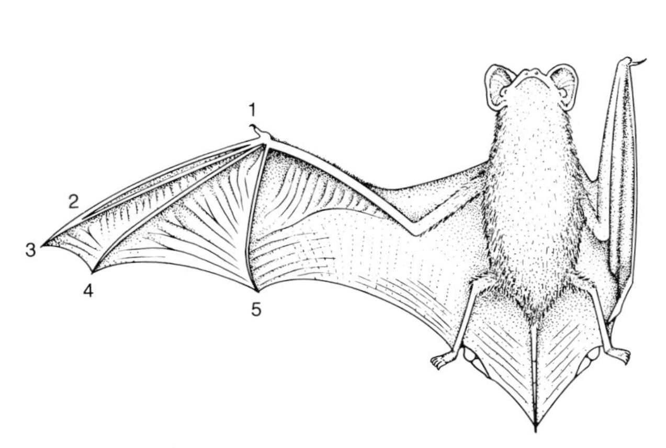

FIGURE 5-9 The extended wing of a bat. The five digits are equivalent to the toes of a land-dwelling mammal and to the fingers of the human hand.

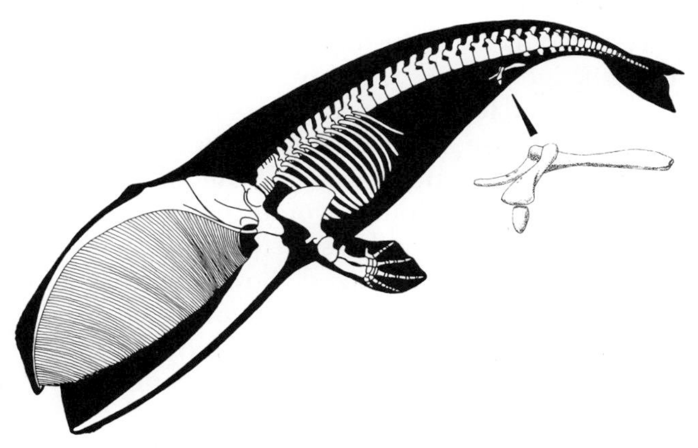

FIGURE 5-10 The skeleton of a baleen whale, the largest living mammal. The enlargement shows the whale's pelvic bones, which resemble those of other mammals but are only weakly developed and serve no apparent function. *(Drawing by Gregory S. Paul.)*

Natural Selection

Darwin also recognized a different type of evidence that pointed to the validity of biological transformation in nature: Animal breeders were known to have produced major changes in domesticated animals by means of selective breeding. If wild dogs could be modified into greyhounds, Saint Bernards, and chihuahuas under domestic conditions, Darwin saw no reason why animals should be anatomically straitjacketed in nature. The question was: What could bring about biological changes under natural conditions?

This question led to Darwin's second great contribution. The first, of course, had been his amassing of an enormous amount of evidence indicating that species had evolved in nature. The second was his conception of a mechanism through which evolution was likely to have taken place. The mechanism Darwin proposed was **natural selection**—a process that operates in nature but parallels the artificial selection by which breeders develop new varieties of domestic animals and plants for human use.

Essentially, artificial selection in domestic breeding involves the preservation of certain biological features and the elimination of others. A breeder simply chooses certain individuals of one generation to be the parents of members of the succeeding generation. Darwin recognized that in nature, many more individuals of a species are born than can survive. Accordingly, he reasoned that success or failure here, as in artificial breeding, would not be determined by accident. In nature it would be determined by advantages that certain individuals had over others—greater ability to find food, for example, or avoid enemies, or resist disease, or deal with any of a number of environmental conditions. By virtue of their longevity, these individuals would tend to produce more offspring than others.

Darwin also recognized, however, that survival was not the only factor influencing success in natural selection, because some individuals with only average life spans were capable of producing more total offspring than others simply because they bore large litters or shed large numbers of seeds. Thus, as long as the members of a breeding population varied substantially in either longevity or rate of reproduction, certain individuals would pass on their traits to an unusually large number of members of the next generation. The kinds of individuals that came to predominate as generation followed generation could then be said to be favored by natural selection.

GENES, DNA, AND CHROMOSOMES

Darwin faced a major obstacle in convincing others that natural selection could operate effectively to produce evolution. Because he lived before the birth of modern genetics, Darwin was not familiar with the mechanisms of inheritance and thus could not explain how an organism could pass along a favorable genetic trait to its offspring. Although the Austrian monk Gregor Mendel had outlined the basic elements of modern genetics only a few years after Darwin published *On the Origin of Species*, Mendel's work was not acknowledged until the turn of the century, two decades after Darwin's death.

Mendel's most significant contribution to modern genetics was the concept of **particulate inheritance**, which explains how certain hereditary factors, which we now call **genes**, retain their identities while being passed on from parents to offspring. Mendel's experiments with pea plants demonstrated that individuals possess genes in pairs, with one gene of each pair coming from each parent. In one of his experiments, Mendel employed a true-breeding white-flowered strain of pea plants and a true-breeding red-flowered strain. (*True breeding* signifies that within the strain, descendants resemble parents throughout a long series of generations.) Mendel's first step was to cross plants of the white-flowered strain with those of the red-flowered strain. The surprising result was that all of the daughter plants had red flowers. When these red plants were crossed with each other, however, they produced both red-flowered and white-flowered descendants.

The most significant aspect of Mendel's work was his discovery that the effect of the gene for white color could surface in the third generation even after its presence had been masked in the second generation. Preservation of genes in this manner—that is, as discrete entities that maintain their identity from generation to generation—constitutes the basis of particulate inheritance, the cornerstone of modern genetics.

Darwin's problem was that he and his contemporaries lived in an era dominated by the erroneous concept of blended inheritance, which held that genetic material is permanently diluted when it is combined with other genetic material through mating. This theory maintained that, barring the influence of nutrition during life, an individual of medium height should result from the mating of a tall individual with a short one. In keeping with this assumption, some of Darwin's opponents argued that any useful new feature that might appear in a species would subsequently be diminished upon being blended with other features as generation followed generation until the feature eventually disappeared. Because he was unaware of the concept of particulate inheritance, Darwin was unable to counter this argument.

Another discovery that was made during the emergence of modern genetics was that genes can be altered. It is now understood that genes are, in fact, chemical structures that can undergo chemical changes, and these changes, or **mutations,** provide much of the variability upon which natural selection operates. Genes are now known to be segments of long molecules of deoxyribonucleic acid, or **DNA**—a compound that carries chemically coded information from generation to generation, providing instructions for growth, development, and functioning of organisms.

Changes in one or more of the hundreds of nucleotides that form genes along a strand of DNA are known as point mutations. These mutations can result from imperfect replication of the DNA strand during cell division, so that one of the new cells formed from the preexisting cell inherits a slightly altered copy of the first cell's genetic material. Point mutations can also occur when an already-existing DNA strand is chemically altered by an external agent, which may be a chemical substance or a dose of radiation such as cosmic radiation or ultraviolet light. In any case, point mutations usually produce changes in the structure of the proteins that are coded by the mutated segments of DNA.

In organisms other than bacteria and their close relatives (Figure 1-5), DNA is concentrated within **chromosomes,** which are elongate bodies found in the nucleus of the cell. Most organisms have chromosomes that are paired, one having been inherited from each parent (Figure 5-11). Chromosomal mutations take the form of changes in the number of chromosomes or in the positions of segments of individual chromosomes. Such changes cause segments of DNA to move with respect to each other. The exact importance of these changes to the ways in which DNA controls the development and operation of an organism remains uncertain.

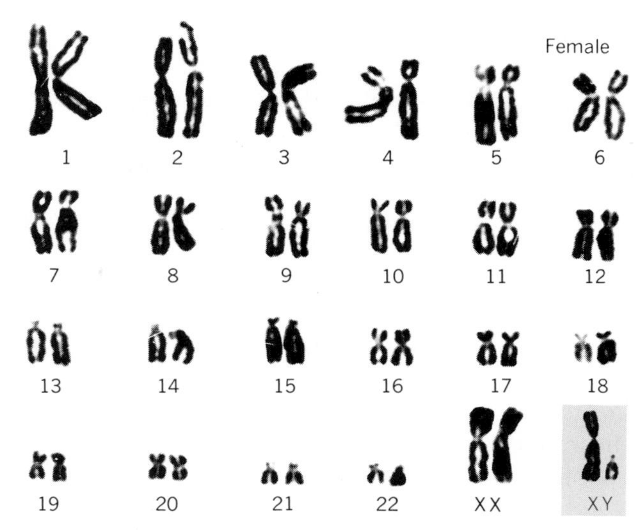

FIGURE 5-11 The complete set of human chromosomes. One member of each of the 23 pairs comes from each parent. The presence of two X chromosomes indicates that this set of chromosomes represents a female; the male condition is determined by the presence of one X and one Y chromosome instead of two X's. *(M. Grumbach and A. Morishima.)*

POPULATIONS, SPECIES, AND SPECIATION

When two animals breed, each normally contributes half of its chromosomes to each offspring by way of a **gamete,** a special reproductive cell that contains only one (rather than a pair) of each type of chromosome. The female transmits this set of chromosomes by way of an egg cell, which is the female gamete, and the male by way of a sperm cell, the male gamete. Similarly, the offspring, if it mates, combines half of its chromosomes with half of those of a member of the opposite sex in order to produce still another generation. The mixing that takes place in this manner, which is known as **sexual recombination,** continually yields new combinations of chromosomes and hence of genes. The process of sexual recombination, in conjunction with occasional point and chromosomal mutations, is responsible for the variability among organisms that provides the raw material for natural selection. Unfortunately, Darwin and his contemporaries had no knowledge of these sources of variability.

In the study of evolution today, the sum total of genetic components of a population, or group of interbreeding individuals, is referred to as a **gene pool.** And as we have seen (p. 7), populations form a species if their members can interbreed. Reproductive barriers between species keep gene pools separate and thus prevent interbreeding. These barriers include differences in mating behavior, incompatibility of egg and sperm, and failure of offspring to develop into fully functioning adults.

Not only can a species as a whole evolve in the course of time, but it can also give rise to one or more additional species. The origin of a new species from two or more individuals of a preexisting one is called **speciation.** Because species are kept separate from one another by reproductive barriers, speciation, by its very definition, entails evolutionary change that produces such barriers. It is widely believed that most events of speciation involve the geographic isolation of one population from the remaining populations of the parent species. This isolated population then follows an evolutionary course that causes it to diverge from the parent species in both form and way of life. Its divergence may result from such phenomena as the occurrence of unique mutations and the guidance of natural selection by unusual environmental conditions. The development of distinct species of finches on the various Galápagos islands (Figure 5-7) illustrates this principle.

EXTINCTION

Fossils provide the only direct evidence that life has changed substantially over long spans of geologic time. They also offer the only concrete evidence that millions of species have disappeared from Earth, or suffered **extinction.**

The idea that a species could become extinct was not widely accepted until late in the eighteenth century. Before that time, fossil forms that seemed no longer to inhabit Earth were thought to live in unexplored regions. In 1786, however, Georges Cuvier, a French naturalist, pointed out that fossil mammoths were so large that any living mammoths could not possibly have been overlooked. Cuvier thus concluded that mammoths were extinct. His argument was well received, and soon the extinction of many species was accepted as fact.

In general, extinction results from particularly extreme impacts of the factors that normally hold populations in check. A limiting factor may be predation on a population, disease, competitive interaction with one or more other species, a restrictive condition of the physical environment, or chance fluctuation in the number of individuals in the population (p. 26). Changes resulting from one or more of these factors have led to the extinction of most of the species of animals and plants that have inhabited Earth; in fact, of all the species that have existed in the course of Earth's history, only a tiny fraction remain alive today.

Species have also disappeared by evolving to the point at which they have been formally recognized as different species. In this process, known as **pseudoextinction,** a species' evolutionary line of descent continues, but its members are given a new name. The point at which the new species comes into being is often arbitrarily designated, because there is no way of determining precisely when members of an evolving group lost the ability to interbreed with its original members.

Mass Extinction

In addition, large numbers of species have vanished throughout the world during intervals of just a few million years in periodic episodes of **mass extinction.** In the chapters that follow, we will review the most conspicuous episodes of mass extinction in Earth's history, including the one in which the dinosaurs disappeared. Although the fossil record is imperfect, it nonetheless reveals that at certain times between 50 and 90 percent of all the species on Earth died out during very brief geologic intervals. All mass extinctions, however, have been followed by at least a partial evolutionary recovery in which the number of species on Earth has increased again. The causes of mass extinction are by no means fully understood, and different mass extinctions may well have had different causes, but all seem to have affected some groups of living things more severely than others.

The kinds of groups that were affected varied from episode to episode. Unfortunately, it appears today that the world is entering a unique interval of mass extinction: one that our own species is bringing about (Box 5-1).

Rates of Origination and Extinction of Taxa

One unique contribution of fossils to biological science is the ability they afford us to assess rates of evolution and extinction. It is only through data derived from the fossil record, for example, that we have been able to measure the rates at which new species, genera, and families have appeared within large groups of animals and plants. We may use the number of species, genera, or families living today as a final number in some calculations, but the geologic record provides essential information about the events that have produced the numbers of these living taxa.

Adaptive Radiation At many times in Earth's history, groups of animals or plants have undergone remarkably rapid evolutionary expansion—that is to say, one or more phyla, classes, orders, or families has produced many new genera or species during brief intervals of time. Rapid expansions of this kind are known as **adaptive radiations.** In this context, the word *radiation* refers to the pattern of expansion from some ancestral adaptive condition to the many new adaptive conditions represented by descendant taxa. The word *adaptive* expresses the idea that the new taxa that develop are functionally related to new ways of life. Figure 5-12 shows how the fossil record permits us to measure the rate at which adaptive radiation has taken place. Here we can see that the number of families of corals increased rapidly during the Jurassic Period. The coral families depicted belong to the group known as the hexacorals, which first appeared in mid-Triassic time, about 215 million years ago. Living species of hexacorals form the beautiful coral reefs of the modern world (see Box 2-1).

Adaptive radiation often occurs in groups of plants or animals within just a few million years of their origin. This pattern is typical because the modes of life of recently formed groups often differ from those of groups that originated earlier. Since the old and new groups occupy different niches, ecologic competition does not restrain the diversification of the new group. In addition, when a group first evolves, predatory animals may not yet have developed efficient methods for attacking the new group's members, and this protected status permits the new group to form many new species in a short period of time.

Sometimes the extinction of one biological group has allowed for the adaptive radiation of another even though the radiating group was not a new one on Earth. Mammals, for example, inhabited Earth during almost all of the Mesozoic Era, but they remained small and relatively inconspicuous until the close of that era, when the dinosaurs suffered extinction. Unrestrained by competition or predation by dinosaurs, mammals then underwent a spectacular adaptive radiation. Their rise to dominance on the land has led paleontologists to label the Cenozoic Era the Age of Mammals. Most of the living orders of mammals, including the order that comprises bats and the one that comprises whales, came into existence within only about 12 million years of the start of the Cenozoic Era. This interval represents only about 2 percent of all Phanerozoic time.

Many episodes of adaptive radiation have been preceded by **adaptive breakthroughs**—the appearance of key features that, along with ecologic opportunities, have allowed the radiation to take place. The rapid growth of the hexacorals' skeletons, for example, has been very important, for it has often allowed these creatures to crowd out other animals that inhabit hard surfaces in shallow marine environments (p. 46). Because their skeletons are porous, hexacorals can quickly assume large proportions without the need for large volumes of calcium carbonate. Thus they have had an edge over slowly growing organisms with which they compete for space on the seafloor. Another adaptive breakthrough for the hexacorals was the development of a symbiotic relationship with algae that live in the tissues of the reef-building species and help the hexacorals develop their skeletons.

A variety of adaptive breakthroughs played major roles in the early Cenozoic adaptive radiation of mammals. The key feature for rodents, for example, may have been gnawing front teeth, which these animals use for eating nuts and other hard seeds. For grazing horses, the development of grinding cheek teeth, mentioned earlier, was a major adaptive breakthrough.

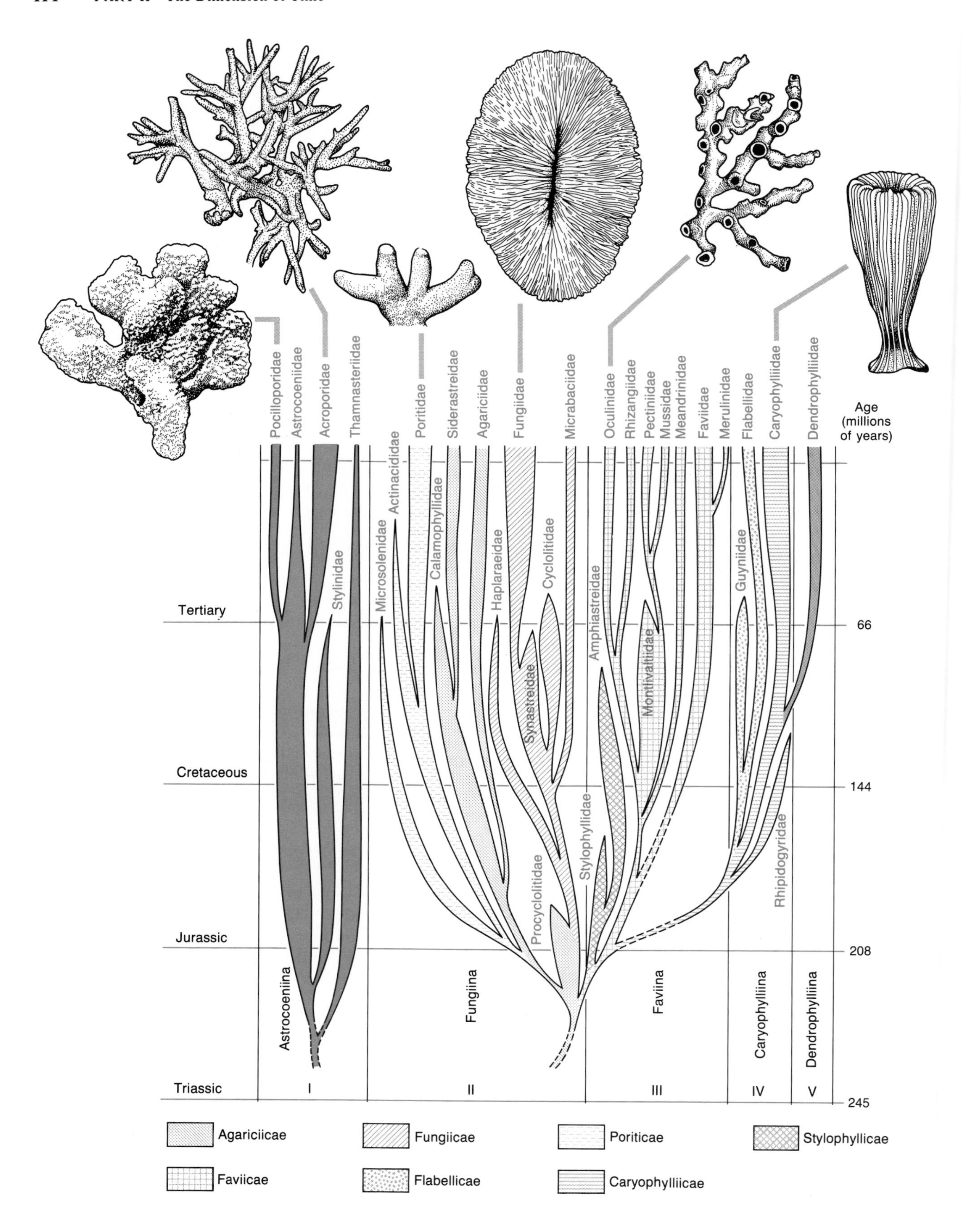

Pocilloporidae
Astrocoeniidae
Acroporidae
Thamnasteriidae
Stylinidae
Actinacididae
Microsolenidae
Poritidae
Siderastreidae
Calamophyllidae
Agariciidae
Haplaraeidae
Fungiidae
Cyclolitidae
Synastreidae
Procyclolitidae
Micrabaciidae
Stylophyllidae
Amphiastreidae
Oculinidae
Rhizangiidae
Pectiniidae
Mussidae
Meandrinidae
Montlivaltiidae
Faviidae
Merulinidae
Flabellidae
Guyniidae
Caryophylliidae
Rhipidogyridae
Dendrophylliidae
Age
(millions
of years)
Tertiary
Cretaceous
Jurassic
Triassic
66
144
208
245
Astrocoeniina
Fungiina
Faviina
Caryophylliina
Dendrophylliina
I
II
III
IV
V
Agariciicae
Fungiicae
Poriticae
Stylophyllicae
Faviicae
Flabellicae
Caryophylliicae

FIGURE 5-12 The pattern of adaptive radiation of the hexacorals since the group originated in mid-Triassic time. This group builds modern coral reefs. Typical representatives of six families are illustrated. All of these colonies are between about 10 and 20 centimeters (4–8 inches) in maximum dimension. Four of the five orders of hexacorals *(roman numerals)* were already present by mid-Jurassic time, and nearly all superfamilies *(patterns)*, but families *(names printed in color)* continued to proliferate during the Cretaceous Period.

The pattern of adaptive radiation seen in Figure 5-12 is typical: Early expansion produced large-scale evolutionary divergence at a very early stage. Note that four of the five modern orders of hexacorals were already present halfway through the Jurassic Period, shortly after the adaptive radiation of hexacorals began. After the new orders of hexacorals became established, however, evolutionary change was more restricted. New families continued to develop at a high rate, but new orders did not. Later, during the Cenozoic Era, few new families evolved, but divergence continued at the genus and species level. This pattern seems to indicate that as a group of animals or plants begins to expand, it quickly exploits any adaptations that its body plan allows it to develop with ease. Later, however, evolution is restricted to the development of variations on the basic adaptive themes that evolved early on. In time, few new families evolve, and eventually only new genera and species evolve.

Local adaptive radiations of the recent past offer special insights into larger adaptive radiations that took place earlier. Many of these local adaptive radiations occurred in isolated environments such as islands or lakes, whose well-defined boundaries prevent the escape of the species within them. When these sites of evolution are of recent origin, they can provide evidence of the remarkably rapid diversification of life. This diversification, in turn, usually reflects the fact that the site of adaptive radiation was uninhabited territory and thus lacked the predators and competitors that might have inhibited the evolutionary diversification.

The Galápagos, where Darwin studied the unique groups of tortoises and finches (Figures 5-6 and 5-7), are the islands most famous as sites of recent adaptive radiation. The Galápagos originated as a result of volcanic eruptions a few million years ago. Many large lakes have also been sites of adaptive radiation in the recent past. In Lake Victoria in Uganda, which came into being no more than a few hundred thousand years ago, about 170 species of fishes of the cichlid group have been identified, and all but three of them can be found nowhere else in the world. Many of these fishes have highly distinctive adaptations; some are specialized for eating insects, others for attacking other fishes, and still others for crushing shelled mollusks (Figure 5-13). Several species of cichlids that still inhabit Lake Victoria look very much like the fishes that seem to have given rise to the great

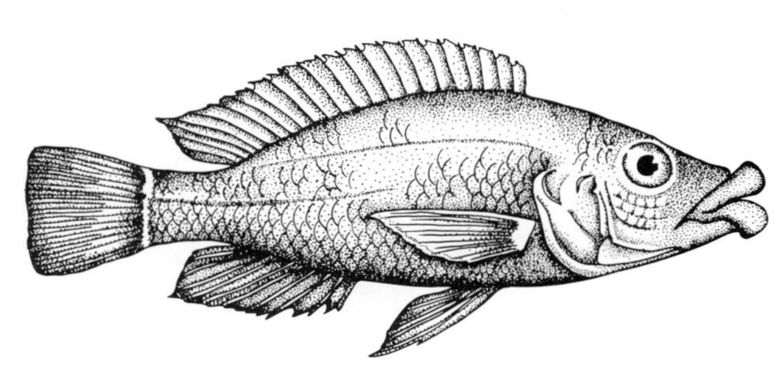

Haplochromis chilotes, a specialized insectivore (62% of actual size)

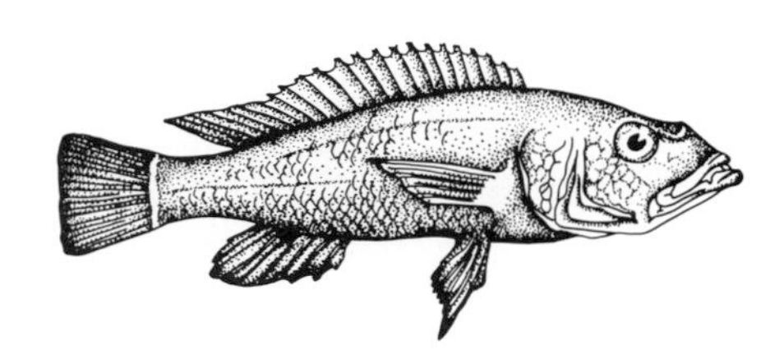

Haplochromis estor, a piscivore (21% of actual size)

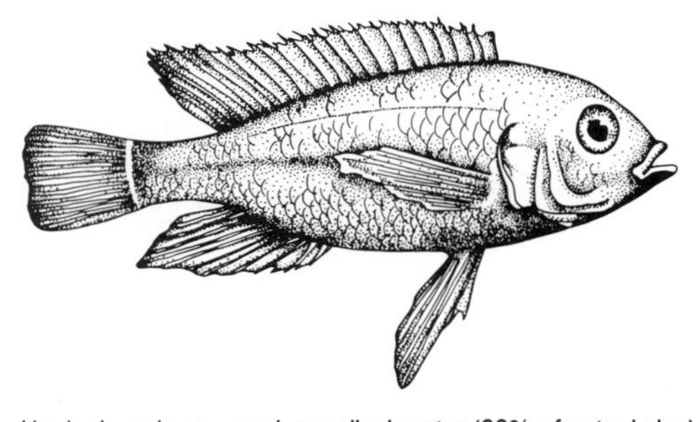

Haplochromis sauvagei, a mollusk eater (60% of actual size)

FIGURE 5-13 Three of the more than 170 species of cichlid fishes that evolved in Lake Victoria, Uganda, within the last few hundred thousand years. The fishes exhibit many different adaptions. *(After P. H. Greenwood, Brit. Mus. [Nat. Hist.] Bull. Suppl. 6, 1974.)*

Box 5-1
The Coming Mass Extinction

The Siberian tiger, a subspecies, is on the verge of extinction. *(Jim Brandenburg/Minden Pictures.)*

The great extinctions of the geologic past provide a unique perspective on the extinctions that humans are now precipitating throughout the world. Comparisons indicate that our activities are triggering a mass extinction that will rival any that the world has known in the past 600 million years.

The fossil record alerts us to look for particular patterns in the crisis we are entering. Large animals, for example, have always tended to suffer relatively high rates of extinction, apparently because their typically small populations leave them constantly at risk. The dinosaurs and mammoths spring most readily to mind.

The fossil record also shows that when the environment deteriorates on a global scale, the percentage of species that are lost tends to be especially great in tropical regions. Tropical species are particularly vulnerable because the great diversity of life in the tropics is packed into complex communities in which many species have specialized ecologic requirements and small populations. A large proportion of these species can exist only in association with certain other species, which provide their habitat or their food. Thus when one species goes, others are sure to follow. The patterns seen in the fossil record are already apparent in the modern-day crisis: Large-bodied and ecologically specialized species have been disappearing most rapidly. (Box 2-1 describes how many species that inhabit coral reefs in present-day tropical seas are imperiled because they have narrow ecologic niches.)

Rates of extinction have varied throughout geologic time, but calculations suggest that during most intervals, only about one species died out each year on the average. These days several species must be dying out somewhere every day. Experts believe that the rate of extinction may climb to several hundred species a day within 20 or 30 years.

The largest number of extinctions can be traced to our destruction of habitats. Of all our depredations, the destruction of tropical forests — most of them rain forests — has the most dire consequences, for two reasons. First, even though these lush habitats occupy less than 10 percent of Earth's land area, they contain most of the world's species. Second, the total expanse of tropical forests being destroyed every year would make an area about the size of West Virginia; and only about 5 percent of the remaining area is protected in parks and preserves. Because their populations fall below critical levels, many species die out in a shrinking forest long before the forest has disappeared altogether. Rain forests can grow again in areas from which they have disappeared, but recovery requires more than a century. The forests are disappearing so rapidly that even a massive restoration program would be too late to preserve vast numbers of species of great beauty, scientific interest, and possible value to humankind as sources of medical drugs and other useful products. Through the natural speciation process, evolution would require several million years to restore the lost diversity. Even then the process would not precisely duplicate any lost species, or even produce species remotely resembling those that belonged to genera or families that had vanished altogether.

Even when we do not attack a habitat with fire, chainsaws, and bulldozers, we can still manage to alter it in ways that make it hostile to many forms of life. Acid rain has devastated vast areas of North America and Europe. The acid forms in the atmosphere from oxides of nitrogen and sulfur emitted by automobiles and power plants. Precipitation and runoff bring it to lakes and rivers. The resulting pollution has killed all the fishes in hundreds of lakes in New York State, and in Nova Scotia it has eliminated the Atlantic salmon from numerous rivers where it once came every year to spawn.

Direct exploitation of animals and plants is another cause of extinction. The African black rhinoceros will be lucky to survive hunters who are eager to cater to people in all parts of the world who cling to the old fantasy that its horn enhances sexual prowess.

After earlier mass extinctions, a few kinds of surviving organisms inherited a world all but free of competitors and predators, and their populations exploded. These were ecologic opportunists—species capable of invading vacant terrain readily and then multiplying rapidly (p. 28). Weeds are not the only opportunists in today's world; so are noisy, aggressive birds such as starlings, the cyanobacteria that form scummy masses in polluted water and then decay to rob their environment of oxygen, and many other forms of life inimical to ours. Thus we not only face the prospect of losing vast numbers of species important to us and many entire ecologic communities in the next few decades; we must also expect their places to be filled by species that impair the quality of human life.

adaptive radiation that has occurred within the lake. In other words, the original species, or descendants very much like them, remain along with much more distinctive products of adaptive radiation.

Given the evidence that dramatic adaptive radiations have taken place in lakes and on islands, it is interesting to consider what may have happened in the aftermath of major extinctions of the past. When the dinosaurs disappeared from Earth at the end of the Mesozoic Era, for example, the great continents of the world must have been the equivalents of large vacant islands that mammals could colonize on a grander scale than had been possible when dinosaurs ruled. For the mammals, which had remained small and inconspicuous in the shadow of the dinosaurs, the opportunity to undergo adaptive radiation must have resembled the evolutionary opportunity that was available to the first cichlids that arrived in Lake Victoria or the first finches that landed on the Galápagos Islands. Opportunities for mammals expanded simply because the entire world, not just a small area, had become available for their occupation.

Thus small-scale adaptive radiations of the recent past seem to offer useful models for understanding larger adaptive radiations of the more distant past. The most basic lesson here is that when preexisting species do not interfere, a small number of founder species can rapidly produce many new species, some of which differ substantially from the original forms.

Rates of Extinction Extinction rates have varied greatly among most large groups of animals and plants in the course of geologic time, and they have varied just as greatly from taxon to taxon. An average mammalian species, for example, has survived for just 1 to 2 million years, which means that the extinction rate for mammals has exceeded 50 percent per million years. In contrast, within many groups of marine life, including bivalve mollusks (clams, scallops, oysters, and their relatives), species have, on the average, existed for 10 million years or more. Under ordinary circumstances, then, only a small fraction of the species within these groups has disappeared every million years.

Groups of animals and plants that are well represented in the fossil record and that have experienced high extinction rates tend to serve well as index or guide fossils (p. 81). The ammonoids, an

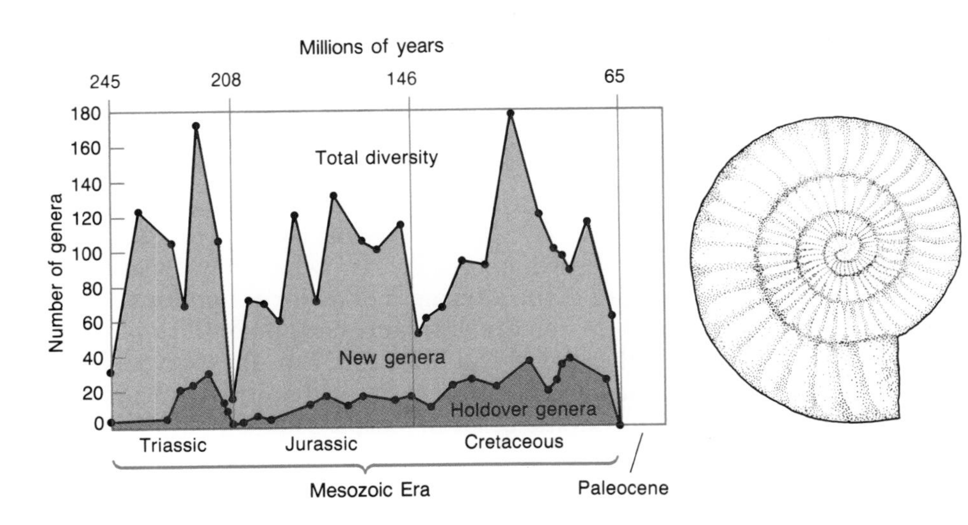

FIGURE 5-14 The appearance and disappearance of ammonoid genera through time. The ammonoids were mollusks related to squids and octopuses but possessed of coiled shells. The turnover rate of genera was high throughout the ammonoids' history. Data plotted for the three stages of the Mesozoic Era show that few genera present during one age were still present in the next. *(After W. J. Kennedy, in A. Hallam [ed.], Patterns of Evolution, Elsevier, Amsterdam, 1977.)*

order of swimming mollusks with coiled shells, meet both requirements. As Figure 5-14 indicates, few ammonoid genera found in any stage of the Mesozoic erathem are also found in the next stage—and yet these genera are usually succeeded by a large number of new ones. Thus the ammonoids experienced both a high rate of extinction of genera and a high rate of formation of new genera.

CONVERGENCE

Convincing evidence that biological form is adaptive is seen in instances of evolutionary **convergence**—that is to say, the evolution of similar forms in two or more different biological groups. This principle is strikingly illustrated by the similarity between many of the **marsupial mammals** of Australia and the other kinds of mammals that live in similar ways on other continents (Figure 5-15). Marsupial mammals, which carry their immature offspring in a pouch, are the products of a radiation that took place on this isolated island continent during the Cenozoic Era. That this radiation has been adaptive is indicated by the fact that these marsupials have *diverged* from each other but simultaneously have *converged*, both in way of life and in body form, with one or more groups of **placental mammals** living elsewhere. (Nearly all nonmarsupial mammals are placentals.) The strong similarities between many Australian marsupial mammals and mammals of other regions must partially reflect the basic evolutionary limitations of the mammalian body plan and mode of development. It would appear, in other words, that certain adaptations are likely to develop under a variety of circumstances, while others are highly unlikely to evolve. The species illustrated in Figure 5-15 have body forms and modes of life that were likely to evolve when primitive mammals were permitted to undergo an extensive adaptive radiation, and accordingly, all developed at least twice.

Almost all adaptive radiations, however, have produced some surprises as well. Judging from what we see elsewhere in the world, we might have predicted, for example, that hoofed, four-footed herbivores resembling deer, cattle, and antelopes would populate the continent of Australia. As it turns out, the Australian equivalents of these large galloping herbivores are kangaroos—animals that hop around on two legs. Apparently it just happened that the breakthrough represented by the kangaroos' hopping adaptation evolved in Australian herbivorous marsupials before an adaptive breakthrough could produce an efficient running apparatus. Once the kangaroos had occupied many regions of Australia, there was simply little opportunity for animals resembling deer, cattle, or antelopes to evolve.

EVOLUTIONARY TRENDS

By examining the evolutionary history of any higher taxon that has left an extensive fossil record, we can observe long-term evolutionary

FIGURE 5-15 Evolutionary convergence between marsupial mammals of Australia and nonmarsupial mammals of other continents. Each of the marsupials is more closely related to a kangaroo than to its counterpart in the other column. *(After G. G. Simpson and W. S. Beck, Life, Harcourt, Brace & World, Inc., New York, 1965.)*

trends — general changes that developed over the course of millions of years. Some of these changes affected form, but others simply affected body size.

Change of Body Size

A general tendency for body size to increase during the evolution of a group of animals is known as **Cope's rule,** after Edward Drinker Cope, a nineteenth-century American paleontologist who observed this phenomenon in his studies of ancient vertebrate animals.

Numerous factors may cause a group of animals to become larger as the group evolves, but all have to do with the tendency of large individuals to produce more offspring than smaller ones. Within species in which males fight for females, for example, larger males tend to win battles and hence to produce a disproportionate number of offspring. Within other species, larger animals may survive to produce more offspring because they are better equipped to obtain food or to avoid predators.

This evolutionary trend cannot continue indefinitely in any animal group, however, because at some point a further increase in body size will inevitably cease to be advantageous. A four-legged animal the size of a large building, for example, could not run or even stand, because its weight would greatly exceed the strength of its limbs. Indeed, many animals could not gather sufficient food or move efficiently if they were appreciably larger than they are.

Given the fact that increases in body size are advantageous only within limits, the great number of animal groups that have evolved toward larger size seems to indicate that most animal orders and families have evolved from relatively small ancestors. Large size tends to impose many adaptive problems for animals, and the specialized adaptations associated with these problems are not easily altered to produce entirely new adaptations. Thus large, highly specialized animals tend to represent evolutionary dead ends.

Some of the problems associated with large size can be seen in the physical adaptations of the elephant. An elephant has such a huge body to feed that it must spend most of its time grinding up coarse food with its molars. The need for constant chewing dictates that an elephant's teeth and jaws must be quite large in relation to its overall size. The elephant's head is therefore large as well—so large that the neck must be quite short to support it (Figure 5-16). To compensate for their consequent short reach, members of the elephant family evolved an enormous trunk from an originally short nose, together with long tusks from originally short teeth. These and other unusual features make it unlikely that modern elephants will ever evolve into very different types of animals.

Manatees (or sea cows), which are blubbery ocean swimmers, are relatives of the elephants (Figure 5-16). Like elephants, manatees are highly specialized animals with limited potential to give rise to substantially different types of animals. The two groups share common ancestors of early Cenozoic age that were small and rodentlike in their general form and adaptations. These small, relatively unspecialized forms easily evolved in a variety of directions, most of which led to larger animals. Elephants and manatees are among the largest.

FIGURE 5-16 The elephant and the manatee. These animal groups differ in form but have a common ancestry within the class Mammalia.

The Structure of Evolutionary Trends

Trends in evolution occur on both small and large scales. A transition from one species to another, for example, represents a simple trend on a small scale. A large-scale trend is one that occurs within a branching limb of the "tree of life." Such a branching limb is known as a **phylogeny.** The change from the oldest known kind of horse to the modern horse represents a large-scale trend in the phylogeny of the horse family (Figure 5-17). The oldest known horse had four toes on each forefoot and three on each hind foot, had relatively simple molars, and was the size of a small dog. The modern horse, in contrast, has a single hoofed toe on each foot, has complex molars, and is a relatively large mammal. Many speciation events separated the ancestral kind of horse from the modern kind. Both the ancestral genus and the modern one have included several species, as have most intermediate genera, and so the transition from one kind to the other represents a complex, large-scale evolutionary trend.

In considering the structure of evolutionary trends, we will focus first on simple trends that cause one species to be transformed into another by a single thread of evolution, and we will then examine complex trends involving many species—trends that consist of many threads of evolution and can thus be viewed as having a fabric.

Some changes from one species to another have occurred in rapid steps of speciation, with the descendant species evolving from the parent species in a relatively short span of time. A probable example is the axolotl, a species whose members remain aquatic throughout life (Figure 5-18). The axolotl evolved from a normal species of salamander—one that is still extant—which underwent metamorphosis from an aquatic juvenile to a terrestrial adult form. The axolotl becomes reproductively mature even though it retains the juvenile body form of its ancestors. The evolutionary transition that produced the axolotl was genetically simple. An axolotl that would normally

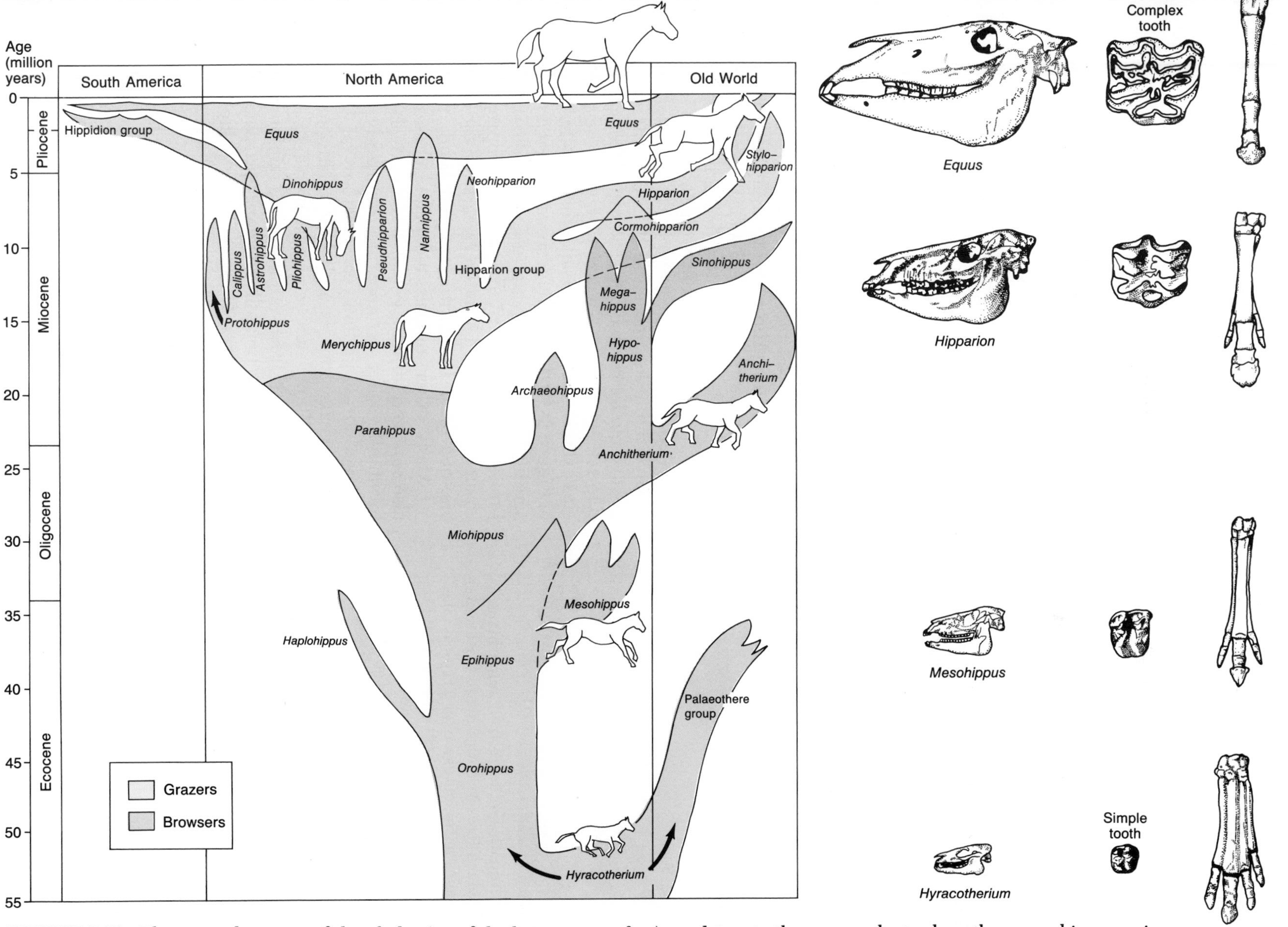

FIGURE 5-17 The general pattern of the phylogeny of the horse family. The surfaces of the molars of some members developed complex cusps that were associated with a transition from browsing on soft leaves to grazing on harsh grasses. The number of toes on the front foot was reduced from four to one. Heads and teeth (but not feet) are drawn to the same scale, to show the general increase in size. *(After B. J. MacFadden, Paleobiology, 11:245–257, with modifications by D. R. Prothero; and G. G. Simpson, Horses, Oxford University Press, New York, 1951.)*

FIGURE 5-18 Aquatic and terrestrial amphibians. *A.* The axolotl, which lives its entire life in fresh water. *B* and *C.* Two stages in the ontogeny of a typical salamander. The aquatic larval stage *(B)* closely resembles the adult axolotl, which in effect never grows up. *(A. After J. Z. Young, Life of Vertebrates, Oxford University Press, London, 1962. B and C. Courtesy of H. Spenser.)*

remain aquatic throughout life can be artificially forced to metamorphose into a terrestrial animal if it is injected with thyroxine, a substance normally produced by the thyroid gland but missing in the axolotl. It thus appears that the axolotl was produced by a speciation event consisting of a simple genetic change that impeded the normal development of the thyroid gland. This change probably occurred quite rapidly on a geologic scale of time. The simple genetic change that produced the axolotl probably spread throughout a single population of the ancestral species. In this respect, however, the axolotl appears to be unusual, because most species that form rapidly from a small ancestral population do not evolve by means of just one genetic change; it appears that several such changes are often entailed.

Other species have evolved slowly by the gradual transformation of an entire species — that is to say, an entire species has changed sufficiently in the course of many generations to be regarded as a new species. Figure 5-19 illustrates this type of evolutionary change in a group of coiled oysters during the Jurassic Period. Recall that the "disappearance" of a species because of evolutionary change is termed pseudoextinction (p. 112).

Like many biologists and paleontologists of the twentieth century, Darwin believed that gradual trends such as that of the coiled Jurassic oysters produced most large-scale evolutionary trends, including those involved in the origin of the modern horse (Figure 5-17). Recently, however, it has been suggested that gradual trends such as those evident in the evolution of the Jurassic coiled oysters are relatively rare. Oysters are bivalve mollusks, and more than 300 species of bivalve mollusks have been identified from Jurassic rocks of Europe — yet very few of these species exhibit gradual trends like the one illustrated in Figure 5-19. In fact, it is estimated that an average bivalve species living in Europe during the Jurassic Period existed without appreciable change for about 15 million years, or for about one-quarter of the entire Jurassic Period.

Many other animals and plant species have also survived for long geologic intervals. Benthic foraminifera, for example, are estimated to have survived for 30 million years; diatoms (p. 43), 25 million years; mosses and their relatives, more than 15 million years; seed-bearing plant species, 6 million years; freshwater fishes, 6 million years; beetles, more than 2 million years; and mammals, about 2 million years. These estimates, which represent a wide variety of living things, suggest that most species evolve very slowly; they indicate that an average animal or plant species is not likely to evolve sufficiently to be regarded as a new species even after it has passed through about a million generations. For comparison, note that the oyster species shown in Figure 5-19 changed enough to be regarded as new species within 2 to 4 million years; even though an individual oyster became

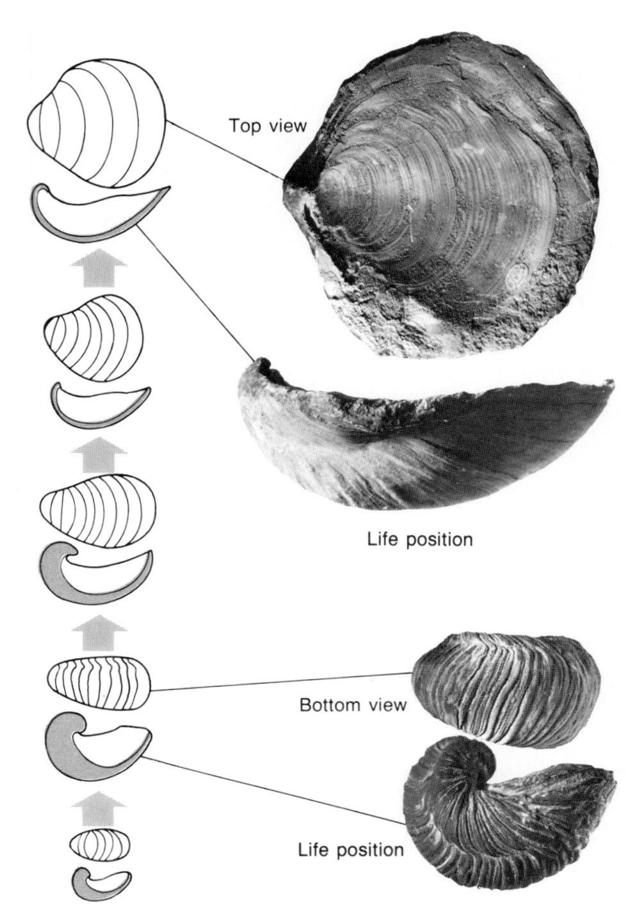

FIGURE 5-19 An apparently gradual trend in a lineage of coiled oysters of the genus *Gryphaea* during an Early Jurassic interval of about 12 million years. During the interval shown here, the shell became larger, attaining the diameter of a small saucer, but it also became thinner and flatter. These animals rested on the seafloor in the orientation labeled "life position." Perhaps the flatter shell was more stable against potentially disruptive water movements. *(After A. Hallam, Philos. Trans. Roy. Soc. London 254B:91–128, 1968.)*

reproductively mature in 2 or 3 years, evolution required a million generations or so to produce a new species. All of these longevities must be viewed in the light of the length of time it has taken for new higher taxa of the same group to develop. Recall that early in the Cenozoic Era, whales evolved from vastly different small, rodentlike mammals in no more than 12 million years. A typical survival time of 2 million years for a single mammal species seems quite sizable in comparison.

According to the traditional, **gradualistic model** of evolution, most evolutionary change takes place in small steps within well-established species. The very slow rate of evolution that characterizes many well-established species has led some paleontologists to oppose the gradualistic model. These paleontologists conclude that most evolution must be associated with speciation—that is, with the rapid evolution of new species from others. This is the **punctuational model** of evolution. Another line of evidence cited in favor of this idea is the evolutionary history that typifies long, narrow segments of phylogeny—segments that undergo little branching but span long intervals of geologic time. If speciation were indeed the site of most evolution, such segments of phylogeny would be expected to exhibit little evolution for the simple reason that they have experienced very little speciation. This is exactly the pattern exhibited by the bowfin fishes (Figure 5-20), a group that has experienced very little speciation and very little evolution during the last 60 million years. The single living bowfin species so closely resembles those of early Cenozoic time that it has been labeled a living fossil. As it turns out, all of the living species that we know to be at the end of long, narrow segments of phylogeny are living fossils. Among other living fossils are the alligator, the snapping turtle, and the aardvark (Figure 5-21). A well-publicized living-fossil plant is the dawn redwood, which was thought to be extinct until it was discovered living in a small area of China in the 1940s.

Even when individual species have evolved slowly, large-scale trends have sometimes developed through steps of change in particular evolutionary directions that have led to distinctive new species. Trends have also developed when species of one type have survived for long intervals while related species of other types have suffered high rates of extinction. The result has been a shift in the biological composition of the genus or family to which the species have belonged.

Bear in mind that although the fossil record is not complete enough to reveal the patterns of many evolutionary trends, it does demonstrate that such trends have occurred: It shows that certain ancient groups of animals or plants were the ancestors of certain modern forms. If we had no fossil record, we would know almost nothing about large-scale evolutionary trends. In fact, in the absence of a fossil record, the idea that large-

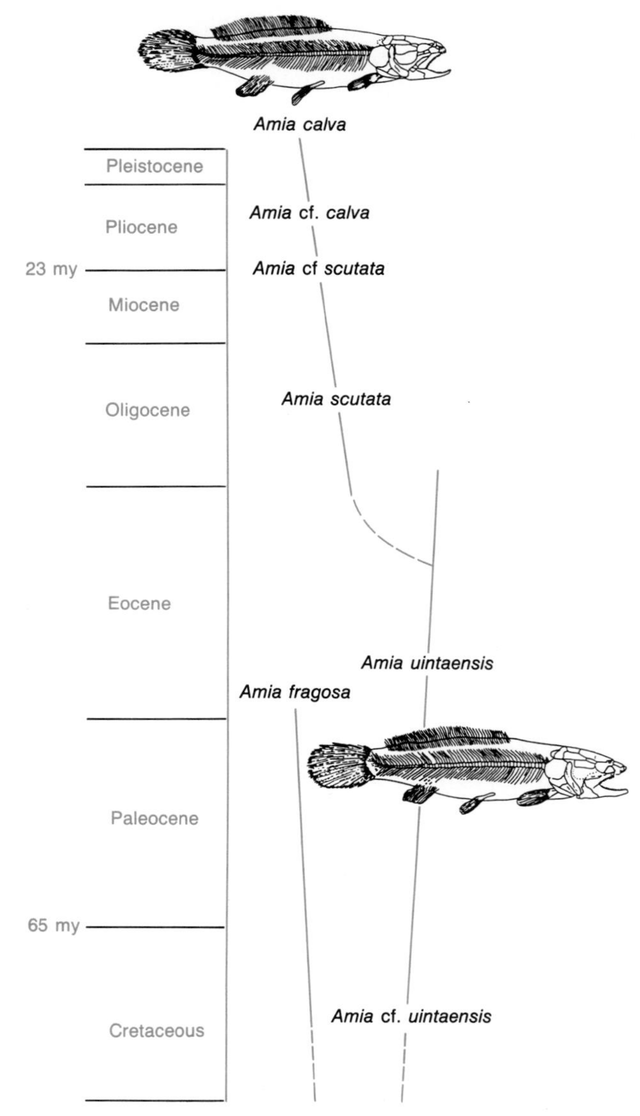

FIGURE 5-20 Living and fossil bowfin fishes. The living bowfin, *Amia calva,* is the product of a long history in which there has been little speciation (branching) and also little evolutionary change. *Amia* grows to a length of more than a meter (~3 feet). *(After J. R. Boreske, Museum of Compar. Zool. Bull. 146:1–87, 1974.)*

scale trends occurred at all would perhaps remain controversial among scientists. The argument that modern forms of life evolved from quite different ancient forms could then be based only on the less direct evidence provided by species of the present world (evidence such as that described on pp. 103–111).

FIGURE 5-21 The aardvark, a living fossil and the only existing species of the order of mammals to which it belongs. Since the beginning of the Miocene Epoch more than 20 million years ago, very few aardvark species have lived at any time. During this interval, aardvarks have evolved very little. *(Silvestris Photoservice/ NHPA.)*

The Irreversibility of Evolution

An evolutionary trend that has resulted from at least several genetic changes is highly unlikely to be reversed by subsequent evolution. This principle is called **Dollo's law,** for Louis Dollo, the Belgian paleontologist who proposed it early in the twentieth century. Dollo's law reflects the fact that it is extremely unlikely that a long sequence of genetic changes in a population will be repeated in reverse order. Thus evolution occasionally produces a species of animals or plants that crudely resembles an ancestor, but it never perfectly duplicates a species that has disappeared. In other words, once a species has been changed by evolution or eliminated by extinction, it is gone forever.

CHAPTER SUMMARY

1 Several lines of evidence convinced Charles Darwin that organic evolution produced the multitudinous species that inhabit the modern world. Among these pieces of evidence were:

The restriction of many closely related groups of species to discrete geographic regions separated by barriers.

Embryological similarities between groups of animals that are dissimilar as adults.

The presence of similar anatomical "ground plans" in animals that live in quite different ways.

The existence in certain animals of vestigial organs that serve no apparent function but resemble functioning organs in other species.

The ability of humans to alter domestic animals and cultivated plants by artificial selection in breeding.

2 Natural selection is a process in which certain kinds of individuals become more numerous in a population because they produce an unusually large proportion of the total offspring. They manage to do so either by surviving a long time or by reproducing at a high rate.

3 The variability that is the basis of natural selection is generated by two mechanisms: genetic mutation and the generation of new gene combinations by sexual reproduction.

4 Speciation is the process by which an existing species gives rise to an additional species; it is believed that, in most cases, the population that becomes the new species is geographically isolated from the remainder of the parent species.

5 Extinction is the dying out of a species. The most important agents of extinction are the ecological factors that normally govern the sizes of populations in nature. Pseudoextinction is the disappearance of a species through its evolutionary transformation into another species.

6 Mass extinction is the disappearance of many species during a geologically brief interval of time.

7 Adaptive radiation is the proliferation of many species from a small ancestral group. Adaptive radiation has usually followed new access to ecological opportunities.

8 Under normal circumstances, the rates of speciation and extinction among groups of animals and plants vary greatly.

9 One of the most convincing kinds of evidence that evolutionary changes are adaptive is evolutionary convergence, or the evolution within two or more higher taxa of species that resemble one another in form and also live in the same way.

10 Long-term evolutionary trends are evolutionary changes that have developed in the course of millions of years. Such trends can develop in many ways, and the relative importance of these ways depends on what percentage of evolutionary change is associated with rapid speciation events and what percentage is represented by the gradual transformation of well-established species.

EXERCISES

1 What geographic patterns suggested to Charles Darwin that certain kinds of species descended from others?

2 What characteristics make a particular kind of individual successful in the process of natural selection?

3 What conditions make it likely that a small group of closely related species will increase to a large number of species by means of rapid speciation?

4 How can evolution proceed by a change in the growth and development of a species?

5 In what ways can an evolutionary trend develop during the history of a genus or a family?

6 What kinds of environmental change can lead to the extinction of a species?

7 How is pseudoextinction related to gradual evolutionary change?

8 Give an example of evolutionary convergence.

ADDITIONAL READING

The Fossil Record and Evolution (Readings from *Scientific American*), W. H. Freeman and Company, New York, 1982.

Gould, S. J., *Bully for Brontosaurus*, W. W. Norton & Company, Inc., New York, 1990.

Stanley, S. M., *The New Evolutionary Timetable: Fossils, Genes, and the Origin of Species*, Basic Books, Inc., New York, 1981.

Stebbins, G. L., and F. J. Ayala, "The Evolution of Darwinism," *Scientific American*, July 1985.

PART III

MOVEMENTS OF EARTH

The most exciting development in the science of geology during the twentieth century has been the theory of plate tectonics. According to this theory, plates of lithosphere form from Earth's mantle and ascend along midocean ridges, slide laterally over the surface of the planet, and descend into the mantle once again. Many plates carry continents with them; the materials that compose continents are of low density, however, and do not descend into the mantle. Movements and interactions of plates deform Earth's crust on a vast scale. They also cause magma to rise through the upper mantle to form igneous rocks on and within the overlying crust.

On continents, mountain ranges are the colossal products of plate motions. Most mountains have formed where two continents on separate plates have collided or where one plate has descended beneath a continent situated on the edge of another plate. Continents grow laterally when mountains rise up along their margins, and they break apart when spreading ridges transect them, producing oceans such as the modern Atlantic.

This complex fold in Israel was produced by mountain-building processes. *(Geological Survey of Israel.)*

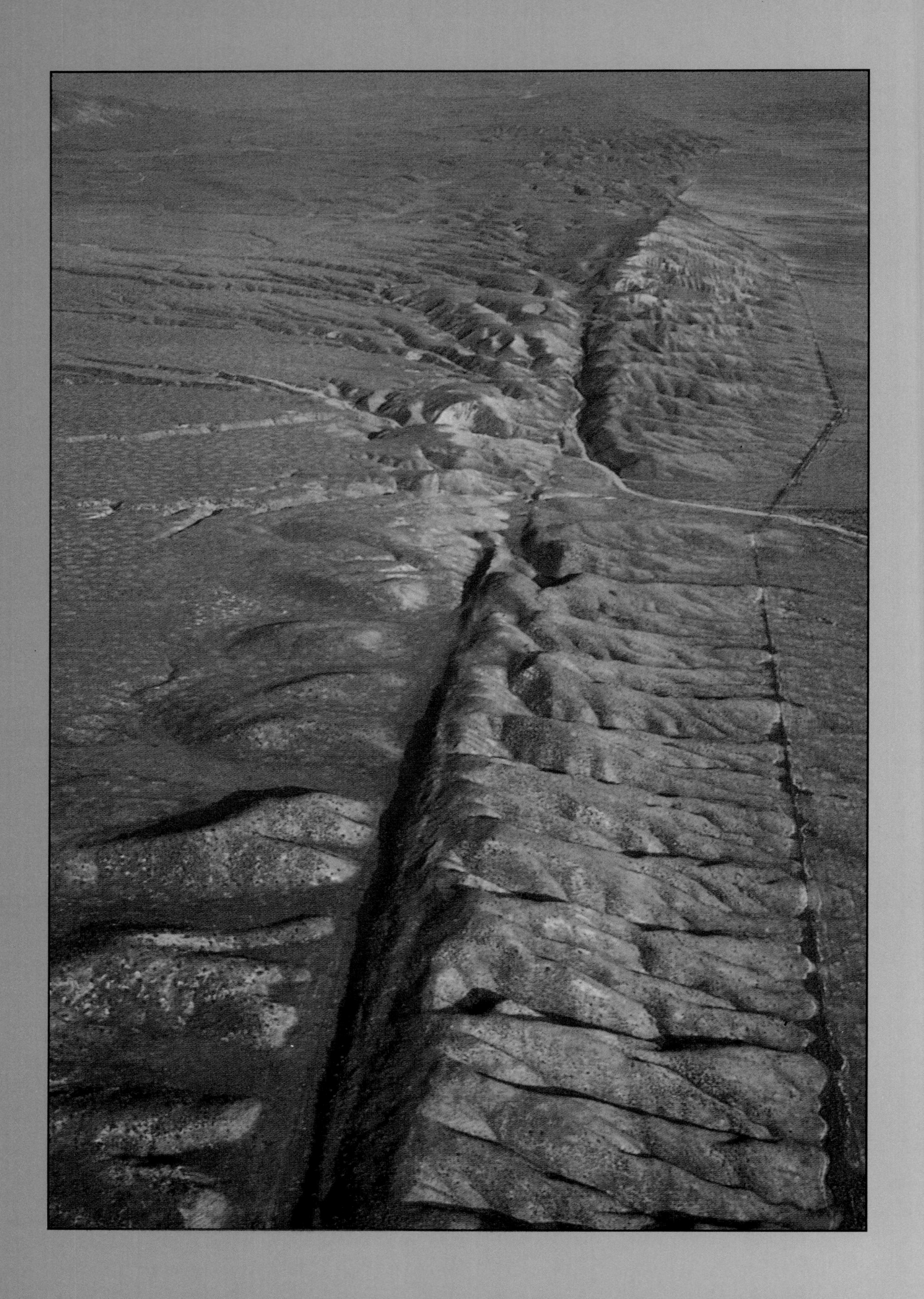

CHAPTER 6

Plate Tectonics

The emergence of the theory of plate tectonics during the 1960s fostered a revolution in the science of geology. **Tectonics** is a term that has long been used to describe movements of Earth's crust. Accordingly, **plate tectonics** refers to the movements of discrete segments of Earth's crust in relation to one another. Whereas continents were once thought to be locked in place by the oceanic crust that surrounds them, the theory of plate tectonics holds that continents move over the surface of Earth because they form parts of moving plates (Figure 6-1). Moreover, continents occasionally break apart or, alternatively, fuse together to form larger continents. The theory of plate tectonics explains why most volcanoes and earthquakes occur along curved belts of seafloor, why mountain belts tend to develop along the edges of continents, and why the present ocean basins are very young from a geologic perspective. Many of the kinds of rock deformation discussed in Appendix II also result from the movements of plates.

The San Andreas Fault in California is a great break within Earth's lithosphere separating large segments, or plates, of the lithosphere. The Pacific plate is on the left, and the North American plate is on the right. The Pacific plate periodically slides northwestward in relation to the North American plate. *(R. E. Wallace, U.S. Geological Survey.)*

THE HISTORY OF OPINION ABOUT CONTINENTAL DRIFT

When the concept of plate tectonics emerged quite suddenly in the 1960s, it resolved many longstanding disputes. For many years, the idea that continents move horizontally over Earth's surface, an idea labeled **continental drift,** failed to receive general support in Europe or North America. In 1944 one prominent geologist went so far as to assert that the idea of continental drift should be abandoned outright because "further discussion of it merely encumbers the literature and befogs the minds of students." Although many geologists may not have read this comment, most agreed with its spirit, and during the 1950s, little attention was given to the possibility that continental drift was a real phenomenon.

When the idea of continental drift first emerged earlier in the twentieth century, it attracted considerable attention, primarily as a result of the arguments of two scientists—Alfred

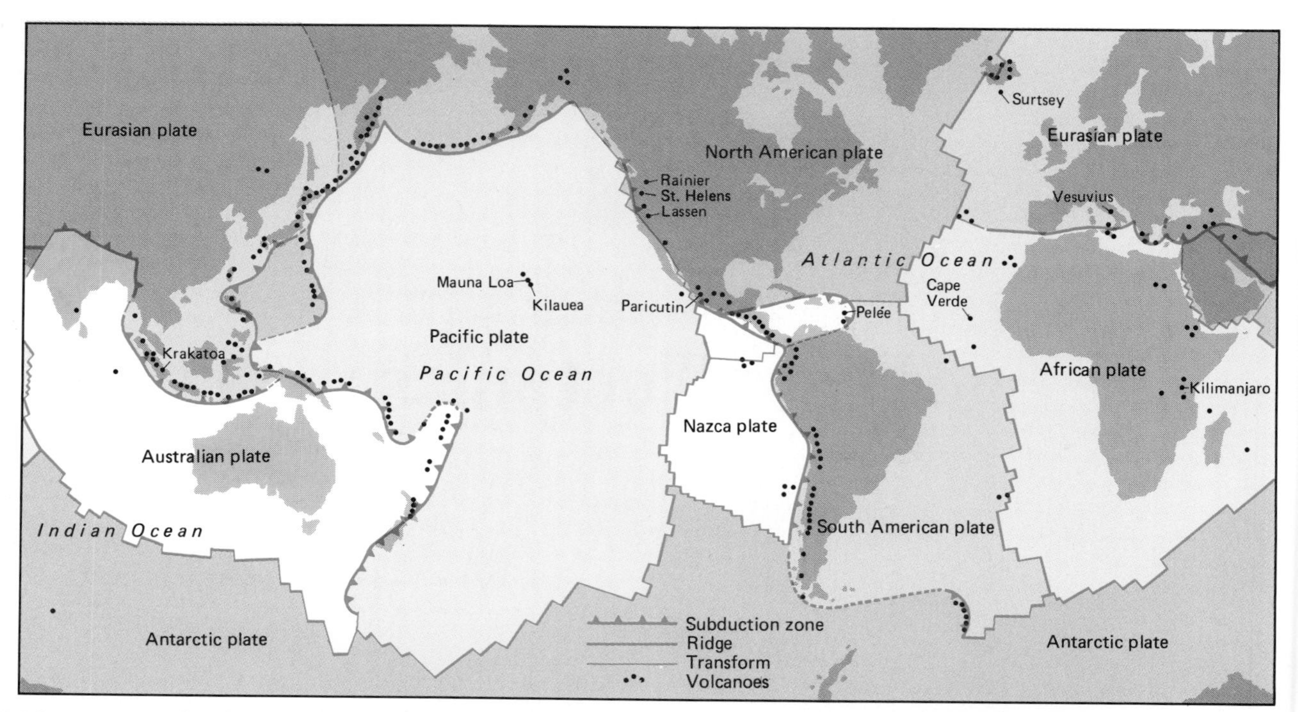

FIGURE 6-1 The distribution of lithospheric plates over Earth's surface. There are eight large plates and several small ones on Earth today. The three kinds of plate boundaries—subduction zones, ridge axes, and transform faults—are shown here (they will be discussed later in the chapter). The locations of volcanoes are also indicated; most lie near subduction zones or ridges. *(F. Press and R. Siever, Earth, W. H. Freeman and Company, New York, 1986.)*

Wegener of Germany and Alexander du Toit of South Africa. We will briefly examine the case that these two men and their followers made and the reasons their arguments were rejected by most of their contemporaries.

Early Evidence

The observation that the coasts on the two sides of the Atlantic Ocean fit together like separated parts of a jigsaw puzzle (Figure 6-2) constituted the first evidence that continents might once have broken apart and moved across Earth's surface. Centuries ago, map readers noted with curiosity that the outline of the west coast of Africa seemed to match that of the east coast of South America. It was not until 1858, however, that Antonio Snider-Pellegrini, a Frenchman, published a book suggesting that a great continent had once broken apart and that the Atlantic Ocean had been formed by powerful forces within Earth.

More traditional geologists clung to the idea that large blocks of continental crust could not move over Earth's surface. Thus, when the distribution of certain living and extinct animals and plants began to suggest former connections between landmasses now separated, most geologists tended to assume that great corridors of felsic rock (the most abundant material in continental crust) had once formed land bridges that connected continents but later subsided to form portions of the modern seafloor. Today it is recognized that this scenario is not realistic, because felsic crust is of such low density that it cannot possibly sink into the mafic rocks that underlie the oceans. Nonetheless, many prominent geologists of the late nineteenth century presented schemes of Earth history that included the concept of felsic corridors.

One phenomenon that led scientists to speculate about ancient land bridges was the similarity between the fauna of the island of Madagascar and that of India, a land that is separated from Mada-

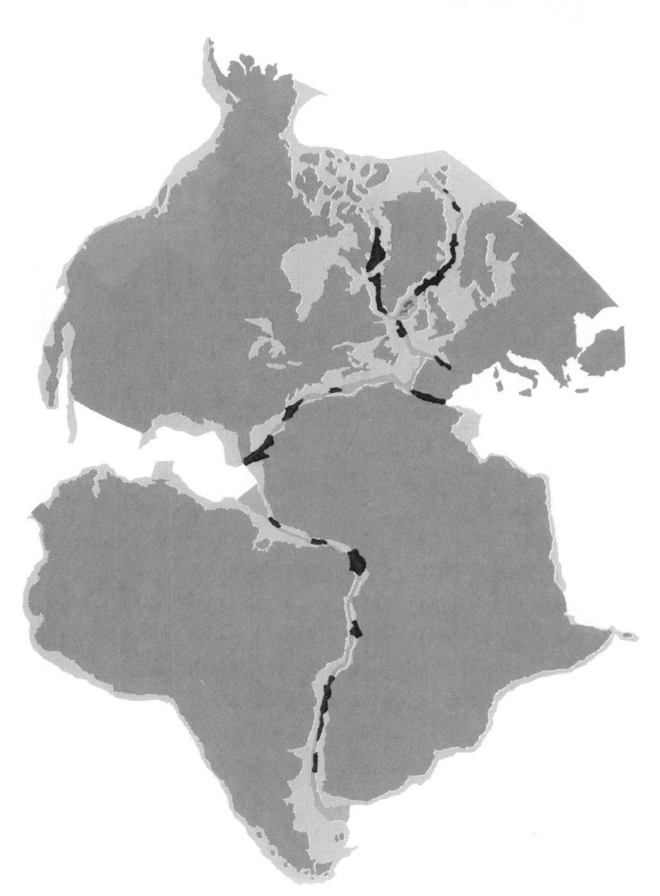

FIGURE 6-2 Computer-generated "best-fit" union of continents that now lie on opposite sides of the Atlantic. This fit was calculated by Sir Edward Bullard and his coworkers at the University of Cambridge. The fit was made along the 500-fathom line of each continental slope. Overlaps of continental margins are shown in black. Gaps between margins are shown in color. *(After P. M. Hurley, Scientific American, April 1968.)*

gascar by nearly 4000 kilometers (~2500 miles) of ocean. Madagascar's mammals are primitive; altogether missing are the zebras, lions, leopards, gazelles, apes, rhinoceroses, giraffes, and elephants that inhabit nearby Africa. In contrast, some of the native animals of India closely resemble those of Madagascar. Neumayr and Suess consequently believed that a now-sunken land bridge had once spanned the western part of the Indian Ocean, connecting Madagascar to India.

A second line of evidence for ancient land connections was found in the fossil record. During the nineteenth century, late Paleozoic coal deposits of India, South Africa, Australia, and South America were found to contain a group of fossil plants that were collectively designated the *Glossopteris* flora, after their most conspicuous genus, a variety of seed fern (Figure 6-3). After the turn of the century, the *Glossopteris* flora was discovered in Antarctica as well. The occurrence of this fossil flora on widely separated landmasses was one of the facts that led Eduard Suess to suggest that land bridges had once connected all of these continents. Suess introduced the name Gondwanaland (Gondwana-Land, in his spelling) to connote the hypothetical continent that consisted of these landmasses and the land bridges that he believed to have connected them. (Gondwana is a region of India where seams of coal yield fossils of the *Glossopteris* flora.)

Then in 1908 the American geologist Frank B. Taylor proposed a new explanation for the connection of ancient landmasses. In essence, Taylor hypothesized that the continents had once lain side by side as components of very large landmasses that eventually broke apart and moved across Earth's surface to their present positions. This general idea is central to modern plate-tectonics theory, although the particulars of Taylor's scheme are no longer accepted.

FIGURE 6-3 A fossil *Glossopteris* leaf from the Permian of India. The leaf is about 12 centimeters (5 inches) long. *(Jim Frazier/Mantis Wildlife Films Pty Ltd.)*

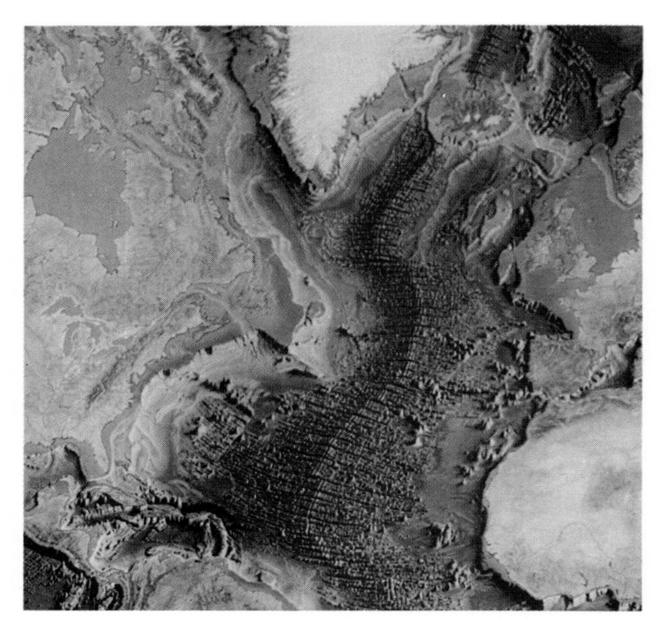

FIGURE 6-4 Artist's representation of the Mid-Atlantic Ridge in the North Atlantic region. Transform faults are conspicuous features oriented at right angles to the ridge axis; some of them offset the ridge slightly. *(Painting by Heinrich C. Berann; Bruce C. Heezen and Marie Tharp, World Ocean Floor, 1977, © Marie Tharp.)*

One of Taylor's most significant contributions was his suggestion that the immense submarine mountain chain now called the Mid-Atlantic Ridge, which is one of several **midocean ridges** of the present-day seafloor, marks the line along which one ancient landmass ruptured to form the Atlantic Ocean (Figure 6-4).

A Twentieth-Century Pioneer: Alfred Wegener

In 1915 Alfred Wegener, a German meteorologist, presented evidence that virtually all of the large continental areas of the modern world were united late in the Paleozoic Era as a single supercontinent, which he labeled Pangaea. The idea that Pangaea existed has been revived with the rise of modern plate-tectonics theory, and today nearly all professional geologists accept it as fact—although, as we will see, they recognize major errors in Wegener's proposed chronology for the fragmentation of this supercontinent (Figure 6-5). Wegener reasoned that Pangaea had

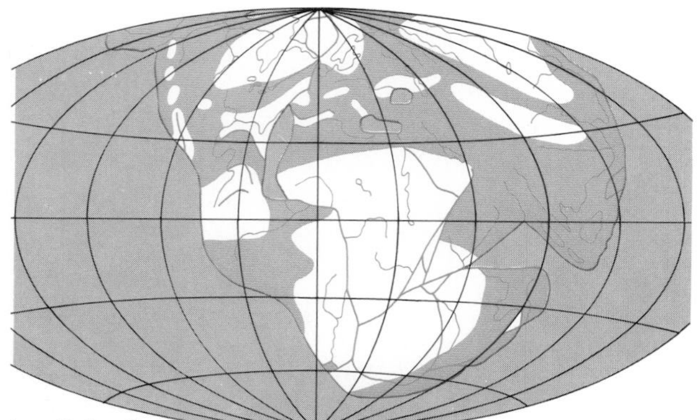

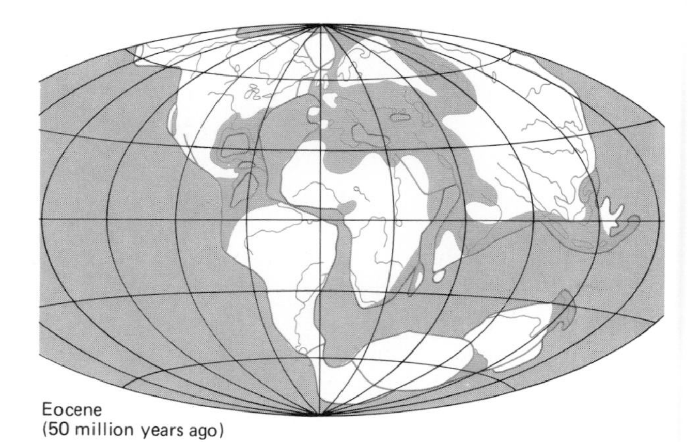

Early Pleistocene
(1.5 million years ago)

FIGURE 6-5 Alfred Wegener's reconstruction of the map of the world for three past times. Africa is placed in its present position as a point of reference. Heavy shading represents shallow seas. Wegener erred in suggesting that Pangaea, the supercontinent shown in the upper map, did not break apart until the Cenozoic Era *(lower two maps)*. *(After A. Wegener, Die Entstehung der Kontinents und Ozeane, Friedrich Vieweg und Sohn, Brunswick, Germany, 1915.)*

broken apart and that the fragments had drifted about. He cited the great rift valleys of Africa as possible newly forming or failed rifts. Wegener's insight was again correct. As we will see, the African rift valleys are now regarded as instances of continental rifting at an early stage.

Wegener supported his theory with several additional arguments. He noted numerous geologic similarities between eastern South America and western Africa, for example, and he also called attention to the many similarities between the fossil biotas of these two widely separated continents. Several extinct groups of animals and plants had been found in the fossil records of two or more Gondwanaland continents, and this led Wegener to argue that these continents must once have lain close together. Even *living* animal and plant groups were shown to exhibit a "Gondwanaland" pattern: A number of individual species were found to be widely distributed among the southern continents.

Wegener's arguments were more fully developed by the South African geologist Alexander du Toit. Du Toit and others introduced a wealth of circumstantial evidence in support of the idea of continental drift—evidence that was publicized both before Wegener's untimely death in 1930 and during the three decades of controversy that followed.

Among the groups of living animals that seemed to support the concept of Gondwanaland were the earthworms. One genus of earthworm, for example, was found to be restricted to the southern tips of South America and Africa, which lay close together in Wegener's Gondwanaland reconstruction. Another genus was encountered only in southern India and southern Australia.

The evidence of the fossil record seemed even more compelling. Wegener plotted a map representing the earlier suggestion that the southern continents once formed a large landmass, most of which had sunk to become the floors of the South Atlantic, Indian, and South Pacific oceans. Wegener's map of this mythical landmass is quite striking. Especially when he added to this map the locations of the *Glossopteris* flora (Figure 6-3), the Gondwanaland continents seemed naturally to belong not where they are today but in a tight cluster (Figure 6-6). Twenty of the twenty-seven species of land plants recognized within the *Glossopteris* flora of Antarctica, for example, have been found as far away as India. It might be suggested that winds could have spread the plants this far by carrying their seeds, but in fact the seeds of the genus *Glossopteris* are several millimeters in diameter—much too large to have been blown across wide oceans.

Animal fossils also played a role in the debate that resulted from Wegener's work. Du Toit noted, for example, that fossils of the reptile *Mesosaurus* (Figure 6-7) occurred at or near the position of the Carboniferous-Permian boundary in both Brazil and South Africa. On both continents, fossils of *Mesosaurus* occur in dark shales along

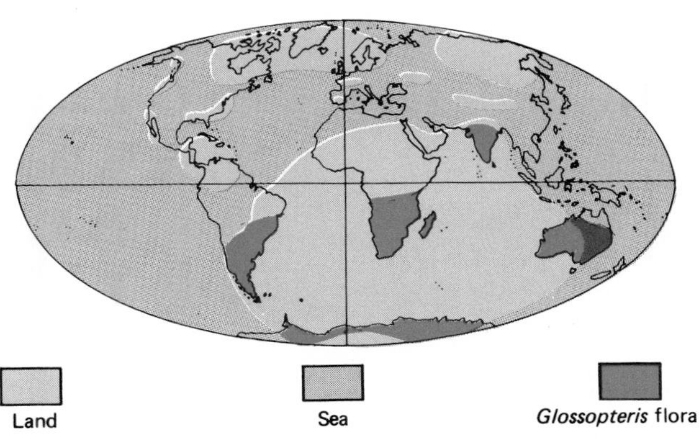

FIGURE 6-6 Distribution of land and seas during the Carboniferous Period as perceived by Alfred Wegener's predecessors, who believed that large areas of the present-day ocean floor then stood above sea level. The distribution of the *Glossopteris* flora strengthens the opposing idea that the Gondwanaland continents were once united.

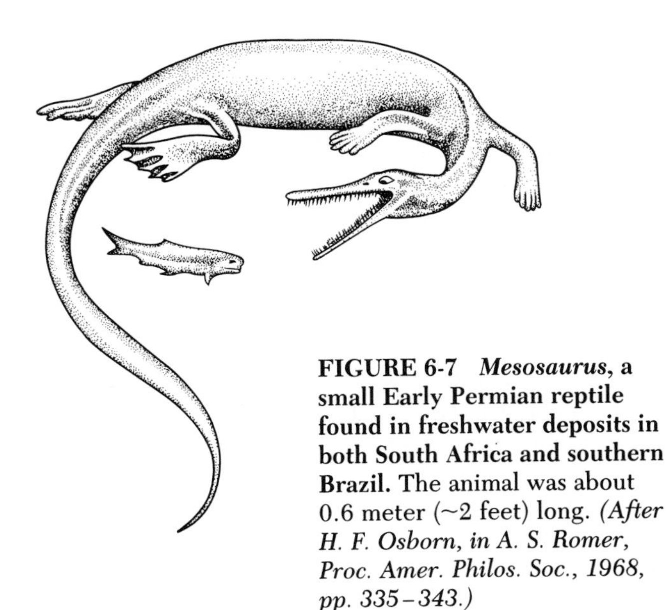

FIGURE 6-7 *Mesosaurus*, a small Early Permian reptile found in freshwater deposits in both South Africa and southern Brazil. The animal was about 0.6 meter (~2 feet) long. *(After H. F. Osborn, in A. S. Romer, Proc. Amer. Philos. Soc., 1968, pp. 335–343.)*

with fossil insects and crustaceans. *Mesosaurus* occupied freshwater and perhaps brackish habitats, and most paleontologists found it difficult to imagine that the animal had somehow made its way across an ocean as broad as the present Atlantic and had then found freshwater depositional settings that were nearly identical to its former habitat.

The Gondwana Sequence

The general stratigraphic context in which the *Glossopteris* flora and *Mesosaurus* are encountered offered further evidence supporting the existence of Gondwanaland. Specifically, Carboniferous and Permian rock units that yield the *Glossopteris* flora form what is known as the Gondwana sequence, which occurs with remarkable similarity in South America, South Africa, India, and Antarctica.

The Gondwana sequence of Brazil bears an uncanny resemblance to the Karroo of South Africa. At the bases of both sequences (Figure 6-8) are glacial tillites that are coarsest at the base and alternate with interglacial sediments, including coals, that yield members of the *Glossopteris* flora. As in South Africa, *Mesosaurus* is found near the base of the Permian in dark shales. Tillites occur only as high as the lowermost Permian, but some members of the *Glossopteris* flora persist into the Upper Permian. Much of the Triassic record consists of dune deposits known as the Botacatú Sandstone, which, like similar dune deposits of South Africa, are succeeded by Jurassic lava flows. Antarctica and India display Gondwana sequences

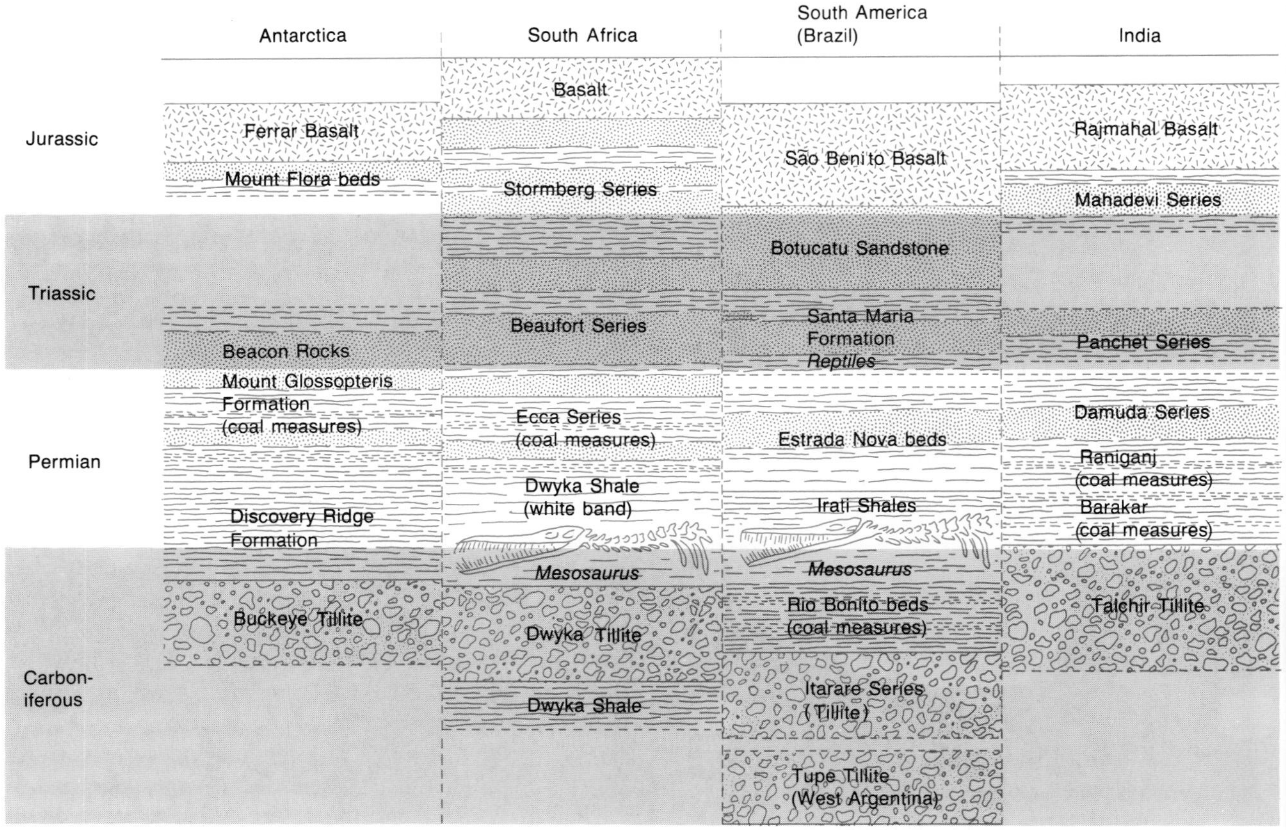

FIGURE 6-8 Correlation of the Gondwana sequences of four continents. In each sequence, glacial tillites are followed by shales and coal beds containing the *Glossopteris* flora. *(Modified from G. A. Duamani and W. E. Long, Scientific American, September 1962.)*

very similar to those of South Africa and South America (Figure 6-8).

Du Toit and other followers of Wegener sought further evidence of ancient land connections in the Gondwana glacial deposits. They measured the orientations of features scoured into underlying bedrock by glaciers and found very interesting patterns (Figure 6-9). They discovered, for instance, that glacial movement in eastern South America was primarily from the southeast, where today no landmass exists that might support large glaciers. In southern Australia, too, there was evidence of glacial flow from the south, where there is now only ocean. Obviously, it would not be at all difficult to account for such movement if the continents had been united as Gondwanaland at the time the glaciers were flowing. Ice flow would then have radiated from the center of a large continent that might be expected to support large glaciers under cold climatic conditions.

Du Toit correctly deduced from geologic evidence that Pangaea did not form until late in the Paleozoic Era. Before Pangaea was formed, Gondwanaland existed as a distinct supercontinent, and the northern continents were united as a second supercontinent called **Laurasia.**

Du Toit also recognized that if South America, Antarctica, and Australia were assembled as Gondwanaland, the mountain belts along their margins would line up. To this rugged feature of Gondwanaland du Toit gave the name Samfrau, which he derived from letters of the names South America, Africa, and Australia. At the same time, du Toit expanded on Wegener's observation that the ancient geologic features of these now-separate continents would match when the continents were placed in their Gondwanaland positions. Du Toit's plot of these structural trends—including regional trends of folds and faults—is shown in Figure 6-10.

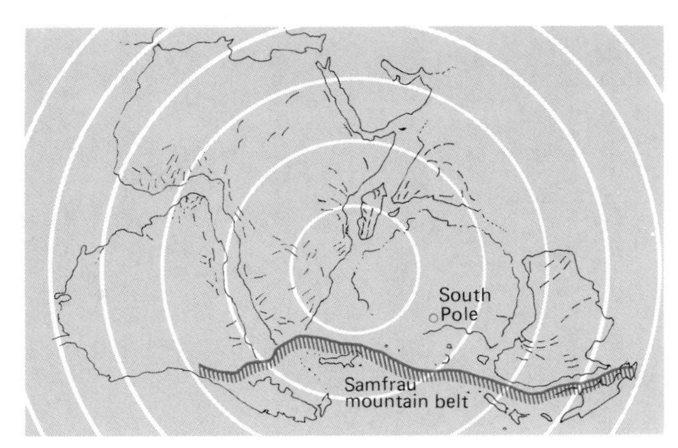

FIGURE 6-10 Alexander du Toit's reconstruction of Gondwanaland. The short lines indicate regional strikes of folds and faults that align well when the continents are assembled to form Gondwanaland. The Andes mountain chain of South America aligns with mountain systems of South Africa, Antarctica, and Australia to form what du Toit labeled the Samfrau mountain belt of Gondwanaland. *(After A. L. du Toit, The Geology of South Africa, Oliver & Boyd Ltd., Edinburgh, 1937.)*

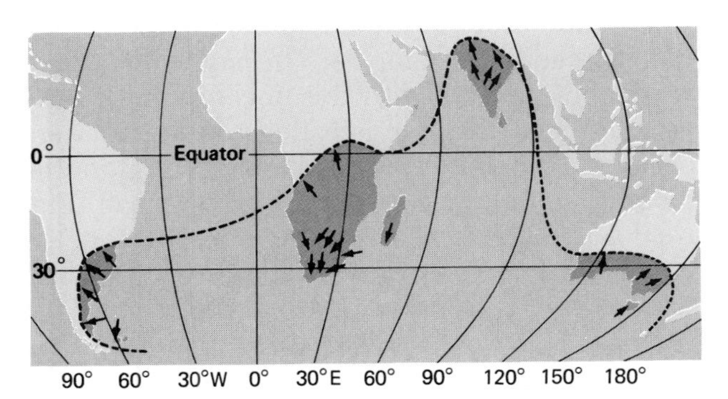

FIGURE 6-9 Locations of late Paleozoic glaciation and the directions in which glaciers flowed. *(After A. Holmes, Principles of Physical Geology, The Ronald Press Company, New York, 1965.)*

The Rejection of Continental Drift

Despite mounting evidence favoring Wegener's and du Toit's ideas, geologists of the Northern Hemisphere continued to view the theory of continental drift with considerable skepticism. The primary source of their dissatisfaction lay in the seeming absence of a demonstrated mechanism by which continents could move over long distances. Geophysicists knew that both continental crust and oceanic crust were continuous above the Mohorovičić (or Moho) discontinuity and hence they could not imagine how continents could be made to move laterally—to plow through oceanic crust (Figure 6-11). In addition, some fossil evidence seemed to contradict the notion of continental drift. Specifically, Wegener had suggested

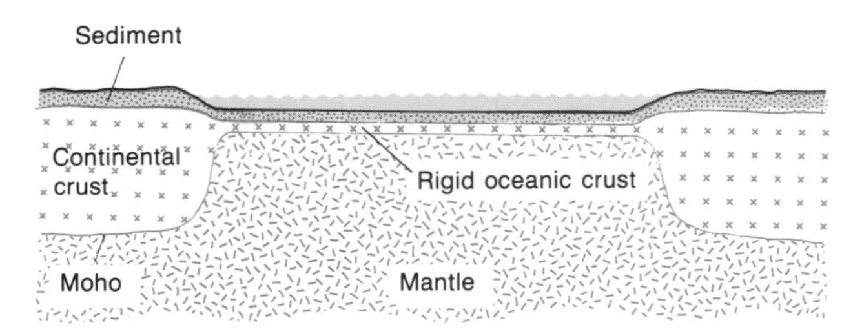

FIGURE 6-11 Cross section of Earth's crust showing the rigid oceanic crust that separates the continents. The Moho is the discontinuity between the crust and the mantle.

a brief timetable for drift in which he proposed that Pangaea, which incorporated virtually all the modern continents, had survived into the Cenozoic Era (Figure 6-5). Thus paleontologists looked for evidence that the world's biotas had evolved into increasingly distinctive geographic groupings since the start of the Cenozoic Era. However, no such evidence was found.

We now know that Wegener made a dating error that misled paleontologists of his time. The rifting of Pangaea had actually begun near the start of the Mesozoic Era—much earlier than Wegener believed. Continents could not have moved far enough during the brief Cenozoic Era to have allowed biotas to diverge greatly; instead, continents have been relatively widely dispersed since the very beginning of this era.

FIGURE 6-12 *Lystrosaurus*, the mammal-like reptile now known from Antarctica as well as from Africa and southeast Asia. *Lystrosaurus* was a herbivorous animal about a meter (3 feet) long, with short legs, beaklike jaws, and a pair of short tusks. *(Painting by Mark Hallett.)*

Faced with so many apparent problems, the idea of continental drift remained highly unpopular in the United States and Europe for decades.

Ironically, after new data supporting plate tectonics had finally brought continental drift into favor, an exciting fossil find was made in Antarctica. This was the discovery in 1969 of the genus *Lystrosaurus*, an animal classified as a member of the mammal-like reptile group, which will be described in Chapter 13. *Lystrosaurus* was a heavyset herbivorous animal with beaklike jaws (Figure 6-12). One has to wonder whether an earlier discovery of *Lystrosaurus* fossils might have revived enthusiasm for continental drift before the advent of plate-tectonics theory.

The Puzzle of Paleomagnetism

Interest in continental movements was renewed in the late 1950s as a result of new evidence derived from **paleomagnetism**—or the magnetization of ancient rocks at the time of their formation. We have already seen how Earth's magnetic field has reversed its polarity on many occasions. During the 1950s, geophysicists attempted to ascertain whether the north and south magnetic poles not only reversed their positions but also wandered about periodically. To explore this possibility, these researchers attempted to determine the previous positions of the magnetic poles by using magnetized rocks as compasses for the past.

As we learned in Chapter 4, a magnetic field frozen into a rock is similar to the magnetic field that is "read" by a compass. The angle that a compass needle makes with the line running to the geographic north pole is called the **declination.** Today the magnetic pole lies about 15° from the geographic pole, so the declination is small (Figure 6-13). A compass needle not only points at the north magnetic pole but, if allowed to tilt in a vertical plane, also dips at a particular angle. As Figure 6-13 shows, the dip of a compass needle varies with the distance of the compass from the magnetic pole. The dip is lowest near the equator, where the lines of force of Earth's magnetic field intersect Earth's surface at a low angle.

Paleomagnetism in a rock also has both a declination and a dip, which serve to indicate the

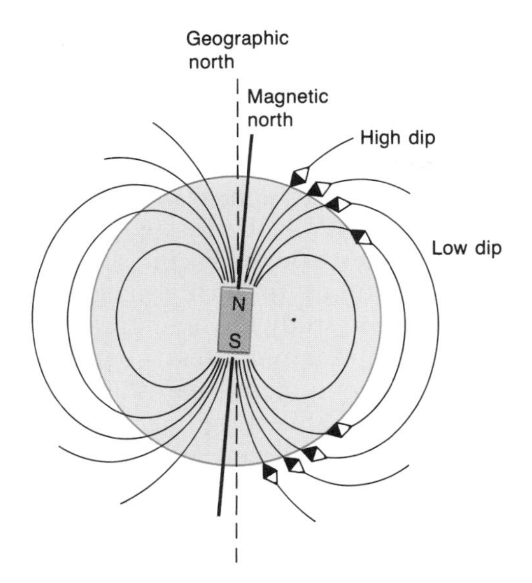

FIGURE 6-13 The structure of Earth's magnetic field. The core has north and south poles and thus behaves like a bar magnet. The north-south axis has a declination of 15° from Earth's north-south geographic axis. Curved lines represent magnetic lines of force. These lines of force have high dips near the poles and low dips near the equator.

apparent direction of the north magnetic pole at the time when the rock was first magnetized, as well as the distance between the rock and the pole. It is important to understand, however, that neither a compass needle nor the magnetism of a rock reveals anything about longitude (position in an east-west direction).

When geologists first began to measure rock magnetism, they found that recently magnetized rocks exhibited a magnetism that was consistent with Earth's current magnetic field. The magnetism in older rocks, however, had different orientations. As data accumulated, it began to appear that Earth's magnetic pole had wandered. A plot of the pole's apparent positions, indicated by rocks of various ages in North America and in Europe, showed that the pole seemed to have moved to its present position from much farther south, in the Pacific Ocean. However, another striking fact emerged to cast doubt on this hypothesis. The path obtained from European rocks differed in detail from that obtained from North American rocks (Figure 6-14*A*). It was recognized that this pattern might actually reflect a history in which

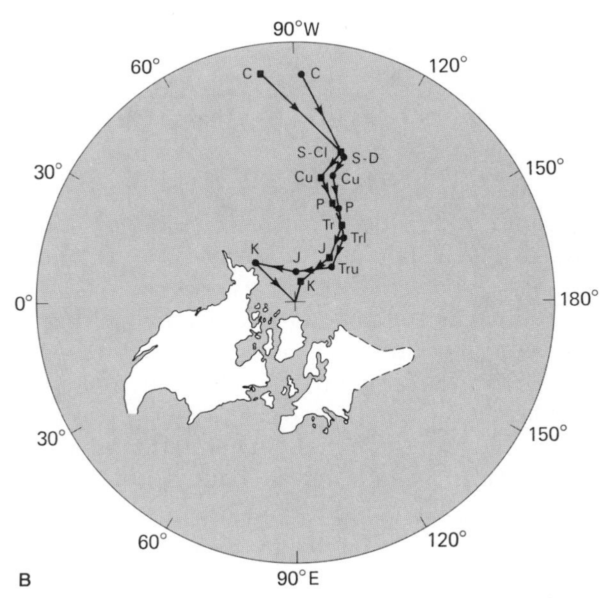

FIGURE 6-14 Apparent polar-wander paths for North America *(circles)* and Europe *(squares)*. *A*. Plot of polar-wander paths based on the assumption that the continents have remained in their present positions. *B*. Plot for North America and Europe juxtaposed, as postulated for Paleozoic time by Wegener and his followers. Here the Paleozoic and Mesozoic apparent polar-wander paths of the two continents nearly coincide, suggesting that the continents were united during the Paleozoic Era. The time-rock units represented are Cretaceous *(K)*, Triassic *(Tr)*, Upper Triassic *(Tru)*, Lower Triassic *(Trl)*, Permian *(P)*, Upper Carboniferous *(Cu)*, Siluro-Devonian *(S-D)*, Silurian to Lower Carboniferous *(S-Cl)*, and Cambrian *(C)*. *(After M. W. McElhinny, Paleomagnetism and Plate Tectonics, Cambridge University Press, London, 1973.)*

the north pole did not wander at all; instead, as Wegener had suggested, the continents of Europe and North America might have moved in relation to the pole and to one another, carrying with them rocks that had been magnetized when the continents were in different positions. This possibility led to the use of the cautious term *apparent polar wander* to describe the pathways that geophysicists plotted.

Tests were conducted to examine the possibility that continents rather than poles had moved. It was hypothesized, for example, that if North America and Europe had once been united and had drifted over Earth's surface together, they should have developed identical paths of apparent polar wander during their joint voyages. The test, then, was to fit the outlines of North America and Europe together along the Mid-Atlantic Ridge to determine whether, with the continents in this position, their paths of apparent polar wander coincided. As Figure 6-14*B* shows, the apparent polar-wander paths of North America and Europe did coincide almost exactly for both Paleozoic and early Mesozoic time. This evidence strongly suggested that the continents had indeed drifted apart, carrying their magnetized rocks with them.

THE RISE OF PLATE TECTONICS

During the late 1950s, these new paleomagnetic data generated widespread discussion of continental drift — especially in Great Britain, where many of the data had been assembled. Even so, most geologists continued to doubt the validity of continental movement. There were two reasons for this continuing skepticism: First, many paleomagnetic measurements were known to be imprecise; and second, the belief persisted that no natural mechanism could move continents against oceanic crust. Then, in 1962, the American geologist Harry H. Hess published a landmark paper proposing a novel solution to this problem.

Seafloor Spreading

In essence, Hess suggested that the felsic continents had not plowed through the dense mafic crust of the ocean at all but that instead the *entire* crust had moved. Hess's ideas were highly unconventional (he labeled his contribution "geopoetry"), but the manner in which he compiled his facts exemplifies the way geologists assemble circumstantial evidence to construct theories. In the following summary of Hess's paper, the critical facts and inferences appear in italics.

During World War II, Hess commanded an American naval vessel in the Pacific, and he seized upon this opportunity to pursue his geologic interests. In order to study the configuration of the ocean floor, for example, Hess kept his ship's echo-sounding equipment operating for long stretches of time. While profiling the bottom in this way, he discovered curious flat-topped seamounts rising from the floor of the deep sea, and he named them **guyots** after the nineteenth-century geographer Arnold Guyot. On the basis of their size and shape, Hess concluded that *guyots were volcanic islands that had been eroded by the action of waves near sea level.* Two decades later, shallow-water fossils of Cretaceous age were recovered from the tops of some guyots, proving that the guyots had indeed once stood near sea level. How the ocean floor on which they sat had subsided to such great depths, however, remained a mystery.

Another piece of evidence that Hess pondered was the apparent youth of the ocean basins. At the time, it was estimated that sediment was being deposited in the deep sea at a rate of about 1 centimeter (~$\frac{1}{2}$ inch) per thousand years — but at this rate, 4 billion years of Earth history would theoretically produce a layer of deep-sea sediment 40 kilometers (~25 miles) thick. In fact, *the average thickness of sediment in the deep sea today is only 1.3 kilometers (less than 1 mile).* Thus, *allowing for some compaction, Hess estimated that the layer of sediment existing in the deep sea represented only about 260 million years of accumulation — a figure that might therefore approximate the average age of the seafloor.* (Hess's calculation was of the right order of magnitude, but we now know that the average age of the seafloor is even younger than 260 million years; in fact, little or none of the seafloor is as old as 200 million years.)

Hess found support for the idea of a youthful seafloor in his observation that there are only about 10,000 volcanic seamounts (volcanic cones and guyots) in all the world's oceans. Hess knew that when a volcano has been eroded to the level at which waves can act on it, it withstands further erosion very effectively — and so he assumed that if the oceans were permanent features, their oldest volcanic seamounts should still be extant. Given the fact that there were only 10,000 volcanic seamounts in modern oceans, Hess further reasoned that if the oceans were nearly as old as Earth—say, 4 billion years old—an average of only one volcano would have formed every 400,000 years or so. The existence of so many obviously young volcanoes indicated to Hess that new volcanoes appear much more frequently—perhaps at a rate of one every 10,000 years. Thus *the relatively small number of volcanic seamounts in modern oceans suggested to Hess that current ocean basins are much younger than Earth.*

Like many earlier workers, Hess noted the central location of the Mid-Atlantic Ridge. He also noted that other midocean ridges tend to be centrally located within ocean basins. (A "best fit" restoration of continents along the Mid-Atlantic Ridge, calculated after Hess's paper was written, is shown in Figure 6-2.) Four other curious facts about these ridges seemed significant to Hess:

1 *They are characterized by a high rate of upward heat flow from the mantle to neighboring segments of seafloor.*

2 *Seismic waves from earthquakes move through the ridges at unusually low velocities.*

3 *Along the crest of each ridge there is a deep furrow.*

4 *Volcanoes frequently rise up from midocean ridges.*

Hess developed a hypothesis that seemed consistent with all of these observations. Essentially, he suggested that midocean ridges represent narrow zones where oceanic crust forms as material from the mantle moves upward and undergoes chemical changes. Hess further maintained that as this material rises, it carries heat from the mantle to the surface of the seafloor. *The expanded condition of the warm, newly forming crust thus accounts for the swollen condition of the crust there—that is, for the presence of a ridge.*

Hess then revived a geophysical concept that had been discussed by a number of earlier researchers—namely, the idea that material within Earth's mantle rotates by means of large-scale thermal convection. The material of the mantle, existing at high temperatures and pressures, must flow like a very thick liquid. As we have seen, the mantle is heated by the decay of radioactive isotopes within it and is cooled from above by loss of heat through the crust. Consequently, the upper part of the mantle, being cool, is more dense than the lower part and thus tends to sink while the lower part tends to rise. In a deep body of liquid, the result is convection, the same circular overturning movement that heat from the sun creates in Earth's atmosphere. Hess proposed that Earth's liquidlike mantle is divided into **convective cells** (Figure 6-15) whose low-density material forms crust as it rises and cools. This crust then bends laterally to become one flank of a ridge (Figure 6-16). *The furrow down the center of many ridges could then be explained as the site at which newly formed crust separates and flows laterally in two directions. Similarly, volcanoes along midocean ridges would represent the rapid escape of mantle material at certain sites, while the low velocity of earthquake waves passing through a ridge would result from the fact that the rocks of the ridge exist at a high temperature and are extensively fractured where they bend laterally to form the basaltic seafloor.*

How do the guyots that Hess discovered fit into this theory? According to his scheme, the seafloor adjacent to a midocean ridge (together with anything attached to this seafloor) moves laterally, away from the spreading center. The volcanoes that frequently form along midocean ridges sometimes grow upward to sea level, as is the case with Ascension Island in the Atlantic. As a

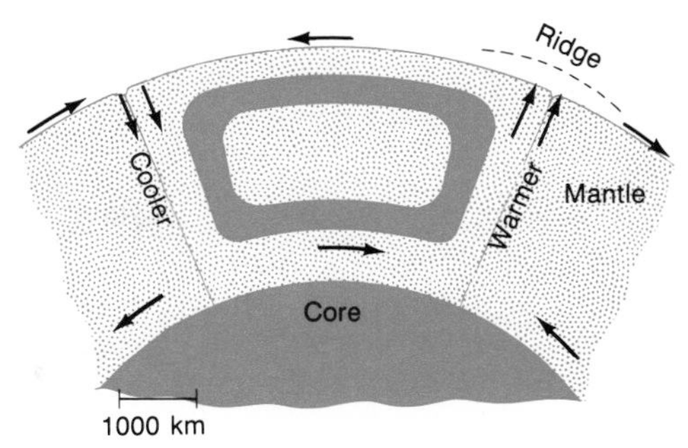

FIGURE 6-15 Convective motion within the mantle as envisioned by Harry H. Hess. Midocean ridges form where the upward-flowing limbs of two adjacent convection cells approach the surface. *(After H. H. Hess, in Petrologic Studies: A Volume in Honor of A. F. Buddington, Geol. Soc. Amer., 1962.)*

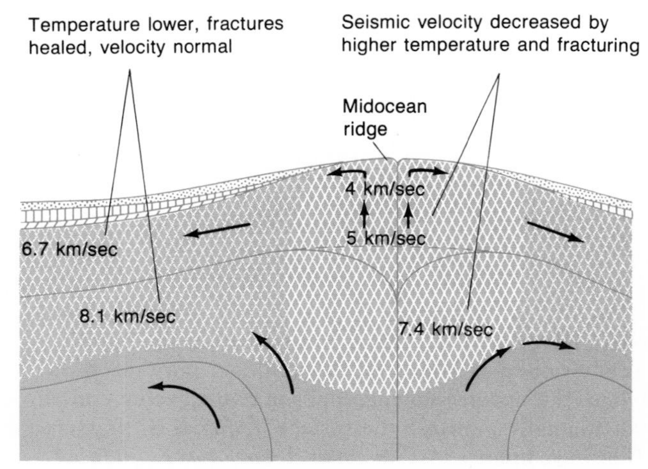

FIGURE 6-16 Hess's model of the structure of a midocean ridge. Arrows show the flow of new crust derived from the convecting mantle. The newly formed crust carries heat from the mantle. This factor, along with fracturing of the rock as it bends laterally, results in low velocities (shown in kilometers per second) for seismic waves passing through the ridge. The elevation of the ridge results from the hot, swollen condition of the newly formed crust. *(After H. H. Hess, in Petrologic Studies: A Volume in Honor of A. F. Buddington, Geol. Soc. Amer., 1962.)*

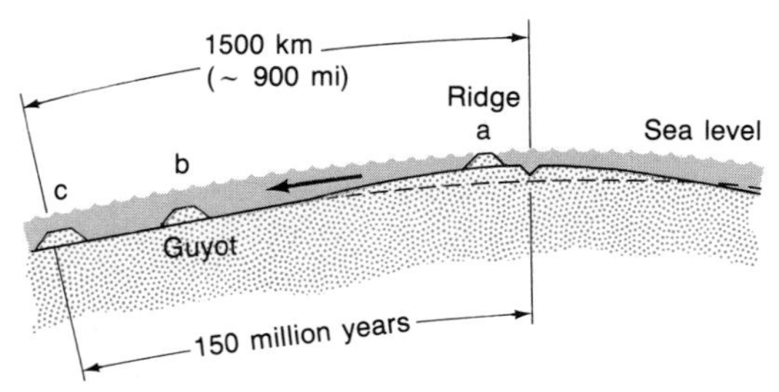

FIGURE 6-17 Hess's interpretation of the way a typical guyot is formed. First, a volcano builds a cone along a midocean ridge. The cone initially stands partly above sea level, and its tip is later planed off by wave erosion *(a)*. The resulting flat-topped structure moves laterally with the spreading crust and is carried gradually downward (*b* and *c*), because the newly formed crust beneath it cools and therefore shrinks as it moves away from the ridge. *(After H. H. Hess, in Petrologic Studies: A Volume in Honor of A. F. Buddington, Geol. Soc. Amer., 1962.)*

volcano moves laterally from the ridge along with the crust on which it stands, it moves away from the source of its lava. It then becomes an inactive seamount, and its tip is quickly planed off by erosion. The seafloor gradually deepens away from midoceanic ridges, because newly formed material of the crust cools and therefore shrinks as it moves laterally away from the ridge. Thus *a truncated seamount is gradually transported out into deep water as if it were on a conveyor belt, and it then becomes a guyot* (Figure 6-17). Assuming that the Atlantic Ocean has developed by seafloor spreading since the end of the Paleozoic, Hess calculated a spreading rate of about 1 centimeter per year.

Continents can be viewed as enormous bodies that float in oceanic crust by virtue of their low density. They would be expected to ride passively along like guyots. Here, then, was Hess's explanation for the fragmentation of continents: He reasoned that when convective cells in the mantle change their locations, the upwelling limbs of two adjacent cells must sometimes come to be positioned beneath a continent. Convective spreading should then **rift** the continent into two fragments and move them apart from the newly formed spreading center. New ocean floor should subsequently form at the same rate on each side of the spreading center. Hess further maintained that the spreading center would continue to operate along the midline of the new ocean basin—and thus persist as a midocean ridge—as long as the convective cells remained in their new location.

If oceanic crust forms and flows laterally without an enormous change in thickness, however, it must disappear somewhere. Hess postulated that it must be swallowed up again by the mantle along the great **deep-sea trenches** that exist at certain places in the ocean floor (Figure 6-18). Movement of crust into the mantle along one side of a trench provided a ready explanation for the fact that *Earth's gravitational field here is unusually weak;* the presence of low-density crustal rock in deep-sea trenches in place of dense mantle rock would be expected to weaken the gravitational force exerted by Earth on objects at or above its surface. *Hess estimated that the formation of new crust along midocean ridges and the simultaneous disappearance of crust into the deep-sea trenches would produce an entirely new body of crust for the world's oceans every 300 or 400 million years.*

Hess's hypothesis of seafloor spreading had two great strengths. First, by asserting that conti-

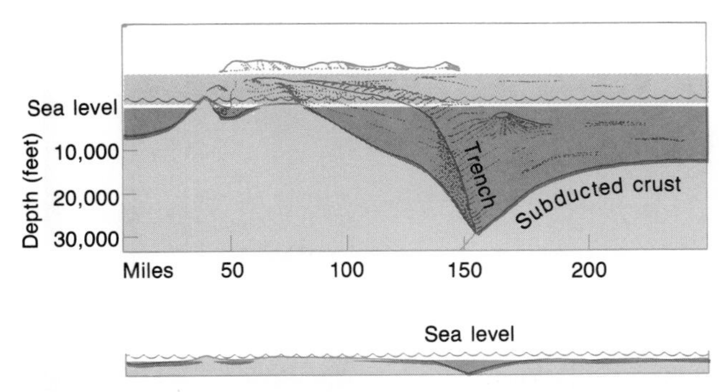

FIGURE 6-18 Section through the Tonga Trench in the Pacific Ocean. The view is northward. In the upper picture, vertical distances are exaggerated by a factor of 10. The lower diagram is drawn without vertical exaggeration. The island toward the left is Kao, a dormant volcano. The oceanic crust to the right (or east) of the trench is moving down into the mantle beneath the oceanic crust to the left. *(Modified from R. L. Fisher and R. Revelle, Scientific American, November 1955.)*

nents move along *with* oceanic crust, it overcame the objection that continents could not move *through* such crust. Second, the hypothesis was consistent with a variety of facts, the most important of which are italicized above. Most of these facts had not previously made sense.

The Triumph of Paleomagnetism

Despite the strong circumstantial evidence in support of Hess's hypothesis, its publication in 1962 created no great stir within the geologic profession. What was needed was a really convincing test of the basic idea. Such a test was soon found. It was based on the well-known fact that Earth's magnetic field has periodically reversed its polarity (p. 96). In 1963 the British geophysicists Fred Vine and Drummond Matthews reported that newly formed rocks lying along the axis of the central ridge of the Indian Ocean were magnetized while Earth's magnetic field was polarized as it is now. This finding came as no surprise, because it was known that other midocean ridges also exhibited "normal" magnetization. It turned out, however, that seamounts on the flanks of the Indian Ocean ridge were magnetized in the reverse way. Vine and Matthews concluded that this pattern might confirm Hess's seafloor-spreading model. They reasoned that if crust is now forming along the axis of any midocean ridge, it must become magnetized with the magnetic field's present polarity as it crystallizes from the molten mantle. In older crust lying at some distance from the ridge, however, reversed polarity should be encountered, and in even older crust farther from the ridge, the polarity should be normal again.

Magnetic "striping" had, in fact, recently been observed on many parts of the seafloor. The stripes are called **anomalies** because their magnetism, if it is normal, adds to Earth's present magnetic field, and if it is reversed, it weakens the magnetic field. Thus the presence of magnetic stripes causes measurements of regional magnetism to be abnormal or anomalous.

The striping patterns were soon put to a more rigorous test. During the 1960s a time scale was developed for late Cenozoic magnetic reversals. This scale was based on measurements of the magnetic polarity of terrestrial rocks of known age. It was assumed that the spreading rate for each midocean ridge had remained reasonably constant over the past 4 or 5 million years. It was then found that the relative widths of seafloor stripes were proportional to the time intervals that these stripes were thought to represent—that is to say, long intervals of normal polarity were represented by broad stripes, short intervals by narrow stripes. Thus the detailed patterns of striping were found to match the known timing of magnetic reversals (Figure 6-19). Because the oceanic crust is not altogether homogeneous, magnetic anomalies do not form perfectly smooth striping patterns on the seafloor. Nonetheless, the patterns can be striking.

An interesting story concerns the misfortunes of a Canadian geologist named L. W. Morley. Morley developed the same model for magnetic anomalies that Vine and Matthews published, but the manuscript in which he outlined his model was rejected by the two journals to which he submitted it in 1963. One reviewer of the manuscript cynically commented that "such speculation makes interesting talk at cocktail parties." Because Vine and Matthews were fortunate enough to have had their paper accepted for publication, they were the ones who ultimately received recognition in the scientific world. Radically new ideas are not easily established in science.

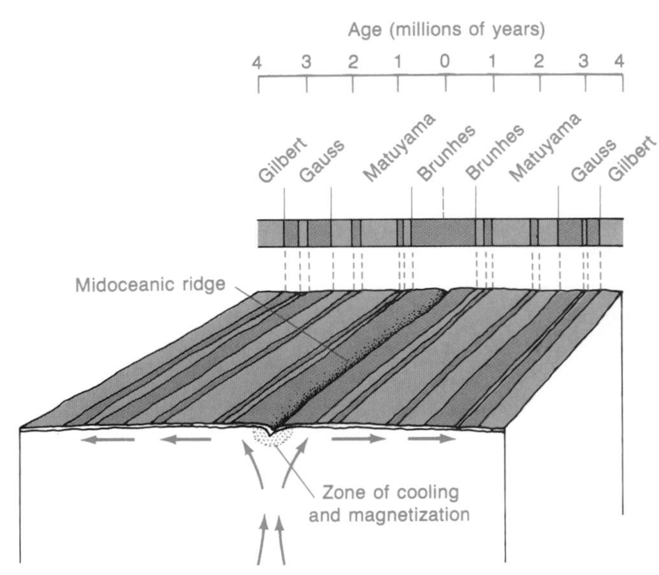

FIGURE 6-19 Magnetic anomaly patterns of the seafloor fit the prediction that they represent magnetic reversals. The time scale shows known magnetic reversals of the past 4 million years. The labels (*Gilbert* through *Brunhes*) represent intervals that are, for the most part, characterized by either normal or reversed polarity, dated by the polarity of terrestrial rocks whose ages are known. The relative lengths of these intervals are remarkably similar to those of the magnetic-anomaly stripes on either side of a midocean ridge. *(After A. Cox et al., Scientific American, February 1967.)*

FIGURE 6-20 Thingvellir Graben in Iceland. This portion of the Mid-Atlantic Ridge is dramatically exposed above sea level. As the rift separates, lava squeezes upward to form new basaltic crust. *(Gudmundur Sigvaldason.)*

What Happens at Ridges

At places in Iceland, the furrow down the center of the Mid-Atlantic Ridge can be seen to be a structural **graben** (Figure 6-20): a valley bounded by **normal faults**—that is, faults whose dip is steeper than 45° and that have allowed a central "fault block" to slip downward (Figure 6-21). Grabens form where the crust is extending—as is the case along a midocean ridge, where crust is forming and moving laterally. As the crust periodically breaks apart along a midocean ridge, lava moves upward to fill the space thus formed, producing new oceanic crust. Extruded lavas have also been recorded along submarine midocean ridges. These lavas sometimes form **pillow basalt,** rock with the hummocky configuration that originates when lava cools underwater (Appendix I).

Midocean ridges and neighboring features of the seafloor have now been mapped in considerable detail. It turns out that these ridges are frequently offset along enormous **transform faults**

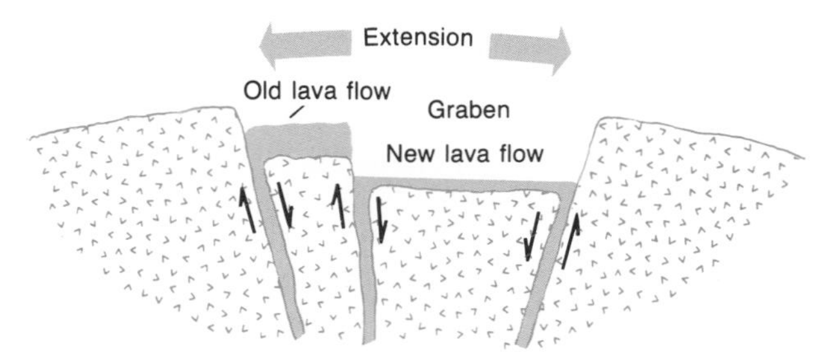

FIGURE 6-21 A graben of the sort that a midocean rift represents. As tension breaks the crust and spreads it laterally, some blocks of crust sink back into the zone of extension along faults. Lavas *(pattern)* moving upward along the faults fill in the space produced by rifting and also flow laterally and solidify on the floor of the graben.

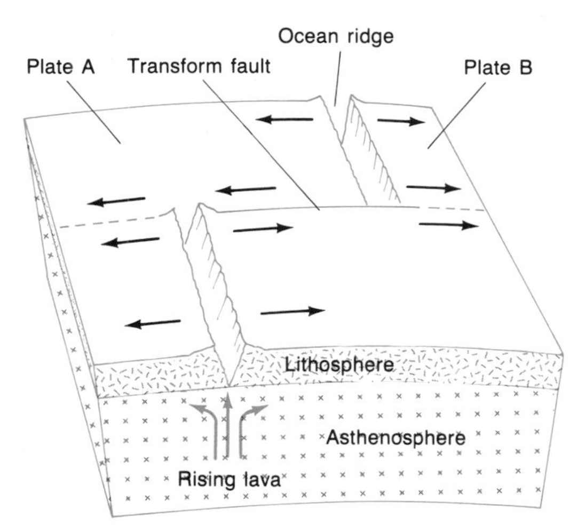

FIGURE 6-22 A transform fault. The central part of this fault is a plate boundary along which two plates slide past one another. Here plate *A* and plate *B* are separating at a rift, and the transform fault offsets this ridge. Arrows indicate the opposite directions of plate motion between the two segments of the ridge. *(After F. Press and R. Siever, Earth, W. H. Freeman and Company, New York, 1986.)*

(Figures 6-4 and 6-22). These fractures of Earth's crust are **strike-slip faults**—that is, high-angle faults along which the rocks on opposite sides move in opposite directions (Appendix II). The famous San Andreas Fault is a transform fault that happens to cut across the edge of the North American continent (p. 128). Earthquakes caused by movement along the San Andreas have caused considerable damage in California.

It is now recognized that the boundary between the crust and the mantle—the Moho (Figure 6-11)—is not the surface along which Earth's "skin" moves. This surface, which lies well below the Moho, is the boundary between the asthenosphere (the partially molten part of the mantle) and the lithosphere (the uppermost mantle and the crust). The asthenosphere-lithosphere boundary is situated closer to the surface beneath midocean ridges, where high temperatures keep mantle material molten even at very shallow depths. Figure 6-23 shows the current configuration of the crust and upper mantle in the vicinity of the Atlantic Ocean, which is still growing by seafloor spreading.

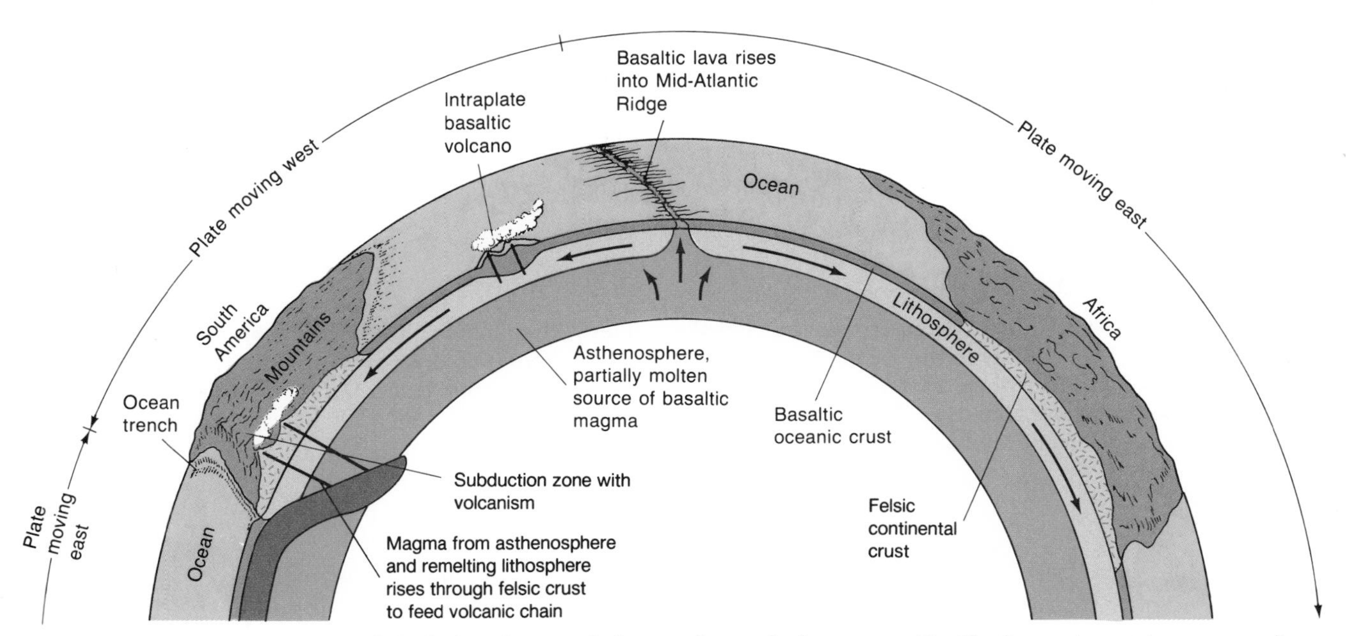

FIGURE 6-23 Cross section of the lithosphere and the asthenosphere in the vicinity of the South Atlantic Ocean. Note that the lithospheric plate that includes South America is moving westward from the Mid-Atlantic Ridge. At the same time, the lithospheric plate beneath the eastern Pacific Ocean is moving eastward; where it meets South America along a deep-sea trench, this oceanic plate moves downward into the asthenosphere. *(After F. Press and R. Siever, Earth, W. H. Freeman and Company, New York, 1986.)*

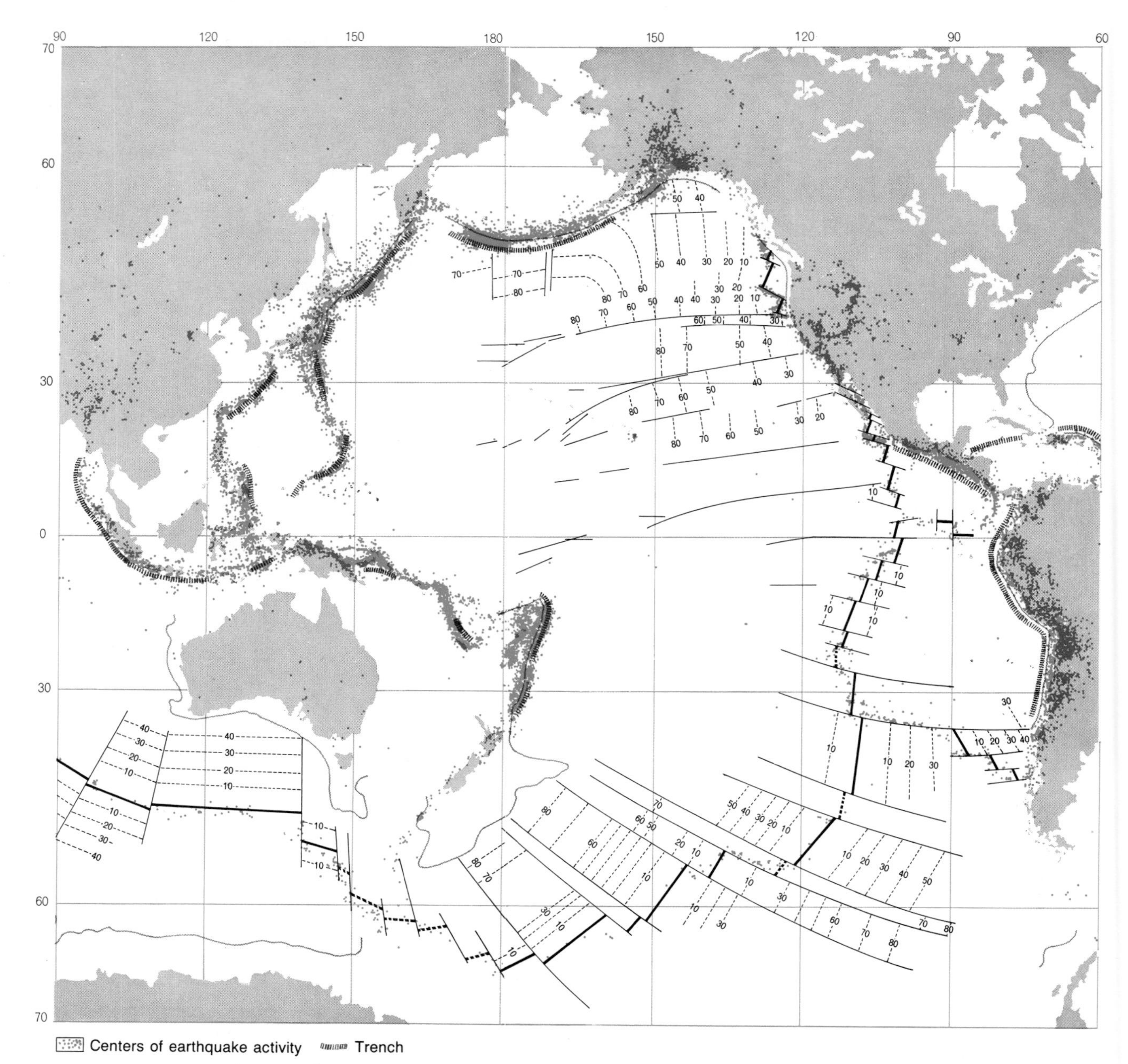

FIGURE 6-24 Distribution of deep-sea trenches and of centers of earthquake activity in the Pacific between 1961 and 1967. Note that when trenches are viewed from above, many can be seen to curve, and that earthquake centers are concentrated along trenches. Dashed lines with numbers indicate segments of seafloor of various ages (millions of years before the present). *(After M. Barazani and J. Dorman, Seismol. Soc. Amer. Bull. 59:369–380, 1969.)*

Subduction at Deep-Sea Trenches

After the publication of Hess's paper, deep-sea trenches attracted much interest. These trenches are the sites of the lithosphere's descent into the asthenosphere—a process now called **subduction.** Most of the trenches of the modern world encircle the Pacific Ocean (Figure 6-24). In the decade before Hess developed his ideas, the geophysicist Hugo Benniof and others had noted that trenches are associated with two other geologic features: volcanoes and **deep-focus earthquakes.** The latter are earthquakes that originate more than 300 kilometers (~190 miles) below Earth's

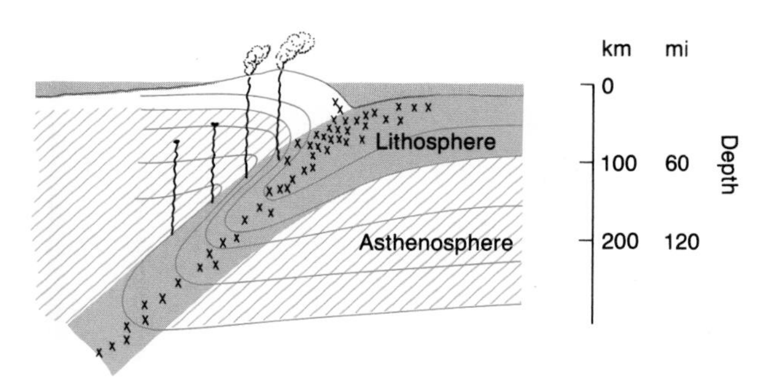

FIGURE 6-25 Volcanic activity produced by subduction. The subducted plate partially melts after it reaches a critical depth. Magma of low density rises and reaches the surface to form volcanoes. Crosses represent earthquake foci. Curved lines are lines of equal temperature. Note that subduction of the lithosphere, which is initially cold, causes these lines to bend downward. *(After S. Uyeda, The New View of the Earth, W. H. Freeman and Company, New York, 1978.)*

surface (p. 13). In areas far from deep-sea trenches, deep-focus earthquakes are rare. Near trenches, however, both shallow- and deep-focus earthquakes are frequent.

The typical spatial relationship between trenches, volcanoes, and earthquake foci is shown in Figure 6-25. As this figure indicates, the earthquake foci fall along a narrow, nearly planar zone that angles down from the trench. A little thought will reveal that the earthquake foci in Figure 6-25 must lie along or within the segment of a plate that is descending into the asthenosphere. Earthquakes must result from the occasional downward movement of a plate. The descent of a plate causes earthquakes to originate deeper than they do elsewhere within the lithosphere, because a descending plate is a portion of lithosphere that is moving down into the asthenosphere. The plate descends because it is cooler and therefore denser than the partially molten asthenosphere, and it produces earthquakes because it occasionally takes a sudden step downward.

Chains of volcanic islands often parallel deep-sea trenches (Figure 6-24). They are positioned in this way because the descending slab of lithosphere undergoes partial melting (Figure 6-25), and any molten material that is less dense than the asthenosphere rises toward the surface as magma. Some of this magma solidifies within the crust to form intrusive igneous bodies that are several kilometers in diameter. The rest reaches the surface to emerge in volcanic eruptions.

A band of subducted lithosphere is called a **subduction zone.** Subduction zones border much of the Pacific Ocean (Figures 6-1 and 6-24). The volcanoes associated with these zones form what is known as the "ring of fire" around the Pacific; Box 6-1 describes cataclysmic volcanism within the ring.

When deep-sea trenches and the chains of volcanoes associated with them are viewed from above, many of them can be seen to have the shape of an arc. The volcanoes that rise above sea level form **island arcs** (Figure 6-26). Lithosphere tends to descend along a curved path for a reason that appears to be quite simple: Because Earth is nearly spherical, lithosphere that is depressed into the less rigid asthenosphere tends to bend downward along a curved line in the way that a circular dent forms in a Ping-Pong ball when you press it with your thumb.

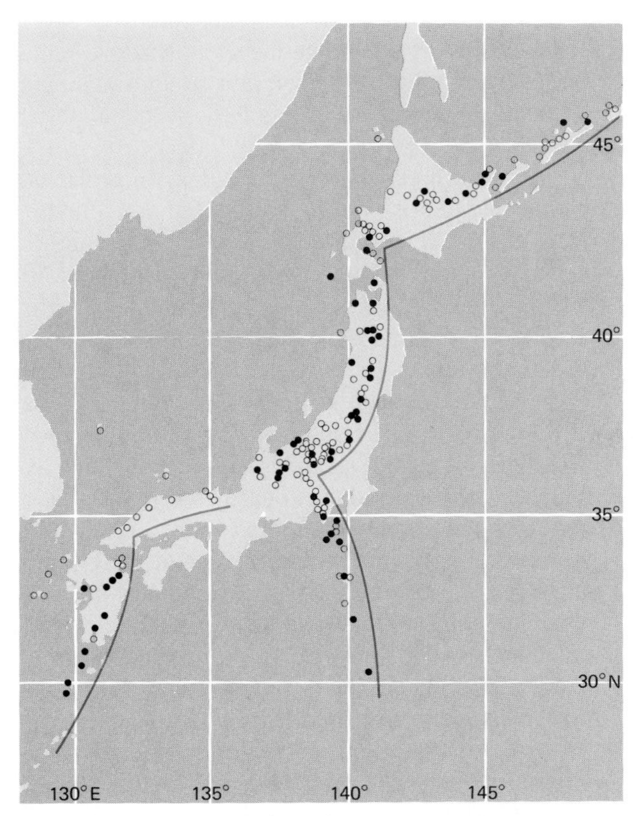

FIGURE 6-26 Island arcs that lie partly along the islands of Japan. Black circles represent active volcanoes; hollow circles represent young but inactive volcanoes.

Box 6-1
The Ring of Fire

In 1928 a small volcano rose above the surface of the sea between the large Indonesian islands of Sumatra and Java. The islanders named it Anak Krakatau, or "child of Krakatau." Nearly half a century before it was born, its parent, a much larger volcanic island, had all but self-destructed in a series of volcanic eruptions that affected the entire world.

Early in 1883, earthquakes repeatedly shook islands near Krakatau, which until then had been quiescent for about 200 years. On May 20, Krakatau's pointed summit began to spew volcanic ash and steam. These events foreshadowed one of the greatest natural catastrophes of recorded history. On August 26, 27, and 28, a series of violent explosions audible to people across 10 percent of Earth's surface sent volcanic ash several tens of kilometers above Krakatau. The resulting airwaves disturbed barographs throughout the world. During these convulsions, Krakatau belched out an estimated 20 cubic kilometers (5 cubic miles) of lava, rock, and ash. By the time the event was over, most of the volcano had collapsed into a huge chamber that had formed beneath it as the volcanic material was expelled.

With each explosion of Krakatau, an avalanche of volcanic debris poured into the sea and a tidal wave surged thousands of kilometers across the ocean. Some of the waves swelled to heights of 40 meters (~25 feet). Altogether they destroyed 165 coastal villages and took more than 36,000 human lives.

When the eruptions of 1883 were over, only a third of the original volume of Krakatau remained above sea level, and steaming debris that it had expelled formed new islands nearby where the sea had previously been 36 meters (~100 feet) deep. In the months after the eruptions, volcanic dust in the atmosphere caused such deep-red sunsets that frightened citizens as far away as New York City mistakenly called out fire engines. The effects on climates throughout the world were longer lasting. Ash that spread throughout the atmosphere screened out sunlght, reducing average global temperatures by as much as $\frac{1}{2}$° Celsius. Only five years later did climates return to normal.

Anak Krakatau. *(Kall Muller/Woodfin Camp.)*

Indonesia, the home of Krakatau, is a veritable nation of volcanoes. With a total of 132, it ranks first in the world in number of volcanoes that have been active within the last 10,000 years. Indonesia lies within the "ring of fire" around the Pacific Ocean, where lithosphere is descending into hot asthenosphere along subduction zones and melting to produce magma. Because of the presence of the ring of fire, more than twenty countries that border the Pacific are vulnerable to tidal waves produced by volcanic eruptions. Most of these countries also face the more direct dangers of volcanism: exposure to flowing lava, flying debris, spreading poisonous gases, and smothering mudslides.

In one way or another, volcanoes have claimed more than a quarter-million human lives during the past 400 years. It is sobering to note that the eruption of Krakatau in 1883 was not an especially large volcanic event. When Tambora, another Indonesian volcano, came to life in 1815, its activity went largely unstudied because the science of geology was still in its infancy, but it is clear that Tambora's eruption dwarfed Krakatau's. The size of Tambora's crater shows that it spewed forth about five times as much material, and its dust screened out so much sunlight that 1816 came to be known as "the year without a summer."

How severely a volcanic event affects humans actually depends more upon the proximity of large populations than upon the size of the eruption. The famous eruption of Mount Vesuvius in A.D. 79 was relatively small, yet it totally destroyed the nearby towns of Pompeii and Herculaneum, trapping thousands of unsuspecting Romans. Ironically, the citizens of Pompeii had worshiped the magnificent smoldering mountain that loomed impressively over them and had built their main street to align with its summit.

We can take various measures to reduce the damage that future volcanic eruptions wreak on human populations. We can construct coastal barriers to tidal waves, for example, or move dwellings inland. More generally, we can establish plans of evacuation for frequently endangered areas, and we can monitor volcanoes to sense movements of subterranean magma so that potential victims can be alerted to the danger of imminent eruptions. Unfortunately, however, millions of humans inhabit risky locations and geologists are monitoring only a handful of menacing volcanoes. Unless efforts are made to improve safety, eruptions will undoubtedly claim thousands of lives during the next several decades.

Relative Plate Movements

Today Earth's lithosphere is divided into eight large plates and several small ones (Figure 6-1). We have already discussed three kinds of plate boundaries. To recapitulate, an actively spreading ridge represents one kind of boundary, and a subduction zone (or trench) represents a second. Plates move away from ridges, along which they form, toward subduction zones, where they are destroyed. As two plates move, they scrape past each other along the third kind of plate boundary, a transform fault (Figures 6-4 and 6-22).

A complex set of forces is responsible for the movement of plates. A plate tends to be dragged along by convective movement of the underlying asthenosphere, but other forces may be more important. For example, a gravitational force tends to slide a plate laterally away from an elevated midocean ridge, where the plate forms (Figure 6-16). In addition, the margin of a plate that sinks into the asthenosphere along a subduction zone (Figures 6-18 and 6-25) pulls the plate toward that subduction zone. Because these gravitational forces that arise at spreading and subduction zones are strong enough to move plates, one cannot assume that all spreading zones or subduction zones necessarily form by movements of the asthenosphere where two convection cells meet.

The configuration of lithospheric plates has changed throughout Earth's history. From time to time, new ridges and subduction zones form and old ones disappear. Although we cannot reconstruct the entire history of plate motions from geologic evidence, much of this history is known for Cenozoic time. The history of spreading along the Mid-Atlantic Ridge is particularly well understood, as are the histories of spreading along the East Pacific Rise west of South America and its continuation south of Australia (Figure 6-27).

The complex history of continental movement described by plate tectonics differs in an important way from the pattern that Wegener envisioned. Wegener thought that the enormous landmass of Pangaea existed as a stable crustal feature for hundreds of millions of years and that it fragmented in a single event. Plate tectonics, on the other hand, entails continuous movement of most landmasses in relation to one another. Because all plates are moving, no piece of lithosphere represents a perfectly immobile block against which the movement of all others can be assessed.

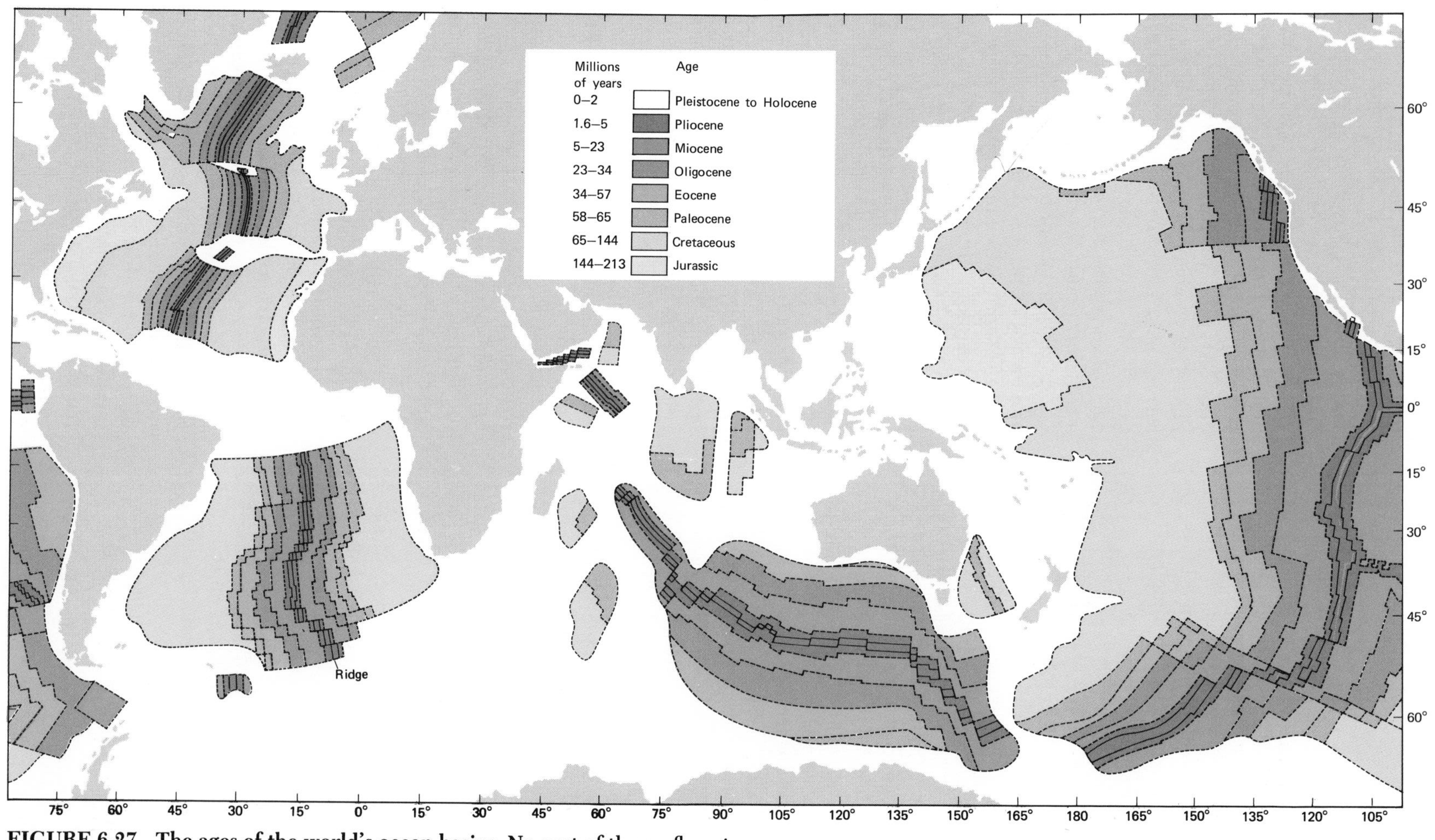

FIGURE 6-27 **The ages of the world's ocean basins.** No part of the seafloor is older than Mesozoic. *(After W. C. Pitman et al., Geol. Soc. Amer. Map and Chart MC-6, 1974.)*

Absolute Plate Movements

Today the enormous Pacific plate is moving to the northwest *in relation to* the plates that neighbor it on the north and northeast (Figure 6-28). How can we measure the *absolute* movement of a plate? Absolute movement is defined as any movement in relation to a fixed feature, such as an immobile point at the surface of Earth's mantle. Nearly immobile points appear to have been discovered in the form of **hot spots.** A hot spot is a small geographic area where heating and igneous activity occur within the crust. A hot spot is located at Yellowstone National Park in Wyoming, where geysers and volcanoes have been present for millions of years. Some hot spots result from the arrival at Earth's surface of a **thermal plume,** or a column of molten material that rises from the mantle. Often a large volcano forms at the surface above a plume that rises through thin oceanic crust. Because most of the mantle is in a state of convective motion, it is remarkable that plumes are nearly stationary. As a plate moves over a plume, its successive positions are commonly recorded as a chain of volcanoes such as the one constituting the Hawaiian Islands (Figure 6-29).

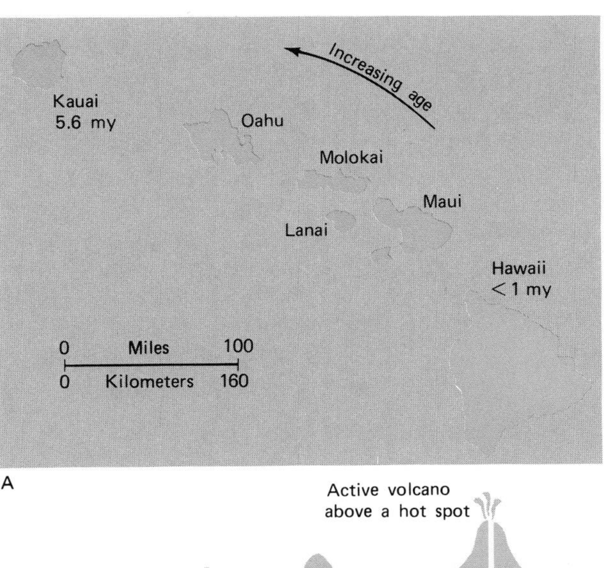

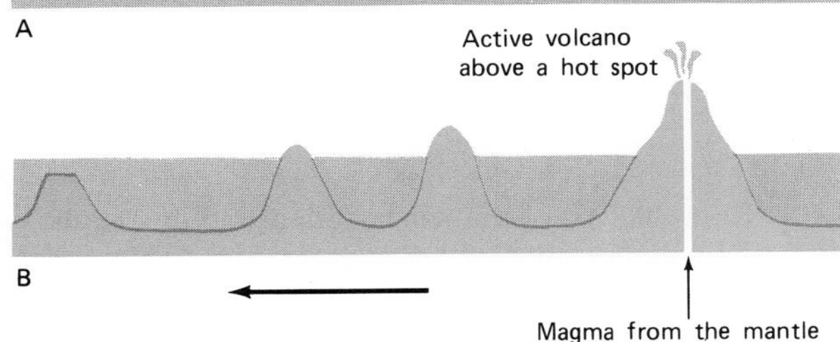

FIGURE 6-29 Formation of a chain of islands as an oceanic plate moves over a hot spot. *A.* The major Hawaiian islands. The islands increase in age toward the northwest. They have formed, one after the other, as the Pacific plate has moved northwestward over a hot spot in the mantle. *B.* Volcanic islands form, one after the other, as an oceanic plate moves over a hot spot.

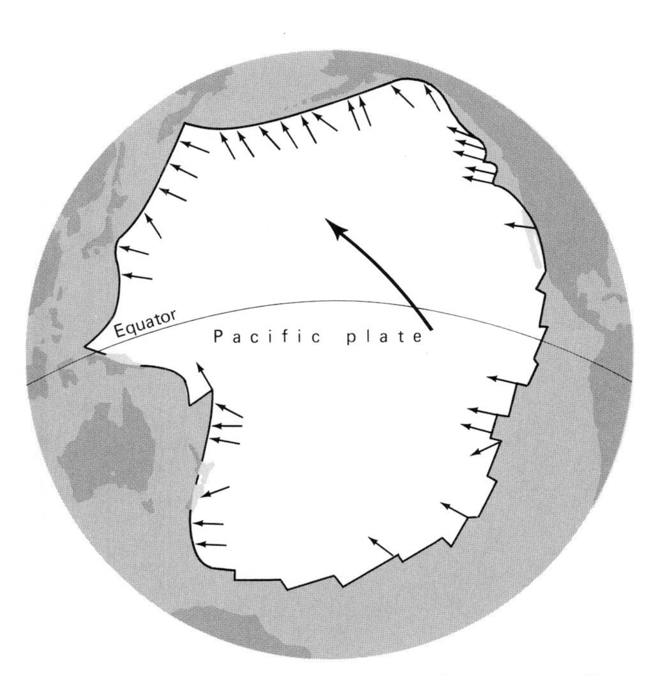

FIGURE 6-28 Movement of the Pacific plate. Small arrows show the directions of movement of the plate in relation to its neighboring plates. The large arrow shows the absolute direction of movement of the Pacific plate. *(After P. J. Wyllie, The Way the Earth Works, John Wiley & Sons, Inc., New York, 1976.)*

Radiometric dating tells us that Hawaii, the largest and easternmost of the islands, is less than 1 million years old, while the small northwestern island of Kauai is about 5.6 million years old, and a long train of even older submarine seamounts extends northwestward beyond Kauai. This age pattern would seem to indicate that the Pacific plate is moving in a west-northwest direction over a stationary hot spot. Note that this direction approximates the one in which the Pacific plate is moving in relation to the plates that border it on the north and northeast (Figure 6-28).

A search of the entire globe has turned up many hot spots that have been active within the last 10 million years (Figure 6-30). Several of them have been used to estimate the absolute

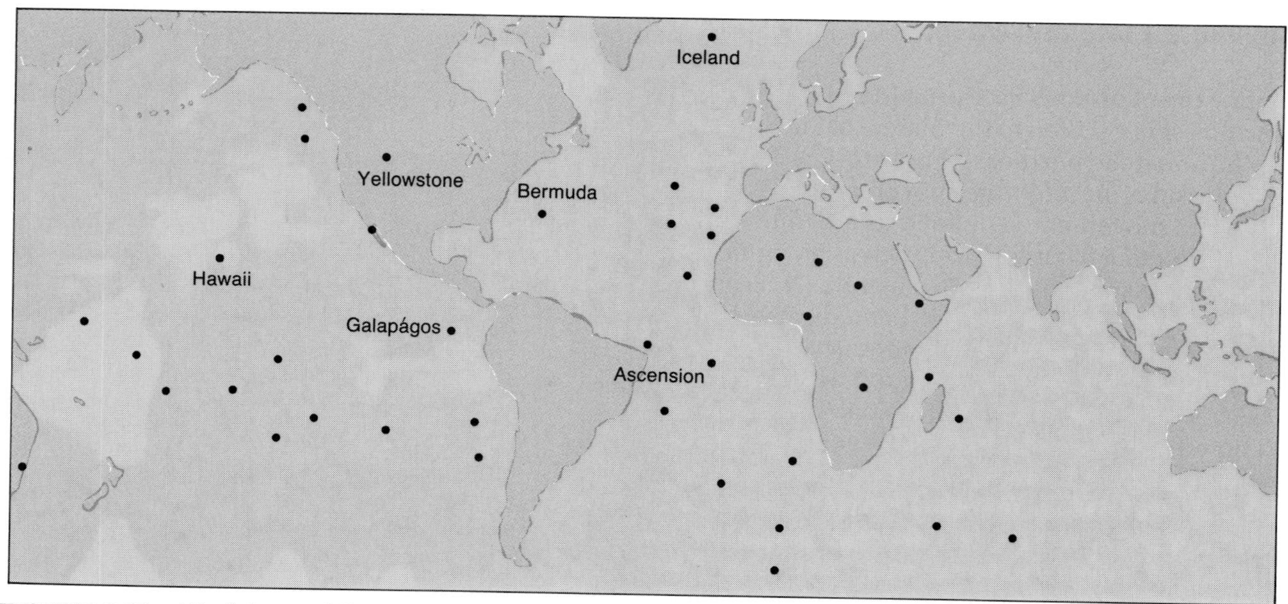

FIGURE 6-30 All of the world's identified hot spots. Note the concentration of hot spots in Africa. Many hot spots, such as Ascension Island, are positioned on or close to midocean ridges. *(After S. T. Crough, Ann. Rev. Earth Planet. Sci. 11:165–193, 1983.)*

directions and rates of plate motion. Many lithospheric plates are moving at a rate of only 5 centimeters (~1 inch) or less per year.

The Rifting of Continents

Eastern Africa provides us with a model of continental rifting in progress. A system of rift valleys, formed during the Cenozoic Era, extends southward from the Red Sea and the Gulf of Aden (Figure 6-31). The rift valleys and the basins harboring the Red Sea and Gulf of Aden are grabens formed by extension and breaking of the continental lithosphere.

When rifts develop, they often begin as three-armed lesions at plate boundaries known as **triple junctions.** Long before the advent of plate tectonics, geologists noticed that at the locations of three-armed grabens, continental crust is frequently elevated into a dome. In the context of plate tectonics, it seems evident that this doming represents the development of a hot spot. In the area of Ethiopia called the Afar Triangle, the Red Sea, the Gulf of Aden, and the north end of the African rift system form a triple junction. Such junctions are common features of Earth's crust (Figure 6-1). The plate boundaries at a triple junction may differ from one another. Each boundary may consist of a spreading zone, a subduction zone, or a transform fault. At the Afar Triangle, the junction happens to involve three spreading zones.

It appears that when a large continent is rifted apart, the jagged line along which it divides often represents a composite structure formed from arms of several three-armed rifts. The rifting that formed the Atlantic Ocean illustrates this phenomenon (Figure 6-32). A three-armed rift usually contributes two of its arms to the composite rift, while the third arm becomes a failed rift—a plate-tectonic dead end. Before it ceases to be active, this third arm forms a graben or a system of grabens that projects inland from the new continental margin formed by the other two arms. Some of the world's largest rivers, including the Mississippi and the Amazon, flow through valleys located in failed rifts that border the Atlantic basin.

It is not uncommon, however, for all three arms of a three-armed rift to develop into segments of plate boundaries. Note, for example, that the Mid-Atlantic Ridge terminates at a triple junction of ridges in the South Atlantic. Similarly,

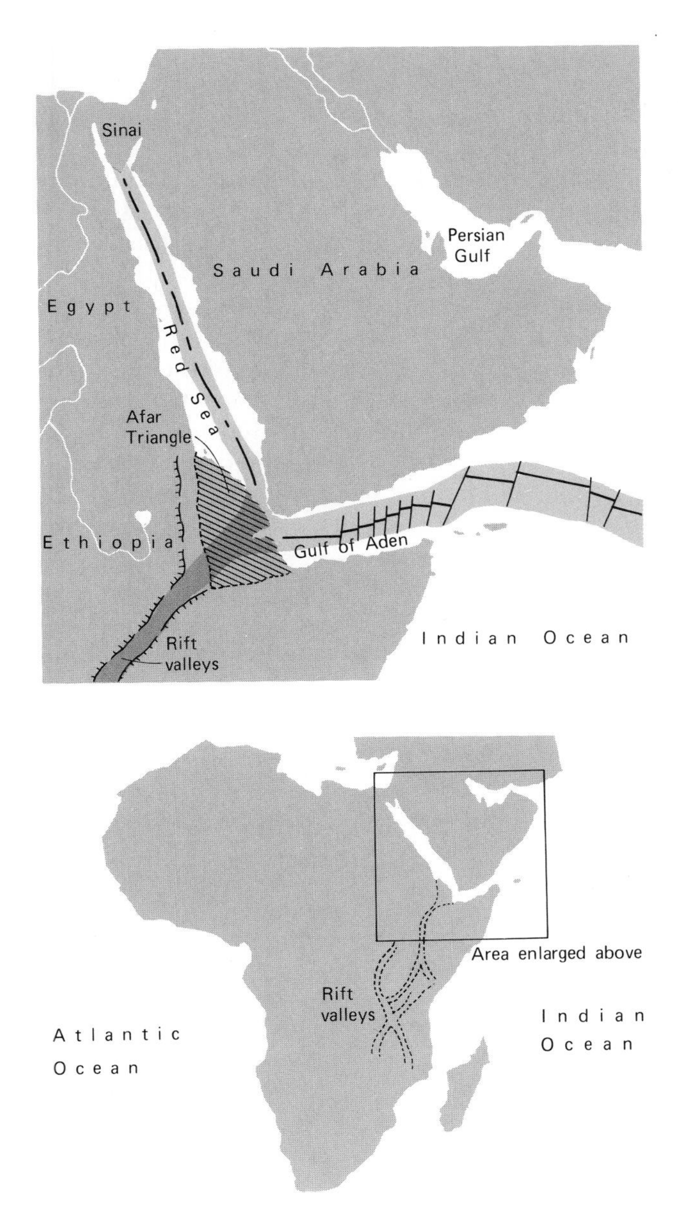

FIGURE 6-31 Three-armed rift along the northeast margin of Africa. Two of the arms represent new oceans: the Red Sea and the Gulf of Aden. The third is beginning to break the continent of Africa apart along Africa's famous rift valleys. The Afar Triangle is a small region of oceanic crust that has been elevated to become land.

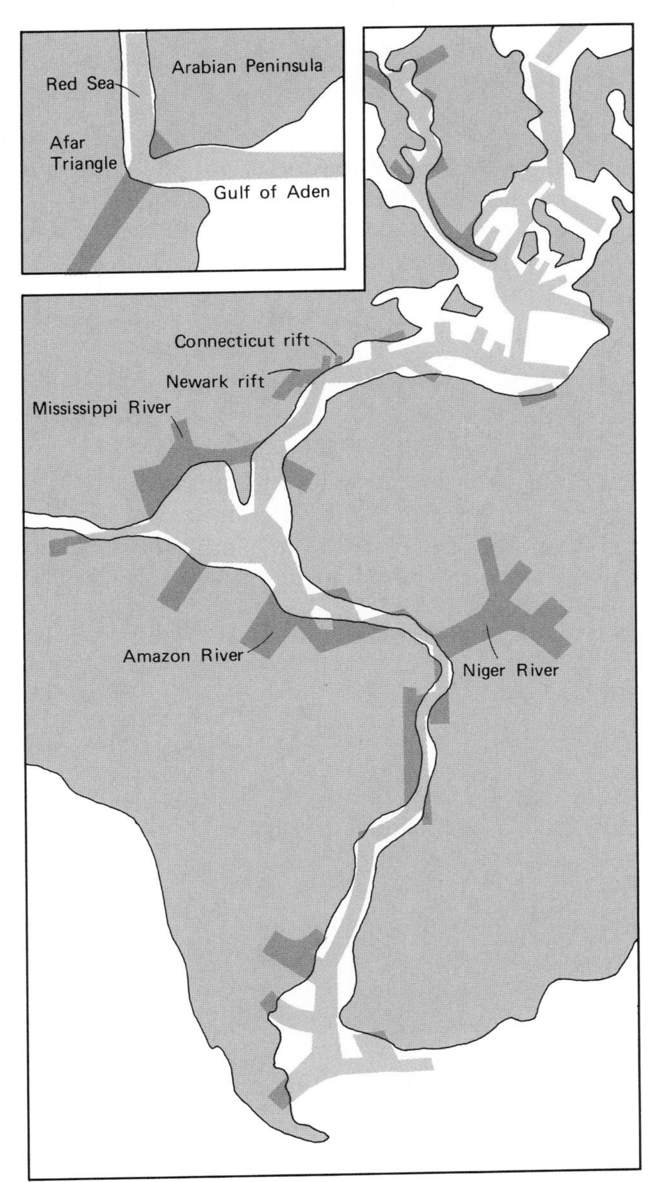

FIGURE 6-32 Ancient three-armed rifts that become apparent when we reassemble continents now bordering the Atlantic Ocean. Many of these rifts contributed two arms to the fracture zone that became the Mid-Atlantic Ridge. *(After K. C. Burke and J. T. Wilson, Scientific American, August 1976.)*

although the rift arm that projects into Africa has not yet divided the continent, it may do so in the future.

It is not surprising that many hot spots are situated on or very near midocean ridges (Figure 6-30). These may be the only surviving hot spots of a larger number that formed three-armed rifts that subsequently became active spreading ridges. The southern portion of the Mid-Atlantic Ridge has shifted to a position slightly to the west of three surviving hot spots that appear to have played a role in its origin.

CLUES TO ANCIENT PLATE MOVEMENTS

A great deal has been learned about plate motions and interactions during Cenozoic time through the study of magnetic striping, the directions and relative rates of plate movement, and hot spots. As we look far back in geologic time, however, these methods cease to be useful, because the evidence on which they are based has been obscured or destroyed by such processes as subduction, metamorphism, and erosion. We must therefore rely on other geologic clues to tell us how plates moved and interacted hundreds of millions of years ago. Even for the Cenozoic Era, complex motions of plates in certain regions (the Mediterranean, for example) can be deciphered only by the study of special clues. Let us examine some of these clues.

Signs of Ancient Rifting

As we have seen, a midocean ridge is often associated with block faulting that is expressed by a graben running along the ridge's midline (Figures 6-4 and 6-20). When a spreading zone first passes beneath continental crust, however, it seldom produces faults that cut cleanly through. Instead, the extension tends to break the thick continental crust into a complex band of fault blocks. Today this process is under way in Africa, where a system of rift valleys passes southward from the Red Sea (Figure 6-31). Each valley is a long, narrow, downfaulted block of crust associated with mafic volcanoes that have welled up from the mantle. (Rifting on the land often produces mafic dikes and flood basalts.) Some of the rift valleys cradle great lakes such as Lake Tanganyika. These rift valleys have been in existence only since early in Miocene time (less than 20 million years).

Before the sea enters a continental rift valley, nonmarine clastic sediments often accumulate rapidly to great thicknesses. The reason for this is that rapid subsidence of a downfaulted basin creates a rugged landscape in which subsequent erosion is rapid as well. The sediments typically include conglomerates derived from the steep valley walls and also **red beds**—sediments whose reddish color is usually attributable to iron oxide cement—and alluvial-plain deposits. Lakes that form within the elongate valleys also leave a sedimentary record. In arid climates, temporary lakes leave accumulations of nonmarine evaporites. If rifting continues long enough, a rift valley becomes so wide and so extended that it opens up to the sea. Because inflow from the sea tends at first to be restricted or sporadic, saline waters within the rift valley evaporate more rapidly than they are renewed. Under such circumstances, marine evaporites form. Waters from the Indian Ocean, for example, only recently gained full access to the currently widening Red Sea. Beneath a thin veneer of marine sediment in the Red Sea, geologists have found evaporites that formed during an earlier time, when there was only a weak connection to the larger ocean, or perhaps even earlier, when arid nonmarine basins were present.

The margins of the Red Sea also exhibit geologic features that typify the early stages of continental rifting. The Afar Triangle (Figure 6-32) was once part of the Red Sea floor. Most of the rocks in this area are basalts similar to those that form oce-

FIGURE 6-33 Tilted fault blocks in the Afar Triangle. The white areas are crusts of evaporite deposits. *(H. Tazieff, Scientific American, February 1970. Photograph by H. Tazieff.)*

anic crust. Much of the topography is the product of block faulting and uplift produced by the high rate of heat flow from the mantle, and in places there are great thicknesses of evaporite deposits (Figure 6-33).

In summary, then, regions of continental rifting are characterized by normal faults, mafic dikes and sills, and thick sedimentary sequences within fault block basins that often include lake deposits, coarse terrestrial deposits, and evaporites followed by oceanic sediments. An ancient example of this suite of features can be found in fault basins of the eastern United States. Two of the basins are labeled in Figure 6-32 as the Newark and Connecticut rifts. These are actually failed rift arms—rifts that never became part of the composite rift that formed the Atlantic Ocean. As we will see in Chapter 13, these rifts passed far inland and never opened wide enough to allow the sea to invade, although they contain other sedimentary sequences that typify incipient continental rifts.

If spreading continues, however, two new continental margins eventually develop and move far from the spreading zone where they formed. Soon they are likely to be flooded by shallow seas, because they move laterally away from the ridge axis, down the slope of the asthenosphere's surface, to regions where heat flow is lower and the asthenosphere is not so swollen (see Figure 6-16). Thus the continental borders, which were tectonically active when they were still close to the spreading zone, become what are termed **passive margins.** Having descended below sea level, these tectonically inactive areas of continental crust accumulate sediment along shallow shelves. Thus, as Chapters 14, 15, and 16 will describe, the Atlantic margin of the United States, after rifting away from Africa early in the Mesozoic Era, soon began to accumulate great thicknesses of sediment and has continued to do so to the present day (Figure 6-34). This passive margin has continuously subsided under the weight of the sediment, making way for more to be laid down.

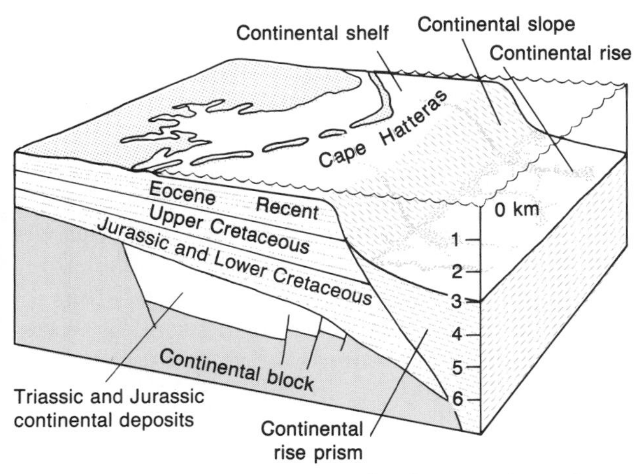

FIGURE 6-34 Accumulation of sediments along the passive margin of the eastern United States since the Jurassic Period. During Triassic and Jurassic time, continental deposits accumulated in fault-bounded basins as the early stages of rifting separated North America and Africa. When the continents separated enough so that seawater could enter to form the Atlantic Ocean, marine deposition commenced along the new passive margins that bordered eastern North America and western Africa.

Evidence of Ancient Subduction Zones

Subduction zones also leave telltale signs in the rock records of continents. One clue is provided by igneous rocks. Some of the magma rising from a subducted plate reaches the surface of the crust and erupts volcanically, while some of it cools within the crust as **intrusions.** Another specific clue is found in the belt between the site of igneous activity and the deep-sea trench, where plate convergence typically creates a zone of intensely deformed rocks (Figure 6-35). Most of the rocks in this deformed belt are deep-ocean

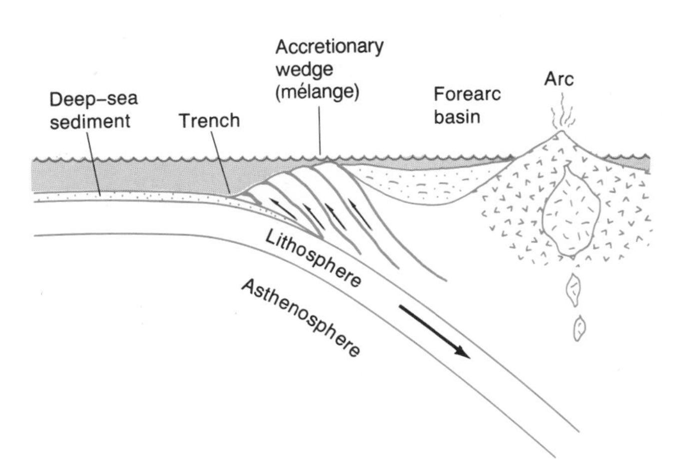

FIGURE 6-35 Cross section of a subduction zone showing the zone of intense deformation between the trench and the volcanic arc. Here sediments are deformed into mélanges and are metamorphosed.

sediments such as dark muds and graywackes, with bits of ocean crust mixed in. Some have been scraped from the descending plate. Others previously descended with the plate to depths as great as 30 kilometers (~20 miles) and rose again, apparently because of their low density. Rocks of subduction zones are characteristically metamorphosed at low temperatures (because the depth at which they are deformed is not great) and at high pressures (because the plates converge with great force). This chaotic, deformed mixture of rocks is called a **mélange** (the French word for mixture). The mélange shown in Figure 6-36 formed when a subduction zone passed beneath the western margin of North America. Rocks of a mélange are also folded and faulted. They are huge slices of material that have broken from the downgoing lithospheric plate. Slice after slice piles up against older material of the mélange, and the slices pile up along thrust faults. The entire body of rocks found in this way constitutes an **accretionary wedge.** Between the accretionary wedge and the igneous arc lies a **forearc basin,** where turbidites and other sediments accumulate in moderately deep water.

FIGURE 6-36 A mélange of the Franciscan sequence in California. Large blocks of exotic material are visible in the dark, metamorphosed deep-sea sediment. This mélange formed during the Mesozoic Era, when deep-sea sediments were pushed against the margin of the continent along a subduction zone. *(J. Schlocker, U.S. Geological Survey.)*

The border of a continent along which subduction occurs, producing igneous activity and deformation, is known as an **active margin.** Sometimes two continental masses converge along a subduction zone. Because they are felsic in composition, neither can descend into the dense mantle. As we will see in Chapter 7, this is one cause of mountain building; over such convergence leads to the formation of enormous faults and extensive folding. Some slices of seafloor usually end up on the surface of the combined continent that is formed by the plate convergence. These segments of seafloor, known as **ophiolites,** consist of pillow lavas that formed oceanic crust and ultramafic rocks that formed the upper mantle. Often present as well are deep-sea sediments such as turbidites, black shales, and **cherts**—rocks that consist primarily of very small quartz crystals precipitated from watery solutions—which originally rested on the seafloor above the lavas and ultramafic rocks. In effect, ophiolites are samples of ancient ocean basins that have been conveniently elevated for our study. Because ophiolites often mark the positions of vanished oceans that once lay between continents, they are key features in our recognition of plate convergence along subduction zones.

Transform Faults in the Rock Record

We have considered the geologic characteristics of *di*verging plate margins and of *con*verging plate margins. Transform faults can be recognized most readily when they have juxtaposed dissimilar geologic **terranes**—that is, geologically distinctive regions of Earth's crust, each of which has behaved as a coherent crustal block. The offset along these faults is commonly very large—tens or hundreds of kilometers—and this kind of strike-slip displacement commonly brings together terranes that developed in different regions and under quite different circumstances.

CHAPTER SUMMARY

1 Large plates of lithosphere cover Earth. The movement of plates over the asthenosphere accounts for much deformation and breakage of rocks in the lithosphere.

2 Similarities of rocks and nonmarine fossils on opposite sides of ocean basins provide strong circumstantial evidence that continents have moved horizontally over Earth's surface. The idea that continents actually drift was strongly opposed for several decades — partly because it was assumed that continents would have to move over or through the oceanic crust that lies between them.

3 It is now recognized that oceanic crust moves right along with the continents. Circular, convective motion within the mantle contributes to a plates' movement.

4 Mantle material is extruded along midocean ridges, spreading laterally in both directions to form the oceanic crust. Oceanic crust is recycled back to the mantle by means of subduction along deep-sea trenches. Geologists can measure the rate of seafloor spreading by observing how rapidly the age of the seafloor increases away from a midocean ridge. The rates of spread are high enough so that all segments of the modern seafloor are of Mesozoic and Cenozoic age. All of the seafloor that formed during the Paleozoic has been consumed along deep-sea trenches, except for the small amount that was preserved because it was attached to continents along subduction zones.

5 Fracturing of continents often begins with the doming of continental crust in several places. Each dome then fractures to form a three-armed rift system. The joining of some of the rift arms then produces a fracture that cuts across the entire continent.

6 Block faulting and deposition of thick siliciclastic sequences and evaporites often mark the beginning of continental fracturing.

7 Continental material is of such low density that it cannot be consumed by the mantle. As a result, when two continents converge along a deep-sea trench, neither becomes subducted; rather, the two become sutured together.

8 When a continent encroaches on a deep-sea trench, a slice of seafloor is frequently pushed up onto the continental surface; when such a seafloor remnant, called an ophiolite, is found within a modern continent, it marks the position of an ancient ocean that disappeared when two continents were united.

EXERCISES

1 List as many pieces of evidence as you can in support of the idea that continents have moved over Earth's surface.

2 What is the geographic extent of the lithospheric plate on which you live?

3 What is apparent polar wander? Draw pictures of Earth showing how the movement of a continent can produce apparent wander of the north magnetic pole.

4 Why are most volcanoes that have been active in the last few million years positioned in or near the Pacific Ocean?

5 How can a hot spot indicate the absolute direction of movement of a plate?

6 What geologic processes occur along midocean ridges?

7 What are failed rifts, and how are they important to our understanding of the breaking apart of continents? (Hint: Refer to Figure 6-32.)

8 What geologic features enable us to recognize ancient continental rifting?

9 What features enable us to recognize ancient subduction zones?

ADDITIONAL READING

Bonati, E., "The Rifting of Continents," *Scientific American,* March 1987.

Gass, I. G., "Ophiolites," *Scientific American,* August 1982.

MacDonald, K. C., and P. J. Fox, "The Mid-Ocean Ridge," *Scientific American,* June 1990.

Nance, R. D., T. R. Worsley, and J. B. Moody, "The Supercontinent Cycle," *Scientific American,* July 1988.

Sullivan, W., *Continents in Motion,* American Institute of Physics, New York, 1991.

Vink, G. E., W. T. Morgan, and P. R. Vogt, "The Earth's Hot Spots," *Scientific American,* April 1985.

CHAPTER 7

Mountain Building

Until recently the process of mountain building was a mystery to geologists. Compounding the puzzle was the fact that, as we saw in Chapter 6, few geologists before the 1960s believed that continents move laterally over Earth's surface. What processes cause mountains to grow, and why do mountains occur in so many different positions on continents? Before the origin of plate-tectonic theory, contradictory ideas emerged. North American geologists, for example, had long been struck by the fact that one chain of mountains, the Cordilleran system, parallels the west coast of their continent and another, older chain, the Appalachian system, parallels the east coast. Before the 1960s, this symmetrical pattern led some North American geologists to conclude that mountains tend to form along the margins of relatively stable continental masses known as **cratons.** Most European geologists, however, pointed to the Ural Mountains, standing between Europe and Asia within the largest landmass on earth, as evidence that long mountain chains could indeed rise up in the center of a continent. Plate-tectonics theory explains both situations: It is true that continuous long mountain chains form only along continental margins; continents can unite, however, and when they do, a mountain chain forms where the margins are welded together within the newly formed large landmass.

The unification of two continents along a subduction zone is termed **suturing.** According to the theory of plate tectonics, subduction zones are the key to the origin of mountain chains on continents. As we will see, however, not all mountain-building events result from continental suturing; a mountain chain also forms when an oceanic plate descends beneath the margin of a solitary continent along a subduction zone. Furthermore, not all mountain chains are formed of continental crust. As we have seen, rifting and swelling of oceanic crust create the great midocean ridges, which are, in effect, submarine mountain chains (Figure 6-4).

Cerro Torre is a mountain peak in the Andes of Argentina. *(Art Wolfe.)*

PLATE TECTONICS AND OROGENESIS

The process of mountain building is known as **orogenesis,** and a particular orogenic episode is termed an **orogeny.** Orogenesis along continental margins is a complex process that is only partially understood. Part of the complexity arises from the fact that every orogenic episode is unique in certain ways. Some orogenies result from the suturing of two large continents, whereas others result when a single continent interacts with an oceanic plate along a subduction zone. In this chapter we will examine features shared by most orogenies, as well as some that are unique to certain episodes.

First, let us consider a mechanism of orogenesis that is especially easy to understand: the collision of two continents. Ultimately, two properties of continental crust—its great thickness and its low density in relation to the density of the asthenosphere—lead to mountain building by collision. Continents are simply too thick and too buoyant to be subducted (Figure 7-1). When a continent encounters a deep-sea trench, its resistance to subduction forces a reversal in the direction of subduction. As a result, the oceanic plate opposite the continent is forced to descend into the mantle. If a second continent is riding on the newly subducting plate, it will eventually collide with the first continent along the subduction zone, and the two continents will be welded together. The juncture formed in this way is called a **suture.**

Continental suturing creates mountain chains because subduction along a trench causes the margin of one continent to wedge beneath the margin of the other (Figure 7-1*D*). The forces of collision cause both continental margins to thicken, and a mountain chain is uplifted along the suture. Often some oceanic crust is pinched up along the suture, forming an ophiolite (p. 154).

It turns out that orogenesis can take place along the margin of a continent that is resting against a subduction zone even when that continent does not collide with another. The central Andes, for example, which fringe the east coast of South America, have risen to great heights without a collision, while an oceanic plate has been subducted beneath them. We will consider how this situation came about after we examine the geologic features of a typical mountain chain.

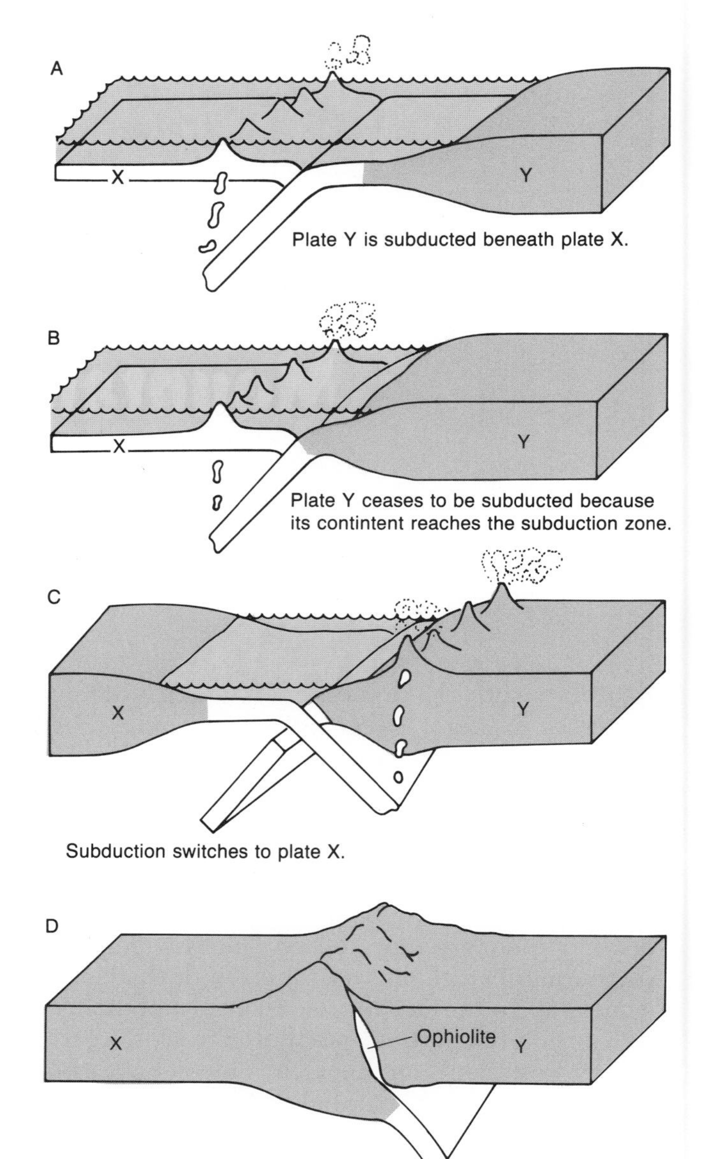

FIGURE 7-1 The processes of suturing and mountain building where two continents *(shown in color)* meet along a subduction zone. *(Peter Kresan.)*

THE ANATOMY OF A MOUNTAIN CHAIN

When a continent encounters a subduction zone, forcing a reversal in the direction of subduction, magma rises into the continental crust that now passes beneath it (Figure 7-1). Some of the magma reaches the surface and forms a chain of volcanoes that elevate the crust, often forming mountain peaks. Some magma also cools within the crust, forming **plutons,** or massive intrusions of igneous rock (Figure 7-2; also see Appendix I).

In keeping with the principle of isostasy (p. 14), the addition of large volumes of low-density igneous rock to the base of the crust causes the crust along the igneous arc to bob upward. This vertical movement contributes to the elevation of the mountain chain. At the same time, a root of crustal rock produced by the igneous arc extends

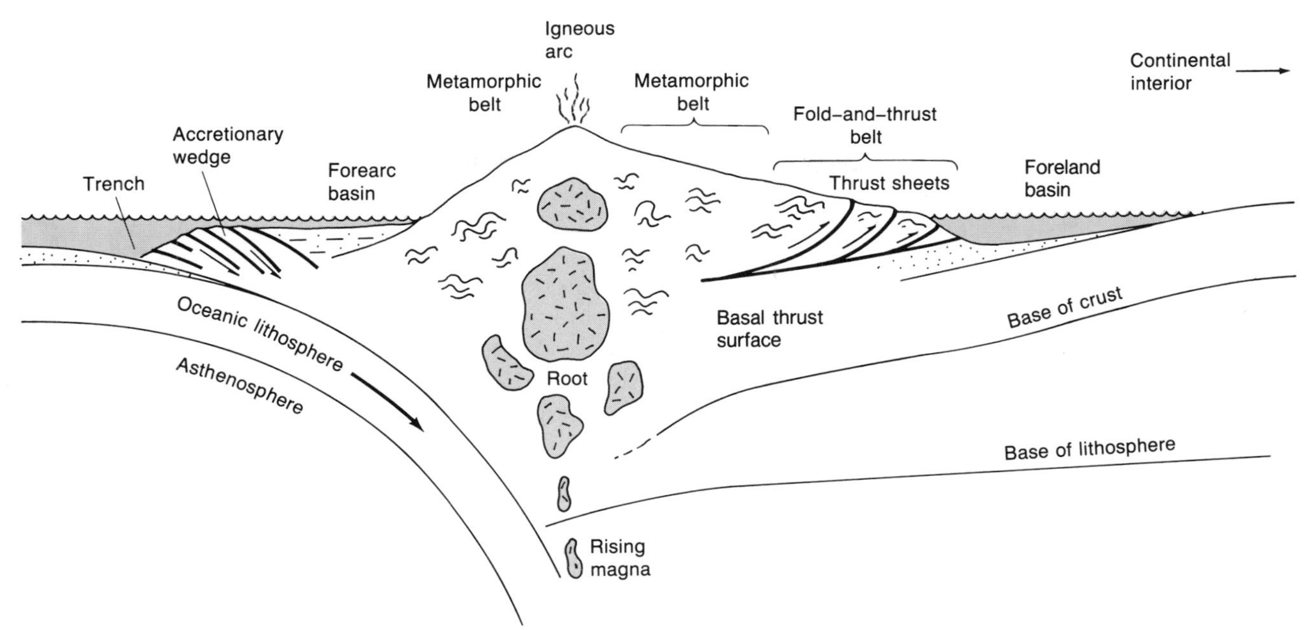

FIGURE 7-2 The configuration of an idealized mountain chain forming where an oceanic plate is being subducted beneath the edge of a continent. This cross section illustrates the general symmetry of the mountain chain. Metamorphism dies out both toward the sea and toward the land from the central igneous arc, and beyond the metamorphic belt, in the direction of the continental interior, is a fold-and-thrust belt. Beyond the inland fold-and-thrust belt, the crust is warped downward to form a foreland basin, where sediments from the mountain system accumulate.

downward beneath the mountain chain, balancing the weight of the mountains by displacing dense rocks in the asthenosphere.

An igneous arc like the one just described forms the core of a typical mountain chain. Extending along either side of this core is a belt of regional metamorphism (see Appendix I). This belt consists of crustal rocks that have been metamorphosed by heat from the core and locally intruded by igneous activity issuing from the core. Rocks of this **metamorphic belt** (Figure 7-2) are also deformed by processes that will be described shortly.

Metamorphism dies out in both directions from the igneous core. Toward the continental interior, the metamorphic belt gives way to a **fold-and-thrust belt** (Figure 7-2). The folds of the fold-and-thrust belt are typically overturned away from the core of the mountain, reflecting the fact that the prevailing forces of deformation come from the direction of the core. Because of their distance from the igneous core, the preexisting sedimentary rocks in the fold-and-thrust area are largely unaffected by metamorphism and are folded less severely than the rocks in the metamorphic belt. As a result, the behavior of these sedimentary rocks during the deformation process is more brittle and less plastic than that of the rocks in the metamorphic belt. Thus the folds are broken by enormous **thrust faults** along which large slices of crust, known as **thrust sheets,** have moved away from the core. The thrust sheets usually slide along a basal thrust surface. A look at Figure 7-2 will reveal how the telescoping action of folding and thrusting above a basal thrust shortens and thickens the crust.

Seaward of the igneous arc, along the subduction zone, rocks are also deformed by folds and thrust faults. This is the position of the accretionary wedge, which forms as material is scraped from the descending oceanic plate and piled along the continental margin (see also Figure 6-35).

MECHANISMS OF DEFORMATION

There has been much debate about the origins of the forces that cause folding and thrusting in mountain belts. One place where the mechanism is easy to understand is the accretionary wedge along the trench (Figure 7-2). Here the descending plate drags newly added slices of accreted material against older ones along thrust faults; the drag forces also fold material within individual thrust slices.

More controversial is the cause of deformation within the metamorphic belts and the fold-and-thrust belt. Two origins have been suggested for the forces that cause large-scale deformation within these belts. The first is simply pressure that the subducted plate applies to the mountain chain, pushing it laterally toward the interior of the continent. The resulting compression takes the form of folding near the igneous arc and folding and thrusting in the more brittle terrain farther toward the continental interior.

The second mechanism proposed as a source of mountain-building forces is less intuitively obvious. Its name, **gravity spreading,** is aptly descriptive, however. This mechanism depends on the fact that rock, although seemingly rigid, can deform under its own weight when that weight becomes great enough. It has been proposed that the necessary weight develops as the igneous arc builds the core of the mountain to a great elevation, while a root develops by isostatic adjustment. When the mountain chain becomes tall enough, it tends to spread laterally, like a mound of pudding heaped too high to remain stable. A mountain chain spreads by deformation along folds and thrust faults. This process can cause thrust sheets to move uphill along the basal thrust, just as pudding that is heaped too high can spread up the gently sloping sides of a dish.

Still controversial is the relative importance of the two kinds of forces that are generated along an actively deforming mountain belt. Some structural geologists view gravitational spreading as the dominant mechanism of deformation. Others believe that its contribution is relatively minor and that the forces applied by the descending plate are the dominant cause of folding and thrusting.

FORELAND BASIN DEPOSITION

The downwarping of the lithosphere beneath an actively forming mountain chain continues for some distance beyond the fold-and-thrust belt. This activity produces an elongate **foreland basin** whose long axis lies parallel to the mountain chain (Figure 7-2). The foreland basin forms rapidly and is usually so deep initially that the sea floods it either through a gap in the mountain chain or through a passage around one end of the chain.

The foreland basin typically deepens so quickly that the first sediments to accumulate within it are shales and turbidite deposits, collectively known as **flysch** (Figure 7-3*A*). These sediments are derived from the mountain chain. The shales represent the slow accumulation of mud in the foreland basin during intervals between episodes of turbidite deposition.

As the mountain system evolves, folding and thrust faulting move progressively farther inland, and mountain building proceeds toward the continental interior. In the process, flysch deposits, too, become folded and faulted. At the same time, the mountain core rises progressively higher, shedding sediments more and more rapidly.

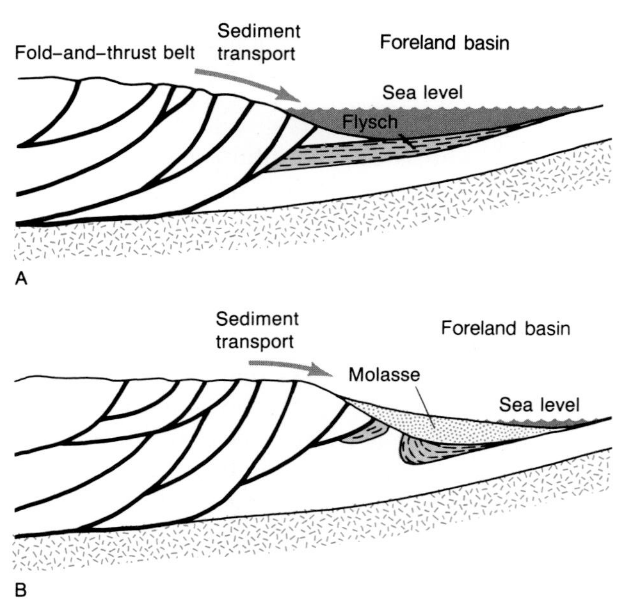

FIGURE 7-3 Evolution of a typical foreland basin. The foreland basin forms as a downwarping of the crust in front of a fold-and-thrust belt. At first the foreland basin receives deep-water flysch deposits *(A)*. As the rate of addition of sediment to the foreland basin becomes greater than the rate of downwarping, the foreland basin fills with sediment and the sea retreats. Deposition of flysch then gives way to deposition of shallow marine and nonmarine molasse *(B)*. Meanwhile, thrusting continues to migrate toward the continental interior, and the flysch deposits are deformed.

Eventually sediment chokes the foreland basin, pushing marine waters out and leaving only nonmarine depositional settings in their place. These areas comprise alluvial fans along the mountain chain and also rivers, floodplains, and related environments farther inland. Collectively, the sediments of all these settings are termed **molasse.** Molasse deposits can accumulate to great thicknesses; as deformation continues, they too can sometimes be folded and faulted. By the time molasse begins to form, the foreland may no longer be a topographic basin but may appear instead as a broad depositional surface sloping away from the mountain front (Figure 7-3*B*). As the foreland subsides beneath the accumulating sediments, however, it remains a *structural* basin (see Appendix II). Molasse deposits pinch out from the margin of a mountain belt toward the interior of the craton. Owing to its prismlike configuration, a thick package of molasse is sometimes referred to as a **clastic wedge.**

The depositional transition from the deep-water flysch to nonmarine molasse occurs during the evolution of most mountain systems. Even molasse deposition comes to an end when orogenic activity stops. Not only does igneous activity cease in the core of the mountain chain, but so do folding and thrusting along the margin. Soon erosion subdues the mountainous terrain and the source of the molasse sediment disappears.

The preceding discussions have dealt with ideal mountain systems. Actual mountain systems, however, frequently lack one or more of the features described here. The following sections describe the histories of some real mountain systems: the Andes, the Himalayas, and the Appalachians. The Rocky Mountains and other mountain ranges of western North America, which form the so-called Cordilleran system, have a history so long and complex that it will be treated in installments in later chapters.

THE ANDES: MOUNTAIN BUILDING WITHOUT CONTINENTAL COLLISION

The Andes Mountains of South America offer an appropriate starting point for our examination of actual mountain belts because they are still forming today: They exhibit orogenesis in progress. Furthermore, Figure 7-2 might be viewed as an idealized representation of the Andean belt as it exists today.

The Andes are associated with the ring of fire. This composite feature, which encircles much of the Pacific Ocean, is in most areas a product of the subduction zones where oceanic plates are colliding (Figure 6-24). In certain segments of the ring, however, subduction is occurring instead along blocks of continental crust. In Japan, for example, volcanic activity is centered on a slice of continental crust that has become separated from the Asian landmass, and in the eastern Pacific Ocean, subduction zones have encroached extensively on continental borders during much of Phanerozoic time. Until recently, subduction and associated igneous activity affected the entire west coast of North America. A subduction zone still operates all along the South American coast, where the Andes Mountains continue to form (Figures 7-4 and 7-5).

The Andean system is the longest continuous mountain chain in the world. Where the southern Andes meet the sea, a transform fault connects them structurally along the seafloor to the Scotia Arc, an island arc that circles back to meet a related active mountain chain in Antarctica. Similarly, where the northern Andes disappear into

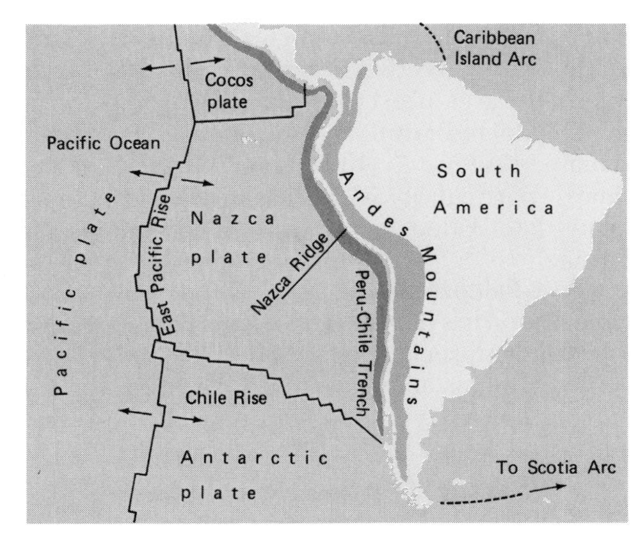

FIGURE 7-4 Map of the Andean mountain chain, showing its position adjacent to a subduction zone, the Peru-Chile Trench. *(After D. E. James, Scientific American, August 1973.)*

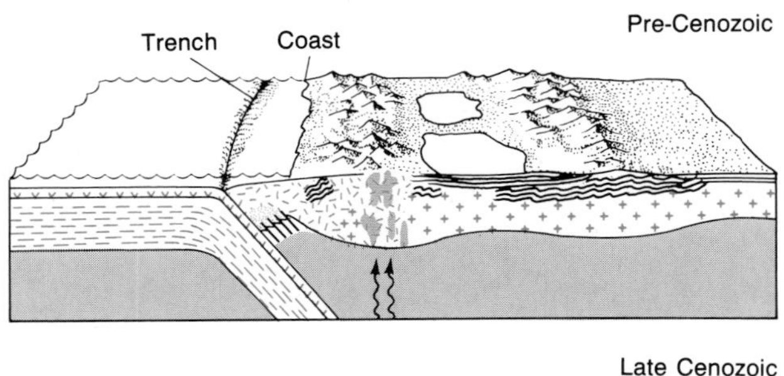

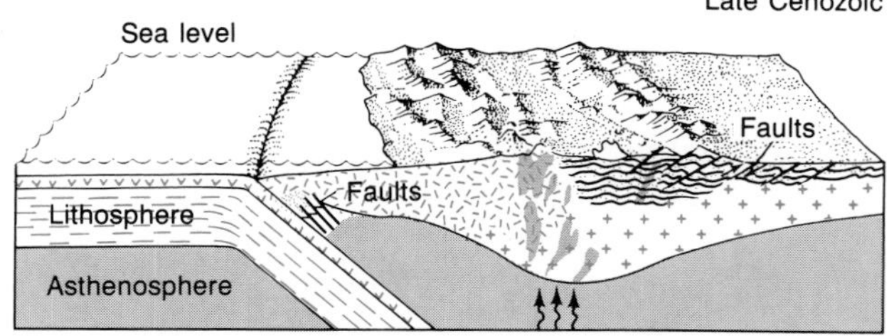

FIGURE 7-5 Formation of the Andes. During pre-Cenozoic time, igneous material was added to the crust from the oceanic plate descending along the marginal trench. During Cenozoic time, igneous activity shifted farther east. Thrust faulting has occurred both east and west of the area of igneous activity. *(After D. E. James, Scientific American, August 1973.)*

the ocean, the igneous arc responsible for their formation loops out into the Caribbean Sea.

The history of mountain building along the west coast of South America extends well back into the Paleozoic Era. During Paleozoic time, considerable thicknesses of marine sediment were laid down along an ancient continental margin, where the Andes now stand, extending the continental margin westward. Widespread unconformities in the area attest to mountain building during the Devonian Period, and the continental margin remained elevated throughout late Paleozoic time—particularly in the eastern Andes, where many late Paleozoic rocks are of nonmarine origin.

Post-Paleozoic mountain building in the Andean region has not caused the continental margin to extend appreciably seaward. Instead, it has involved igneous activity and uplift within the Paleozoic boundary of the continent. The present pattern of mountain building began early in the Mesozoic Era, when a subduction zone came to lie along the margin of South America (Figure 7-5). Enormous volumes of igneous rock have since risen from the subducted oceanic plate and have been added to the Andean crust, thickening it in some places to more than 70 kilometers (~45 miles). When Charles Darwin sailed around the world on the *Beagle,* he noted the presence of Cenozoic marine fossils at high altitudes in the Andes. These fossils offered proof that the Andes had been elevated greatly during recent geologic time. Darwin also saw at firsthand that the movements occurred in pulses: He witnessed earthquakes during which land along the seacoast was suddenly raised several feet. From a distance Darwin also observed the eruption of Andean volcanoes. We now know that for about the last 200 million years the Andean crust has been thickened by the addition of igneous material below and also has been bobbing up isostatically. At the same time, volcanic rocks have been piled on top. We also understand why most major earthquakes originate in major orogenic belts such as the Andes (see Box 7-1).

The details of Andean mountain building are complex and only partially understood, but certain general patterns are of interest. For example, igneous activity has steadily shifted toward the continental interior during Mesozoic and Cenozoic time; in other words, magma has ascended at positions farther and farther inland (Figure 7-5). Probably because today the subducted plate descends at a low angle, the zone where volcanoes are formed and magma cools below the surface to form intrusive rocks is now centered about 200 kilometers (~125 miles) inland from the coast. Earlier the subducted plate probably descended at a steeper angle, thus reaching the depth of partial melting nearer the coast.

THE LOFTY HIMALAYAS

The Himalayas, having formed primarily during the Neogene Period, are a relatively youthful mountain system. Partly because of their youth, the Himalayas are the tallest mountain range on Earth. The Himalayan front rises abruptly from the flat Ganges Plain; not far from the front, Mount Everest, the tallest mountain on Earth, towers to 8848 meters (~5.5 miles) above sea level. Even the broad Tibetan plateau, which lies to the north of the Himalayan front (Figure 7-6), stands at an average elevation of about 5 kilometers (~3 miles) above sea level—higher than any mountain peak in the 48 contiguous United States.

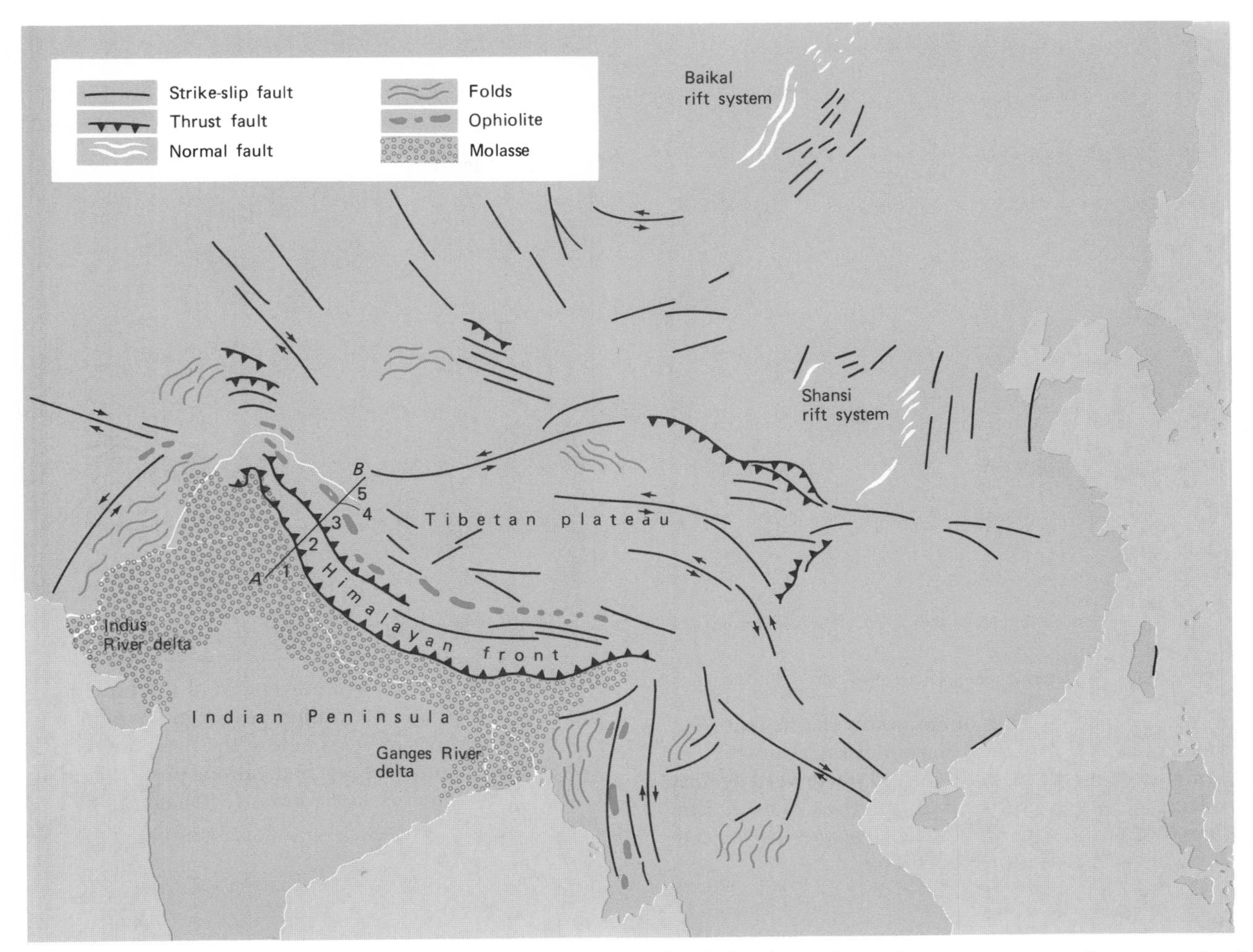

FIGURE 7-6 Geologic features of the Himalayan region. The high-standing Tibetan plateau is bounded by thrust faults, especially in the south, and molasse is being shed southward from the plateau. Numerous strike-slip faults throughout the region seem to have permitted the Asian crust to squeeze eastward, as the Indian peninsula has pushed northward. *(Modified from P. Molnar and P. Tapponier, Scientific American, April 1977.)*

Plate Movements

The Himalayas form part of a great series of mountain chains of Cenozoic origin that stretch from Spain and North Africa to Indochina (Figure 7-7). All of these chains formed as a consequence of the northward movement of fragments of Gondwanaland, the immense southern continent that broke apart during the Mesozoic Era (p. 133). The Alps and other Cenozoic mountains of the Mediterranean region formed as the African plate moved northward against the Eurasian plate.

Recall that the Indian peninsula, which projects southward from the Himalayas, was originally a fragment of Gondwanaland. By late in the Mesozoic Era, this fragment was moving northward as an island continent within the large Australian plate (Figure 7-8). The collision of this Indian craton with Eurasia resulted in the uplifting of the Himalayas.

When did the Indian craton arrive? During Eocene time, shallow seas covered much of the Indian craton, and limestones were laid down over large areas. Coarse sediment derived from mountainous terrain was first deposited on top of the limestones in late Miocene time. Apparently it was not until shortly before this time that mountain building began. Sediments in the Indian

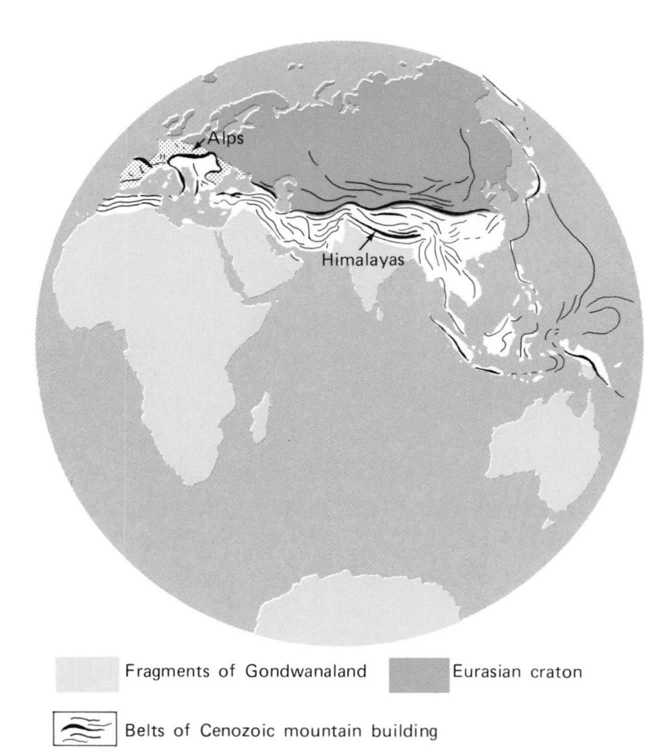

FIGURE 7-7 The series of mountain chains that formed along the southern margin of Eurasia when fragments of Gondwanaland moved northward against the large northern continent during the Cenozoic Era. *(After H. Cloos, Einführung in die Geologie, Verlag von Gebruder Borntraeger, Berlin, 1936.)*

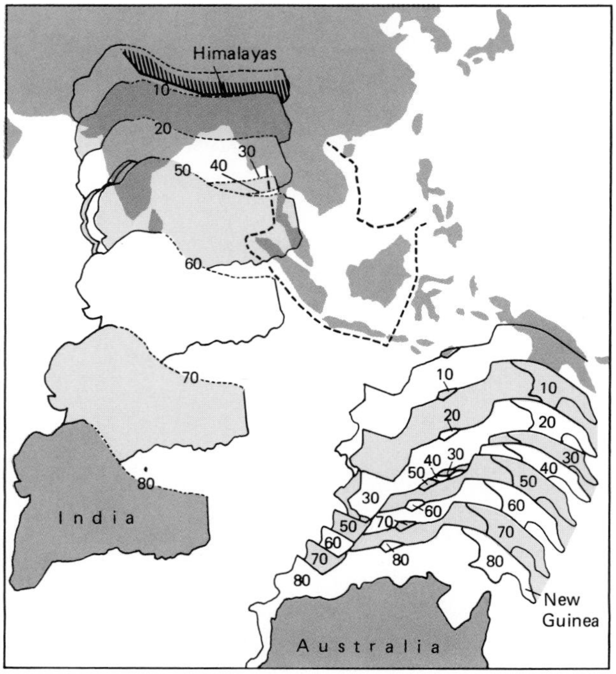

FIGURE 7-8 Northward movement of the Indian craton between 80 and 10 million years ago. Numbers represent the millions of years before the present when geographic boundaries reached various positions. *(After C. McA. Powell and B. D. Johnson, Tectonophysics 63:91–109, 1980.)*

Ocean provide additional evidence of the timing of orogenic activity. The oldest deep-sea turbidites deposited offshore from the Indus and Ganges rivers (Figure 7-6) date to the middle Miocene. The rivers themselves cannot be much older, and they came into being when the Himalayas began to form, perhaps 20 million years ago.

Indeed, much of the Himalayan chain has been uplifted during the last 5 million years.

Pattern of Orogenesis

Figure 7-9 shows in greater detail how the Himalayas formed. When India was approaching Eurasia, riding on the Australian plate, the northern margin of this plate was being subducted beneath Eurasia (Figure 7-6). When India arrived, being a continental mass, it could not be subducted. As a result, subduction ceased, and so did the associated igneous activity along the southern margin of Tibet. Convergence of the Australian and Eurasian plates continued, however, and about 20 million years ago India began to wedge beneath the southern margin of Tibet without descending into the asthenosphere (Figure 7-9*B*).

Sediments of the forearc basin that had bordered Tibet were squeezed up along the suture, along with material of the accretionary wedge and solid oceanic crust, to form an ophiolite. At some unknown time another dramatic event took place: The northern margin of India, consisting of sediments and underlying continental crust, broke away from the rest of the Indian continent. The remaining Indian continent then slid beneath the margin of the continent for at least 100 kilometers (about 60 miles) along a huge thrust fault, which is known as the Main Central Thrust. This fault can be seen today in many areas of the Himalayas,

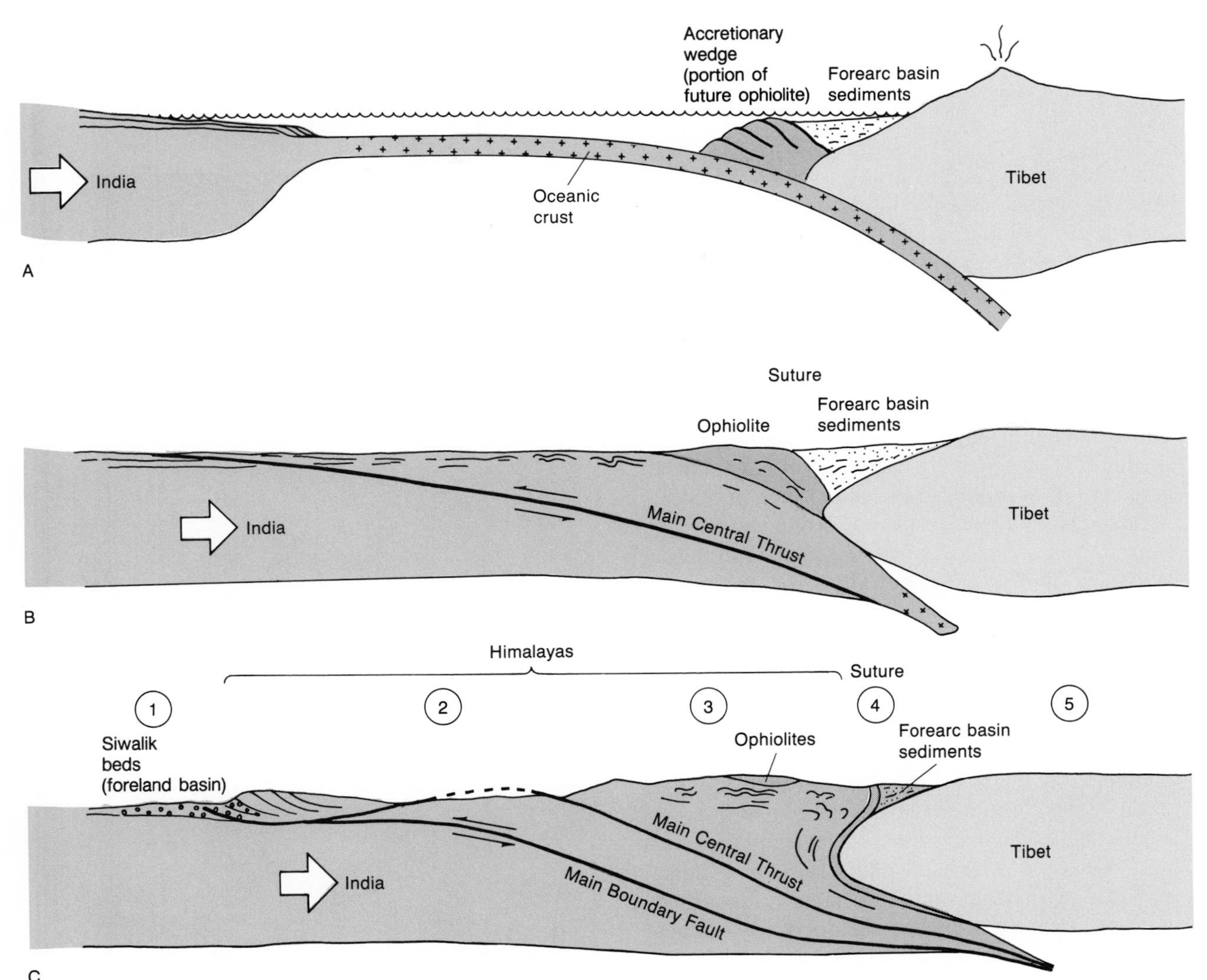

FIGURE 7-9 Cross section showing how the Himalayas may have formed when the Indian craton wedged beneath the margin of Eurasia. *A*. Slightly before 20 million years ago the Indian craton was rafted toward Eurasia as part of a plate being subducted beneath Tibet. *B*. About 20 million years ago the Indian craton began to wedge beneath Tibet and fractured along the Main Central Thrust. Movement along this fault thickened the crust, forming mountainous terrain. Compression in the suture zone uplifted and deformed the accretionary wedge that had bordered Tibet, producing ophiolites. *C*. Today movement has shifted to a new thrust fault, the Main Boundary Fault, and the crust has further thickened. Molasse is being shed southward, and older molasse deposits are being deformed in the vicinity of the Main Boundary Fault. (Numbers refer to zones identified in Figure 7-6 along line *AB*.) *(Adapted from P. Molnar, American Scientist 74:144–154, 1986.)*

where valleys have cut deep into the mountains. Movement along the Main Central Thrust ceased sometime before 10 million years ago and a new fault, the Main Boundary Fault, developed below it (Figure 7-9*C*). Movement along this fault has continued to the present day. The result of the movement along the two faults has been a great thickening of the Indian crust, as the margin of India has underthrust the slices of crust that have broken from it. Presumably, as the Indian subcontinent continues to press against Tibet, a new thrust fault will form sometime in the future so that yet another slice of crust will be added to the Himalayan region.

Box 7-1
Where Earth Shakes

The Richter scale is the universal yardstick by which geologists measure the magnitude of earthquakes. An earthquake that scores 3 on the scale will be felt indoors by some people but will do no damage to buildings; one that reaches 6 will topple chimneys and weak walls; and one of magnitude 8 will destroy virtually all human structures. Most major earthquakes are sudden movements of earth at the boundaries of lithospheric plates. Since 1906, all continental earthquakes that have exceeded 8 on the Richter scale have originated in mountain belts produced by the collision of plates. A subducted plate moves downward in sudden steps. When the massive plate takes an unusually large step, it produces a very large earthquake, and where the subduction is beneath a continent, the continent shakes violently. Similarly, collision of continents along a subduction zone causes episodic earthquakes until the subduction zone shifts to a new location.

Two regions have been the sites of all large continental earthquakes of recent decades: mountain belts that form segments of the ring of fire around the Pacific Ocean and the chain of mountain belts that stretches from southern Europe to China. Individual earthquakes have taken tens of thousands of lives in densely populated sectors of these zones, especially in China and in Central and South America. The 1976 earthquake centered beneath the city of Tangshan, China, stands as the most disastrous of modern times. It leveled the city and killed about 240,000 people. In contrast, the much larger 1964 Alaskan earthquake, which originated beneath a bay near Anchorage, caused only 131 deaths. Eleven of the victims lost their lives as far away as Crescent City, California, where a huge tidal wave crashed ashore. This wave had spread from the coast of Alaska, where a pulse of subduction had elevated the continental margin and disturbed shallow seas. Few lives were lost in Alaska because the population near the earthquake's center was sparse.

The famous San Francisco earthquake of 1906 occurred when the Pacific plate suddenly moved northwestward against the North American plate along the San Andreas fault (p. 128). Although this earthquake was of magnitude 8.3, it caused little damage outside California because its shock waves were damped by the soft sediments that form much of the terrane nearby. Nonetheless, the San Francisco earthquake and the fires it triggered, claimed about 700 lives. If an earthquake of similar magnitude struck the same area today, it would take an even heavier toll because the population of the region has exploded since 1906. The Loma Prieta earthquake, caused by movement along the San Andreas fault in 1989, was only of magnitude 7.0. It was, nonetheless, the most costly natural disaster in the history of the United States,

The Borah Peak earthquake in 1983 produced this fault scarp in the Rocky Mountains. *(Paul Karl Link, Idaho State University.)*

causing more than $10 million in damages in California as well as more than 2400 human injuries.

Only rarely does an earthquake that registers higher than 8 on the Richter scale originate in the middle of a continent. In 1911 and 1912, however, earthquakes of this magnitude shook a vast area of the United States. They were named the New Madrid earthquakes, after a town close to their place of origin in Missouri. Because these earthquakes shook the rigid craton, their shock waves traveled great distances, ringing church bells and cracking pavement as far away as the District of Columbia. Fortunately, so few towns had been built west of the Appalachians by 1811 that the New Madrid earthquakes caused no more than about ten human deaths. Even though the New Madrid earthquakes originated far from any plate boundary, they also resulted from plate movements. They were caused by movement along faults that formed long ago, in Proterozoic time, along a failed rift that sliced into the southern flank of North America. The Mississippi River follows the axis of the ancient rift, where Earth's crust remains thin and weak even though the faults lie buried beneath Phanerozoic sediments. Every year many tiny earthquakes vibrate the land near New Madrid without attracting the average resident's notice. An earthquake that rivals those of 1911 and 1912 would damage human constructions across a large region of the United States. Such an event may not occur for many decades or centuries, but long-range predictions are impossible. Episodic movements along the ancient faults, some small and some large, apparently result from stresses produced by North America's movement over the asthenosphere.

Only early warnings, safety instruction, and construction of resilient buildings can protect populations against earthquakes. Less than two years before the catastrophic 1976 earthquake in China, earth movements damaged hundreds of buildings in the Chinese city of Haicheng. Weeks before the crisis began, however, sensitive instruments had detected subtle but ominous geologic changes, including a slight tilting of the land. The instruments' warning led to evacuation before Earth began to shake, and casualties were relatively light.

A fold-and-thrust belt has formed above the Main Central Thrust and Main Boundary faults where these faults approach the surface at the southern margin of the Himalayas (Figure 7-9*C*). In the foreland basin to the south of this belt, a huge body of sediment, the Siwalik beds, has formed from material that has eroded from the mountains. The Siwaliks, which are famous for having yielded large numbers of fossil mammals of Neogene age, constitute molasse that has accumulated in a foreland basin that formed where the crust has been depressed by the adjacent mountain chain. This foreland basin has never been deep enough to admit the ocean, however, so it has received only nonmarine sediments. As the fold-and-thrust belt has continued to advance, it has deformed some of the Siwalik strata.

It has been suggested that as the Indian craton pushes northward into the Eurasian landmass, it is squeezing the larger landmass toward the east. According to this notion, Asia can be compared to a tube of toothpaste being squeezed by a thumb (India) toward the tube's open end (the Pacific Ocean). Slices of crust move past one another along strike-slip faults. Earthquakes still rumble through the Himalayan region as a result of movement along faults, and there is every reason to believe that mountain building here is far from over.

THE APPALACHIANS: AN ANCIENT MOUNTAIN SYSTEM

The Appalachians are a much older system than either of the two youthful, rugged mountain belts we have examined so far. Their internal structures were formed mainly during the Paleozoic Era, more than 200 million years ago. The heavily eroded surface of the Appalachians gives them a look of antiquity. In fact, even the modest topographic relief that the mountain chain displays is largely of Cenozoic origin; by late Mesozoic time, the original Appalachians had been worn down. The hills and mountains that we now call the Appalachians were formed by subsequent rejuvenation of the region — a process consisting of gentle uplift followed by slow erosion of resistant bodies of rock and more rapid erosion of weak ones. Much of the heavily eroded Appalachian system now lies beneath thick Mesozoic and Cenozoic deposits of the Atlantic Coastal Plain and the continental shelf. This buried portion of the Appalachians can be studied only through the use of complex and costly techniques such as seismic analysis and drilling.

In our examination of the Appalachian mountain system, we will review the system's geologic provinces, the three episodes of mountain building documented in the sedimentary record, and the plate movements that caused the intervals of mountain building.

Provinces of the Appalachians

Eastern North America is divisible into numerous discrete geologic provinces (Figures 7-10 and 7-11). Along the east coast, sediments of the Coastal Plain overlap the heavily eroded Appalachian mountain belt, which includes more than one province. In the Chesapeake Bay region, these sediments overlap the Piedmont Province. The Piedmont rocks typify the central and southern Appalachians in that they are metamorphosed, intensely deformed, and intruded by scattered igneous plutons. A less deformed, relatively unmetamorphosed zone, the Valley and Ridge Province, lies closer to the continental interior than does the Piedmont. Still farther inland are the slightly uplifted and only mildly deformed Appalachian plateaus.

The gradient from intense deformation in the Piedmont to weak deformation in the plateaus resembles the gradients in the Andes (Figure 7-5) on the inland side of the axis of igneous activity; thus it would appear that the exposed Appalachians represent the inland flank of a mountain system. We know little about the structure of the central axis and the other flank, however, because for the most part they lie buried beneath sediments of the Coastal Plain and the submerged continental shelf. Before delving into the history of the Appalachians, let us take a closer look at the present structure of the Valley and Ridge and Piedmont provinces as well as the narrow Blue Ridge Province, which lies between them.

The Valley and Ridge Province is so named because of its many structural folds and faults; the Piedmont is much less rugged terrain (Figure 7-11). The Valley and Ridge Province, which has been described as the most elegant folded mountain belt on Earth, is what many nongeologists think of as the Appalachian Mountains. In fact, this

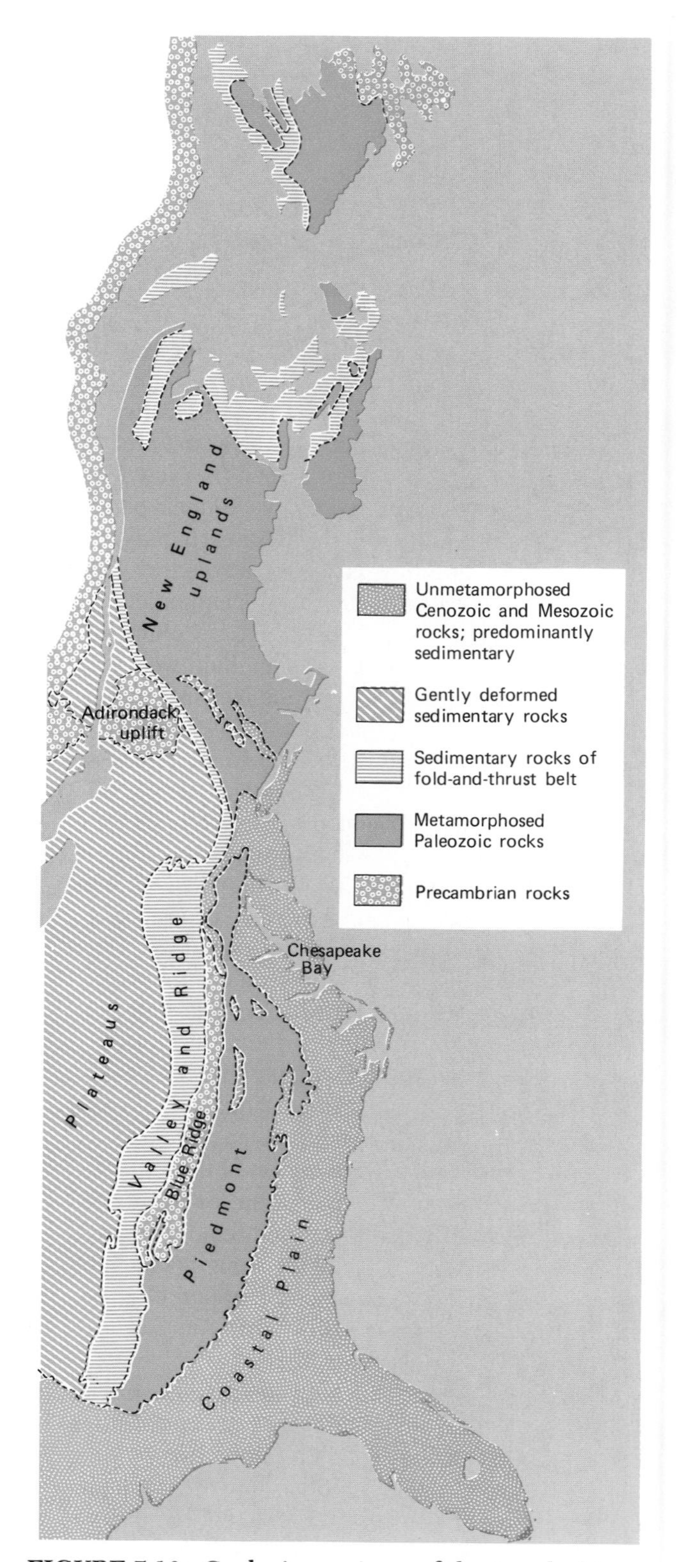

FIGURE 7-10 Geologic provinces of the Appalachian region. A portion of the original Appalachian mountain system has been leveled by erosion and buried beneath the Coastal Plain. *(Modified from G. W. Colton, in G. W. Fisher et al. [eds.], Studies of Appalachian Geology: Central and Southern, John Wiley & Sons, Inc., New York, 1970.)*

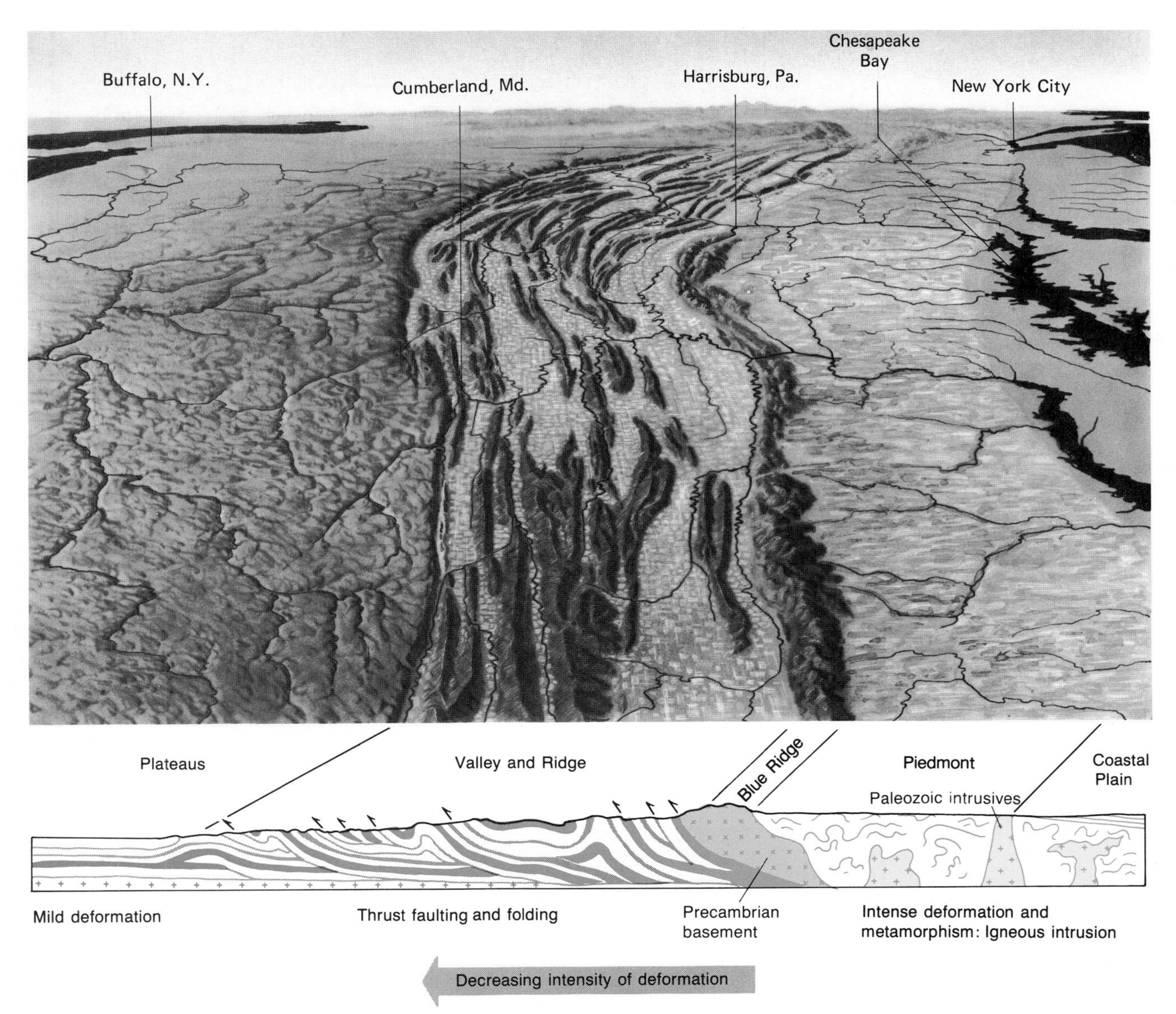

FIGURE 7-11 Aerial view and idealized cross section of the Appalachian region. The aerial view is toward the northeast. Sediments of the Coastal Plain lap up on the worn-down eastern portion of the Appalachian system. To the west of the Coastal Plain, the low-lying Piedmont Province is separated from the Valley and Ridge Province by the conspicuous Blue Ridge Province. *(Modified from J. S. Shelton, Geology Illustrated, W. H. Freeman and Company, New York, 1966.)*

province constitutes the western fold-and-thrust belt of the Appalachian system. Sedimentary rocks have been folded and transported toward the continental interior as immense thrust sheets. In the southern Appalachians, many thrust faults are exposed at the surface. Strata above the basal thrust in many places have been telescoped to half their original horizontal span or less. In the north, fewer thrust faults reach the present erosion surface, and folding is a more conspicuous feature of the landscape. Many of the thrust faults descend to a basal thrust that slopes uphill in the direction of thrusting. It is clear from this evidence that the lowest Appalachian thrust sheets moved uphill toward the interior of the continent.

The Piedmont Province includes the zone of intense deformation to the southeast of the fold-and-thrust belt. As we will see, many segments of Piedmont terrane were formed from blocks of crustal rock that were welded to the North American craton during the orogenic events that produced the Appalachians. Because of their proximity to the eastern margin of the craton during orogenesis, these blocks suffered intrusion from igneous arc activity and also experienced metamorphism and intense folding. Rocks of the Piedmont behaved in a less brittle fashion than those of the nearby fold-and-thrust belt; their plasticity under high temperatures and pressures contorted them into complex patterns.

The Piedmont is separated from the Valley and Ridge by the narrow Blue Ridge Province. Geographers have applied the name Blue Ridge to a conspicuous chain of hills in Virginia (Figure 7-12); however, the Blue Ridge Province also includes the Great Smoky Mountains farther south and the Green Mountains of New England. This province consists primarily of a band of crystalline rocks of the Precambrian basement that moved upward along thrust faults during the formation of the Appalachian system.

Three Orogenic Episodes

The Appalachian system was produced not by a single orogenic episode but by three successive episodes, all of which occurred during the Paleozoic Era.

The stratigraphic record in the Valley and Ridge Province reflects a history of sediment deposition that reveals much about the sequence of mountain building in the Appalachians. The most obvious feature of the block of sedimentary rocks within the Valley and Ridge Province is its extreme thickness along the southeastern margin of the province; the block thins toward the continental interior (Figure 7-13). Within this block of sediments are two large packages, one resting on top of the other. Each of these packages represents a **tectonic cycle** of deposition consisting of three units. The first of these units represents stable, preorogenic conditions, and the second and third represent orogenic conditions.

More specifically, the basal unit of the tectonic cycle contains shallow-water sedimentary deposits consisting primarily of sandstones and carbonates laid down on a shallow platform. This sequence formed during a period of tectonic qui-

FIGURE 7-12 The Blue Ridge Mountains of Virginia. *(Peter Kresan.)*

escence in the region, when there was little mountainous terrain to supply siliciclastic sediments. The sequence was deposited when the eastern margin of North America was a passive margin.

The second unit consists of deep-water flysch deposits that record the existence of a foreland basin where the carbonate bank once stood. This dramatic change in depositional setting reflects the onset of an episode of mountain building to the southeast, because it is in this direction that the flysch deposits become thicker and more coarsely grained.

The third unit of the tectonic cycle of deposition contains shallow marine and nonmarine siliciclastics, including some red beds. These molasse deposits began to form when the influx of sediments from mountain building increased, pushing back the waters of the foreland basin (Figure 7-3).

The First Tectonic Cycle of Deposition in the Appalachians

Let us look in greater detail at the events recorded by the first tectonic cycle in the central Appalachians. This cycle began during latest Precambrian time with the deposition of siliciclastic sediments derived from the continental interior along the stable margin of the North American craton. During the Early Cambrian Epoch, seas transgressed progressively farther over the craton, and by the end of the period, the siliciclastics along the eastern continental margin had given way to carbonate deposits (Figure 7-13), producing an immense carbonate platform that stretched from Newfoundland to Alabama (Figure 7-14). We know that in many places this platform formed a steep edge on the continental shelf because deep-water deposits to the southeast contain blocks that tumbled down from the shelf (Figure 7-15). Stromatolites, mud cracks, and other indications of intertidal or shallow subtidal conditions indicate that the platform itself maintained a position near sea level; in this respect, it resembles the modern Bahama banks, which are, however, much smaller structures (Figure 3-29). As the Cambrian Period progressed, the seas transgressed farther and farther over the continental interior; still, the carbonate platform formed the shelf edge throughout Cambrian and part of Ordovician time.

During mid-Ordovician time, the depositional framework along the east coast changed drastically. Carbonate deposition ceased, and flysch deposits such as the Martinsburg Formation and other deep-water units were laid down. Deposition of black shale predominated at the outset of this activity; later, turbidites became prevalent (Figure 7-13). That all these deposits derived from an eastern source is indicated by sole markings in the Martinsburg turbidites (Figure 3-33).

The orientation of the foreland basin and the arrival of siliciclastic sediments from the southeast indicate that by this time mountain building had begun to the southeast. Radiometric dating of igneous rocks in the Piedmont Province confirms the existence of a Middle and Late Ordovician episode of mountain building now known as the **Taconic orogeny.**

Eventually sediment was being supplied by the eastern area of uplift faster than the mountains were subsiding. Near the end of the Ordovician Period, flysch gave way to molasse in the form of shallow marine and nonmarine clastics, some of which were red beds. Thus, in the central Appalachians, the Juniata and Tuscarora formations consist of coarse shallow marine and nonmarine deposits in a clastic wedge or molasse sequence that tapers out toward the northwest. Cross-bedding and other features reveal that this was indeed the primary direction of transport. This clastic wedge represents the final interval of the first tectonic cycle of deposition in the Appalachians. During the Silurian Period, coarse sediments ceased to arrive from the east, signaling that the Taconic orogeny had ended and that the Taconic Mountains had been effectively leveled by erosion.

Plate Movements Responsible for the Taconic Orogeny

Since the advent of plate tectonics, there has been much interest in reconstructing the movements of lithospheric plates that were responsible for Appalachian mountain building.

One difficulty in reconstructing plate movements is that a key form of evidence, paleomagnetism, gives information only about the latitudinal position and orientation of ancient continents; it tells us nothing about continental movements in an east-west direction (p. 137). This limitation is

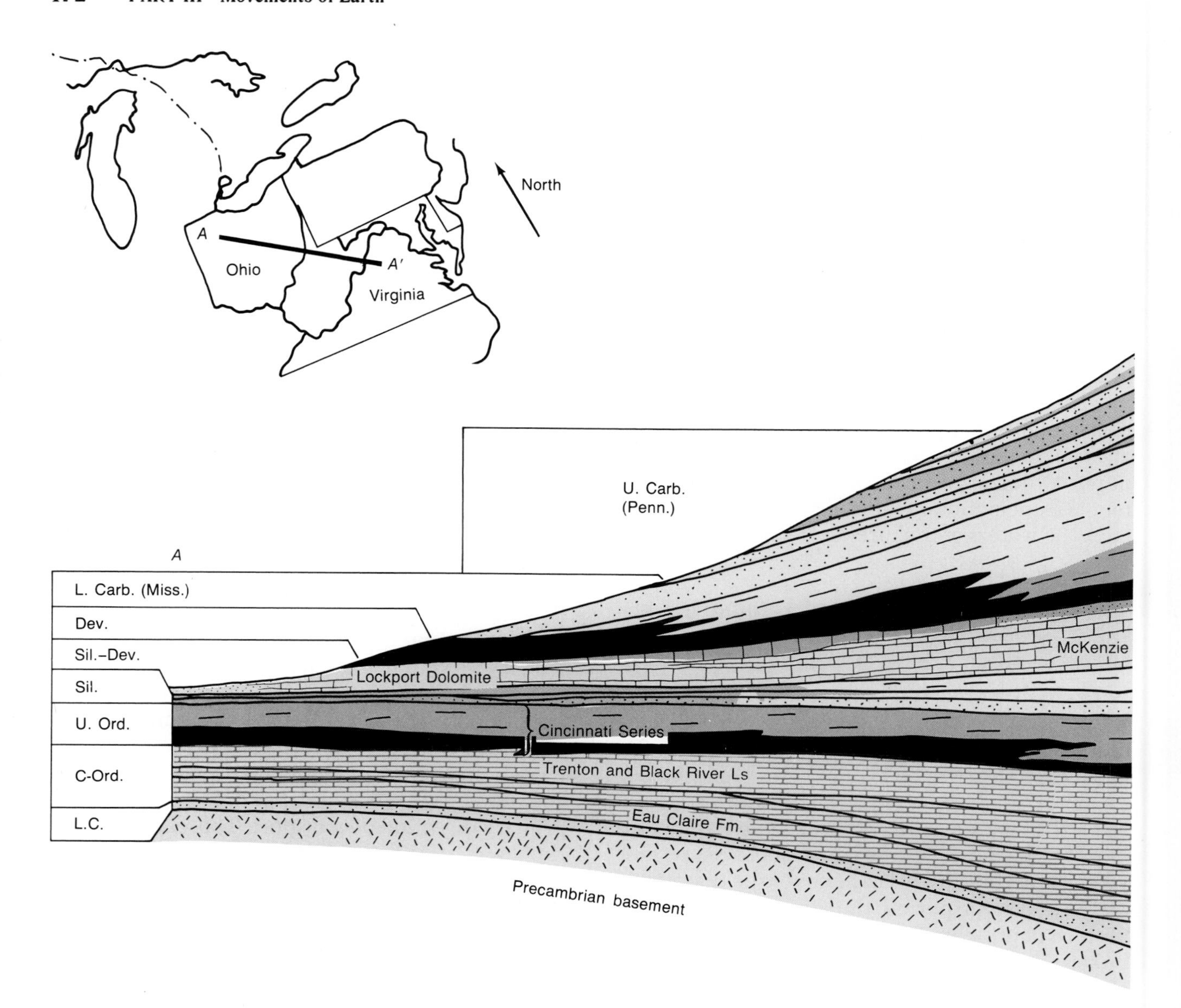

FIGURE 7-13 Stratigraphic cross section through the central Appalachians west of the Blue Ridge Province, with vertical exaggeration. Folds and faults are not shown. The thickest deposits lie to the southeast, in the Valley and Ridge Province. The thinnest deposits lie to the northwest, in the plateau region west of the Appalachians. This package of Paleozoic strata represents two tectonic cycles of deposition. Each cycle contains a sedimentary record of mountain building to the east. In the origin of each cycle, flysch and then molasse were deposited in a foreland basin that formed where a shallow shelf previously existed. The orogenies that produced the cycles are shown on the right. *(Modified from G. W. Colton, in G. W. Fisher et al. [eds.], Studies of Appalachian Geology: Central and Southern, John Wiley & Sons, Inc., New York, 1970.)*

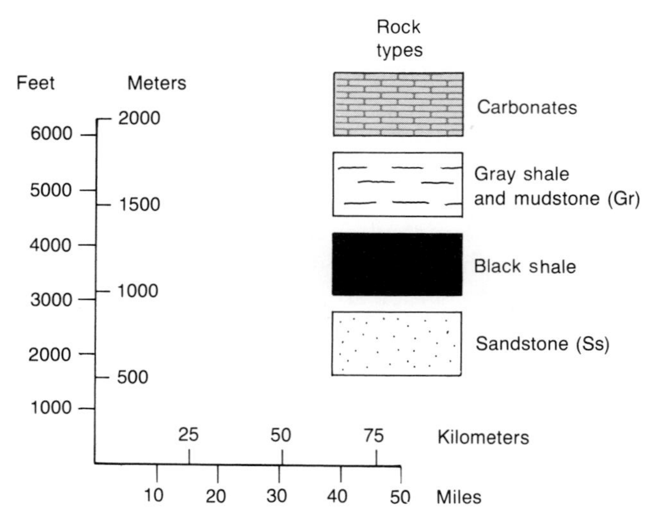

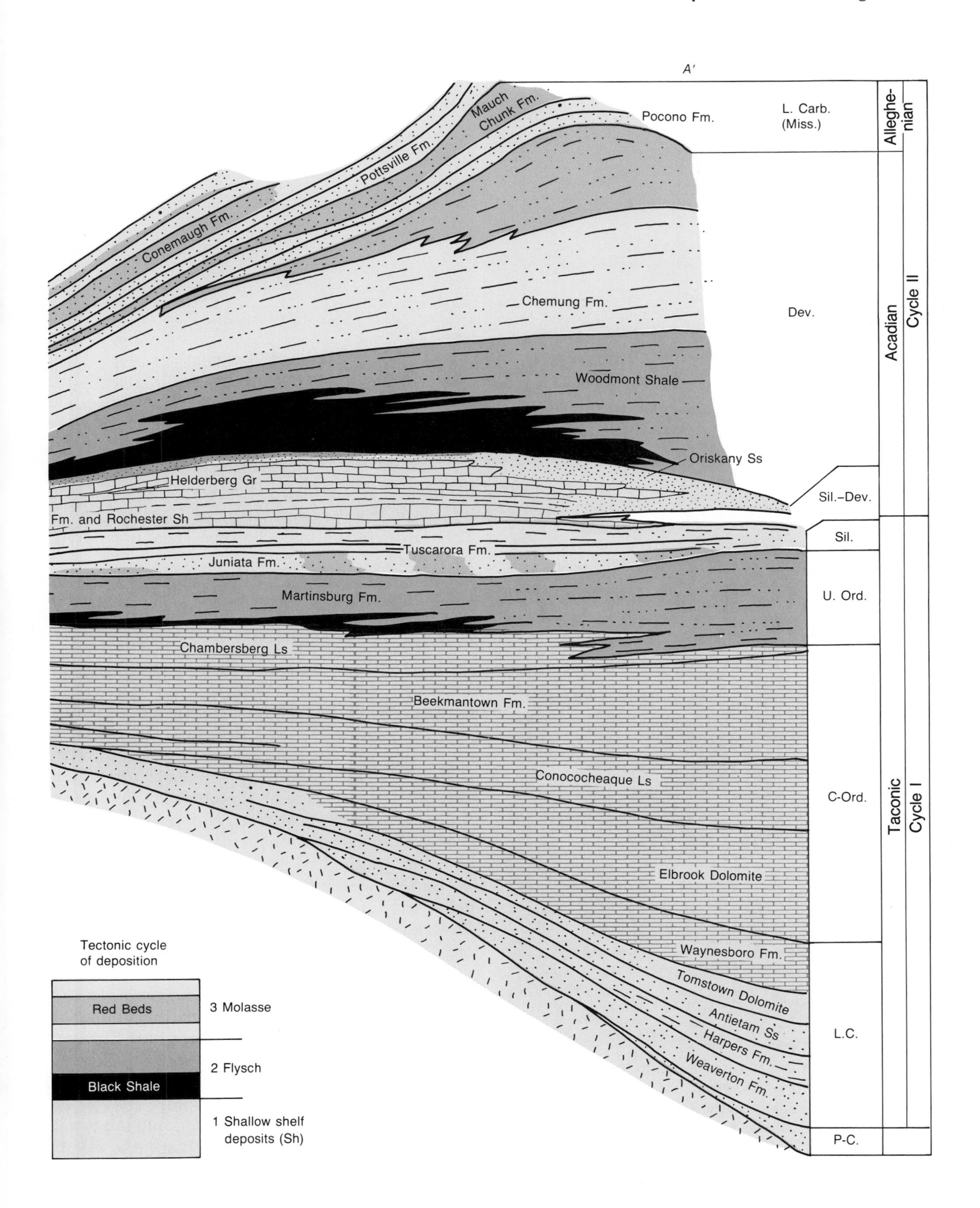
A′
Pocono Fm.
Mauch Chunk Fm.
Pottsville Fm.
Conemaugh Fm.
Chemung Fm.
Woodmont Shale
Oriskany Ss
Helderberg Gr
Fm. and Rochester Sh
Tuscarora Fm.
Juniata Fm.
Martinsburg Fm.
Chambersberg Ls
Beekmantown Fm.
Conococheaque Ls
Elbrook Dolomite
Waynesboro Fm.
Tomstown Dolomite
Antietam Ss
Harpers Fm.
Weaverton Fm.
L. Carb. (Miss.)
Dev.
Sil.-Dev.
Sil.
U. Ord.
C-Ord.
L.C.
P-C.
Allegheinian
Acadian
Taconic
Cycle II
Cycle I
Tectonic cycle of deposition
Red Beds
3 Molasse
2 Flysch
Black Shale
1 Shallow shelf deposits (Sh)

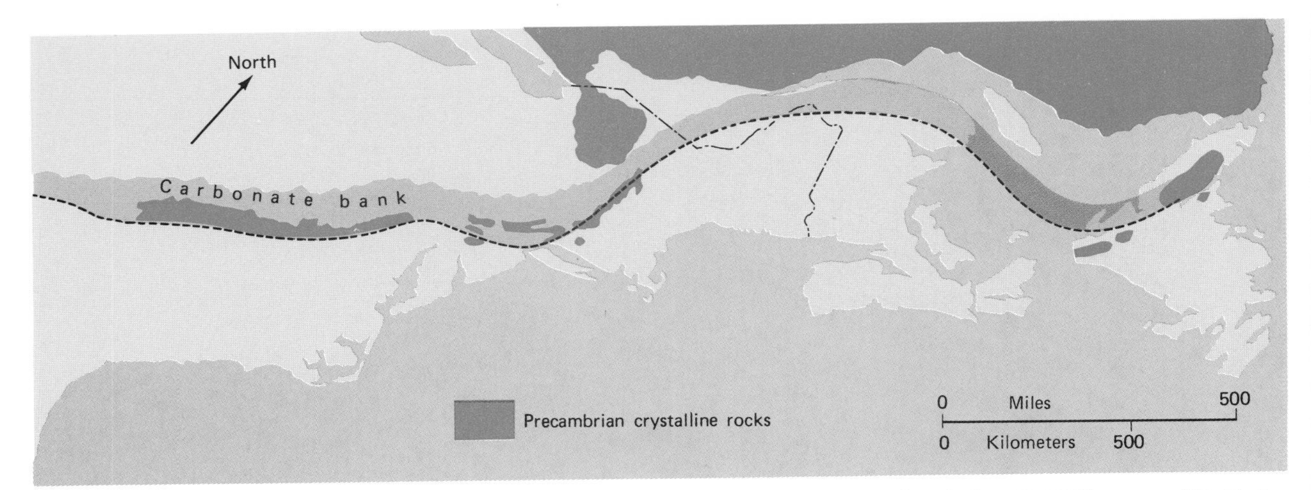

FIGURE 7-14 Shallow-water Cambro-Ordovician deposits representing an ancient carbonate bank along the eastern margin of North America. The dashed line shows the easternmost outcrops of those deposits. In some places erosion has removed the carbonates, exposing Precambrian rocks. *(After H. Williams and R. K. Stevens, in C. A. Burk and C. L. Drake [eds.], The Geology of Continental Margins, Springer Publishing Company, Inc., New York, 1974.)*

FIGURE 7-15 A bed of breccia within the Frederick Limestone of central Maryland. The clasts of the breccia are pieces of shallow-water limestone that slid over the edge of the Cambrian carbonate platform into deep water. A hammer rests on the outcrop. *(R. Demicco.)*

evident when we consider plate positions during Late Ordovician time, when the Taconic orogeny took place.

Paleomagnetic data do indicate that early in the Paleozoic Era, North America was separated from Europe and Africa by what has become known as the Iapetus Ocean. North America was at that time part of a large continent known as Laurentia, which lay to the northwest of northern Europe, which in turn formed a separate continent that geologists call Baltica (Figure 7-16). Greenland was also part of Laurentia and was attached to North America, while England, which was not yet attached to Scotland, lay far to the south. The border of eastern North America consisted of the long carbonate platform on which the sediments of the first Appalachian cycle were deposited. As we have seen, that carbonate platform collapsed during the Ordovician Period and a foreland basin was formed. What plate movements led to this change?

Considerable narrowing of the Iapetus Ocean between Laurentia and Baltica is suggested by the composition of shallow-water marine faunas of Ordovician age in North America and Europe. While Lower Ordovician faunas of the two continents are quite different, Upper Ordovician faunas are similar. Apparently the Iapetus Ocean was narrow enough by late Ordovician time to permit many species to cross from one side to the other.

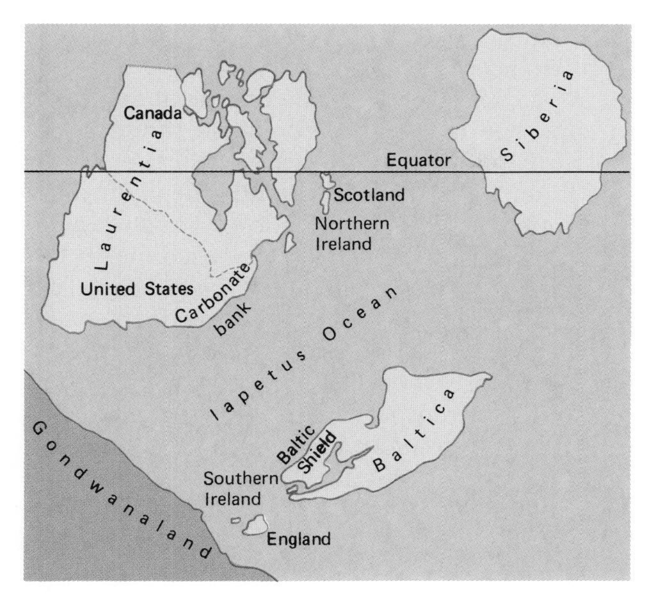

FIGURE 7-16 Positions of landmasses and the Iapetus Ocean in Middle Ordovician time. *(After C. R. Scotese et al., Jour. Geol. 87:217–268, 1979.)*

Paleomagnetic data also reveal that during Ordovician time, Baltica was moving northward toward North America. For this to happen, oceanic lithosphere between the two continents had to be disappearing along a subduction zone. The question is: Did the two continents eventually collide and become sutured together? Paleomagnetic data are not precise enough to settle the question. The answer is provided, however, by rocks that became attached to the margin of North America at this time. These rocks are now situated in New England and adjacent areas of Canada, where they constitute **exotic terranes.** An exotic terrane is a block of lithosphere that has been sutured to a much larger continent. Figure 7-17

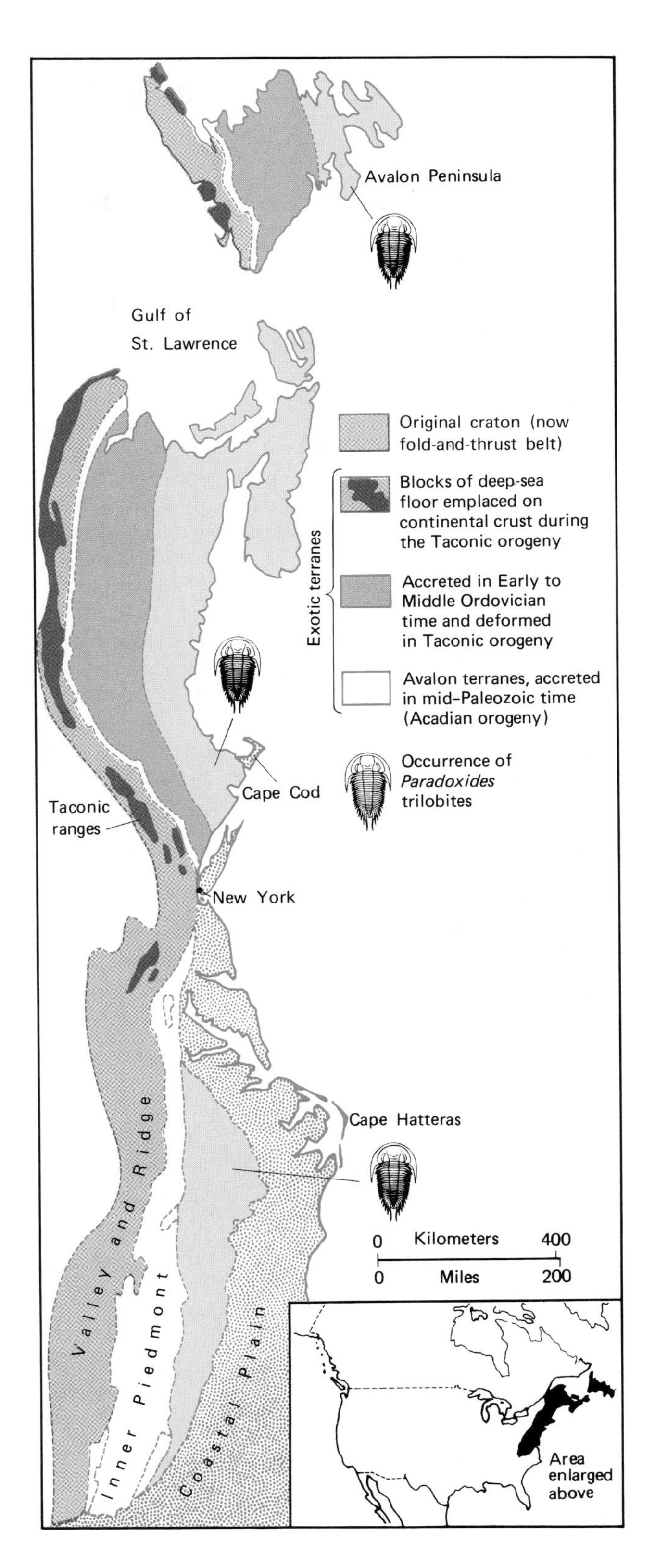

FIGURE 7-17 Exotic crustal blocks in the Appalachians. As would be expected, the older blocks (those accreted to the craton before the Late Ordovician Taconic orogeny) lie inland from the Avalon blocks, which were accreted during the mid-Paleozoic Acadian orogeny. The Cambrian trilobite *Paradoxides,* which is found in the Avalon exotic blocks, is a genus that is also found in Europe but not in rocks of the original North American craton. The Taconic ranges and other small exotic blocks that rest on rocks of the original craton were thrust from the east during the Taconic orogeny (Figure 7-19). *(Modified from H. Williams and R. D. Hatcher, Geology 10:530–536, 1982.)*

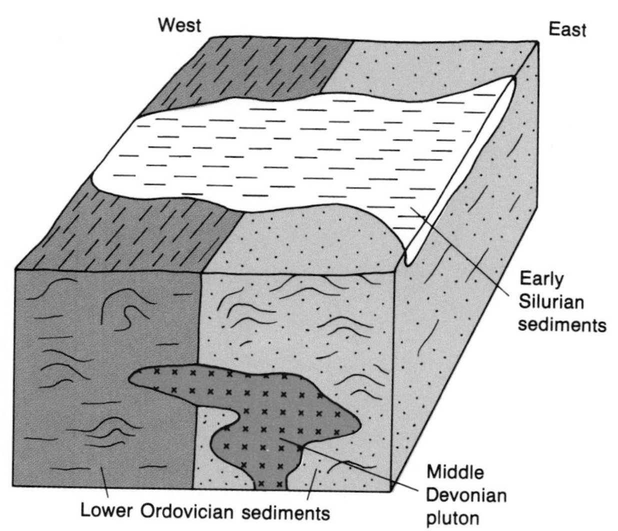

FIGURE 7-18 Recognition of exotic blocks of crust in the Appalachians. In this hypothetical illustration, two adjacent blocks can be distinguished from each other because they are characterized by different kinds of Lower Ordovician sediments that do not intergrade (shale is in the western block and sandstone is in the eastern block). The Early Silurian sediments that blanket the two blocks of crust and the Middle Devonian pluton that cuts through both of them are not deformed in the manner of the Lower Ordovician sediments. This pattern indicates that the two blocks were deformed and welded together during the Late Ordovician Taconic orogeny.

displays these terranes and others that were attached to North America during a subsequent orogenic interval. Figure 7-18 indicates how geologists identify an exotic terrane and estimate the time at which it was sutured to a host continent.

The exotic terrane that was accreted to northeastern North America during the Ordovician Period contains large volumes of rocks produced by the igneous activity of island arcs. These igneous rocks indicate that the Taconic orogeny was initiated during Middle Ordovician time by the collision of North America with an island arc, which constituted a microcontinent, not by collision with Baltica. During this collision, the island arc overrode the North American continental shelf, depressing the carbonate platform and forming a chain of mountains along its margin (Figure 7-19). The downward flexure of the continental crust produced a foreland basin to the west of the mountains, and sediment from the mountains began to accumulate here as flysch.

As the island arc and the continent continued to converge into Late Ordovician time, folding and thrusting moved progressively farther inland. The flysch that was shed from the newly forming mountains was caught up in the deformation along the suture zone along with deep-sea deposits that had originally accumulated on the continental rise of North America. Eventually these deep-sea deposits moved along thrust faults over depressed rocks of the carbonate platform (Figure 7-19). Today these transported deep-sea deposits form the Taconic mountain ranges of New York State and nearby areas (Figure 7-17).

Near the end of the Ordovician Period, the collision of the arc and the continent proceeded until sediment was shed very rapidly from the new mountain chain, choking the foreland basin and expelling the sea. Thus began the interval of molasse deposition depicted in Figure 7-13: The first tectonic cycle of deposition entered its final phase.

The Acadian and Alleghenian Orogenies

The second and third mountain-building episodes that contributed to the Appalachian system were the **Acadian orogeny** of the Devonian Period and the **Alleghenian orogeny** of the Late Carboniferous Epoch. These events, like the Taconic orogeny, left their mark in the form of rocks still visible in the Appalachian region. Also like the Taconic, the two later orogenic episodes have been explained in terms of major movements of lithospheric plates. Unlike the Taconic orogeny, however, the Acadian and Alleghenian orogenies were associated with major continental collisions—the suturing of North America first to northern Europe and later to Africa.

The Second Tectonic Cycle of Deposition in the Appalachians The Silurian Period was an interval of renewed quiescence along the Atlantic coast of North America. Orogenesis presumably ended when the margin of North America, wedged beneath the igneous arc, blocked further subduction so that both convergence and igneous activity ceased.

Well before the end of Silurian time, the mountains that had formed during the Taconic

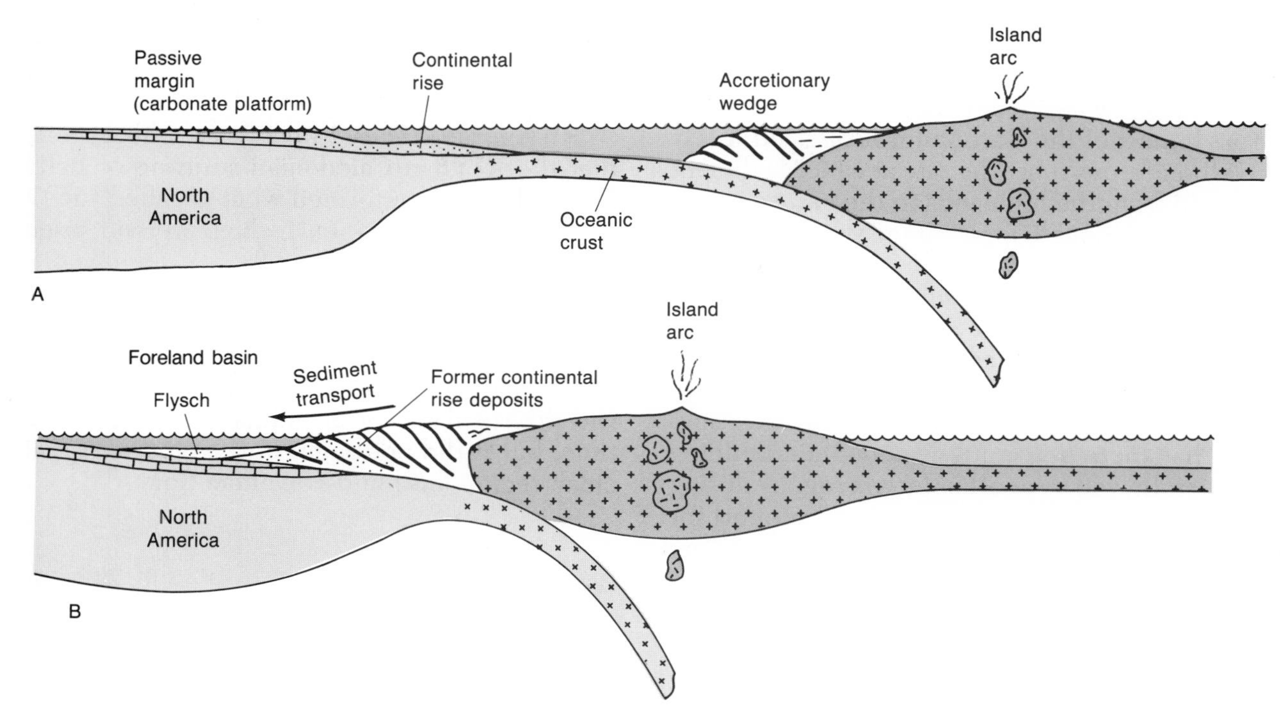

FIGURE 7-19 The plate movements that in mid-Ordovician time transformed the passive margin of northeastern North America *(A)* into a foreland basin *(B)*. When the passive margin, which supported a carbonate bank, encountered an island arc, the accretionary wedge bordering the island arc was thrust over the continental margin, as were deep-water deposits that had accumulated along the continental rise of North America, in front of the carbonate bank. The island arc thus joined North America as an exotic terrane.

orogeny were leveled by erosion. Seas then flooded the new, passive continental margin. During the remainder of the Silurian and part of the Devonian, shallow-water carbonates of the Helderberg Group accumulated over a broad shelf along this margin. These sediments are succeeded in places by a thin body of clean, well-sorted quartz sand. This unit, known as the Oriskany Sandstone (Figure 7-13), was deposited along the beach that fringed the Devonian sea.

The shallow-water deposits at the base of the second tectonic cycle of deposition, like those at the base of the first cycle, were succeeded by flysch deposits, including the Woodmont Shale of Figure 7-13, during Early Devonian time. This activity signaled the origin of a new foreland basin as well as the onset of the Acadian orogeny. As in the first cycle, flysch deposition was characterized by the initial accumulation of mud throughout most of the foreland basin and later by the predominance of turbidite deposition.

Near the beginning of Late Devonian time, as the Acadian orogeny intensified, clastic wedges built westward, pushing back the waters of the foreland basin. Thus began the molasse phase of the second tectonic cycle of deposition. The great thickness of the Late Devonian clastic wedge in the northern and central Appalachians (Figure 7-13) indicates that the Acadian orogeny elevated very large mountains to the east.

In fact, as we will see shortly, these mountains were uplifted by the collision of North America and northern Europe. During the Late Devonian phase of molasse deposition, sands accumulated in nonmarine environments to the east and spread to marine environments of the foreland basin farther west. In Late Devonian time, these sediments formed what is often called the Catskill delta, the thickest accumulation of which is in New York State. In fact, the Catskill was not a single delta but a complex of alluvial, deltaic, and submarine environments. Some of the meandering-river deposits are red beds resembling some of the molasse units of the first Appalachian tectonic cycle. In places meandering-river sequences are piled on top of one another to great thicknesses (Figure 3-15).

Remarkably, the molasse phase of the second tectonic cycle of deposition continued more than 50 million years, from Late Devonian time through most of the Carboniferous Period. Early in Carboniferous time, the rate at which sediment was shed from the highlands to the east seems to have been reduced for a time, suggesting that the Acadian highlands may have been subdued by erosion. The rate of influx increased again, however, leading to the accumulation of thick siliciclastic units such as the Mauch Chunk and Pottsville formations (Figure 7-13), which are meandering-river deposits. This evidence suggests that there was a lull in mountain-building activity followed by another large orogeny during the Carboniferous Period. Additional support for this inference is supplied by dating of igneous and metamorphic rocks in the Piedmont, which reveals a pulse of igneous activity centered in Late Carboniferous time and perhaps extending into the Permian Period. This pulse is called the Alleghenian orogeny.

Thus the thick molasse sequence of the second tectonic cycle represents sediment shed from mountains that formed during two successive orogenic episodes, the Acadian and the Alleghenian. The second followed so closely upon the first that, although the supply of siliciclastic sediment slowed in the interim, it never fully stopped. It seems evident, therefore, that some highlands remained in the region of orogenic activity and that the continued accumulation of molasse right up to the start of the Alleghenian orogeny prevented the foreland basin from becoming deep enough to permit flysch to be deposited when the Alleghenian orogeny began. As we will see in greater detail shortly, the second pulse of orogeny resulted from the collision of North America with Africa. Before considering this event, we must examine the cause of the Acadian orogeny.

Plate Movements and the Acadian Orogeny
Although the Taconic orogeny was not associated with a collision of large continents, there is clear evidence of such a collision during the middle Paleozoic Acadian orogeny. The Acadian event is known as the Caledonian orogeny in Europe, where it affected Scandinavia and Great Britain (Figure 7-20).

Many nonmarine deposits of the Old Red Sandstone Formation of northern Britain resemble sediments of the Upper Devonian Catskill clastic wedge of New York and Pennsylvania (Figure 7-13); likewise, similar fossil freshwater fishes of Devonian age are found in Britain and eastern North America. These similarities reflect the fact that the Acadian/Caledonian suturing of Baltica to North America formed what is called the Old Red Sandstone continent, which we will discuss in more detail in Chapter 11.

Dates of igneous activity, folding, and foreland basin deposition all reveal that Acadian/Caledonian orogenic activity began in the north and spread southward. This activity was under way in Greenland and Scandinavia by mid-Silurian time, but it did not begin in the central Appalachian region until mid-Devonian time.

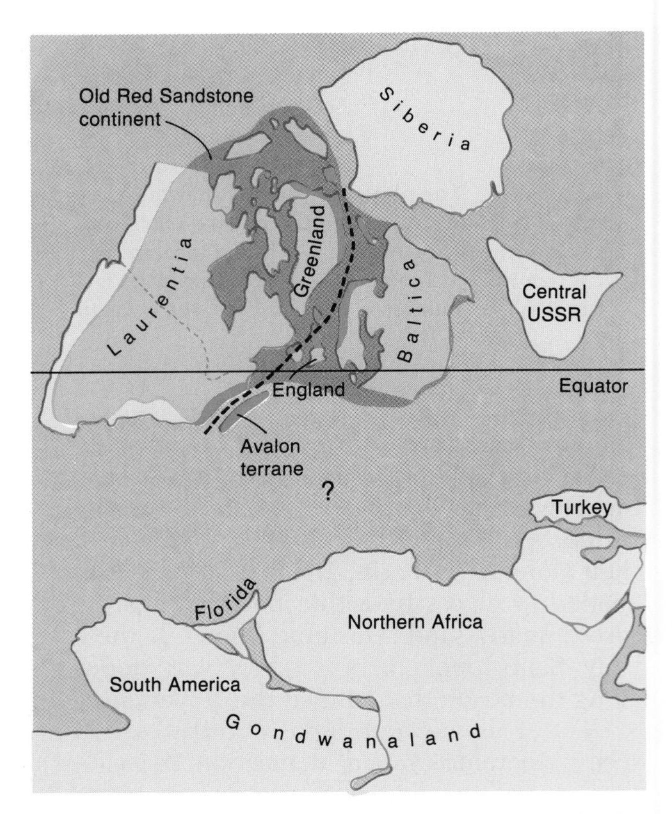

FIGURE 7-20 World geography during Acadian/Caledonian mountain building. The suture *(broken line)* marked the disappearance of the Iapetus Ocean by Early Devonian time. Newly formed mountains (the Appalachians and Caledonides) are shown in color. The question mark between Gondwanaland and the Old Red Sandstone continent indicates that there is uncertainty whether there was a broad sea here or whether Gondwanaland lay closer to North America, perhaps even taking part in the Caledonian orogeny. *(Modified from C. R. Scotese et al., Jour. Geol. 87:217–268, 1979.)*

As it turns out, there was more to the Acadian orogeny than the collision of North America with Baltica, which united Baltica with northern Canada and Greenland. To the south of this region there was also orogenesis, but of a different type. Here, after the Old Red Sandstone continent formed, North America collided with an island arc, probably along the same subduction zone that caused continental suturing to the north. Part of this island arc survives in the Appalachians today as exotic blocks known as the Avalon terrane.

In New England and southern Canada, the Avalon terrane was accreted to exotic terranes that had been attached to North America during the earlier Taconic orogeny (Figure 7-17). Blocks of the Avalon terrane range in age from Precambrian to Ordovician. They include mafic and ultramafic bodies of rock that have been interpreted to represent ophiolites, and they also comprise large volumes of extrusive igneous rocks of types produced by island arcs. Thus the history of the Avalon terrane appears to resemble that of the terrane added to North America during the Taconic orogeny: The Avalon terrane was an island arc that was sutured to eastern North America during the Acadian orogeny after Baltica and Laurentia collided. Perhaps this arc extended southward from the southern end of Baltica as this continent approached North America to form the Old Red Sandstone continent (Figure 7-20).

Long ago it was noted that Cambrian trilobite faunas of the Avalon and related terranes contain such genera as *Paradoxides* that are also found in European faunas (Figure 7-17). It seemed strange that typical American faunas should be found in western Newfoundland, just a short distance inland from these exotic faunas. Plate tectonics explains the puzzle with the revelation that the Avalon terrane became attached to North America during the Acadian orogeny. It brought with it fossil trilobites related to those of Baltica—the continent with which it traveled westward to North America.

Plate Movements and the Alleghenian Orogeny In Early Carboniferous time it lay farther south, separated from the northern continent by a tropical seaway. The closure of this seaway and the attachment of Gondwanaland to the Old Red Sandstone continent in Late Carboniferous time were associated with the Alleghenian orogeny in the Appalachian region. This was the primary suturing event in the creation of Pangaea, and it produced mountains not only in eastern North America but also along the Gulf Coast (the Ouachita Mountains) and in southern Europe and northwestern Africa. Later Europe became attached to Asia, and this suturing produced the elevation of the Ural Mountains (Figures 7-21 and 7-22).

The geologic structures produced by the Alleghenian orogeny are very much in evidence in the Appalachians today, because this was the final Appalachian orogeny, and its effects have not been masked by later episodes of mountain building.

In the southern Appalachians, the segment of the Piedmont lying inland of the Avalon-like block is of uncertain origin, but seismic studies suggest

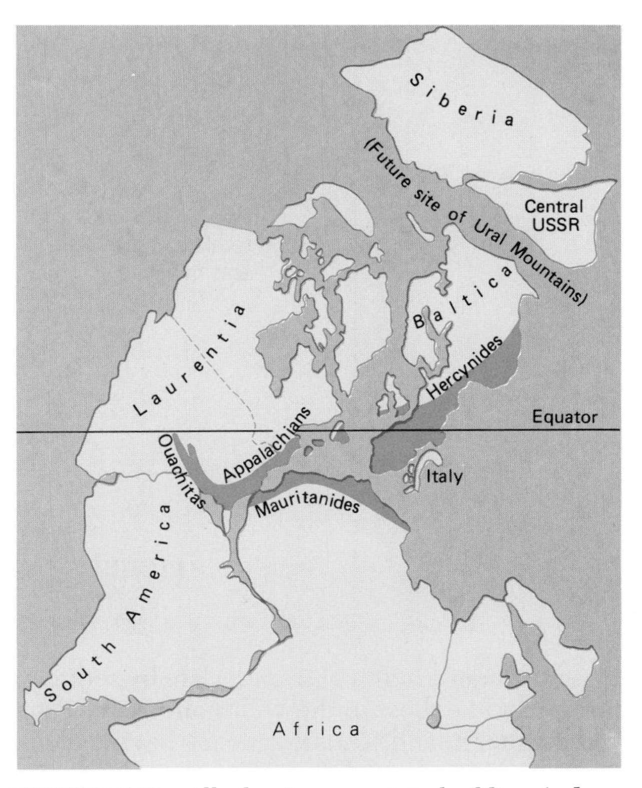

FIGURE 7-21 Alleghenian mountain building *(colored area)* during Late Carboniferous time. Northward movement of Gondwanaland late in Paleozoic time closed off the western end of the Tethys Sea. Mountain building occurred not only here, where Gondwanaland was sutured to North America, but also to the east, along the southern margins of Great Britain and Eurasia. *(Modified from C. R. Scotese et al., Jour. Geol. 87: 217–268, 1979.)*

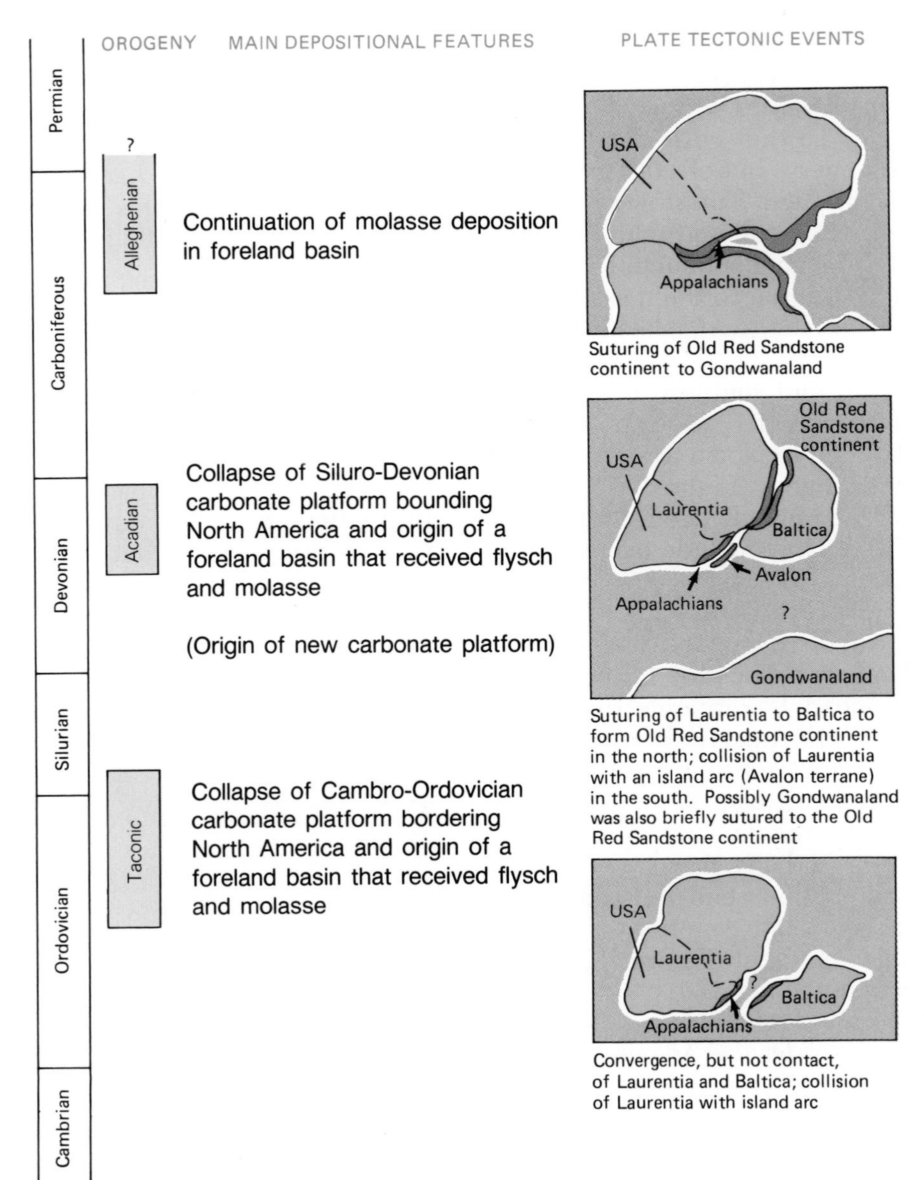

FIGURE 7-22 Summary of Appalachian orogenic events.

that this inner Piedmont is underlain by sedimentary rocks like those of the Valley and Ridge Province further inland. Thus it appears that the highly metamorphosed and deformed rocks of the southern inner Piedmont have been thrust hundreds of kilometers inland over rocks of the fold-and-thrust belt. This movement seems to have occurred during the Alleghenian orogeny, when igneous plutons were also emplaced in the Piedmont. At the same time, farther inland, rocks of the Valley and Ridge Province, which range in age from latest Precambrian to Carboniferous, underwent the folding and thrusting that give them their present configuration. These features of the Appalachian mountain belt indicate that the collision of the Old Red Sandstone continent with Gondwanaland had profound effects.

As we will see in later chapters, the Appalachians were largely leveled by erosion during the Mesozoic Era, when an episode of rifting separated North America from Europe and Africa, creating the Atlantic Ocean. The moderate elevation of the Valley and Ridge Province that we see today has resulted from secondary uplift, probably caused by isostatic adjustment during the Cenozoic Era. It was also during the Mesozoic Era that Pangaea broke apart to form new oceans, including the Atlantic. The Atlantic Ocean basin formed by the origin of a great rift system along which the basin is still expanding (Figure 6-4). When rifting was initiated, it did not precisely follow the ancient margin of the North American craton, however—the margin along which the great carbonate platform had stood in Cambrian and Early Ordovician time (Figure 7-14). Instead, Mesozoic fracturing followed a line slightly farther to the east, leaving attached to North America the slivers of exotic terrane that today reveal so much about the history of Appalachian mountain building (Figure 7-17).

CHAPTER SUMMARY

1 A mountain chain forms when two continental blocks converge along a subduction zone. The Himalayas have formed in this way.

2 A mountain chain also forms when the edge of a moving continent encounters a subduction zone. This situation is causing the elevation of the Andes today.

3 The Himalayas have been elevated because the crustal block that now forms peninsular India has been forced beneath the margin of Asia, doubling the thickness of the crust.

4 The Appalachian mountain system displays a sedimentary sequence characteristic of mountain building. The sequence begins with shallow-water sediments that have been deposited along a continental border. When mountain building began along the border, a foreland basin was created by a downwarping of the crust inland from the orogenic activity. This foreland basin became the site of deposition for flysch in the form of shales and graywackes of deep-water origin. Eventually the foreland basin became choked with sediment from the young mountains and deep-water flysch deposits gave way to shallow marine and nonmarine clastic sediments, known as molasse.

5 In the Appalachians, three episodes of mountain building are evident in the sedimentary record. The first destroyed an enormous early Paleozoic carbonate platform that had extended along most of the eastern margin of North America. The second and third were associated with the coalescing of continents to form the supercontinent Pangaea. As the Appalachians formed, small blocks of crust were added to the margin of North America.

EXERCISES

1 Where does a foreland basin develop in a typical mountain chain?

2 What is flysch and where does it form?

3 What is molasse? Why does it normally accumulate after flysch?

4 Why are the Andes Mountains taller than the Appalachians?

5 How can mountain chains form without continental collision?

6 Why do mountains have roots?

7 Are the ophiolites found within a mountain chain episode normally older or younger than the foreland basin deposits?

8 Examine a world map or globe to locate mountain chains that are not discussed in this chapter. Then locate these chains on the plate-tectonic map of the world (Figure 6-1). See if you can figure out how the presence of each mountain system might relate to plate-tectonic processes. (Some of the answers will appear in the chapters that follow.)

ADDITIONAL READING

Burchfiel, B. C., "The Continental Crust," *Scientific American*, September 1983.

Cook, F. A., L. D. Brown, and J. E. Oliver, "The Southern Appalachians and the Growth of Continents," *Scientific American*, October 1980.

James, D. E., "The Evolution of the Andes," *Scientific American*, August 1973.

Molnar, P., "The Geologic History and Structure of the Himalaya," *American Scientist* 74:144–154, 1986.

PART

IV

THE PRECAMBRIAN WORLD

Earth came into being about 4.6 billion years ago, forming — with the other planets of our solar system — from a whirling cloud of dust. During the first billion years or so of Earth history, continents were small, volcanism was widespread, and life consisted of bacteria and their lowly relatives. During this Archean Eon, radioactive elements were abundant in the lithosphere and released heat at a high rate. By the beginning of the Proterozoic Eon, however, Earth had cooled, larger continents were forming, and modern plate-tectonic processes were in operation. In the course of Proterozoic time, cells more advanced than bacteria arose within aquatic environments, and from them multicellular plants and animals evolved.

Lightening and eruptions from a volcano in Japan illuminate the night sky. Early in Precambrian time, in the absence of higher life, physical processes like these played a conspicuous role at Earth's surface. *(Nishiinoue/Sipa Press.)*

CHAPTER

8

The Archean Eon of Precambrian Time

Since the last century, the interval of Earth history that preceded the Phanerozoic Eon has been known as the Precambrian. Although the term *Precambrian* has no formal status in the geologic time scale, it has traditionally been employed as though it did. The Precambrian includes nearly 90 percent of geologic time, ranging from 4.6 billion years ago, when Earth formed, to the start of the Cambrian Period about 4 billion years later.

Two eons are formally recognized within the Precambrian: the Archean and the Proterozoic. The Archean Eon includes about 45 percent of Earth's history—the interval from about 4.6 to 2.5 billion years ago. During the Archean Eon, Earth underwent enormous physical changes and life developed on its surface. Many details of Archean history, however, remain unknown or poorly understood.

In fact, much less is known of both the Archean and Proterozoic eons than of the succeeding Phanerozoic Eon. One reason for this discrepancy is that, although the Precambrian constitutes such a large span of geologic time, Precambrian rocks form less than 20 percent of the total area of rocks exposed at Earth's surface (Figure 8-1). Erosion has destroyed many Precambrian rocks, and metamorphism has so altered others that they can no longer be dated and therefore cannot be recognized as Precambrian. Still other Precambrian rocks lie buried beneath younger sedimentary and volcanic rocks. Index fossils are seldom found in Precambrian rocks because primitive organisms without durable skeletons predominated until the end of Precambrian time. As a result, stratigraphic correlation of Precambrian rocks has been based largely on radiometric dating.

Most geologic information about the Precambrian has been derived from cratons, those large portions of continents that have not experienced tectonic deformation since Precambrian or early Paleozoic time. All the continents of the present world include cratons that consist primarily of Precambrian rocks (Figure 8-2). A **Precambrian shield** is a largely Precambrian portion of a craton that is exposed at Earth's surface. The largest is

In northwestern Canada the Canadian Shield is a largely barren Precambrian terrane that was scoured by glaciers during the recent Ice Age and is now dotted by lakes. This photograph shows outcrops of numerous dikes that formed when magma forced its way into giant cracks in the Precambrian crust. *(R. S. Hildebrand, Geological Survey of Canada.)*

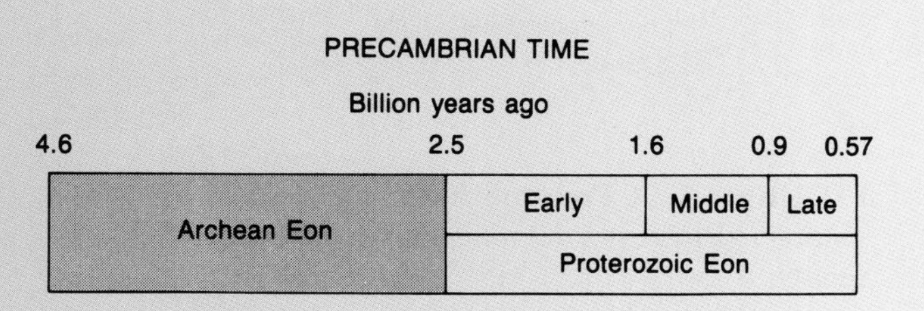

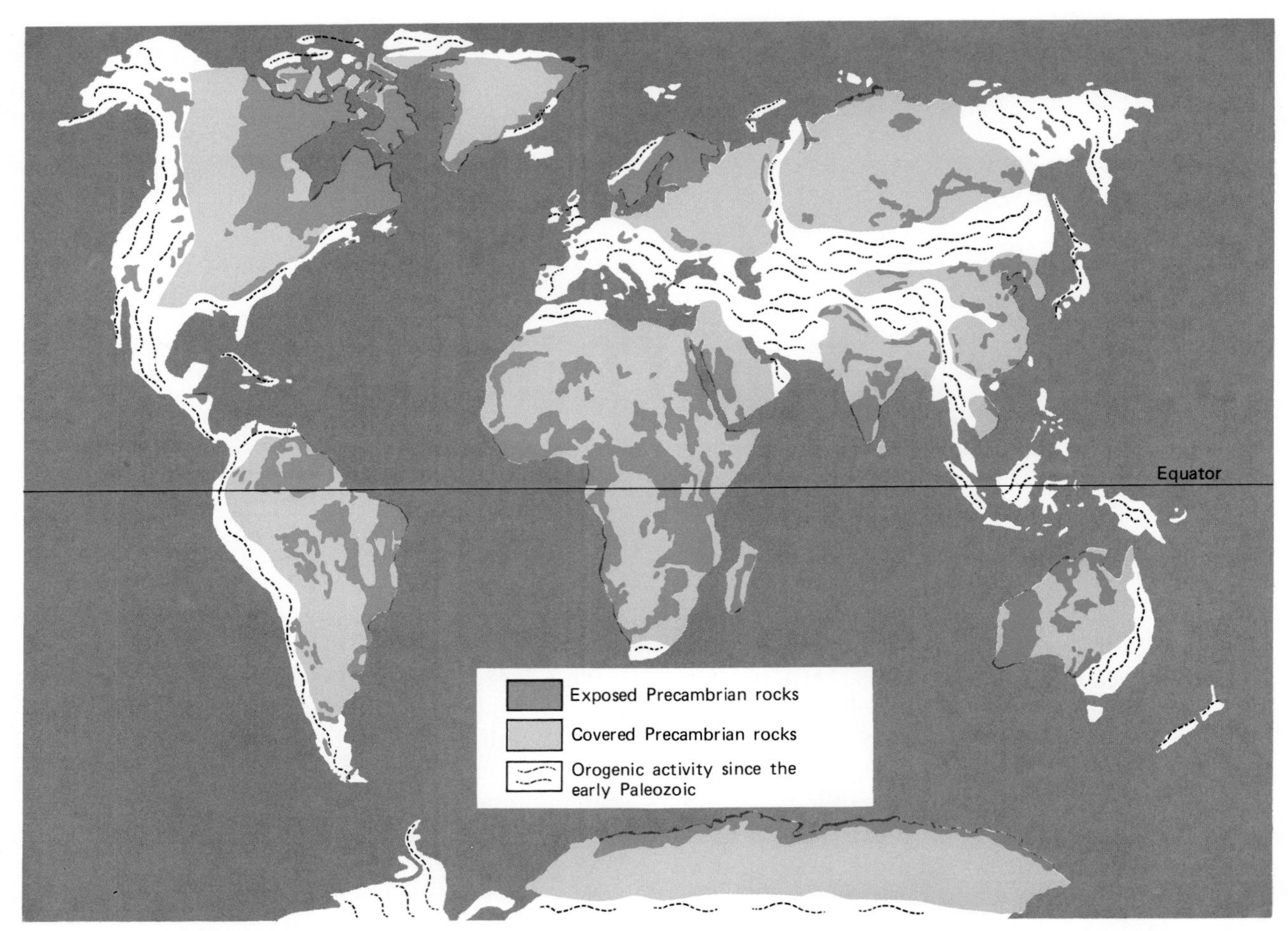

FIGURE 8-1 Distribution of Precambrian rocks in the modern world. Note that these rocks form or underlie most of the cratonic area of the modern world. *(After A. M. Goodwin, in B. F. Windley [ed.], The Early History of the Earth, John Wiley & Sons, Inc., New York, 1976.)*

the vast Canadian Shield, which has recently (geologically speaking) become more fully exposed by the action of Pleistocene glaciers (p. 184). Although shields contain some sedimentary rocks, they consist primarily of crystalline (igneous and metamorphic) rocks. As we will see, mountain belts that formed during Precambrian time left traces that can be recognized within Precambrian shields, but their elevated topography was destroyed long ago by erosion. Today the only Precambrian rocks that stand at high elevations within mountain ranges are those that have been uplifted by Phanerozoic orogenies.

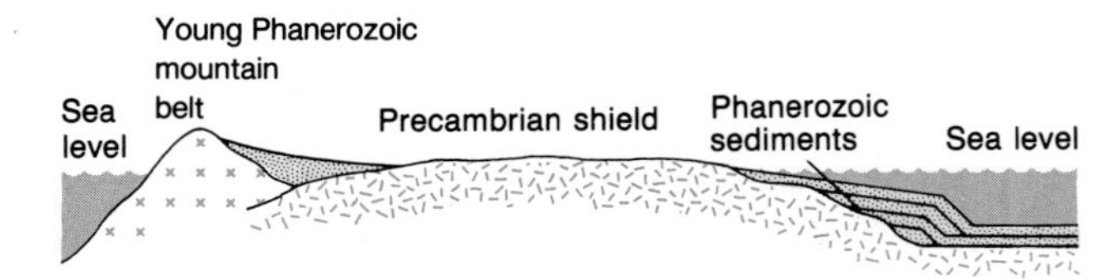

FIGURE 8-2 A Precambrian shield flanked by younger (Phanerozoic) mountain belts and sediments.

It was during the Archean Eon that Earth acquired its basic configuration, with the mantle and crust surrounding the core (p. 14). Early in Archean time, Earth's crust seems to have differed from what it is today. Earth was much hotter than it is now, and apparently there were no large felsic cratons. By the end of Archean time, however, large cratons had begun to form, and plate-tectonic processes were modifying these cratons in the same way that they do now. The transition to this modern style of tectonism ushered in the Proterozoic interval of Precambrian time (discussed in Chapter 9). In the absence of useful biostratigraphic data, the boundary between the Archean and Proterozoic intervals is defined by its absolute age of 2.5 billion years before the present.

We will begin our review of Archean events by considering when and how Earth and other planets of the solar system may have formed. We will then review evidence that suggests how Earth changed during the remainder of Archean time.

AGES OF THE UNIVERSE AND THE PLANETS

All seriously considered theories of the origin of our solar system are based on the premise that its planets formed almost simultaneously. Part of the evidence for simultaneous origin is the observation that the planets' orbits around the sun lie in nearly the same plane. Had the planets formed independently, we would expect their orbits to occupy different planes.

Precisely when the planets came into being has been a more difficult issue for scientists to resolve. Astronauts report that the most beautiful object they see from space is Earth, whose surface is partially blanketed by swirling white clouds set against the blue background of extensive oceans (Figure 8-3). But while Earth's water is aesthetically pleasing and necessary for life, its abundance near the planet's surface makes rapid erosion inevitable. Continuous alteration of the crust by erosion and also by igneous and metamorphic processes makes unlikely any discovery of rocks nearly as old as Earth. Thus geologists have had to look beyond this planet in their efforts to date Earth's origin. Fortunately, we do have samples of rock that are generally believed to represent the primitive material of the solar system. These samples are **meteorites** — extraterrestrial objects that have been captured in Earth's gravitational field and have subsequently crashed into our planet (Figures 8-4 and 8-5).

Some meteorites consist of rocky material and, accordingly, are called **stony meteorites**

FIGURE 8-3 Earth viewed from space. The blue regions are oceans and the swirling white masses are clouds. *(NASA.)*

FIGURE 8-4 Barringer Crater, near Flagstaff, Arizona. This fresh crater, which is slightly more than 1 kilometer ($\frac{3}{5}$ mile) in diameter, was created by the impact of a meteorite only 25,000 years ago. Numerous fragments have been collected nearby. *(Meteor Crater Enterprises, Inc.)*

(Figure 8-5). Others are metallic and have been designated **iron meteorites** even though they contain subordinate amounts of elements other than iron. Still others consist of mixtures of rocky and metallic material and thus are called **stony-iron meteorites.** Meteorites come in all sizes, from small particles to the small planets known as asteroids; no asteroid, however, has struck Earth during recorded human history. Many meteorites appear to be fragments of larger bodies that have undergone collisions and broken into pieces.

Meteorites have been radiometrically dated by means of several decay systems, including rubidium-strontium, potassium-argon, and uranium-thorium (p. 86). The fact that the dates thus derived tend to cluster around 4.6 billion years suggests that this is the approximate age of the solar system. After many meteorites had been dated, it was gratifying to find that the oldest ages obtained for rocks gathered on the surface of the moon also approximated 4.6 billion years: This must, indeed, be the age of the solar system. Ancient rocks can be found on the moon because the lunar surface, unlike that of Earth, has no water to weather and erode rocks and is characterized by only weak tectonic activity.

Determining the age of the universe, which turns out to be more than three times as old as our solar system, has been more complicated. Stars in the universe tend to be clustered into enormous disklike galaxies (Figure 8-6). The distance is increasing between our galaxy, known as the Milky Way, and all others. In fact, all galaxies are moving away from each other, so that the universe is expanding. It is not the galaxies themselves that are expanding, but the space between them. What is happening is analogous to inflation of a balloon with small coins attached to its surface (Figure 8-7). The coins behave like galaxies: Although

FIGURE 8-5 A stony meteorite. This particular stone fell in a meteorite shower near Plainville, Texas. *(R. A. Oriti.)*

A

B

FIGURE 8-6 The configuration of a spiral galaxy. *A.* This is galaxy NGC 6744. *B.* In this edge-on view, the galaxy's arms spiral outward from the central bulge. *(A. Anglo-Australian Telescope Board. B. NOAO.)*

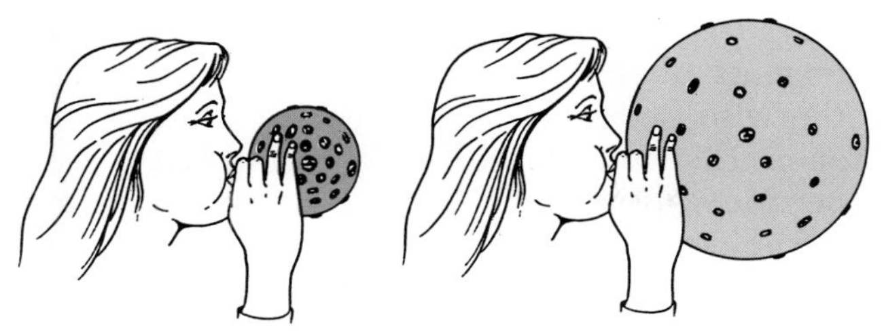

FIGURE 8-7 An analogy for the expansion of the universe. The space between galaxies expands like the balloon, but the galaxies themselves, like the coins, do not expand but only move farther apart. *(Adapted from C. Misner, K. Thorne, and J. Wheeler.)*

they do not expand, the space between them does. Before the galaxies formed, matter that they contain was concentrated with infinite density at a single point, from which it exploded in an event called the **big bang.** Even after it became assembled into galaxies, matter continued to spread in all directions from the site of the big bang.

The evidence that the universe is expanding makes it possible to estimate its age. This evidence, called the **redshift,** is an increase in the wavelengths of light waves traveling through space — a shift toward the end of the spectrum of wavelengths where visible light is red. Expansion of the space between galaxies causes this shift by stretching light waves as they pass through it. The redshift is taking place in light waves emanating from all galaxies, and some of these light waves reach Earth. The farther light waves have traveled through space, the greater the redshift they have undergone. For this reason, light waves from distant galaxies have larger redshifts than those from nearby galaxies. Calculations based on these redshifts indicate that between about 15 and 18 billion years ago all of the galaxies would have been at one spot, the site of the big bang. This, then, is the approximate date of the big bang and the age of the universe.

ORIGIN OF THE SOLAR SYSTEM AND EARTH

We know more about the origin of distant stars than we do about the origin of our own solar system. Our solar system formed long ago, and we have no other young solar systems to observe at close range. Scientists can train their telescopes on multitudes of large stars in various stages of development — but unfortunately, these stars are too far away to allow us to observe how planets may be forming in association with them.

Galaxies, which are enormous clusters of stars, form by the gravitational collapse of dense clouds of gas (mainly hydrogen) into stars. Our galaxy, which originated less than 10 billion years ago, is made up of approximately 250 billion stars. And even after a galaxy is formed by the establishment of some stars, secondary stars are continually born within its spiral arms, where galactic matter is concentrated and the pressure varies (in time and space) as matter contracts and expands.

Today, to explain the origin of our solar system, most experts favor some version of the **solar nebula theory,** which holds that the solar system formed from a cloud of particulate matter called cosmic dust. The details of this theory, however, are widely debated.

The Sun

Although all stars form in essentially the same manner, their histories vary according to their size. Our sun, for example, apparently formed from material remaining from a star that imploded, or collapsed violently, to form heavy elements. After this collapse, a **supernova** — or an exploding star that casts off matter of low density — was formed. What remained was a dense cloud that condensed as it cooled after the explosion. This dense cloud, or **nebula** (Figure 8-8), is assumed to have had some rotational motion when it formed and must then have rotated more and more rapidly as it contracted, just as ice skaters automatically spin more rapidly (conserving angular momentum) as they pull in their arms.

The Planets

It is possible that the sun formed alone from a nebula and then captured a cloud of cosmic dust that formed the planets. Alternatively, the sun and the planets may have formed from the same dust cloud. What does seem certain, however, is that the planets formed either during or soon after the

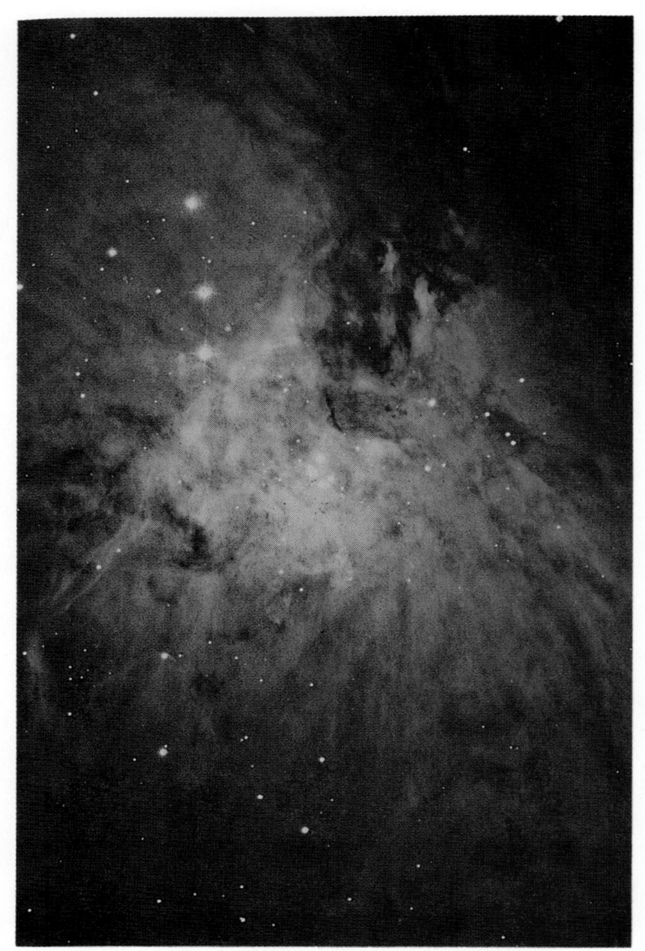

FIGURE 8-8 A nebula in which stars are being born within the constellation Orion. Nebulas such as this are the visible features that outline the arms of spiral galaxies. *(Anglo-American Telescope Board.)*

birth of the sun. In fact, the sun and the planets appear to have originated during an interval of time no longer than about 100 million years. This has been deduced from the excessive concentration in some meteorites of the stable isotopes xenon 129 and plutonium 244. ("Excessive" in this case means present in an amount greater than is normal for meteorites.) Excessive amounts of these isotopes in a meteorite indicate that some of the short-lived parent isotopes were originally present in the meteorite material and then decayed to xenon 129 and plutonium 224. Thus the planet or planetlike body of which such a meteorite is a fragment must have formed soon after the elements of the solar system came into being; otherwise, the short-lived parent isotopes of xenon 129 and plutonium 244 could not have been incorporated into the material of meteorites in a sufficiently high concentration to leave excessive amounts of their daughter products. The conclusion is that the sun cannot be very much older than the meteorites, the moon, and the planets, which have ages of about 4.6 billion years.

Planets that are positioned far from the sun formed largely from volatile (i.e., easily vaporized) elements. It would appear that these elements were expelled from the hot inner region of the nebula to solidify in colder regions far from the sun. More dense materials tended to be left behind, and these materials formed the inner planets, including Earth.

Several steps led to the formation of the planets. When the rotating dust cloud reached a certain density and rate of rotation, it flattened into a disk, and the material of the disk then segregated into rings, which later condensed into the planets (Figure 8-9). Each planet began to form by the aggregation of material within one of these rings. The aggregates eventually reached the proportions of asteroids, which are commonly about 40 kilometers (~25 miles) in diameter, and coalesced to form planets.

After the planets formed, the rocky debris left behind that we refer to as asteroids remained in orbit around the sun. Some of these asteroids survive to this day, but most have collided with larger planets to become part of them. Others have undoubtedly had their motions so severely disturbed by near collisions that they have passed out of the solar system.

The origin of the moon, which circles Earth, is still debated. Was the moon an independent body that was captured by Earth's gravitational field? Was it formed when a collision between Earth and another planet separated a large chunk of Earth? Or was it accreted from matter in orbit around our primitive planet? Lunar rocks obtained in the Apollo space program exhibit an isotopic composition of oxygen that is remarkably similar to that of terrestrial rock. This suggests that the materials that constitute the moon formed near Earth from the same portion of the solar nebula. (Meteorites, in contrast, which are presumed to form in a variety of regions, show widely varying isotopic compositions of oxygen.)

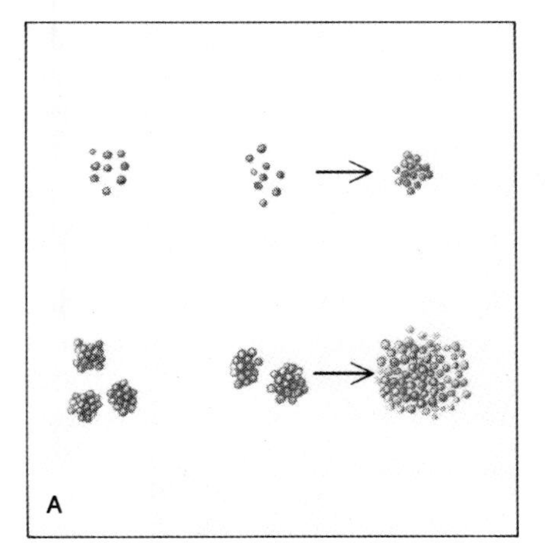

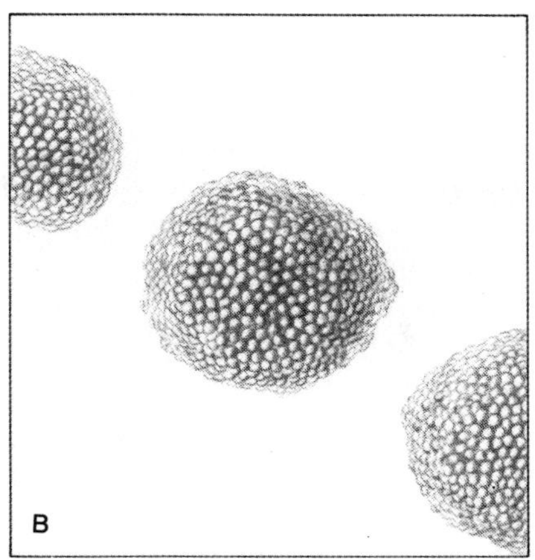

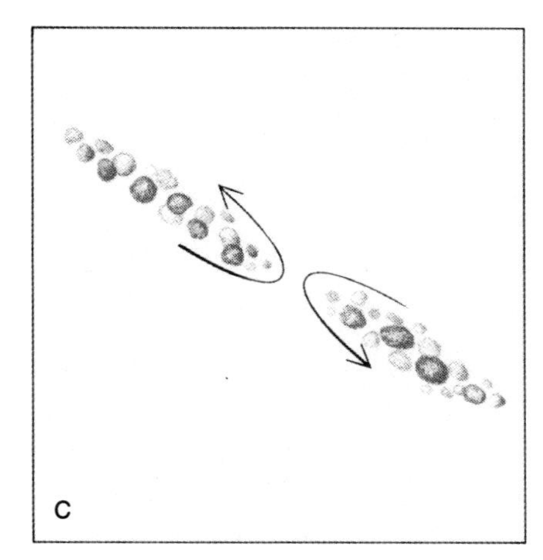

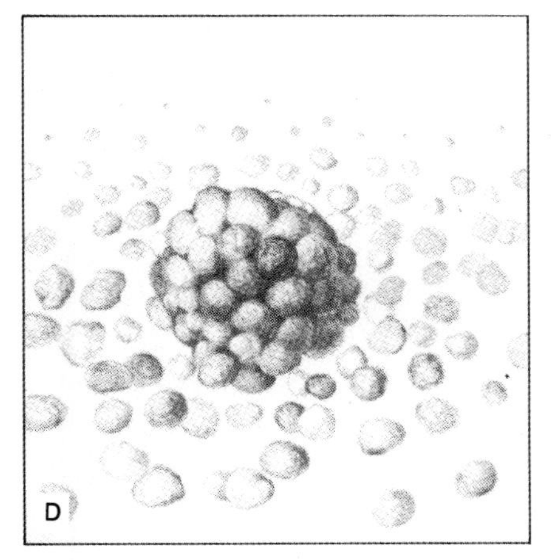

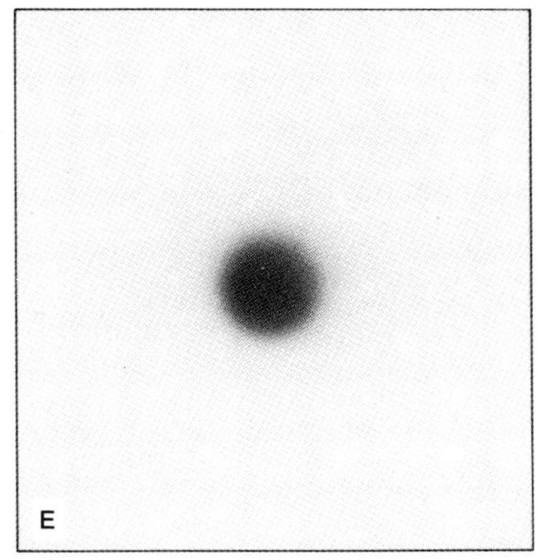

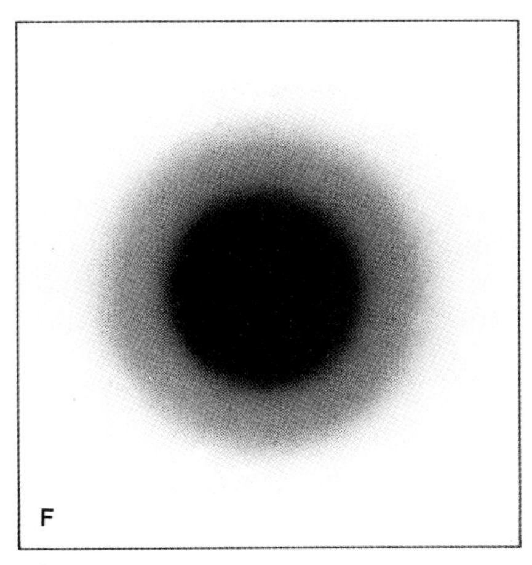

FIGURE 8-9 How a planet forms. The process begins when dust grains collide and stick together, forming larger and larger clumps *(A)*. The clumps move toward the central plane of the nebula and collect into bodies the size of asteroids *(B)*. These bodies gravitate into clusters *(C)*, and collide *(D)* to form the nucleus of a planet *(E)*. The planet grows and may attract gas from the nebula. If the planet is large enough, it may attract so much gas and draw it in so closely that the gas forms a dense shell representing most of the planetary mass *(F)*. *(From A. G. W. Cameron, "The Origin and Evolution of the Solar System." Copyright © 1975 by Scientific American, Inc. All rights reserved.)*

HOW EARTH AND ITS FLUIDS BECAME CONCENTRICALLY LAYERED

The ideas described earlier in this chapter explain the manner in which our planet may have come into being, but they do not account for many of Earth's particular features. How, for example, did the concentric structure of the solid planet and the fluids above it develop? How did metal and rock become segregated into core, mantle, and crust? Have there always been continents and ocean basins? How did the oceans and the atmosphere develop? These are questions that we will consider in this section.

The Core, Mantle, and Crust

Earth's concentric layers, from the core to the atmosphere, are arranged according to density, with the least dense layer, the atmosphere, on the outside. The explanation would be simple if the primordial planet were known to have been a molten mass, in which case the high-density materials would automatically have sunk toward the center while less dense components would have floated toward the surface and formed the crust. The mantle, which is of intermediate density, would then have remained between crust and core.

Was Earth initially molten? Rapid accretion would concentrate a great deal of heat, but slow accretion would allow components to cool as they fell into place. Many experts currently believe that Earth accreted components in a solid state (Figure 8-9)—but even if this was the case, Earth must have developed a liquid interior when it heated to a critical temperature upon reaching a certain size. At this point, gravitational collapse would inevitably have moved heavy elements (primarily iron) to the center, forming Earth's core. Such contraction would then have released an enormous amount of energy—enough, some claim, to raise Earth's temperature by as much as 1200°C! Under such conditions, the mantle would also have segregated in a liquid state, and radioactive elements (especially isotopes of uranium, thorium, potassium, and rubidium) would then have moved to positions near the surface as a consequence of their tendency to combine chemically with other elements in low-density silicate minerals.

The Atmosphere

The asteroids that coalesced to form Earth were too small for their gravitational fields to have held gases around them as atmospheres. We can thus conclude that Earth did not inherit its atmosphere from these ancestral bodies; instead, the gases that the primitive planet retained as an atmosphere must have been emitted from within Earth after it formed, while it was in a liquid state, so that they could easily escape to the surface. If the core, mantle, and crust became differentiated when Earth first became liquefied, extensive **degassing**—or loss of gases to Earth's surface—would have accompanied this differentiation.

Degassing by way of volcanic emissions has continued to the present, albeit at a much lower rate than early in Earth's history. The chemical composition of gases released from modern volcanoes provides a general picture of what the early atmosphere contained: primarily water vapor, hydrogen, hydrogen chloride, carbon monoxide, carbon dioxide, and nitrogen. In the modern world, plant photosynthesis is responsible for most of the oxygen in the atmosphere. In the absence of plants, less oxygen entered or formed within the early Archean atmosphere. The early atmosphere would thus have been inhospitable to animal life.

Hydrogen and helium are the only elements of low enough density to have escaped from Earth's gravitational field during the period of rapid degassing. Noble gases (neon, argon, and their chemical relatives)—are also relatively rare in Earth and its atmosphere today in comparison with their abundance elsewhere in the solar system. Although they are rare, carbon, nitrogen, and water are more abundant in Earth than the noble gases, and it seems evident that they were preferentially retained as a result of being partially locked up in solid compounds. Noble gases rarely combine with other elements.

The Oceans

The rapid degassing of the liquid planet released hot clouds of water vapor. Initially, Earth's great heat would have kept the water in a gaseous state. Only when the planet cooled sufficiently would water have fallen as rain and remained on Earth as lakes, rivers, and oceans.

Like modern rain, the rains that formed the earliest oceans are assumed to have contained few salts. Salts accumulated in early seawater by the reaction of the water, and of carbon dioxide dissolved within it, with natural minerals such as clays and carbonates. Calculations show that seawater should have become approximately as saline early in Archean time as it is now. Since then, salts have been precipitated on the seafloor as evaporite sediments about as rapidly as they have been added to the oceans, and thus the salinity of seawater has not varied greatly. At the same time, the global hydrological cycle has moved water but not salts from the oceans to the atmosphere and back again (Figure 2-7).

A HOTTER EARTH AND SMALLER CONTINENTS

During Archean time, heat must have flowed upward through Earth's lithosphere more rapidly than it does today, because Earth's radioactive "furnace" was hotter. Recall from Chapter 1 that

Earth's heat source is constantly diminishing as radioactive isotopes decay without being renewed (p. 18). Because these decay rates are constant, geologists can calculate the approximate difference between the rate at which Earth produces heat today and the rates of times past. The total rate of heat production was perhaps twice as high near the end of Archean time as it is now (Figure 8-10), and earlier it was even higher. As a consequence, hot spots should have been numerous and the lithosphere should have been fragmented into many small plates separated by numerous rifts, subduction zones, and transform faults. As we shall see, the nature of Archean rocks confirms this prediction.

Geologists still debate how rapidly the total volume of Earth's continental crust increased during Archean time through the operation of partial melting. There is no question, however, that what felsic crust existed even late in the Archean was divided into relatively small continents, sometimes termed protocontinents.

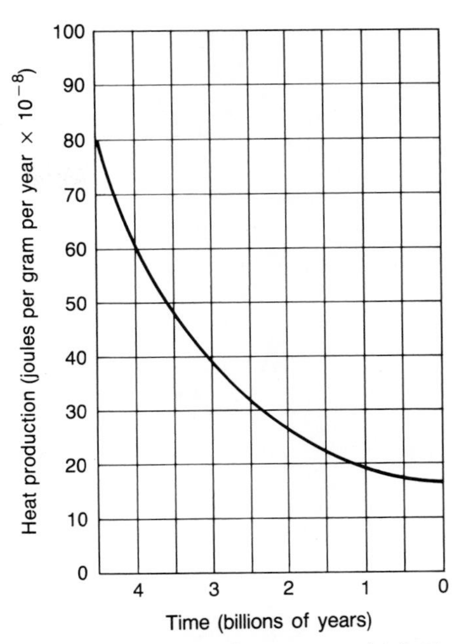

FIGURE 8-10 Decline in the rate at which Earth has produced heat through time. *(After W. H. K. Lee, Ph.D. thesis, University of California at Los Angeles, 1967.)*

THE GREAT METEORITE SHOWER

Before moving to a general discussion of Archean rocks, we will consider what must have been one of the most remarkable episodes in Earth's history: the pelting of Earth by large numbers of meteorites and asteroids over a period of several hundred million years. Given the rarity of early Archean rocks, our only evidence that this extraterrestrial shower took place is provided by other planets and particularly by the moon. These other bodies of the solar system have not undergone the weathering and erosion that constantly alter Earth's surface.

Impacts on the Moon

Earthbound observers have long commented on the moon's pockmarked appearance. The moon's craters, which are known as **maria** (singular, mare), the Latin word for "seas," were first sketched by Galileo. It is only from manned lunar exploration and from photographs provided by artificial satellites, however, that we have gained detailed knowledge of these enormous craters. The maria that face the earth have an average diameter of approximately 200 kilometers (~125 miles) and are distributed in a crescent-shaped pattern (Figure 8-11). At first it was not known whether the maria were craters produced by volcanoes or were indeed the impact scars of huge meteorites, but meteoritic origin has now been established by detailed study of the maria and neighboring lunar terrane. The entire surface of the moon is, in fact, scarred with craters, most of which are much smaller than the enormous maria. The lunar highlands surrounding the maria consist of rock fragments that testify to the pulverization of the lunar crust by the impact of falling meteorites. The maria facing Earth are floored by immense flows of dark volcanic basalts, which give the maria their dusky appearance. These basalts, which formed as a result of the heat generated by impacts, are themselves scarred by smaller craters, and the relative ages of successive lava flows can be deduced from the density of cratering.

Dating of associated rocks has shown that most large lunar craters are quite old, ranging in age from slightly less than 4.0 billion years to about 4.6 billion years. It has been estimated that during the early cataclysmic interval of lunar history, meteorite impacts were more than 1000 times more frequent than they are now.

FIGURE 8-11 Crescent-shaped arrangement of dark maria on the side of the moon that faces Earth. *(Lick Observatory.)*

Impacts on Earth

We can see that craters of all sizes also formed on planets of the solar system whose surfaces have not been so heavily altered as Earth's. These craters formed early in the solar system's history, when it was, in effect, being cleared of many small remaining pieces of solid material that had consolidated from the solar nebula but had not been assimilated into planets or captured as planetary satellites. Thus the conclusion seems inescapable that Earth was subjected to the same kind of meteorite bombardment as its neighboring moon. The composition and structure of Earth's crust must have been altered by the material contributed by this enormous shower of meteorites.

Unfortunately, the rock record that might have provided us with details of the Archean meteorite shower has been all but destroyed by igneous and metamorphic activity on Earth, as well as by weathering and erosion. Geologists have identified numerous post-Archean craters, despite the decline in the rate of meteorite impacts — a decline that has resulted from the continual loss of meteors through collision with planets. Some of the largest post-Archean meteorites appear to have devastated life on Earth (Box 8-1).

THE ORIGINS OF CONTINENTAL CRUST

As we learned in Chapter 1, continents consist primarily of thick felsic crust and are surrounded by the thinner mafic crust of ocean basins. An important question is how the two kinds of crust came into being. To understand the answer, it is important to recognize that Earth's interior cooled from its fully molten state primarily by means of convection, which carried hot material to the surface. The primitive crust probably formed rapidly during this brief interval of cooling as magma flowed to the surface at a high rate. Most of the early crust was composed of mafic material that rose from the denser ultramafic mantle to form relatively thin oceanic crust, like that of the modern world.

How, then, did the felsic crust of continents come into being? It developed secondarily, mostly from oceanic crust that descended into the mantle and released felsic components whose low density caused them to rise toward the surface as magma. More specifically, the mafic oceanic crust, containing water, descended into the mantle. Some portions of this crust melted and lost water. Chemical reactions between the products of melting and nearby lithospheric materials produced felsic magmas. These low-density magmas rose toward the surface, leaving ultramafic material behind. This process is called **partial melting** because felsic materials are melted away from ultramafic materials, which melt only at higher temperatures.

Probably most Archean felsic magmas were derived from basalts along midocean rifts. Support for this idea comes from modern Iceland, where basalts produced along the Mid-Atlantic Ridge stand above sea level (Figure 6-23). In this area, basalts have piled up to such great thicknesses that they have sunk isostatically, in the way that a root forms beneath a mountain (Figure 1-17). These basalts have descended to depths at which they have undergone partial melting, which has produced felsic magma that has risen to form both intrusive and extrusive rocks.

Continental crust was probably present earlier than 4 billion years ago, but the oldest well-dated rocks are metamorphic rocks. Some of these are found in a small area near the southern tip of Greenland. This is one of many relatively small

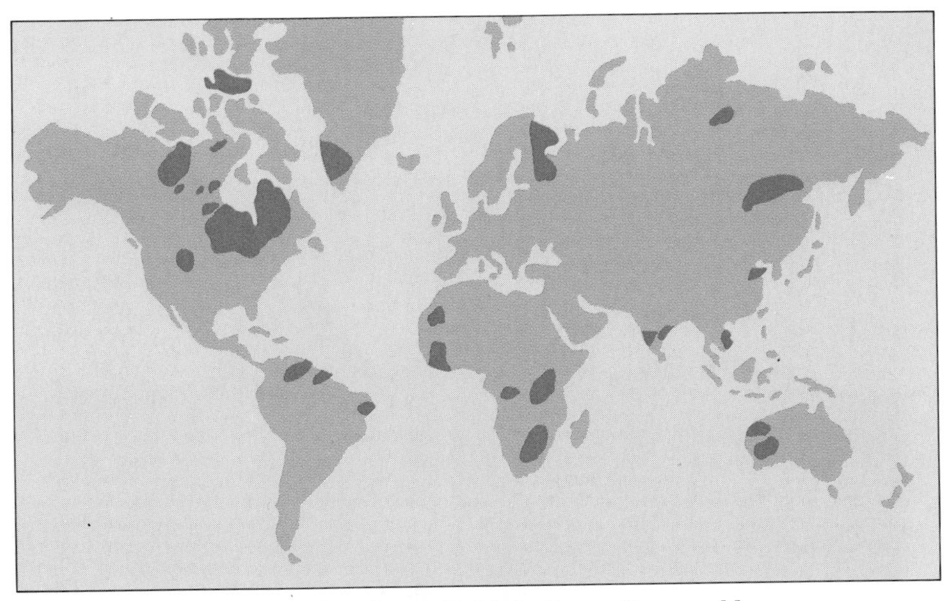

FIGURE 8-12 Locations of Archean shields in the modern world.

Archean shield areas scattered across the modern world (Figure 8-12). Among the Archean rocks of Greenland are rocks that were metamorphosed about 3.8 billion years ago—after they had accumulated as sediments sometime earlier (Figure 8-13). In fact, metamorphism has undoubtedly altered many bodies of rock that are older than 3.8 billion years and reset their radiometric clocks to give younger ages.

Grains of the mineral zircon from various regions offer evidence that continental crust was present on Earth 4.1 to 4.2 billion years ago. At that time, zircon grains originated, as indicated by uranium-lead dating (p. 86). The zircon grains were preserved because, being especially resist-

FIGURE 8-13 Some of the oldest known rocks on Earth. These banded iron formations from the Isua area of southern Greenland were deposited as sediments at least 3.8 billion years ago. Individual laminae, which are deformed and weakly metamorphosed, are a few millimeters thick. Laminae of chert (fine-grained quartz) alternate with darker laminae that contain iron minerals. *(Preston Cloud and David Pierce.)*

Box 8-1
The Threat from Outer Space

This artist's conception shows a marine landing for a large bolide colliding with Earth. *(Painting by William K. Hartmann.)*

During the past few years, geologists have gathered evidence that a large extraterrestrial object struck Earth about 65 million years ago and killed off the dinosaurs. This evidence has heightened human interest in the threat that asteroids and comets may pose to life on Earth.

Nearly all of the meteorites that have struck Earth in recent times have come from the belt of asteroids that circles the sun between Mars and Jupiter. Jupiter is such a large planet that its movement disturbs these bodies, and occasionally one is tossed into an orbit that crosses that of Earth. Sometimes the result is collision. Comets that strike Earth have traveled farther than meteorites—from far beyond the solar system, where they are concentrated in two huge clouds. A few, however, have orbits that bring them close to the sun, and the sun's intense heat vaporizes some of the ice that forms them. The tail we observe when a comet streaks across the night sky as a "shooting star" is the water vapor streaming away behind it. Some comets that have lost all of their ice remain in solar orbits as solid masses; most of these comets are indistinguishable from asteroids. When a comet loses mass as a result of vaporization, its orbit changes, and occasionally one crashes to Earth.

Just how likely is it that an asteroid or comet will strike Earth in the near future? We have two ways of estimating how often extraterrestrial objects of various sizes crash into our planet. One is to count the objects of various sizes whose orbits cross that of Earth. Knowing the configurations of the orbits as well, scientists can calculate probabilities of collision. A second approach is to measure and date craters excavated during the past 3 billion years. Radiometric dating of metamorphic rocks formed during impact gives the age of a crater.

An object must weigh about 350 tons or more to form an impact crater. Smaller objects simply break apart and lodge in Earth's surface. Stony meteorites are so brittle that they break into small pieces when they encounter Earth's atmosphere.

In general, only metallic meteorites remain large enough to form craters when they strike Earth.

Studies of orbiting bodies and craters indicate that extraterrestrial objects have struck Earth at a roughly constant rate during the past 3 billion years. By the beginning of this interval, the planets had swept up most of the debris left over from the formation of the solar system. During the past 3 billion years, the deaths of comets have supplied some of the new bodies that have threatened Earth; the rest have been asteroids that have escaped from the belt between Mars and Jupiter. These new bodies have appeared about as rapidly as other bodies have disappeared in planetary collisions.

Telescopes reveal that today there are about 1000 bodies larger than 1 kilometer (~0.6 mile) in diameter whose orbits cross that of Earth. Few of these bodies exceed 20 kilometers (~12 miles) in diameter, and the sizes of craters discovered on Earth suggest that few objects larger than this have, in fact, struck our planet during the past 3 billion years. On the other hand, it is estimated that an object as large as 10 kilometers in diameter has struck Earth about once every 40 million years.

Most extraterrestrial bodies that have collided with Earth have made marine landings because oceans have always covered most of the planet. Unfortunately, geologic processes have destroyed many craters that extraterrestrial bodies have left on continents. Even so, recognized craters tell interesting stories. The two largest, each about 140 kilometers (~85 miles) in diameter, are the Sudbury crater in Ontario, Canada, and the Vredefort crater in South Africa. Each was formed nearly 2 billion years ago by the impact of an object that must have been in the neighborhood of 10 kilometers in diameter. The Popigai crater in Siberia is the third largest yet discovered, with a diameter of about 100 kilometers (~60 miles). Its age, somewhere between 50 and 30 million years, places the time of impact within the Age of Mammals, and the object that landed was probably slightly less than 10 kilometers in diameter.

Collision with a very large extraterrestrial object—one slightly larger than 10 kilometers—should have serious consequences for environments on Earth. The cloud of particles blasted into the atmosphere should plunge the planet into total darkness for at least a few months. Lack of sunlight would produce subfreezing temperatures and perhaps accumulation of deep snow on continents, even at low latitudes. Even after the dust began to settle and the sun peaked through the haze, the reflectance of light from widespread snow would probably leave Earth cool for years. The energy of the explosion would convert atmospheric nitrogen to acidic nitrous oxides. As a result, atmospheric moisture would fall to Earth as acid rain. Acid rain could damage many forms of life. It would also attack carbonate rocks and liberate carbon dioxide, producing a greenhouse effect and eventually reversing the initial climatic trend, perhaps warming the planet to unusually high temperatures.

What forms of life would survive such changes? How many humans would die? We have no certain answers; we can only wonder whether the arrival of an extraterrestrial object 10 or 15 or 20 kilometers in diameter may someday bring an end to our species' reign on Earth.

This aerial view shows Manicouagan crater, an impact structure in the Canadian Shield of Ontario that is about 70 kilometers (~43 miles) in diameter and is partly occupied by lakes. The impact occurred in Late Triassic time, more than 200 million years ago. *(EROS.)*

ant to weathering, they survived the destruction of their parent rock to become detrital components of sedimentary rocks. The mineral zircon forms by the metamorphism of felsic rocks. Thus the ancient zircon grains prove that felsic crust existed as early as 400 or 500 million years after Earth formed.

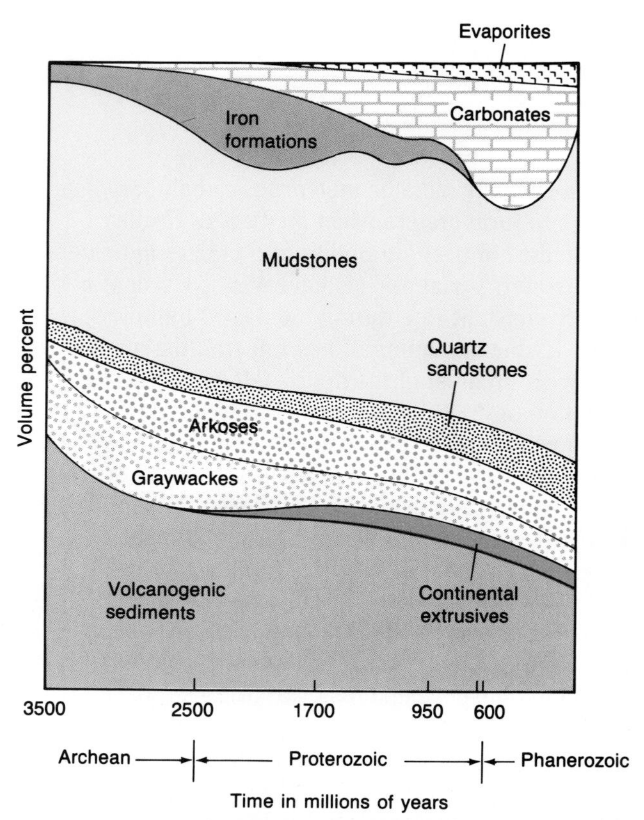

FIGURE 8-14 Changes in the relative proportions of rock laid down on cratonic surfaces in the course of geologic time. *(Modified from F. T. Mackenzie, in J. P. Riley and G. Skirrow [eds.], Chemical Oceanography, Academic Press, London, 1975. Based on data from A. B. Ronov.)*

ARCHEAN ROCKS

Archean rocks differ in average composition from younger rocks. This difference reflects the configuration of the Archean continental crust, which was relatively thin and divided into numerous small continents that lay along subduction zones. During Archean time, numerous volcanic arcs produced large volumes of dark igneous rocks. In addition, large bodies of dark sedimentary rocks formed from the erosion of these volcanic rocks. The thin nature of the Archean continental crust is preserved in the modern world: Areas of existing cratons that are of Archean age extend less far downward from the surface, on the average, than do cratons that have formed by continental accretion since Archean time.

General Features of Sedimentary Rocks

It is a striking fact that most Archean sediments are of deep-water origin. They include graywackes, mudstones, iron formations, and sediments derived from volcanic activity. In contrast, sediments deposited in terrestrial and shallow marine environments are relatively uncommon in the Archean record. These include carbonates and quartz sandstones (Figure 8-14). We have seen that in more recent times, crustal subsidence has caused shallow marine sediments such as those of the Bahamas (p. 70) and nonmarine sediments such as those of rift valleys (p. 152) to be buried quite deep in very short spans of time. Why, then, were no similar sequences preserved in Archean time? The apparent answer is that there are no widespread continental or continental-shelf deposits in the Archean record simply because there were no large continents during most of Archean time.

As the next section will explain, the configuration of rocks within Archean terranes also points to a predominance of small protocontinents.

Greenstone Belts

The characteristic configuration of Archean terranes is evident in satellite photographs of shield areas (Figure 8-15). Over broad areas, podlike bodies known as **greenstone belts** sit in masses of high-grade metamorphic rocks of felsic composition (**gneisses**, for example). Rocks of the greenstone belts themselves are generally weakly metamorphosed, and the green metamorphic mineral chlorite gives them their name.

FIGURE 8-15 Satellite photograph of greenstone belts in the Pilbara Shield of Western Australia. These belts are dark bodies of rock between circular bodies of light-colored crystalline rock that represent felsic crust of Archean protocontinents. The felsic body at the top is about 40 kilometers (~25 miles) across. *(CSIRO Division of Exploration Geoscience, Australia.)*

FIGURE 8-16 Archean pillow basalt, Sioux Lookout area, northwestern Ontario. The "pillows" have been planed off by erosion. *(F. J. Pettijohn.)*

Igneous rocks of greenstone belts, before they were metamorphosed, were mostly mafic and ultramafic volcanic rocks of the kind extruded along volcanic arcs, with felsic volcanic rocks present in smaller volumes. Many of the volcanics display pillow structures, which indicate that the lava that formed them was extruded underwater (Figure 8-16). Not surprisingly, many of the sedimentary rocks of greenstone belts, though now metamorphosed, can be seen to have originally formed from detritus eroded from dark volcanic rocks. Metamorphosed turbidites are common, as are dark mudstones, now mostly metamorphosed to slate. These sediments were apparently deposited in forearc basins and other environments situated along subduction zones (p. 153).

Cherts and iron-rich sedimentary rocks known as **banded iron formations** are also found in the Archean sedimentary belts. These rocks are occasionally widespread but are seldom very thick. They were far more prevalent during the Proterozoic Eon, so they will be discussed at some length in Chapter 9. It is notable, however, that the banded iron formation at Isua, in southern Greenland, represents one of the oldest known bodies of rocks on Earth. Like other banded ironstones of the Precambrian, it consists of iron-rich layers alternating with quartz layers (Figure 8-13). The Isua rocks are believed to have originated by chemical precipitation in marine basins, and the quartz within them is thought to have existed initially as chert precipitated from seawater. Submarine volcanic eruptions were the source of the dissolved silica that was precipitated as chert.

Other characteristic but less abundant Archean detrital sediments are coarse conglomerates containing large, rounded cobbles that suffered considerable stream or beach abrasion (Figure 8-17). These conglomerates are not cross-bedded like stream gravels, however; they seem instead to have been dumped into place, perhaps

FIGURE 8-17 Coarse conglomerate of the Archean Manitou Series, Mosher Bay, northwestern Ontario. The felsic pebbles, cobbles, and boulders in these rocks testify to the presence of continental crust nearby. *(F. P. Pettijohn.)*

by enormous turbidity flows traveling down steep slopes. It is therefore not surprising that the conglomerates tend to occur adjacent to or folded into greenstone belts; they seem to be nearshore facies. The conglomerates contain pieces of greenstone as well as pieces of felsic crystalline rocks eroded from protocontinents.

The rocks that today form greenstone belts came to lie within Archean protocontinents through continental accretion (Figure 8-18). Having formed along subduction zones, they were wedged between protocontinents that collided and became sutured together. Given the small size of the protocontinents and the abundance of subduction zones, greenstone belts formed frequently and came to constitute a large proportion of Archean terranes.

LARGE CRATONS APPEAR

Radiometric dating shows that during early Proterozoic time, large bodies of magma were intruded into cratonic rocks. It can therefore be concluded, in accordance with the principle of intrusive relationships, that cratons of moderate proportions had come into being very late in Archean time. The hardened magma often forms upright tabular structures, or **dikes** (see Appendix I), that range in age from about 2.1 to 2.5 billion years. Large mafic dikes of this age, often occurring in swarms associated with other mafic intrusives, have been identified in North America, Greenland, the Baltic Shield, South America, southern Africa, India, and Australia. Evidence also indicates that in many parts of the world, major metamorphic episodes occurred slightly earlier, about 2.7 to 2.3 billion years ago. Geolo-

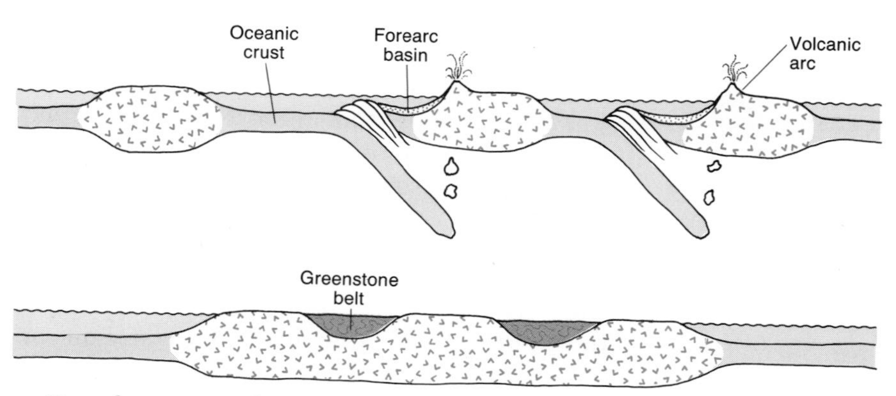

FIGURE 8-18 Formation of greenstone belts. Forearc basin sediments, deformed oceanic crust, and arc volcanics along the margins of protocontinents *(top)* become squeezed between protocontinents during suturing to become podlike greenstone belts *(bottom)* in a larger protocontinent.

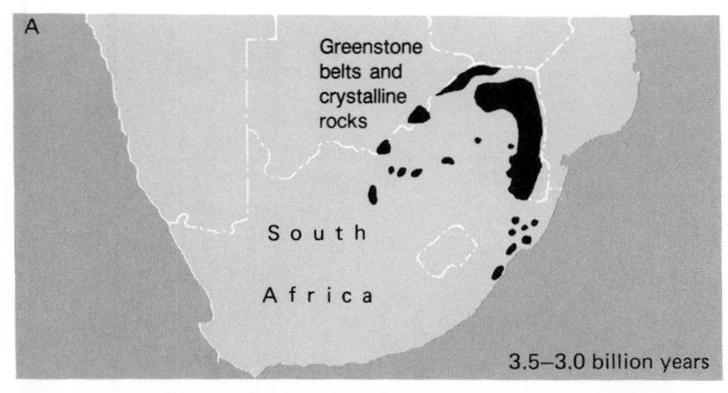

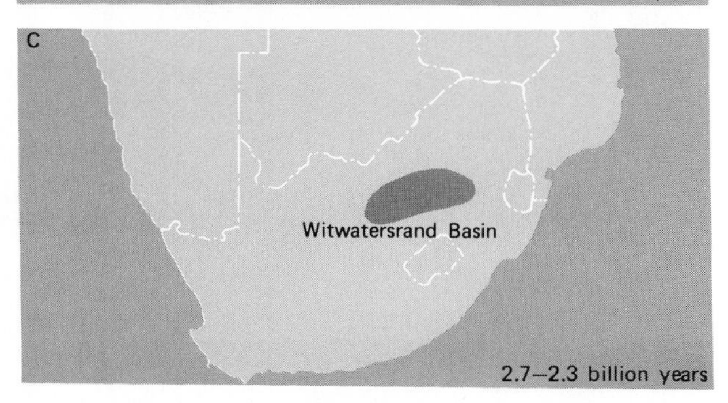

FIGURE 8-19 The distribution of some of the world's oldest known extensive cratonic rocks, in the Archean of southern Africa. Siliciclastic deposits of the Pongola Basin *(B)* and Witwatersrand Basin *(C)* accumulated on a broad crust composed partly of greenstone sequences *(A)*. (See Figure 8-15.) *(Adapted from C. R. Anhaeusser, Philos. Trans. Roy. Soc. London A273:359–388, 1973.)*

gists do not yet understand the origins of this metamorphic interval, but they do know that it resulted in the resetting of many radioactive clocks and in the consolidation of many crustal elements into sizable cratons.

There is also evidence, however, that "cratonization" did not occur simultaneously throughout the world. In most areas, typical Archean greenstone associations formed until approximately 2.5 billion years ago, but in southern Africa a large craton was already present about a half billion years earlier (Figure 8-19). Here a greenstone sequence more than 3 billion years old is immediately overlain by a substantial body of clastic sediments known as the Pongola Supergroup. The Pongola, in turn, is followed by the Witwatersrand sequence, which is famous for the gold deposits that accumulated as detrital material within it (Figure 8-19). The Pongola Supergroup consists of deposits that are 3 billion years old but are nonetheless strikingly similar to intertidal sequences of younger portions of the stratigraphic record (Figure 8-20). The presence of a broad intertidal belt during Pongola deposition indicates that a sizable landmass existed 3 billion years ago in southern Africa.

The overlying Witwatersrand sequence, with sediments ranging in age from about 2.5 to 2.8 billion years, consists of sediments that cover about 40,000 square kilometers and have a maximum thickness of nearly 8 kilometers (~5 miles). These deposits accumulated in nonmarine environments on the surface of a craton. The coarsest sediments, which are widespread conglomerates, are most common in the upper division of the Witwatersrand, and it is from these sediments that gold is mined. Gold is a very dense metal and thus tends to settle from moving water with much larger particles. During Witwatersrand deposition, bits of gold accumulated with larger silicate pebbles. The gold-bearing gravels apparently formed bars within braided streams that seem to have washed enormous quantities of sedimentary debris from highlands across alluvial fans into one or more lakes. Lake environments were the sites where mud and occasionally sand accumulated.

Even at the close of Precambrian time, the large continents remained unusual in one important respect: They were barren of advanced forms of life. Long before the first large cratons existed, however, living cells had begun to populate the marine realm, where they remained at a primitive stage of development for perhaps a billion years or more.

ARCHEAN LIFE

Of all the planets in our solar system, only Earth is well suited to life as we know it. One of the reasons is that its size is right. On a much larger planet, the

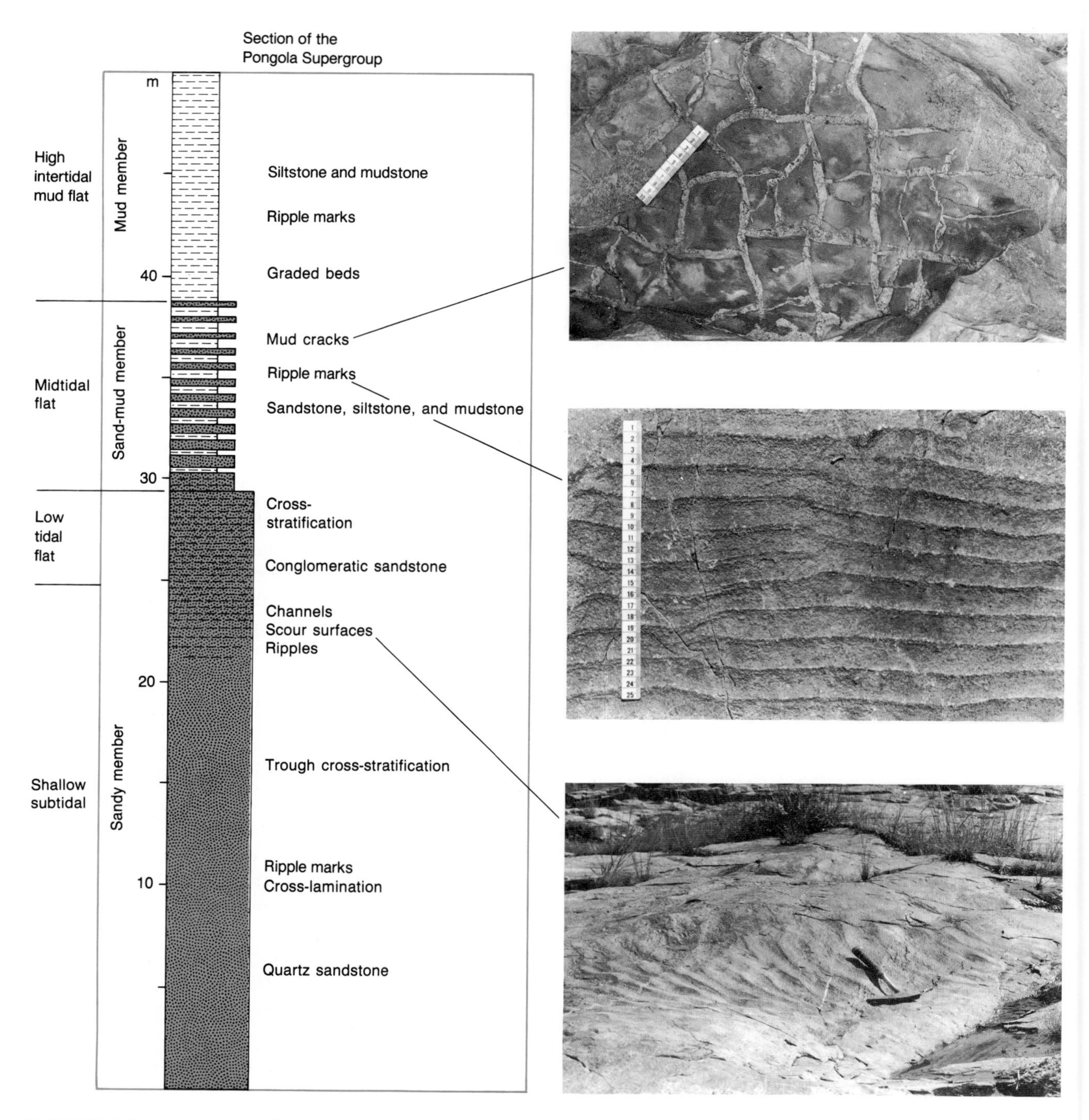

FIGURE 8-20 Depositional environments represented within a stratigraphic section of the Pongola Supergroup of southern Africa. The section includes a 3-billion-year-old regressive sequence in which a tidal flat prograded seaward over subtidal environments. In the lower, sandy member, cross-stratification and symmetrical ripples are common. Some of these ripples are double-crested, reflecting the ebb and flow of tides. Tidal channel deposits floored by pebbles are present, particularly in the upper part. Above them, the sand-mud member bears ripples and mud cracks, which suggests a shallower, midtidal environment. The uppermost mud member bears smaller mud cracks as well as mud chips and appears to represent a high intertidal mud flat that lay landward of the zones where sand settled from tidal waters. *(After V. von Brunn and D. K. Hobday, Jour. Sedim. Petrol. 46:670–679, 1976.)*

gravitational pull on the atmosphere would be so great that the resulting atmospheric density would exclude sunlight, which is the fundamental source of energy for life. A much smaller planet, on the other hand, would lack sufficient gravitational attraction to retain an atmosphere with life-giving oxygen. In addition, Earth's temperatures are such that most of its free water is liquid, the form that is essential to life. Even Venus, our nearest neighbor closer to the sun, is much too hot to allow water to survive in a liquid state. Mars, our nearest neighbor farther from the sun, has a cooler surface, but an atmosphere so thin that liquid water would evaporate from the planet's surface almost immediately.

Fossil Evidence

We can never hope to possess clear fossil evidence of the origin and early evolution of life on Earth, because cells break down easily and even chemical components that might be indicative of past events can quickly deteriorate beyond recognition. Nevertheless, certain facts suggest that life is at least as old as the oldest rocks now known. For one thing, graphite, a mineral consisting of pure carbon, is present even in the oldest known sedimentary rocks, the banded iron formations of Isua, Greenland (Figure 8-13)—perhaps reflecting concentration of carbon by organisms. In fact, carbon found in sedimentary rocks of Archean age tends to have nearly the same ratio of isotopes (^{13}C to ^{12}C) that characterizes biological systems today.

More direct evidence of very early cellular life is provided by stromatolites, the internally layered structures shaped like domes, mounds, or pillars that were described in Chapter 3. Today cyanobacteria form stromatolites along the margins of warm seas, but they are much less abundant than they once were. Recall that within a well-preserved stromatolite, layers of carbonate sediment rich in organic matter alternate with layers of purer carbonate sediment. Stromatolites usually grow side by side in large numbers; some stromatolites, especially those of the geologic past, have attained heights of several meters.

Because structures resembling stromatolites sometimes form by deposition of layered sediment in the absence of cyanobacteria, we cannot always identify fossil stromatolites with certainty. The oldest structures that are thought to be stromatolites occur in the Pilbara Shield of Australia in rocks 3.4 to 3.5 billion years old (Figure 8-21). Stromatolites are known to occur in slightly younger rock units, such as the Bulawayan Group of Rhodesia, for which indirect dating gives an age of approximately 2.8 billion years, and the 3-billion-year-old Pongola Supergroup. They are also found in a few other Archean units, but in general they are much less common here than in Proterozoic sediments. The rarity of Archean stromatolites probably reflects the rarity of Archean shelf deposits. We cannot be certain, however, that the organisms that formed Archean stromatolites were photosynthetic, and therefore we cannot draw definite conclusions concerning

A

B

FIGURE 8-21 One of the oldest known geologic structures thought to be stromatolites (*A*, top view; *B*, cross section). This layered structure from the Warawoona Group of the Pilbara Shield of Western Australia is between 3.4 and 3.5 billion years old. *(D. R. Lowe.)*

the environments in which these stromatolites formed. Certain threadlike bacteria, which are not cyanobacteria, form stromatolitelike structures today in hot springs such as those of Yellowstone National Park, but most of those bacteria are not photosynthetic; that is to say, they do not use light as a source of energy to synthesize food. Thus the possibility exists that similar bacteria formed the stromatolitelike structures found in Archean rocks.

Actual fossils of cyanobacteria and other bacteria have also been tentatively identified in Archean rocks. Among the oldest cells yet known are filaments from a rock at North Pole, Western Australia, where the apparent stromatolites of Figure 8-21 are also found. These filaments, which are thought to be approximately 3.5 billion years old (Figure 8-22), are about the same size as those of modern cyanobacteria (Figure 8-23) and display similar transverse partitions. Somewhat younger spheroidal (nearly spherical) structures resembling other types of cyanobacteria occur in the Fig Tree Group of South Africa. These structures, which are about 3 billion years old, are preserved in what appear to be various stages of cell division (Figure 8-24).

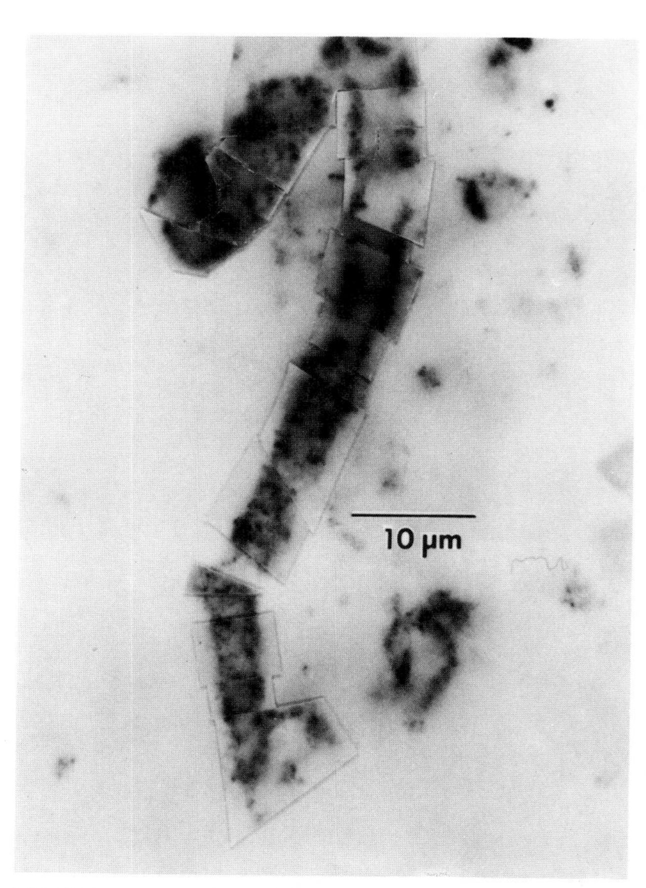

FIGURE 8-22 A fossil filament from North Pole, Western Australia. (The scale bar represents 10 microns.) This and associated filaments display distinctive transverse partitions and may be between 3.4 and 3.5 billion years old, in which case they may be the oldest known fossil cells of cyanobacteria. *(S. M. Awramik.)*

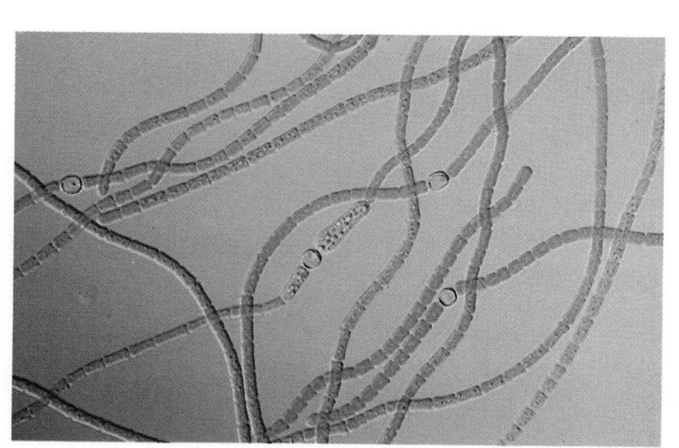

A

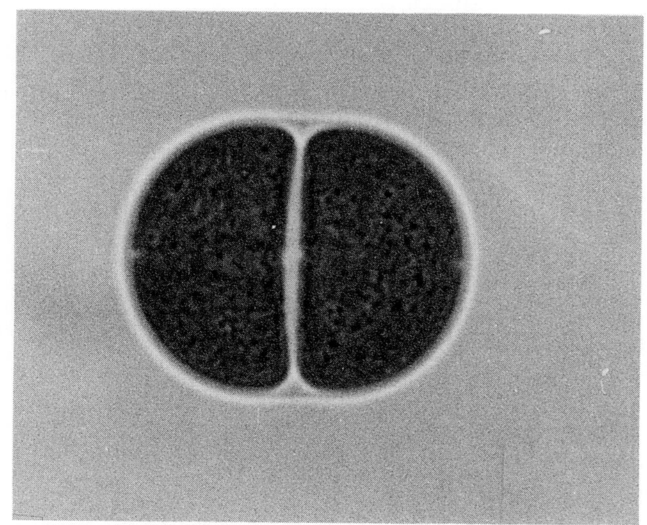

B

FIGURE 8-23 Living filamentous and spheroidal (nearly spherical) cyanobacteria. The spheroidal cell is in the process of dividing. *(A. S. Barns and N. Pace. B. M. Potts.)*

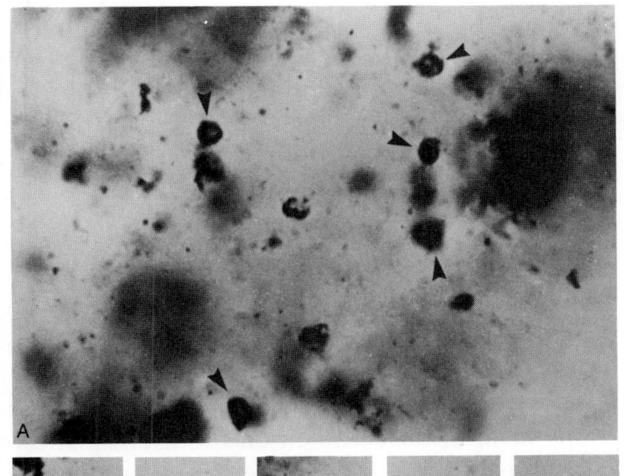

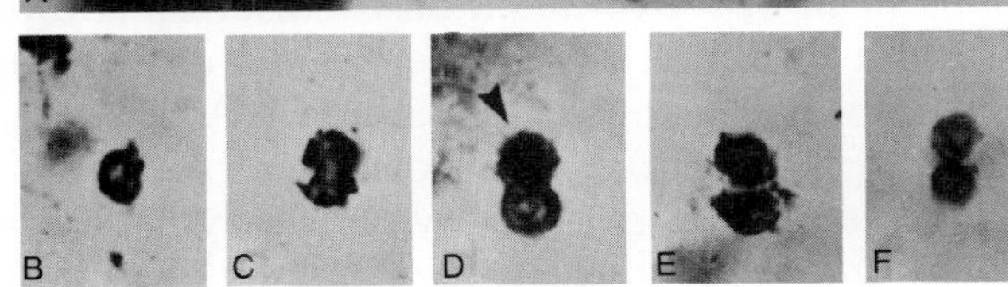

FIGURE 8-24 Microfossils older than 3 billion years from the lowest part of the Fig Tree Group of South Africa. Many of the cells are in some stage of division (*B* through *F*). They may represent cyanobacteria or bacteria. *(A. H. Knoll and E. S. Barghorn, Science 198: 396–398, 1977.)*

Bacteria, including cyanobacteria, seem to be the only forms of life represented in Archean rocks. Thus it is significant that bacteria differ markedly from all other groups of cellular life in the modern world; they are placed by themselves in the kingdom Monera (see Figure 1-5). Recall that Monera are single-celled organisms characterized by a primitive kind of cell that does not have a nucleus and whose DNA is not clustered into discrete chromosomes. This type of cell, which is known as a **prokaryotic cell,** also lacks certain other internal organlike structures (organelles) that are present in more advanced forms of life. Forms of life that do possess chromosomes as well as nuclei and other organelles are known as **eukaryotes.** Of all the cellular forms of life in the world today, only bacteria and cyanobacteria are **prokaryotes.** All other cellular taxa are eukaryotes, and fossils of these organisms are unknown from the Archean and early portion of the Proterozoic record. Thus, with respect to the history of life on Earth, the Archean Eon might well be labeled the Age of Prokaryotes.

Chemical Evidence

Although the fossil record will never tell us a great deal about the earliest stages of organic evolution, such information has been derived from other sources. In the early 1950s, for example, researchers found that they could readily produce amino acids in the laboratory by sending electrical sparks through sealed vessels containing ammonia, methane, hydrogen, and steam (Figure 8-25). These sparks may duplicate what lightning did in the Archean world. Similar laboratory results have since been obtained with simple starting

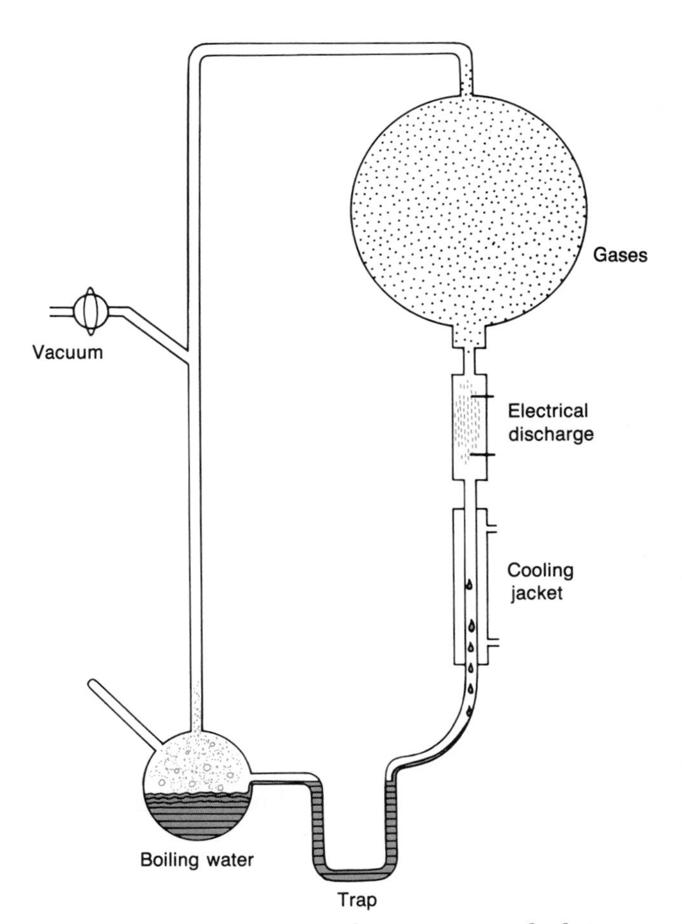

FIGURE 8-25 Experimental apparatus in which S. L. Miller produced amino acids by circulating ammonia (NH_3), methane (CH_4), water vapor (H_2O), and hydrogen past an electrical discharge. Amino acids accumulated in the trap. *(From G. Wald, "The Origin of Life." Copyright © 1954 by Scientific American, Inc. All rights reserved.)*

mixtures such as carbon monoxide, nitrogen, and hydrogen. In some cases, ultraviolet light has been substituted for electrical sparks as a source of energy. **Amino acids** are the building blocks of **proteins,** and proteins are important compounds in living systems. Thus, experiments demonstrating the generation of amino acids under conditions that might well have characterized primitive Earth are highly significant. It has also been found that the heating of amino acids drives off water and links them into chains called **polypeptides,** which resemble proteins but are less complex. This, too, could have occurred in many parts of primitive Earth.

While amino acids and proteins are essential features of terrestrial life, they do not account for a basic aspect of living systems: the capacity for self-replication. On Earth this capacity resides within the DNA and RNA of cells. The genetic messages of chromosomes are chemically encoded in DNA; these messages are transcribed by ribonucleic acid, or RNA, which is similar in structure to DNA, and are then carried by RNA to regions of the cell where, according to the chemical prescription, proteins are constructed from amino acids. RNA and DNA are nucleic acids. Nucleic acids are not complex chemical structures, and it does not seem at all preposterous to imagine that the first nucleic acid formed in nature by the assembly of sugars, phosphates, and nucleotide bases; but we do not know exactly how this happened or precisely how the codes for protein synthesis came to reside in DNA.

Another problematic step was the origin of cell-like bodies that have the ability to build true proteins. Although such bodies have not been created in the laboratory, simpler spherical bodies have been produced from proteinlike compounds in water or salt solutions.

The earliest forms of cellular life that employed nucleic acids to build proteins and to reproduce may have obtained their nutrition in either of two ways: They may have been animal-like consumers (**heterotrophs**) or plantlike producers (**autotrophs**) (Figure 8-26). Consumers obtain food directly from the environment, whereas producers manufacture their own food from simple raw materials that they obtain from the environment (p. 27). We will consider two scenarios for early Archean evolution — one in which the earliest single-celled organisms are animal-like and the other in which they are plantlike. Modern bacteria live in many different ways, and as very primitive organisms, some of which may have survived from Archean time with little modification, they offer clues about evolution during Archean time.

Cellular Beginnings?

It is not clear whether the earliest cells were autotrophic or heterotrophic — that is to say, whether they required food and energy from their environment in order to maintain themselves and to reproduce. Like all modern cells, early cells probably employed a compound known as ATP (adenosine triphosphate) as a source of energy. Unlike modern cells, however, early cells may have obtained the ATP they needed from their environment — by eating it, in effect. ATP is easily produced from simple gases in the laboratory, and it may well have formed inorganically in the Archean world. Even very primitive animal-like cells must have had at least a limited ability to synthesize other essential compounds, although they fed on amino acids, sugars, and other molecules necessary for their existence. It has often been assumed that these organisms lived in a lake or an ocean filled with a sort of natural "soup" of essential molecules.

Certain cells developed the ability to ferment organic compounds — that is, to break down such compounds into simpler ones and to use the energy liberated in this way to build some of the compounds they needed. In other words, they became plantlike (Figure 8-26). Many bacteria employ fermentation today. Most of them ferment sugar or cellulose (the material that forms the cell walls of plants) into products such as ethanol (ethyl alcohol). Energy liberated by fermentation is stored as ATP until it is used to build compounds essential to the operation of the bacterial cell.

Some bacteria that are fermenters also obtain energy by converting compounds that contain sulfate into others that contain sulfide — that is to say, they remove oxygen atoms that have been attached to sulfur atoms. These bacteria cannot tolerate oxygen, and most therefore live in the muds of swamps, ponds, or lagoons. The hydrogen sulfide that these bacteria liberate into their muddy environments generates the characteristic

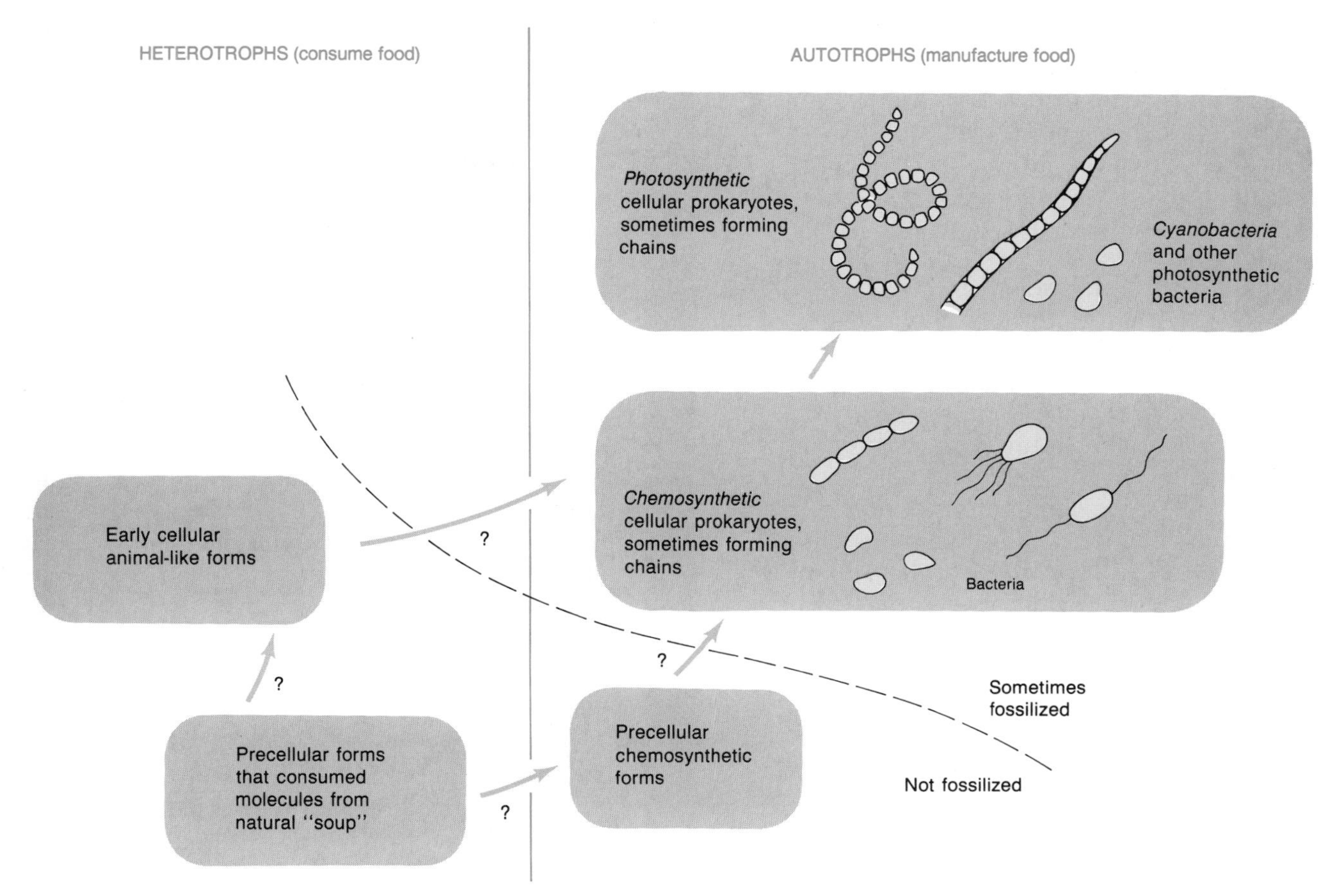

FIGURE 8-26 Possible sequence of nutritional changes in the early evolution of life. The earliest chemosynthetic forms might have been either cellular or precellular.

"rotten egg" smell of those settings. If, as is widely believed, Earth's atmosphere contained little oxygen during Archean time, such **sulfate-reducing bacteria** may have occupied a much wider range of environments than they do today.

In recent years it has been discovered that a few kinds of bacteria are distinct from all others. These have been labeled **Archaebacteria.** This group includes the methane-producing bacteria. Living abundantly in marine and freshwater sediments and in sewage, these bacteria feed on a few simple organic compounds. Methane, sometimes referred to as marsh gas, is a waste product of their metabolism. Some experts view Archaebacteria as being so different from other bacteria that they belong in a separate kingdom. The fact that they, like sulfate-reducing bacteria, can live only in the absence of oxygen makes it possible that they were quite widespread early in Archean time, when little or no free oxygen was present in the atmosphere.

Bacteria that conduct fermentation or reduce sulfate are said to engage in **chemosynthesis.** The origin of the energetically more efficient process of photosynthesis in living organisms was an important breakthrough that altered the ecosystem profoundly. Photosynthesis, as we saw in Chapter 2, is the process by which single-celled algae and multicellular plants employ the green pigment chlorophyll to transform the energy of sunlight into chemical energy. This energy is then used to transform carbon dioxide and water into energy-rich sugar. When it is released from the sugar, the energy also fuels essential chemical reactions.

Presumably cyanobacteria were not the only Archean Monera that conducted photosynthesis.

Other **photosynthetic bacteria** probably existed as well. Such organisms exist today in the form of purple and green bacteria that inhabit moist areas lacking free oxygen. Like cyanobacteria, these other photosynthetic bacteria employ sunlight to produce sugar.

The evolution of the cyanobacteria, which release free oxygen, had a profound effect on the global ecosystem. The oxygen that was liberated by these organisms accumulated in the atmosphere to form the reservoir from which humans and other animals were later able to breathe. In Chapter 9 we will examine the current controversy over the time it took for Earth's atmosphere to accumulate.

Whatever the details of the early history of life may have been, one thing seems certain: Missing from the Archean world of prokaryotic life were animals and advanced animal-like cells that feed upon bacteria and cyanobacteria. The origin and early evolution of these advanced consumers will be discussed in Chapter 9.

CHAPTER SUMMARY

1 Radiometeric dating has revealed that stony meteorites, which represent the primitive material of the solar systems, are 4.6 billion years old, as are the most ancient moon rocks. This, then, is the apparent age of Earth and the other planets of the solar system.

2 Earth originated by condensation of material that had been part of a rotating dust cloud. Earth became stratified into crust, mantle, and core because material of high density accreted before material of low density or because dense material sank toward the center of the young liquid Earth.

3 The Archean interval of Earth history extended from the time of the planet's origin to approximately 2.5 billion years ago.

4 Between the time it formed and slightly later than 4 billion years ago, Earth was pelted by large numbers of meteorites. During the same interval, meteorites produced most of the large craters that are still visible on the moon, whose surface is less active than Earth's.

5 The formation of large continents was inhibited in early and middle Archean time by the abundance of radioactive elements whose decay produced heat at a high rate; under conditions of high heat flow, Earth's crust was divided into small protocontinents.

6 The most readily studied Archean rocks occur in greenstone belts. Greenstones are metamorphosed dark volcanic and sedimentary rocks. They formed along subduction zones adjacent to small continents, from which were derived dark mudstones and graywackes associated with the volcanics.

7 Large continental landmasses apparently did not form until late in Archean time. The oldest of those now recognized are in South Africa, where shallow marine and nonmarine siliciclastic sediments were spread over sizable landmasses between 3 and 2 billion years ago.

8 Cyanobacteria formed stromatolites during Archean time. Bacteria, including cyanobacteria, are also represented in Archean rocks by fossil cells. These groups represent the most primitive forms of cellular life on Earth today, lacking cell nuclei and chromosomes.

9 The earliest forms of life presumably consumed molecules from their environment. Later, with the evolution of chemosynthesis and photosynthesis, organisms developed the ability to manufacture their own food. Cyanobacteria appear to have been among the earliest photosynthetic organisms.

EXERCISES

1 What is a Precambrian shield? Where is one located in North America?

2 What reasons are there to believe that Earth was pelted by vast numbers of meteorites early in its history?

3 Why might we expect Earth to be nearly the same age as its moon and the material that forms meteorites?

4 What geologic features characterize greenstone belts and how did these belts form?

5 What types of sedimentary rocks were rare in the Archean Eon? What does this suggest about the nature of cratons during Archean time?

6 Why did magma rise from the mantle to Earth's surface at a higher rate during Archean time than it does today?

7 What features make Earth a more hospitable place than other planets for life as we know it?

8 What are stromatolites? From what we know of their formation today, why might we expect them to have been present early in the history of Earth?

9 What are some of the ways in which bacteria obtain their nutrition? How may the modes of life of certain living bacteria shed light on early evolution?

ADDITIONAL READING

Cloud, P., *Oasis in Space: Earth History from the Beginning,* W. W. Norton, New York, 1988.

Grieve, R. A. F., "Impact Cratering on the Earth," *Scientific American,* April 1990.

Margulis, L., *Early Life,* Science Books International, Boston, 1982.

Nisbet, E. G., *The Young Earth: An Introduction to Archean Geology,* Allen & Unwin, Boston, 1987.

Silk, J., *The Big Bang: The Creation and Evolution of the Universe,* W. H. Freeman and Company, New York, 1988.

Windley, B. F., *The Evolving Continents,* John Wiley & Sons, New York, 1984.

Woese, C. R., "Archaebacteria," *Scientific American,* June 1981.

CHAPTER 9

The Proterozoic Eon of Precambrian Time

The Proterozoic Eon, which succeeded the Archean Eon 2.5 billion years ago, was in many ways more like the Phanerozoic Eon, in which we live. We have already seen a foreshadowing of this difference between the Proterozoic and Archean eons in the origin of large cratons late in Archean time. The persistence of large cratons throughout the Proterozoic Eon produced an extensive record of deposition in broad, shallow seas — a pattern that differed substantially from the Archean record of deep-water deposition, which is now largely confined to greenstone belts and adjacent areas. In addition, more Proterozoic than Archean sedimentary rocks remain unmetamorphosed and are therefore accessible for study.

The extensive deposits of Proterozoic age document ancient mountain-building events that are strikingly similar to those of the Appalachians and other younger orogenic belts, and they reveal records of major intervals of glaciation, at least one of which seems to have affected most of the world. Also present in Proterozoic rocks is a fossil record of organic evolution that reveals a transition from the simplest kinds of single-celled organisms at the start of the Proterozoic Eon to more advanced single-celled forms and, finally, to multicellular plants and animals, some of which belonged to modern phyla.

This array of global events is the subject of this chapter. We will also view the Proterozoic world on a regional scale and learn how the modern continents began to take shape.

Stromatolites are present in Archean rocks but are much more abundant in Proterozoic rocks, where they are the most conspicuous fossils. Large stromatolites are rare today, but examples in Shark Bay, Western Australia *(above)*, closely resemble 2-billion-year-old examples *(below)*. The long axes of the living forms parallel the prevailing wave direction. The Proterozoic stromatolites are from the Great Slave Lake area of Canada. Their long axes must have lain parallel to the prevailing wave direction and perpendicular to the ancient shoreline. *(P. F. Hoffman.)*

A MODERN STYLE OF OROGENY

As we saw in Chapter 8, cratons of modern proportions first began to form about 3 billion years ago, late in Archean time. This was also the time when the oldest known sedimentary deposits

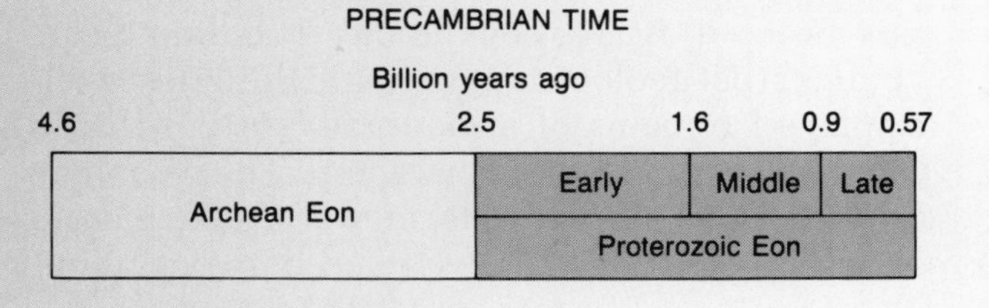

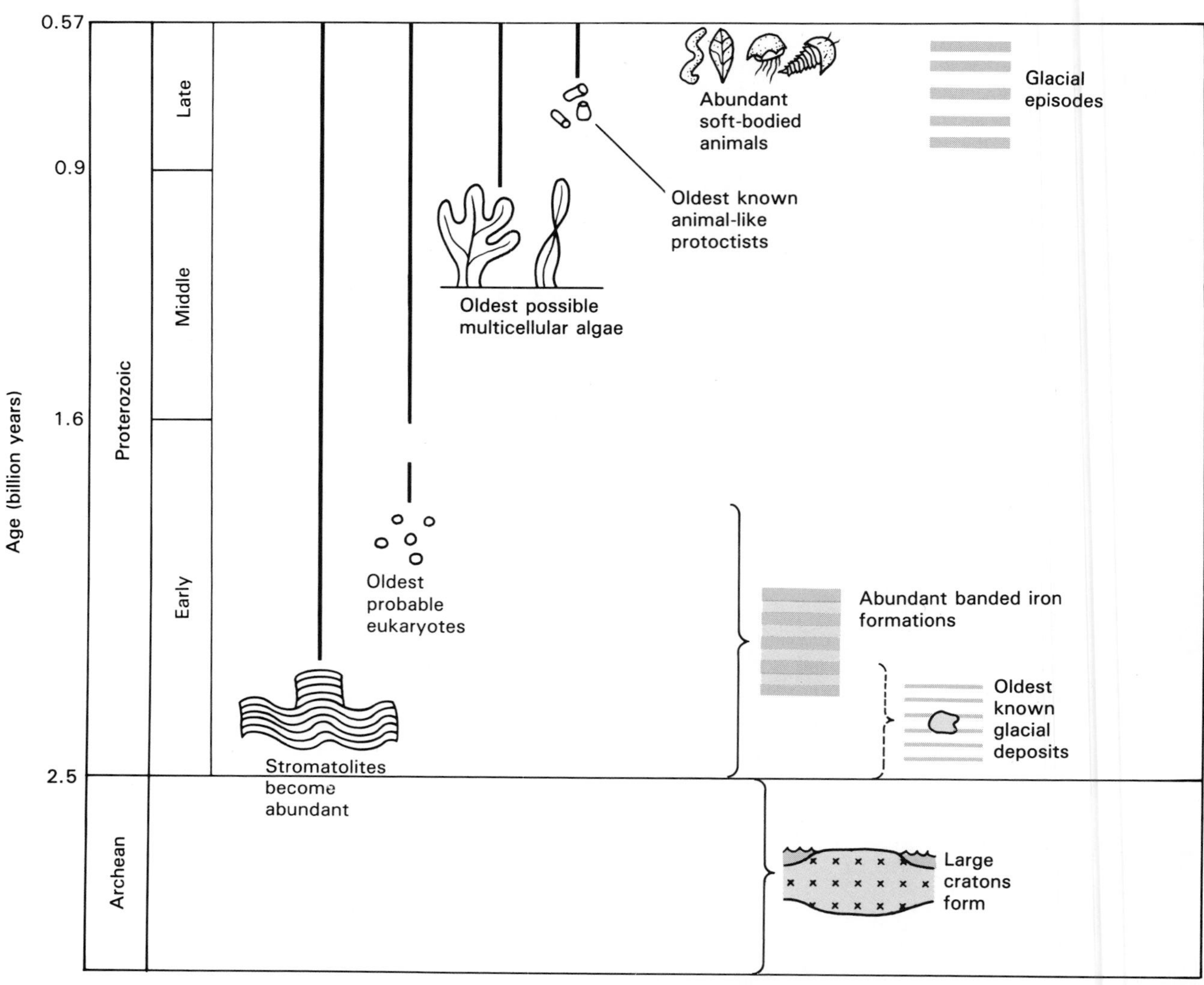

Iron formations continued to appear in abundance early in Proterozoic time, and the oldest known glacial sediments were deposited. Perhaps in response to the presence of broad continental margins, stromatolites became more plentiful. The eukaryotic cell apparently evolved between 1.8 and 1.6 billion years ago. The oldest fossils of multicellular algae are only slightly younger, but complex animals did not evolve until latest Proterozoic time, after glaciers spread across many areas of the world.

were laid down over substantial continental areas in southern Africa (p. 201). Although mountain-building processes resembling those of the Phanerozoic world were undoubtedly in operation by this time, it is in rocks about 1 billion years younger that geologists have found the oldest well-displayed remains of a mountain system that is thoroughly modern in character. The Wopmay orogen of Canada, which formed along the margin of an early continent, developed between about 2.1 and 1.8 billion years ago, over a large area that is now approximately 100 kilometers (~60 miles) to the west of Hudson Bay. Today remarkably well-preserved sedimentary rocks of this orogen are exposed along the low-lying surface of the Canadian Shield as a result of continental glaciation that has repeatedly scoured the orogenic belt over the past 2 million years.

The present Epworth Basin, which is the site of the Wopmay orogen, lies along the western

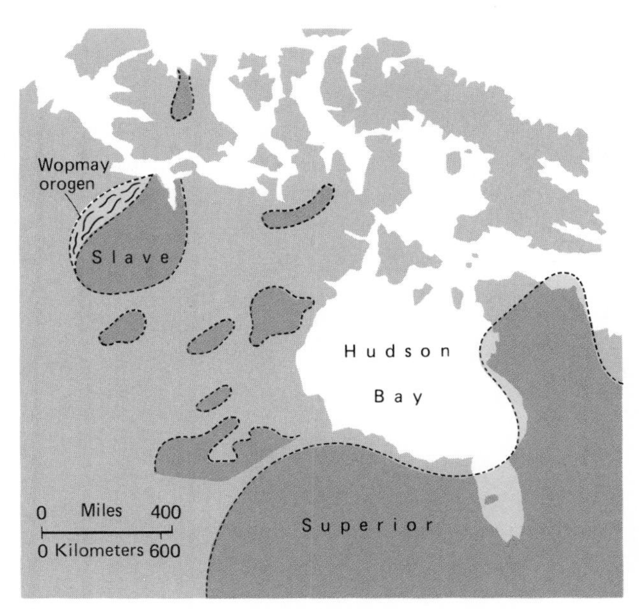

FIGURE 9-1 The Wopmay orogen, which formed about 2 billion years ago along the margin of the Slave Province of northwestern Canada. The Slave Province and other regions of Archean terrane are shown in brown. Figure 8-12 shows their locations in North America. *(After P. F. Hoffman, Philos. Trans. Roy. Soc. London A233:547–581, 1973.)*

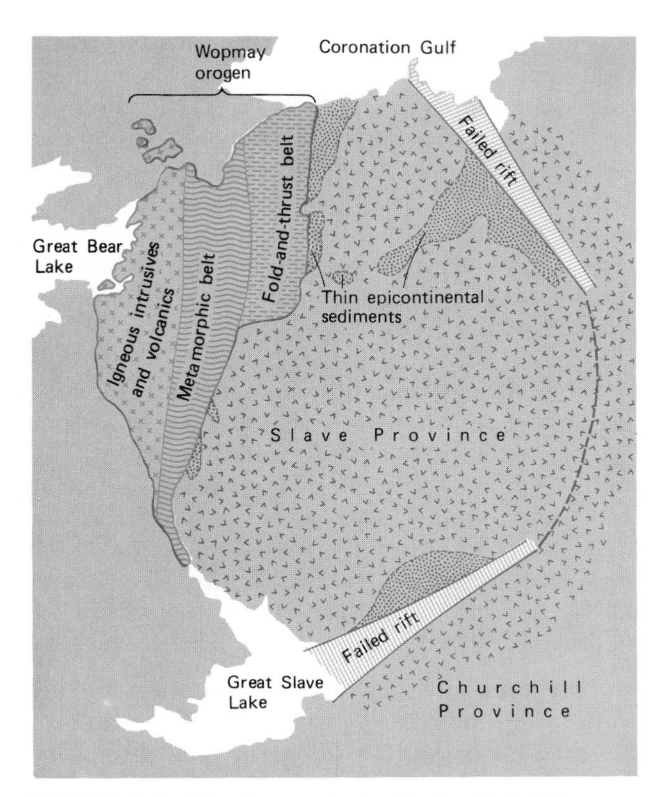

FIGURE 9-2 The three principal belts of the Wopmay orogen (shaded) and associated features. While sediments accumulated along the margin of the Slave craton where the orogen later developed, thinner sequences accumulated inland on the craton and thick sequences in failed rifts that cut into the craton. *(After P. F. Hoffman, Philos. Trans. Roy. Soc. London A233:547–581, 1973.)*

margin of the geologic region known as the Slave Province (Figures 9-1 and 9-2) and is floored by an ancient fold-and-thrust belt (Figure 9-3). Although it has long been planed off by erosion, this ancient zone of deformed rocks bears a striking resemblance to the younger belts described in Chapter 7. In the Wopmay orogen, thrusting was toward the east, and igneous intrusions associated with the deformation now lie primarily within the Bear Province to the west. A belt of metamorphism lies between the igneous belt and the fold-and-thrust belt. To the east, epicontinental sedimentary rocks continuous with those of the fold-and-thrust belt are flat-lying except where they lie adjacent to local upwarps of underlying rock. Like sedimentary deposits of younger fold-and-thrust belts, those of the Wopmay belt show a clear relation to tectonic history. Near the end of Archean time, before the orogen was formed, most of what is now called the Slave Province existed as a discrete craton. Then, early in the Proterozoic Eon, the sedimentary sequences

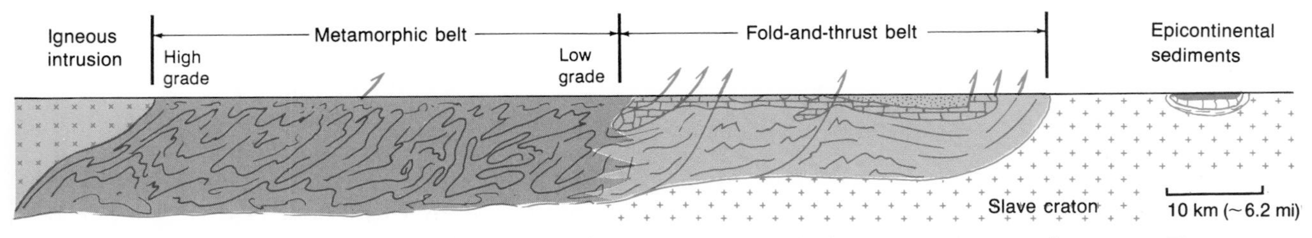

FIGURE 9-3 A cross section of the northern part of the Wopmay orogen (depicted with vertical exaggeration). Compare the three belts as shown here in cross section with their appearance on the map in Figure 9-2. *(After P. F. Hoffman, Philos. Trans. Roy. Soc. London A233:547–581, 1973.)*

developed on the interior of this craton, and thick shelf deposits accumulated along its western margin. As in younger mountain belts, the shelf deposits were succeeded by flysch and then by molasse deposits. In fact, the thick sequence of deposits in the Wopmay fold-and-thrust belt closely resembles that found in each of the tectonic cycles of the Appalachian orogen (Figure 7-13). The Wopmay sequence has the following characteristics:

1 The first thick deposit, which formed along the edge of the continental shelf, is a quartz sandstone that prograded toward the basin (Figure 9-4). This unit resembles the sequence of Lower Cambrian shelf sands at the base of the first Appalachian tectonic cycle. The quartz sandstones of the Wopmay orogen grade westward into deep-water mudstones and turbidites that now lie within the metamorphic belt.

2 Next come rocks in which stromatolites and dolomite predominate. These rocks formed along a Proterozoic carbonate platform that resembled the Cambro-Ordovician platform of the first Appalachian tectonic cycle. In the Proterozoic platform, sedimentary cycles record repeated progradation of tidal flats across a shallow lagoon. Laminated dolomite that formed in the lagoonal environment is at the base of each cycle, while at the top are oolitic or stromatolitic deposits that must have formed in environments fringing the lagoon on its landward side (Figure 9-5). Enormous stromatolite mounds grew to the west, along the shelf margin. These mounds formed a persistent barrier, behind which the fine-grained deposits of the lagoon were trapped. Thus stromatolites bounded the lagoon on both its landward and seaward margins. The present metamorphic zone consists of a thinner sequence of mudstones that represent deeper environments beyond the shelf edge, together with beds of dolomite breccia that contain blocks as large as 50 meters (~ 165 feet) in length. It is obvious that these beds were transported down the steep slope in front of the shelf edge by catastrophic flows of submarine debris, much like the flows that produced the breccia shown in Figure 7-15.

3 The carbonate platform deposits give way to transitional mudstones, which reflect a downwarping of the platform as a foreland basin was formed.

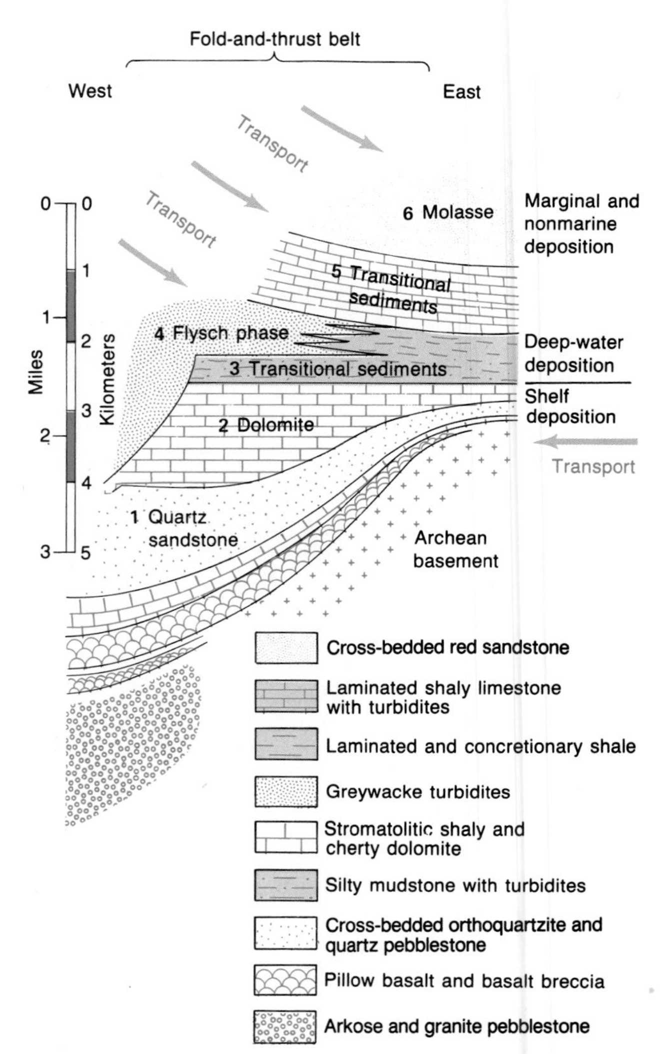

FIGURE 9-4 The sequence of development of sediments in the fold-and-thrust belt of the Wopmay orogen. Numbers refer to depositional units described in the text. Units 1 and 2 represent marine deposition along a shallow continental shelf. Units 3 and 4 are deep-water deposits, including flysch, that accumulated when the shelf foundered as mountain building began to the west. Unit 5 consists of shallow-water deposits transitional between flysch below and molasse above. Unit 6, the molasse phase of deposition, followed the exclusion of marine waters by a heavy influx of sediment from the west. *(After P. F. Hoffman, in M. R. Walter [ed.], Stromatolites, Elsevier Publishing Company, Amsterdam, 1976.)*

4 The mudstones are followed by flysch deposits (turbidites) similar to Upper Ordovician flysch of the first Appalachian tectonic cycle. A westward thickening of the Wopmay flysch, and the inclusion within it of particles derived from plutonic

FIGURE 9-5 Stromatolites within shelf deposits of the Wopmay orogen. *(P. F. Hoffman, Geological Survey of Canada.)*

rocks to the west, indicate that the source area of siliciclastics now lay seaward of the orogenic belt, just as it did in the development of each Appalachian tectonic cycle, when the foreland basin formed. Thus it is clear that the Wopmay foreland basin formed when plutonism and orogenic uplift began in the offshore area to the west.

5 The deep-water turbidites grade upward into beds containing mud cracks and stromatolites, both of which formed in shallow-water environments and thus point to a shallowing of the foreland basin.

6 The influx of sediments eventually pushed marine waters from the Wopmay foreland basin, and the Wopmay cycle, like younger tectonic cycles, ended with an interval of molasse deposition. The Wopmay molasse consists largely of river deposits in which cross-bedding is conspicuous.

Two kinds of evidence suggest that the Proterozoic Wopmay orogen had the same pattern of formation as a modern orogenic system. First, the parallel igneous, metamorphic, and fold-and-thrust belts resemble similarly arranged belts of younger mountain ranges (Figure 9-3). Second, within the fold-and-thrust belts, shallow-water shelf deposits are succeeded by flysch deposits that give way to molasse deposits.

Another indication that normal plate-tectonic processes were operating in northern Canada about 2 billion years ago is evidence of continental rifting. East of the Wopmay orogen, slicing into the platform from the northwest and southwest, are deep, narrow troughs containing thick sedimentary sequences (Figure 9-2). These troughs represent failed rifts—that is, rift systems that ceased to be active before they cut across the entire Slave craton.

It seems likely that the westward-facing edge of the Slave craton, along which the Wopmay orogen formed, came into being more than 2 billion years ago, when a rift system broke apart a slightly larger continental mass. More or less simultaneously, the two rifts that later failed began to form at right angles to the newly forming shelf edge. Shortly thereafter, about 2 billion years ago, the shelf edge must have foundered when it was rafted up against a subduction zone. Then, as we have seen for younger orogens, igneous activity elevated the crust seaward of the shelf edge and mountain building began.

PROTEROZOIC GLACIAL DEPOSITS

The fact that the Slave terrane along the Wopmay orogen behaved like rigid continental crust when it was rifted and deformed indicates that by 2 billion years ago Earth was markedly cooler than it had been 3 billion years ago, when magmas were pushing up from the mantle to the surface throughout much of what is now the Canadian Shield. That climates in this region were also quite cool approximately 2 billion years ago is shown by evidence that glaciers spread over the land.

Early Proterozoic Glaciation

Just to the north of Lake Huron in southern Canada are some of the most spectacularly exposed ancient glacial deposits: those of the Gowganda Formation, which forms part of the Huronian Supergroup. Well-laminated mudstones in this formation consist of varves that formed in standing water in front of glaciers. In Chapter 3 these ancient deposits were compared to the strikingly similar glacial varves that formed nearby, where Toronto is now located, just a few thousand years ago (Figure 3-6). Some of the laminated Gowganda mudstones contain drop-stones—pebbles and cobbles that appear to have fallen to the bottom of a sea or lake from ice that melted as it floated out from a glacial front (Figure 3-7). Whether the mudstones accumulated in an ocean or in a lake is not clear, but it is evident that they alternate with tillites, which seem to have been deposited when glaciers encroached on the body of water. Some of the pebbles and cobbles of these tillites are faceted or scratched from having slid along at the bases of moving glaciers.

Although the exact age of the Gowganda deposits has not been determined, it is known that the unit is slightly more than 2 billion years old, because it rests on 2.6-billion-year-old crystalline rocks and is intruded by rocks that are 2.1 billion years old. Tillites of similar age are found elsewhere in Canada as well as in Wyoming, Finland, southern Africa, and India. Thus it would appear that there was an interval of extensive continental glaciation not long after the transition from Archean to Proterozoic time.

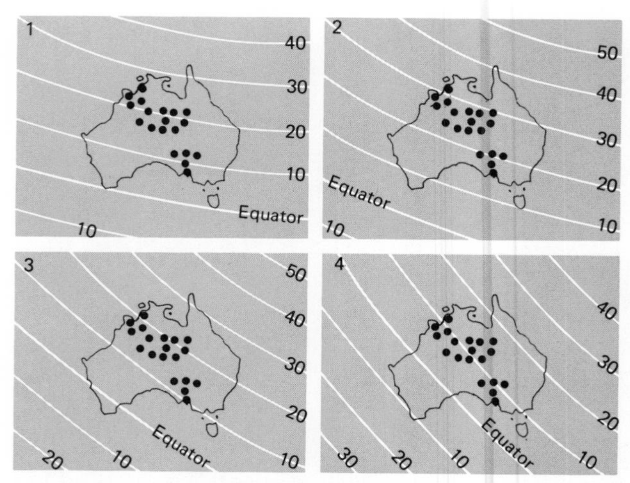

FIGURE 9-6 Latitudinal positions of Australia at four successive times during the late Proterozoic interval in which it was subjected to glaciation. The glaciated regions lay within 30° of the ancient equator at all times. *(After M. C. McWilliams and M. W. McElhinny, Jour. Geology 88:1–26, 1980.)*

Late Proterozoic Glacial Episodes

Rocks ranging in age from less than 1 billion years to about 2.3 or 2.4 billion years show little evidence of glacial activity, indicating that after the early Proterozoic period of glaciation, continental glaciers were rare or disappeared altogether for about 1.5 billion years. Then, between about 850 and 600 million years ago, in late Proterozoic time, a series of glacial episodes occurred. Tillites and other glacial deposits of late Proterozoic age can be found on all major continents of the world except Antarctica. In North America, for example, late Proterozoic tillites are widely distributed along both the Cordilleran and the Appalachian orogens.

Although correlation from one part of the world to another is difficult, the presence of two or more discrete tillites at each of several localities indicates that the world experienced several glacial episodes during late Proterozoic time. At least two and possibly three or more of these episodes are known to have occurred in Africa, for example. The last took place about 600 million years ago, when glacial deposits accumulated in many parts of the world.

The global distribution of late Proterozoic tillites is puzzling because it is difficult to comprehend how glaciers could have spread over so many continents. Indeed, the limited paleomagnetic data now available suggest that even regions lying close to the equator experienced some degree of continental glaciation during this time. Almost all of Australia, for example, seems to have lain within 30° of the equator throughout the latter part of Proterozoic time (Figure 9-6), and yet glaciation in Australia was extensive. It is possible that most of Earth was at times quite cold late in Proterozoic time.

ATMOSPHERIC OXYGEN

In Chapter 8 we learned that Earth's primitive atmosphere, which formed by the degassing of the planet, contained little or no free oxygen. It is not known precisely when oxygen reached its present level in the atmosphere, but it is now widely agreed that a moderate level was reached about 2 billion years ago, early in Proterozoic time. Thus the presence of very low levels of atmospheric oxygen seems to be another characteristic that distinguishes the Archean world from all subsequent times except the earliest part of the Proterozoic Eon. Before we look at the evidence that supports this conclusion, we must first consider why the atmosphere serves as a reservoir of free oxygen today.

It would appear that the concentration of oxygen in Earth's atmosphere has not varied substantially over the past few hundred million years, even though oxygen is continuously removed from the atmosphere by such processes as oxidation of minerals at Earth's surface and respiration. (Respiration is the process by which organisms employ oxygen to obtain energy from their food; in effect, organisms use oxygen to burn up their food in the same manner that fire oxidizes a flammable material, releasing energy in the form of heat.) The atmospheric reservoir of oxygen persists because oxygen is continuously returned to it by photosynthesis and, to a lesser extent, by the breakdown of water in the upper atmosphere by the sun's rays (Figure 2-6).

A number of factors prevent the concentration of atmospheric oxygen from increasing significantly. Were oxygen to build up much beyond its present level, for example, chemical weathering of sediments and rocks would become more intense and would consequently "soak up" excess oxygen. If plants became markedly more abundant on Earth and thus began to liberate more oxygen, animals and simpler respiring organisms, including bacteria, would also become more abundant because more plants and plant debris would be available for them to feed on. The resulting increase in the rate at which oxygen was consumed by respiration would then offset the increase in the amount of oxygen produced by plants.

With these factors in mind, let us now examine the evidence supporting the assumption that low levels of atmospheric oxygen existed until early Proterozoic time.

The Case for Early Anaerobic Evolution

One piece of evidence suggesting that Earth's earliest atmosphere was anoxic (oxygen-free) lies in the fact that the chemical building blocks of life could not have formed in the presence of atmospheric oxygen. Indeed, chemical reactions that yield amino acids from simpler compounds in the laboratory (p. 205) are inhibited by even smaller amounts of oxygen than those found in Earth's atmosphere.

Furthermore, oxygen prevents the growth of the most primitive living bacteria, including purple photosynthetic bacteria and bacteria that obtain energy from fermentation. As we have seen, photosynthetic bacteria as well as methane-producing bacteria and bacteria that derive energy from fermentation rather than from respiration are currently restricted to anoxic habitats such as swamps, ponds, and lagoons (p. 206). For this reason, these bacteria are called *anaerobic.* The fact that the simplest living cellular organisms are anaerobic suggests that Earth's first forms of cellular life had similar metabolisms, and it also suggests that these organisms may have evolved in this way because there was virtually no oxygen in the atmosphere when they came into being. If this was the case, the subsequent buildup of atmospheric oxygen limited these bacteria to the environments they now occupy.

The Oxidation State of Minerals

Also widely cited as evidence of an Archean atmosphere containing relatively little oxygen is the distribution of uranium and iron minerals in Precambrian rocks. Under the atmospheric conditions of the modern world, the uranium oxide mineral uraninite (UO_2) is readily oxidized further and quickly dissolves from rocks. The iron sulfide mineral pyrite (FeS_2), known as "fool's gold," also disintegrates readily when it is exposed to the modern atmosphere. Today uraninite and pyrite are seldom found in the sediments of rivers and beaches, and they are similarly rare in ancient siliciclastic sediments younger than about 2 billion years. In contrast, both minerals are relatively abundant in buried nonmarine and shallow marine siliciclastic deposits that are older than 2 billion years; uraninite, for example, has been found in such rocks on five continents.

Significantly, red beds (see Appendix I) display the opposite pattern: They are never found in terrains older than 2.2 or 2.3 billion years. Hematite, a highly oxidized iron mineral, gives red beds their color. Often the hematite found in red beds has formed secondarily by oxidation of other iron minerals that accumulated with the sediments. During Phanerozoic time, oxygen has been plentiful in Earth's atmosphere, so that this secondary oxidation has often occurred within a few millions or tens of millions of years after the sediments were deposited. It would appear that oxidation of this type did not occur early in Earth's history.

Some researchers have argued that red beds cannot be found in Archean sedimentary sequences because these sequences formed mainly in deep marine basins, whereas most red beds are of nonmarine origin. The Huronian and Witwatersrand sequences are especially significant in this regard because they formed more than 2 billion years ago on sizable cratons (p. 201). Although both of these sequences comprise large volumes of nonmarine sediments and would therefore be expected to include some red beds under modern atmospheric conditions, both lack red beds but contain substantial quantities of detrital pyrite and uraninite. This condition has been widely cited as evidence that oxygen concentrations remained low until 2 billion years ago or later.

Banded iron formations are other sedimentary rocks whose distribution may reflect the history of Earth's atmosphere. Rocks of this type are rare except in Precambrian terranes. Although banded iron formations are present in Archean terranes and, as we have seen, are among the oldest known rocks on Earth (Figure 8-13), most accumulated early in the Proterozoic Era, between about 2.5 and 1.8 billion years ago. The term *banded iron formation* refers to a bedding configuration in which layers of chert that is sometimes contaminated by iron alternate with layers of other minerals that are richer or poorer in iron than the chert (Figure 9-7). The iron in these formations may take any of a variety of forms, including iron oxide, iron carbonate, iron silicate, and iron sulfide. However, the present chemical form of the iron is not necessarily the form in which it was originally deposited, and in many cases there is evidence that the mineralogy of the iron has altered over time. Banded iron formations account for most of the iron ore that is mined in the world today. One reason for their economic value is that many contain iron in the form of magnetite (Fe_3O_4), whose iron-to-oxygen ratio is higher than that of hematite (Fe_2O_3).

Banded iron formations accumulated in offshore waters. Many are associated with turbidites. Both the iron and the silica in these sediments appear to have come from hot, watery emissions from the seafloor associated with igneous activity. The kind of layer that was deposited at any time probably depended on the chemical composition of nearby watery emissions.

The fact that the iron in banded iron formations was at least weakly oxidized when these structures first formed implies that some oxygen was present in the environment at this time.

FIGURE 9-7 A weakly metamorphosed banded iron formation in northern Michigan that is about 2 billion years old. *(Bruce Simonson, Oberlin College.)*

Precambrian soils offer additional testimony. These thin units, although rare, when well preserved reveal the chemical nature of weathering during the time the soil was formed. The concentration of oxygen in the atmosphere influences this weathering. Studies of the minerals that constitute Precambrian soils indicate that atmospheric oxygen had reached at least 15 percent of its present level by 2 billion years ago—and may even have reached its present level by this time.

Photosynthesis and Oxygen

It is widely agreed that photosynthesis caused atmospheric oxygen to build up during Precambrian time. Before such a buildup could occur, however, natural reservoirs known as **oxygen sinks** had to be filled. Essentially, oxygen sinks are chemical elements and compounds that combine readily with oxygen and that are believed to have been present in the crust or atmosphere immediately after Earth formed. Sulfur and iron are two of the most important oxygen sinks. (Note how iron that we extract from naturally occurring compounds rusts when it is exposed to the oxygen in the atmosphere.)

Stromatolites and Oxygen

Today photosynthesis liberates oxygen at such a high rate that it could refill the atmosphere within just a few thousand years. Given the probable presence of cyanobacteria by at least 3.5 billion years ago, how could it have taken more than a billion years for atmospheric oxygen to fill major sinks and then approach a high level? The answer may well be that there were simply not enough stromatolites to liberate oxygen at a high rate until 2.3 or 2.2 billion years ago, when the geologic record reveals that stromatolites became very abundant (Figure 9-8; Major Events, p. 212). Their earlier rarity may be attributable to certain geologic factors. Perhaps the absence of large continents and continental shelves during Archean and Early Proterozoic time restricted the growth of stromatolites to the steep margins of small Archean landmasses, which offered little space for colonization.

It is also likely that nutrients were not easily elevated from the deep sea to shallow water for use by stromatolites during Archean time. Along the margins of continents today, you will recall, currents drag surface waters away from the shore, and nutrient-laden waters from great depths replace the surface waters (p. 45). During Archean time, however, such upwelling may have been weak and nutrient supplies sparse because there were no large continents. Then, as continents grew, more extensive upwelling may have accelerated the rate of photosynthesis and oxygen production.

LIFE OF THE PROTEROZOIC EON

It appears that even as late as the Early Proterozoic interval, after an estimated 1.5 billion years of evolution, Earth was populated exclusively by single-celled prokaryotic forms of life. In the course of Proterozoic time, however, a great expansion of life issued from these simple forms. Major Events (p. 216) outlines what we now know of this expansion and depicts other Proterozoic

FIGURE 9-8 A large reeflike mound built by stromatolites in Proterozoic rocks of Victoria Island, Arctic Archipelago, Canada. Such structures are common in Proterozoic rocks. This mound, which is 6 meters (~20 feet) in diameter, has been exposed by erosion. *(G. M. Young, Precambrian Res. 1:13–41, 1974.)*

events described earlier. We will soon examine what is known about advanced forms of life that evolved before the start of the Cambrian Period, but first let us review the fossil evidence supporting the continued dominance of prokaryotes early in Proterozoic time. This evidence takes the form of stromatolites and molds of microscopic cells.

Stromatolites

We know that stromatolites first became abundant in the fossil record about 2.3 or 2.2 billion years ago, perhaps because the area covered by continental shelves increased. Stromatolites remained abundant throughout the rest of Precambrian time and did not decline until early in the Paleozoic Era, when animals diversified greatly and (as we shall see in Chapter 10) inhibited the growth of stromatolites. Proterozoic rocks contain stromatolites of many shapes, some of which form reeflike structures (Figure 9-8). Attempts to use these fossils for stratigraphic correlation, however, have been hindered by the fact that filamentous cyanobacteria of a single type often produce stromatolites that have different shapes in different environments (p. 210 and Figure 9-8). Similar ecologic influences on the form of these organisms during Precambrian time have made it difficult to identify evolutionary trends that might make fossil stromatolites useful for dating purposes.

Fossil Prokaryotic Cells

One of the most interesting fossil discoveries since the middle of this century has been the identification of cell remains in Precambrian rocks. Such remains were first identified in the Gunflint Chert of Ontario and northern Minnesota. The Gunflint fossils, which are about 1.9 billion years old, display a wide variety of shapes (Figure 9-9) and appear to represent bacteria or cyanobacteria, as do the members of other microscopic fossil assemblages of early Proterozoic age. Analysis of these fossils has revealed not only that prokaryotic forms of life persisted from Archean time into the Proterozoic Eon, but also that many of the forms found in Precambrian rocks closely resemble forms that are alive today.

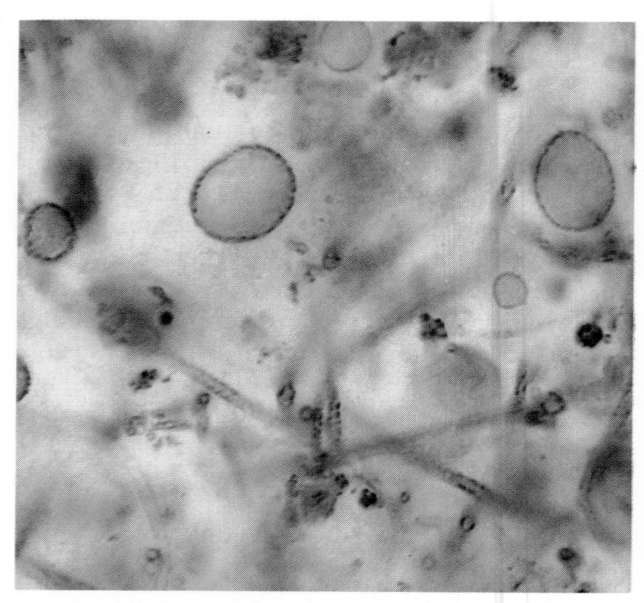

FIGURE 9-9 Fossil cells of the Gunflint Formation of the Lake Superior region. Both filamentous and spheroidal forms are present. The large spherical forms are about 10 μm in diameter. *(Andrew H. Knoll.)*

The Earliest Eukaryotes

All forms of life except cyanobacteria and bacteria are eukaryotes — that is, forms whose cells contain chromosomes, nuclei, and other advanced internal structures. Eukaryotes apparently evolved about 1.8 billion years ago. Thick-walled spherical cells that were apparently eukaryotes are abundant in rocks between 1.8 and 1.6 billion years old, as are chemical compounds that occur in eukaryotes but not in prokaryotes.

Fossil eukaryotic cells called **acritarchs** first appear in slightly younger rocks. These near-spherical or many-pointed forms are the dominant group of algal plankton found in the Paleozoic fossil record (Figure 9-10). Some acritarchs are believed to have been the resting stages (or cysts) of dinoflagellate cells. Dinoflagellates constitute one of the most important groups of planktonic algae today (p. 43) and are known to be eukaryotic. We do not know for certain, however, that all acritarchs were dinoflagellates. In fact, some experts believe that Proterozoic acritarchs were not dinoflagellates but instead belonged to the group

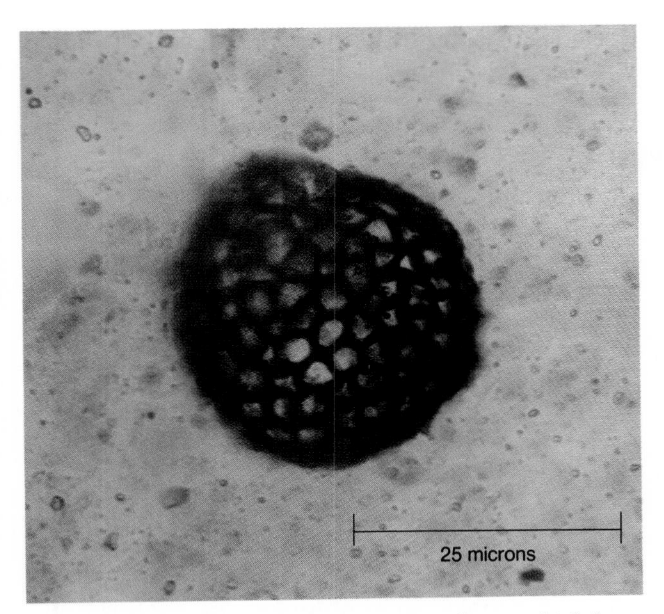

FIGURE 9-10 The acritarch *Dictyotidium*, which has a thick, complex wall structure that is unknown in prokaryotes. This fossil is about 750 million years old and has a diameter of about 35 μm. *(Nicholas Butterfield.)*

known as green algae. In any event, the large size of many Proterozoic acritarchs, together with the chemical composition of their cell walls, suggests that acritarchs were eukaryotic. It is also significant that the cell walls of some Proterozoic acritarchs exhibited complex patterns similar to those that today are restricted to eukaryotes. Interestingly, the fossil record of acritarchs, like that of large cells in general, begins in rocks about 1.4 billion years old.

Despite the antiquity of the acritarch fossil record, acritarchs are rarely found in rocks older than about 850 or 900 million years. In contrast, they are both abundant and diverse in younger rocks. In fact, the fossil record reveals that acritarchs underwent a rapid adaptive radiation between 900 or 850 and 700 million years ago. For this reason, acritarchs are very useful for dating rocks of the latest Precambrian age.

It is interesting to note that acritarchs suffered a mass extinction about 600 million years ago, when glaciers spread across the world (p. 216). Only a few simple spheroidal types survived this crisis, which may have resulted from worldwide cooling.

Early Evolution of Eukaryotes

Because the fossil record of single-celled eukaryotes lacking skeletons is very poor, most of our ideas about the early evolution of this group are based on evidence derived from living organisms. The most remarkable of these ideas, which was not widely accepted until the 1970s, is that the eukaryotic cell arose from the union of two or more prokaryotic cells, at least one of which came to reside within another. According to this theory, the prokaryotic cell that first came to live inside another was altered in minor ways to form a structure called a **mitochondrion,** one or more of which is present in nearly all eukaryotic cells (Figure 9-11). Mitochondria are, in fact, the structures that allow cells to derive energy from their food by means of respiration. In mitochondria, complex compounds are broken down by oxidation, which yields energy and, as a by-product, carbon

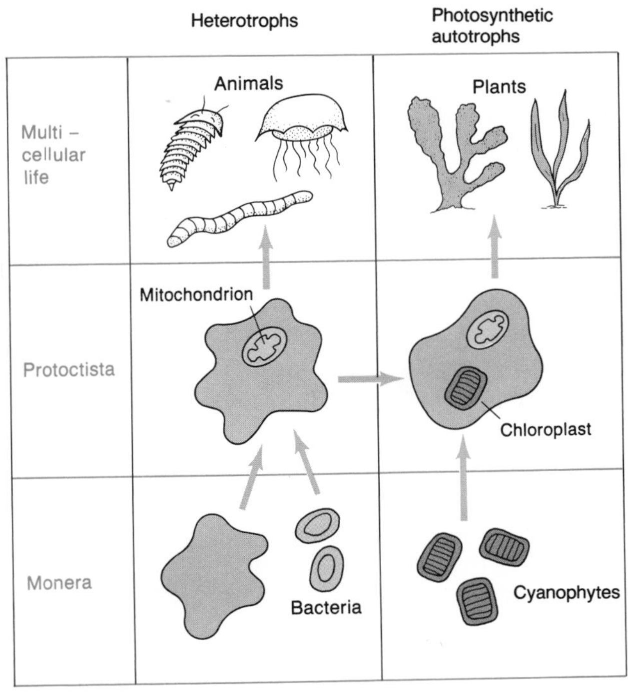

FIGURE 9-11 The probable sequence of major events leading from Monera to multicellular animals and plants. The first protoctist apparently evolved when one moneran engulfed but failed to digest another, which then became a mitochondrion. A plantlike protoctist evolved when an animal-like protoctist engulfed but failed to ingest a cyanobacterium cell, which then became a chloroplast.

dioxide. A mitochondrion is apparently the slightly modified evolutionary descendant of a small bacterium that became trapped within a larger one, as evidenced by the presence in mitochondria of DNA and RNA, both of which are essential to the independent existence of any cell. It is assumed that the smaller cell that became a mitochondrion was eaten by the larger one but proved resistant to the digestive processes of the predator cell.

Single-celled eukaryotes and a few simple multicellular organisms constitute the kingdom Protoctista (Figure 1-5), and single-celled animal-like members of this kingdom (that is, consumers rather than producers) are called **protozoans.** It is believed that the earliest protoctists were protozoans, because their immediate ancestors were consumers that had eaten bacteria, which later became mitochondria. Among the more familiar living protozoans are amoebas and ciliates (Figure 9-12). A minority of protozoan groups, including foraminifera (p. 80), have skeletons that can be fossilized.

When did the protozoans first evolve? The oldest known protozoans in the fossil record, a group of organisms with rigid, vase-shaped skeletons, are no older than about 0.8 billion years (Figure 9-13). The earlier existence of protozoans without these durable skeletons can be inferred from the fact that protozoans were apparently the first eukaryotes and, as we have seen, eukaryotes probably existed as early as 1.8 billion years ago.

It is widely agreed that plantlike protoctists, like protozoans, evolved as a result of the union of two kinds of cells. In this major step in the evolution of the Protoctista, a protozoan consumed and retained a spherical or oblong cyanobacteria cell. This cell then became an intracellular body known as a **chloroplast** (Figure 9-11), which served as the site of photosynthesis both in plantlike protoctists and in higher plants, which evolved from them. The similarity between cyanobacteria and chloroplasts is striking. In both, for example, the pigment chlorophyll, which absorbs sunlight and permits photosynthesis, is located on layered membranes.

In many kinds of protoctists, photosynthesis is conducted within chloroplasts, and it is generally believed that plantlike protoctists evolved many times when protozoans retained within their cells the cyanobacteria they had eaten. Thus mobile

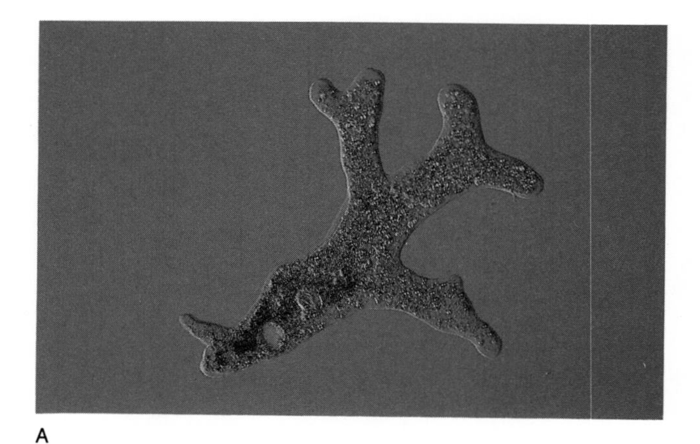
A

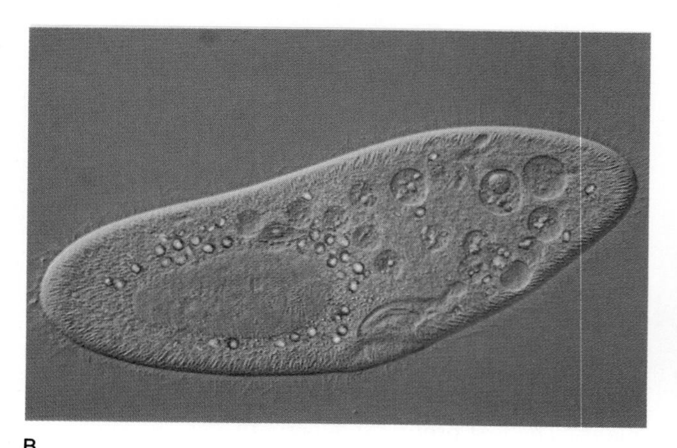
B

FIGURE 9-12 Living protozoans. *A.* An amoeba, which is a living animal-like protoctist with a changeable shape. *B.* The ciliate *Paramecium*, whose surface is covered with cilia. *(M. Walker/NHPA.)*

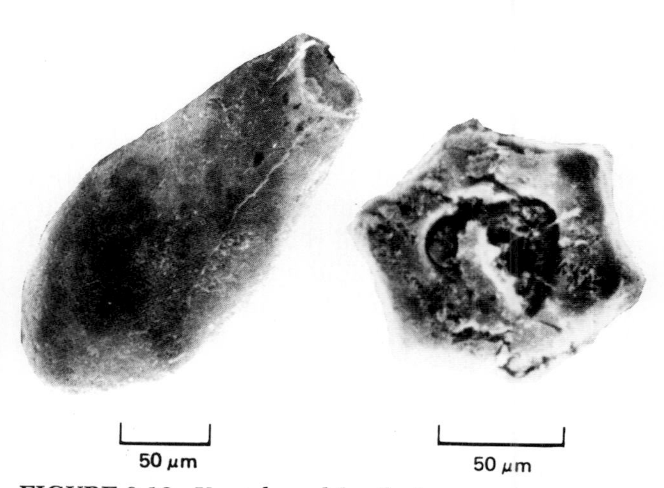
50 μm   50 μm

FIGURE 9-13 Vase-shaped fossils that appear to represent animal-like protoctists from the late Proterozoic Chuar Group of the Grand Canyon. *(B. Bloeser.)*

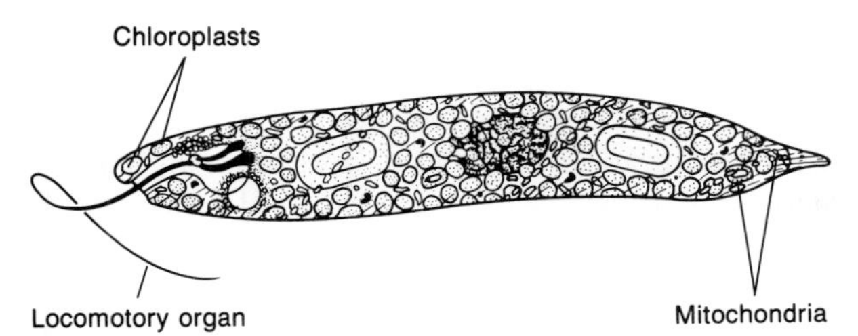

FIGURE 9-14 ***Euglena*, a plantlike protoctist, which, like many animal-like protoctists, is mobile.** This cell, which is about 60 μm long, contains numerous chloroplasts.

protozoans may have evolved into mobile photosynthetic forms such as *Euglena* (Figure 9-14), while immobile protozoans evolved into immobile photosynthetic forms.

Multicellular Algae

Algae (singular, alga) is a term that encompasses a variety of forms. Photosynthetic protoctists, for example, are called algae, as are the prokaryotic cyanobacteria and a large group of multicellular plants. The multicellular plants that are known as algae differ from more advanced land plants in that they lack multicellular reproductive structures to protect their eggs and embryos.

Today algae that are composed of many connected cells can be found in both freshwater and marine environments. Some groups form lettucelike carpets along rocky seashores, and others comprise the great kelps that rise up from some seafloors in twisted ribbons tens of meters long. Even these large forms, however, are fleshy structures that decay easily and thus are unlikely to be fossilized. Some ribbonlike fossils of Proterozoic age are probably segments of cyanobacteria mats that only superficially resemble multicellular algae. U-shaped fossils from the 0.8- or 0.9-billion-year-old Little Dal Group of northwestern Canada, however, have smooth, nearly symmetrical outlines (Figure 9-15); they and a few other Proterozoic fossils as old as 1.1 or 1.2 billion years appear to represent multicellular algae.

FIGURE 9-15 An apparent fossil specimen of multicellular algae from northwestern Canada. The smooth outline of this carbonaceous fossil could not easily have formed by the fragmentation of mats of cyanobacteria. The scale bar represents 10 millimeters (0.39 inch). *(M. R. Walter and H. J. Hoffman.)*

Multicellular Animals

Multicellular animals evolved from animal-like protists rather than from multicellular plants. It is unlikely that any kind of stationary multicellular alga would evolve into an animal that must gather food. On the other hand, it is likely that certain mobile, predatory animal-like protists evolved into higher animals simply by developing multicellular body forms.

Before there were single-celled or multicellular eukaryotic consumers, Earth's ecosystem was relatively simple. Because the primary photosynthetic producers — cyanobacteria and eukaryotic algae — did not suffer predation, their proliferation in aquatic settings was limited only by the supply of nutrients essential to their growth. Seas and lakes were, in effect, saturated with algae — a situation that may have slowed evolution by leaving very little room for the origin of new species. Although the first organisms to feed on algae must have been animal-like protists, today these forms are feeble in this role in comparison with multicellular animals, which, because of their size, can

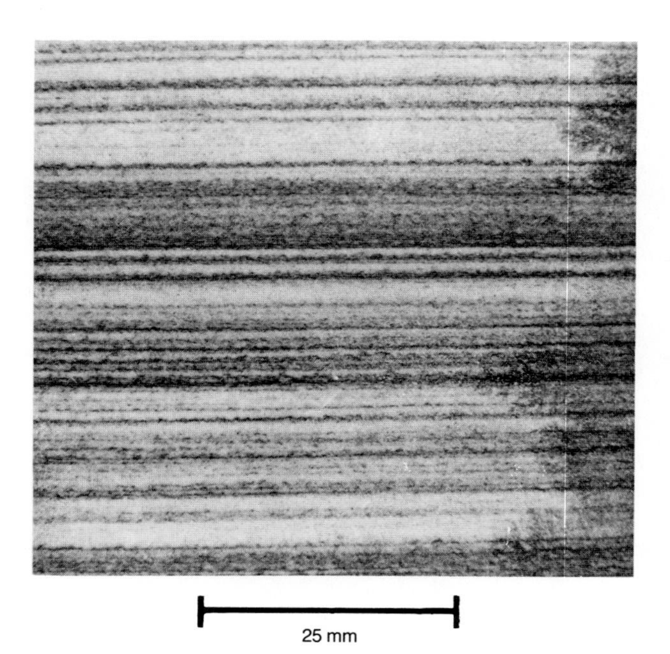

FIGURE 9-16 Laminated siltstone from the 1.3-billion-year-old Greyson Shale (Belt Supergroup) in Montana. Here, as in other rocks older than late Proterozoic, we see no evidence of burrowing by invertebrate animals. *(C. W. Byers.)*

consume algae rapidly. Nonetheless, the origin of eukaryotic consumers added a new nutritional level to many aquatic ecosystems (p. 27). Exactly when multicellular animals evolved remains uncertain, but we now have evidence that it was not until late in Proterozoic time that these organisms first appeared. For many years paleontologists have searched well back into the Precambrian interval for skeletons of multicellular animals, but their search has been unsuccessful. The only skeletons they have found have come from rocks situated close to or above the Precambrian–Cambrian boundary, which, as we will see, cannot be recognized precisely in most areas.

Trace Fossils Even during the last century, it was acknowledged that fossil skeletons appear in the stratigraphic record quite suddenly near the base of the Cambrian System. The complexity and variety of fossilized Cambrian life gave rise to speculation that multicellular animals had a long Precambrian history, during which they lacked hard parts and therefore left no fossil record. One highly effective means of testing this hypothesis is based on the assumption that soft-bodied multicellular animals—those that lack hard parts—have crawled over the seafloor or burrowed into it throughout their existence, and in doing so have often left trace fossils in the form of tracks, trails, and burrows in sedimentary rocks (p. 11). If soft-bodied invertebrate animals had existed for a long interval of Proterozoic time, scientists reasoned, some of them would have left such trace fossils. A search for trace fossils in Precambrian rocks has since turned up a striking pattern—such fossils have been found only in rocks less than about 600 million years old. For example, 1.3-billion-year-old sedimentary rocks of the Belt Supergroup of Montana exhibit no tracks or trails. As Figure 9-16 indicates, Belt mudstones are often strikingly well layered in comparison with younger deposits, in which layers of sediment are often disrupted or destroyed by burrowing animals.

Precambrian trace fossils display a general evolutionary pattern. The oldest trace fossils are geometrically simple structures; often they are tubes made by wormlike animals that burrowed through the sediment. In several regions of the world, as stratigraphic sections progress upward toward and into the Cambrian System, simple trace fossils are replaced by more and more complex types whose shapes also vary increasingly

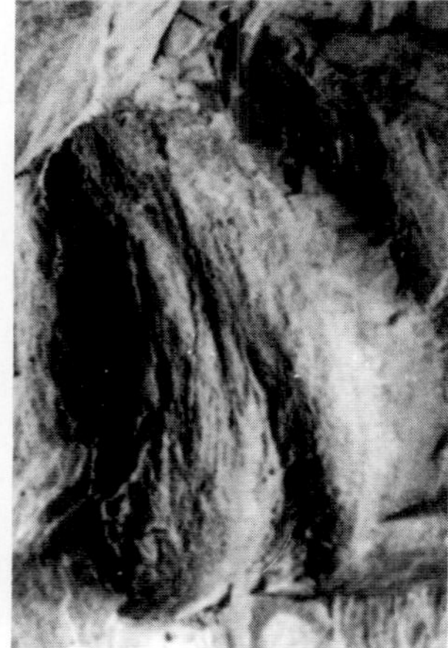

A B

FIGURE 9-17 Undersurfaces of sandstones with fillings of relatively complex burrows in Norway. *A.* Filling of a feeding burrow. *B.* Filling of a shallow burrow on which can be seen scratch marks left by the legs of the animal that dug it. *(Courtesy of N. L. Banks.)*

FIGURE 9-18 Representatives of the late Precambrian Ediacara fauna of Australia. *A.* A problematic flat, segmented form (× 1). *B.* An animal that appears to be intermediate in form between a segmented worm and an arthropod (× 1.7). *C.* A strange form of uncertain biological relationships (× 1.35). *D.* An animal that may be a jellyfish (× 0.7). *E.* An animal that may be a sea pen (× 0.6). *F.* A form that may be related to the echinoderms (× 1.2). *(M. F. Glaessner.)*

(Figure 9-17). This pattern of increase in both complexity and variety seems to represent the initial evolutionary diversification of multicellular animals in the world's oceans. Nowhere in the world have undoubted trace fossils been found below the youngest Precambrian glacial sediments. It appears, then, that little multicellular animal life existed before the final glacial episode.

Imprints of Soft-Bodied Animals Additional evidence supporting the presence of multicellular life in the latest Precambrian interval has come to light since 1950. In many parts of the world, imprints of soft-bodied animals have been found in sandstone below the oldest rocks containing fossil skeletons (Figure 9-18). Many of these fossils have been thought to represent jellyfishes or sea pens, but these identifications have been questioned. Jellyfishes and sea pens are living groups that belong to the phylum **Coelenterata** (the same phylum as corals). Whereas jellyfishes float in the water, sea pens are stalked creatures that stand upright on the seafloor. The Ediacara fauna of Australia is the most famous of late Precambrian "soft-bodied" faunas and was also the first to be recognized. In addition to the possible coelenterates, the Ediacara fauna includes several kinds of animals that cannot be related with certainty to any living group. Fossils of soft-bodied animals such as those of the Ediacara fauna are rare in rocks of Phanerozoic age. It seems likely that the late Precambrian faunas that were preserved escaped destruction only because few predators or scavengers at that early time were capable of devouring their carcasses quickly.

The Delayed Diversification of Animal Life The diversification of animal life near the end of the Proterozoic Eon happened about a billion years after the origin of the eukaryotic cell. For many decades, geologists assumed that some external factor, such as a low level of atmospheric oxygen, accounted for this delay. There is no evidence that atmospheric oxygen had failed to reach a suitable level much earlier in Proterozoic time, however, or that any other condition of Earth's environment thwarted the evolution of animals.

In fact, the delayed diversification of animals may simply have resulted from the time that evolution required to produce key adaptations, such as the nerve cell—a complex message-carrying cell without which coordinated muscular locomotion and advanced modes of feeding are impossible. A single kind of nerve cell characterizes all animals more complex than sponges. There is no comparable kind of cell in sponges or protozoans. Very primitive multicellular forms that lacked neuromuscular systems may have existed long before the end of the Proterozoic without leaving a fossil record.

Plants, in fact, experienced a similar delay. Although fleshy multicellular algae apparently existed 1.1 or 1.2 billion years ago, as we will see in Chapter 11, advanced plants did not invade the land until nearly 400 million years ago. This invasion was delayed by the time required for evolution to produce essential adaptations, including the hollow cells that conduct fluids in all land plants except mosses and their relatives.

FIGURE 9-19 A living annelid worm. Like most marine annelids, this worm has a pair of small, leglike flaps on each of its many segments, but it lacks true legs. *(L. Margulis and K. V. Schwartz, Five Kingdoms, W. H. Freeman and Company, New York, 1987.)*

It is clear that animals more advanced than coelenterates were well established before the end of Proterozoic time. Among these advanced groups were the segmented worms known as **annelids,** which include modern earthworms as well as many kinds of marine and freshwater species (Figure 9-19). Annelids undoubtedly formed many of the tubelike fossil burrows of Late Proterozoic age. Also present were early members of the phylum **Arthropoda,** which includes modern crabs, lobsters, insects, and spiders. Arthropods have external skeletons and jointed legs, a set of which presumably made the scratches visible in the burrow shown in Figure 9-17*B*. Coelenterates, annelids, and arthropods not only continued to flourish into Paleozoic time but have remained important up to the present time.

PROTEROZOIC CRATONS: FOUNDATIONS OF THE MODERN WORLD

Although geologists have long attempted to determine how the continents of the modern world originated, they have managed to trace the histories of these continents only into Proterozoic time. Indeed, the configurations and relative positions of most older, Archean microcontinents will probably never be known. At the same time, uncertainties remain even about the histories of large Proterozoic cratons. The difficulty is that depositional patterns and structural trends of rocks more than half a billion years old are often obscured by erosion, metamorphism, or burial; paleomagnetic data are also sparse and difficult to interpret. Nonetheless, geologists have reconstructed partial histories for most large blocks of Proterozoic crust and have thus gained some knowledge about how they became part of the Phanerozoic world that forms the subject of the final seven chapters of this book.

In this section we will learn how North America grew episodically during Precambrian time and how this continent became part of a vast supercontinent that fragmented at the very end of the Proterozoic Eon. We will also review the early histories of the landmasses that came to constitute Gondwanaland and Baltica. These continents, too, formed part of the supercontinent before it broke apart.

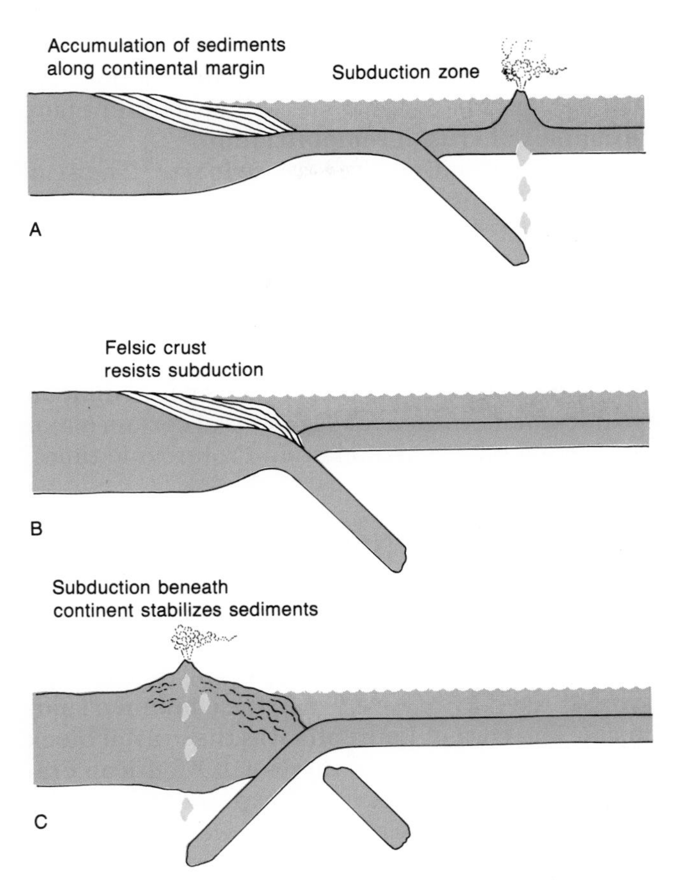

FIGURE 9-20 Stabilization of sediments that have accumulated along the margin of a continent *(A)*. The continent comes to rest along a subduction zone. Because the continent is of low density and resists subduction, the direction of subduction is reversed *(B)*. Igneous activity then adds rock to the continental margin and metamorphoses the sediments that have collected there *(C)*.

Continental Accretion

Before we discuss the histories of individual cratons, let us consider how cratons undergo changes in size. As we have seen, the sizes of cratons increase greatly when major cratons become sutured together along a subduction zone, and this process is usually accompanied by mountain building in the vicinity of the suture (p. 157).

Cratonic growth on a smaller scale, which is known as **continental accretion,** also entails mountain building, but this process occurs at the margin of a single large craton. As we noted earlier, marginal accretion can result either through the suturing of a **microplate** to a large craton along a marginal subduction zone (p. 200) or from the compression and metamorphism of sediments that have accumulated along a continental shelf. The latter process is sometimes referred to as **orogenic stabilization,** because it thickens the crust and hardens both unconsolidated sediments and soft sedimentary rocks (Figure 9-20).

Stabilization is a cannibalistic process inasmuch as some of the sediment that is deposited and stabilized along a continental margin is derived from the interior of the continent by erosion. On the other hand, limestone that accumulates along continental margins is precipitated from seawater or is secreted by organisms and thus represents an external contribution to the mass of the continent—as do the igneous rocks and oceanic crust that become welded to a continental margin resting along a subduction zone.

Orogenic processes do not simply add material to continents; they also alter preexisting crust. Regional metamorphism, for example—which sometimes operates in conjunction with structural deformation—remobilizes continental crust. As we saw in Chapter 4, metamorphism often alters the character of preexisting rocks beyond recognition and resets their radiometric clocks so that the age of the crust can no longer be determined. In reviewing the Precambrian history of individual Proterozoic cratons, we will encounter many examples of crustal **remobilization.**

How do continents decrease in size? They can shrink by erosion, but this process operates so slowly that it has little overall significance. Far more important is the process of continental rifting, which operates on many scales. It can remove a small sliver of crust, or it can divide a large craton in half. Although continental rifting that took place more than half a billion years ago is difficult to document, evidence suggests that certain major rifting events occurred late in Proterozoic time.

The Assembly of North America

Greenland today is a continent in its own right, but during Proterozoic and most of Phanerozoic time it was attached to North America. Recall that Laurentia is the name given to the combined landmass. The core of Laurentia was the crustal block that now forms most of the North American craton. This ancient block is well exposed today as the largest Precambrian shield in the world: the Canadian Shield (p. 184).

FIGURE 9-21 **Major geologic features of North America.** The Canadian Shield ends where sediments of the interior lowlands lap over it on the south and west. The Cordilleran, Ouachitas, and Appalachian orogens flank the North American craton on the west, south, and east.

The Canadian Shield constitutes a large portion of the North American craton, including a small part of the northern United States (Figure 9-21). Precambrian rocks also underlie the interior of the continent to the south, where they are overlain by a relatively thin veneer of Phanerozoic sedimentary rocks. Rocks obtained from wells that penetrate the Phanerozoic cover have provided a good picture of the general distribution of buried Precambrian rocks. These rocks, together with the exposed rocks of the Canadian Shield and Precambrian rocks that have been elevated by Phanerozoic mountain building in the American West, reveal that in the course of Proterozoic time, Laurentia gained territory by continental accretion.

Evidence that Laurentia was growing by accretion during Proterozoic time began to appear decades ago, when the recognition of structural trends and regional occurrences of rock units permitted geologists to recognize natural geologic provinces within the Canadian Shield. More recently, reliable radiometric dates for rocks of the Canadian Shield and for subsurface rocks bordering the shield have yielded a much more detailed picture. Uranium-lead techniques now yield dates with precision within about 10 million years for rocks that are about 2 billion years old. These dates are obtained from crystalline rocks, and thus they represent episodes of igneous and metamorphic activity.

North America also grew rapidly during Proterozoic time by becoming sutured to other cratons. In fact, as we will see shortly, near the end of the Proterozoic Eon it was united with nearly all of Earth's other landmasses to form a vast supercontinent only slightly smaller than Pangaea.

Suturing of Archean Cratons The first stage in the formation of North America, before it became part of a supercontinent, was the assembly of at least six microcontinents into a sizable craton (Figure 9-22). This amalgamation took place within only about 100 million years, between 1.95 and 1.85 billion years ago. Each of the microcontinents that were combined consisted of litho-

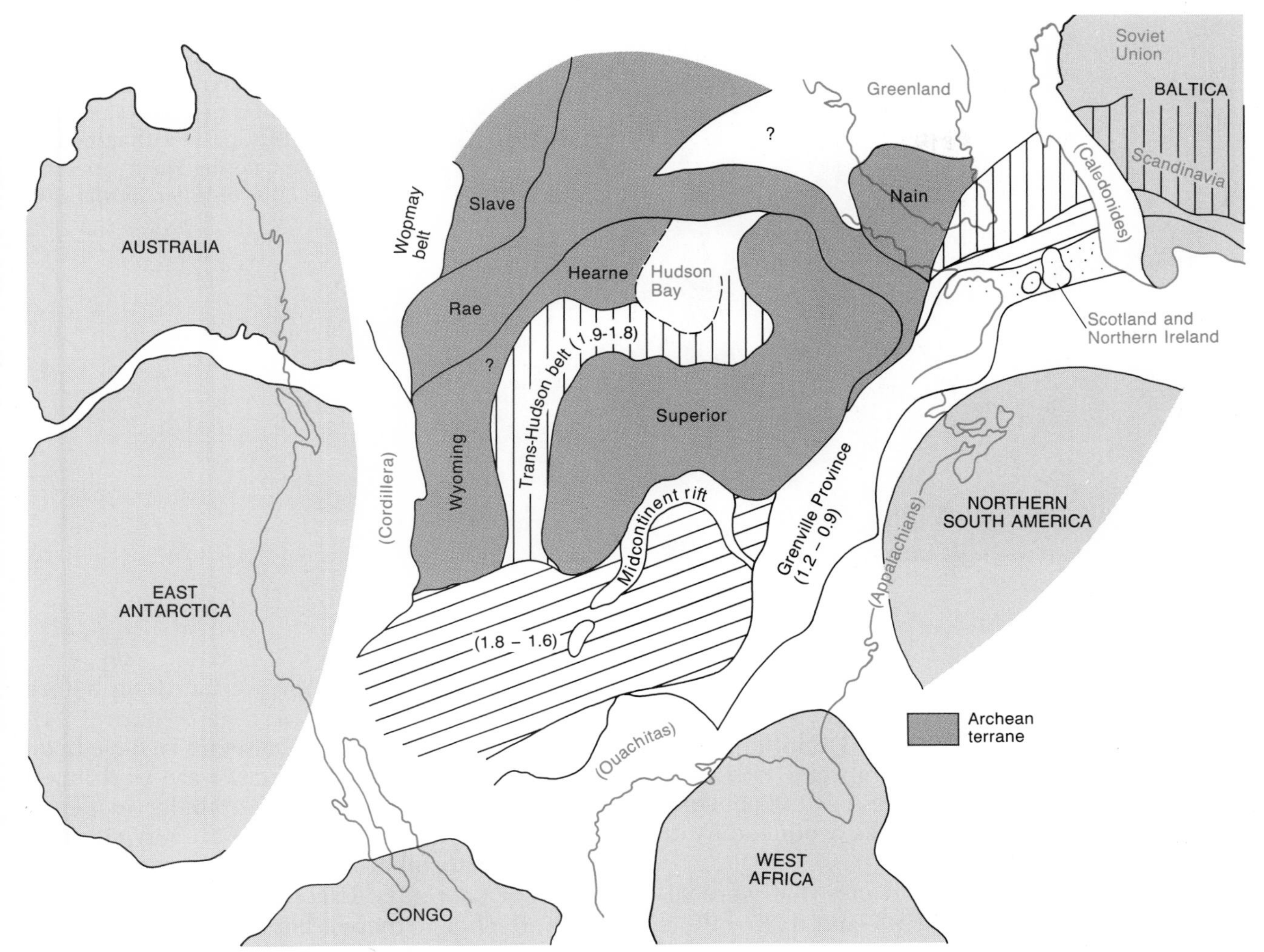

FIGURE 9-22 Geologic provinces of North America late in Proterozoic time, when this continent was attached to other landmasses (see Figure 9-24). Numbers in parentheses give times of origin in billions of years. Provinces of Archean terrane, in the north, represent Archean microcontinents that were amalgamated 1.95–1.85 billion years ago. The Wyoming and Hearne provinces may constitute a single terrane. The Trans-Hudson Belt consists of newly formed crust that was caught between the Superior terrane and the Archean terrane to the west. The origin of the broad province of the central United States 1.8–1.6 billion years ago resulted in substantial continental accretion toward the south. The Grenville Province formed when North America was sutured to Baltica and landmasses that later became portions of Gondwanaland. *(Derived from P. F. Hoffman, Ann. Rev. Earth and Planet. Sci. 16:543–603, 1988.)*

sphere that formed during Archean time. Today the former microcontinents represent Archean terranes, which lie mostly within the Canadian Shield (Figure 9-21). The largest of these Archean terranes is the Superior Province, which crops out as far south as Minnesota. The Wyoming and Hearne provinces may actually have formed a single microplate; in the narrow zone of contact between the two, the geologic evidence is inconclusive. The Wyoming Province is exposed south of the Canadian Shield in mountainous uplifts in Wyoming (Figure 9-23) and also in the Black Hills, a blisterlike structure in South Dakota, whose gold deposits are discussed in Box 9-1.

Most of the Archean terranes were sutured directly together, but the Superior Province is

FIGURE 9-23 A U-shaped glaciated valley in the Beartooth Mountains of Wyoming. Archean rocks form the core of the Beartooth uplift. The glacial scouring took place within the last 2 million years, during Earth's most recent ice age. *(Martin Miller/Earth Lens.)*

separated from the Wyoming and Hearne provinces by a broad zone of rocks that formed about the time of the suturing, 1.9 to 1.8 billion years ago (Figure 9-22). This zone comprises both deep-sea sediments squeezed up between the converging cratons and crystalline rocks produced by an igneous arc.

South of the Archean terranes that were sutured together between 1.95 and 1.85 billion years ago is a broad zone of crust that formed shortly thereafter, between about 1.8 and 1.6 billion years ago. The rocks of this zone are exposed in uplifts from southern Wyoming to northern Mexico, and they have been sampled by drilling through the sedimentary rocks that blanket most of the central and western United States. The composition of these rocks suggests that they formed by island arc activity and by associated sedimentation. The rate of continental accretion was very rapid by Phanerozoic standards. In the central and western United States, the continental margin expanded southward about 800 kilometers (~500 miles) in 200 million years.

Thus far we have considered only regions that remain parts of North America and Greenland. In fact, early in Proterozoic time the combined landmass called Laurentia appears to have been part of a larger craton. Geologic similarities between the Wyoming Province and both eastern Antarctica and eastern Australia suggest that these regions were attached to one another long before the start of late Proterozoic time (Figure 9-24). Thus western North America seems to have been connected to cratons that today are positioned in the Southern Hemisphere. Similar evidence points to a connection between the terranes that now constitute Siberia and the northern Canadian region of Laurentia. Exactly when Laurentia became attached to these other landmasses is not yet known.

Middle Proterozoic Rifting in Central and Eastern North America The growth of the landmass that now constitutes North America was threatened in mid-Proterozoic time by the greatest disturbance of the central North American craton during the last 1.4 billion years. This was an episode of continental rifting that took place between about 1.2 and 1.0 billion years ago. The episode was characterized by faulting and by mafic igneous activity in which lavas poured into downwarped basins along a belt that extended from the Great Lakes region to Kansas (Figure 9-25). Had the crescent-shaped zone of rifting extended to the margins of the craton, the eastern United States would have drifted away as a separate small craton. This did not happen, however; the rifting failed.

The rocks that formed within the failed midcontinent rift include hardened lavas known as

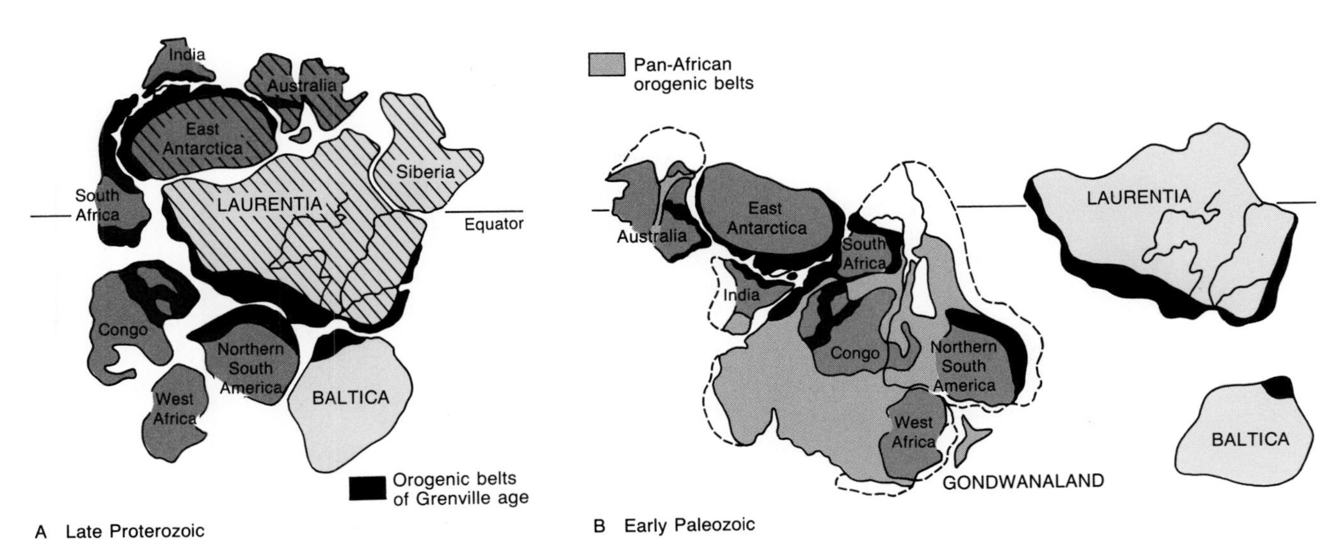

FIGURE 9-24 Changing paleogeographic patterns between late Proterozoic and early Paleozoic time. *A.* The likely configuration of the late Proterozoic supercontinent, in which North America occupied the central position. The lined portion of the supercontinent represents the smaller craton thought to have included North America earlier in the Proterozoic. Continental areas shown in dark brown represent future elements of Gondwanaland. The supercontinent was assembled by suturing along orogenic belts of Grenville age. *B.* The assembly of Gondwanaland along Pan-African orogenic belts after Laurentia and Baltica broke away. *(After P. F. Hoffman, Science 252:1409–1412, 1991.)*

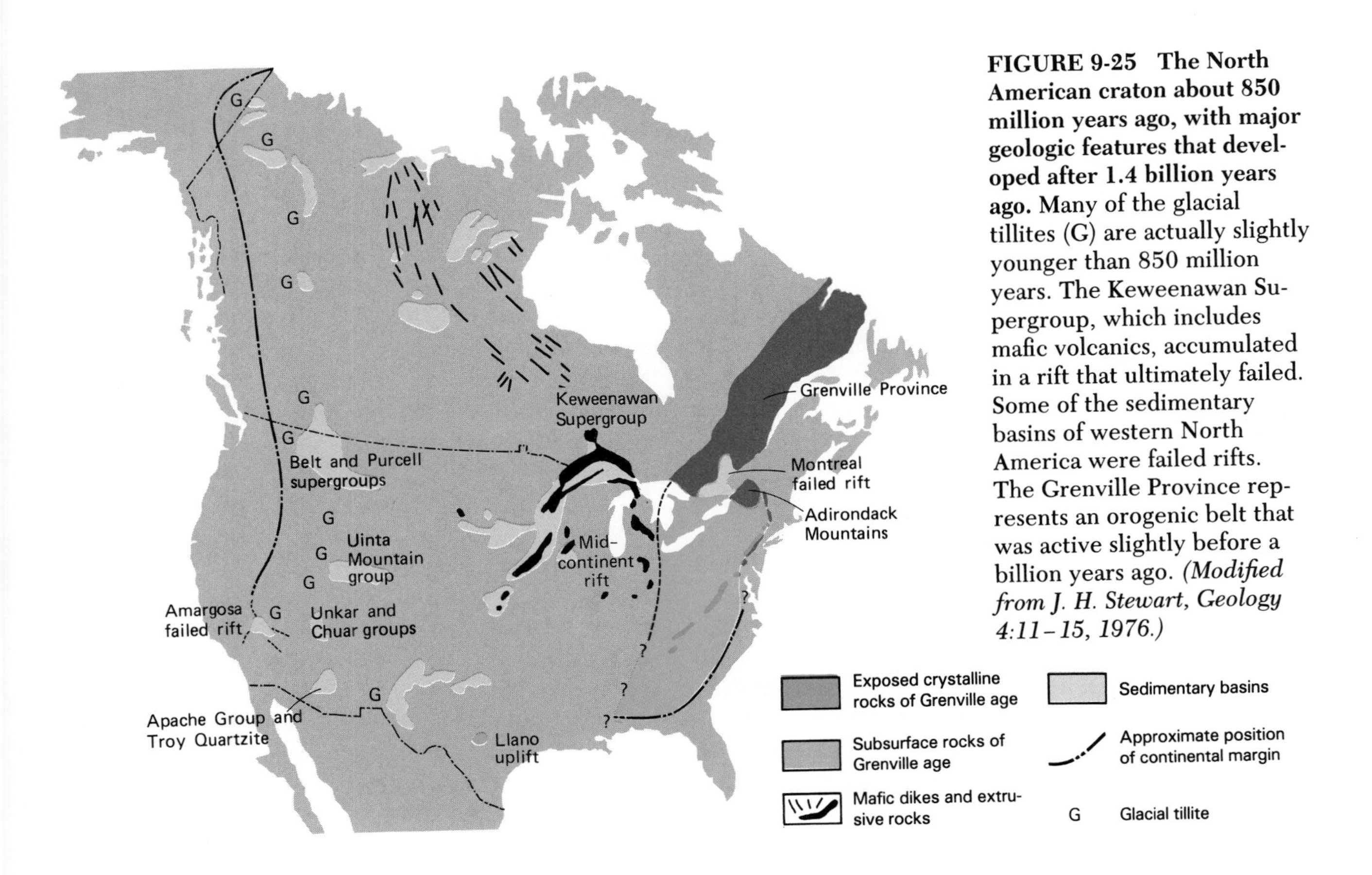

FIGURE 9-25 The North American craton about 850 million years ago, with major geologic features that developed after 1.4 billion years ago. Many of the glacial tillites (G) are actually slightly younger than 850 million years. The Keweenawan Supergroup, which includes mafic volcanics, accumulated in a rift that ultimately failed. Some of the sedimentary basins of western North America were failed rifts. The Grenville Province represents an orogenic belt that was active slightly before a billion years ago. *(Modified from J. H. Stewart, Geology 4:11–15, 1976.)*

Box 9-1
A Mountain of Gold

The craggy peaks of the Black Hills stand high above the Great Plains *(Paul Horestead.)*

After more than a century of operation, the Homestake mine in the Black Hills of South Dakota ranks first among all gold mines, with total profits in excess of $1 billion. The forested Black Hills, a dome-shaped outlier of the Rocky Mountains, stand conspicuously above the surrounding prairie. Their gold comes from Archean rocks that were metamorphosed early in Proterozoic time and are now exposed in the center of the dome. The gold was probably emplaced there by hot watery solutions, perhaps along an Archean midocean rift. Then, about 1.6 billion years ago, hot fluids from regional metamorphism concentrated much of the gold in veins, along with quartz and other minerals. The age of the deposits is not unusual. Most of the gold in Earth's crust is in Precambrian rocks. Perhaps this heavy metal has come largely from the dense mantle. We know that early in the planet's history mantle material moved upward in great volumes along numerous midocean ridges.

Keweenawan basalts, which are exposed near the southern border of the Canadian Shield (Figure 9-26). They contain ore deposits of native copper, a mineral consisting of elemental copper uncombined with other elements. Similar basin basalts lie to the southwest beneath the sedimentary cover of the Midwest, as has been revealed both by the examination of rock cuttings taken from deep wells and by the detection from Earth's surface of strong magnetism. Because these basalts are rich in iron and magnesium and are therefore of high density, their presence is also associated with a feature known as the Midcontinental Gravity High, which is a local increase in Earth's gravitational field as measured from the surface. While the basalts were forming, numerous basic dikes were also emplaced across the Canadian Shield to the north (Figure 9-25 and p. 184).

The quest for gold in the Black Hills in many ways epitomizes the nineteenth-century conquest of the American West. Meriwether Lewis and William Clark heard of the Black Hills in 1804, during their epic journey to the Pacific, but did not visit them. Beginning in the 1820s, other explorers and prospectors ventured into the mountains, but at great risk. Ezra Kind was the last survivor of a party of six who rode into the Black Hills in 1833, and before he too died, he scraped a final message on a slab of sandstone: "Got all of the gold we could carry our ponys all got by the Indians I have lost my gun and nothing to eat and Indians hunting me."

Gold excites prospectors not only because of its value but also because even in nature it is usually as shiny as a wedding ring. As a so-called noble metal, it does not form compounds with other elements, and it flashes its purity to the naked eye. Sporadic reports of gold in the Black Hills tantalized adventurers for years. Congress nonetheless ratified a treaty in 1868 that included the Black Hills in a new reservation for the Sioux. The treaty banned prospectors from the region, but dreams of wealth led many to violate the law, and the actions of a reckless general named George Armstrong Custer only added to the incentive. Custer was ordered to lead a combined force of soldiers and civilians to reconnoiter the Black Hills region in 1874. With gold clearly on his mind, Custer hired a geologist to take part in the expedition. Once in the Black Hills, Custer diverted the group's activities from exploring to prospecting. His party found only a few particles of the noble metal, but apparently he wanted recognition as the man who first discovered the rich gold deposits that were widely anticipated. Newspapers throughout the country reported on Custer's expedition, and when the general failed to correct exaggerated stories about an abundance of gold, thousands of prospectors flocked into the Black Hills. The treaty with the Sioux became a worthless piece of paper.

The flurry of activity quickly bore fruit. The first rich deposits to be found were stream sediments, but soon prospectors tracked down the veins of quartz that supplied this detrital gold, and mining of these veins yielded even greater riches. The great Homestake mine opened in 1876. That same year General Custer, still seeking fame, led his soldiers into the famous massacre at the Little Big Horn, in which he and 269 of his men lost their lives. The Sioux's victory did them more harm than good; increased hostility toward them added to the economic incentives to terminate the treaty of 1868. In 1877, faced with the government's threat to cut off supplies to their reservation, the Sioux reluctantly signed a new agreement that removed them from the Black Hills, a region that for generations they had regarded as sacred.

Legitimate at last, mining camps in the Black Hills expanded into towns. Deadwood, which sprang up close to the Homestake mine, became the town most infamous for lawlessness. Calamity Jane earned her nickname there, and a bullet in the back of the head took Wild Bill Hickock's life in a Deadwood saloon. Civilization has subdued the town, but the Homestake mine continues to churn out thousands of tons of ore every day and still has larger gold reserves than any other American mine.

The configuration and composition of Keweenawan rocks and their subsurface counterparts indicate the presence of a failed rift in the eastern United States (p. 152). The Keweenawan volcanics, for example, are associated with red siliciclastic rocks and alluvial-fan conglomerates in what appear to be downfaulted troughs—configurations that tend to occur in newly forming continental rifts. Thus it would appear that about 1.3 billion years ago, a spreading center formed beneath the late Precambrian craton and began to rift it apart. It is possible that this spreading center intercepted the eastern margin of the ancient craton, but this remains uncertain. In any event, it is obvious that the Keweenawan rifting was abortive. Rifting ceased before the continent was split but left its mark in the enormous volumes of mantle-derived lavas that were disgorged along a

FIGURE 9-26 Columnlike joints in the Edwards Island flow, a body of volcanic rock within the Keweenawan Supergroup of northern Michigan. *(N. K. Huber, U.S. Geological Survey.)*

belt more than 1500 kilometers (~900 miles) long and 100 kilometers (~60 miles) wide!

The Grenville Orogenic Belt While the midcontinent rifting was in progress, an episode of mountain building took place along the east coast of North America. This was the Grenville orogeny, which spanned the interval from about 1.2 to 1.0 billion years ago. This orogeny constituted another step in the accretion of the North American continent, adding a belt of terrane that stretched from northern Canada to the southeastern United States (Figures 9-22 and 9-25). The Grenville orogeny stabilized a large volume of sediments that had accumulated along the margin of eastern North America before about 1.2 billion years ago (Figure 9-27).

The igneous and metamorphic activity of the Grenville orogeny ended about 1 billion years ago. The resulting crystalline rocks and others are best exposed in the Canadian portion of the Grenville Province (Figure 9-22). To the south, most crystalline rocks of Grenville age are buried, but some crop out here and there—especially in the Adirondack uplift of New York State and, as we noted earlier, in the Blue Ridge Mountains and other uplifts lying between the Valley and Ridge Province and the Piedmont. In the Gulf Coast region, most of the Grenville Belt lies beneath Phanerozoic cover, but in central Texas crystalline rocks of Grenville age appear at the surface as a small, isolated prominence known as the Llano uplift (Figure 9-25).

The relationship between the Grenville orogeny and the midcontinent rift remains a puzzle. What is clear is that the Grenville event entailed the collision of eastern North America with another large craton. The other continent involved in the Grenville collision apparently included a landmass that later became a major component of Gondwanaland—the component that included Africa and South America. Forming another segment of the colliding landmass was Baltica (Figure

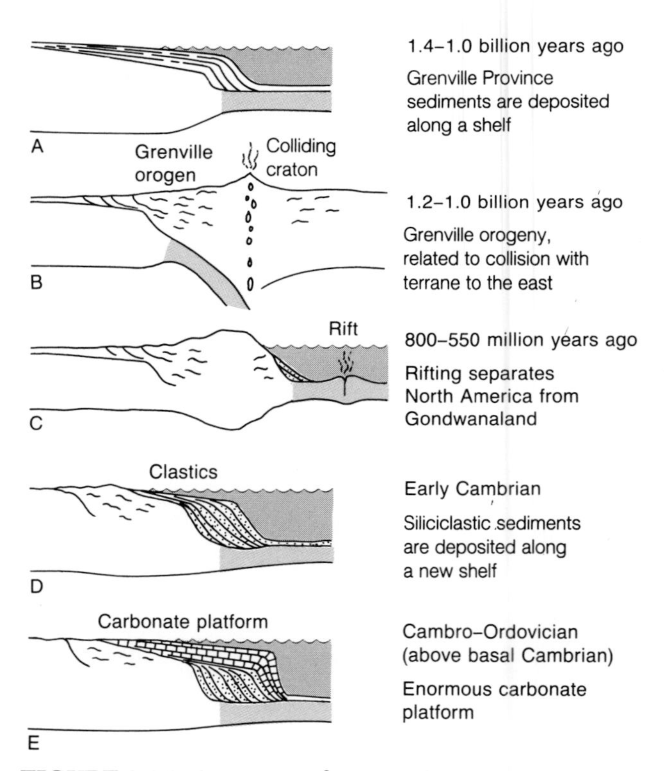

FIGURE 9-27 Sequence of events along the eastern part of the North American craton from middle Proterozoic time until early Paleozoic time. Sediments of the Grenville Province were deposited before 1 billion years ago along a shelf margin *(A)*. The Grenville orogeny took place when one or more large cratons were sutured to eastern North America *(B)*. Rifting in latest Proterozoic and earliest Phanerozoic time *(C)* produced a new continental margin, along which siliciclastics accumulated *(D)* and then a carbonate platform developed *(E)*.

9-24). The evidence for this pattern of collision is seen in regions of Africa, South America, and Europe, where there are remnants of mountain systems that are the same age as the Grenville orogenic belt (Figure 9-22).

In fact, the Grenville collision was part of a long zone of tectonic suturing that encircled much of the continent to which North America belonged. Orogenic belts in southern Africa and the Indian peninsula suggest that these regions became attached to eastern Antarctica at this time. If these reconstructions are correct, landmasses that would later become Gondwanaland formed a crescent that wrapped around North America (Figure 9-24*A*). The landmass thus formed rivaled the Phanerozoic supercontinent Pangaea in total size. Pangaea was more elongate in a north-south direction, however, nearly spanning the globe from pole to pole (Figure 6-5, top). Paleomagnetic data indicate that the late Proterozoic supercontinent also straddled the equator, but it was a compact landmass confined to low latitudes.

The Great Rifting Event During the Proterozoic–Phanerozoic transition, a remarkable tectonic event took place: Laurentia was, in effect, expelled from the Proterozoic supercontinent after a massive episode of rifting; Baltica also broke away (Figure 9-24*B*). Evidence of rifting in North Carolina suggests that it began there about 800 million years ago and continued into Early Cambrian time. This evidence takes the form of sedimentary rocks and radiometrically dated volcanic rocks that are younger than those affected by the Grenville orogeny. The volcanics are primarily in the lower portion of the stratigraphic sequence that accumulated after the Grenville event.

The Birth of North America's West Coast In western North America, the final rifting that carried away future components of Gondwanaland was preceded by tectonic episodes that produced failed rifts from northern Canada to southern Arizona. The resulting basins, shown in Figure 9-25, received large volumes of sediment. In the northern United States, the largest of the basin sequences is the Belt Supergroup (Figure 9-28), which ranges in age from about 0.9 to 1.5 billion years. The northern extension of this formation is known in Canada as the Purcell Supergroup, but for simplicity we can apply the name Belt to the entire sequence. In general, the Belt thickens toward the west, sometimes reaching a thickness of 16,000 meters (~53,000 feet).

The Belt formed in a northwesterly trending failed rift (Figure 9-25). Sandstones increase in abundance toward the western part of the sequence, while limestones increase toward the east, where sediments accumulated in the

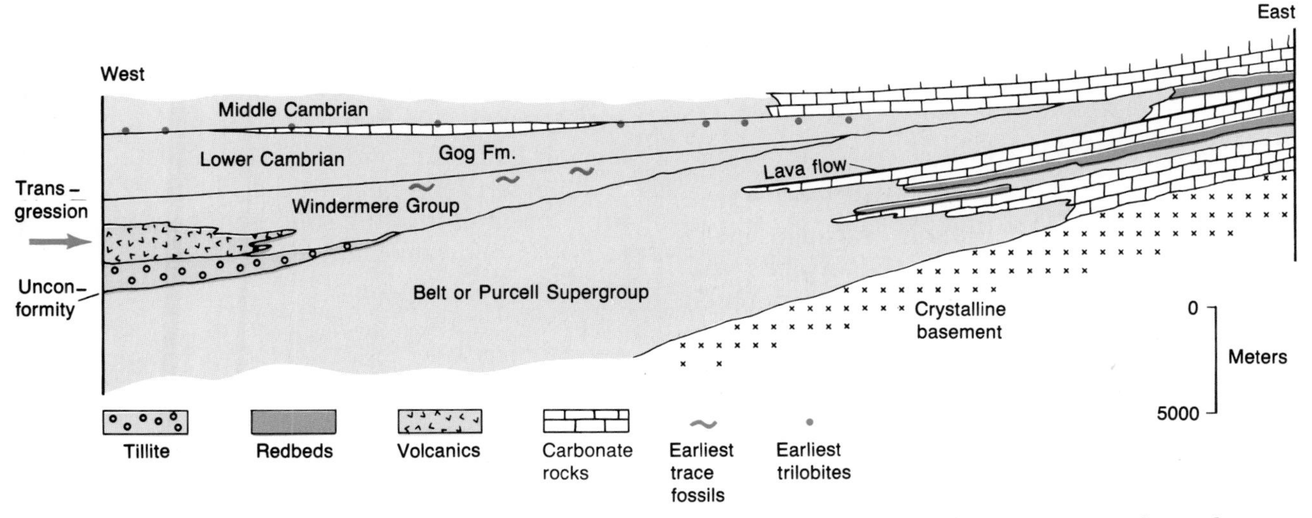

FIGURE 9-28 Stratigraphic sequence in Canada just north of Montana, ranging from the middle Proterozoic through the middle Cambrian. The Belt Supergroup accumulated in a failed rift trending to the northwest (see Figure 9-25). *(Modified from P. B. King, The Evolution of North America, Princeton University Press, Princeton, New Jersey, 1977.)*

shallower water. In general, however, mudstones predominate. The Belt apparently formed as a result of the accumulation of sediments in very shallow water during rapid subsidence. Salt crystals and mud cracks, both of which indicate a drying up of shallow bodies of water, are present in Belt sediments, and shallow-water stromatolites are also abundant in the limestones.

The final global glaciation of the Proterozoic, described earlier, took place about 600 million years ago, shortly after the great rifting event. In the region occupied by the Belt Supergroup, and in fact from one end of western North America to the other, tillites accumulated. Whether the continental breakup may somehow have triggered the major ice age remains a matter of speculation. In any case, after the rifting event, the western border of North America remained a passive margin for hundreds of millions of years. In fact, it has never since been sutured to another large craton. As we will see in later chapters, however, it has grown westward through suturing events that have added small bodies of crust. It has also, of course, been the scene of extensive mountain building and igneous activity that have added crust during the latter part of Phanerozoic time. Thus, at the start of the Phanerozoic Eon, the passive western margin of North America lay well to the east of its present position. As we will see in Chapter 10, this margin soon became the site of a vast carbonate platform, much like the one that developed along the similar, newly formed passive margin bordering the east coast (Figures 7-14 and 9-27).

CHAPTER SUMMARY

1 At least as early as 2 billion years ago, plate-tectonic processes formed mountain belts with characteristics similar to those of Phanerozoic mountain belts.

2 Continental glaciers spread over parts of Canada and other regions more than 2 billion years ago and also covered many parts of the world between 1 billion and 600 million years ago.

3 The presence of red beds in sedimentary sequences younger than 2.2 or 2.3 billion years, together with the rarity of the easily oxidized minerals pyrite and uraninite in such sequences, suggests that atmospheric oxygen had reached a moderate level early in Proterozoic time.

4 Banded iron formations, which formed in the presence of oxygen, accumulated in great abundance within marine basins between about 2.5 and 1.8 billion years ago. They may have formed when the concentration of oxygen in the atmosphere was lower than it is today.

5 Stromatolites first became abundant about 2.2 or 2.3 billion years ago. Their success at this time, which may have resulted from the growth of continental shelves, probably led to a buildup of atmospheric oxygen.

6 Cells and chemical compounds that probably represent some of the oldest eukaryotes are found in rocks that are 1.8 to 1.6 billion years old.

7 Acritarchs are fossils of single-celled planktonic algae that almost certainly were eukaryotic. They underwent an adaptive radiation between 800 and 700 million years ago and then suffered a mass extinction at the time of the last major Precambrian glacial episode, about 600 million years ago.

8 The oldest fossils that appear to represent predatory protists are about 800 million years old. The oldest unquestioned fossils of multicellular animals are younger than about 600 million years.

9 The modern North American craton was assembled from smaller Archean cratons early in Proterozoic time. Later in the Proterozoic, it became part of a huge supercontinent that contained nearly all of Earth's landmasses.

10 Near the time of the Proterozoic–Phanerozoic transition, the supercontinent fragmented, leaving Laurentia and Baltica as isolated supercontinents. Very early in the Phanerozoic Eon, other fragments of the supercontinent coalesced to form Gondwanaland.

EXERCISES

1 Compare the basic features of the Wopmay orogenic belt with those of the Appalachian orogenic belt described in Chapter 7.

2 What kinds of geologic evidence suggest that glaciers were present on Earth more than 2 billion years ago?

3 List as many differences as you can between the Archean world and the world as it existed 1 billion years ago.

4 What arguments favor the idea that little atmospheric oxygen existed on Earth until 2 billion years ago?

5 What reasons do we have for believing that multicellular animals did not exist 2 billion years ago?

6 How does the history of North America illustrate continental accretion?

7 How does the Appalachian orogenic belt of North America relate to the Grenville orogenic belt?

8 Using a world map, locate the modern positions of the various landmasses that make up the supercontinents shown in Figure 9-24*A* and *B*.

ADDITIONAL READING

Cloud, P., *Oasis in Space: Earth History from the Beginning,* W. W. Norton, New York, 1988.

Knoll, A. H., "End of the Proterozoic Eon," *Scientific American,* October 1991.

Margulis, L., *Early Life,* Science Books International, Boston, 1982.

McMenamin, M. A., "The Emergence of Animals," *Scientific American,* April 1987.

Vidal, G., "The Oldest Eukaryotic Cells," *Scientific American,* February 1984.

Windley, B. F., *The Evolving Continents,* John Wiley & Sons, Inc., New York, 1984.

PART V

THE PALEOZOIC ERA

An evolutionary explosion of invertebrate animals with skeletons ushered in the Paleozoic Era, or "interval of ancient life." Before long, fishes also evolved, and midway through the era, some of them developed jaws. Multicellular plants invaded the land, soon to be joined by scorpions, insects, and amphibians (descended from fishes), and later by reptiles (descended from amphibians). Mass extinctions punctuated the history of Paleozoic life, the final one striking at the end of the era. Paleozoic time saw the formation of several mountain ranges, including the Appalachians, together with their counterparts in Europe, and early mountain ranges in the American West. At the close of the Paleozoic Era, nearly all of Earth's landmasses were united into a single supercontinent that stretched from pole to pole.

Eurypterids were among the many arthropod groups that evolved early in the Paleozoic Era. These aquatic relatives of scorpions were predatory swimmers, some of which had large claws. This relatively small Silurian specimen was the size of a large carrot, although flatter. *(Chip Clark.)*

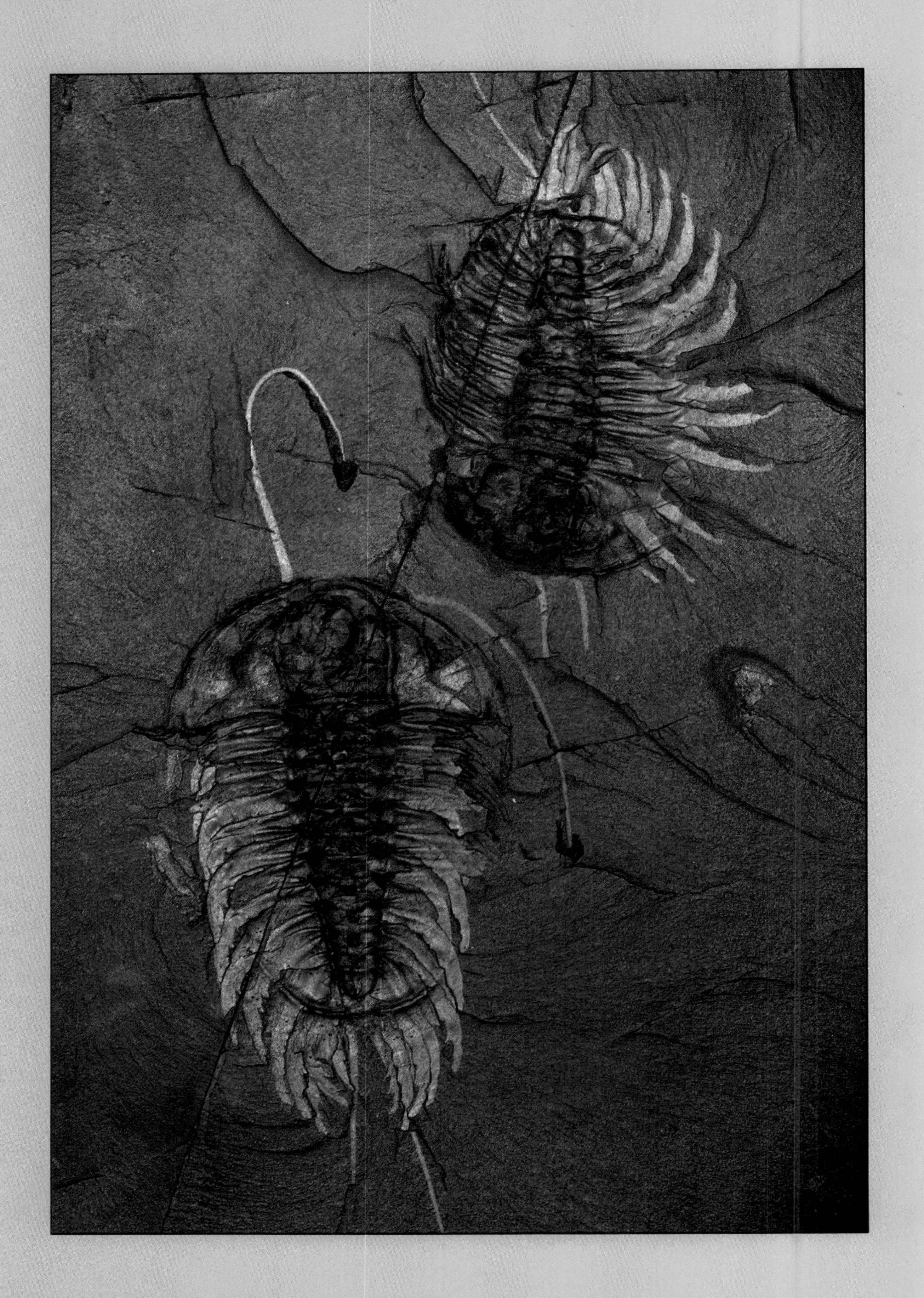

CHAPTER

10

The Early Paleozoic World

Chapter 1 described how the Cambrian and Ordovician systems of the Paleozoic Erathem were established more than a century ago in what is now Wales. The rock record of these early Paleozoic systems documents the Taconic orogeny of the Appalachian mountain belt, discussed in Chapter 7, as well as widespread mountain building in Gondwanaland, mentioned in Chapter 9. Early Paleozoic rocks of marine origin are also well displayed on the broad surfaces of cratons, reflecting the fact that, with brief interruptions, sea level rose in the course of the Cambrian Period and remained high during most of Ordovician time. Among the revelations of the resulting rock record is that life in the oceans diversified rapidly at the end of Proterozoic time. Despite brief episodes of mass extinction during Cambrian time, about as many families existed in the marine ecosystem near the end of the Ordovician Period as during any subsequent interval in the Paleozoic Era.

These fossil trilobites were found in a slab of Burgess Shale from western Canada. Vestiges of the antennae and also of the limbs, which became splayed out from beneath the body during the process of burial, are preserved. (Approximately life size.) *(Chip Clark.)*

LIFE

The story of early Paleozoic biotas is essentially one of life in the sea. It is presumed that certain simple kinds of protists and fungi had made their way into freshwater habitats by this time, but no fossil record of early Paleozoic freshwater life is known. The terrestrial realm, too, was barren of

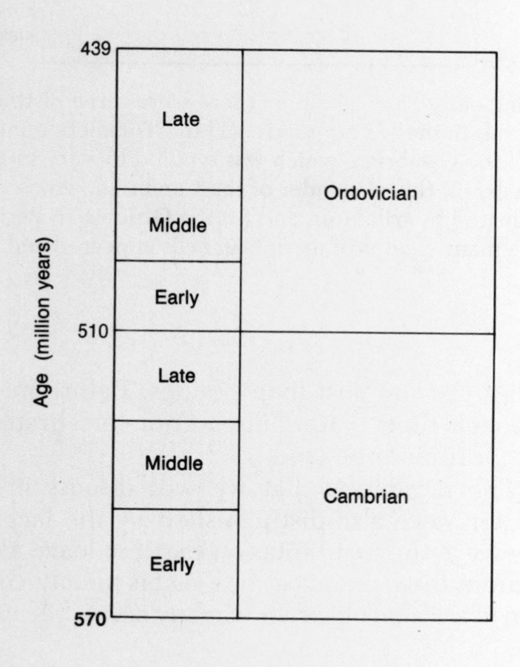

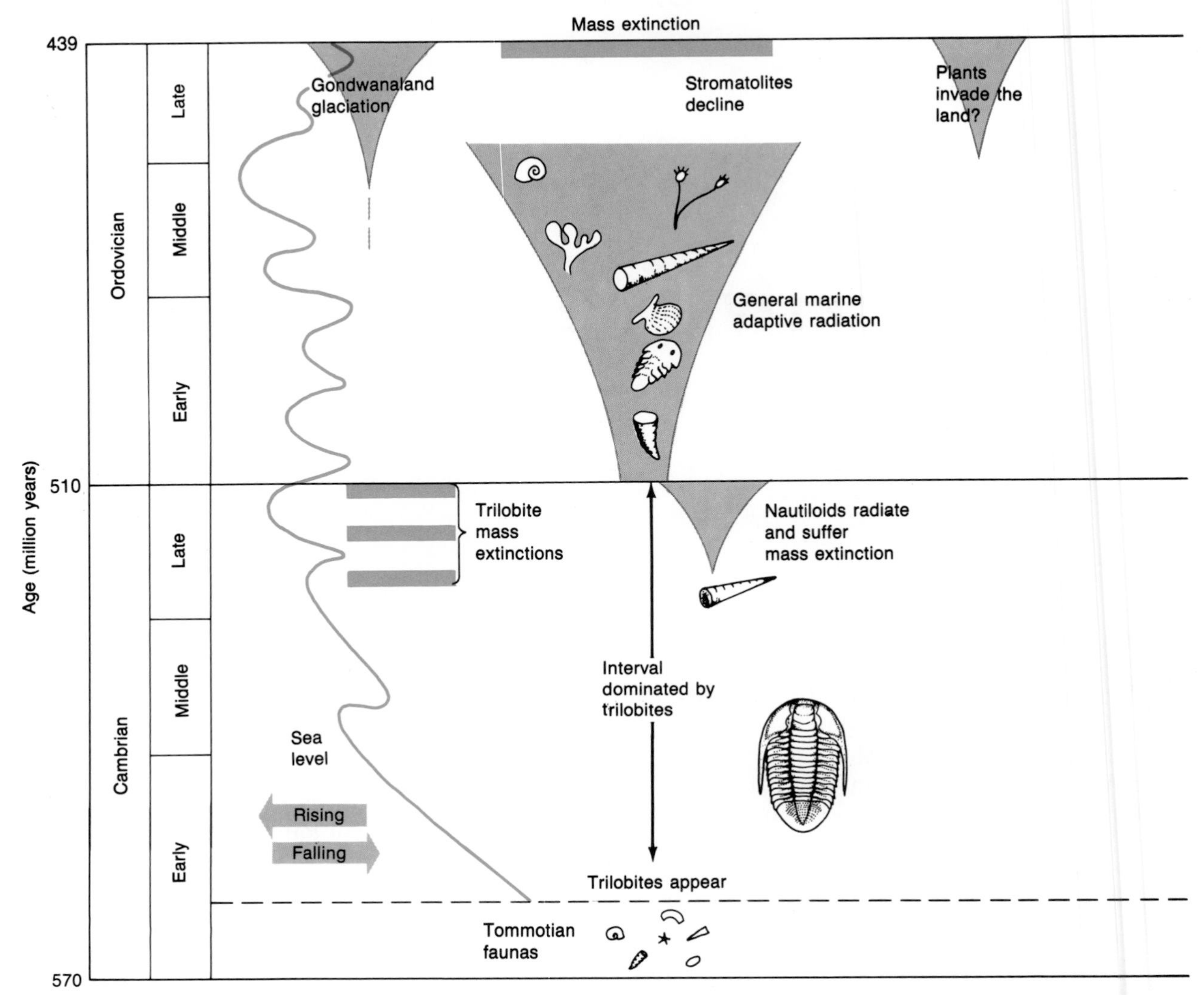

During early Paleozoic time there were three distinctive intervals in the history of life: (1) the Tommotian interval of the Cambrian, which was typified by very small animals; (2) the remainder of the Cambrian, which was dominated by trilobites; and (3) the Ordovician Period, when many groups of marine animals appeared and stromatolites declined. Note that several large drops of sea level failed to cause mass extinctions and that two of the Late Cambrian mass extinctions did not coincide with a lowering of sea level. The mass extinction at the end of the Ordovician coincided with the climax of glaciation in Gondwanaland.

all but the simplest living things. Before middle Paleozoic time, neither insects nor vertebrate animals occupied the land.

The organisms that we will discuss in this chapter were also distinguished by the fact that they were the first biotas on Earth to leave a conspicuous fossil record — one that is plainly visible even to a casual observer in many areas, because it includes a great variety of shells and other kinds of skeletons composed of durable minerals.

In Chapter 9 we learned that the first major adaptive radiation of multicellular marine animals occurred during the last few tens of millions of years of the Proterozoic Eon. Nearly all of the creatures that emerged during this time, however, were soft-bodied; many, for example, were naked jellyfishlike animals or worms. During the earliest segment of Cambrian time, the seas became populated by a different kind of fauna — one consisting of small shelled animals, some of which failed to survive into later Cambrian time. This fauna was the world's oldest diverse group of shelled animals; then a more conspicuous and enduring Cambrian biota of shelled animals was followed by a still more highly diversified biota in the Ordovician Period.

The Tommotian Biota: Early Skeletons

The first diverse biotas of animals with skeletons are found in rocks of the Tommotian Stage, an interval that spanned about 15 million years. This stage was not formally added to the base of the Cambrian System until the 1970s, when its fossil record was brought to the attention of the general scientific community. Before this time, the exceedingly small Tommotian fossils had been overlooked by all but a few paleontologists, but they have since been found on many continents. Some belong to groups of animals that still live in the ocean, such as sponges and mollusks (Figure 10-1). Also found in Tommotian rocks, however, is a host of strange skeletal elements that cannot be assigned to any living phylum and that show no apparent relation to any group of fossils found in rocks younger than Cambrian age.

The development of the types of skeletons that characterize Tommotian faunas constituted a major evolutionary event. Although skeletons are known to support soft tissue and to facilitate locomotion, such adaptive functions cannot explain why so many different kinds of skeletons developed suddenly in the early part of Tommotian time. It has been suggested that a chemical change within the oceans triggered the production of these skeletons, but this hypothesis does not explain why some skeletons were composed of calcium carbonate and others of calcium phosphate — two compounds with quite different chemical properties. The rapid evolution of various kinds of external skeletons is at least partly attributable to the presence of enemies; the first multicellular animals must have fed on single-celled creatures and might also have fed on larger plants. The effective predation of some animals on others marked a change in the basic structure of ecosystems.

FIGURE 10-1 Fossils that represent the Tommotian fauna, the oldest diverse skeletonized fauna on Earth. All of the specimens shown here are small; none exceeds a few millimeters in length. *A.* A fossil that appears to represent primitive mollusks with coiled shells. *B* to *E.* None of these specimens can be assigned to a familiar group of animals. *(A, C, and D. From S. C. Matthews and V. V. Missarzhevsky, Jour. Geol. Soc. London 131:289–304, 1975. B and E. S. Bengston.)*

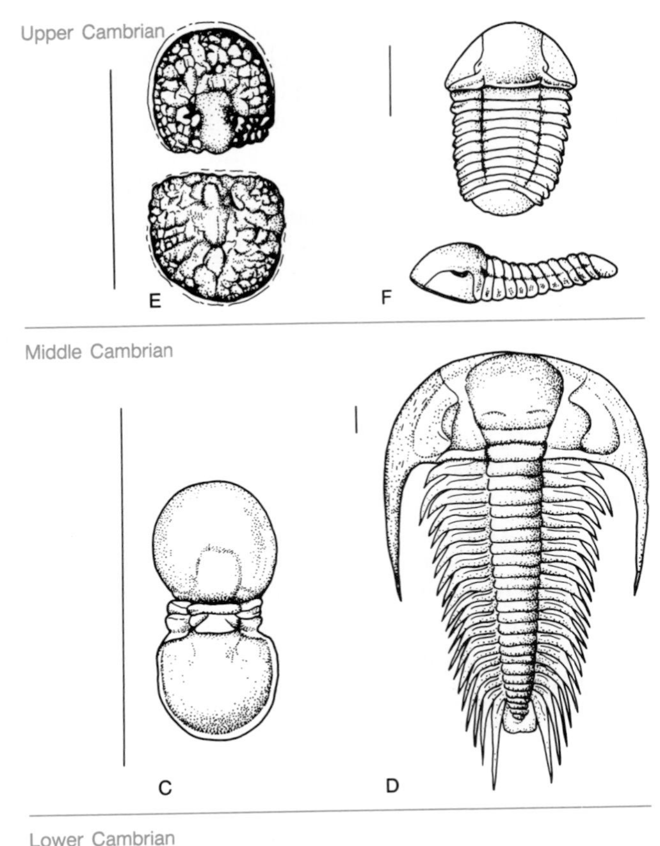

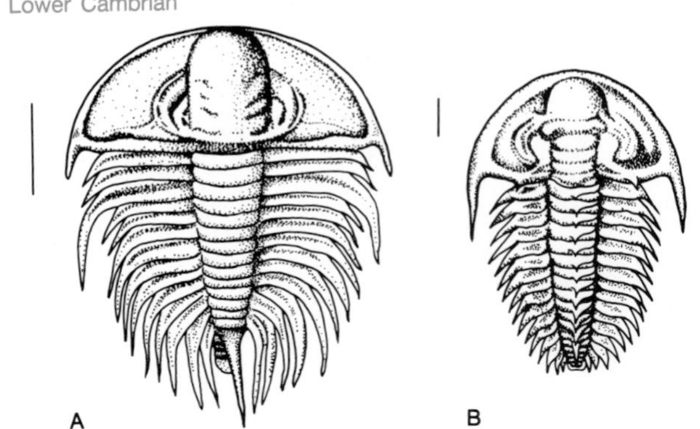

FIGURE 10-2 Typical Cambrian trilobites. Trilobites were arthropods (invertebrate animals with segmented bodies and jointed legs). The soft body and many legs were positioned beneath the flexible, jointed skeleton. Trilobites had mouthparts for chewing small pieces of food. Most species crawled over the seafloor, but some burrowed in sediment, and a few small species, including those labeled *C* and *E*, were planktonic. *A. Olenellus. B. Holmia. C. Lejopyge. D. Paradoxides. E. Glyptagnostus. F. Illaenurus.* (Scale bars represent 1 centimeter [~$\frac{3}{8}$ inch].) *(After Treatise on Invertebrate Paleontology, Part O, R. C. Moore [ed.], Geological Society of America and the University of Kansas Press, Lawrence, 1959.)*

Later Cambrian Marine Life

The brief Tommotian interval of Cambrian time was followed by the evolution of many larger marine animals with hard parts.

Bottom Dwellers The most conspicuous animals moving over Cambrian seafloors were the **trilobites** (Figure 10-2). A few types of trilobites may actually have lived during Tommotian time, but none have yet been found in association with typical Tommotian faunas. Many Cambrian trilobite species survived for only short spans of geologic time — 1 million years or less. For this reason, and because most trilobite species have proved easy to identify, trilobites have served as the principal index fossils for Cambrian strata. Most trilobites crawled or swam along the seafloor, frequently leaving conspicuous trace fossils. Some of these traces resulted from scratching or digging by the trilobites' many appendages, while others record trilobites' paths as they crawled about (Figure 10-3). Apparently other small trilobites either swam or floated in the water (Figure 10-2*C*, *E*).

Algal stromatolites were also more abundant during the Cambrian and Ordovician intervals than in later times.

Double-valved **brachiopods,** shelled animals that fed on organic matter suspended in the water, are abundant in Cambrian strata as well, but their representatives are small and of limited variety (Figure 10-4). These organisms, which are also

FIGURE 10-3 A Cambrian trilobite track preserved as a cast on the underside of a bed of sediment. Scratches made by the appendages of the trilobites are clearly visible. *(T. P. Crimes.)*

FIGURE 10-4 **Cambrian brachiopods.** The articulate genus *Eorthis,* which lived on the surface of the sediment *(left)*, and the inarticulate genus *Lingula*, a burrowing form that survives in modern seas *(right)*. *(Chip Clark.)*

found in modern seas, resemble bivalve mollusks but are not related to them. **Mollusks** are also common in Cambrian strata, but they, too, are small, and many advanced molluscan groups that became highly significant later in Paleozoic time were still inconspicuous or absent. **Echinoderms** were represented in the Cambrian by a remarkable variety of classes (Figure 10-5), but none closely resembled modern echinoderms such as starfishes, sea urchins, and sea cucumbers. A few other groups of fossils that are well represented in younger Paleozoic rocks also occur in Cambrian strata, but in small numbers. Among these fossils are **conodonts,** which are toothlike structures made of bone that belonged to a group of swimming animals that were the earliest known vertebrates (Figure 10-6), and **ostracods,** a group of bivalved arthropods that survives today (Figure 10-7).

It can be assumed that many important groups of soft-bodied animals flourished during the Cambrian Period without leaving fossil records. One indication of the variety of such forms is provided by the unique sample of animals preserved in the

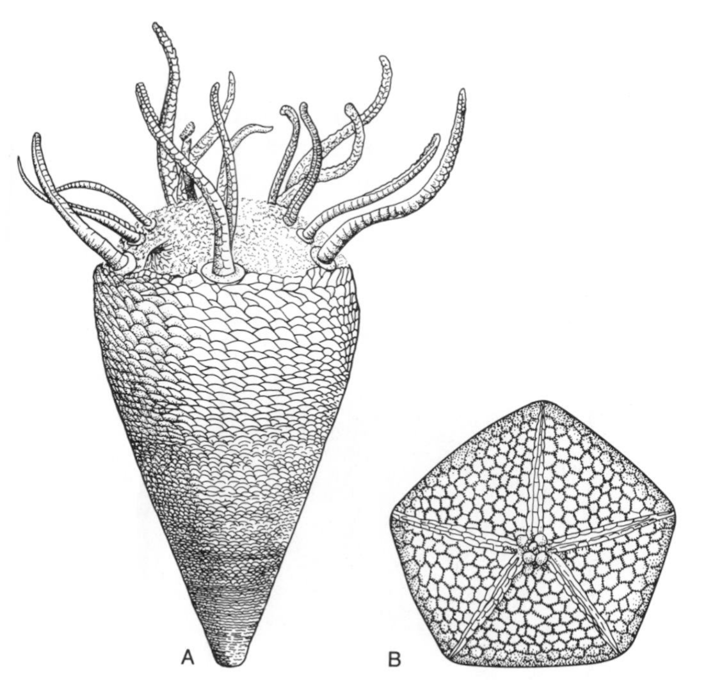

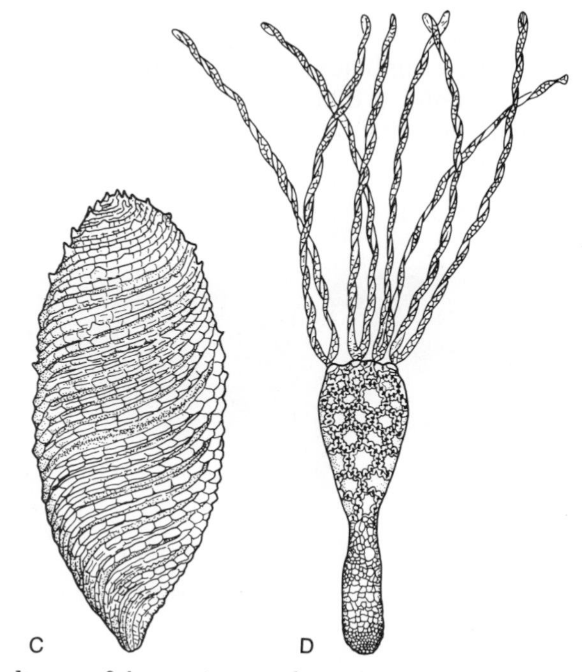

FIGURE 10-5 **Strange Cambrian echinoderms that show no close relation to any younger group.** *A, B,* and *D*. Attached forms that apparently fed on organic matter suspended in the water. *C*. A flexible form that probably burrowed in sediment. Most living echinoderm groups, including starfishes and sea urchins, display a fivefold radial symmetry similar to that which can be seen in *B*. *(After Treatise on Invertebrate Paleontology, Part U, R. C. Moore [ed.], Geological Society of America and the University of Kansas Press, Lawrence, 1966.)*

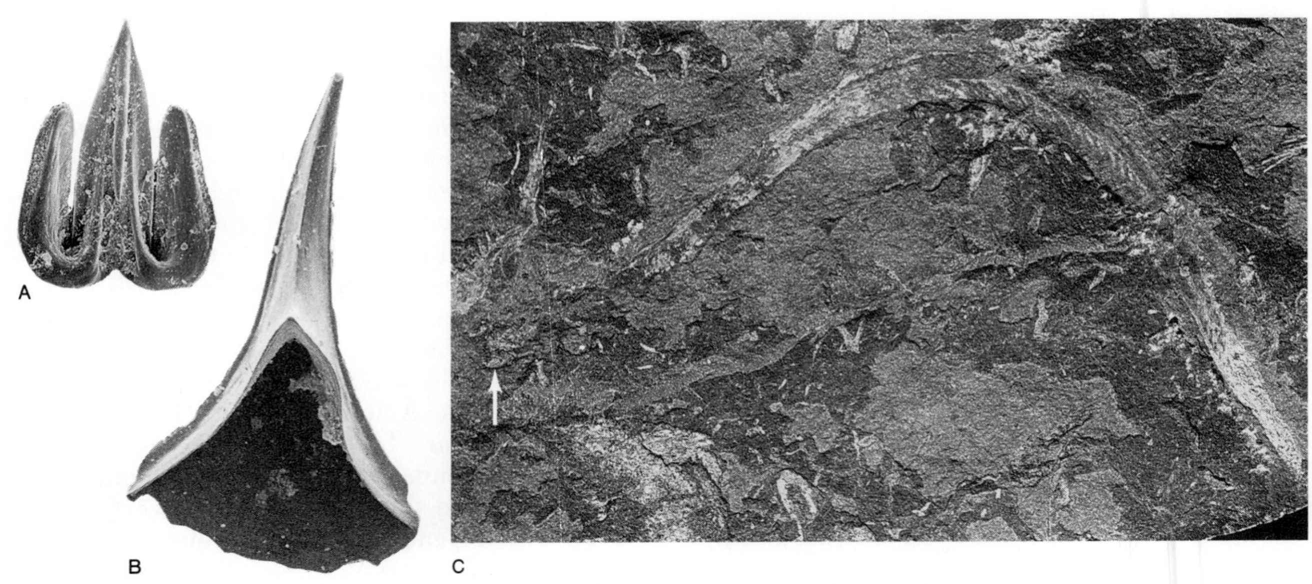

FIGURE 10-6 Cambrian conodonts. *A. Westergaardodina* (×75). *B. Furnishina* (×45). *C.* Only in the 1980s were the remains of an entire conodont animal found in the fossil record. This unique fossil is a late Paleozoic impression of an eel-like swimming vertebrate (×3). Conodonts, which were teeth, are found in the head region *(arrow)*. *(A and B. K. J. Müller. C. From D. E. G. Briggs et al., Lethaia 16:1–14, 1983.)*

Middle Cambrian Burgess Shale, which lies in the Rocky Mountains of British Columbia. Later we will examine the environment in which the Burgess Shale formed, but at this point it is sufficient to note that this black shale accumulated in a deep-sea environment that was virtually free of oxygen, so that the soft-bodied animals preserved there were not exposed to predators or to the types of bacteria that normally cause soft tissue to decay. Many of them are preserved as impressions and as carbonizations of soft tissues (Figure 10-8). Most impressive among the Burgess Shale species is the array of nontrilobite **arthropods**, which are distributed among 30 genera or so — more than

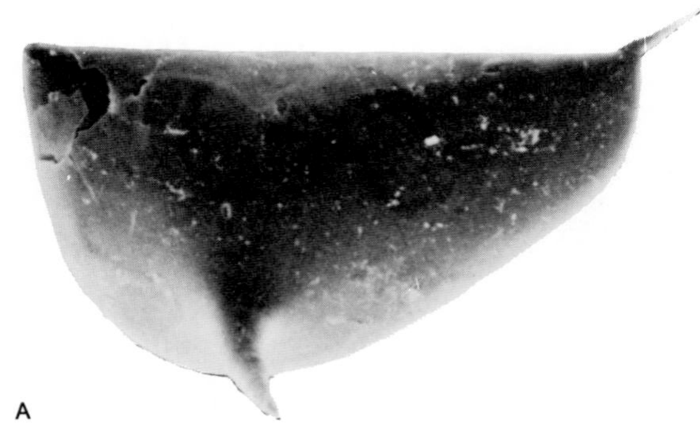

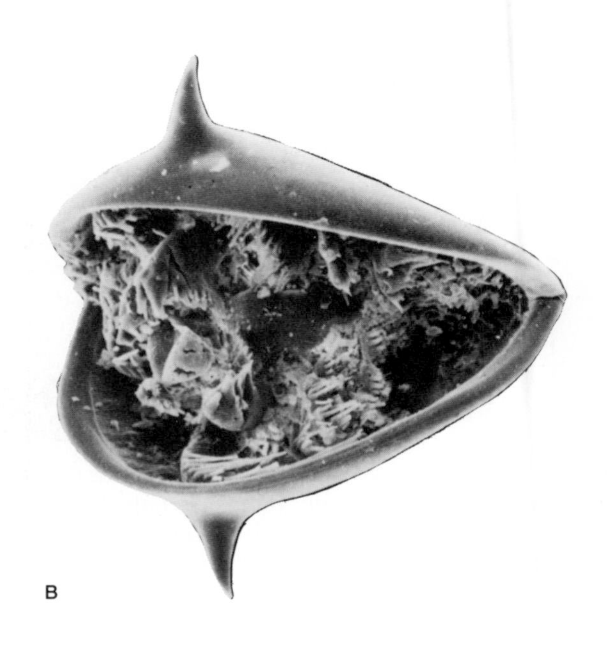

FIGURE 10-7 The Cambrian ostracod *Vestrigothia*. *A.* Lateral view of the bivalved shell (×110). This shell is the part of the ostracod that is usually fossilized. *B.* This remarkable specimen is a shell that is gaping open to reveal many appendages that were well preserved, because soon after the animal died they were coated with phosphatic material (×100). *(From K. J. Müller, Lethaia 12:1–27, 1979.)*

A

B C

FIGURE 10-8 Animals without durable shells from the Burgess Shale of British Columbia. *A.* An arthropod specimen related to the trilobites. *B.* A polychaete worm. *C.* An especially important animal that is intermediate in form between polychaete worms and arthropods; it had a wormlike body but possessed walking legs that resembled those of the arthropods. *A* and *B* ×3, *C* ×4. *(Smithsonian Institution.)*

twice the number of trilobite genera in the Burgess fauna. Second in variety in this unusual fauna are the **annelid worms;** six families, all of which survive today, have been found in the Burgess Shale, an indication that these worms were already highly diversified in Cambrian time. In fact, this important group probably produced many of the burrows found in rocks of latest Precambrian age.

Early Plankton Most of the Cambrian animals described above were herbivores that fed on algae. The abundance of algal stromatolites on the seafloor was noted earlier, and in the waters above were single-celled planktonic algae that must also have been important producers in Cambrian seas. Modern forms of planktonic algae, however, were not yet present; instead, the phytoplankton known as acritarchs flourished during the Cambrian Period and persisted throughout the entire Paleozoic interval (Figure 10-9). The cysts in which acritarchs were encased during resting stages are all that remain of these organisms in the fossil record. As we saw in Figure 9-10, acritarchs were conspicuous in the late Precambrian record as well. It is possible that other types of phytoplankton also played major roles in early Paleozoic seas without leaving fossil records.

Primitive Predators What carnivores existed in Cambrian time? Some of the wormlike animals and arthropods of the Burgess Shale fauna were equipped with small pincers or specialized mouth-parts that were adapted for killing or chewing up small animals. Trilobites and conodont animals also had small jawlike structures

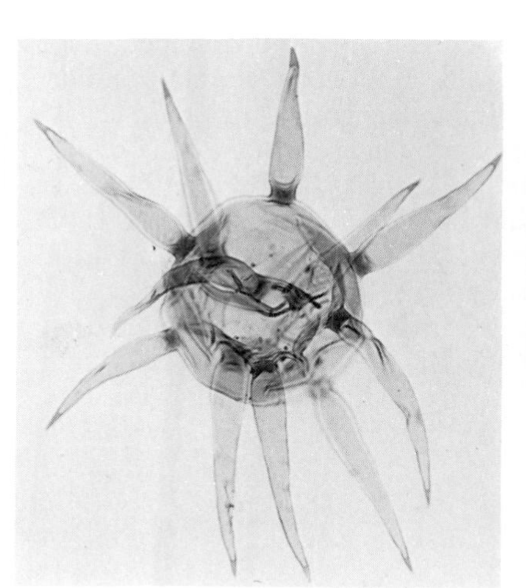
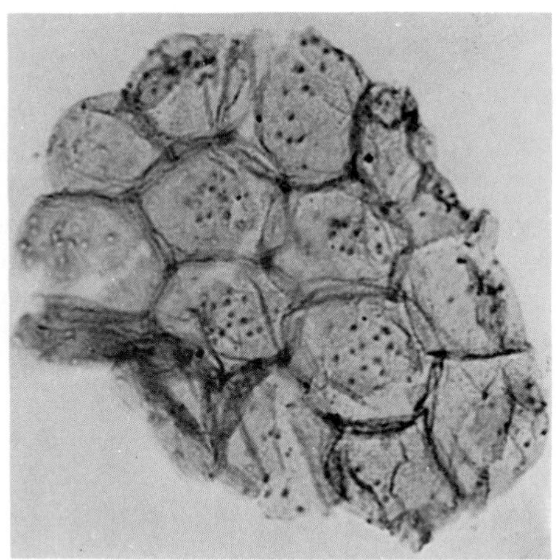

FIGURE 10-9 Acritarchs from Ordovician deposits of Oklahoma. A single-celled spiny variety *(left)* contrasts with the cluster of cells shown *(right)*. *(H. Tappan, The Paleobiology of Plant Protists, W. H. Freeman and Company, New York, 1980.)*

that were capable of crushing tiny species of prey. Jellyfishes must also have captured small victims with their stinging cells. One group of predators that arose very late in the Cambrian Period were the **nautiloids** (see Figure 10-13). Like the modern mollusks with which they are united in the class **Cephalopoda,** nautiloids were predators that used their tentacles to grasp prey and their beaks to tear the prey apart. Numerous nautiloid species have been found in very late Cambrian strata in China, but only a few have been found elsewhere in the world. Cambrian nautiloids were quite small; most measured between 2 and 6 centimeters (~1.0 and 2.5 inches) in length.

Entirely missing from Cambrian seas, as far as we know, were large carnivores—animals that could break large shells or tear apart large prey in the way that crabs and jawed fishes do in the modern world. Bony plates of very small fishes have recently been found in Cambrian rocks (Figure 10-10), but they did not belong to biting fishes; they are the remains of a large group of jawless fishes that, as we will see in Chapter 11, burrowed in mud for small items of food during both Cambrian and middle Paleozoic time.

The Earliest Reefs During Early Cambrian time the **archaeocyathids** flourished. These cone-shaped creatures of uncertain biological relationship were attached to the substratum by the tips of their skeletons (Figure 10-11). The vaselike shape and sedentary habits of archaeocyathids suggest that these animals fed on organic matter suspended in the water and that they at least superficially resembled sponges. Archaeocyathids were partially responsible for the origin, in Early Cambrian time, of the world's first organic reefs. The main framework builders were archaeocyathids, but algae and other organisms of unknown biological relationships also contributed to the solid structures. These early reefs, which stood above

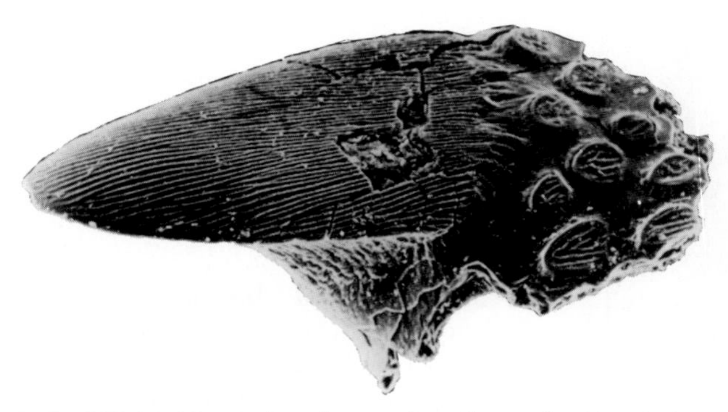

FIGURE 10-10 A tiny, bony plate from the Cambrian System of Wyoming, believed to have belonged to a small jawless fish. *(U.S. Geological Survey.)*

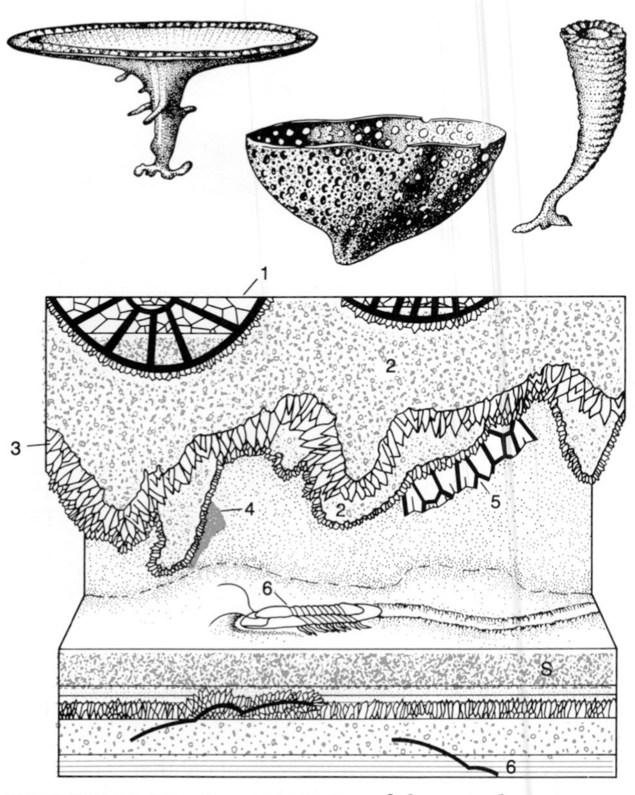

FIGURE 10-11 Reconstruction of three archaeocyathid species ***(above)*** **and one of the world's oldest organic reefs, located in the Lower Cambrian Series of Labrador** ***(below).*** All three archaeocyathid species were vase- or bowl-shaped. It seems likely that these animals were similar to sponges in that they pumped water through their porous walls, but it is not certain whether they were closely related to any group of modern organisms. The diagram shows the composition of the reef, which was constructed by several kinds of organisms, the most important of which were archaeocyathids (1) and calcareous algae (2). Cavities were encrusted with crystals of the mineral calcite (3), which were precipitated from seawater, and by organisms whose biological relationships are uncertain (4, 5). Trilobites (6) left tracks on sediment flooring cavities in the reef and also left fossil remains within the sediment. *(Archaeocyathid drawings from D. R. Kobluck and N. P. James, Lethaia 12:193–218, 1979. Diagram after I. T. Zhuravleva, Akad. Nauk. U.S.S.R., Geol. Geofiz. Novosibirsk, 2:42–46, 1960.)*

the seafloor as low mounds, ceased to form before the end of Early Cambrian time, when archaeocyathids became extinct. All that remained until mid-Ordovician time were very small, inconspicuous reeflike structures formed by the organisms that had been secondary contributors to the archaeocyathid reefs.

Patterns of Adaptive Radiation The trilobites originated near the end of Tommotian time or shortly thereafter and immediately underwent a remarkable adaptive radiation. Of some 140 families of trilobites recognized in Paleozoic rocks, more than 90 have been found in Cambrian strata.

Many of the types of marine animals that appeared during the Cambrian Period were strange groups that included only a few genera and species; indeed, some are classified as discrete classes or even phyla. The early Paleozoic history of the phylum Echinodermata illustrates this phenomenon. Today this phylum includes only a few large groups, such as starfishes and sea urchins, but quite a number of bizarre echinoderm classes evolved during Cambrian and Ordovician time (Figure 10-5). None included many species or genera, and most survived only a short time. As discussed in Box 10-1, many body plans were "tried out" in this manner, but only a few succeeded. This pattern is sometimes referred to as evolutionary "experimentation," with the understanding that what happens is not a planned event but rather a development produced blindly by nature. Of the many groups of invertebrates that appeared during Tommotian and later Early Cambrian time, only a few—such as sponges, snails, brachiopods, and trilobites—flourished long afterward.

The Cambrian adaptive radiation of marine animals with skeletons was not without interruption. During the latter part of Cambrian time, several mass extinctions eliminated most of the trilobite species in North America and in at least some other regions of the world. We will examine these extinctions later in this chapter, in the context of paleogeography.

The Ordovician Adaptive Radiation

The last of the Cambrian mass extinctions, at the very end of the Cambrian Period, eliminated large numbers of nautiloid and trilobite genera. This was a crisis from which the trilobites never fully recovered. Trilobites are found in many Ordovician strata (Figure 10-12) but not in abundances or diversities comparable to those of many Cambrian limestones. The Ordovician Period was instead characterized by the adaptive radiation of many other groups of animals. Some of them almost certainly originated in Cambrian time but failed to diversify markedly until the Ordovician, while others probably did not appear until Ordovician time. The Ordovician adaptive radiation populated the seas with many classes and orders of animals that continued to flourish in later Paleozoic periods. Indeed, it is possible that all animal phyla from subsequent geologic intervals were present in the seas by the end of Ordovician time.

FIGURE 10-12 Ordovician trilobites. The spiny genus is *Ceraurus*, an animal about 4 centimeters (1.6 inches) long that crawled over the surface of the sediment. *Calimene*, which was also a surface crawler but frequently rolled up for defense against predators, was about 5 centimeters (2 inches) long. *Isotelus (right)* was a smooth form that burrowed in the sediment, as depicted in the right foreground of Figure 10-13. This half-grown specimen is about 8 centimeters (3 inches) long. *(Chip Clark.)*

An interesting aspect of the Late Ordovician fauna is that most of the skeletonized members were animals that lived on the surface of the sediment rather than within it (Figure 10-13). It is difficult to move about and obtain oxygen within sediment, and in Ordovician time relatively few kinds of animals had developed methods of coping with these problems. Let us take a closer look at life in Ordovician seas.

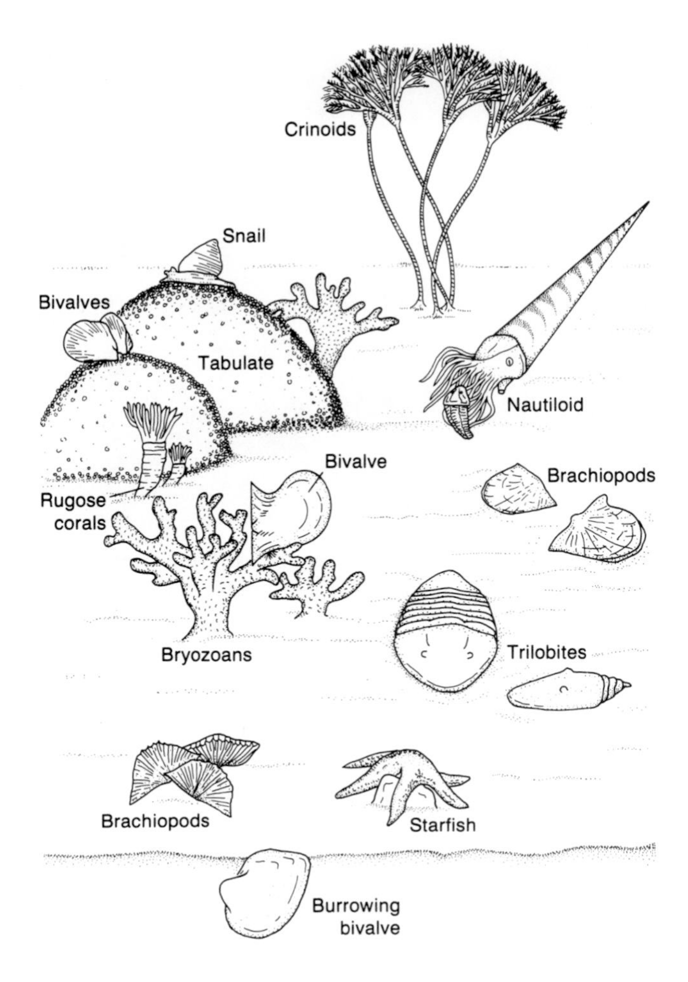

FIGURE 10-13 Life of a Late Ordovician seafloor in the area of Cincinnati, Ohio. Fossils of many of the groups of animals represented here can be seen in Figures 10-12 and 10-14 through 10-20. Note that at this early stage of Phanerozoic evolution, relatively few animals lived within the sediment. At the left a snail crawls over a large tabulate colony, and two bivalve mollusks are attached to another tabulate colony by threads that give the bivalves stability. Another bivalve is similarly attached to the branch of a bryozoan colony. Two solitary rugose corals, lodged alongside colonies, have their tentacles outstretched for food. Stalked crinoids are waving about in the center of the picture, feeding on suspended matter with their arms. To their right, a large nautiloid prepares to eat a trilobite that it has trapped in its tentacles; below the nautiloid's eye is a spoutlike siphon that is used to expel water for jet propulsion. Two kinds of suspension-feeding brachiopods live on the seafloor. In the right foreground are trilobites of a type that left trace fossils, indicating a burrowing mode of life. In the central foreground a starfish prepares to devour a bivalve by prying apart the shell halves with its sucker-covered arms; then, by extruding its stomach, the starfish can digest the bivalve within its opened shell.

Three groups, which were all present in Cambrian times but were not highly diversified, provide especially important Ordovician index fossils: the articulate brachiopods, the graptolites, and the conodonts.

Articulate brachiopods (Figure 10-14) are the most conspicuous group of well-preserved fossils both in Ordovician rocks and in all younger Paleozoic systems as well. These animals were immobile suspension feeders that either rested on sediment, were partially buried in sediment, or attached themselves to solid objects.

Small Swimmers and Floaters Graptolites were especially common in Ordovician and Silurian times (see Figure 4-2). They are most frequently found in black shales — partly because they were too fragile to be easily preserved in sand and partly because many of them were oceanic plankton that sank to muddy deep seafloors after death. Because most individual species existed for less than a million years, fossil graptolites (like fossil trilobites) are especially useful for correlation.

The wide distribution of conodonts suggests that these toothlike structures also represent elements of creatures that floated or swam. The recent discovery of the first carbonized impression of the conodont animal reveals the presence of fins, which suggests that they were swimmers (Figure 10-6). The broad distributions and short stratigraphic ranges of individual conodont species make these fossils ideally suited for correlation. Geologists can also extract conodonts from limestones in large numbers simply by dissolving the rock with acid — a process that does not attack the phosphatic material of which these fossils are composed.

Attached Animals Joining the brachiopods as important sedentary animals of Ordovician seafloors were the **rugose corals,** which are sometimes known as horn corals because of their shape, and the **crinoids,** which are sometimes called sea lilies, even though they were animals rather than plants (Figure 10-15). Primitive corals and crinoids had existed during the Cambrian Period, but at such low diversities that they were insignificant members of the ecosystem.

Three groups of colonial animals with sturdy skeletons attained importance on Ordovician seafloors. Of these, the **bryozoans,** or moss animals (Figures 10-13 and 10-15), were the most con-

FIGURE 10-14 Articulate brachiopods of Ordovician age. The large form in back on the right is shown reclining on an ancient seafloor in Figure 10-13. *(Chip Clark.)*

FIGURE 10-15 Ordovician animals that stood upright on the seafloor. *A.* One of the oldest kinds of rugose corals. Like modern hexacorals, this primitive coral probably used tentacles to capture its prey. *B.* An early crinoid (sea lily); animals like this were suspension feeders attached to the seafloor by a flexible stalk formed of disclike elements of calcite that interlocked like poker chips and were encased in flesh. A modest variety of similar crinoids inhabit modern seas. *C.* A colony of bryozoans exhibiting finger-sized branches in which a number of brachiopods are nestled. (Figure 10-13 shows Ordovician rugose corals, crinoids, and bryozoan colonies as they grew on the seafloor.)

FIGURE 10-16 Ordovician reef builders. The massive form with internal layering is a stromatoporoid colony about 23 centimeters (9 inches) across, and the colonies with vertical tubes are tabulate corals. *(Chip Clark.)*

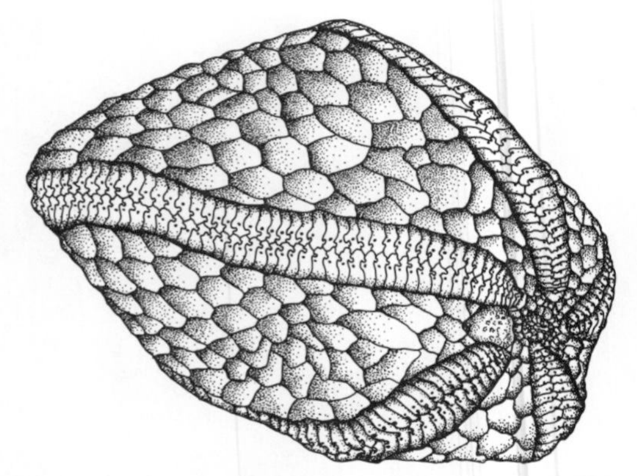

FIGURE 10-18 An Ordovician sea urchin (×2) that differed from living echinoids in its flexibility. In life the globular skeleton was covered with spines, which served for locomotion. The animal probably fed on algae or plant debris. Its mouth was positioned underneath (at the "south pole") and the anus was positioned near the "north pole." *(After E. W. McBride and W. F. Spencer, Philos. Trans. Roy. Soc. London 229B:91–136, 1939.)*

spicuous. The others, the **stromatoporoids** and the **tabulates** (Figure 10-16), attained their greatest importance as reef builders in middle Paleozoic times. The recent discovery of close living relatives of stromatoporoids reveals that these fossil forms were sponges with dense skeletons. Some tabulates may also have been sponges, although most of them were probably coral-like coelenterates. Because tabulates became extinct at the end of the Paleozoic Era, we may never be certain of their biological relationships.

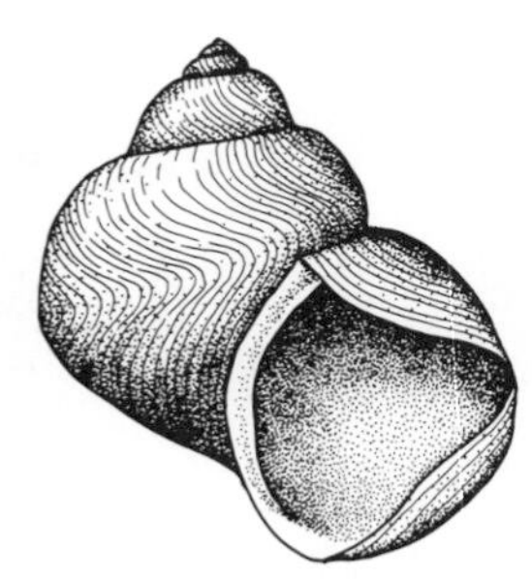

FIGURE 10-17 Snails (gastropods) that inhabited Ordovician seas. The shell at the left probably belonged to an animal that was largely stationary and rested on the seafloor in the position shown here. The shell at the right belongs to a group of crawling snails represented in modern seas by a few surviving species. Figure 10-13 shows this type of snail moving over a tabulate colony. *(After Treatise on Invertebrate Paleontology, Part I, R. C. Moore [ed.], Geological Society of America and the University of Kansas Press, Lawrence, 1960.)*

Bottom Dwellers That Flourish Today In addition to trilobites, the mobile epifauna of Ordovician time included new varieties of snails (or gastropod mollusks) as well as the first **echinoids** (or sea urchins), which differed from modern echinoids in that they had flexible bodies (Figures 10-17 and 10-18). The ostracods on Ordovician seafloors (Figure 10-7) were also more diverse than those of Cambrian time.

The bivalve mollusks, which apparently were represented by only a few tiny species during the Cambrian Period, radiated during Ordovician time to develop a variety of forms adapted to modes of life on or within the substratum (Figure 10-19). During Ordovician time, burrowing bivalves attained the position that they hold today—they became the most successful group of burrowing suspension feeders with shells.

Ordovician Predators Jawless fishes persisted from Cambrian into Ordovician time but continued to contribute only fragmentary fossil remains to the rock record, so we know little about their shapes. During early Paleozoic time, before fishes acquired jaws, only invertebrates preyed on large animals. As Figure 10-13 indicates, two groups seem to have dominated here—the starfishes (Figure 10-20) and the nautiloids. All five of the living orders of starfishes were already present in Ordovician time.

FIGURE 10-19 Bivalve mollusks of Ordovician age. The two species on the left were attached to the surface of the substratum by threads that they secreted. The species on the right lived in the sediment. *(U.S. Geological Survey.)*

New Levels of Biological Diversity The rapidity of adaptive radiation during this second period of the Paleozoic Era is illustrated by the early history of the bryozoans. Very few bryozoans are found in Lower Ordovician rocks, but Middle Ordovician rocks, which represent a period of some 20 million years, record the rapid development of diverse bryozoan faunas. Within the Simpson Group of Oklahoma, for example, early Middle Ordovician strata contain just five or six bryozoan species. Successively younger strata yield increasing numbers of species. In the uppermost parts of the Simpson Group, the variety of species rivals that of diverse Late Ordovician bryozoan faunas.

The adaptive radiation of many groups of marine invertebrates during the Ordovician Period is shown in Figure 10-21, where the number of marine invertebrate families is plotted against time. The darkest portion of the graph represents fami-

FIGURE 10-20 An Ordovician starfish—one of the first of this class of animals. Figure 10-13 shows a starfish feeding on a bivalve. *(Smithsonian Institution.)*

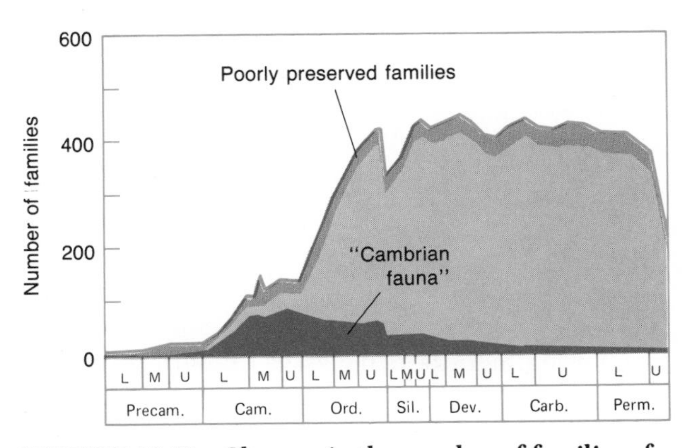

FIGURE 10-21 Changes in the number of families of marine invertebrates through the Phanerozoic Era. The "Cambrian fauna" consists of families that are found only in the Cambrian System or are best represented there. The expansion of life that produced the Cambrian fauna was followed by a new adaptive radiation in the Ordovician Period, and this Ordovician expansion, despite a mass extinction at the end of the period, produced a general level of diversity that was maintained until the end of the Paleozoic Era. *(After J. J. Sepkoski, Paleobiology 4:223–251, 1978.)*

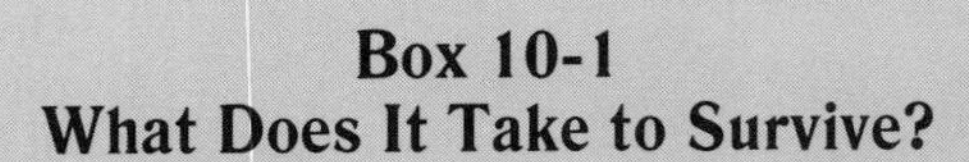

Box 10-1
What Does It Take to Survive?

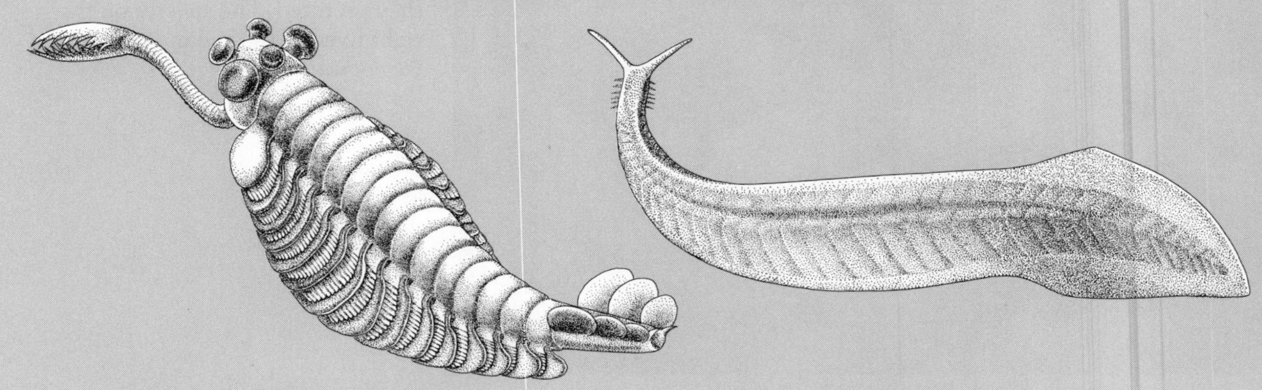

Many animals that were uniquely preserved in the Burgess Shale display unusual features. *Opabina (left)* had five eyes and food-gathering pincers at the end of a strange nozzle. *Pikaia (right)* was a swimming chordate. *(Drawings by Marianne Collins, © by W. W. Norton & Company, Inc., reprinted by permission from S. J. Gould, Wonderful Life.)*

The diversification of animal life in Late Proterozoic and Cambrian time was a veritable evolutionary explosion (p. 224). When humans reflect on this episode, they tend to focus on the many successful groups of animals it produced—the groups that have survived—and to ignore the groups of very early animals that cannot be assigned to living phyla. Many of these animals, such as the strange-looking one shown above on the left, died out soon after they appeared.

When we assemble the full cast of characters found in these early faunas, we can hardly avoid asking why some groups survived long after Cambrian time whereas others soon vanished. Were the victims of early extinction biologically inferior to the groups that survived? Were they less effective in competing with other taxa or in avoiding predators? Or were they simply the unlucky victims of catastrophic extinctions that struck down species without regard to their ecological abilities in a healthy

lies belonging either to classes that are restricted to the Cambrian System or to classes that are better represented here than in younger systems. These families can be said to form the Cambrian fauna, although even this fauna is divisible into the strange, early Tommotian fauna and the succeeding Cambrian fauna, dominated by trilobites.

What is most striking about the post-Cambrian fauna is that it consisted of slightly more than 400 known families by the end of the Ordovician Period—approximately the same number that characterized all subsequent intervals of Paleozoic time. In contrast, diversity in the Cambrian Period had leveled off at about 150 families. We can draw two conclusions from these observations: First, the great adaptive radiation of the Ordovician Period greatly augmented the diversity of the marine ecosystem; and second, while some old families died out and other new ones appeared later in the Paleozoic Era, something seems to have held the number of families in check from Ordovician time until the end of the era. At least three factors may have been operating here: (1) Environments may have become filled with life

environment? Such accidental death might have come in the mass extinctions that decimated the trilobites during the latter part of Cambrian time, for example, or the global crisis that brought the Ordovician Period to a close.

Not only did the very ancient groups of animals that have survived to the present escape these catastrophes, but for hundreds of millions of years thereafter they coped successfully with predators and with taxa that vied with them for food and other resources. Some of the taxa that failed to survive appear to have been ecologically deficient. Some, for example, were weakly armored against predators and probably could no longer flourish after advanced predators evolved early in the Paleozoic Era. It is clear that external skeletons, which evolved in many taxa in earliest Cambrian time, played an important role in thwarting predators. Some of these skeletons, however, were weak and flexible. A multitude of fossil plates from the skeletons of eocrinoids shows that these stalked ancestors to modern sea lilies were very abundant, yet their fossil skeletons are rarely intact; they normally fell to pieces soon after the animals died. Some of their relatives, however—cystoids, blastoids, and crinoids—had robust skeletons that abound intact in the fossil record, and these animals flourished long after Cambrian time. Although we have an incomplete picture of Paleozoic predators, we know that one important group, the nautiloids, evolved at the end of Cambrian time. Rapid adaptive radiation of these beaked predators produced numerous families early in Ordovician time, and the impact on other marine animals, including some of the kinds found in the Burgess Shale, must have been profound.

All of the early groups of animals were at risk soon after they evolved simply because they included few species. Diversification increases chances of survival, and many groups of animals that failed to survive beyond Cambrian time are known from only a few fossil species.

Among the many strange animals of the Burgess Shale fauna was a rather ordinary-looking creature that belonged to a group with a great future. This was the genus *Pikaia,* which had a structure known as a notochord. The notochord evolved into the vertebrate backbone, and it identifies the animal as a chordate—a member of the phylum to which humans and all other vertebrates belong. Probably *Pikaia* itself was not our ancestor, but one of its Cambrian relatives certainly was—some form undiscovered in the fossil record. It is sobering to observe that if accidental extinction had swept away the early chordates before some of them evolved into fishes, no four-legged vertebrate, to say nothing of a human being, would ever have walked the land.

to the point where new forms could no longer easily evolve; (2) the evolution of increasingly effective predators may have made it increasingly difficult for new forms to evolve; and (3) most of the animals that existed may have been too specialized to give rise easily to totally new types. We cannot at present assess the relative importance of these factors.

Reefs of a New Type By Middle Ordovician time, adaptive radiation had produced a number of new reef-building animals. Thus, some 40 million years after the archaeocyathid reefs of the Cambrian had ceased to grow, a new era of reef building began. Some of the first Middle Ordovician reefs were built by bryozoans in northeastern North America. Stromatoporoids and tabulates also contributed to later reef building and later expanded to dominate organic reefs during Silurian and Devonian time. Although many Ordovician reefs were small mounds or patch reefs similar to those of the Cambrian Period (Figure 10-11), others exceeded 100 meters (~300 feet) in length and 6 or 7 meters (~20 feet) in height.

Many Ordovician reefs, in addition, display several stages of development. The community that inhabited and added to a reef at each developmental stage was followed by a different one. Massive stromatoporoids were often dominant in the final stage (Figure 10-22).

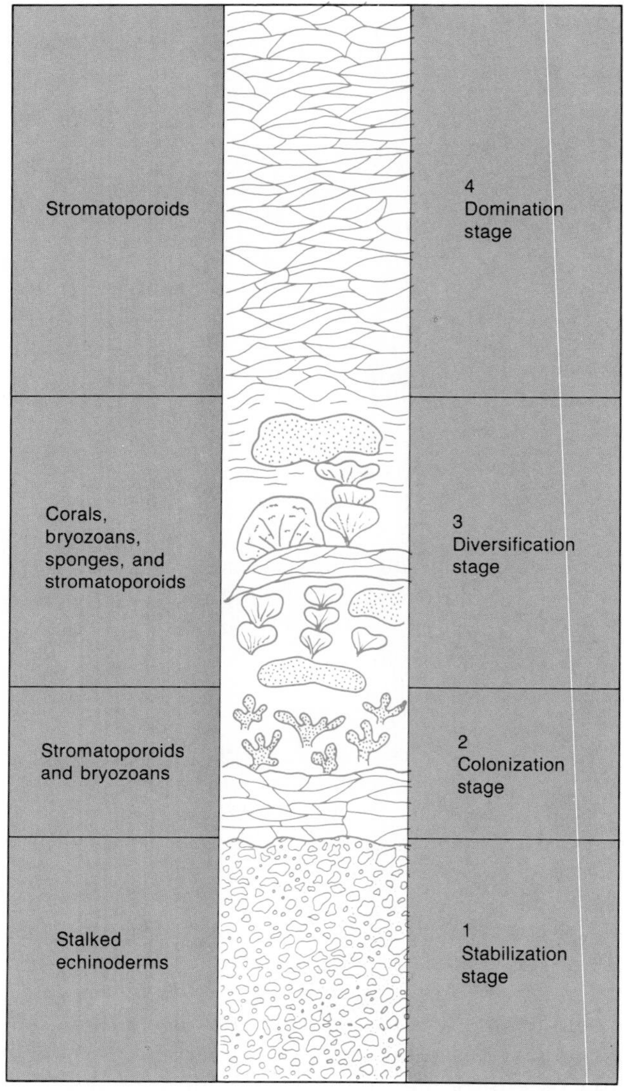

FIGURE 10-22 Stratigraphic section through a reef in the Crown Point Formation of Vermont. Four stages of development are shown. First, stalked echinoderms stabilized the seafloor with the holdfasts by which they attached themselves (stage 1). Then the stabilized seafloor was colonized by stromatoporoids and by branching bryozoans (stage 2). These were later joined by a variety of reef-building forms (stage 3). Eventually massive stromatoporoids dominated the reef (stage 4). *(After L. P. Alberstadt et al., Geol. Soc. Amer. Bull. 85:1171–1182, 1974.)*

Animal Life and the Decline of Stromatolites Of all the Phanerozoic periods, only the Cambrian and Ordovician were characterized by abundant stromatolites. This abundance was carried over from late Precambrian times. Stromatolites continued to blanket enormous areas of the seafloor during Cambrian time (Figure 10-23), but by the end of the Ordovician interval large stromatolites were rare.

The few areas where stromatolites grow in the modern world offer some clues as to what happened to stromatolites during the Ordovician Period. The types of cyanobacteria that form stromatolites occur widely in modern seas, but only in areas usually above high tide and in hypersaline lagoons do they prosper well enough to form conspicuous stromatolites. Marine animals are largely absent from both of these kinds of habitats. In more normal marine environments, animals burrow through algal mats and also eat them; their ravages prevent the mats from forming stromatolites. Experiments have shown that when animals are excluded from small areas of seafloor in tropical climates, algal mats flourish, just as they did long ago. It seems evident that the great adaptive radiation of Ordovician life (Figure 10-21) produced a variety of animals that tended to prevent stromatolites from developing in all but unusual habitats. As a result, the character of shallow seafloors was forever altered.

Did Plants Invade the Land? Chapter 11 will reveal that in Silurian time, small multicellular plants were well established in moist terrestrial environments. Although similar plants probably invaded the land during the Ordovician Period, the evidence is not yet conclusive. It consists of fossilized sheets of cells similar to those that cover the surfaces of modern land plants as well as structures that closely resemble the spores released by primitive (non-seed-bearing) land plants of the modern world (Figure 10-24). If land plants did evolve during the Ordovician Period, it seems likely that they were restricted to moist habitats, as mosses are in modern times.

Terminal Ordovician Mass Extinction

The Ordovician Period concluded with one of the greatest mass extinctions in all of Phanerozoic time. Considerably more than half of the species

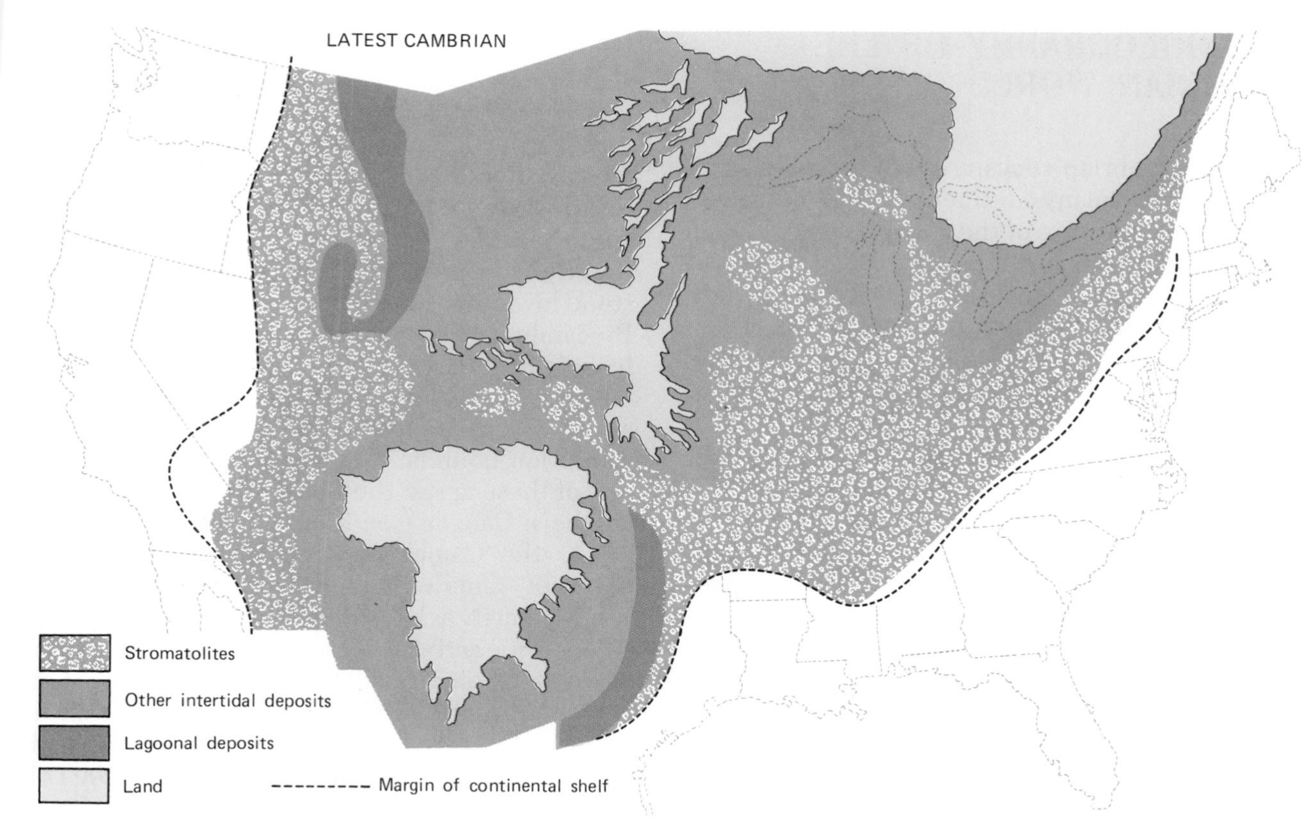

FIGURE 10-23 The widespread occurrence of stromatolites in North America during the latest part of the Cambrian Period, when shallow seas spread across much of the continent. These stromatolites occupied vast intertidal and supratidal areas. *(Modified from C. Lochman-Balk, Geol. Soc. Amer. Bull. 81:3191–3224, 1970.)*

of brachiopods and bryozoans in North America died out, and on a broader geographic scale, 100 families of Ordovician marine animals failed to survive into the Silurian Period. A mass extinction is a geographic phenomenon, striking biotas over a large area of the globe. For this reason, we will examine the causes of the terminal Ordovician mass extinction—together with the causes of the Cambrian mass extinctions—in the context of wordwide paleogeography.

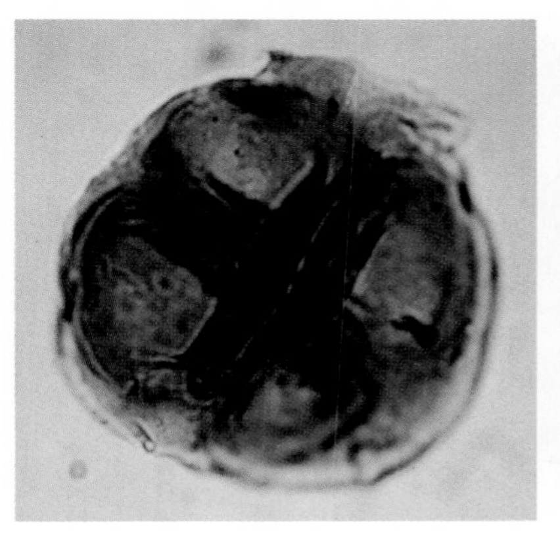

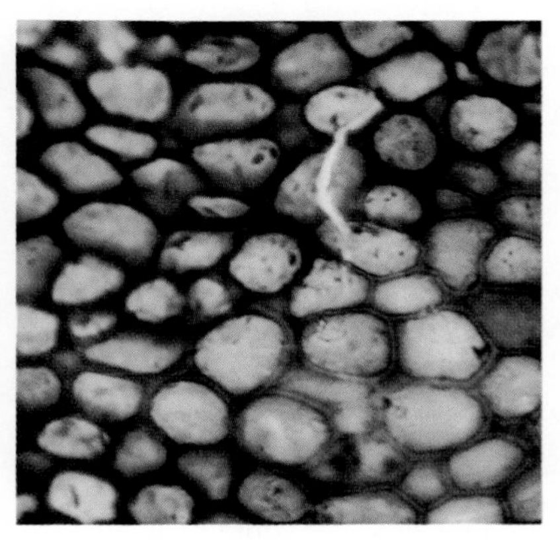

FIGURE 10-24 Late Ordovician fossils that may represent plants that lived on the land. *A.* Spores that resemble those of modern land plants (×670). *B.* A sheet of fossil cells that resemble those covering the surfaces of some land plants (×384). *(From J. Gray et al., Geology 10:197–201, 1982.)*

PALEOGEOGRAPHY OF THE CAMBRIAN WORLD

Because Cambrian rocks are so widespread and are found on so many continents, geologists have a much clearer picture of the Cambrian world than of the Late Proterozoic world. This paleogeographic framework gives us considerable insight into the repeated mass extinctions of Cambrian marine life.

Continents of the Cambrian World

As we saw in Chapter 9, rock magnetism and other geologic evidence strongly suggest that most cratons were fused into one giant supercontinent near the end of Precambrian time. The arrangement of continents late in Cambrian time was strikingly different. By that time, Gondwanaland and several smaller landmasses occupied equatorial zones (Figure 10-25), and no continent lay near either pole. This positioning of continents helps explain why Cambrian limestones are common on many cratons—most shallow-water deposits of Cambrian age accumulated in tropical or near-tropical climatic zones.

The Cambrian Period was notable for the progressive flooding of continents. The stage for this trend was set near the end of Precambrian time, when most of Earth's cratons stood largely exposed above sea level. In fact, the oceans probably stood lower in relation to the cratons during latest Precambrian time than at any time during the Paleozoic Era. As a result, only scattered local areas on modern continents yield a continuous record of shallow-water deposition across the Precambrian-Cambrian boundary. In Chapter 9 we discussed one of these areas: the Rocky Mountain region of southern Canada (see Figure 9-28).

As the Cambrian Period progressed, many parts of Gondwanaland remained above sea level—partly as a result of regional uplifts caused by orogenic activity between 800 and 400 million years ago. Some smaller cratons, however, show evidence of continued encroachment of Cambrian seas until little of their total area remained exposed late in Cambrian time (Figure 10-25). This

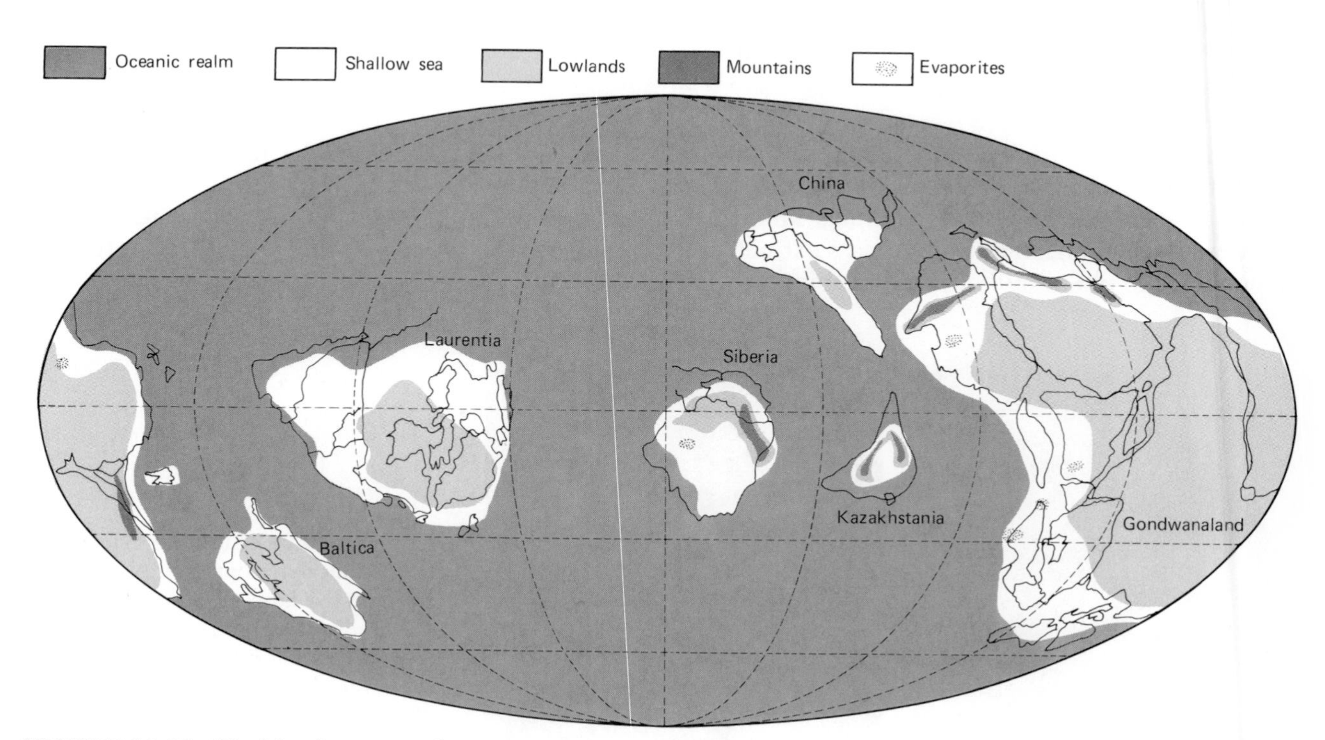

FIGURE 10-25 World paleogeography in Late Cambrian time. Continents were positioned at low latitudes, and many were inundated by shallow seas. *(After R. K. Bambach et al., American Scientist 68:26–38, 1980.)*

flooding represented one of the largest and most persistent sea-level rises of the entire Phanerozoic Eon. It was interrupted in North America only by a modest regression in Middle Cambrian and by another during Late Cambrian time.

A Characteristic Pattern of Deposition

Slightly before the beginning of Cambrian time, as the seas began to encroach on broadly exposed continents, siliciclastic sediments were eroded from the continents and accumulated around the continental margins. Examples of this pattern, which were described in earlier chapters, are the Windermere Formation of western North America (Figure 9-28) and the sandstones and shales that form the first tectonic cycle in the eastern Appalachians (Figure 7-13).

When the seas encroached farther over most continents during Middle and Late Cambrian times, a characteristic sedimentary pattern emerged. To understand the nature of this pattern, let us examine the geography of Laurentia, the landmass that included North America, Greenland, and Scotland (shown in Figure 10-25).

At all times during the Middle and Late Cambrian, some part of central Laurentia stood above sea level (Figure 10-26). Around the margin of the land, belts of marine deposition were arranged in a concentric fashion. Siliciclastic sediments derived from the craton were deposited in the innermost belt. This belt was essentially the same as the marginal siliciclastic belt that surrounded the continent during earliest Cambrian time, but it had shifted inland along with the shoreline. Seaward of this belt were broad carbonate platforms that were often fringed by reefs or stromatolites and that, as we have seen for eastern North America, terminated along a steep slope. Muds and breccias derived from the platform accumulated in deep water near the base of its steep slope (Figure 7-15), and subduction zones that lay close to cratons contributed volcanic rocks to the deep-water belt from the opposite direction.

Trilobites, the dominant skeletonized animals of Middle and Late Cambrian oceans, were distributed around continents in a pattern corresponding to the arrangement of the sedimentary belts. Certain groups of these trilobites are found primarily in deep-water deposits. They include small, blind forms that apparently lived as plankton and also the genus *Paradoxides*, which characterizes the exotic blocks of oceanic volcanic terrane that became attached to eastern North America early in Paleozoic time (Figure 7-17). Other groups were largely restricted to the broad carbonate-floored shallow seas of continents; trilobites of this type are found over large areas of the North American craton—lowlands that were extensively flooded by warm tropical seas during Late Cambrian time (Figure 10-25).

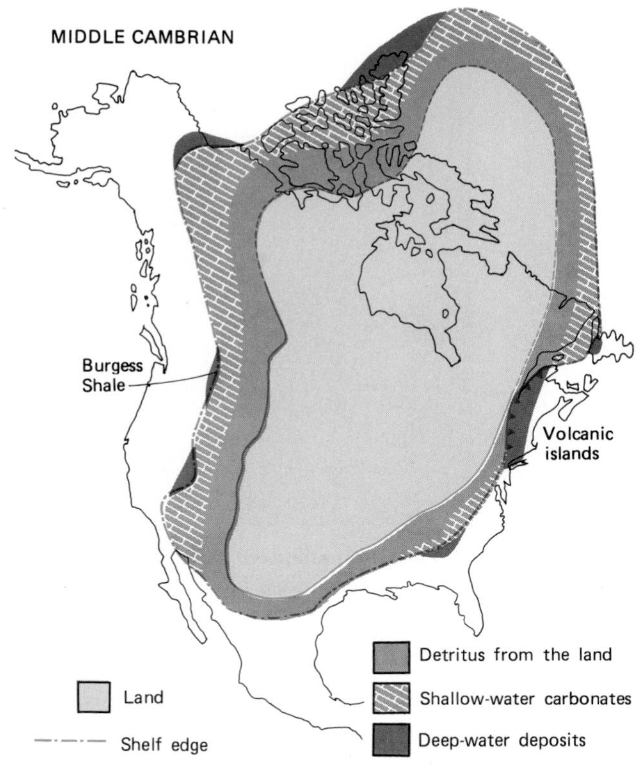

FIGURE 10-26 Concentric pattern of sediment deposition around the margin of Laurentia during Middle Cambrian time. Note the location of the Burgess Shale, renowned for its fauna of soft-bodied invertebrates, at the base of the Middle Cambrian continental shelf in western Canada. *(After A. R. Palmer, American Scientist 62:216–224, 1974.)*

Periodic Mass Extinctions of Trilobites

Trilobite species that inhabited tropical seas—including some small species thought to have been planktonic—were the ones that suffered in the

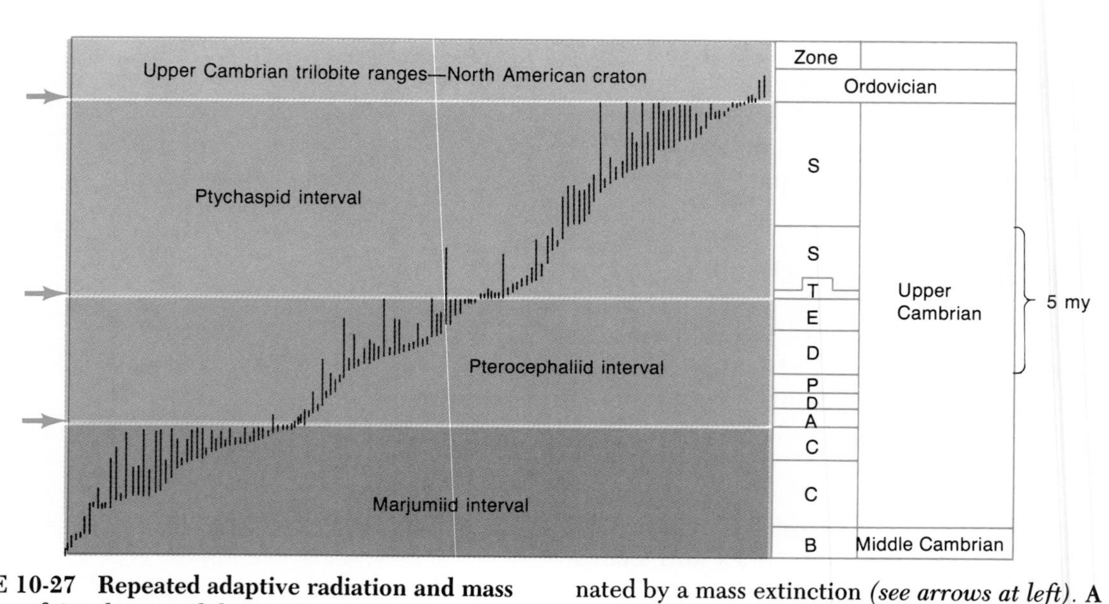

FIGURE 10-27 Repeated adaptive radiation and mass extinction of Cambrian trilobites. The vertical bars indicate the stratigraphic ranges of important species in Middle and Upper Cambrian deposits of North America. The ranges form clusters and thus delineate three successive adaptive radiations, each of which was terminated by a mass extinction *(see arrows at left)*. A stratigraphic interval representing approximately 5 million years is shown on the right. Note that many trilobite species survived for less than 1 million years. *(After J. H. Stitt, Oklahoma Geol. Surv. Bull. 124:1–79, 1977.)*

repeated mass extinctions of the Cambrian Period. Each mass extinction was followed by an adaptive radiation that restored the diversity of shallow-water trilobites to a high level (Figure 10-27). These events are well documented for North America but thus far have been recorded elsewhere only in Australia.

Each adaptive radiation of Cambrian trilobites occupied several million years, but each extinction was quite sudden (Figure 10-27). The fossil record reveals that each extinction took place during the deposition of a thin layer of sediment; thus it can be inferred that each must have occurred over an interval of no more than a few thousand years. The transition from one adaptive radiation to another followed a characteristic pattern that is illustrated in Figure 10-28. The mass extinction is recorded in a layer of limestone or shale just a few centimeters thick. Above this "disaster" layer, in sedimentary beds just a meter or so thick, a variety of new trilobite genera join species that survived the mass extinction. These beds accumulated during a brief time of biotic adjustment in which opportunistic species flourished for a short time and then dwindled in the face of competition from new, more successful forms. The beds above contain a different fauna that consisted of about half as many species. From this fauna there issued a new adaptive radiation that lasted several million years—until another mass extinction started the cycle again.

What led to the periodic mass extinction of trilobites? Elimination of shallow-water habitats by lowering of sea level can be ruled out, because the extinctions did not all coincide with regression of seas from most areas of Laurentia. In fact, a sizable regression occurred in Late Cambrian time, between two mass extinctions (Major Events, p. 242). Because the adaptive radiation that preceded each mass extinction took place in association with the deposition of tropical limestones, it has been suggested that a sudden, temporary cooling of the seas was the agent of the trilobites' periodic massive death. This idea gains support from evidence that the adaptive radiation following each mass extinction issued from a group of trilobites that lived offshore in cool, deep waters marginal to the continent. These offshore

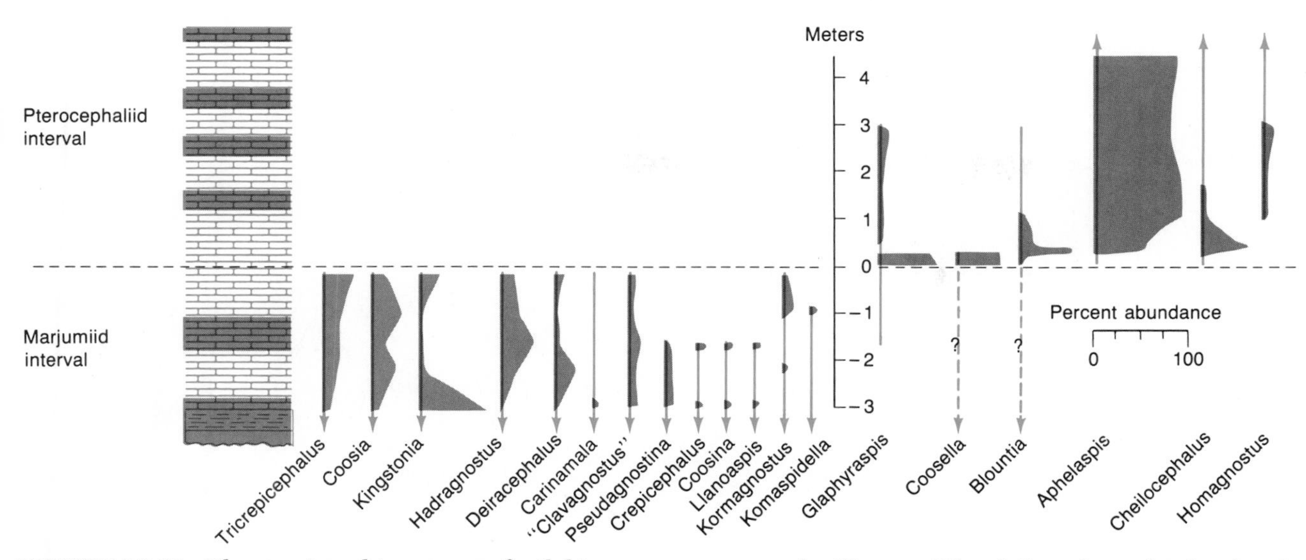

FIGURE 10-28 The stratigraphic pattern of trilobite mass extinction recorded in limestones at the top of the Cambrian Bonanza King Formation of Nevada. The gray band above each genus shows how the relative abundance of the genus changed through time. Many genera of the Marjumiid interval (see Figure 10-27) disappear at the "0 meter" level. Just above this level, only a few forms are found, and, in the absence of other genera, these forms were very abundant. Arrows indicate that species are also found below or above the occurrences displayed in the diagram. *(After A. R. Palmer, Alcheringa 3:33–41, 1979.)*

trilobites did not suffer in the mass extinction. Such evidence is only circumstantial, however, and the case for temperature as an agent of mass extinction remains unproved.

PALEOGEOGRAPHY OF THE ORDOVICIAN WORLD

After its general rise during the Cambrian Period, sea level stood high during much of Ordovician time, flooding broad cratonic areas (Major Events, p. 242). Widespread regressions during that period do not appear to have had any major effect on marine life. There is abundant evidence, however, that a cooling of climates contributed to the major marine extinction that took place at the close of the Ordovician Period. Before discussing the probable cause of this extinction, let us examine the profound local climatic consequences of the movement of two major continents, Gondwanaland and Baltica.

Baltica Moves Northward

As late as Middle Ordovician time, the center of Baltica lay far south of the equator. The Ordovician temperature gradient from equator to pole was nonetheless gentle enough to allow diverse marine faunas to occupy the shallow seas of Baltica. During the latter half of the Ordovician Period, Baltica moved toward the equator. Migrating north along with Baltica were England and southern Ireland, each of which had either been attached to the North African margin of Gondwanaland or had lain nearby (Figure 10-29). It was this movement that brought Baltica close to the eastern margin of Laurentia when the first Paleozoic tectonic cycle of the Appalachian region ended with orogeny and the deposition of flysch and clastic wedges (Figure 7-13). As Baltica and England moved toward Laurentia, the brachiopods, trilobites, and graptolites of these two northward-moving landmasses became increasingly similar to those of Laurentia. As Chapter 11 will reveal, it was not until Siluro-Devonian time, however, that the Iapetus Ocean closed and

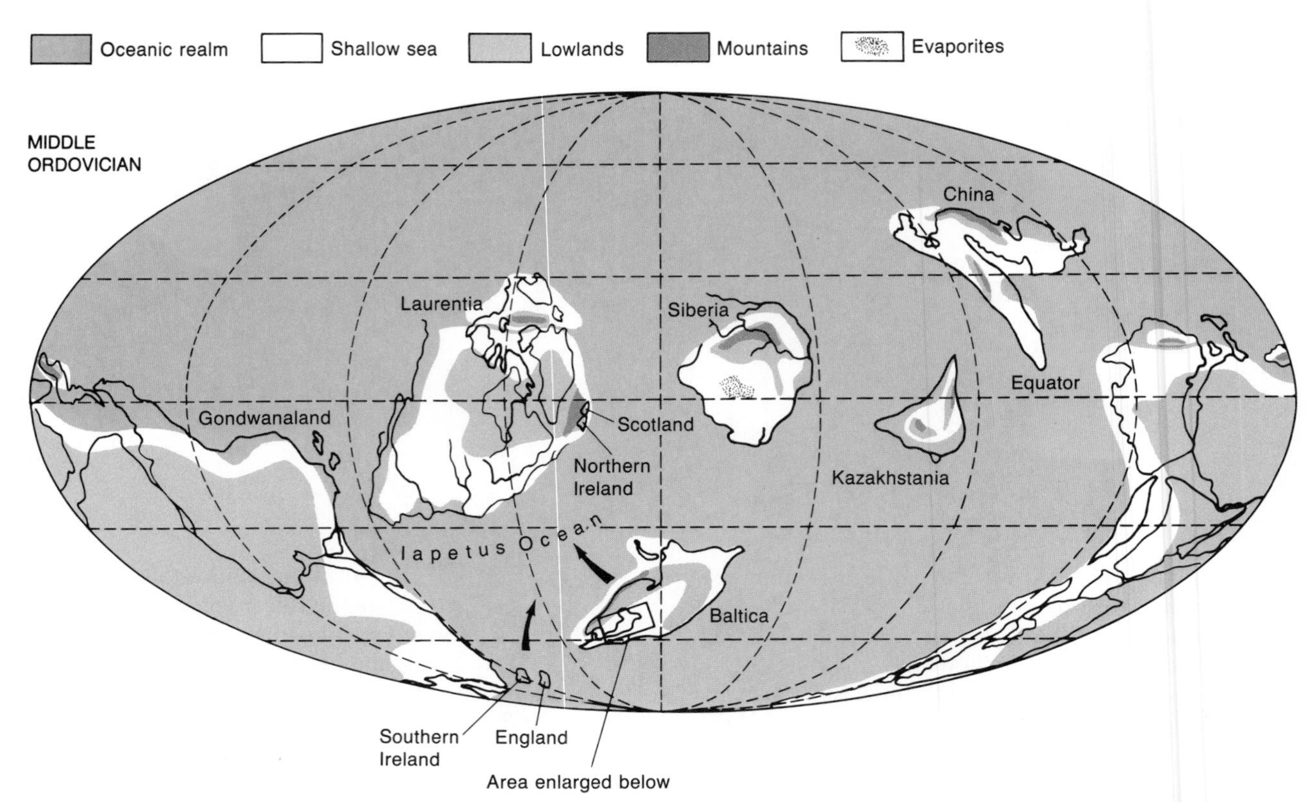

FIGURE 10-29 Movement of Baltica northward during the Ordovician Period. Arrows show the directions in which Baltica, southern Ireland, and England moved during Late Ordovician time. *(After R. K. Bambach et al., American Scientist 60: 26–38, 1980.)*

Baltica was united with Laurentia, England with Scotland, and the northern part of Ireland with the southern part.

Glaciers in Gondwanaland and Marine Mass Extinction

The mass extinction at the close of Ordovician time was one of the most severe ever to strike life in the oceans, eliminating about a hundred families of marine animals. It devastated the tropical reef community, which by this time was dominated by bryozoans, tabulates, and stromatoporoids. Also hard hit were trilobites, nautiloids, brachiopods, crinoids, and bryozoans. Plate movements seem to have played a major role in triggering the crisis.

While Baltica moved toward the equator, Gondwanaland moved poleward. Thus, while Baltica became warmer, parts of Gondwanaland became much cooler. Gondwanaland lay squarely on the equator late in Cambrian time (Figure 10-25), but by mid-Ordovician time, only its northern margin was equatorial (Figure 10-29). Several million years before the end of the Ordovician Period, glaciers grew in and around the south polar region of Gondwanaland, and as the period came to a close, the glacial episode reached a climax that was accompanied by a mass extinction in the marine realm.

Because the bulk of the evidence for extensive glaciation lies in remote desert regions of northern Africa, it was not discovered until the 1970s. At that time three or four levels of glacial deposits as well as a remarkable variety of glacial features, including ancient moraines (p. 54), were found in the central Sahara Desert together with scratch marks on Upper Ordovician rocks and ice-rafted boulders (Figure 10-30). These features and less extensive ones on other continents point to widespread glaciation late in Ordovician time. Possible tillites in North and South America may be of Silurian age; confirmation of the age and glacial ori-

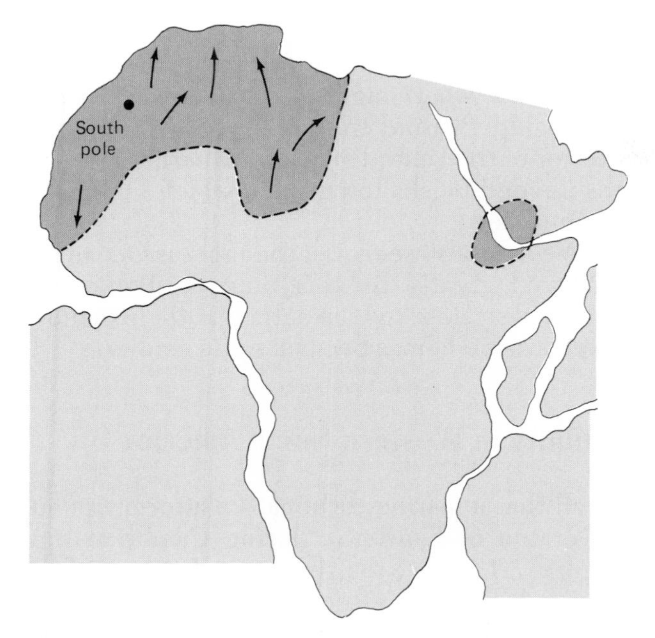

FIGURE 10-30 Late Ordovician glacial activity in northern Africa. The map shows the glaciated area (shaded), directions of ice flow (arrows), and the position of the Late Ordovician South Pole. The photograph is of glacial sediments in the Arabian Peninsula; the arrow points to a cavity where an ice-transported boulder about 1 meter (~1 yard) in diameter has weathered out of the tillite. *(Courtesy of H. A. McClure.)*

gin of these deposits would indicate a continuation of the glacial episode beyond Ordovician time. In any event, this episode seems to have peaked near the close of the Ordovician Period, causing sea level to drop rapidly.

Without question, the movement of Gondwanaland over the South Pole contributed to the Ordovician glacial episode. In general, oceans tend to remain warmer than the land in cold regions, in part because the albedo (the reflectance of sunlight) is usually higher for land than for water (p. 31). Moreover, ocean waters of cold regions usually mix with waters from warmer regions; so a glacier may accumulate on a large body of land in a cold polar region while an ocean in the same vicinity remains ice-free.

It has been suggested that the lowering of sea level at the end of the Ordovician Period contributed to the mass extinction of marine life by reducing the area of shallow seafloor, but the connection is doubtful. Large numbers of modern species survive on small areas of seafloor: the shallow seafloor around the small, isolated Hawaiian islands today supports about a thousand species of shelled mollusks. Moreover, episodes of sea-level lowering earlier in Ordovician time failed to cause mass extinction, even though they were of the same magnitude as the drop at the end of the period (Major Events, p. 242).

Several patterns suggest that cooling of the seas played an important role in the Late Ordovician mass extinction. For one thing, extinction was heaviest in the tropics, as we would expect if seas cooled on a global scale. Under such circumstances, temperate climatic zones would be expected to migrate equatorward, and faunas adapted to these zones would also be expected to migrate. On the other hand, cooling of the tropics would leave species adapted to warm tropical conditions with no place to go. The fossil record not only reveals the predicted heavy extinction at low latitudes, as evidenced by the destruction of the tropical reef community, but also shows that faunas that had lived at high latitudes migrated toward the equator as the mass extinction progressed during Late Ordovician time. Both life of the seafloor and planktonic graptolites migrated in this way. Further evidence of the cooling of equatorial seas is the fact that the deposition of limestone, which is favored by warm conditions, slowed in earliest Silurian time.

REGIONAL EXAMPLES

Now that we have viewed early Paleozoic history from a distance, let us focus on some interesting regional developments, first in eastern North

America late in Ordovician time, when the Iapetus Ocean narrowed or closed altogether and mountains were uplifted, then in western and central North America during Cambro-Ordovician time.

The Border of a Narrowing Iapetus Ocean

As we have seen, the first Paleozoic tectonic cycle of the Appalachians ended in Late Ordovician and Silurian time with the deposition of clastic wedges that were shed westward from a newly forming mountain belt produced by the Taconic orogeny (Figure 7-13). Figure 10-31 shows the regional geography when the latest Ordovician clastic wedges were forming. As this figure indicates, the Juniata and Queenston formations spread westward from what might be called the Taconic Mountains. Beyond the clastic belt, the central interior of the United States (southern Laurentia) was flooded by shallow seas in which limestones accumulated.

We have also seen that the Iapetus Ocean had narrowed greatly by this time, as Baltica approached so close to Laurentia that their shallow-water marine faunas became quite similar.

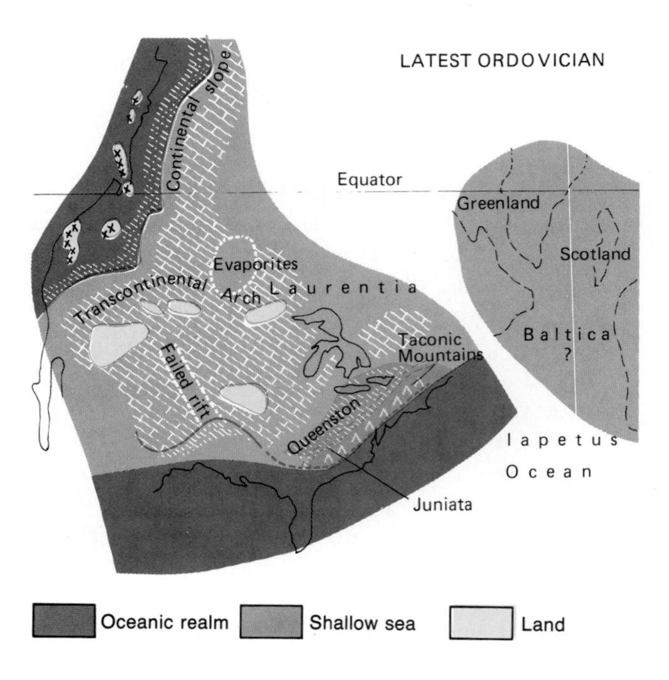

FIGURE 10-31 Pattern of deposition in North America during latest Ordovician time, when glaciers were spreading over parts of Gondwanaland. The Iapetus Ocean narrowed as Baltica and Laurentia converged. In eastern North America, the Juniata and Queenston formations formed as clastic wedges of sediment shed inland from newly formed mountains (see Figure 7-13). Carbonate sediments were laid down in the shallow seas that inundated almost all of the craton of North America. These deposits blanketed the failed rift in southern Oklahoma that had extended inland from the Gulf Coast in the early Paleozoic. Several low islands remained along the Transcontinental Arch. The continental margin bordering western North America remained stable, but volcanic islands lay offshore.

Stability in Western North America

Recall that a marine carbonate platform rimmed the craton of Laurentia during Cambrian time (Figure 7-14), and that in the east this platform persisted into Late Ordovician time, when orogeny caused it to sink and give way to environments of flysch and then molasse deposition (Figure 7-13). The marginal carbonate platform survived much longer in western North America. Parts of it sank into deep water during Silurian time, but orogeny did not strongly affect the western margin of the craton until the end of the Devonian Period.

Thus throughout Cambro-Ordovician time a stable continental shelf bounded western North America, passing diagonally across what is now southern California (Figure 10-31). Coarse, poorly sorted sediments derived from shallow-water environments accumulated at the base of a steep platform edge, just as they did in the east (Figure 7-15). In many places there are great thicknesses of these limestones, composed in part of debris from shallow-water stromatolites and invertebrates (Figure 10-32). Deposits that accumulated on deep seafloors beyond the platform include black limey mudstones and limestones.

To the north, in British Columbia, the continental rise at the base of the carbonate bank was the depositional setting of the famous Burgess Shale, where an amazing fauna of soft-bodied animals was preserved. It is said that this fauna was discovered by a curious accident. In 1909 Charles Walcott, the secretary of the Smithsonian Institution and an expert on Cambrian fossils, was riding along a narrow mountain trail when his horse stumbled over a block of shale in which Walcott caught a glimpse of a spectacularly well-preserved fossil. Careful examination of the strata

FIGURE 10-32 An outcrop of the Cambro-Ordovician Hales Limestone in Nevada, representing the ancient continental slope of western North America (see Figure 10-31). The photograph shows a graded bed of coarse fragments of shallow-water limestone that were transported down the continental slope. *(H. E. Cook and M. E. Taylor, U.S. Geological Survey.)*

FIGURE 10-33 A fossil-collecting party from the Smithsonian Institution working at their quarry in the Burgess Shale in British Columbia. Figure 10-8 shows specimens from the quarry. *(Smithsonian Institution.)*

above the trail turned up the layer, 2 meters (~7 feet) thick, from which the fossil-bearing block had fallen. Walcott organized a quarrying operation and removed nearly all of the fossil-bearing material (Figure 10-33).

Since Walcott's day the Burgess assemblage has been studied extensively. The fact that the fauna is so well preserved indicates that it was entombed in an oxygen-free environment from which destructive bacteria and scavenging animals were excluded. Stratigraphic evidence further indicates that the Burgess Shale was deposited at the foot of the steep carbonate shelf. In fact, the escarpment is still preserved in cross section in the mountainside some 200 meters (~650 feet) above the beds that preserve the fossils (Figure 10-34). Presumably the carbonate bank stood close to sea level, so this figure of 200 meters approximates the depth at which the Burgess fauna was preserved. The Burgess fauna was collected from a series of turbidite beds. Within each bed, calcareous siltstone grades upward into fine-grained mudstone. The beds apparently formed when turbid flows descended the escarpment from one or more channels in the carbonate bank. Most of the animals found in the Burgess Shale probably lived along the continental margin and were swept farther down the steep continental slope by the turbid flows. It is possible that they were preserved in the absence of oxygen because they were buried very rapidly, but because several flows produced the same result, it seems more likely that the entire site was an oxygen-free basin near the foot of the continental slope—a depression filled with stagnant water from which oxygen had become depleted. The Santa Barbara Basin off the coast of California may represent a modern analog. In any case, we must be grateful for this spectacular glimpse of the soft-bodied marine life of Middle Cambrian time.

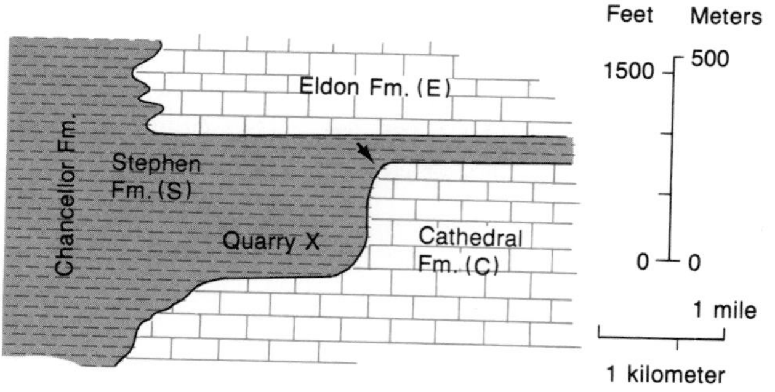

FIGURE 10-34 Location of the Burgess Shale Quarry within the Stephen Formation in British Columbia. The formations labeled in the diagram are identified in the photograph by letters. The arrow in the diagram points to the edge of the ancient continental shelf. The Burgess Shale accumulated at the foot of the steep slope below this shelf edge. *(Photograph from W. H. Fritz, Proc. N. Amer. Paleont. Conv. J:1155–1170, Allen Press, Lawrence, Kansas, 1969.)*

CHAPTER SUMMARY

1 The early Paleozoic world was populated by three successive faunas of marine invertebrates: first the earliest Cambrian Tommotian fauna of small animals, many of which are not known from later intervals; then the main Cambrian fauna, which was dominated by trilobites but included the earliest vertebrates; and finally the Ordovician fauna. The latter resembled faunas of later Paleozoic time in the diversity of its components, which included many kinds of brachiopods, mollusks, echinoderms, and other marine invertebrates.

2 During the Cambrian Period, trilobites suffered periodic mass extinctions, the last of which occurred at the close of the period.

3 During Cambro-Ordovician time, stromatolites declined, apparently because of the grazing and burrowing activities of new groups of animals.

4 During the Ordovician Period there developed a highly successful reef community that was dominated by tabulates, stromatoporoids, and colonial rugose corals. This community went on to thrive throughout almost all of middle Paleozoic time.

5 Early in the Cambrian Period, many continents stood unusually high above sea level; but as the period progressed, the continents were increasingly flooded. Siliciclastic deposits fringed the land, and carbonates were laid down across the expanding continental shelves and along marginal platforms.

6 Late in the Ordovician Period, the Iapetus Ocean narrowed as Laurentia and Baltica moved closer together. At the same time, subduction led to mountain-building episodes in eastern North America.

7 Whereas the carbonate platform bordering eastern North America was destroyed by Late Ordovician mountain building, the carbonate platform that bordered western North America remained intact into middle Paleozoic time.

8 Late in the Ordovician Period, after an interval during which the seas had stood unusually high throughout the world, global sea level dropped as continental glaciers spread over the part of Gondwanaland that had moved toward the South Pole. At the same time, a mass extinction eliminated many groups of marine life.

EXERCISES

1 Why do geologists know more about the life that colonized early Paleozoic seafloors than about the life that floated and swam above those seafloors?

2 What fossil evidence suggests that there were more effective predatory animals during the Ordovician Period than in Cambrian time?

3 In which continental regions is the climate likely to have been warmer in Late Cambrian time than it is today? (Compare Figure 10-25 with a map of the modern world.)

4 Give two reasons why limestone was more widely deposited in shallow waters during the Late Cambrian Period than it is today. (Question 3 supplies one hint, and Figures 10-25 and 10-23 provide another.)

5 How did the Cambro-Ordovician history of the eastern margin of North America differ from that of the western margin? (Your answer will be more complete if you reexamine the discussion of the origin of the Appalachian Mountains in Chapter 7.)

6 Where was Gondwanaland at the start of the Paleozoic Era?

7 What is the significance of the Burgess Shale? In what geographic region and environmental setting did it form?

8 Describe the general paleogeography of Laurentia during mid-Cambrian time.

ADDITIONAL READING

Bruton, D. L. (ed.), *Aspects of the Ordovician System: A Handbook.* Universitetsforlaget, Oslo, 1984.

Gould, S. J., *Wonderful Life: The Burgess Shale and the Nature of History,* W. W. Norton, New York, 1989.

McMenamin, M. A., "The Emergence of Animals," *Scientific American,* April 1987.

Morris, S. Conway, and H. B. Whittington, "The Animals of the Burgess Shale," *Scientific American,* July 1979.

Palmer, A. R., "Search for the Cambrian World," *American Scientist* 62:216–224, 1974.

CHAPTER

11

The Middle Paleozoic World

The oceans of the world stood high during most of Silurian and Devonian time, leaving a widespread sedimentary record on every continent. Marine deposition was interrupted in one region, however, by the most profound plate-tectonic event of middle Paleozoic time: the suturing of Baltica to Laurentia along a zone of mountain building. As we saw in Chapter 7, this orogenic event produced the Caledonide mountain belts of Europe and the Acadian orogen of the Appalachian region.

In the northern British Isles, Silurian rocks were tilted by the Caledonian orogeny, and thus an angular unconformity separates them from the overlying Devonian sediments. It was farther south, in Wales, however, that in 1835 Roderick Murchison founded the Silurian System, along with the Cambrian. Five years later, Murchison and Adam Sedgwick formally recognized the Devonian System, naming it for the county of Devon, along the southern coast of England. They recognized that the fossils in this system were intermediate in character (we would now say intermediate in evolutionary position) between those of the Silurian System below and those of the Carboniferous System above. (The Carboniferous System, though younger than the Devonian, had been recognized earlier in the century.)

The broad, shallow epicontinental seas of Silurian and Devonian time teemed with life. In the tropical zone, a diverse community of organisms built reefs larger than any that had formed during early Paleozoic time. More advanced predators were also on the scene, including the first jawed fishes — a few of which were the size of large modern-day sharks. The Devonian Period was also distinguished by the progressive colonization of land habitats by new forms of life. Plants, for example, were restricted to marshy environments in

The Old Red Sandstone forms cliffs along the coast of the Orkney Islands off northern Scotland. *(John Forbes/PEP.)*

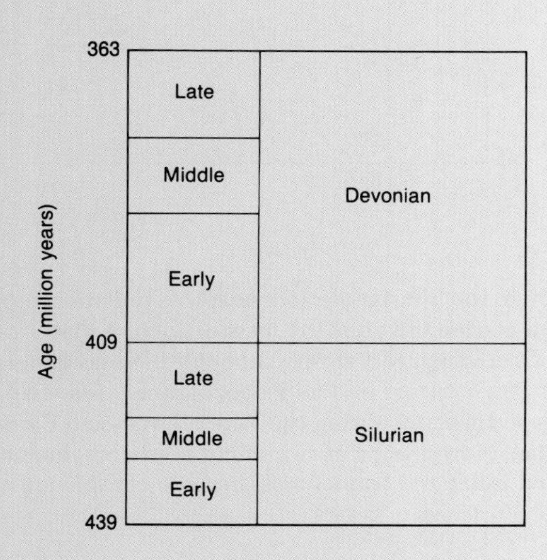

MAJOR EVENTS OF MIDDLE PALEOZOIC TIME

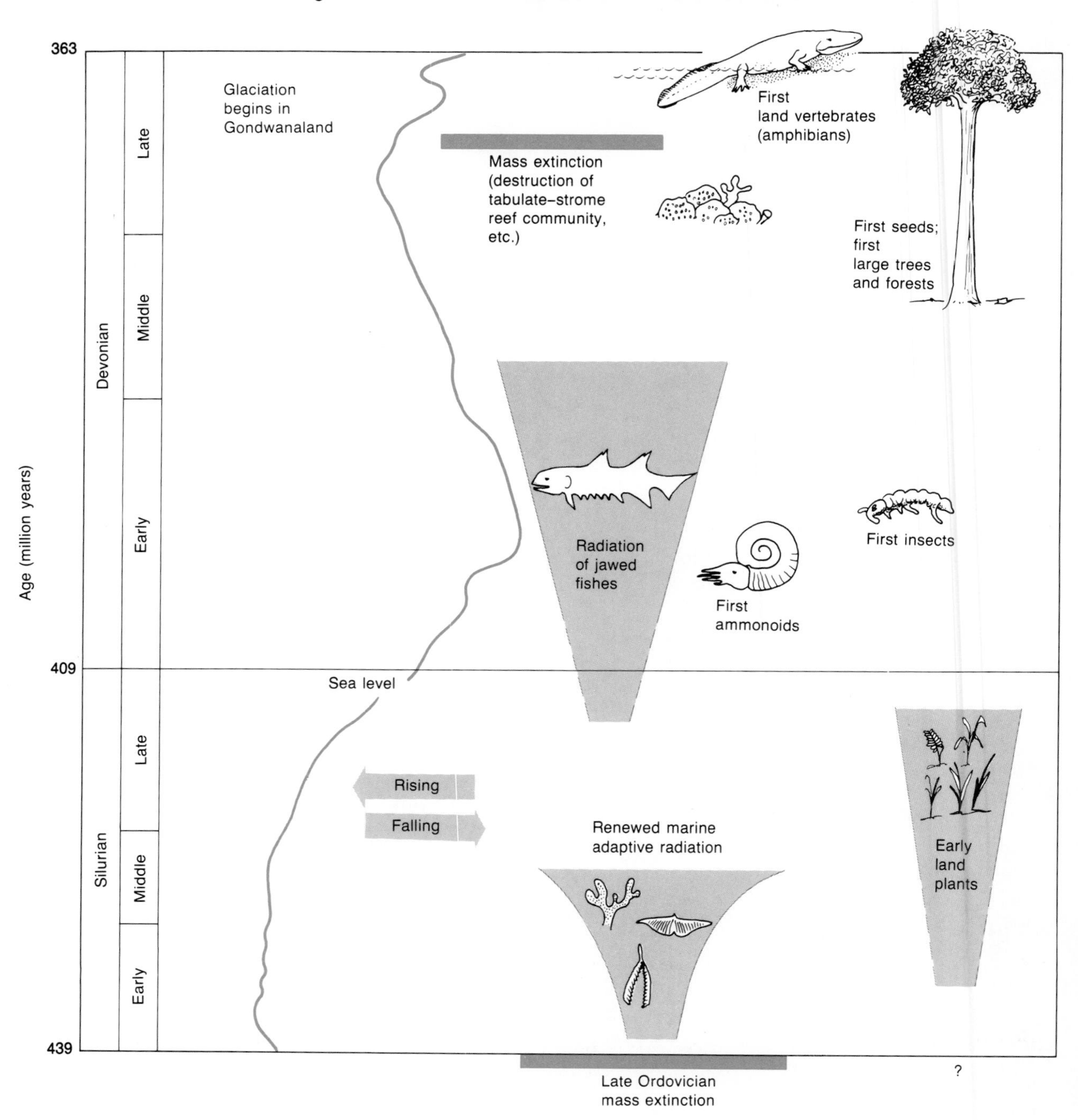

Early in the Silurian Period, adaptive radiation replenished marine life after the mass extinction that brought the Ordovician to a close. Although the seas generally stood high during middle Paleozoic time, sea level dropped sharply during the Late Silurian and Early Devonian. New groups of swimming predators, including jawed fishes and ammonoids, became conspicuous early in the Devonian Period, when arthropods also invaded the land and the first insects appeared. During Late Devonian time, continental glaciers formed in Gondwanaland and a mass extinction struck marine faunas. Vertebrates moved onto the land shortly after the first large trees evolved there and the first large forests developed.

Silurian time but were forming large forests by Late Devonian time. The oldest known insects are also of Early Devonian age; and near the end of the Devonian Period, the first vertebrate animals crawled up onto the land, the fins of their ancestors having been transformed into legs. Shortly before the end of the Devonian Period, however, a wave of mass extinction swept away large numbers of aquatic taxa, leaving an impoverished biota in their place during the final 3 or 4 million years of the Devonian Period and the earliest segment of late Paleozoic time.

LIFE

After the great mass extinction at the close of the Ordovician Period, many of the decimated taxa diversified once again. Their recovery surpassed the Ordovician adaptive radiation, yielding superior reef builders and swimming predators. Plants, too, expanded their ecologic role on the land, and animals invaded the terrestrial realm.

Aquatic Recovery

Most of the marine taxa that had flourished during the Ordovician Period rediversified after the terminal Ordovician mass extinction to become prominent members of the Silurian and Devonian marine biota. The trilobites failed to recover fully, however, and were less conspicuous in middle Paleozoic than in early Paleozoic seas. For other groups, recovery took the form of renewed adaptive radiation. As Figure 11-1 shows, all of the important Paleozoic articulate brachiopods were well represented in middle Paleozoic seas.

Other marine groups that recovered during middle Paleozoic time were the bivalve and gastropod mollusks (Figure 11-2). The bivalves, in fact, expanded their ecologic role by invading nonmarine habitats; some of the oldest known freshwater bivalves are found in the Upper Devonian strata of New York State. On the surface of the seafloor, bryozoans (Figure 10-13) rediversified after the crisis that ended the Ordovician, and crinoids (Figure 10-15) and rugose corals (Figure 11-3) increased in variety. Acritarchs were the dominant group of fossil phytoplankton in middle Paleozoic time, just as they had been in late Precambrian and early Paleozoic time (Figures 9-10 and 10-9). One of the most spectacular Early Silurian adaptive radiations, however, was that of the graptolites, which had nearly disappeared at the end of the Ordovician Period (as Figure 4-2 shows). The number of species of graptolites known in the British Isles increased from about 12 to nearly 60 during just the first 5 million years or so of Silurian time.

Luxuriant Reefs Most of the Silurian radiations of marine life did not vastly alter marine ecosystems; instead they refilled niches that were vacated by the mass extinction at the end of the Ordovician. Builders of organic reefs, however, did diversify in new ways, and in some shallow waters they produced much larger reefs than any of Cambro-Ordovician age. In Chapter 10 we saw that a new kind of organic reef developed in mid-Ordovician time: While the earliest reefs of the Middle Ordovician were formed entirely by bryozoans, later Middle Ordovician reefs were more complex, with tabulates and stromatoporoids playing dominant roles in their construction. Reef communities of this second general type, which we can call **tabulate-strome reefs,** diversified and persisted for about 120 million years, until late in the Devonian Period. The success of these reefs was a result of mid-Paleozoic adaptive radiations of tabulates, colonial rugose corals, and stromatoporoids (Figure 10-16).

During the Silurian Period, tabulate-strome reefs occasionally attained heights of 10 meters (~35 feet) above the seafloor. During Devonian time, however, tabulate-strome reefs assumed enormous proportions. We will examine some of these large reefs later in this chapter. In areas subjected to strong wave action, the growth of tabulate-strome reefs followed a characteristic ecologic succession (Figure 11-4). First, sticklike tabulates and rugose corals colonized an area of subtidal seafloor. A low mound was then formed when these fragile forms were encrusted by platy and hemispherical tabulates and colonial rugose corals. Finally, as the mound grew up toward sea level, stromatoporoids and algae encrusted the seaward side, forming a durable ridge. Tabulates and colonial rugose corals occupied a zone of quieter water behind the ridge, and beyond them

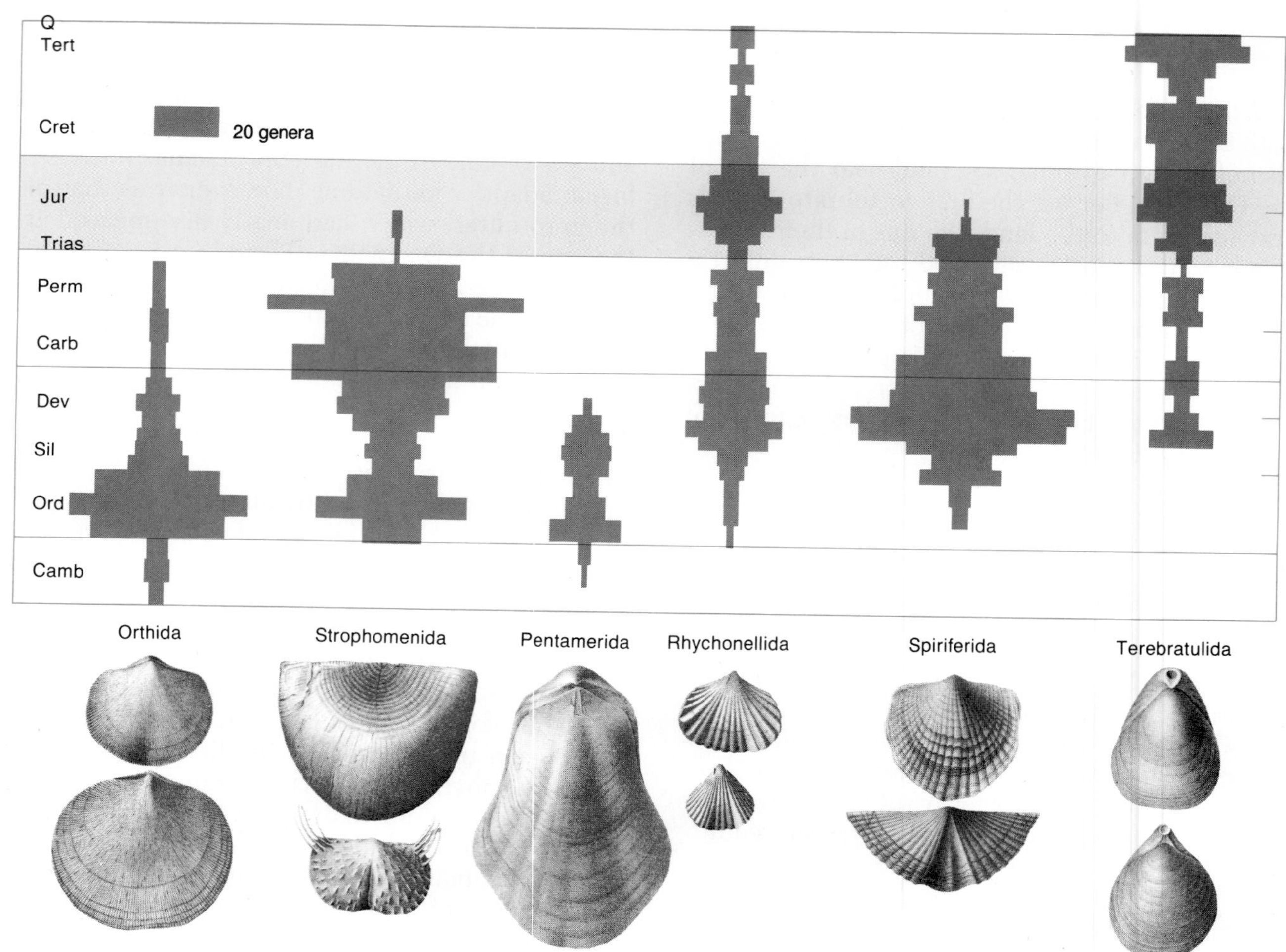

FIGURE 11-1 Middle Paleozoic articulate brachiopods. *Above:* The diversity (numbers of genera) of the six major orders of Paleozoic brachiopods, all of which were well represented in middle Paleozoic (Silurian and Devonian) time. *Below:* Typical middle Paleozoic representatives of each brachiopod order. *(Diversity data from Treatise on Invertebrate Paleontology; brachiopod illustrations from James Hall's volumes of the New York State Natural History Survey [1862–1894].)*

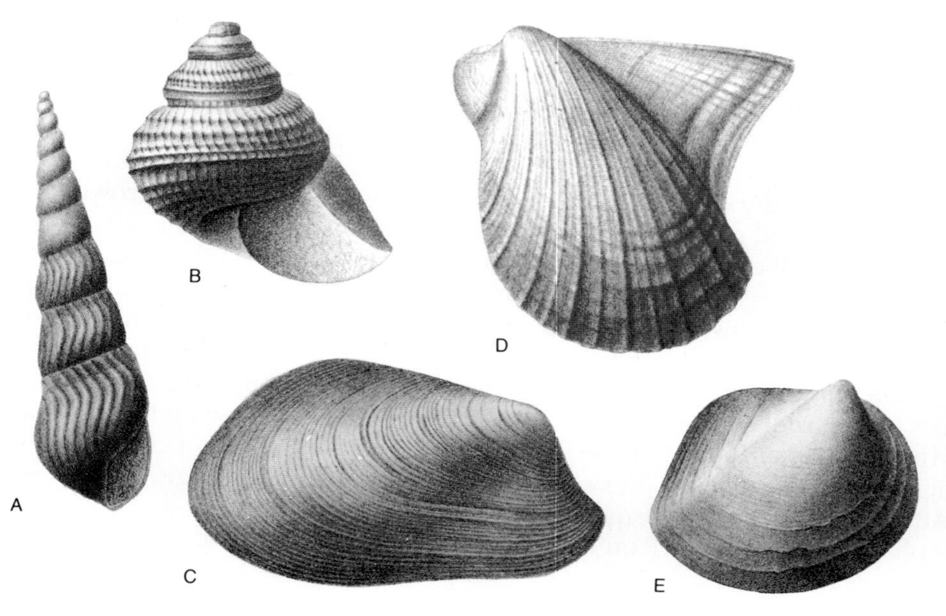

FIGURE 11-2 Typical mollusks that inhabited middle Paleozoic seafloors. *A* and *B*. Two snails. *C* and *D*. Two bivalves that attached to the substratum by threads. *C* closely resembles certain modern species that live partially buried in sediment, and *D* is related to modern wing shells that are attached to objects standing upright on the seafloor. *E* is a species that burrowed in sediment. *(From James Hall's volumes of the New York State Natural History Survey [1879–1885].)*

FIGURE 11-3 Rugose corals of Devonian age. The polyp of each solitary coral sat on the bladelike septa at the top of the cone-shaped skeleton, which was about the size of a radish. *(Chip Clark.)*

FIGURE 11-4 Ecologic succession of a typical Devonian reef. *(1)* The pioneer community consisted of fragile, twiglike rugose corals and tabulates. *(2)* Broad and moundlike tabulate colonies dominated during the intermediate stage of development. *(3)* In the mature stage, the reef grew close to sea level, and waves broke against a ridge of massive, encrusting stromatoporoids; behind them were species that were adapted to quieter water, and leeward of these was a lagoon populated by fragile, twiglike species. The leeward side of the lagoon was bounded by a small stromatoporoid ridge. At left is a photo of a moundlike reef formed by tabulates and rugose corals. The fossils were collected in Michigan (see Figure 11-26) and were then reassembled at the Smithsonian Institution to recreate the reef. This moundlike reef represents the intermediate stage (2). *(Diagram after P. Copper, Proc. Second Internat. Coral Reef Symp. 1:365–386, 1975. Photograph by the author.)*

FIGURE 11-5 Reconstruction of an Upper Devonian reef in New York State. Numerous kinds of corals are present. In the right foreground is a huge, spiny trilobite that measured about 45 centimeters (~18 inches) in length. *(Carnegie Museum of Natural History.)*

was a lagoon in which mud-sized sedimentary grains accumulated along with coarser skeletal debris from the reef. Pockets of fossils preserved in reef rock reveal that a wide variety of invertebrates inhabited tabulate-strome reefs; brachiopods and bivalve mollusks attached themselves to a typical reef, snails grazed over it, and crinoids and lacy bryozoans reached upward from its craggy surface. Although its fauna would look unusual to us today, a living, fully developed tabulate-strome reef (Figure 11-5) would certainly seem as colorful and spectacularly beautiful as the coral reefs that flourish in the modern tropics (p. 47; Box 2-1).

New Swimming Invertebrates Perhaps the greatest change in the nature of aquatic ecosystems during middle Paleozoic time resulted from the origin of new kinds of nektonic (swimming) animals, many of which were predators. The most important of these swimmers among the invertebrates were the ammonoids. These coiled cephalopod mollusks evolved from a group of straight-shelled nautiloids during Early Devonian time (Figure 11-6). After giving rise to the ammonoids, the nautiloids persisted at low diversity. The ammonoids, in contrast, diversified rapidly, and because their species were distinctive, widespread, and relatively short-lived, they serve as important

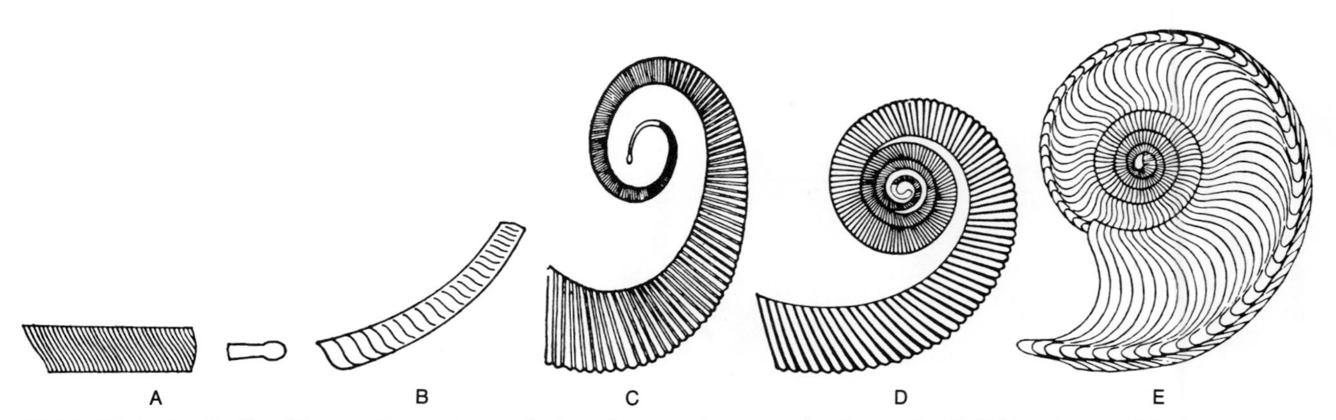

FIGURE 11-6 Shells of Lower Devonian cephalopod mollusks from the Hunsrück Shale of Germany reveal the apparent evolutionary sequence leading from nautiloids to early ammonoids. *A* and *B*. Fragments of nautiloids of the group that evolved into ammonoids. *C* through *E*. Early ammonoid species representing various degrees of coiling. The bulblike shape of the earliest part (tip) of each shell, together with other shared features, suggests that these species are closely related to one another; the coiling sequence displayed here apparently represents the evolutionary sequence. *(After H. K. Erben, Biol. Rev. 41:641–658, 1966.)*

FIGURE 11-7 Reconstruction of a Late Silurian eurypterid. This animal was about 0.5 meters (~20 inches) long. The appendages beneath the head of this species bore sharp spikes for stabbing prey. *(Chip Clark.)*

guide fossils in rocks ranging in age from Devonian to latest Mesozoic. (Ammonoids died out along with the dinosaurs at the end of the Mesozoic Era.)

The eurypterid arthropods were a second important group of invertebrate predators that proliferated during middle Paleozoic time. These distant relatives of scorpions were swimmers, and many had claws. Although the eurypterids appeared in the Ordovician Period and survived until Permian time, their most conspicuous fossil record is in middle Paleozoic rocks (p. 238 and Figure 11-7). Unlike ammonoids, eurypterids ranged into brackish and freshwater habitats.

Jawless Fishes Other swimmers that were adapted to both marine and freshwater conditions were the fishes. The major groups are illustrated in Figure 11-8. Whereas only fragments of fish

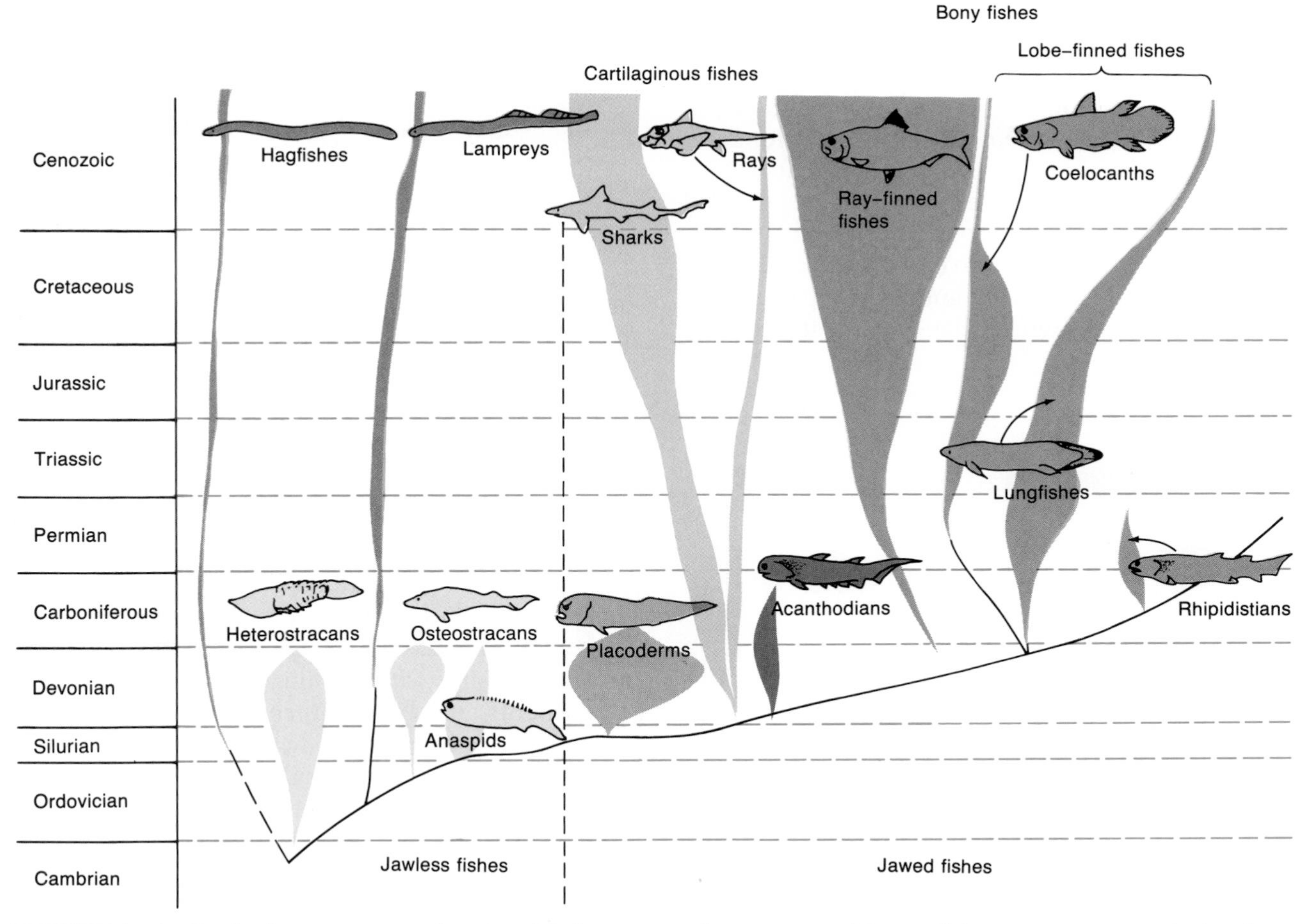

FIGURE 11-8 Geologic occurrence of various kinds of fishes. By Devonian time, all the major groups living today were in existence; no ostracoderms and few placoderms survived beyond the Devonian Period. *(Modified from E. H. Colbert, Evolution of the Vertebrates, John Wiley & Sons, Inc., New York, 1980.)*

skeletons have been found in early Paleozoic sediments (Figure 10-10), the Silurian and Devonian systems have yielded diverse, fully preserved fish skeletons, many of which were found in freshwater deposits of lakes and rivers.

We do not know when fishes first occupied fresh water. The fact that all known Cambro-Ordovician fish remains have been found in marine deposits does not prove that fishes evolved in the ocean, but it does lend support to the idea. Most Silurian fish remains, unlike most Cambro-Ordovician fish fossils, come from freshwater deposits. One of the most conspicuous groups of these fishes were the **ostracoderms,** whose name means "bony skin." Ostracoderms were small animals with paired eyes like those of higher vertebrates. Lacking jaws and covered by bony armor, ostracoderms did not resemble modern fishes. In addition, ostracoderms had small mouths and could have consumed only small items of food. Many of these fishes, such as *Hemicyclaspis* (Figure 11-9), also had flattened bellies that apparently were adapted to a life of scurrying along the bottoms of lakes and rivers. The upper fin of the asymmetrical tail of *Hemicyclaspis* was elongate; when the animal wagged it back and forth for swimming, this structure would have pushed its head downward rather than upward. In contrast, the ostracoderm known as *Pteraspis* (Figure 11-10) had a curved belly and an elongate lower fin in its tail that would have lifted it upward, suggesting a life of more active swimming well above lake floors and river bottoms. Ostracoderms not only lacked jaws for chewing but also lacked highly mobile fins, which provide more advanced fishes with stability and control of their movements, and bony internal skeletons; it is assumed that ostracoderms had cartilaginous internal skeletons. These animals continued to thrive throughout most of the Devonian Period but disappeared at the end of that interval.

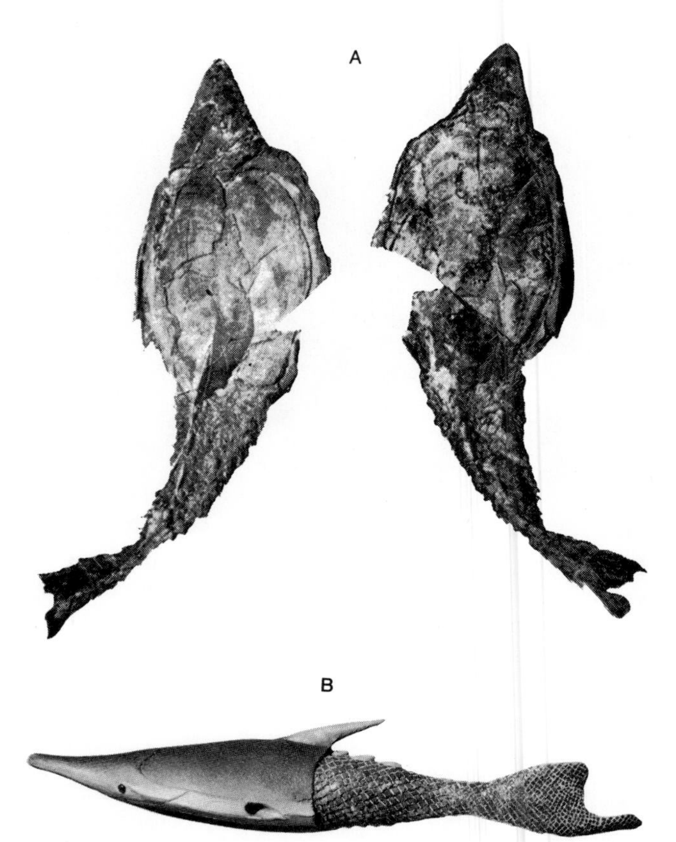

FIGURE 11-10 ***Pteraspis*, a Devonian ostracoderm (jawless fish) that was about 6 centimeters (~$2\frac{1}{4}$ inches) long.** Note the difference between the asymmetry of the tail here and in *Hemicyclaspis* (Figure 11-9). *A.* Upper and lower surfaces of a well-preserved specimen from the Old Red Sandstone of Wales. *B.* Side view of a restoration. *(A. E. I. White, Philos. Trans. Roy. Soc. London B225:381–457, 1935. B. Field Museum of Natural History.)*

FIGURE 11-9 Reconstruction of *Hemicyclaspis*, a Devonian ostracoderm (jawless fish) that was about 20 centimeters (~8 inches) long. The upper picture shows the asymmetrical tail, and the lower picture shows the flattened belly; both of these features probably relate to a life of scurrying along the floors of lakes and rivers. *(Field Museum of Natural History.)*

Fishes with Jaws Late in Silurian time a second, quite different group of small marine and freshwater fishes made their appearance. These were the **acanthodians**—elongate animals with nu-

FIGURE 11-11 Restoration of *Euthacanthus*, a member of the most primitive group of jawed fishes, which are known as acanthodians. This animal was about 20 centimeters (~8 inches) in length. *(Field Museum of Natural History.)*

merous fins supported by sharp spines (Figure 11-11; see Box 11-1). The acanthodians appear to have been the first fishes to possess several features that were passed on to more advanced, modern fishes: Their fins were paired; scales rather than bony plates covered their bodies; and, most important, they had jaws. With the origin of jaws, a wide variety of new ecologic possibilities opened up for vertebrate life—possibilities that related primarily to the ability to prey on other animals. Unlike ostracoderms, many acanthodians must have been predators that fed on small aquatic animals. As Box 11-1 indicates, it seems evident that the jaws of acanthodians evolved from the gill supports of ancestral fishes.

Acanthodians declined near the end of Devonian time, but they left an evolutionary legacy of great ecologic significance. We do not know precisely how more advanced groups of fishes were related to acanthodians, but during the Devonian Period, a great adaptive radiation of descendant jawed fishes added new levels to the food webs of both freshwater and marine habitats. Soon very large fishes were feeding on smaller fishes, which, in turn fed on still smaller fishes. At the top of this food web were the largest representatives of the group, the **placoderms.** These heavily armored jawed fishes made their appearance during the Devonian Period but almost disappeared before the beginning of Carboniferous time. A few placoderms are known from uppermost Silurian and Lower Devonian freshwater deposits, and a wide variety of freshwater species existed by mid-Devonian time. Only secondarily did placoderms make their way into the oceans, and they were not highly diversified until Late Devonian time. During their brief stay on Earth, however, this remarkable group of fishes expanded to include a wide variety of species. *Dunkleosteus,* a Late Devonian marine genus, attained a length of some 10 meters (~30 feet). Like other placoderms, it had armorlike bone protecting the front half of its body (Figure 11-12), but its unarmored tail was exposed to attack, remaining flexible for locomotion. Figure 11-13 depicts *Dunkleosteus* pursuing *Cladoselache,* a shark commonly found with it in black shales of northern Ohio.

Sharks were, in fact, among the most important groups of fishes in Devonian seas. Unknown in rocks older than mid-Devonian, the sharks may have been the last major group of fishes to evolve. Devonian sharks were primitive forms, and few greatly exceeded 1 meter (~3 feet) in length.

Also making an appearance in Devonian time were the **ray-finned fishes.** These jawed forms, which attained only modest success during the Devonian Period, went on to dominate Mesozoic and Cenozoic seas. They include most of the familiar modern marine and freshwater fishes, such as trout, bass, herring, and tuna. The term *ray-finned* refers to the fact that the fins of these fishes are supported by thin bones that radiate from the body—the bones that can be seen through the transparent fins of living fishes. The oldest mid-

FIGURE 11-12 The massive, armored skull of *Dunkleosteus*, a placoderm fish of Late Devonian age. This formidable species of jawed fish attained a length of more than 10 meters (~30 feet). Note the bony teeth and the armor protecting the eye. *(Chip Clark.)*

FIGURE 11-13 The giant placoderm *Dunkleosteus*, shown in pursuit of the Late Devonian shark *Cladoselache*. (See Figure 11-12.) *(Drawing by Z. Burian under the supervision of Professor J. Augusta.)*

Devonian ray-finned fishes, such as *Cheirolepis* (Figure 11-14), differed from modern representatives in that they had asymmetrical tails and diamond-shaped scales that did not overlap.

Cheirolepis and many other freshwater fishes of Devonian age have been found in deposits that accumulated around the Old Red Sandstone continent, a land area formed, as we will see later in this chapter, by the closing of the Iapetus Ocean. This ancient landmass included parts of modern North America, Greenland, and western Europe.

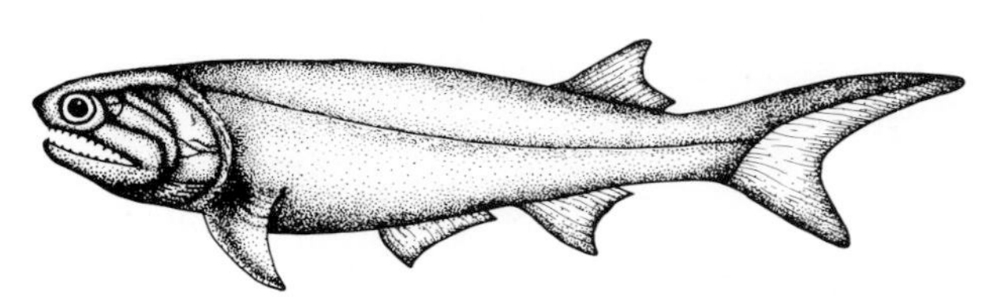

FIGURE 11-14 *Cheirolepis*, a primitive ray-finned fish of mid-Devonian age. The tail of this animal was strongly asymmetrical, and the small, diamond-shaped scales did not overlap. The animal was about 10 centimeters (~4 inches) long. *(After A. S. Romer, Vertebrate Paleontology, University of Chicago Press, Chicago, 1966.)*

The origin of the ray-finned fishes was an event of great significance, but so was the origin of a related group of jawed fishes—one that included the lungfishes and the lobe-finned fishes.

Fishes with Lungs The Devonian Period was the time of greatest success for the **lungfishes,** which survive today via only three genera—one in South America, one in Africa, and one in Australia. (Presumably this fragmented distribution reflects the Mesozoic breakup of Gondwanaland.) The Australian genus, *Neoceratodus,* so closely resembles the Triassic genus *Ceratodus* that it is commonly referred to as a "living fossil." The surviving lungfishes are named for the lungs that allow them to gulp air when they are trapped in stagnant pools during the dry season. Such lungs presumably served a similar function in Devonian time.

FIGURE 11-15 The lobe-finned fish *Eusthenopteron*, a large animal that exceeded 50 centimeters (~20 inches) in length. This unusually well-preserved specimen is from Upper Devonian deposits at Scaumenac Bay, Canada. (See Figure 11-23.) *(Swedish Museum of Natural History.)*

Lungfishes belong to a group known as **lobe-finned fishes** (Figure 11-15). These fishes derive their name from their paired fins, whose bones are not radially arranged, as in ray-finned fishes, but instead are attached to their bodies by a single shaft. Most lobe-finned fishes inhabited fresh water, but one unusual group, the coelacanths, invaded the oceans. A single coelacanth genus survives today in deep waters off the southeastern coast of Africa. Lobe-finned fishes declined after the Devonian period but left a rich evolutionary legacy. As we will see, lobe-finned fishes are the ancestors of all terrestrial vertebrates, including humans; their lungs were the predecessors of our own.

The Impact of Swimming Predators The great diversification of jawed fishes and, to a lesser extent, the expansion of ammonoids and eurypterids must have had a profound effect on many relatively defenseless aquatic animals. These predators may have contributed to the middle Paleozoic decline in the diversity of trilobites. About 80 families of trilobites are known from the Ordovician (far fewer than are known from the Cambrian), but only 23 families have been found in Silurian deposits. It seems likely that the weakly calcified external skeletons of trilobites offered little resistance to the jaws of fishes, and certainly trilobites had no mechanism for rapid locomotion. In Chapter 10 it was suggested that nautiloids, which evolved at the very end of the Cambrian Period, suppressed the diversification of trilobites during the Ordovician Period. Continuing with this line of reasoning, we must suspect that more advanced cephalopods and jawed fishes were largely responsible for the further decline of trilobites in mid-Paleozoic time. The small, apparently defenseless ostracoderms, which died out late in the Devonian Period, must also have served as easy prey for jawed fishes. Ostracoderms even lacked the ability to burrow in sediment, which at least some of the trilobites could do (Figures 10-3, 10-12, and 10-13).

Plant Life: Invasion of the Land

It is difficult to imagine how the landscape looked in Precambrian and early Paleozoic times, before there were conspicuous terrestrial plants. Certain terrestrial environments must have been populated by algae and other simple plants and plant-like organisms, but there were no forests or meadows, and there must have been large areas of barren rock and soil with little or no humus (decayed organic matter). Thus one of the most important events revealed by the fossil record of Silurian and Devonian life was the invasion of terrestrial habitats by higher plants.

The basic requirements for the terrestrial existence of large multicellular plants are very different from those of plants that live in water. Unlike water, air is a fluid that has a much lower density than the tissues of a plant; thus, if a plant is to stand upright in air, it must have a rigid stalk or stem. A tall plant must also be anchored by a root system or by a buried horizontal stem, which serves the further indispensable function of collecting water and nutrients from the soil.

The first upright plants to make their way onto the land lacked the roots, leaves, and efficient means to transport nutrients that made their descendants so successful. Essentially, these plants were simple rigid stems. Fragments of such early plants have been found in Silurian rocks. Silurian plants seem to have been pioneers that lived near bodies of water, and they may actually have been semiaquatic marsh dwellers rather than fully terrestrial plants.

Vascular Plants Most large plants of the modern world are **vascular;** that is, their stems have one set of special tubes to carry water and nutrients upward from their roots and another to distribute the food that the plants manufacture for themselves. Most large modern plants also bear leaves, which serve to capture the sunlight that assists in the manufacture of food.

Box 11-1
Jaws, Evolution, and Genetic Engineering

The origin of jaws teaches us an important lesson about evolution. The process of natural selection must work with the materials at hand. Genetically based variability within populations is the raw material on which natural selection must work, but this variability is limited. New biological structures must evolve from old ones, not from raw materials derived from sources outside the organism. Even some potentially useful modifications of animals and plants that simple genetic changes might produce have never actually occurred simply because, by chance, the genetic changes have never taken place. Mutations, after all, are genetic accidents.

The evolution of jaws transformed the marine ecosystem. It also added a new dimension to life on land when jawed vertebrates emerged from aquatic environments near the end of the Devonian Period. Both in the sea and on the land, the existence of jaws permitted sophisticated predators to evolve. Ultimately, evolution produced human jaws by remodeling those of fishes. Of course, there were many stages of development along the way, and other lines of evolution produced jaws of various other types.

Where did fishes get their jaws? Evolution did not grow them from tiny beginnings. Instead, as evolution has so often done in "building" new biological structures, it produced jaws from other features that were already well developed. Jaws evolved during the Devonian Period from bars that supported the gills of primitive fishes. Fossil sharks provide much of the evidence.

In sharks, as in other primitive vertebrates, skeletal bars lie on either side of the throat between the gill slits—the openings that allow water to pass through the gills. Each bar has an upper and lower

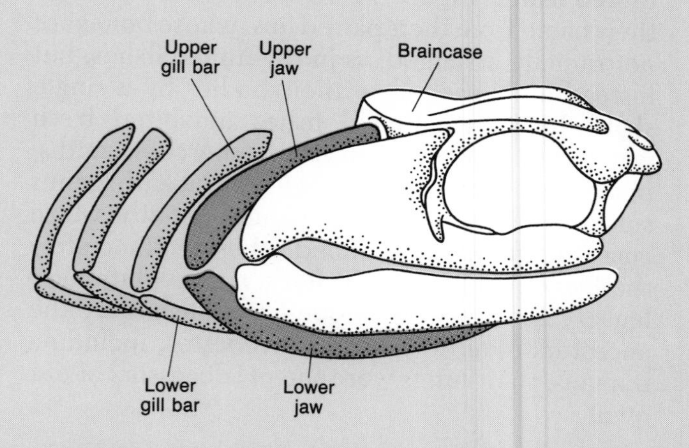

This diagram of the brain case, jaws, and gill supports of a Carboniferous shark illustrate the primitive configuration of jaws in fishes. *(After R. Zangrel and M. E. Williams, Paleont. 18:333–341, 1975.)*

A major adaptive breakthrough for life on land, before the evolution of roots and leaves, was the origin of vascular tissue. Two kinds of tubes developed—one for the transport of water and nutrients and another for the transport of manufactured food. Figure 11-16 shows the tubes within a stem of the Early Devonian genus *Rhynia*. A few kinds of vascular plants are found in nonmarine deposits of latest Silurian age (Figure 11-17). These plants had branched leaves as well as bulbous organs that shed spores.

Spore-Bearing Plants **Spores** are reproductive structures that can grow into new adult plants when they are released into the environment. Ferns are familiar spore-bearing plants in the modern world. The fossil record of spores resembling modern ones extends well back into the Ordovician System (Figure 10-24), but while these older spores suggest that upright land plants existed much earlier than the Late Silurian, they may in fact represent aquatic or semiaquatic species. In some Early Devonian forms, solitary spore-bearing organs stood atop upright stalks, while other species displayed clusters of spore-bearing organs in similar positions; and in still other species, organs were arrayed along the upright stalks (Figure 11-16).

part that connect to form a backward-pointing V. The jaws of some primitive Devonian sharks resemble these bars in both shape and orientation. Unlike our jaws, those of primitive sharks were not attached to the skull: They were positioned directly in front of the gill bars and were aligned with them. In fact, some fossil sharks' jaws give the appearance of being the first gill bars in the series, although they differ slightly in shape from the bars and they bear teeth. It is difficult to avoid the conclusion that the jaws amount to modified gill bars. A strong similarity between the muscles that operate the jaws and those that close the gill slits reinforces this conclusion. Fossils of primitive sharks found in black shales in northeastern Ohio have been magnificently preserved under anoxic conditions similar to those that produced the Burgess Shale fauna of soft-bodied animals (p. 264). These fossils display traces of the jaw muscles of the sharks, along with many other anatomical features.

The similarity between gill bars and primitive jaws extends even to the teeth. Small, pointed structures called denticles toughen the skin of sharks, giving it the texture of sandpaper. The teeth that line the jaws of primitive sharks resemble the denticles of their skin. Thus it appears that evolution produced primitive teeth along the jaws simply by enlarging the denticles in the skin that overlay the ancestral gill bars.

An engineer with the power to design and build a shark would produce a better jaw than the primitive one that evolved from the gill bar. The skeleton of the early jaw consisted of soft cartilage rather than bone, and it floated loosely in soft tissue, without connection to the braincase. The primitive teeth also remained too much like denticles to function as effectively as those that an engineer might design or those that evolution eventually produced.

When humans interfere with the natural process of evolution by breeding domestic animals and plants selectively to improve their value to society, they are at the mercy of the variability that appears in the populations they breed. They cannot produce any kind of cow, chicken, or corn plant that they desire. Genetic engineering—the process in which biologists artificially redesign organisms by restructuring their genetic codes—carries human interference further, but even genetic engineering has limitations. Only certain genetic recombinations are possible, and many of them produce organisms that will not develop properly or function well.

It is easy to see why evolution has produced many organisms that seem highly imperfect and why it has failed to produce all kinds of organisms that our imaginations can conjure up. Chance has limited its productions, and so has the requirement that it create new forms of life from ones that already exist.

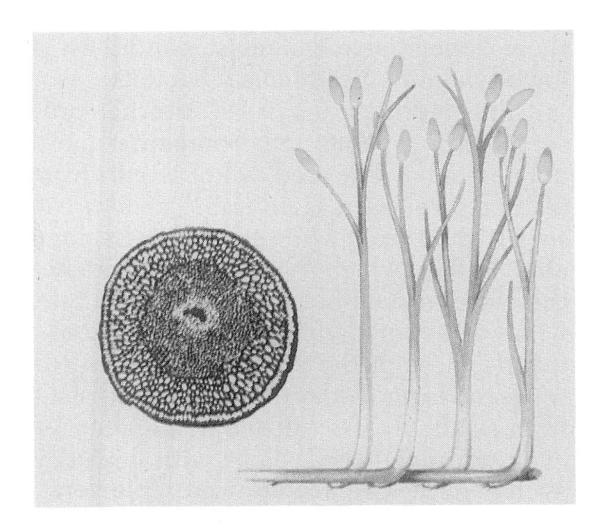

FIGURE 11-16 The Early Devonian vascular plant *Rhynia*. The reconstruction shows how, in this primitive form, a simple horizontal stem served the function of a root system and there were no leaves. The vascular tissue that transported water is seen as a small dark area in the cross section of the stem; around this, and visible as a narrow, light-colored ring in the cross section, is vascular tissue that transported food. (*From M. E. White, The Flowering of Gondwana, Reed Books Pty. Ltd., 1990.*)

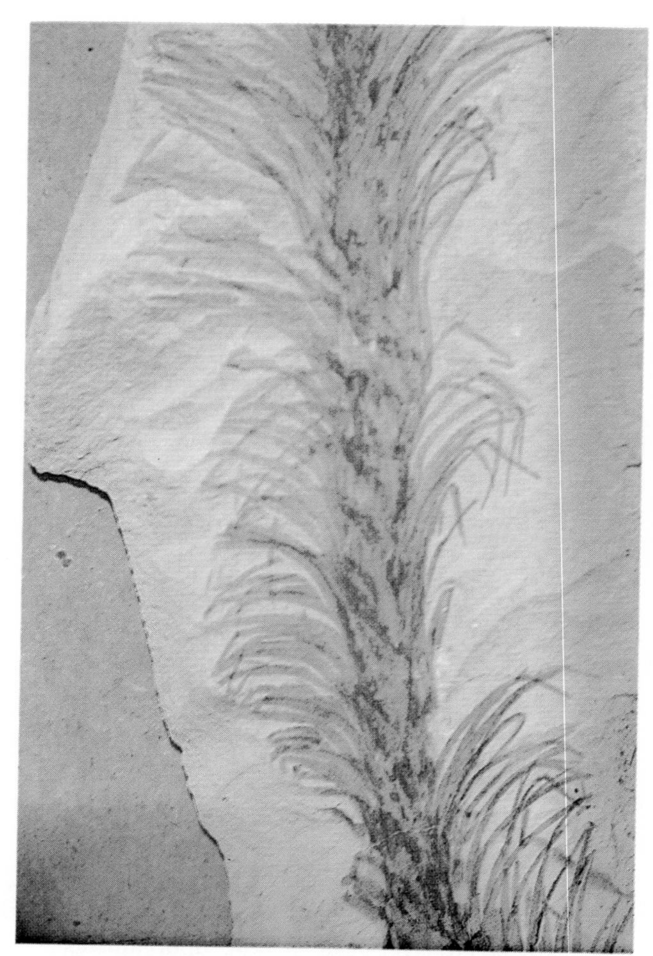

FIGURE 11-17 One of the oldest known vascular plants, a specimen of the genus *Baragwanathia* from the Upper Silurian of Victoria, Australia. This plant was about 2.5 centimeters (~1 inch) in diameter. *(Francis M. Haeber, Smithsonian Institution.)*

Whether or not land plants existed much before latest Silurian time, it was apparently near the end of the Silurian Period that vascular tissues evolved. As a result of this physiological breakthrough, a great adaptive radiation took place in Early Devonian time. The Devonian plants that resulted were still relatively low, creeping forms that lacked well-developed roots and leaves, but during Early and Middle Devonian time, more complex plants evolved. The vascular tissues of early vascular plants such as *Rhynia* were confined to a narrow zone of the stem (Figure 11-16) and so were mechanically weak and inefficient at conducting liquid. By late Devonian time, however, some plants had developed vascular tissues that occupied a larger volume within the stem and were therefore mechanically stronger and more efficient transporters of nutrients. Plant groups with these useful traits also evolved roots for support and for effective absorption of nutrients as well as leaves for capturing large quantities of sunlight. These plants seem to have competitively displaced such plants as *Rhynia,* which were less efficient at obtaining nutrients, synthesizing food, and growing to a large size.

Certain of the small plants that arose during Early and Middle Devonian times are classified as **lycopods.** This group includes the tiny club mosses of the modern world (Figure 11-18), but as we will see in Chapter 12, some late Paleozoic lycopods grew to the proportions of trees, and their petrified remains supply much of modern society's coal. These large forms vanished by the end of the Paleozoic Era, however, and only tiny creeping lycopods resembling the primitive types of the Early Devonian have survived to this day.

Because of the limitations imposed by their mode of reproduction, these Early Devonian land plants must have formed low marshes along bodies of water (Figure 11-19). Like living ferns and other present-day spore-bearing plants, they have a complex reproductive cycle that would have restricted them to habitats that were damp at least part of the year. This reproductive cycle requires not only a conspicuous spore-bearing plant, such as a fern, but also a tiny, inconspicuous plant over whose surface a sperm must travel to fertilize an egg. The sperm requires moist conditions to make its journey. As the Devonian Period progressed, however, the appearance of a second adaptive "innovation," the **seed,** liberated land plants from their dependence on moist conditions and allowed them to invade many habitats.

Since fertilization is an internal process in seed-bearing plants, environmental moisture is not necessary. The seed, which results from fertilization, is released as a durable structure that can sprout into a plant when conditions become favorable. Pollen, which represents a different part of the cycle in seed-bearing plants, also tolerates a wide variety of environmental conditions. Pollen travels through the air to fertilize eggs so that seeds can form. Today most large land plants grow from seeds. As we will see in Chapter 14, however, advanced seed plants with flowers did not evolve until Cretaceous time. Flowers attract

FIGURE 11-18 Reconstructions of *Protolepidodendron (A)* and *Asteroxylon (B)*, primitive Devonian forms of lycopods. *Lycopodium (C)* is a small form like the Devonian fossils, growing only a few centimeters tall. In contrast to these oldest and youngest representatives, many lycopods of late Paleozoic time were large trees. *(C. D. Klees/PEP.)*

FIGURE 11-19 Reconstruction of an Early Devonian landscape, where many of the first land plants still bordered bodies of water. These plants were generally less than 1 meter (~3 feet) tall. *(Chase Studio, Inc.)*

insects and birds, which effect fertilization by unknowingly carrying pollen from flower to flower. More primitive, flowerless seed-bearing plants lack this sophisticated mechanism of fertilization, relying instead on less efficient agents—primarily wind—to carry pollen from plant to plant.

Flowerless seed plants originated in Late Devonian time and soon became important elements of late Paleozoic terrestrial floras (Figure 11-20), opening up a new world to the plant kingdom. During Late Devonian time, for the first time, dry land was invaded on a vast scale. Seed plants soon grew into large trees with strong, woody stems (Figure 11-21), which changed the face of Earth. Trees formed the world's first forests, where there had been only barren land before.

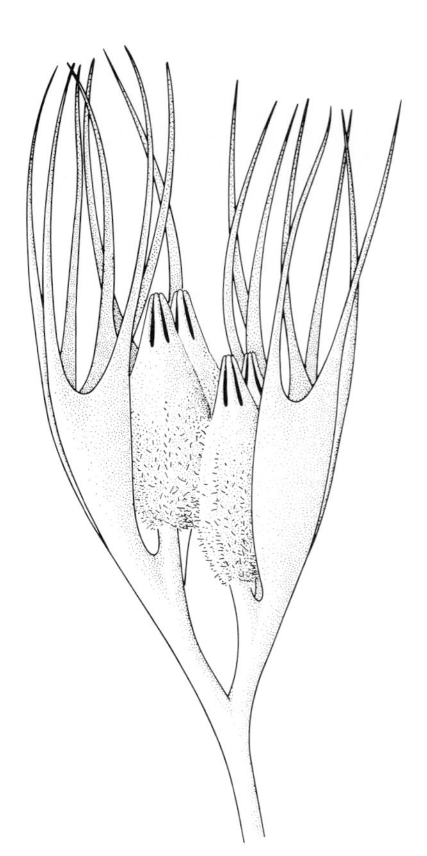

FIGURE 11-20 Reconstruction of Upper Devonian plants that bore some of the oldest known seeds at the ends of their branches. Individual seeds, four of which are shown, were about 1 centimeter ($\sim\frac{1}{4}$ inch) in length. *(J. M. Pettitt and C. B. Beck, University of Michigan Paleont. Contrib. 22:139–154, 1968.)*

One of the consequences of the spread of terrestrial vegetation was that, for the first time in Earth's history, plants carpeted the soil and gripped it with their roots, thereby stabilizing it against erosion. Braided-river deposits, which reflect rapid erosion (p. 59), characterize the clastic wedges of Precambrian and early Paleozoic age. Only in middle Paleozoic time, when vegetation first stabilized the land, did rivers begin to meander and to deposit sediment in orderly cycles (p. 60) on a large scale.

Animals Move onto the Land

The Rhynie Chert of Scotland is a Lower Devonian formation that has yielded not only a variety of beautiful fossil plants but also some of the oldest known nonmarine arthropods, including scorpions and flightless insects.

Arthropods probably invaded dry land in Late Silurian time, before some of their terrestrial representatives were preserved in the Rhynie Chert—but it was not until the Late Devonian interval that vertebrate animals made a similar transition. Anatomical evidence indicates that the four-legged vertebrates most closely related to fishes are the **amphibians**—frogs, toads, salamanders, and their relatives. That amphibians represent the most primitive four-legged vertebrates is also suggested by the fact that these animals are legless and aquatic early in life. They hatch from eggs in water, spend their juvenile existence there, and then usually metamorphose into air-breathing, land-dwelling adults. Living amphibians are small animals that differ substantially from the large fossil amphibians found in upper Paleozoic rocks. Their evolutionary history began late in the Devonian Period.

In eastern Greenland, the remains of unusual vertebrate animals have been found in uppermost Devonian rocks of the Old Red Sandstone continent. These fossils, many of which are assigned to the genus *Ichthyostega*, represent creatures that are strikingly intermediate in form between lobe-finned fishes and amphibians: The lobe fin itself is formed of an array of bones resembling that found in amphibians; similarly, the complicated teeth of lobe-finned fishes closely resemble the teeth of early amphibians (Figure 11-22). These features

FIGURE 11-21 ***Archaeopteris*, one of the very oldest kinds of large tree.** *A.* A segment of a branch with leaves. *B.* Drawing of the entire tree. *(After C. R. Beck, Biol. Rev. 45:379–400, 1970.)*

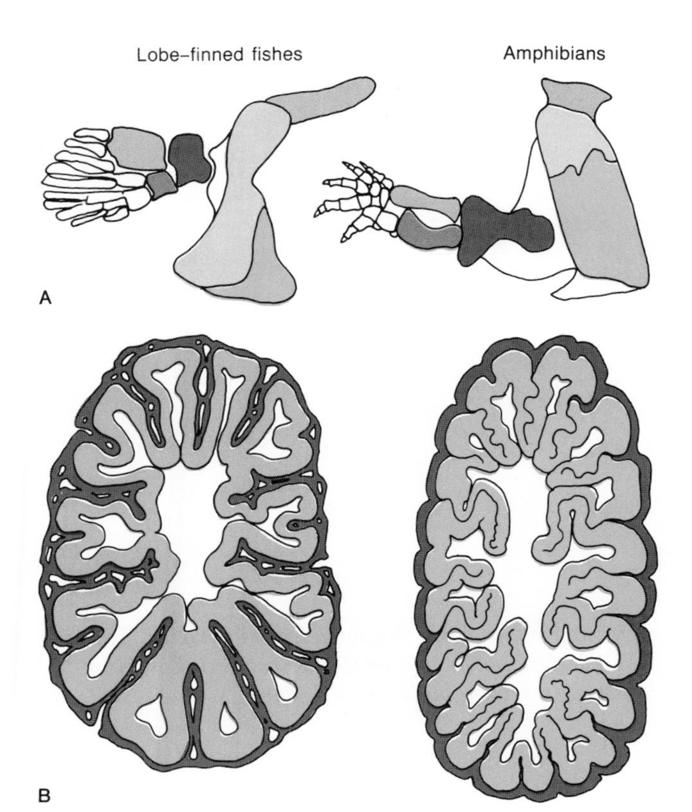

FIGURE 11-22 Lobe-finned fishes *(left)* and amphibians *(right)*. *A.* The shoulder and limb bones; shading identifies particular bones that the two groups have in common. *B.* Cross sections of teeth, showing the unusual, complex structure found in both groups.

alone strongly suggest that amphibians evolved from lobe-finned fishes, but additional features make the derivation a certainty. *Ichthyostega* had four legs, as do amphibians, but its skull structure was remarkably like that of a lobe-finned fish. The creature also had a fishlike tail — one feature of its ancestors that was probably not useful on land (Figure 11-23). Because of this intriguing combination of features, *Ichthyostega*, which was not discovered until the present century, represents what is commonly termed a "missing link."

The lung, which was occasionally used by early fishes to breathe air, was available for exploitation long before amphibians evolved. In the way the lung became a full-time supplier of oxygen we can see yet another example of the "opportunism" of evolution. Unlike the gill supports, which evolved into jaws earlier in vertebrate evolution, the lungs developed from a preexisting structure that required very little evolutionary modification to open up an entirely new mode of life.

More than 80 million years passed between the time when vascular plants appeared on land (Late Silurian or earlier) and the time when the first amphibians evolved (latest Devonian). It is not surprising that vascular plants colonized the land before vertebrate animals, because a food web must be built upward from the base; animals

FIGURE 11-23 The primitive amphibian *Ichthyostega*, whose legs contrast with the fins of the approximately contemporary lobe-finned fish *Eusthenopteron* (left) and the lungfish *Rhynchodipterus* (right). The trunk and branches belong to the tree fern *Eospermatopteris*. *(Drawing by Gregory S. Paul.)*

cannot live on land in the absence of an adequate supply of edible vegetation.

Amphibians evolved so late in Devonian time that they played no significant role in the ecosystem of that period. It was the Carboniferous and Early Permian that might be called the Age of Amphibians. Nevertheless, the dominant animals of that interval descended from the Late Devonian *Ichthyostega* or from similar taxa that remain undiscovered.

PALEOGEOGRAPHY

In general, the Silurian and Devonian were periods when sea level stood high in relation to the surfaces of major cratons (Major Events, p. 270). Early in Silurian time, sea level rose from its low position at the end of the Ordovician Period—a rise that is thought to have resulted from the melting or partial melting of the extensive polar glaciers that had formed late in Ordovician time. Simultaneously, many marine invertebrate taxa underwent the adaptive radiations discussed earlier in this chapter.

The widespread occurrence of organic reefs, carbonates, and evaporites strongly suggests that middle Paleozoic climates were relatively warm. Because they were especially warm during the Devonian Period, it has been suggested that shallow-water tropical marine communities were more widespread at this time than at any other time in the Phanerozoic Eon. Climates not only were warm but in many areas were relatively dry; in fact, larger volumes of evaporite deposits accumulated in middle Paleozoic than in early Paleozoic time. Note that evaporites of Silurian age lie within 30° or so of the ancient equator but that some Devonian evaporites formed farther north and south, apparently reflecting the widespread distribution of warm climates during the Devonian Period (Figure 11-24).

Continents and Oceans

Figure 11-24 displays the positions of continental areas during Middle Silurian time and during late Early Devonian time, about 65 million years later.

An important new geographic feature to appear during Devonian time was the Old Red Sandstone continent, named for a well-known, largely Devonian sandstone unit of the British Isles (p. 268). This continent of high relief formed when Laurentia and Baltica were welded together as the Iapetus Ocean disappeared. Earlier we saw that this collision began in the north during the Silurian Period and progressed southward, ending late in Devonian time (p. 178). Paleomagnetic evidence reveals that by the end of the Devonian Period, the gap between the Old Red Sandstone continent and Gondwanaland had narrowed. The proximity of these continents explains the fact that North America, Europe, and North Africa share at least 80 percent of their Late Devonian genera of marine invertebrates.

Biogeographic Provinces of the Devonian Period

During Silurian time, conspicuous marine groups, such as the brachiopods and graptolites, were remarkably cosmopolitan in that a number of species and genera inhabited the shallow seas of many continents. At this time the continents formed a relatively tight cluster. By Early Devonian time, however, mountain building associated with the formation of the Old Red Sandstone continent left an embayment that projected from the south far into the interior of what is now North America (Figure 11-24). During the first half of the Devonian Period this area became the Appalachian province, and its fauna was distinctive. Further transgression of the seas connected it with the west coast, however, so that the Appalachian marine faunas were blended again with others.

At the same time, possibly because of a cooling of polar regions, a discrete marine province formed along the margin of southern Gondwanaland. This southern realm, which is called the Malvinokaffric Province, was populated by a fauna that appears to have lived in cool water. Within this province early in Devonian time, the Paraná Basin (now central South America) was partly barred from the open ocean by newly formed mountains in the Andean region (Figure 11-25). The center of the Paraná Basin lay perhaps only 15° from the South Pole, so it is not surprising that, though stromatoporoids, tabulates, and rugose corals formed reefs throughout most of the shallow-water areas in the world, they were not found in this basin. Also missing were bryozoans and ammonites. Burrowing bivalves formed a large percentage of the marine species in the Paraná Basin, just as they do in polar regions today.

Late Devonian Mass Extinction

One of the most devastating mass extinctions of marine life in all of Phanerozoic time took place near the end of the Devonian Period. Geologists divide the Upper Devonian Series into two stages—the Frasnian Stage and the Famennian

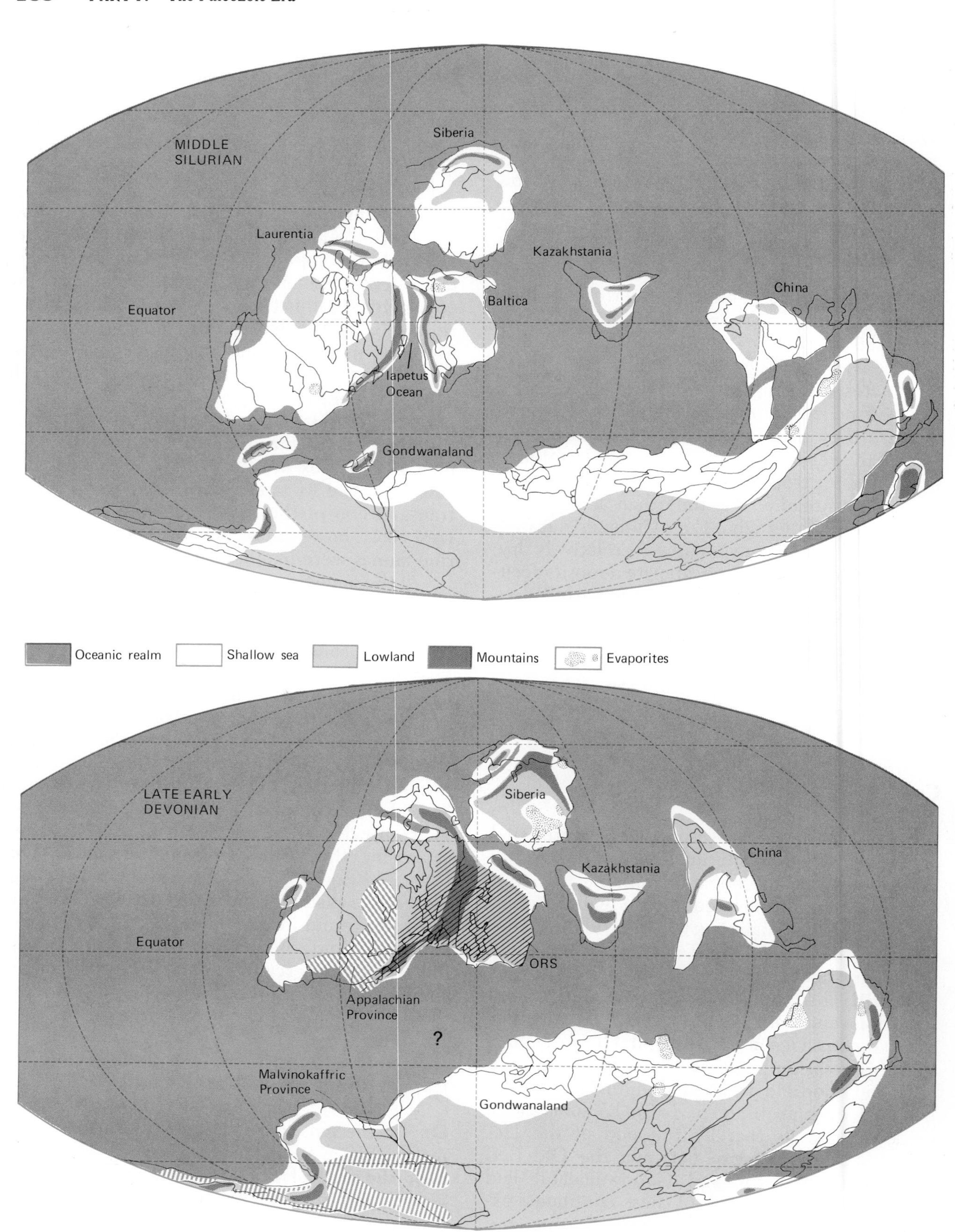
MIDDLE SILURIAN
Siberia
Laurentia
Kazakhstania
China
Equator
Baltica
Iapetus Ocean
Gondwanaland
Oceanic realm
Shallow sea
Lowland
Mountains
Evaporites
LATE EARLY DEVONIAN
Siberia
Kazakhstania
China
Equator
ORS
Appalachian Province
?
Malvinokaffric Province
Gondwanaland

FIGURE 11-24 World geography during middle Paleozoic time. During this interval a broad tropical seaway developed north of Gondwanaland. Note how the Old Red Sandstone continent (ORS) was formed during this interval by the union of Baltica and Laurentia; the Appalachian biogeographic province became isolated to the west. Shallow seas near the South Pole formed the Malvinokaffric Province, where species were apparently adapted to cold temperatures. In the lower map, the question mark in the tropical seaway between the Old Red Sandstone continent and Gondwanaland signifies the possibility that this seaway may have closed during the Devonian Period. *(Modified from R. K. Bambach et al., American Scientist 68:26–38, 1980.)*

Stage. The great extinctions occurred late in Frasnian and early in Famennian time.

On the land, vascular plants appear to have been unaffected by the Late Devonian crisis, but it is difficult to interpret the meaning of their persistence, because they differed so much from modern plants.

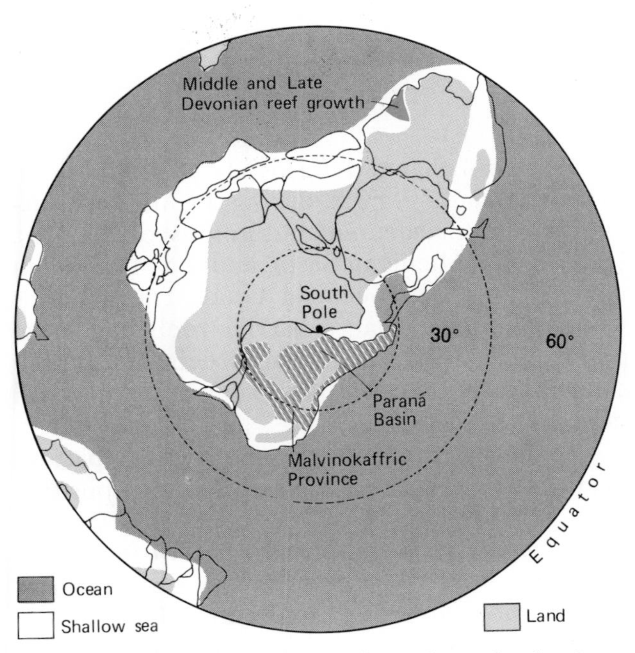

FIGURE 11-25 The position of Gondwanaland in late Early Devonian time. The Paraná Basin, which lay near the south pole on what is now the continent of South America, supported cold-water marine faunas. Later in Devonian time, reefs grew extensively near the equator along what is now the western margin of Australia. *(Modified from C. R. Scotese et al., Jour. Geol. 87:217–277, 1979.)*

In the marine realm, brachiopods were hard hit; only about 15 percent of Frasnian brachiopod genera are found in Famennian strata. Ammonoids experienced a similar decline, and many types of gastropods and trilobites disappeared as well. Two complex communities seem to have suffered almost total collapse: the tabulate-strome reef community and the pelagic community (plankton and nekton). The tabulate-strome community had enjoyed its greatest faunal diversity during Middle Devonian time, 100 million years or so after it came into existence. During much of the Frasnian Age, reef species had already diminished somewhat, but at the end of Frasnian time, the reef community suffered truly catastrophic extinctions. Tabulates, stromatoporoids, and rugose corals are rarely seen in Famennian strata, and tabulate-strome reefs are virtually nonexistent both here and in younger Paleozoic strata. Within the pelagic system, acritarchs, the only group of phytoplankton with an extensive Devonian fossil record, suffered heavy losses, and placoderms, the dominant pelagic carnivores, almost disappeared.

The demise of the middle Paleozoic reef community exemplifies an important geographic pattern of the Late Devonian mass extinction: Tropical taxa were most severely affected. In contrast, the polar communities of the Malvinokaffric Province of South America (Figure 11-25) were largely unaffected. This bias against tropical species suggests that an episode of global cooling may have led to the mass extinction. Fossil evidence in New York State supports this hypothesis. Here glass sponges, which today live in cool waters, experienced an evolutionary expansion while many previously diverse groups of marine animals were on the decline. After the mass extinction, when the groups that had suffered heavy losses began to rediversify, the glass sponges dwindled, perhaps in response to a return of warm climatic conditions.

Supporting these ideas is another pattern: After the mass extinction, the rate at which limestones accumulated in shallow seas declined.

The Start of a New Glacial Interval

The glacial episode that began in Late Ordovician time, when Gondwanaland migrated over the South Pole, appears to have ended during the Silurian Period. Because no glacial deposits of Early

Devonian age are known in Gondwanaland, it has been suggested that Gondwanaland migrated away from the South Pole in Late Silurian time, so that its southern region became less frigid. In any event, near the end of the Devonian Period, glacial deposits were laid down again in Gondwanaland. The location was northern South America, then near the South Pole, which lay in eastern Africa, according to paleomagnetic data (Figure 11-25). Thus it appears that an episode of polar cooling and glaciation began in Late Devonian time, about the time the mass extinction struck marine life in the tropics. This association strengthens the idea that climatic cooling caused the extinction.

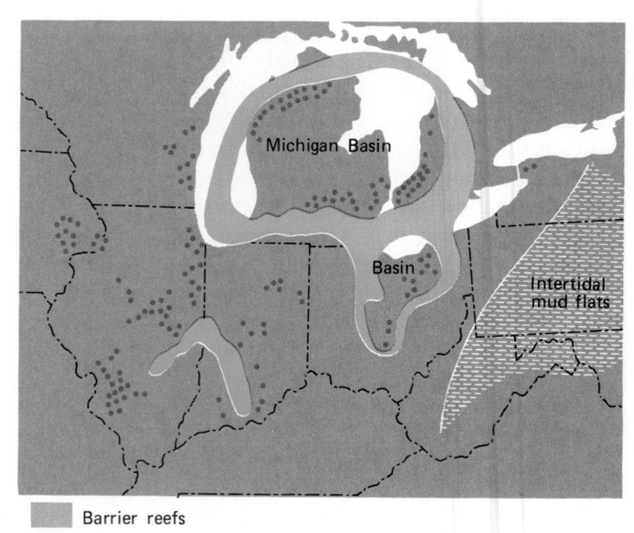

FIGURE 11-26 Middle Silurian reefs of the Great Lakes region. Barrier reefs encircled the Michigan Basin and a smaller basin in Ohio. They also flourished in southern Indiana and Illinois. Extensive mud flats now lay to the east in the Pennsylvania region, in contrast with the environments of coarse clastic deposition that occupied this area in Early Silurian time. *(Modified from K. J. Mesolella, Amer. Assoc. Petrol. Geol. Bull. 62:1607–1644, 1978.)*

REGIONAL EXAMPLES

The Silurian and Devonian periods were times of widespread reef development and carbonate deposition, but they were also times of orogeny for the continents that bordered the Iapetus Ocean. While eastern North America was being transformed from an Early Silurian highland to a Middle Silurian carbonate shelf, reefs and evaporite deposits were forming farther to the west. Later in the Devonian Period, the Old Red Sandstone continent formed with the closure of the Iapetus Ocean, and in western North America reefs continued to grow and mountains rose up.

Eastern North America: Carbonates, Reefs, and Evaporites

The first tectonic cycle of the central Appalachians ended with the deposition of clastic wedges of sediment shed westward from the Taconic Mountains, which formed along eastern Laurentia late in the Ordovician Period (Figure 7-13). The Silurian Period began with a continuation of this pattern. As the eastern highlands were subdued by erosion, however, the site of clastic deposition became more broadly flooded by shallow seas, and late in Silurian time the deposition of shallow-water carbonates initiated the second tectonic cycle of the central Appalachians.

To the west, tabulate-strome reefs dotted shallow epicontinental seas (Figure 11-26). Here, however, the pattern of sedimentation and reef development changed drastically from Middle to Late Silurian time. First, during the Middle Silurian, two basins accumulated muddy carbonates. One was the Michigan Basin, and the other lay in what is now north-central Ohio. These basins were bounded by large barrier reefs and were populated by scattered pinnacle reefs. At this time siliciclastic muds were still accumulating on broad tidal flats to the east.

As the Silurian Period progressed, this pattern changed. To the east, siliciclastic mud deposition gave way to carbonate sedimentation. Central Pennsylvania and neighboring areas were now the sites of accumulation of supratidal, intertidal, and shallow subtidal carbonates (Figure 11-27). At the same time, the Michigan and central Ohio basins came to be only weakly supplied with seawater and thus turned into evaporite pans in which dolomite, anhydrite, and halite were precipitated. The resulting deposits are a major source of rock salt today. It appears that marginal reefs in the Michigan Basin grew so high during Middle Silurian time that they restricted the flow of water into the

FIGURE 11-27 Mud cracks in intertidal deposits of the Middle Silurian Wills Creek Formation in western Maryland. A broad area of intertidal deposition lay to the east of the reef growth in the Great Lakes region. *(From F. J. Pettijohn and P. E. Potter, Atlas and Glossary of Sedimentary Structures, Springer-Verlag, New York, 1964.)*

basin. In time, evaporation and possibly a slight lowering of sea level led to the exposure and consequent death of the reefs. Although a weak flow of seawater into the basin replenished the water that had been lost by evaporation, the rate of evaporation was so high that evaporite minerals were precipitated around the margins of the basin and even at considerable depths within it. At first the center of the basin was moderately deep, but as evaporites accumulated, the water there grew progressively shallower until eventually the sea was excluded altogether.

Because the conditions within the evaporite basins were so inhospitable during the Late Silurian, reefs grew only in the southwest, in Indiana and Illinois. Here a number of reefs have been studied in detail. The most famous is the Thornton Reef of northern Illinois (Figure 11-28). The structure of this reef indicates the direction of the prevailing winds at the time it was growing: The stromatoporoid ridge obviously faced waves advancing from the southwest.

As we saw in Chapter 7, carbonate deposition of the second tectonic cycle in the central Appalachians was interrupted by the onset of orogenic

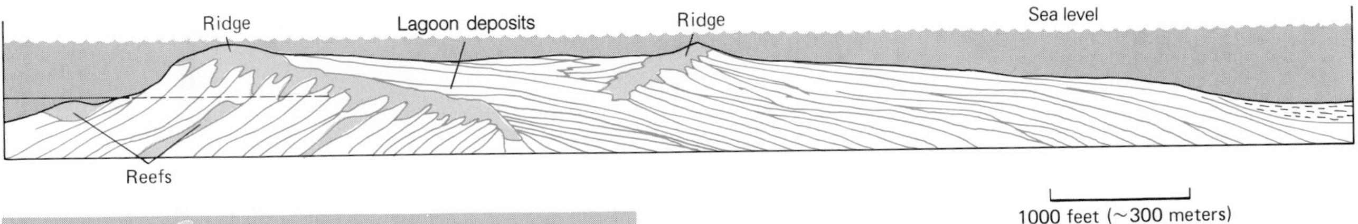

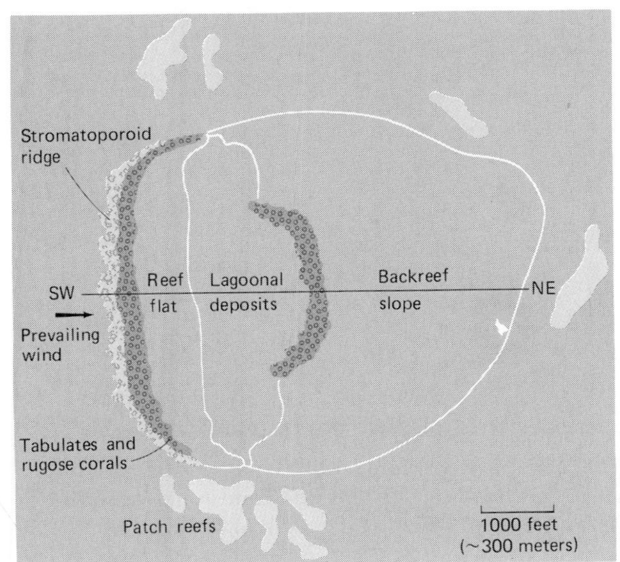

FIGURE 11-28 The Thornton Reef of northern Illinois. This circular reef is exposed in a limestone quarry. The upper diagram represents a cross section of the reef along the axis indicated in the map view below. A durable stromatoporoid ridge faced the prevailing direction of wind-driven waves. Behind the ridge was a zone of tabulates and rugose corals, and beyond were a reef flat and a lagoon. A weaker stromatoporoid ridge bounded the lagoon on the leeward side. *(After J. J. C. Ingels, Amer. Assoc. Petrol. Geol. Bull. 47:405–440, 1963.)*

activity and continental suturing during the Devonian Period. The Old Red Sandstone continent was formed by this suturing.

The Old Red Sandstone Continent

One of the classic angular unconformities in the geologic record occurs in Scotland between Devonian beds of the Old Red Sandstone and the nearly vertical Silurian beds on which they rest. It was at the locality shown in Figure 11-29 that James Hutton recognized the meaning of stratigraphic unconformity in 1788. Thus it came to be understood that the Old Red Sandstone was deposited after a Silurian episode of mountain building—as we now know, after the mountain building that occurred during the suturing of Baltica and Laurentia. The Old Red crops out over large areas of Scotland. Chunks of its distinctive red sandstones can be found in parts of Hadrian's Wall, built by the Roman emperor Hadrian across northern England in the second century.

Acadian Suturing The Old Red includes not only rocks of Early, Middle, and Late Devonian age but also rocks representing Late Silurian and earliest Carboniferous times. For a long time geologists found it puzzling that such a large volume of sediment could have been shed from highlands in the British Isles when most of the isles' area formed a depositional basin. Now the puzzle has been solved within the framework of plate tectonics by the reassembly of the landmass that formed when Laurentia, the British Isles, and Baltica were united during mid-Paleozoic time. Portions of this landmass that stood above sea level formed what has been called Laurasia, or the Old Red Sandstone continent (Figure 11-24). The British Isles, which lay near the southeastern margin of this landmass, were the site of extensive freshwater molasse deposits derived from mountains in the area that is now Greenland (Figure 11-30). Recall that the orogenic activity associated with the formation of this continent proceeded from north to south (p. 178). Molasse deposition, which resulted from tectonic activity known as Caledonian

FIGURE 11-29 Angular unconformity between the Old Red Sandstone *(top left)* and Silurian rocks *(bottom left)* at Siccar Point, Berwickshire, Scotland. The Silurian rocks were tilted and folded when Baltica collided with Laurentia to form the Old Red Sandstone continent (see Figure 11-24). The Old Red Sandstone was subsequently deposited near the margin of this continent. *(Institute of Geological Sciences, British Crown copyright.)*

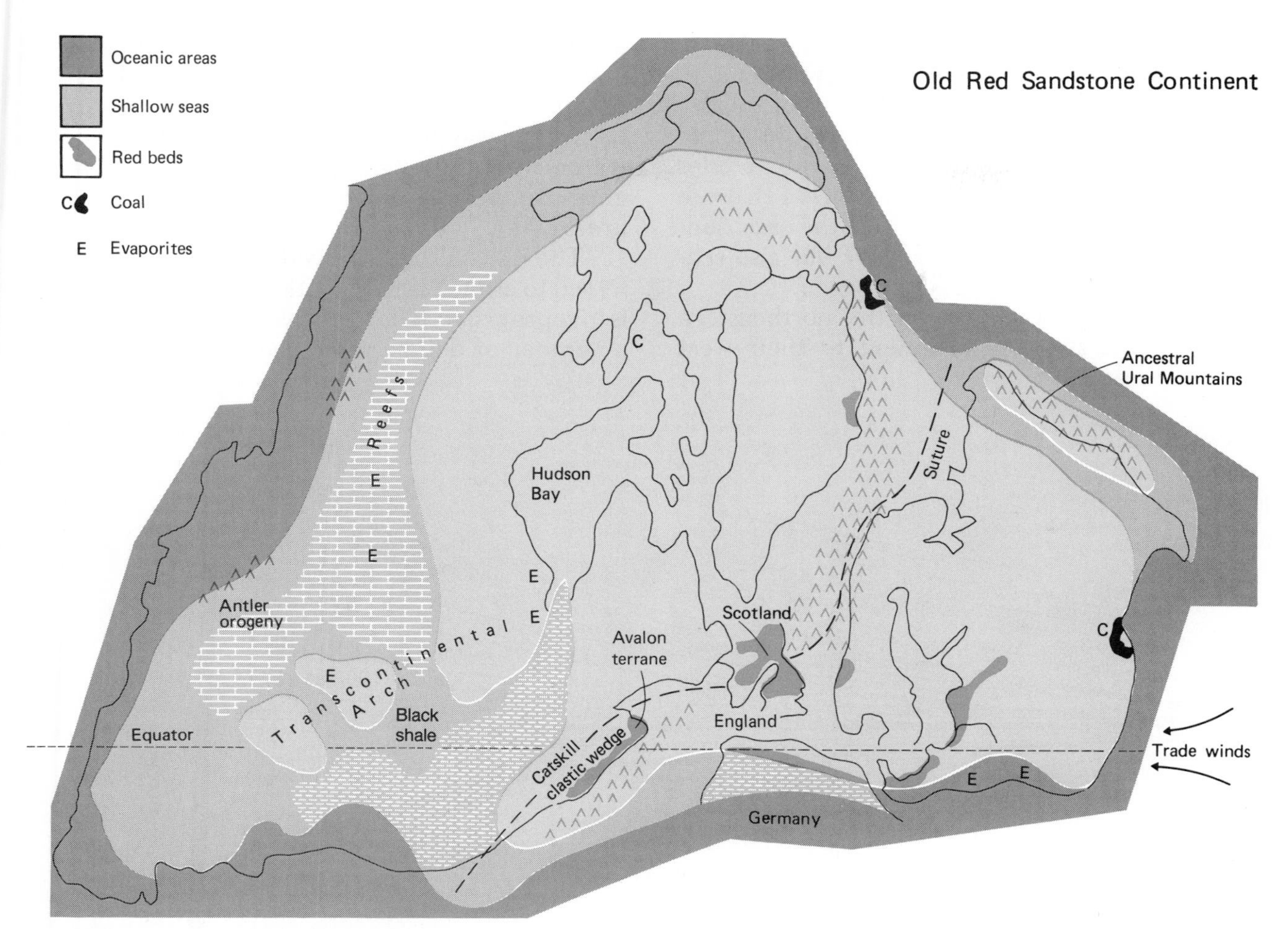

FIGURE 11-30 The Old Red Sandstone continent during Late Devonian time. The central mountain belt running from north to south formed as Baltica converged with Laurentia (see Figure 11-24). Red beds were concentrated in the south, near the equator. Deep-water deposits accumulated in what is now central Germany. Shales accumulated in North America west of the Catskill clastic wedge, and limestones, reefs, and evaporites formed farther west. The Antler orogeny affected the western margin of the continent.

in Britain, was under way in northern Britain early in Devonian time, but in the northeastern United States, the oldest molasse (clastic-wedge) deposits, which resulted from the interval of deformation known as Acadian, are of mid-Devonian age (Figure 7-13). The proximity of Britain and eastern North America during the latter half of the Devonian Period explains another longstanding problem as well—the similarity of freshwater fishes and early land plants of the two regions.

Interior and Eastern North America During much of Devonian time, an arm of land may have extended southwestward across what is now the western interior of North America, where the Transcontinental Arch persisted from early Paleozoic time (Figure 11-30) and little or no sediment accumulated. The Devonian equator passed through the southern part of the Old Red Sandstone continent. Prevailing trade winds must have blown from the east, as they do today (Figure 2-10). Minor coal deposits, which formed from early land plants, are found in the east, where moist air must have risen as it passed landward and dropped moisture that it gained as it passed over seas to the east (Figure 11-30). Although evaporites are found here and there near the eastern margin of the continent, they are best developed in the rain shadow to the west, in the area of North America where the Rocky Mountains now stand. The climate was at least intermittently hot and dry enough along the east and west coasts of the Old Red Sandstone continent so that caliche nodules formed in abundance in low-lying areas that are now represented by ancient soils. In the east, rainfall was probably not only heavier but also seasonal, as in the tropics today.

The eroding mountains in the northeastern United States supplied sediment to their west much more rapidly than the crust subsided. The result was a westward regression, recorded in the enormous Catskill clastic wedge, which thins to the west, away from the source area. Nonmarine red beds in the east give way to finer-grained marginal marine and marine deposits to the west (Figure 11-31).

The Catskill clastic wedge is sometimes referred to as the Catskill delta, but the designation is inappropriate. The Catskill wedge does include a number of deltaic sequences that characteris-

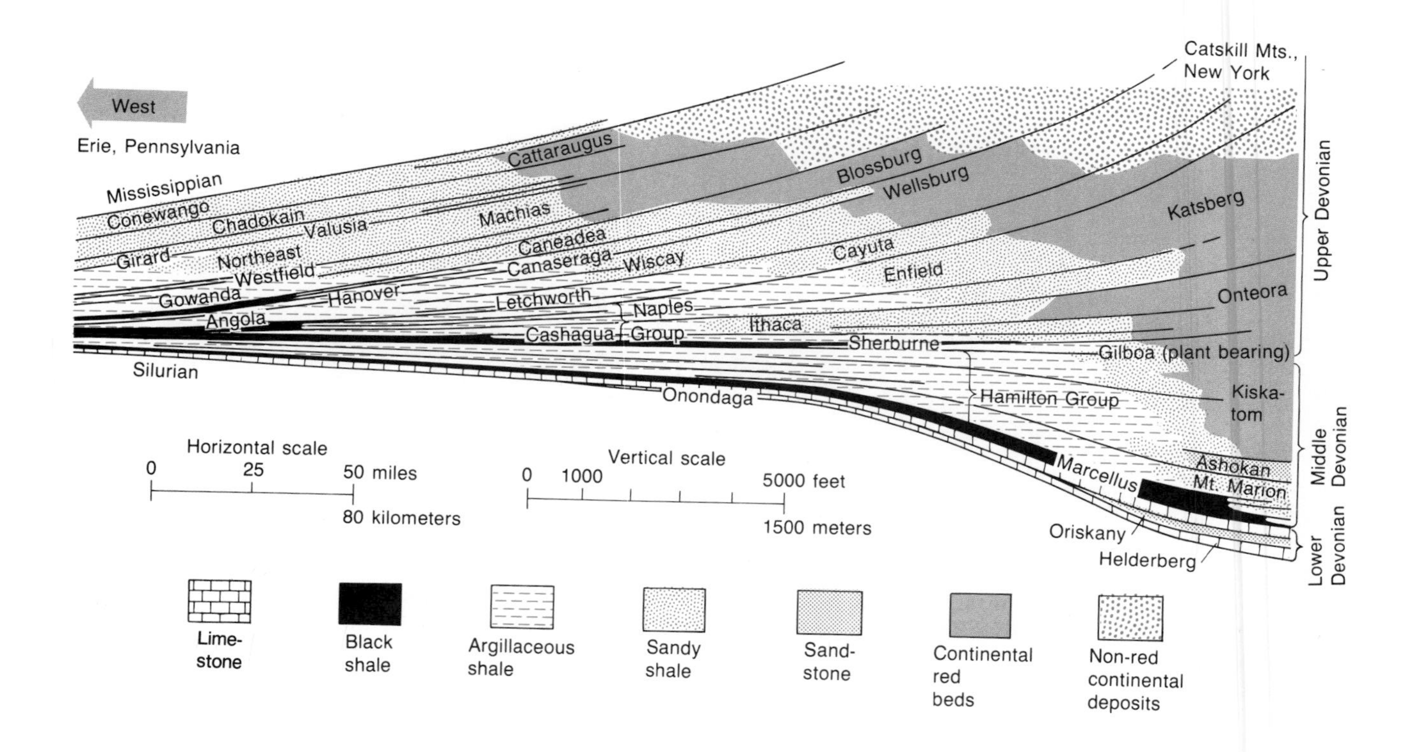

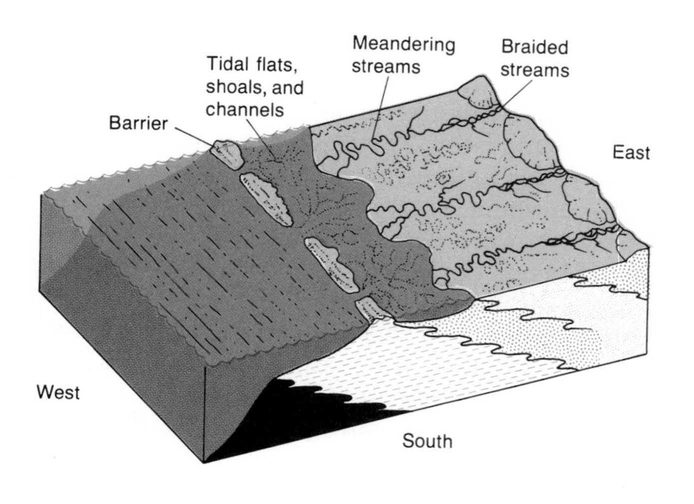

FIGURE 11-31 Devonian sedimentary rocks of New York State. *Above:* Stratigraphic cross section of the formations. Coarse deposits shed from mountains to the east constitute a clastic wedge that is thinning to the west. *Below:* Environments of deposition of the Catskill clastic wedge and associated deposits of New York State. Braided streams meander seaward from the foot of mountains to the east. Eventually they empty into tidal channels behind barrier islands. Muds are deposited offshore. *(Upper diagram after P. B. King, The Geological Evolution of North America, Princeton University Press, Princeton, New Jersey, 1977; lower diagram after J. R. L. Allen and P. F. Friend, Geol. Soc. Amer. Spec. Paper 106:21–74, 1968.)*

tically display coarsening-upward deltaic cycles (Figure 3-19), but it is really a composite of these sequences and others, including meandering-river deposits in which fine-grained sediments occur near the top (Figure 3-15). Sandstones and conglomerates were laid down in braided streams near highlands, meandering-river deposits formed downstream, and tidal-flat deposits with marine faunas accumulated along the shoreline. Offshore were barrier islands and sandy shelves, and farther offshore, muddy seafloors.

Late in the Devonian Period a muddy seaway lay between the Old Red Sandstone continent and the Transcontinental Arch, extending northward to the Hudson Bay area (Figure 11-30). The giant placoderm *Dunkleosteus* flourished in this sea, together with other fishes now preserved in the black shales of northern Ohio (Figures 11-12 and 11-13). Near the end of Devonian time, the deposition of black muds extended westward, leaving a remarkably large area of eastern and central North America blanketed with these sediments.

Reef Building and Orogeny in Western North America

Along the continental shelf west of the Old Red Sandstone continent in what is now western Canada, tabulate-strome reef complexes developed during the latter half of the Devonian Period (Figure 11-32). Reefs here took the form of elongate barriers, atolls with central lagoons, and platforms. Many of the reefs that formed in this area now lie buried beneath the surface, and their porous textures have created traps for petroleum. The Devonian reefs of western Canada varied in form and organic composition, but typically they displayed the sequence of development illustrated in Figure 11-4. Reef growth began with the development of a meadow of sticklike corals on a seafloor of burrowed mud. These corals were then overgrown by colonial tabulates, which in turn gave way to massive stromatoporoids. Behind the reef, carbonate mud accumulated in quiet water. South of the belt of reef growth, carbonates were deposited in shallow seas, just as they had been early in Paleozoic time (Figure 11-30).

During Frasnian time, subsidence of the reef area allowed deep-water muds to encroach on the shelf where reefs had flourished late in Middle

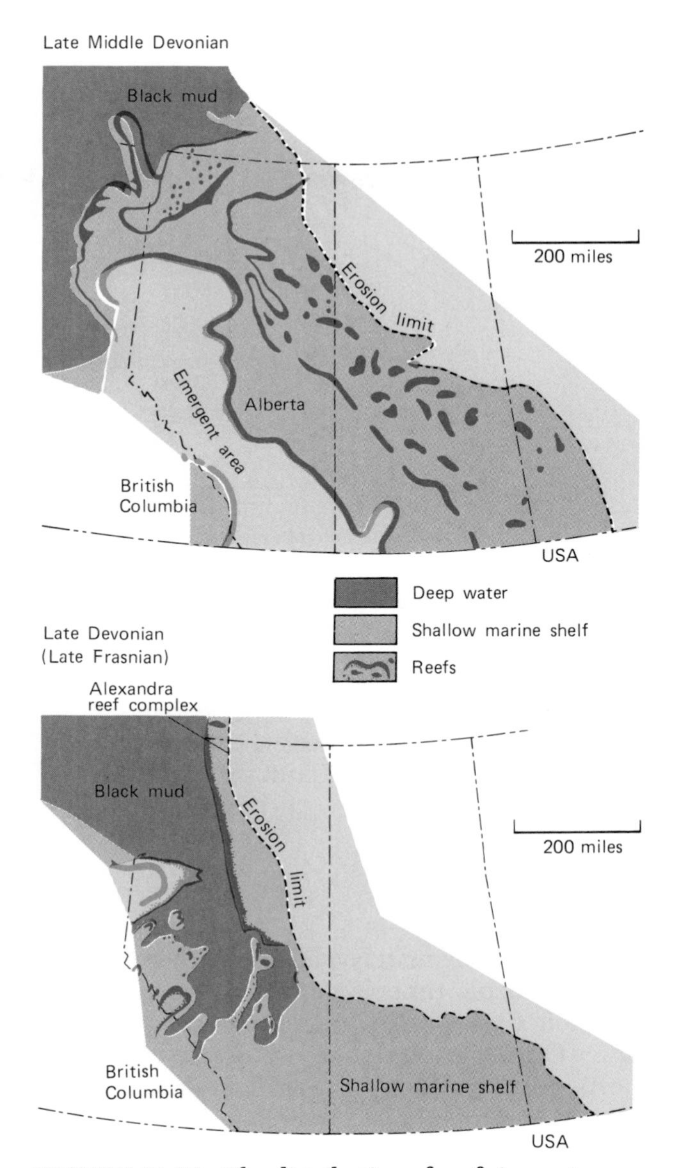

FIGURE 11-32 The distribution of reefs in western Canada during the latter part of Devonian time. Late in Frasnian time, reef growth ceased, and black mud spread onto the continental shelf. Northeast of the erosion limit, Middle and Late Devonian rocks have been eroded away. *(After E. R. Jamieson, Proc. N. Amer. Paleont. Conv. J:1300–1340, 1969.)*

Devonian time (Figure 11-32). Then, at the end of Frasnian time, the worldwide faunal crisis discussed earlier brought an end to tabulate-strome reef building throughout the world.

Through all of Silurian and nearly all of Devonian time, the western margin of North America

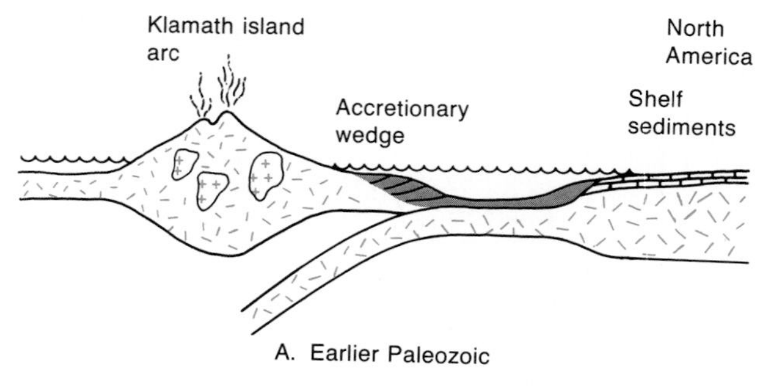

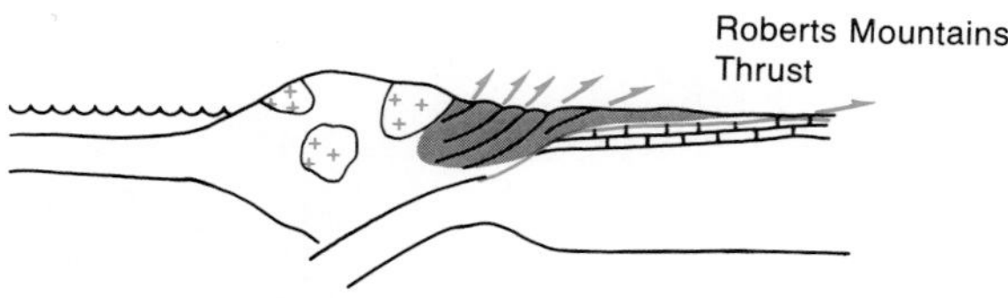

FIGURE 11-33 The likely mechanism by which the Klamath Arc was added to the North American continent by the Antler orogeny during late Devonian and Mississippian time. The basin between the craton and the Klamath Arc *(A)* closed. As the continental crust was thrust beneath the volcanic crust of the Klamath Arc, deep-sea sediments slid onto shallow-water carbonates along the Roberts Mountains Thrust *(B)*.

remained approximately where it had been during early Paleozoic time (Figure 11-30). In middle Paleozoic time, however, an island arc stood offshore. Ophiolite sequences (graywackes, shales, cherts, and volcanics) in the Klamath Mountains and in the Sierra Nevada of present-day northern California record the presence of this Klamath Arc (Figure 11-33*A*). Rocks in Nevada show that this simple geographic picture became more complex between Middle Devonian and Early Mississippian time; they reveal closure of the basin between the Klamath Arc and the craton. In central Nevada, deep-sea deposits like those of northern California can be seen to have been thrust as far as 160 kilometers (~ 100 miles) onto the craton (Figure 11-33*B*). The principal thrust fault along which this movement occurred is called the Roberts Mountains Thrust.

The collision of the arc and continental margin that produced the Roberts Mountains Thrust is known as the **Antler orogeny**. This was the first sizable episode of mountain building in the Cordilleran region of North America during Phanerozoic time. The remainder of the Cordilleran story, which will appear in the chapters that follow, has mountain building as its dominant theme.

CHAPTER SUMMARY

1 Many important middle Paleozoic groups of marine life, depleted during the great mass extinction at the end of the Ordovician Period, reexpanded at the start of the Silurian Period. Other forms, such as the ammonoids and the jawed fishes, were new. Jawed fishes and mollusks occupied freshwater as well as marine habitats.

2 The Silurian Period witnessed the invasion of the land by vascular plants, followed in Devonian time by the invasion of arthropods (scorpions and insects) and vertebrate animals (amphibians). During the Devonian Period, spore-bearing plants were joined by seed-bearing plants, which did not require moist habitats for reproduction and thus were not restricted to the fringes of aquatic habitats.

3 During most of middle Paleozoic time, world climates were relatively warm, and the seas stood high in relation to the surfaces of continents.

4 The Old Red Sandstone continent was formed by the unification of Laurentia and Baltica with the closure of the Iapetus Ocean along the chain of mountains that arose in the Acadian/Caledonian orogeny. Freshwater sediments of the new landmass yield well-preserved fish fossils.

5 Tabulate-strome reefs flourished throughout middle Paleozoic time and were especially abundant in the Great Lakes region and near the western continental margin of North America.

6 Shortly before the end of the Devonian Period, a great mass extinction eliminated many forms of marine life, including bottom-dwelling animals such as those that formed the tabulate-strome reef community and most acritarchs, which were floating algae. Species that occupied cold regions seem to have survived preferentially, suggesting that changing climatic conditions may have been the cause of the extinction.

EXERCISES

1 In what important ways did invertebrate life change between Ordovician time and Devonian time?

2 What animals have the oldest extensive fossil record in freshwater sediments?

3 In what way did terrestrial environments of the Late Devonian Period look different from those of Early Devonian time?

4 What evidence do fossil bones and teeth provide that amphibians evolved from fishes?

5 Where was Greenland in relation to North America and Europe toward the end of the Devonian Period?

6 Where was Gondwanaland located at the end of Devonian time?

7 How do reefs of middle Paleozoic age illustrate ecological succession?

8 Buried ancient reefs are commonly porous structures that serve as traps for petroleum. If you wanted to drill for oil in Devonian reefs, what geographic regions would seem most promising?

9 What caused large quantities of sediment to accumulate in the south-central part of the Old Red Sandstone continent during the Devonian Period?

ADDITIONAL READING

Benton, M. J., *Vertebrate Paleontology,* Unwin Hyman, London, 1990.

Dineley, D. L., *Aspects of a Stratigraphic System: The Devonian,* John Wiley & Sons, Inc., New York, 1984.

Taylor, T. N., "Reproductive Biology in Early Seed Plants," *Bioscience* 32:23–28, 1982.

Thomas, B. A., and R. A. Spicer, *The Evolution and Palaeobiology of Land Plants,* Croom Helm, London, 1987.

CHAPTER

12

The Late Paleozoic World

The late Paleozoic interval of geologic time included the Carboniferous Period, when new groups of animals and plants influenced the accumulation of sediments, and the subsequent Permian Period, when many of these organisms died out in the greatest mass extinction in all of Phanerozoic time. Before this event, during Carboniferous time, skeletal debris from Carboniferous marine organisms accumulated to form widespread limestones, and spore-bearing trees stood in broad swamps, contributing their wood to the formation of coal when they died.

The late Paleozoic world was marked by major climatic changes that today are reflected in the distribution of rocks and fossils. Glaciers, for example, spread over the south polar region of Gondwanaland during the Carboniferous Period and then receded during Permian time. A general drying of climates at low latitudes during the Permian Period led to a contraction of coal swamps and to the extinction of spore-bearing plants and amphibians, both of which required moist conditions. At the same time, seed plants and mammal-like reptiles inherited the earth, and evaporites accumulated in many areas.

In addition to the great mass extinction, another major event took place near the end of the

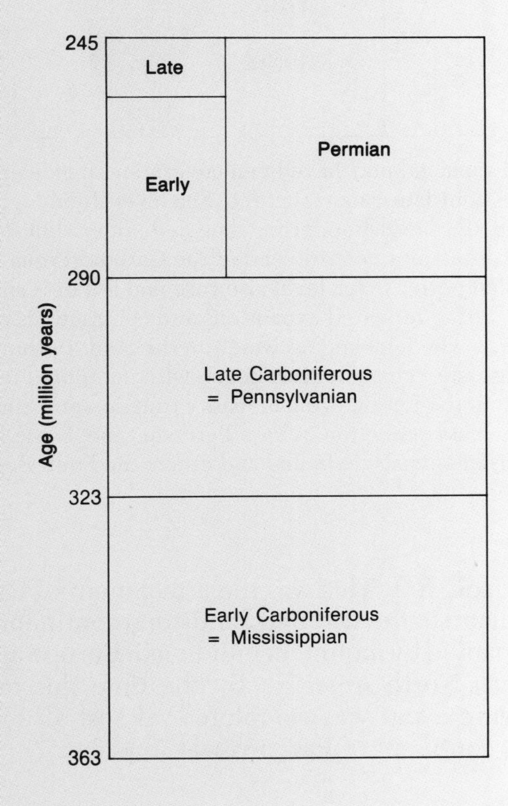

During Late Carboniferous time, logs of spore-bearing trees accumulated in broad swamps to form coal deposits that are now exploited by modern societies. Here ferns and seed ferns form the undergrowth beneath lycopod trees. On the fallen log on the right is a cockroach, one of the many kinds of insects that evolved during Late Carboniferous time. *(Field Museum of Natural History, Chicago, neg. 75400c.)*

MAJOR EVENTS OF LATE PALEOZOIC TIME

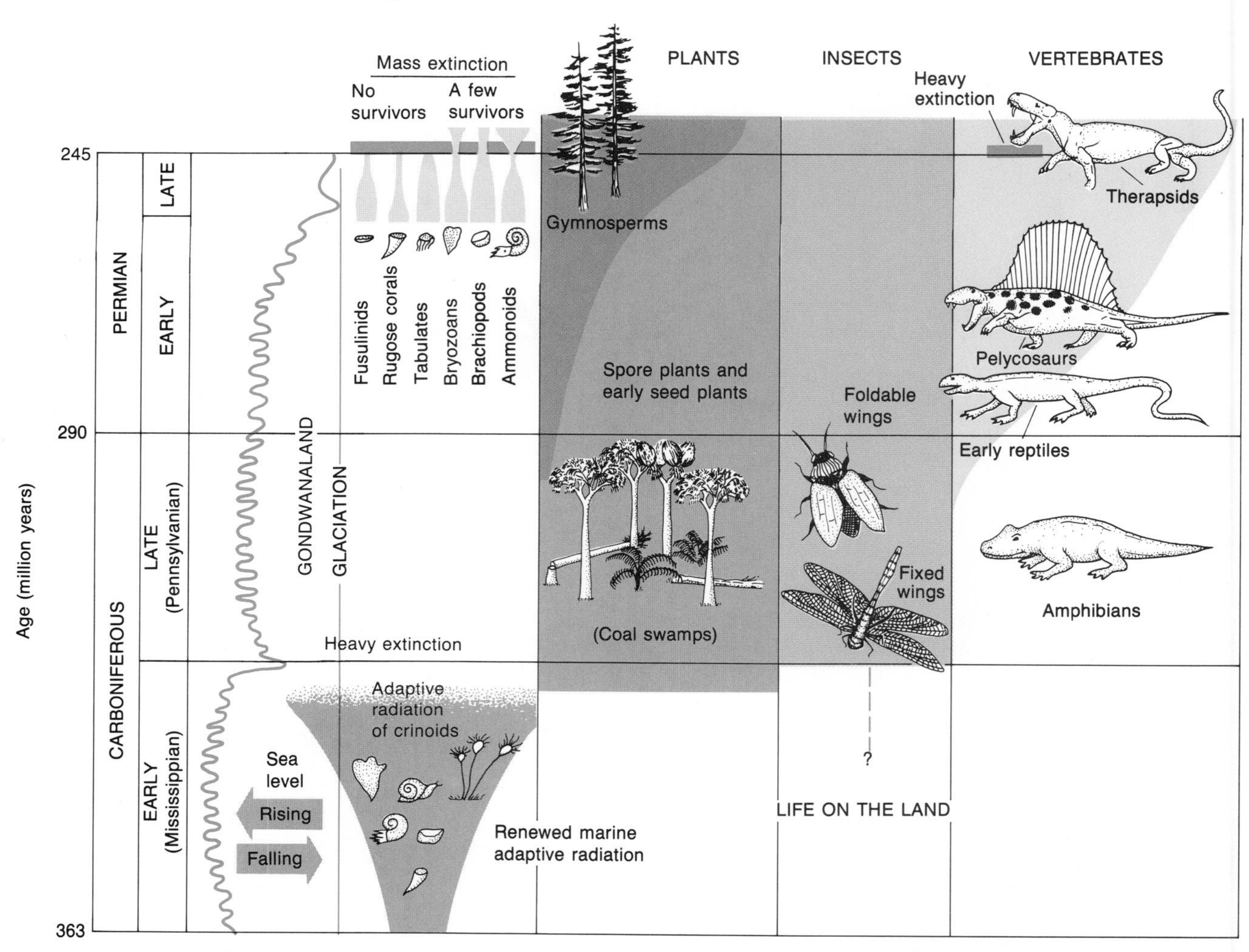

Gondwanaland may have been covered with glaciers throughout late Paleozoic time. Sea level stood highest in mid–Early Carboniferous time and underwent its biggest declines near the end of the Carboniferous and Permian periods. Sea level also rose and fell on a smaller scale during repeated expansions and contractions of glaciers, which locked up water on the land. In the oceans, late Paleozoic time began with continued recovery from the Late Devonian mass extinction and ended with a mass extinction in Late Permian time. Some large groups of animals dwindled and others died out. Beginning in Late Carboniferous time, climates became increasingly arid, and gymnosperms replaced spore-bearing plants as the dominant plants on the land. Late in Carboniferous time, insects with the ability to fold their wings evolved for the first time, and during the Permian Period, reptiles of increasingly high levels of adaptation replaced amphibians as the dominant large land animals. The most advanced mammal-like reptiles, the therapsids, underwent rapid adaptive radiation during Late Permian time and then suffered heavy losses at the close of this epoch.

Paleozoic Era. This was the attachment of Gondwanaland to the Old Red Sandstone continent, accompanied by mountain building in Europe and in eastern North America. By the time this major suturing event was completed, almost all of the supercontinent of Pangaea was in place.

The Carboniferous System was formally recognized in Britain in 1822, early in the history of modern geology. Its name was chosen to reflect the system's vast coal deposits, which had long been mined for fuel. Actually, it is only the upper part of the Carboniferous System that harbors enormous volumes of coal; the lower part contains an unusually large percentage of limestone. Recognizing this distinction, American geologists, late in the nineteenth century, began to refer to the lower, limestone-rich Carboniferous interval as Mississippian because of its excellent exposure along the upper Mississippi River Valley and to the upper, coal-rich interval as Pennsylvanian because of its widespread occurrence in the state of Pennsylvania. Soon the Mississippian and Pennsylvanian were informally recognized as separate systems in North America, and in 1953 the United States Geological Survey officially granted them this status.

Although this difference between the upper and lower parts of the Carboniferous System is evident elsewhere in the world, most geologists in Europe continue to recognize just one system: the Carboniferous. Because this book covers world geology, we will follow the European practice, but occasionally it will be helpful to reiterate that the Lower and Upper Carboniferous systems are equivalent to the Mississippian and Pennsylvanian, respectively.

Roderick Murchison, who established the Silurian System and coestablished the Devonian, recognized and named the Permian System in 1841. Permian rocks were later identified in Britain and in other regions, but it was in Russia that Murchison founded the system. He named it after Perm, a town on the western flank of the Ural Mountains, where an expedition had taken him in 1840.

LIFE

Marine life of the late Paleozoic interval did not differ markedly from that of Late Devonian time except for the absence of several groups of marine organisms that died out in the Late Devonian mass extinction. As we shall see, the changes that took place on land were far more profound. Numerous insects of remarkably modern appearance evolved during this time, and many new kinds of spore-bearing trees colonized broad swamps, where their remains formed coal deposits. These plants later dwindled and were replaced by seed-bearing trees, which eventually dominated the land. In parallel fashion, amphibians, which were tied to water for reproduction—like spore-bearing plants—initially dominated terrestrial habitats but were later replaced by more fully terrestrial reptile groups. By the end of Permian time, the terrestrial reptiles displayed a variety of adaptations for feeding and locomotion, many of which resembled those of mammals.

Marine Life

Some groups of marine life never recovered from the mass extinction of Late Devonian time. Tabulates and stromatoporoids, for example, never again played a major ecologic role. The ammonoids, on the other hand, rediversified quickly and once again assumed an important ecologic position; indeed, ammonoid fossils are widely employed to date late Paleozoic rocks (Figure 12-1). Also persisting from Devonian time as mobile predators were diverse groups of sharks and ray-finned bony fishes. Gone shortly after the start of Carboniferous time, however, were the armored placoderms that had ruled Devonian seas. The absence of armored placoderms and of similar fishes after earliest Carboniferous time reflected a general trend: Although the late Paleozoic is not known for vast changes in the composition of marine life, heavily armored taxa of nektonic (swimming) animals tended to give way to more mobile forms. Apparently, as the Paleozoic Era progressed the ability to swim rapidly became a near necessity, probably because of the increasingly effective predators that inhabited the seas during this interval. After Devonian time, armored fishes never again dominated marine habitats, and heavy-shelled nautiloids also declined in number. In contrast to these heavy, awkward forms, the swimmers that thrived in late Paleozoic time—the ammonoids and, especially, the sharks and ray-finned fishes—were highly mobile.

We know little about the groups of algae that floated in late Paleozoic seas alongside fishes and ammonoids. Phytoplankton are not well represented in the late Paleozoic fossil record, although some groups must have prospered without leaving recognizable fossils. The record of the readily preserved acritarchs shows that this group persisted but never reexpanded after the great extinction that led to its decline near the end of the Devonian Period.

After the decline of the tabulate-strome community (p. 289), organic reefs remained poorly

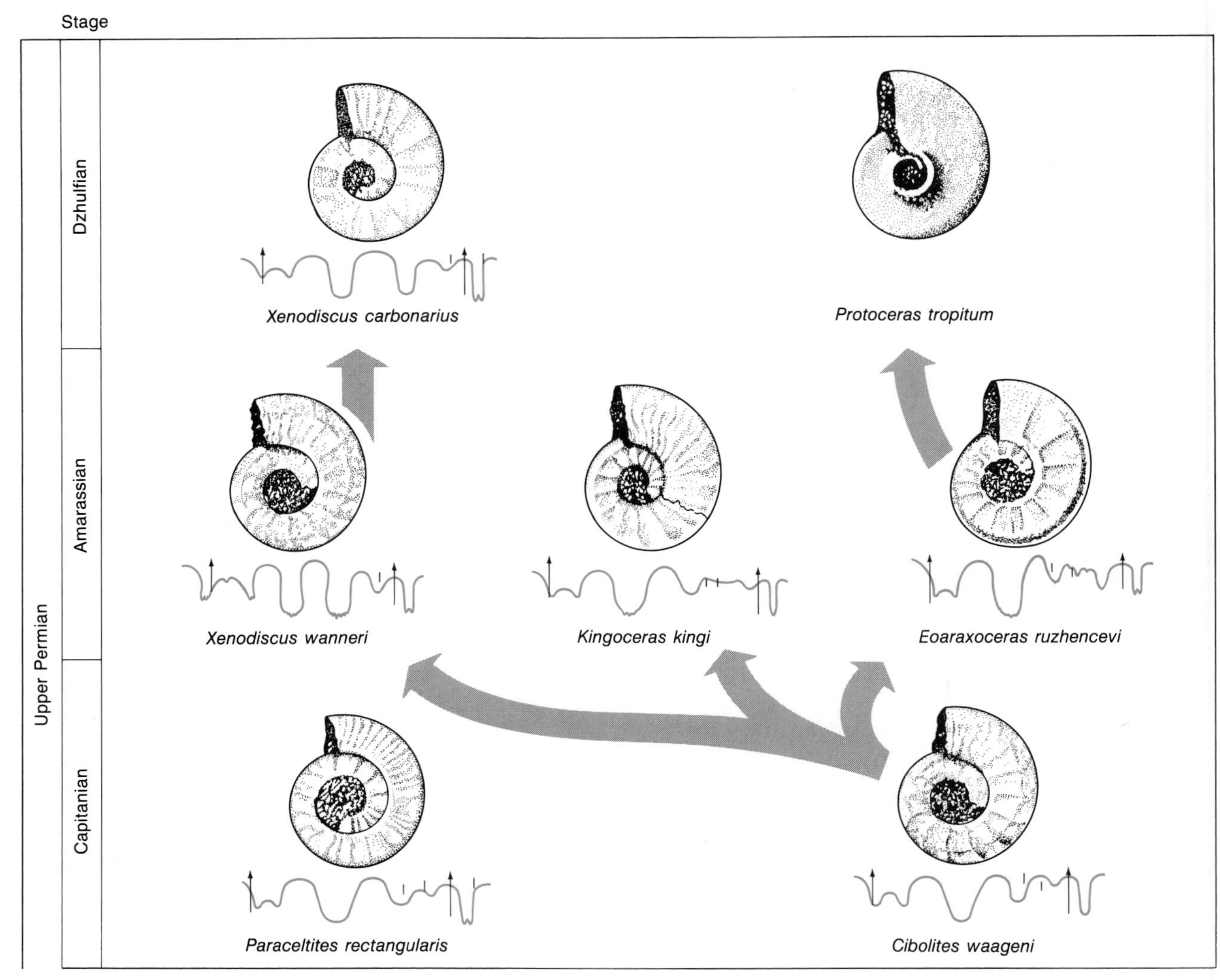

FIGURE 12-1 Species of the ammonoid family Xenodiscidae, arranged according to their evolutionary relationships. A few species of this type were virtually the only ammonoids that survived into the Triassic Period. *(After C. Spinosa et al., Jour. Paleont. 49:239–283, 1975.)*

developed throughout Paleozoic time because of the scarcity of effective frame-building organisms. Corals of the type that build modern reefs did not evolve until the Triassic Period. In the interim, for the most part only low banks and relatively small mounds were formed. Among the groups of animals and plants responsible for these structures were sponges, bryozoans, and calcareous algae. Certain groups of animals—bryozoans, crinoids, and foraminifera—also contributed vast amounts of skeletal debris to the formation of bedded limestones.

Brachiopods rebounded from the Late Devonian mass extinction to resume a prominent ecologic role. A group of spiny brachiopods known as productids enjoyed particular success. Some immobile productids employed their spines to anchor or support themselves in sediment, and a group of Permian productids developed cone-shaped shells that were attached by spines to the frameworks of solid reefs (Figure 12-2). Like the brachiopods, burrowing and surface-dwelling bivalves continued to thrive in late Paleozoic time, as did gastropods.

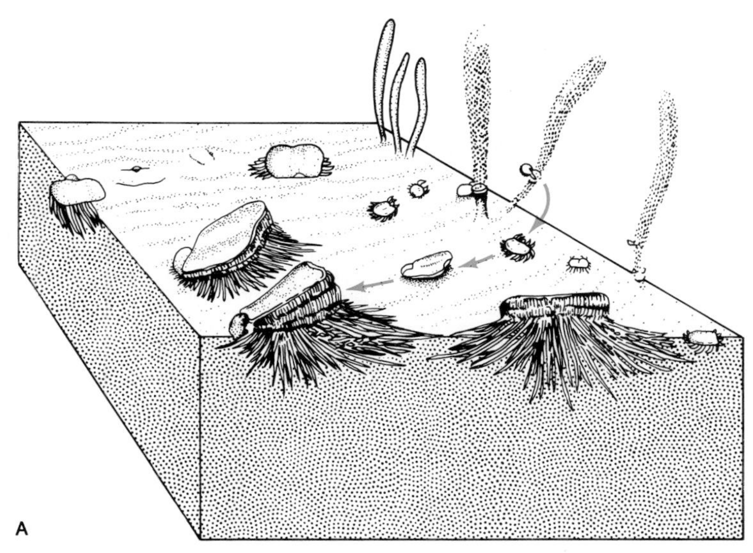

FIGURE 12-2 Modes of life of late Paleozoic spiny brachiopods of the productid group. *A.* Reconstruction showing changes in the life habits of a mud-dwelling species during its lifetime *(arrows)*. The juvenile brachiopods appear to have been attached to stalks of algae by curved spines. Then the algae died, and the small brachiopods came to rest on fine-grained sediment. As the brachiopods grew, their long spines served as "snowshoes," preventing the animals from sinking into the sediment. Thus the brachiopods could pump water in and out between the two halves of their shells in order to obtain food and oxygen without the danger of being clogged by mud. *B.* A group of Permian brachiopods of the genus *Prorichthofenia.* The lower halves of the shells of these coral-like animals were cone-shaped rather than cup-shaped, and throughout their lives their spines were attached to hard objects—in this case, the shells of neighboring brachiopods. The upper halves of the shells were flattened lids. *(A. After R. E. Grant, Jour. Paleont. 40:1063–1069, 1966. B. Smithsonian Institution.)*

FIGURE 12-3 Reconstruction of an Early Carboniferous (Mississippian) meadow of crinoids (sea lilies). Sharks, which were also well represented in Early Carboniferous seas, cruise above the crinoids. *(Chase Studios, Inc.)*

Crinoids—animals that were attached to the seafloor and captured floating food that came within reach of their waving arms as they floated by—expanded to their greatest diversity early in the Carboniferous Period, forming meadows in many areas of the seafloor (Figure 12-3). During this time, these organisms contributed vast quantities of carbonate debris to the rock record (Figure 12-4), leading to widespread limestone deposition during the Early Carboniferous (Mississippian) Period. Other animal groups also contributed to the formation of Carboniferous limestones. Prominent among them were lacy bryozoans (Figure 12-5) and the fusulinid foraminifera (Figure 12-6), whose late Paleozoic adaptive radiation made them the primary constituents of some limestones (Figure 12-7).

Lacy bryozoans were sheetlike, colonial animals that stood above the seafloor and fed on suspended organic matter. These organisms not only contributed skeletal debris to limestones but also trapped sediment to form reeflike structures. During Early Carboniferous time, lacy bryozoans

FIGURE 12-4 Early Carboniferous limestone composed largely of skeletal debris from crinoids. The particles that are shaped and stacked like poker chips are segments of crinoid stalks, most of which were flexible. The largest of the stalks shown here are about as thick as a pencil. *(Institute of Geological Sciences, British Crown copyright.)*

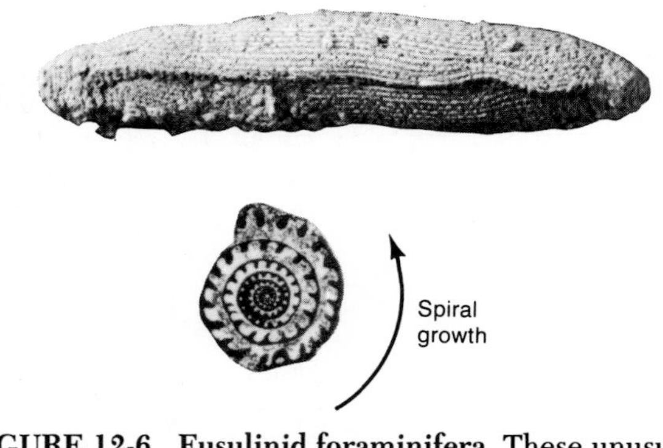

FIGURE 12-6 Fusulinid foraminifera. These unusually large, single-celled creatures secreted skeletons that were commonly shaped like grains of wheat, which many resembled in size. The lower picture is of a cross section, showing the spiral mode of growth. Fusulinids are highly useful for dating late Paleozoic rocks. *(External photograph. Smithsonian Institution. Cross section photograph. R. C. Douglass, U.S. Geological Survey.)*

grew locally in great profusion, trapping fine-grained sediment and growing upward through it to form mounds that usually stood well below sea level in quiet water (Figure 12-8). Around the flanks of the mounds, crinoids often grew in large meadows.

The **fusulinids,** a group of large foraminifera that lived on shallow seafloors, included only a few genera in Early Carboniferous time but underwent an enormous adaptive radiation during the Late Carboniferous and Permian. Some 5000 species have been found in Permian rocks alone. Although they were single-celled, amoebalike creatures with shells, fusulinids included species that exceeded 10 centimeters (~4 inches) in length. Owing to their abundance and rapid evolution, fusulinids are important guide fossils for Upper Carboniferous and Permian strata.

During Late Carboniferous time, certain types of calcareous algae, shaped somewhat like

FIGURE 12-5 Lacy, fanlike bryozoans of late Paleozoic age. In life, these colonies stood upright and were the size of a small fern. *(Smithsonian Institution.)*

FIGURE 12-7 Fusulinids forming the bulk of a specimen of Upper Carboniferous limestone. This species reached a length of about 1 centimeter (~0.4 inch). Compare this illustration with Figure 12-6. *(Chip Clark.)*

FIGURE 12-8 A Lower Carboniferous carbonate mound in New Mexico. The mound, which is about 100 meters (~328 feet) thick, grew on the seafloor, where lacy, fanlike bryozoans trapped carbonate mud. Bedded, coarse-grained limestones on the flanks of the mound were produced by meadows of crinoids. *(L. Pray.)*

large cornflakes, assumed a dominant role in trapping fine-grained sediment to form carbonate mounds on the seafloor. During the Permian Period, other kinds of algae joined sponges and lacy bryozoans to form reeflike banks. The most spectacular of these banks was a structure in west Texas that will be described later in this chapter. The large size of this bank was unusual for late Paleozoic time, when, as we noted earlier, there were few effective frame builders.

Plant Life on Land

Plants gave the Carboniferous Period its name, and in no other geologic interval are plant fossils more conspicuous; low-grade coal (Appendix I) from this period typically contains recognizable stems and leaves. Coal deposits developed chiefly in lowland swamps, where fallen tree trunks accumulated in large numbers. Because it takes several cubic meters of wood to make one cubic meter of coal, it is evident that the vast coal beds of Late Carboniferous age represent an enormous biomass of original plant material. Wetlands were far more extensive than they are today (Box 12-1).

Early Carboniferous floras, which formed little coal, resembled floras of Late Devonian time. They included a wide variety of plants, some of which were early representatives of the groups that became conspicuous in Late Carboniferous coal swamps. It appears that considerable evolutionary "experimentation" took place early in the Carboniferous Period, and what emerged as the highly successful late Paleozoic flora of the coal swamps and adjacent habitats was a small number of genera, each represented by large numbers of species. The most important coal-swamp genera were *Lepidodendron* and *Sigillaria,* two types of lycopod trees that contributed many of the logs that were buried and compressed to form coal (Figure 12-9). As we have seen, the lycopod group had been present during Early and Middle Devonian time, but only as small plants (Figure 11-18). Like smaller lycopods, *Lepidodendron* and *Sigillaria* were spore plants that were confined to swampy areas. *Lepidodendron* was the more successful genus; some of its species grew more than 30 meters (~100 feet) tall and measured 1 meter (~3 feet) across at the base.

At the feet of the treelike plants of the Upper Carboniferous was an undergrowth that consisted primarily of a wide variety of ferns and fernlike plants. Although some of them were spore plants like the modern ferns, many others were so-called seed ferns, which, as their name suggests, reproduced by means of seeds (Figure 12-10). Because seed ferns are difficult to distinguish from spore-bearing ferns on the basis of their foliage alone, they were not recognized as a separate group until 1904. Many seed ferns were small, bushy plants, but others were large and treelike.

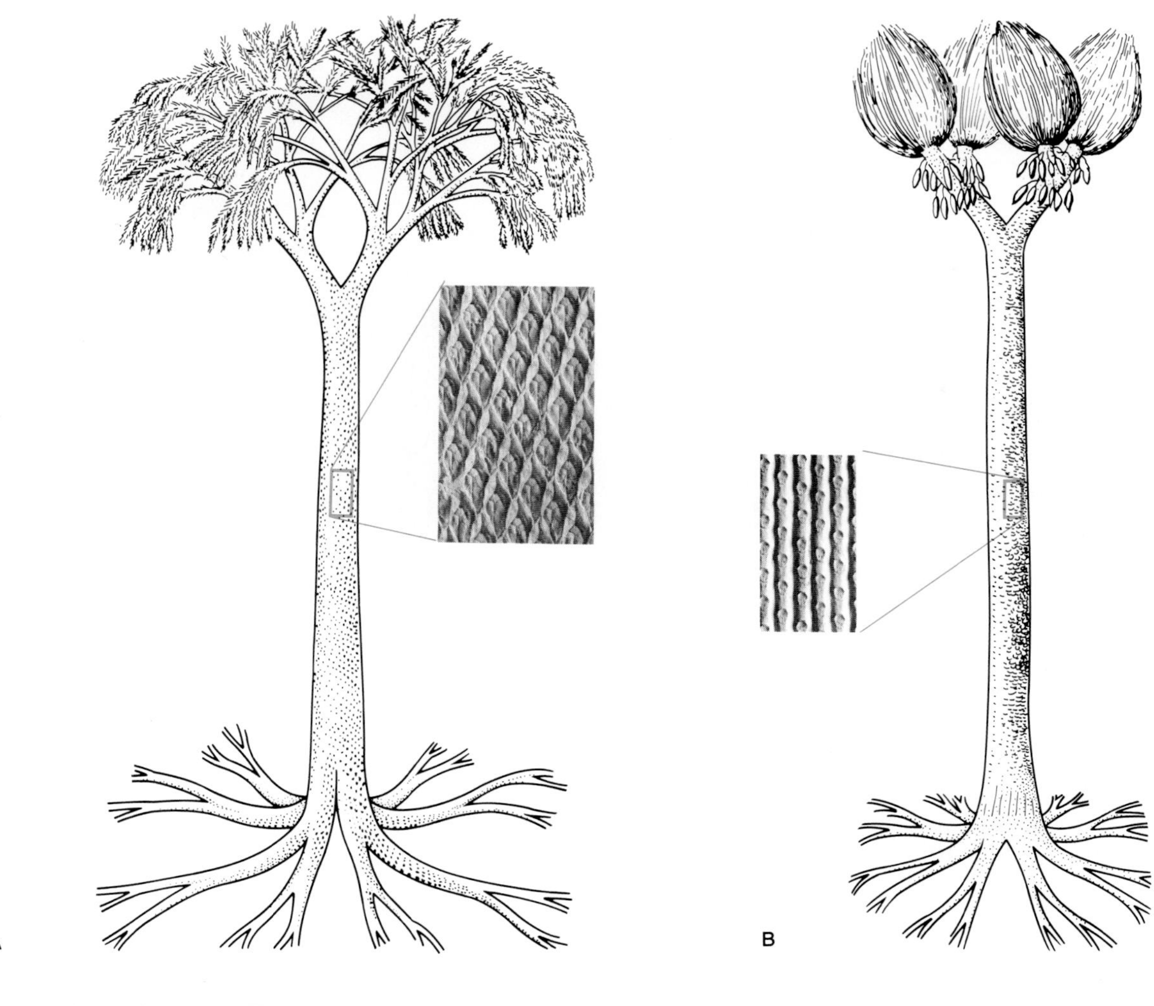

FIGURE 12-9 The dominant trees of Carboniferous coal swamps, belonging to the lycopod group. The genus *Lepidodendron (A)* has a trunk with a spiral pattern of leaf scars (positions where leaves were formerly attached). Note the similar spiral arrangement of branches in the early vascular plant *Baragwanathia* (Figure 11-17); this small, simple plant may have been an ancestor of the treelike lycopods. The trunk of *Sigillaria (B)* has vertical columns of leaf scars. The roots of lycopod trees are often preserved as fossils. Those shown here *(C)* are in Scotland. *(A and B. Photographs of leaf scars, Field Museum of Natural History, Chicago. C. Institute of Geological Sciences, British Crown copyright.)*

Glossopteris, the famous plant so abundant in Gondwanaland, was a seed fern. Although it had tonguelike leaves, familiar to most geologists (Figure 6-3), the entire plant had a treelike appearance, with its leaves arranged in clusters (Figure 12-11).

Not all Late Carboniferous vegetation occupied coal swamps. In fact, seed ferns and spore plants, called sphenopsids, were more abundant on higher ground. Fossils of these groups are more common in sands and muds that accumulated along levees and floodplains of rivers than in

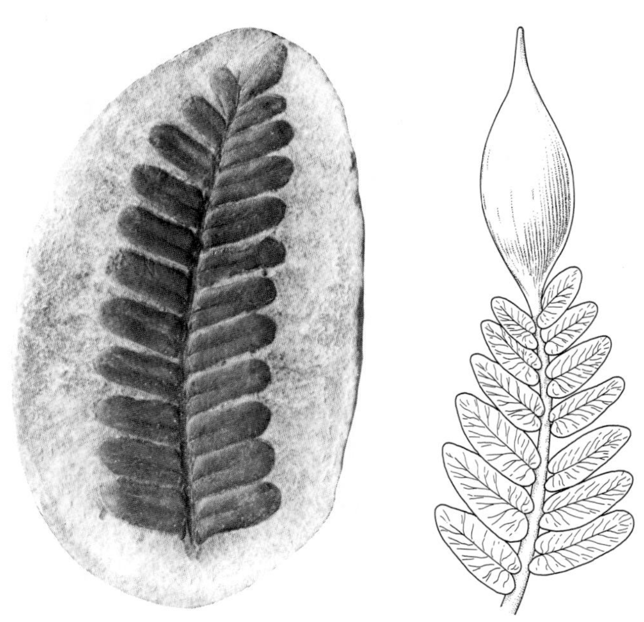

FIGURE 12-10 Late Carboniferous (Pennsylvanian) seed ferns of the genus *Neuropteris*. A fossil leaf from the Mazon Creek Formation of Illinois and a drawing showing the seed pod at the end of a branch. (See p. 298.) *(Field Museum of Natural History, Chicago.)*

coals that accumulated in permanently wet coal swamps. Late Carboniferous sphenopsids were similar in general form to horsetails, which are small sphenopsids of the modern world: They were characterized by branches that radiated from discrete nodes along the vertical stem (Figure 12-12). They also had horizontal underground stems that bore roots. Some Late Carboniferous sphenopsids, such as members of the genus *Calamites,* were tree-sized plants.

Another important group of Late Carboniferous plants that occupied high ground was the **cordaites,** a group of tall trees that often reached 30 meters (~100 feet) in height (Figure 12-13). As seed plants, cordaites were liberated from moist habitats and formed large woodlands that resembled modern pine forests. In fact, cordaites

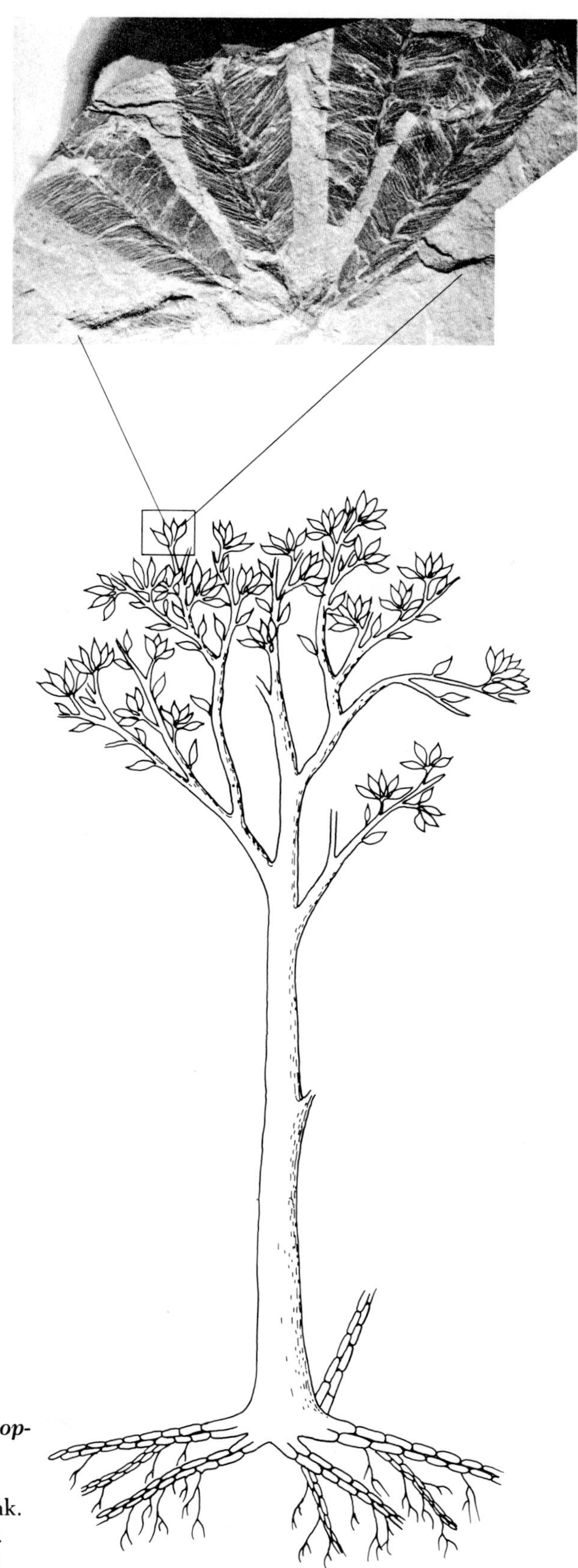

FIGURE 12-11 The famous Gondwanaland seed fern *Glossopteris*. The name means "tongue leaf," and the tongue-shaped leaves, which are sometimes found preserved in the clusters in which they grew, were positioned at the top of a large trunk. This is one of many treelike genera of seed ferns. *(After D. D. Pant and R. S. Singh, Palaeontographica 147[B]:42–73, 1974.)*

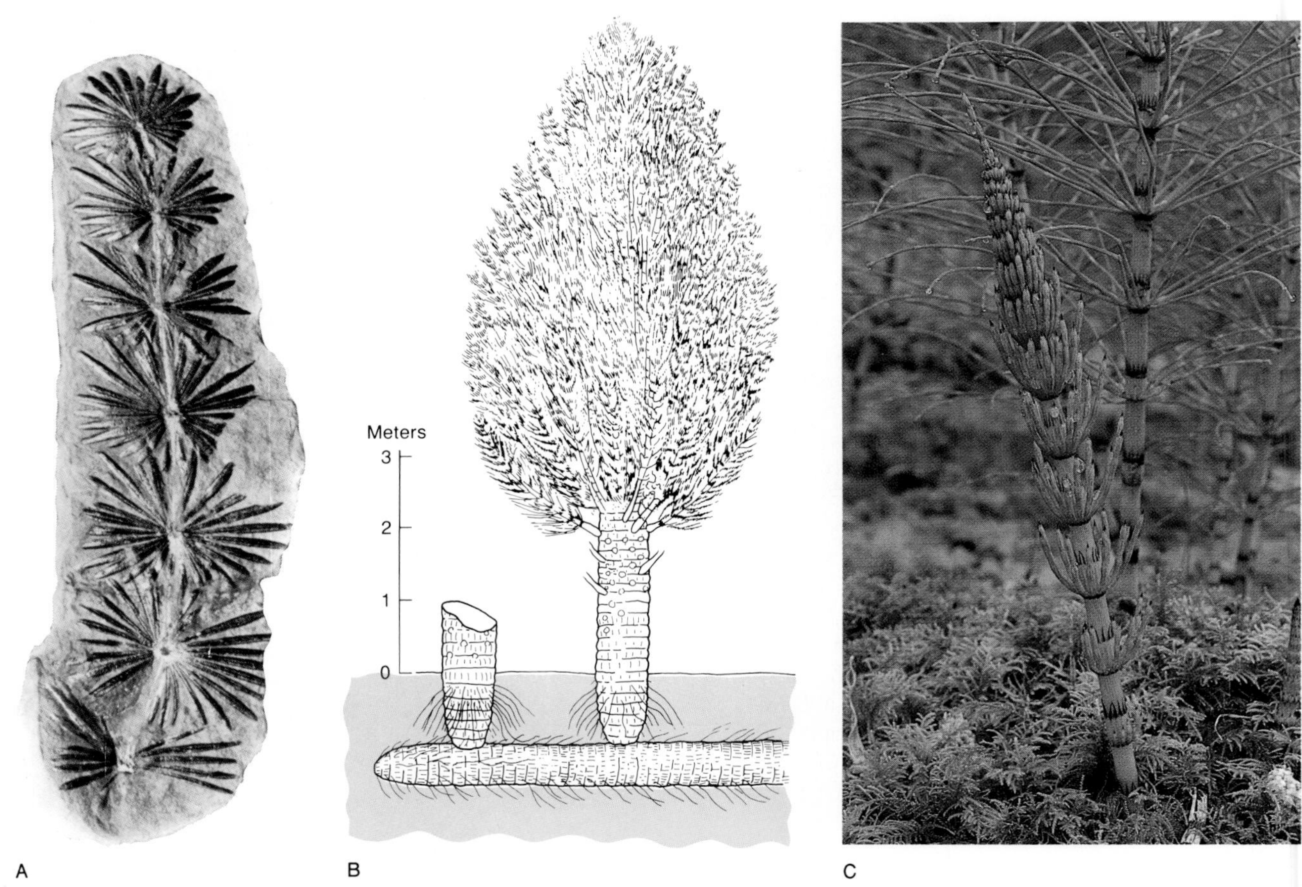

FIGURE 12-12 The Late Carboniferous sphenopsid plant *Calamites*. Branches such as the one preserved in the fossil *(A)* were positioned at intervals along the segmented tree trunk *(B)*, just as they are on the segmented stalk of living horsetails *(C)*, which grow to only about a meter (~3 feet) in height. *(A. Field Museum of Natural History, Chicago. C. Jeff Foott/Survival Anglia.)*

belonged to the group known as **gymnosperms** ("naked-seed plants"), which include the living **conifers,** or cone-bearing plants (pines, spruces, redwoods, and their relatives). The seeds of these plants are lodged in exposed positions on cones or on other reproductive organs. Gymnosperm seeds thus differ from the covered seeds of flowering plants, a group that did not emerge until the Mesozoic Era.

The floras that flourished in Late Carboniferous time continued to dominate into the Early Permian Period but subsequently declined. By the end of Permian time, few lycopods or sphenopsids the size of trees remained on Earth, while the cordaites, which were all treelike plants, disappeared altogether. It is interesting to note that nearly all of the lycopods and sphenopsids that survived the Paleozoic Era were small, inconspicuous creeping forms, some of which persist today as "living fossils" (Figures 12-12 and 11-18). During the Permian Period, gymnosperms, including conifers, took over terrestrial environments. Figure 12-14 shows that the foliage of one of these conifers, *Walchia*, resembled the needled branches of certain living conifers. In *Walchia*, as in other conifers, seeds were borne nakedly on cones. Gymnosperm floras, having expanded in Late Permian time, prevailed throughout Triassic, Jurassic, and Early Cretaceous time (Figure 12-15) and thus are often thought of as representing Mesozoic vegetation. Mesozoic vegetation, however, had a head start on other life of the new era. We will learn more about this kind of vegetation in Chapter 13, which describes the first portion of the Mesozoic Era — a time when not only gymnosperms but also dinosaurs came to dominate the land.

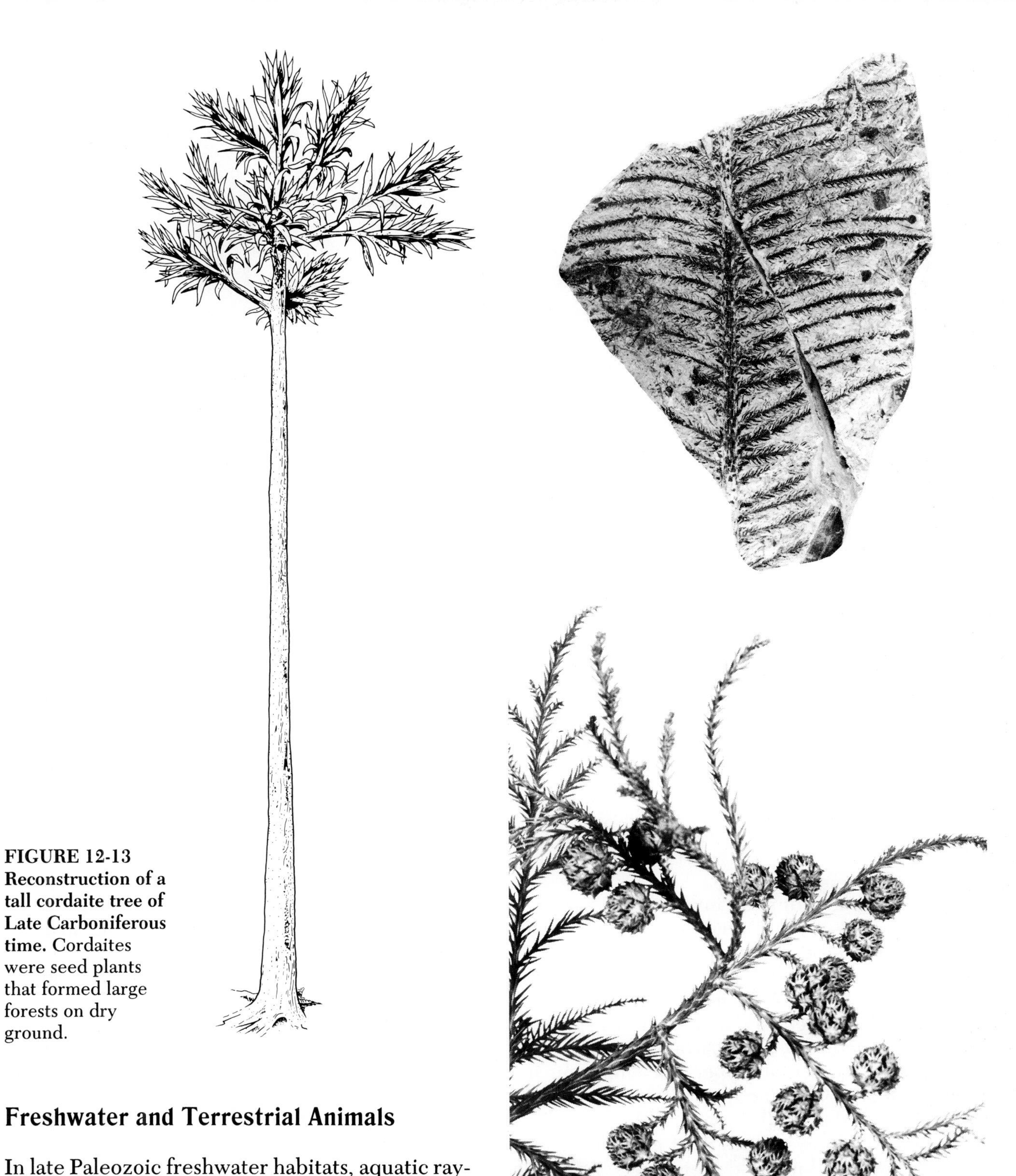

FIGURE 12-13 Reconstruction of a tall cordaite tree of Late Carboniferous time. Cordaites were seed plants that formed large forests on dry ground.

FIGURE 12-14 The early conifer *Walchia* of Late Carboniferous age *(top)* and a living species of conifer that is related to the redwoods *(bottom)*. Like living conifers, *Walchia* had needled branches and reproduced by means of cones. *(From E. B. Blazey, Palaeontographica 146[B]:1–20, 1974; photograph by J. E. Canright.)*

Freshwater and Terrestrial Animals

In late Paleozoic freshwater habitats, aquatic ray-finned fishes continued to diversify and were joined by freshwater sharks that have no close modern relatives. For the first time, mollusks also became conspicuous in freshwater environments; shells of many species of clams are found in freshwater and brackish sediments associated with coal deposits.

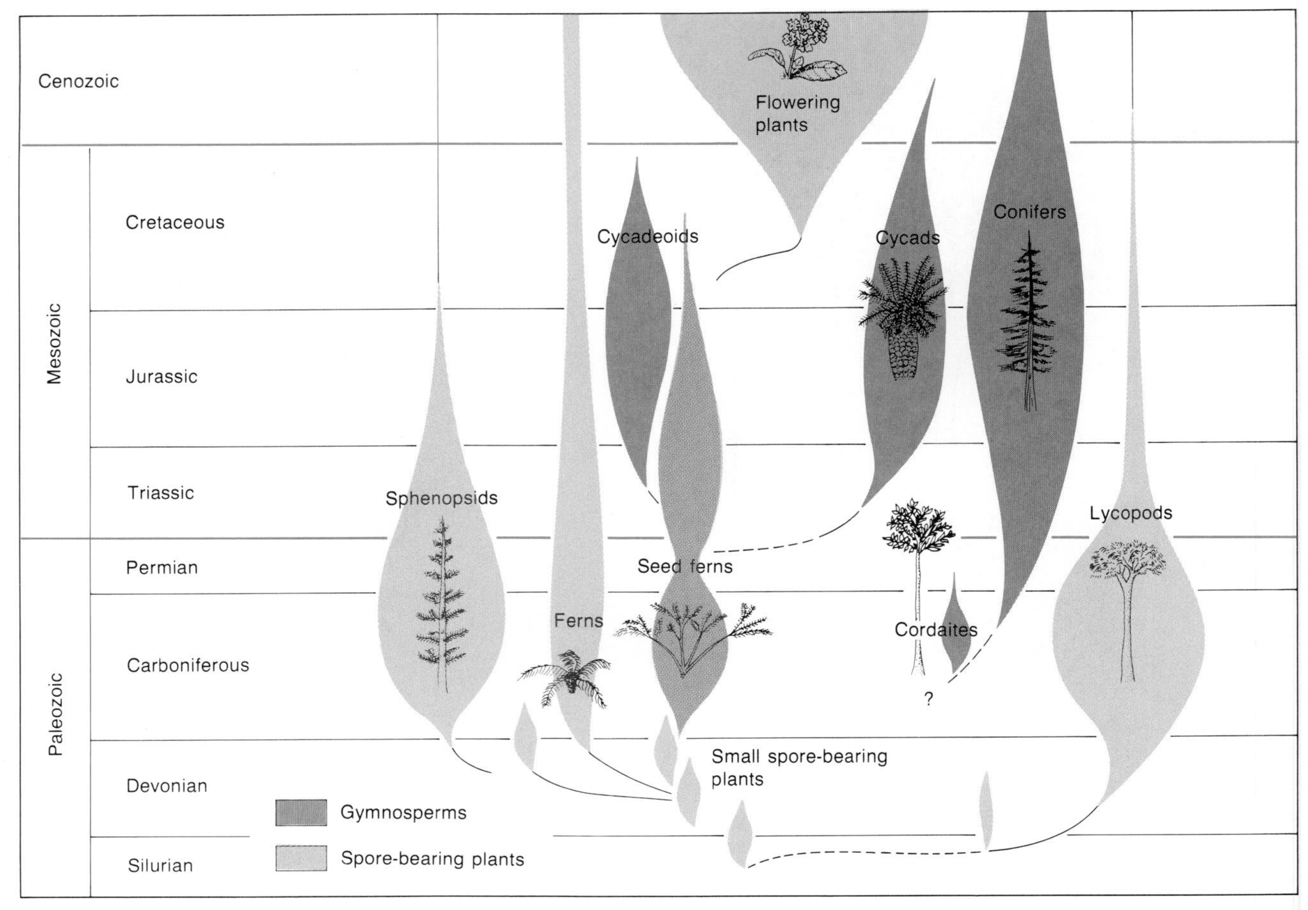

FIGURE 12-15 History of major groups of swamp- and land-dwelling plants. Spore-bearing plants dominated Silurian, Devonian, and Carboniferous floras. Seed ferns are the oldest known seed plants. Gymnosperms (naked seed plants) dominated Mesozoic floras, but the early conifers, which belonged to this group, diversified greatly during the Permian Period, while spore-bearing groups declined. *(Modified from A. H. Knoll and G. W. Rothwell, Paleobiology 7:7–35, 1981.)*

On land, a group of invertebrate animals, the insects, assumed a very important ecologic role — one that they have never relinquished. The oldest known insects are of Early Devonian age, but they were wingless forms. Although no Lower Carboniferous insect fossils are known, insects had evolved wings by Late Carboniferous time; in fact, a wide variety of Upper Carboniferous insects have been identified, primarily by means of their preserved wings (Figure 12-16). The earliest flying insects differed from most modern species in that they could not fold their wings back over their bodies; the only two living orders of insects that lack this ability are the dragonflies and the mayflies, both of which occur in Upper Carboniferous deposits along with several other orders. Giant dragonflies — animals with wingspans of nearly half a meter (~18 inches) — found in the Carboniferous of France have given rise to the false impression that Carboniferous landscapes were populated with many kinds of huge insects. In fact, only one giant Carboniferous species is known. The rest were of normal size by modern standards.

The fact that advanced insects with foldable wings are found in younger Upper Carboniferous deposits indicates that the insects underwent an extensive radiation before the beginning of the Permian Period. Like many modern insect species, some of these Carboniferous forms had eggs that hatched into caterpillar-like larvae, while others possessed specialized egg-laying organs or mouthparts that were adapted to sucking juices from plants. The legs of still other species were highly modified for grasping prey, leaping, or running. Indeed, many of these insects appear to have been as highly adapted for particular modes of life as are insects of the modern world.

FIGURE 12-16 Wings of a Late Carboniferous insect from the Mazon Creek Formation of Illinois. The color pattern is beautifully preserved in this remarkable specimen. *(F. M. Carpenter, Proc. N. Amer. Paleont. Conv. [I]:1236–1251, 1969.)*

FIGURE 12-17 *Eryops*, a large amphibian from the Lower Permian Series of Texas. This animal, which reached a length of about 2 meters (~6 feet), was probably semiaquatic, occupying the margins of rivers and lakes and preying on fishes. *(Field Museum of Natural History, Chicago.)*

It is sometimes difficult to distinguish between aquatic and terrestrial vertebrates of late Paleozoic time because many four-legged animals lived along the shores of lakes, rivers, shallow seas, or swamps and divided their time between the land and the water. The general history of terrestrial and semiaquatic vertebrate animals of late Paleozoic time and the history of plants, insects, and marine life are summarized in Major Events (p. 300).

The only vertebrates populating the Early Carboniferous landscape were amphibians, many of which retained aquatic or semiaquatic habits throughout their lives. Carboniferous amphibians, however, did not closely resemble their modern relatives. The frogs, toads, and salamanders that comprise most living species of Amphibia are small, inconspicuous creatures; they seem to be the only kind of animals belonging to this class that can thrive in the modern world in the face of competition and predation by advanced mammals and reptiles. Carboniferous and Early Permian amphibians, in contrast, had the world largely to themselves and thus displayed a much broader spectrum of shapes, sizes, and modes of life; some superficially resembled alligators (Figure 12-17), others were small and snakelike, and a few were lumbering plant eaters. Some Carboniferous amphibians measured 6 meters (~20 feet) from the ends of their snouts to the tips of their tails. The species that were fully terrestrial as adults were covered by protective scales.

The Rise of the Reptiles The oldest known **reptiles** are found in deposits near the base of the Upper Carboniferous (Pennsylvanian) System. Most of the skeletal differences between the earliest reptiles and their amphibian ancestors were minor, relating to such features as the roof of the mouth, the back of the skull, the inner ear, and the vertebrae.

Reptiles also differ from amphibians in their mode of reproduction. The key feature in the origin of the reptiles was the **amniote egg,** which is also employed by modern reptiles and birds. This egg provides the embryo with a nutritious yolk and two sacs: one (the amnion) to contain the embryo and the other to collect waste products. A durable outer shell protects the developing embryo. The amniote egg allowed vertebrates for the first time to live and reproduce away from bodies of water. The oldest specimen suspected of being a reptile egg is of Early Permian age (Figure

Box 12-1
Wetlands, Then and Now

(Bates Littlehales.)

Coal swamps of the Carboniferous Period were among the most extensive wetlands of all time. Wetlands are continental areas that are normally moist at least part of the year and that support luxuriant growths of vegetation. Some have standing water throughout the year; the soil of others is constantly moist; and still others are flooded only about once a year. Wetlands were widespread during Carboniferous time because the seas stood high enough to spread across large areas of cratons. Along the margins of these epicontinental seas were broad, nearly flat lowlands where brackish and fresh water tended to accumulate and coal swamp floras flourished. Today, because continents stand higher above sea level, wetlands form only narrow fringes along the coasts and occupy small areas of the interiors. Altogether, wetlands cover only 6 percent of the surfaces of modern continents, and they are shrinking rapidly under the impact of human activities. When the United States was founded in 1776, its wetlands spread across 215 million acres, but only about 100 million of these acres remain.

Most of the wetlands of the modern world are informally termed marshes or swamps. Tall, grass-like plants that can tolerate brackish conditions form coastal marshes in relatively cool climates. In frostfree areas these plants give way to mangroves, which are shrublike trees that are able to grow in standing salt water. Plants such as grasses and cattails also grow in freshwater marshes, many of which have open water in the center or stretch along rivers. In contrast to marshes, swamps support abundant trees and shrubs, such as willows and cypresses, that tolerate moist soil or even permanent standing water.

Among the most interesting freshwater wetlands are peat bogs, where low-growing plants, including mosses, produce an environment that is so acidic and so depleted of oxygen that it is hostile to the kinds of microorganisms that would ordinarily decompose dead plant material. In the absence of these organisms, the plant material accumulates as peat. Peat bogs are concentrated in moist regions at high latitudes—the northern Great Lakes area and New England in North America, for example, as well as large areas of Canada and Alaska. Peat is widely used as fuel. Peat harvesters in eastern Europe have recovered about 2000 almost perfectly preserved human bodies from ancient bogs. The absence of oxygen in the soggy peat that floors bogs preserves the flesh of any animal that sinks into it: no bacteria, no decay. One of the oldest bodies found, that of a man who lived in Denmark about the time of Christ, was found with a rope around the neck.

Wetlands perform such important functions in the global ecosystem that their recent decline is alarming. They replenish groundwater, for example, by trapping fresh water and allowing it to percolate into the soil instead of flowing quickly to the sea. The plants that flourish in wetlands trap sediment and organic matter rich in nutrients. Many of these nutrients would otherwise be lost to the sea, or they would pollute rivers, lakes, and lagoons by promoting the growth of cyanobacteria. When scummy masses of cyanobacteria die, their decay robs water of oxygen and excludes other forms of life. The abundant vegetation of wetlands releases a large quantity of oxygen to the atmosphere every year. It also yields organic detritus that supports dense populations of animals. About 90 percent of the commercial fish species of Florida, for example, depend on the mangrove ecosystem for food or shelter.

Humans are currently destroying about one-third of a million acres of wetlands every year. The Everglades of Florida, the largest and most famous freshwater marsh of North America, are now less than half their original size. The Everglades are known as "the River of Grass" because their broad waters actually flow southward very slowly from Lake Okeechobee to the sea. Artificial dikes and levees now restrict the flow of water from the lake, and some of it is diverted for human use. As a result of human interference, many parts of the Everglades have dried up or contain so little water during the dry season that their biotas are shrinking. About 60,000 wading birds occupied this great marsh in 1940. Today there are only about one-quarter as many.

Prairie potholes—small basins scoured by glaciers—contain valuable inland marshes that are also disappearing at a rate that threatens many forms of life. Prairie potholes are scattered over a large area of the northern Great Plains, where they support a great variety of animals and serve as breeding grounds for half of the ducks of North America. Farmers have drained or filled in many of these small marshes, vastly reducing the total area available to waterfowl and other animals.

Unfortunately, humans have tended to regard wetlands as nuisances—places where alligators lurk or mosquitoes breed. These soggy areas have seemed so unattractive that "swamp" has become a metaphor for a place not to be. Swamps, it has seemed, are best drained or filled in to produce solid land. Only recently have humans begun to value wetlands even though they are not the sort of place where most of us would want to live. And today our valuable bogs, swamps, and marshes look fragile indeed when we compare their shrunken remains with the vast coal swamps that cloaked lowlands during Carboniferous time.

12-18), but it is generally assumed that the amniote egg originated in Carboniferous time, when reptiles evolved.

Because the amniote egg was in essence a self-contained pond, it eliminated the need for early life in water and thus enabled reptiles to exploit the land more fully. There is an interesting parallel here with the evolution of the seed in plants. As we have seen, spore plants, like amphibians, require environmental moisture during part of their life cycle. The origin of the more advanced groups—the seed plants and reptiles—represented a transition to a fully terrestrial existence.

Later reptiles developed yet another feature of great importance: an advanced jaw structure that could apply heavy pressure upon closing and could slice food by means of bladelike teeth. Carboniferous amphibians and early reptiles could close their jaws quickly with a snap, but they could apply little pressure. Moreover, they had pointed teeth that could kill prey by puncturing them but that could not slice or tear food apart; therefore, these animals were forced to swallow their meals whole.

Despite the origin of reptiles in Late Carboniferous time, amphibians continued to prosper into Early Permian time. During the Permian Period, however, reptiles diversified and apparently began to replace amphibians in various ecologic

FIGURE 12-18 A specimen alleged to be the world's oldest known fossil egg, from the Lower Permian Series of Texas. It is not known exactly what kind of reptile might have laid this egg, and some experts question its authenticity. *(A. S. Romer.)*

FIGURE 12-19 Skeleton of *Dimetrodon*, a fin-backed mammal-like reptile of the pelycosaur group from the Lower Permian Series of Texas. The fin, which was supported by long vertebral spines, served an uncertain function. Some workers believe that skin stretched between the spines was used to catch the sun's rays, allowing the animal to raise its temperature to a level above that of its surroundings. From snout to tail, this predator exceeded 2 meters (~6 feet) in length. *(Field Museum of Natural History, Chicago.)*

roles, probably because the reptiles had more advanced jaws and teeth as well as greater speed and agility. Permian rocks of Texas have yielded large faunas of amphibians and reptiles that reveal this pattern. By Early Permian time, the **pelycosaurs, finback reptiles** and their relatives, had become the top carnivores of widespread ecosystems. Their stratigraphic occurrence suggests that many lived in swamps and that some may have been semiaquatic. *Dimetrodon* (Figure 12-19) was one such carnivore. It was about the size of a jaguar and had sharp, serrated teeth. While even the Permian carnivorous amphibians, such as the alligator-like *Eryops*, were forced to swallow small prey whole, *Dimetrodon* could tear large animals to pieces (Figure 12-20).

Dimetrodon belonged to a group known as the **mammal-like reptiles.** The skull structure of these animals in some ways resembled that of mammals, which evolved from them. One particular group of mammal-like reptiles that evolved in mid-Permian time was especially similar to mammals. These were the **therapsids** (Figure 12-21), whose legs were positioned more vertically beneath the body than were the sprawling legs of primitive reptiles or even pelycosaurs. In addition, the jaws of therapsids were complex and powerful, and the teeth of many species were differentiated, somewhat like those of a dog, into frontal incisors for nipping, large lateral fangs for puncturing and tearing, and molars for shearing and chopping food.

A New Level of Metabolism Many experts believe that the therapsids were **endothermic,** or warm-blooded—which means that by virtue of a high metabolic rate, they maintained their body temperatures at relatively constant levels that usually exceeded the temperature of their surroundings. Hair similar to that of modern mammals may have insulated therapsids' bodies (Figure 12-21). Even if they were endothermic, however, therapsids may not have kept their body temperatures at levels as constant as those of mammals. In any case, the upright postures and complex chewing apparatuses of advanced Permian therapsids show that these active animals approached the mammalian level of evolution in anatomy and behavior.

The endothermic condition is especially significant in that it allows animals to maintain a sustained level of activity—to hunt prey or to flee from predators with considerable endurance. **Ectothermic** (or cold-blooded) reptiles, in contrast, must rest frequently in order to soak up heat from their environments. Endothermic metabolism, along with advanced jaws, teeth, and limbs, may account not only for the success of the therapsids during Permian time but also for the decline of the pelycosaurs, which were probably ectothermic. In fact, while pelycosaurs declined to extinction during Late Permian time, therapsids underwent a spectacular adaptive radiation. More than 20 families of these advanced animals seem to have evolved in just 5 or 10 million years, and they were the dominant groups of large animals in Late Permian terrestrial habitats. Therapsids seem to have represented an entirely new kind of animal—one so advanced that it was able to diversify very quickly. While it is tempting to assume that the therapsids caused the pelycosaurs to decline throughout the world, it is only in the Karroo beds of South Africa that the therapsid record is sufficiently complete for a large enough geologic interval to suggest such a pattern. The sizable Permian therapsid fauna of central Russia, however, includes families that have not been found in South Africa, suggesting that a great variety of therapsids populated the Late Permian world. Unfortunately, elsewhere in the world the Permian fossil record of therapsids is rather poor.

FIGURE 12-20 Early Permian scene beside a body of water. *Dimetrodon* (see Figure 12-19) threatens *Eryops* (Figure 12-17). Early insects are in the foreground, and the small reptile *Araeoscelis* is climbing a tree of the genus *Cordaites*. The vine is *Gigantopteris* and the small plants are *Lobatannularia*. *(Drawing by Gregory S. Paul.)*

PALEOGEOGRAPHY

During late Paleozoic time the major continents moved closer and closer together until by early Mesozoic time they seem to have fused together as the supercontinent Pangaea. Even early in the Carboniferous Period, however, the continents were rather tightly clustered (Figure 12-22A). As

FIGURE 12-21 Late Permian scene in the South African part of Gondwanaland. Therapsids are shown along an ice-covered stream in a snowy environment, where they may have been able to live by virtue of being endothermic. In this reconstruction, they are shown to have hair, which is associated with endothermy in mammals. The largest animal is *Jonkeria;* in the background is *Trochosaurus;* animals at lower right are *Dicynodon;* and the very small form is *Blattoidealestes. (Drawing by Gregory S. Paul.)*

FIGURE 12-22 World geography in Carboniferous time. *A.* The major continents were rather tightly clustered on one side of the globe in Early Carboniferous (Mississippian) time. The Old Red Sandstone continent (ORS) persisted from Devonian time north of the Tethys Sea. At low latitudes, coal deposits were formed along the eastern portions of continents, and limestones and evaporites accumulated in many areas, especially in the Midcontinent Province to the west of the Old Red Sandstone continent. Enormous glaciers spread over Gondwanaland near the South Pole. *B.* In Late Carboniferous time the Tethys Sea was closed in association with Hercynian mountain building. Coal deposits formed over a larger total area at this time than at any other period in Earth history; some formed quite far north. In Gondwanaland, continental glaciers spread to remarkably low latitudes and were separated from tropical coal swamps (formed by the Euramerian flora) by steep temperature gradients. The Gondwana and Siberian floras flourished under cooler conditions. *(Modified from R. K. Bambach et al., American Scientist 68:26–38, 1980.)*

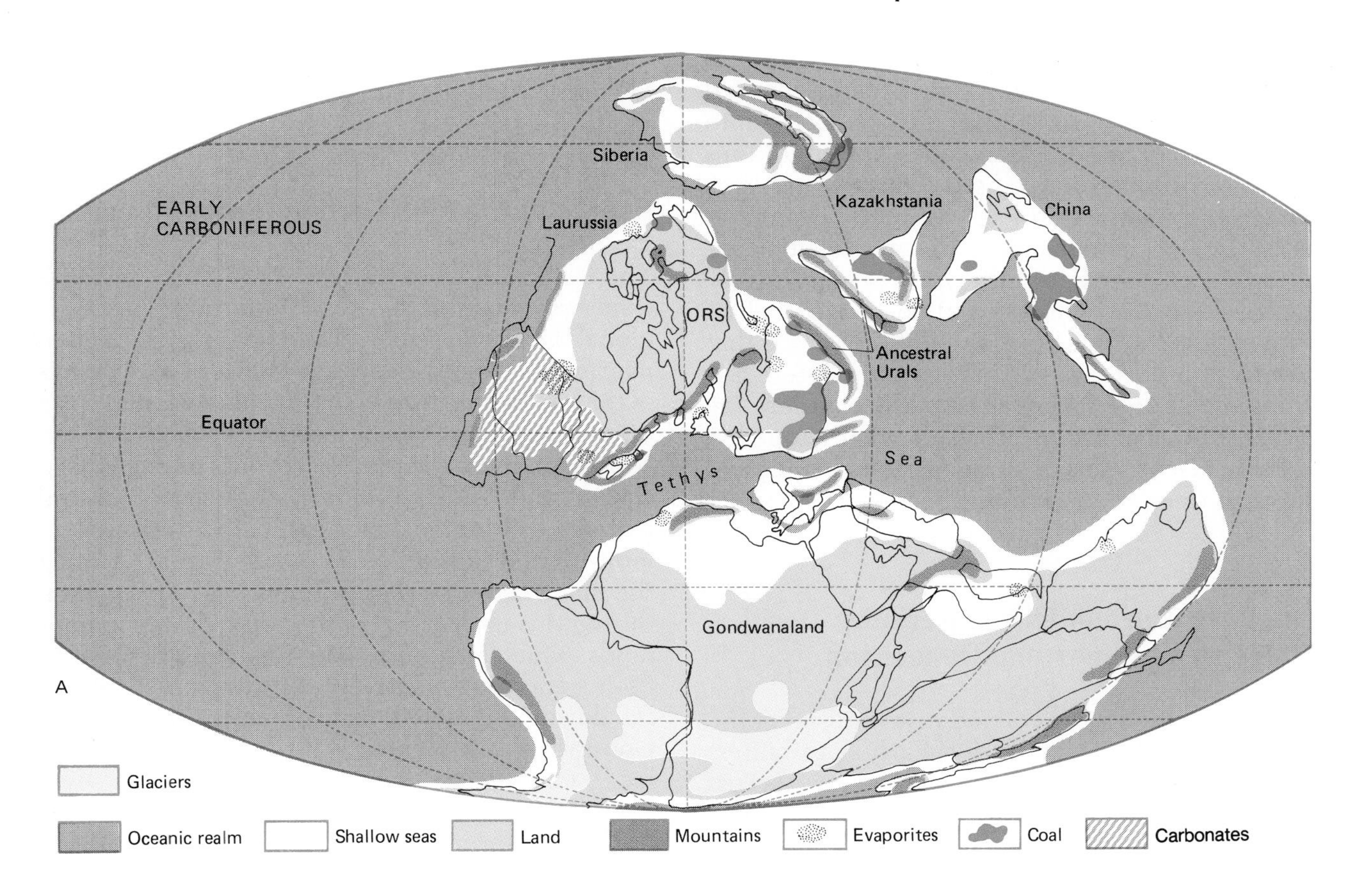
Siberia
EARLY CARBONIFEROUS
Laurussia
Kazakhstania
China
ORS
Ancestral Urals
Equator
Tethys
Sea
Gondwanaland
A
Glaciers
Oceanic realm
Shallow seas
Land
Mountains
Evaporites
Coal
Carbonates

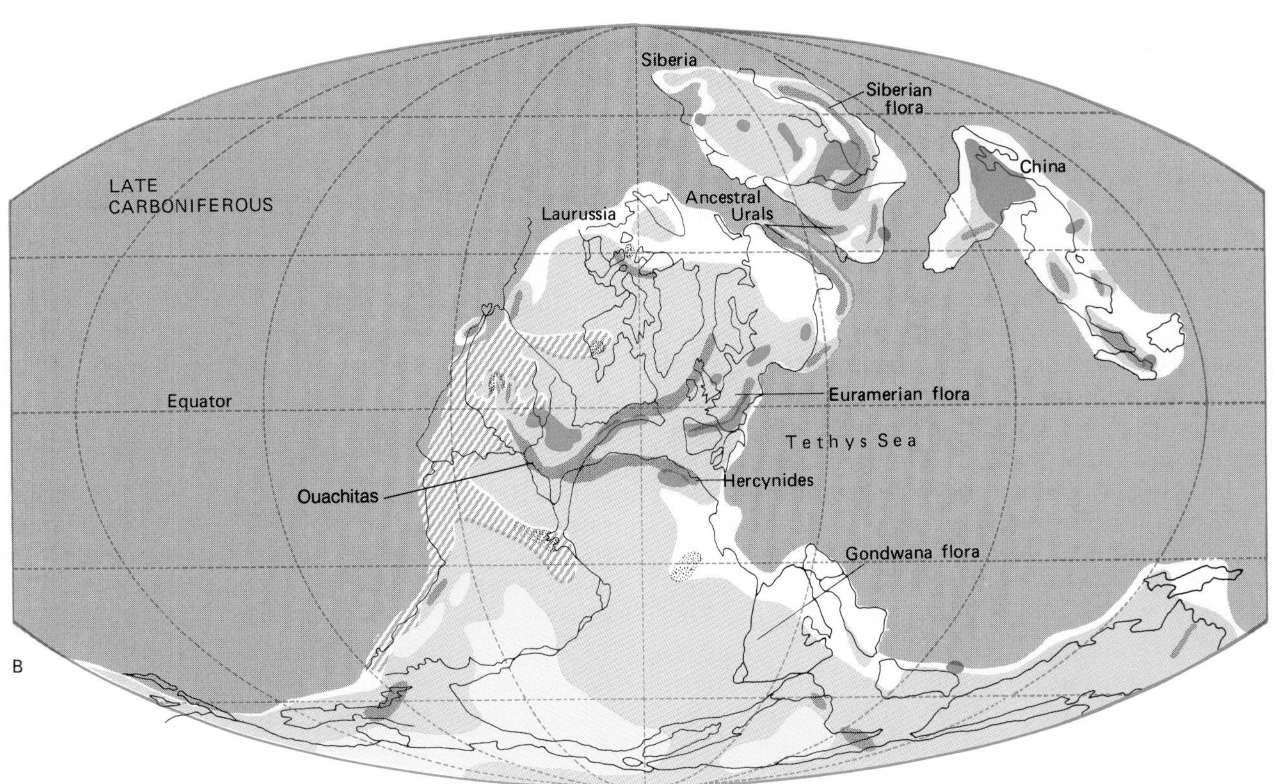
Siberia
Siberian flora
China
LATE CARBONIFEROUS
Laurussia
Ancestral Urals
Equator
Euramerian flora
Tethys Sea
Ouachitas
Hercynides
Gondwana flora
B

the period progressed, the sector of Gondwanaland that lay over the South Pole became covered by a large continental glacier that persisted into the Permian Period. Meanwhile, hot conditions prevailed in equatorial regions. Thus, not only was the late Paleozoic a time of vast biotic change; it was a time of changing and extreme climatic conditions. Coal deposits formed most extensively during Late Carboniferous time, accumulating both at low latitudes (for example, in areas that now form central North America and western Europe) and at high latitudes near the southern ice sheets. Let us consider precisely what happened in Early Carboniferous, Late Carboniferous, and Permian time.

The Early Carboniferous Period: Widespread Limestone Deposition

Sea level, which had declined near the end of the Devonian Period, rose during Early Carboniferous time, so that warm, shallow seas spread across broad continental surfaces at low latitudes. As a result, limestones accumulated over large areas, often with crinoid debris as their most important component.

Early in Carboniferous time, limestones accumulated in many areas of Gondwanaland, but nearer the South Pole temperatures were strikingly cooler. Tillites reveal that throughout all or nearly all of Carboniferous time, large areas of Gondwanaland were blanketed by ice sheets. The earliest of the ice sheets may have been the same ones that formed in Late Devonian time (p. 290), or they may have developed after the Devonian glaciers melted. Uncertainty remains because preserved tillites provide an incomplete record of glaciation; also, many tillites are poorly dated.

Warm, moist conditions prevailed in some continental areas nearer the equator. Thus, coal-swamp floras which first became established early in Carboniferous time, flourished along the northeastern margin of the Old Red Sandstone continent, which survived from Devonian time (Figure 12-22*A*). Trade winds must have continued to bring this continent moisture from the oceans to the northeast, but the western part of the continent was in the rain shadow of the Caledonides and the Appalachian Mountains. Here, across what is now central and western North America, evaporites and limestones accumulated in broad, shallow seas.

Events at the Mid-Carboniferous Boundary

The transition from Early to Late Carboniferous time was marked by two important events: a global decline in sea level and heavy extinction of marine life (Major Events, p. 300). In many parts of the world, the drop in sea level is evidenced by a disconformity in shallow marine deposits. Thus, in North America, the marine records of the Mississippian and Pennsylvanian systems are separated by a disconformity that is estimated to represent more than 4 million years in some areas. Among the marine groups that suffered heavy extinction during this interval were the crinoids and ammonoids, which lost more than 40 and 80 percent of their genera, respectively. Presumably sea level fell during this crisis because glaciers expanded in Gondwanaland, locking water up on the land. It has been suggested that the drop in sea level caused the extinction of life in shallow seas, but, as we have seen, comparable episodes of sea-level lowering at other times had little effect on marine life. Perhaps, then, the mid-Carboniferous extinctions resulted from cooling of the seas that was associated with expansion of glaciers.

The Later Carboniferous Period: Continental Collision and Temperature Contrasts

During the middle portion of Carboniferous time, the northward movement of Gondwanaland caused that continent to collide with the Old Red Sandstone continent. The mountains thus formed in southern Europe are known collectively as the Hercynides, and the orogeny as a whole is known as the Hercynian (or Variscan). Hercynian mountains also formed in northwestern Africa, where they became known as the Mauritanides. In North America the Hercynian orogeny, known here as the Alleghenian, in effect continued where the Caledonian orogeny left off, extending the Appalachian mountain chain southwestward and forming the adjacent Ouachita Belt in Oklahoma and Texas (Figure 12-22*B*).

On the land, latitudinal temperature gradients steepened during Late Carboniferous time—that is to say, there were extreme differences in temperature between the equator and the poles. Continental glaciers pushed northward to within nearly 30° of the ancient equator, a latitude where subtropical conditions have prevailed during most of Phanerozoic time. It seems amazing that tropical coal swamps flourished in North America and western Europe not much farther north than the northernmost Carboniferous glaciers (Figure 12-22*B*).

Recall that coal deposits formed in frigid Gondwanaland as well. Nonetheless, the *Glossopteris* flora that produced the coal deposits in Gondwanaland differed substantially from the so-called Euramerian flora of the equatorial region, which was named for Europe and North America. *Lepidodendron* and *Sigillaria*, the dominant Euramerian elements, were present in Gondwanaland, but many of the plants of Gondwanaland (Figure 6-3) are unknown from northern continents. The *Glossopteris* flora was adapted to the cool climates of the glacial regime in the south. Siberia, which lay near Earth's other pole, also had a distinctive flora adapted to cold conditions.

There is compelling evidence that the fossil floras of Gondwanaland and Siberia grew under cold conditions. Many cold climates are strongly seasonal, and seasonal growth of wood produces distinctive rings in the cross sections of tree trunks. The Upper Carboniferous floras of Gondwanaland in the south and of Siberia in the north are known for their distinctive tree rings (Figure 12-23). In contrast, the Euramerian fossil trees that grew near the Carboniferous equator were of the tropical type: They lacked seasonal rings.

The composition of fossil floras in North America and Europe reveals that tropical climates changed significantly in the course of Late Carboniferous time. About two-thirds of the way through Late Carboniferous time, lycopods, sphenopsids, and seed ferns declined in terrestrial plant communities, while spore-bearing ferns assumed a correspondingly larger role. This change appears to have signaled a shift to drier climatic conditions. Coal continued to form from decaying vegetation in swamps, but lycopods were no longer the primary contributors. This change foreshadowed even greater climatic changes during Permian time.

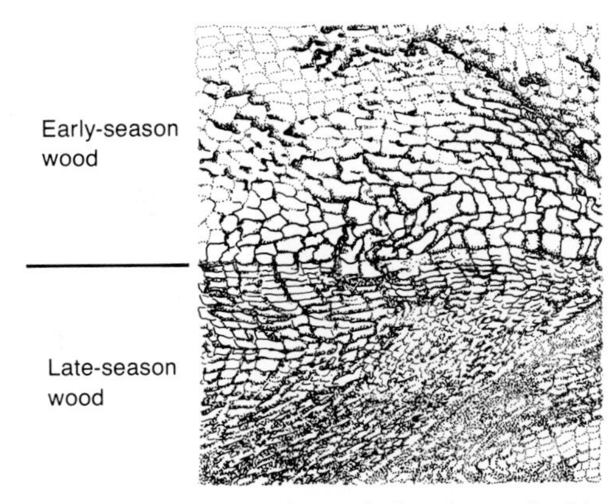

FIGURE 12-23 Section through fossil wood of late Paleozoic or Triassic age from Antarctica, showing the details of the boundary between two growth rings. Each space in the woody tissue was occupied by a single cell. When the late-season wood of the earlier (lower) layer formed, the cells were small and the tree grew slowly. Growth was interrupted during winter, but spring then stimulated the growth of early-season wood, resulting in large cells and rapid growth. The number of growth rings indicates the age of a tree, but rings are not well developed in tropical climates. In late Paleozoic time, high-latitude climates were strongly seasonal, producing growth rings in Siberia near the North Pole and in such regions as Antarctica near the South Pole. *(After a photograph by J. M. Schopf.)*

The Permian Period: Climatic Complexity

As a result of complex topographic conditions and steep climatic gradients, the floras of Permian time were more provincial than those of any other Phanerozoic period—with the possible exception of the most recent interval, when continents have been widely dispersed and the waxing and waning of continental glaciers have produced much geographic differentiation.

The Permian floras remained distinct even though they were not separated by vast oceans. In Permian time, the suturing of Siberia to eastern Europe along the Ural Mountains resulted in the nearly complete assembly of Pangaea (Figure 12-24). Southeast Asia remained as the only separate landmass of large size, and it would be attached during the Mesozoic Era. Contributing to the climatic contrasts of the late Paleozoic globe were several mountain chains, including those of the Hercynian, which formed during the suturing

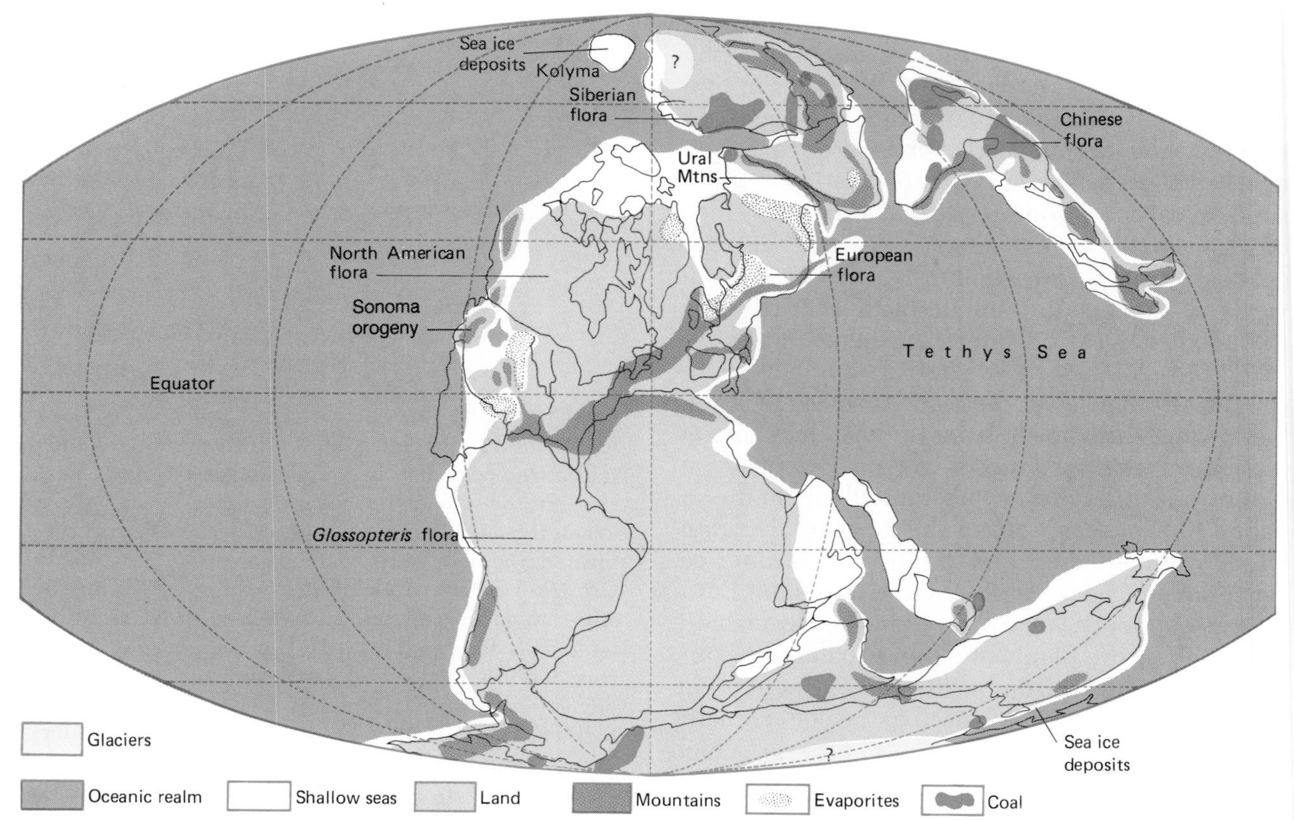

FIGURE 12-24 World geography in Late Permian time, when the ocean separating Europe from Asia was closing along the Ural Mountains to form the supercontinent of Pangaea. The Permian landmasses had a complex geography. Many mountain ranges stood high above lowlands. Five distinctive floras are labeled on the map. Floras produced coal only at high latitudes, while equatorial regions were dry. Glaciers persisted in Antarctica and also formed in Siberia, which now encroached on the North Pole. *(Modified from R. K. Bambach et al., American Scientist 68:26–38, 1980.)*

of Gondwanaland to the Old Red Sandstone continent. Furthermore, polar regions remained quite cold and equatorial regions quite hot. As a result of these contrasts, the Late Permian floras of low latitudes remained distinct from the *Glossopteris* flora to the south and also from the flora of Siberia, a continent that, although now attached to Europe, remained near the North Pole (Figure 12-24). Southeast Asia was still a separate continent, and by now its flora had become unique. The Euramerian flora had also now broken down into more local floras separated by barriers. The Permian floras east and west of the Appalachians, for example, differed significantly.

The various Permian floras had one thing in common: They changed dramatically in the course of Permian time. The floras of the United States and southern Europe, which were in the tropical zone, were altered in mid-Permian time. At higher latitudes—Asia in the north and Australia in the south—the floral transition came later, close to the end of the period. As we noted in Chapter 2, terrestrial plants are highly sensitive to climatic conditions, and it is believed that the Permian flora changed in response to climatic changes. It appears that, in general, plants adapted to moist conditions gave way to ones favored by drier habitats. In the north, the coal-swamp floras were replaced by plant communities dominated by conifers such as *Walchia* (Figure 12-14). Many late Paleozoic conifers apparently resembled conifers of the modern world: They could thrive under dry conditions. In the south, the *Glossopteris* flora of Gondwanaland, which was adapted to moist conditions, gave way to the *Dicroidium* flora, named for a genus of gymno-

FIGURE 12-25 Cross-bedded dune deposits of the Permian Coconino Sandstone, which crops out in the vicinity of the Grand Canyon in Arizona. *(Tom Bean.)*

sperms with forked, leaf-bearing branches. *Dicroidium* appeared first in tropical areas and then migrated into southern regions of the Gondwanaland region of Pangaea, apparently tracking shifting climates. The general trend in Permian time toward drier climates and floras adapted to them represented a continuation of the changes that, as we have seen, altered floras during Late Carboniferous time.

As Figure 12-24 indicates, the highly arid conditions of Permian time resulted in the deposition of great thicknesses of evaporites in the southwestern United States and in northern Europe. There is, in fact, a greater concentration of salt deposits in the Permian than in any other geologic system. In addition, dune deposits are unusually common in the Permian System, where they record the locations of ancient deserts (Figure 12-25).

THE TERMINAL PERMIAN EXTINCTION

The Paleozoic Era ended with what may have been the greatest mass extinction in all of Earth history. This was the first great mass extinction to strike Earth's biota after vertebrate animals had invaded the land on a grand scale, and terrestrial vertebrates were among its primary victims. Nearly 20 families of Permian therapsids failed to survive into Triassic time, but the extinction of therapsids was not a single event in latest Permian time. Rather, there were several pulses of extinction, with taxa of large body size suffering the heaviest losses each time and new taxa evolving from smaller survivors. It has been suggested that the extinctions resulted from climatic changes that altered the terrestrial plant communities at the base of the therapsids' food web. As we have seen, the Permian was indeed a time of floral change, but the details remain unclear.

In the marine realm, the Permian crisis entirely swept away the fusulinids, which had been highly successful in mid-Permian seas, and also the rugose corals, tabulates, and trilobites—although the latter two groups were already very much on the decline before the crisis began. The ammonoids hung on by a thread: Only a handful of their species survived into the Triassic Period. The brachiopods, bryozoans, and stalked echinoderms suffered heavy losses, and the bivalve and gastropod mollusks were struck moderately hard.

A Drop in Sea Level It is difficult to determine the pattern of the marine extinction in time and space, because marine stratigraphic sections that span the Permo-Triassic boundary without interruption are rare. Disconformities represent the crucial interval in nearly all continental areas, because global sea level fell dramatically in Late Permian time, causing continental surfaces to experience erosion rather than deposition (Major Events, p. 300). The most complete stratigraphic sections across the Permo-Triassic boundary are located in southern China, where paleontologists recognize a final Permian interval, the Changxingian Stage, which is not well represented by strata on other modern continents.

In at least 20 sections in southern China, at the top of the Chiangxingian Stage there is an important sedimentary unit known as the Transition Bed (Figure 12-26). This unit is typically about 1 meter (~3 feet) thick, and most of its fauna consists of taxa that survived from the Permian Period. Also present, however, are elements known elsewhere only from Triassic beds. The principal macroscopic constituents of the Transition Bed are bivalve, gastropod, and ammonoid mollusks.

FIGURE 12-26 The terminal Permian boundary at Shangsi, China, where the strata have been tilted by tectonism. In this view, the Transition Bed is positioned above the monument that has been erected to mark the boundary. *(D. W. Boyd.)*

Nonetheless, because of heavy extinction before the deposition of the Transition Bed, many Late Permian members of these surviving taxa are absent from it.

The last Chinese representatives of other major taxa—the fusulinids, rugose corals, tabulates, and trilobites—are found in rocks positioned 2 meters or less below the base of the Transition Bed. The pulse of extinction that victimized these taxa was so strong that some stratigraphers argue that the upper boundary of the Permian System should be placed at the base of the Transition Bed.

The Question of Cause As we have seen in previous chapters, patterns of extinction suggest causes. As it turns out, the Late Permian mass extinction shared certain patterns with earlier crises of the Paleozoic Era. Mimicking the patterns of the earlier crises, this one took its heaviest toll at low latitudes, devastating the reef-building community and other tropical marine groups. The geographic ranges of several Permian taxa contracted toward the equator before the taxa died out. Their final refuge was the Tethyan Sea, which sat astride the equator (Figure 12-24). These patterns as well as other evidence suggest that climatic cooling played a major role in the crisis. One of these factors is that the rate of limestone deposition was low in the aftermath of the mass extinction—in earliest Triassic seas—just as it was after the Ordovician and Devonian crises. Furthermore, two important types of Permian reef builders, which are unknown from the Lower Triassic record, reappear as reef builders in mid-Triassic rocks. These are the calcareous alga *Tubiphytes* and the calcareous sponge *Girtyocoelia* (Figure 12-27). It is difficult to imagine what temporary environmental change other than oceanic cooling might have suppressed the growth of these simple organisms for so long and then allowed them to flourish again.

Despite the circumstantial evidence that climatic cooling contributed to the Late Permian crisis, we have no clear picture of what might have caused this kind of environmental change. One fact that may be pertinent is that, although glaciers in the Southern Hemisphere were generally on the decline, both of Earth's poles were frigid in Late Permian time. We know this from the presence in marine sediments of large dropstones, which were released from floating ice. In the north, dropstones occur in Late Permian mudstones that are widely distributed within a terrane known as Kolyma. Today Kolyma forms most of the northeastern extremity of Siberia, but in Late Permian time it was a separate landmass, probably close to the North Pole (Figure 12-24). In the south, dropstones are found in Late Permian rocks of southeastern Australia. They were apparently rafted there on icebergs that broke from glaciers in Antarctica.

The huge supercontinent that included nearly all of Earth's landmasses was moving northward in Late Permian time, stretching virtually from pole to pole. Perhaps its encroachment on the north pole caused large glaciers to form here for the first time in the Paleozoic Era. This may also have been

FIGURE 12-27 A Permian calcareous sponge of the genus *Girtyocoelia*. This reef builder had pea-sized chambers that were interconnected. Water that passed through the chambers was expelled through holes after food particles were filtered from it. *(Chip Clark.)*

the first time that both of Earth's polar regions were icy at the same time. In addition, sea level dropped substantially in Late Permian time (Major Events, p. 300), eliminating the warm, shallow seas that had flooded the continents. If cold waters had spread toward the equator as sea level dropped, only narrow shelf areas, bathed in relatively cool water, would have been available to forms of life that had adapted to shallow seas. Perhaps this explains the heavy extinction of species that required tropical conditions. This is a hypothesis, however, not a well-supported model. The cause of the crisis that ended the Paleozoic Era is still a subject of debate.

REGIONAL EXAMPLES

Let us begin our discussion of specific regions of the late Paleozoic world by examining the economically important coal deposits in North America and western Europe; and then we will review tectonic events in the American Southwest, the history of an enormous reef complex in west Texas (the site of large petroleum reservoirs), and Permian orogenic activity along the Pacific margin of North America.

Coal Deposits in North America and Europe

In Late Carboniferous (Pennsylvanian) time, while the Michigan and other rivers flowing from the Appalachians continued to form molasse deposits in eastern North America, coal swamps spread over the floodplains of these rivers and over the margins of epicontinental seas. In North America these swamps extended far to the west of the mountains over much of the nearly flat midcontinent. Some coal basins that are now separate were probably connected at the time they were formed. The Michigan Basin, however, formed in isolation, and the basins in New England and eastern Canada may have done so as well (Figure 12-28). We now recognize that most of the coal beds of western Europe developed as molasse deposits equivalent to those of North America, but at the other end of the Hercynian mountain chain (Figure 12-22*B*).

Two Kinds of Cycles One of the most striking aspects of Late Carboniferous coals is the cyclical nature of their stratigraphic occurrence. Cycles that included coal were of two types: those formed by meandering rivers and those formed by changes in sea level that apparently resulted from the expansion and contraction of continental glaciers in Gondwanaland.

In cycles produced by meandering rivers, coal beds represent overbank deposits. In other words, a single coal bed was produced by a forest growing in a swamp that was separated from a river channel by a natural levee. Cyclical coal-bearing deposits of this kind are found in Pennsylvania, where they formed along large rivers that drained the youthful Appalachians.

Farther west, in the nearly flat midcontinent region of the United States, Late Carboniferous

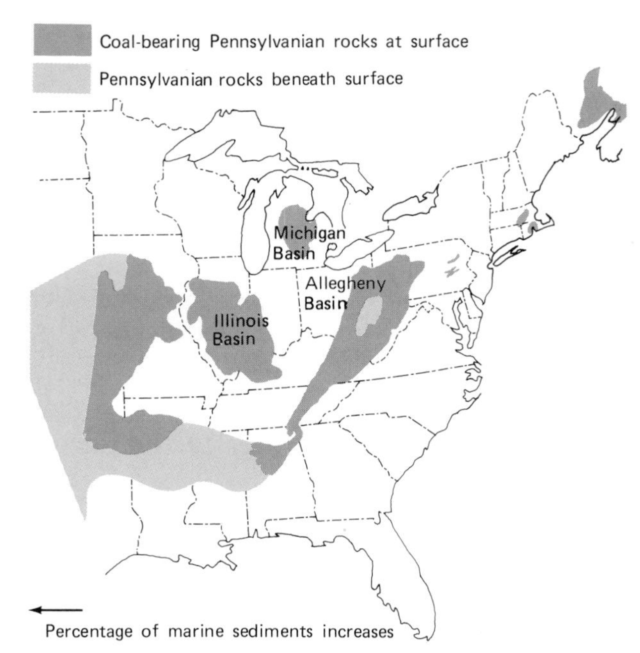

FIGURE 12-28 Distribution of coal-bearing cyclothems of Pennsylvanian (Upper Carboniferous) age in eastern North America. Before experiencing erosion, the cyclothems were even more extensive in the areas east and south of the Illinois Basin. The main area of coal formation extended from the margin of the early Appalachian Mountains on the east to an area of fully marine deposition to the west.

(Pennsylvanian) cycles are of a different type. The coal beds of these cycles formed in swamps that bordered shallow seas. Here and in similar settings on other continents, coal beds are thin but widespread, occurring within cycles that include marine deposits. Dozens of similar cycles are commonly found superimposed on one another. Such cycles in coal beds are known as **cyclothems** in North America and as **coal measures** in Britain.

How Cyclothems Formed The fact that so many marine and nonmarine habitats are often represented in just a few vertical meters of stratigraphic section indicates that the depositional gradient was very gentle. It appears that only a slight vertical movement of the sea or of Earth's crust accounted for substantial advance or retreat of the water with related shifting of environments.

The coal now found within a cyclothem began to form as peat within a coal swamp. The coal swamps seem to have occupied lowland areas neighboring the sea. They were fed by the rivers whose deposits lie beneath them (Figure 12-29). It is possible that the entire swamp was, in effect, a broad river into which inland streams emptied—

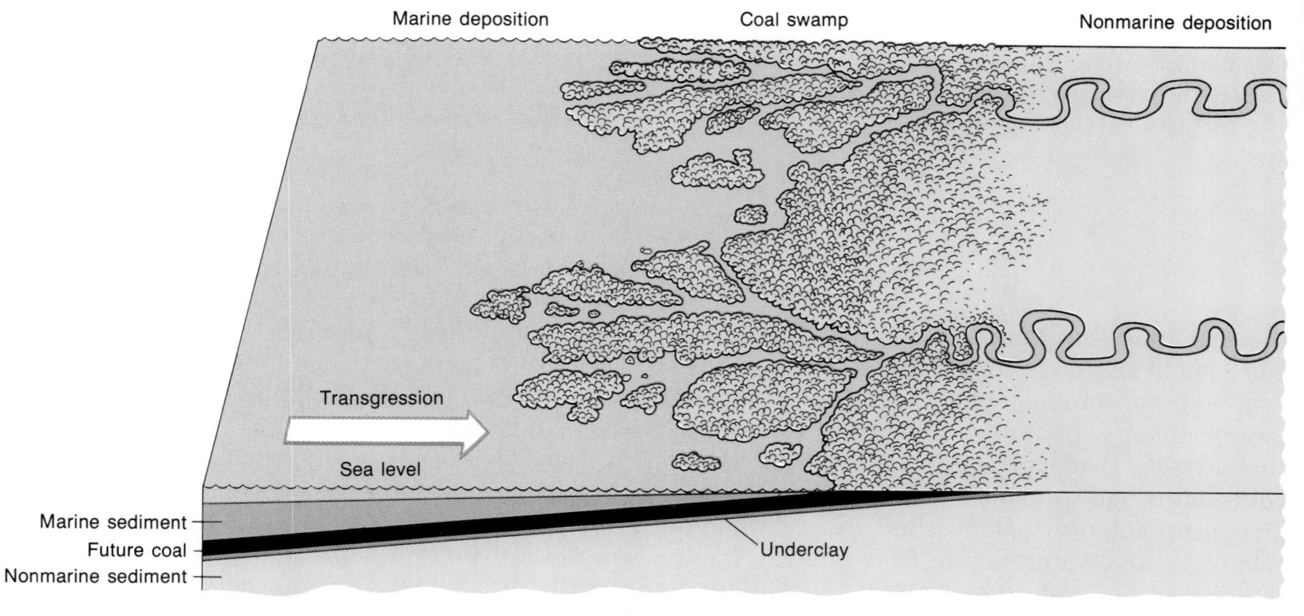

FIGURE 12-29 A block diagram showing the development of a transgressive sequence that is forming as sea level rises and the shoreline shifts inland. A transgression produces the lower part of a cyclothem, where marine sediments are superimposed on nonmarine sediments. Underclay is the nonmarine material upon which coal swamp plants grew and then died to produce the peat that ultimately would become coal.

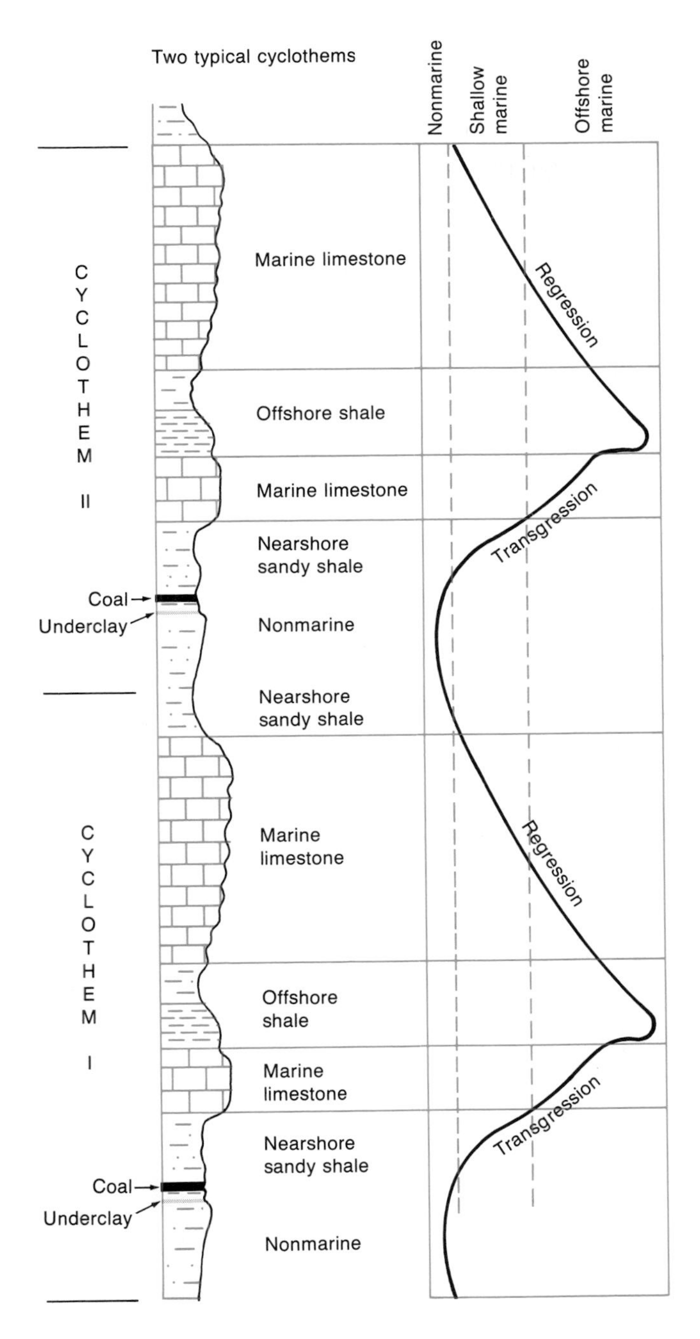

FIGURE 12-30 Two idealized cyclothems. The coal swamps migrated over these nonmarine deposits as the transgression proceeded. Above the coal are deltaic and other marginal marine sediments, which were deposited above the coals as the sea spread inland. The marginal marine deposits are succeeded, in turn, by marine limestones that represent fully marine conditions and then by black shales that represent deep environments that were present in the region at the time of maximum transgression. In time, sea level began to fall again, and the depositional sequence was reversed.

one that flowed so slowly that its movement could not have been observed with the naked eye. This is what the Everglades swamp of Florida is today. The Everglades "river" also flows over a very flat region—the southern part of the Florida peninsula, which is being partially drowned by the rising sea. The water of the Everglades remains fresh except near the edge of the sea.

Cyclothems were formed by alternating transgressions and regressions of shallow seas. A transgression resulted in the deposition of marginal marine peat (future coal) on top of nonmarine deposits and of marine sediment on top of the peat (Figure 12-29). Regression reversed the sequence, burying marine deposits beneath peat and then nonmarine sediments (Figure 12-30).

Glaciers and Sea Level

The cause of the fluctuations in sea level that produced the Carboniferous cyclothems has long been debated. It is doubtful that tectonic movements of the crust caused these frequent transgressions and regressions, because such crustal movements should be highly localized, and many cyclothems are widespread (some, for example, can be traced from Pennsylvania to Kansas). Furthermore, there is no apparent reason why the Carboniferous Period alone should have been characterized by such tectonic movements. The most likely explanation for the rapid transgressions and regressions is that the world's oceans rose and fell as the Gondwanaland glaciers repeatedly melted and re-formed. Why, then, do we not find similar cycles representing the Pleistocene interval of the past 1.6 million years, when continental glaciers have expanded and retreated many times? The explanation seems to be that in recent times—even during interglacial periods of high sea level such as the one in which we live—the continents have remained relatively emergent. The seas are rising and falling over steeply sloping continental margins; they are not invading and receding from vast, almost flat interior lowlands as they did when they formed the cyclothems of Kansas, Illinois, and neighboring regions.

Sedimentary records show that glaciation ceased altogether in South America long before glaciers disappeared from Australia. This observation suggests that glaciers did not wax and wane simultaneously in all parts of Gondwanaland. Thus the rise or fall of the world's oceans at any time must have depended on the averaging of glacial events within the entire glaciated area. The glacial hypothesis for the control of cyclical deposition gains support from the fact that the interval during which this kind of deposition took place coincided with the mid-Carboniferous to mid-Permian interval of glaciation in Gondwanaland.

Earth Movements in the Southwestern United States

The late Paleozoic was also a time of mountain building along a zone that extended from Utah across Oklahoma and Texas to Mississippi (Figure 12-31). Here the Ouachita Mountains formed as a westward continuation of the Appalachians. Today traces of the two mountain chains meet at right angles beneath flat-lying younger deposits, and although the zone of contact is not well understood, the exposed segment of the Ouachitas is a fold-and-thrust belt resembling the Appalachian Valley and Ridge Province (Figure 12-32). One difference is that the folded rocks of the Ouachitas, which range in age from Ordovician to Middle Pennsylvanian, consist of deep-water black shale and flysch deposits that have been thrust northward against shelf-edge carbonates of similar age. In other words, deformation took place offshore from the continental margin (Figure 12-31), and after it began, the rate of deposition in the adja-

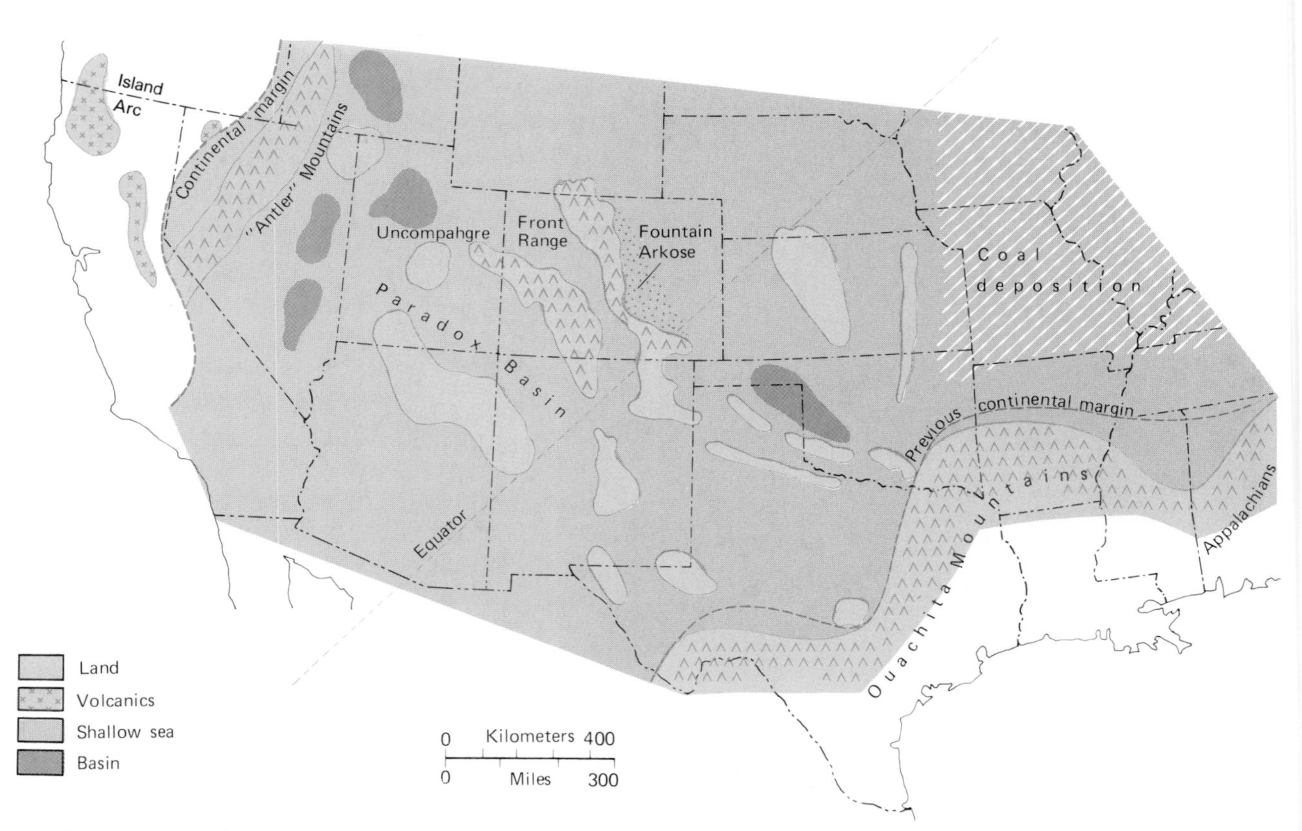

FIGURE 12-31 Paleogeography of the southwestern United States during Late Carboniferous (Pennsylvanian) time. Partway through the Late Carboniferous interval, Hercynide deformation formed the early Ouachita Mountains at the southern end of the Appalachians. Most of this deformation occurred offshore, uplifting oceanic sediments and welding them onto the previous continental margin. Coal-bearing cyclothems formed in marginal marine environments in the midcontinent region. To the west, shallow seas covered most of the craton, but many uplifts and basins developed from Texas to eastern Nevada, apparently in association with the Ouachita orogeny. The highest uplifts, the Front Range and the Uncompahgre, are together known as the Ancestral Rocky Mountains. Farther west, mountains produced by the earlier Antler orogeny still bordered the continent, and an island arc now lay along a subduction zone offshore in what is now California and Oregon.

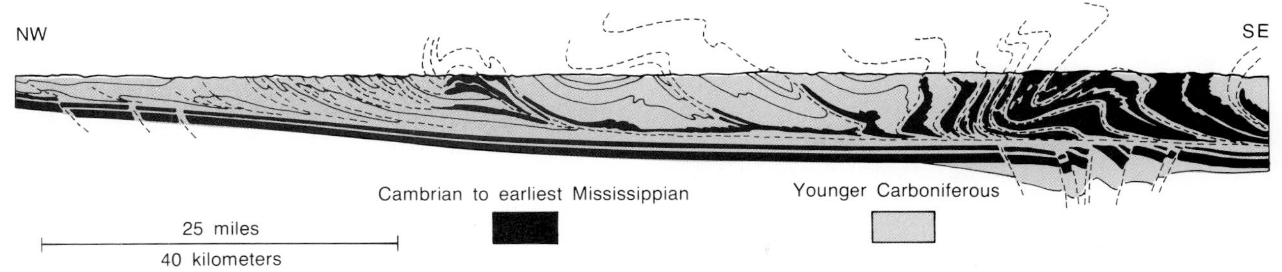

FIGURE 12-32 Cross section through the Ouachita Mountains as they exist today in the southeastern corner of Oklahoma. Here large volumes of deep-water sediment have been thrust northwestward. Note the general similarity between the style of deformation here and in the Appalachian fold-and-thrust belt (Figure 7-11). *(After J. Wickham et al., Geology 4:173–180, 1976.)*

cent basin increased. In fact, the deformed region seems to have behaved as an unusually deep foreland basin in which enormous volumes of deep-water Carboniferous deposits continued to accumulate on top of already-deformed older deposits. These younger, thicker deposits were, in turn, folded and thrust northward. Most of this deformation was completed before the start of the Permian Period.

The Ouachita deformation was part of the general Hercynian sequence of events that united Gondwanaland with the Old Red Sandstone continent (Figures 12-22*B* and 12-24). Although plate-tectonic events in the vicinity of the Ouachita system were complex and remain poorly understood, it is known that several microplates in this region were similar to those involved in the origin of the Alps. Some of the microplates south of the Ouachita system eventually became parts of Central America (Figure 12-33).

The craton to the north and west of the Ouachita deformation also underwent important tectonic movement. It is not clear just how these cratonic movements were related to the Ouachita orogeny, but they were largely vertical; enormous areas in what is now the southwestern United States became transformed into a series of uplifts and basins (Figure 12-31). Many of these structural features are bounded by high-angle faults. The basins accumulated Late Carboniferous (Pennsylvanian) and, in some cases, Permian deposits. Clastic debris shed from the uplifts was deposited rapidly in nearby basins as coarse arkose. **Arkose** is a sedimentary rock that consists

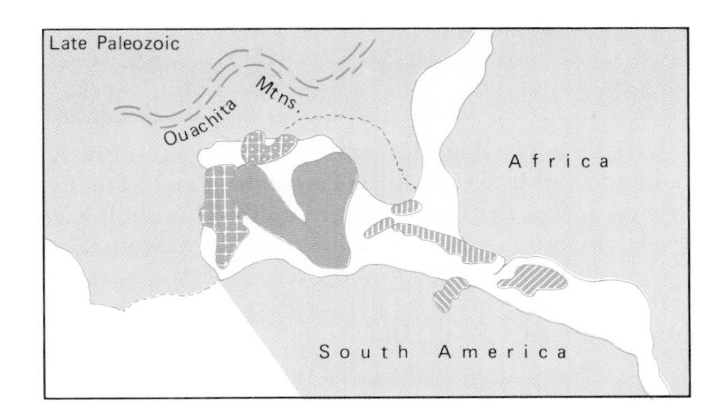

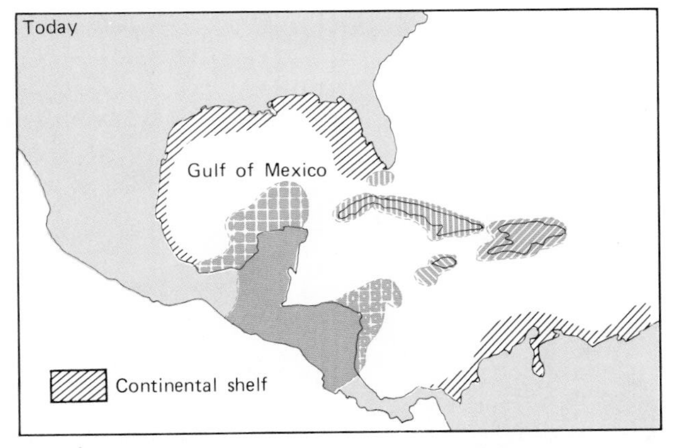

FIGURE 12-33 Approximate positions of microplates south and east of the Ouachita fold-and-thrust belt late in Paleozoic time *(above)* and today *(below)*. Since late Paleozoic time some microplates have shifted southward, leaving the Gulf of Mexico in their place. *(Modified from A. G. Smith and J. C. Briden, Mesozoic and Cenozoic Paleocontinental Maps, Cambridge University Press, Cambridge, 1977.)*

FIGURE 12-34 Black Canyon of the Gunnison River, Colorado, cut into crystalline Precambrian rocks that formed part of the Uncompahgre uplift. This uplift was leveled by erosion during the first half of the Mesozoic Period, and Upper Jurassic and Cretaceous rocks, seen here in the distance, were deposited on top of the eroded surface. *(Martin Miller.)*

largely of feldspar, a mineral that weathers to clay if it is not buried rapidly (see Appendix I).

Two of the uplifts, the Front Range and Uncompahgre uplifts, are commonly referred to as the Ancestral Rocky Mountains. The Ancestral Rockies developed during Late Carboniferous time; they were elevated and then subdued by erosion in an area where portions of the Rocky Mountains stand today. It is estimated that the Uncompahgre uplift rose to between 1.5 and 3.0 kilometers (~ 1 or 2 miles) above the surrounding seas, which flooded much of western North America. This elevation is comparable to that of the modern Rockies above the Great Plains to the east. The Front Range uplift is named for the Front Range of the modern Rockies, which now extends slightly farther east than the late Paleozoic uplift. Growth of the Ancestral Rockies elevated Precambrian rocks beneath, which later were leveled by erosion. The Precambrian roots of the Ancestral Rockies can still be observed where more recent secondary uplift has caused rivers to cut deep gorges (Figure 12-34).

At places in the basin lying between the Uncompahgre and Front Range uplifts, arkosic sands and conglomerates accumulated to thicknesses exceeding 3 kilometers (~ 2 miles). The Fountain Arkose, which formed along the eastern flank of

FIGURE 12-35 Northward view along the Front Range of the Rocky Mountains near Morrison, Colorado. The core of the Rockies lies to the left. Tilted upward along its margin are the so-called Flatirons. These are formed of Foundation Arkose, which consists of sediment shed from the Ancestral Rocky Mountains that lay slightly to the west in Late Carboniferous time. The Fountain Arkose was later tilted upward when the modern Rockies formed. *(T. S. Lovering, U.S. Geological Survey.)*

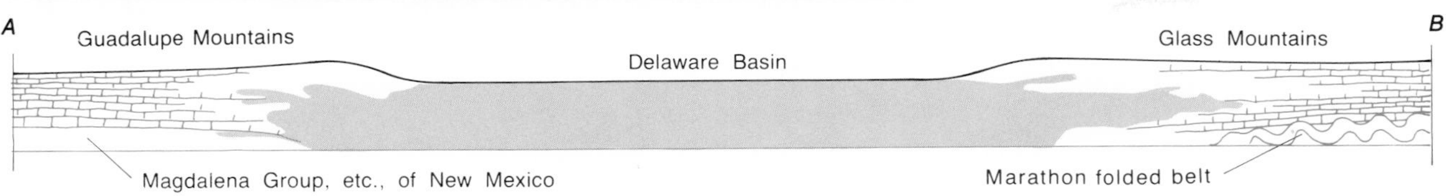

FIGURE 12-36 Aerial view of the Guadalupe Mountains from above the Delaware Basin. The Capitan reef Limestone of the Guadalupe Mountains rims the basin. The cross section below, from northwest to southeast, shows the configuration of the basin. (Figure 12-38 shows the location.) *(National Park Service.)*

the Front Range uplift, was later upturned along the front of the modern Rockies when they were uplifted. Here, through differential erosion, the Fountain stands out in central Colorado as a series of spectacular ridges (Figure 12-35).

The Ancestral Rockies lay close to the Late Carboniferous equator, where easterly equatorial winds must have prevailed. It is therefore understandable that the ancient mountains seem to have produced a rain shadow to their west. Here, in the Paradox Basin (Figure 12-31), great thicknesses of evaporites—primarily halite (Figure AI-1)—accumulated.

The Permian System of West Texas

The Delaware Basin of Texas and New Mexico is one of the most famous geologic structures in the world, both because of its economic importance and because it offers spectacular scenery. Although it has not been occupied by the sea for more than 200 million years, it remains a topographic basin. A person can stand in its center today and view ancient carbonates that formed as banks or reefs around the margin during the Permian Period (Figure 12-36). Earlier, during the latter part of Late Carboniferous (Pennsylvanian) time, the shallow seas that had shifted back and forth over the coal basins of the central United States withdrew westward, never to return. In earliest Permian time they remained only in Texas and in neighboring areas, where they were connected to the seas that still flooded the western margin of North America (Figure 12-37). The Ancestral Rockies still stood high, as a mountainous island, and the young Ouachita Mountains bordered the southern margin of North America as what must have been a rugged and imposing mountain range. To the northwest of this range, detrital material came to rest in a marginal foreland basin.

During Early and Middle Paleozoic time, marine deposits accumulated in the area that now forms west Texas, which was a broad, shallow basin on the continental shelf. During the uplift of the Ouachita range and other Carboniferous uplands, a small fault block rose up within the west Texas basin, dividing it into the Delaware Basin and the Midland Basin (Figure 12-37). Both of these basins subsequently received large thicknesses of sediment that have yielded enormous quantities of petroleum.

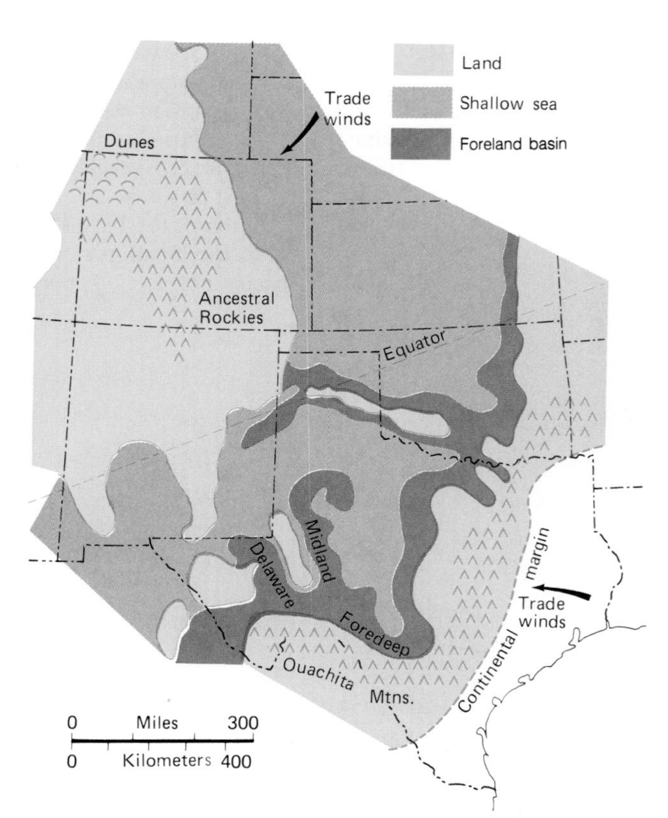

FIGURE 12-37 Paleogeography of Texas and neighboring regions in earliest Permian time. Shallow seas now flooded only parts of western North America, including this region between the Ouachita Mountains and the Ancestral Rockies. Mild deformation of the craton northwest of the Ouachitas continued from Late Carboniferous time, and the Delaware and Midland basins formed in western Texas as part of the Ouachita foreland basin. *(Modified from J. M. Hills, Amer. Assoc. Petrol. Geol. Bull. 56:2303–2322, 1972.)*

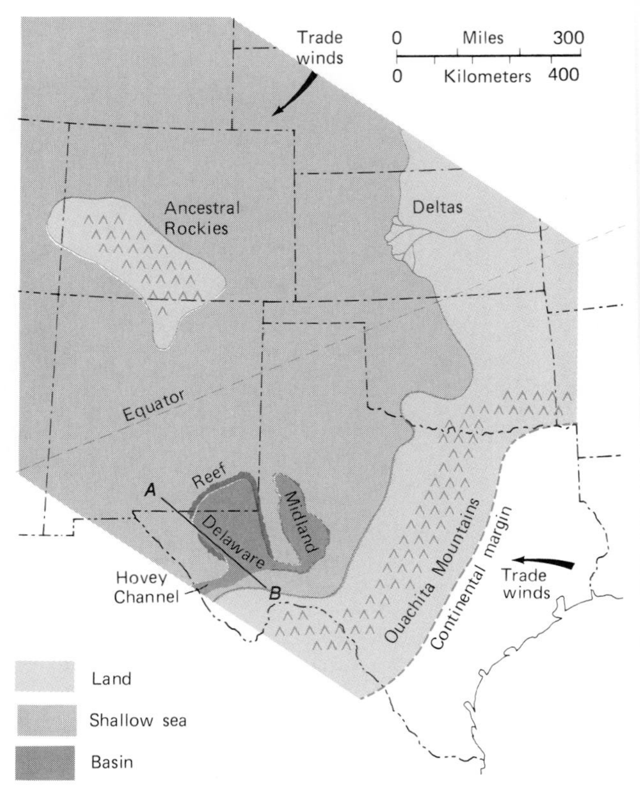

FIGURE 12-38 Paleogeography of Texas and neighboring regions when reefs encircled the Delaware Basin in Late Permian time. During this interval a narrow passageway (Hovey Channel) connected the Delaware Basin with the open ocean to the west, but the Midland Basin was eventually filled with sediment. Line *AB* shows the location of the cross section in Figure 12-36.

Reef Growth While reefs grew upward around the Delaware Basin, the Midland Basin to the east became filled with sediment (Figure 12-38). Along with surrounding areas, it was then flooded by shallow seas. Although by this time, the Ancestral Rockies had been lowered by erosion, they still formed a large island to the northwest. The Delaware Basin lay very close to the Permian equator, and the Ouachita chain must have left the basin in the rain shadow of equatorial winds blowing from the east. The shallow seas that surrounded the basin were the sites of carbonate and evaporite deposition in what was obviously an arid climate.

As time passed and sea level rose, the reef grew upward more rapidly than the Delaware Basin filled in, and eventually the reef stood high above a basin that was some 600 meters (~2000 feet) deep (Figure 12-39). Although the waters have long since withdrawn, this is the configuration that remains today (Figure 12-36). Early in its history, when the Delaware Basin was relatively shallow, its floor was inhabited by snails, deposit-feeding bivalves, sponges, and brachiopods. Later, when the basin had deepened, these animals decreased in number; the only abundant marine fossil remains in the younger basin deposits are conodonts (Figure 10-6), radiolarians (Figure

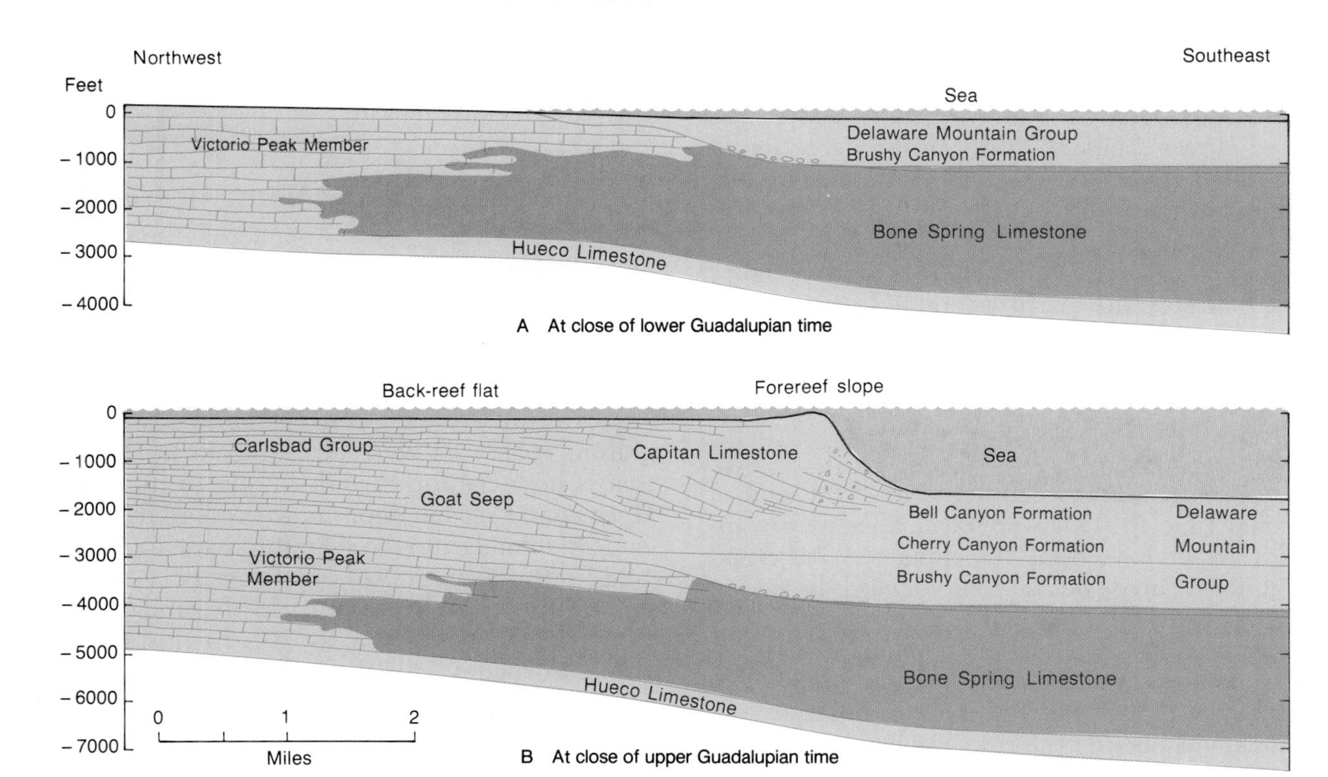

FIGURE 12-39 Profiles of the reef that now forms the Guadalupe Mountains early in Guadalupian time *(A)* and later in Guadalupian time *(B)*, when the reef had grown rapidly upward while the basin deepened. (The Guadalupian Stage was the next-to-last stage of the Permian Period.) Note that the reef advanced toward the basin center as it grew upward. *(After N. D. Newell et al., The Permian Reef Complex of the Guadalupe Mountains Region, Texas and New Mexico, W. H. Freeman and Company, New York, 1953.)*

AI-18), and ammonoids (Figure 12-1), all of which lived high in the water column and sank to the bottom when they died. Also present are plant spores that were blown into the basin from the land. We can conclude that when the Capitan Limestone formed (Figure 12-39), the floor of the deep basin below was poorly oxygenated; oxygen used up in the decay of organic matter was not replenished, and few bottom-dwelling animals could survive.

The reeflike structure that rims the Delaware Basin was built during the Guadalupian Age, the second-to-last age of the Upper Permian. The reef was formed primarily by sponges (Figure 12-27), algae, and lacy bryozoans (Figure 12-5). The crest of the reef was covered by shallow water that also bathed an extensive back-reef flat (Figure 12-39). Rubble from the reef periodically tumbled down the forereef slope into the basin. The ancient talus slope is present today in bedding that dips at angles as high as 40°. In the rubble of the forereef slope are many beautifully preserved fossils whose originally calcareous hard parts have been replaced by durable silica; among them are sponges (Figure 12-27) and shells of fusulinid foraminifera that lived in shallow habitats but periodically washed down to lodge in the slope rubble. Some were swept into the basin by turbidity flows that left conspicuous graded beds in the rocks of the basin; these beds constitute the Delaware Mountain Group. Most sediments of this unit are dark sands and silts that periodically washed into the basin, apparently when sea level was low and the reef surface was exposed to erosion.

When the older bedding surfaces of the carbonates that ring the Delaware Basin are traced laterally, a different configuration becomes apparent. These older bedding surfaces show that the early reefs, known as the Goat Seep Formation, stood in much lower relief above the basin

(Figure 12-39). From its earliest days until late in the Permian, the Delaware Basin was connected with the open sea to the southwest through what is called Hovey Channel. Early in the evolution of the basin, when the reefs were low, the connection and the resulting pattern of water circulation permitted oxygen-rich waters to reach the basin floor so that animals could live there (Figure 12-40). Later the basin deepened, but Hovey Channel remained shallow, a conformation that caused the bottom waters of the basin to become stagnant and poor in oxygen, excluding almost all forms of life.

Death of the Reef Eventually, near the end of Permian time, the Delaware Basin filled with evaporites. As we have seen, the climate of Texas and neighboring areas became arid toward the end of the Permian, and the rate at which waters evaporated from the Delaware Basin may have increased. Hovey Channel may also have become so constricted that the rate of evaporation there occasionally exceeded the rate at which new water was supplied. In any case, the reef stopped growing and the basin ultimately filled with evaporites; distinct layers in some of these evaporite deposits extend over many hundreds of square kilometers (Figure 4-9). It is possible that this layering reflects seasonal changes similar to those responsible for glacial varves.

The Delaware Basin evaporites remained in place for a long period, protecting the magnificent geologic record along the walls and floor of the basin. Fresh water later dissolved the evaporites in many areas, exposing the ancient reef-encircled structure (Figure 12-36).

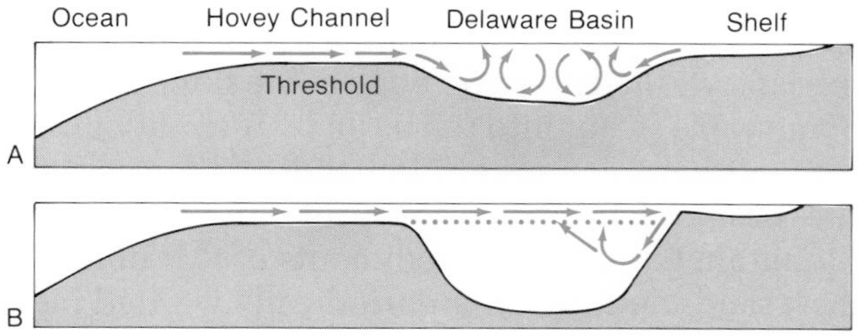

FIGURE 12-40 Patterns of water circulation in the Delaware Basin. Early in its history, the basin was shallow enough that well-oxygenated surface water reached its floor *(A)*. Later, when the basin had deepened, good circulation was restricted to the upper waters, and the bottom waters thus became stagnant *(B)*. *(After N. D. Newell et al., The Permian Reef Complex of the Guadalupe Mountains Region, Texas and New Mexico, W. H. Freeman and Company, New York, 1953.)*

The Western Margin of North America

What was happening to the west of the Delaware Basin and the Ancestral Rocky Mountains? During late Paleozoic time the western margin of the North American craton passed through what is now northwestern Nevada (Figure 12-31). Hundreds of kilometers from the margin of the craton, volcanoes were active again in the area that is now California and slightly to the north. Here coarse clastic deposits were shed from volcanic highlands into the surrounding seas. An orogenic episode in Nevada in latest Permian and Early Triassic times was remarkably similar to the Antler orogeny (Figure 11-30). In this second, **Sonoma orogeny,** as in the Antler, marine deposits were thrust upward over the continental margin. The Sonoma orogeny was of great significance in that it entailed the complete closure of the basin between the volcanic arc and the North American continent. While some of the deep-sea deposits of the basin were thrust onto the continent, others were welded onto the continental margin along with the volcanic terrane of the arc. The result was a considerable westward growth of the North American crust.

CHAPTER SUMMARY

1 Marine life of late Paleozoic time in many ways resembled life of the middle Paleozoic, but the tabulate-strome reef community was gone. In addition, four groups that expanded enormously contributed large volumes of skeletal debris to limestones: first crinoids and lacy bryozoans, later flakelike algae and fusulinid foraminifera.

2 In Carboniferous time the coal-swamp floras, which were dominated by trees of the genera *Lepidodendron* and *Sigillaria*, played a major ecologic role, as did seed ferns and, on drier land, sphenopsids and cordaites. During the Permian Period, however, climates in the Northern Hemisphere became warmer and drier, and these plant groups gave way to conifers and other gymnosperms.

3 Carboniferous coal swamps produced coal beds that commonly occur within depositional cycles, some of which represent meandering rivers and others alternating transgressions and regressions of shallow seas.

4 Throughout nearly all of late Paleozoic time, continental glaciers blanketed the south polar region of Gondwanaland and large areas of this great southern continent were populated by the cold-adapted *Glossopteris* flora.

5 Insects originated during the Carboniferous Period and underwent a great adaptive radiation. During the Permian Period, amphibians were displaced from terrestrial habitats by early mammal-like reptiles, including finbacks, but these soon gave way to more advanced mammal-like reptiles, the therapsids.

6 In mid-Carboniferous time Gondwanaland became sutured to the Old Red Sandstone continent, forming the Hercynide mountain chains.

7 The latter part of the Permian Period was a time of hot, dry conditions and widespread evaporite deposition in equatorial regions. The Permian Period—and thus the Paleozoic Era as well—ended with an enormous extinction that appears to have eliminated most species of invertebrates in the ocean as well as many species of vertebrates on the land.

8 In the United States west of the Appalachians, the Hercynian orogenic episode affected the central and southern Appalachians and also created the Ouachita Mountains in Oklahoma and Texas. Uplifts and basins formed north and west of the Ouachitas. The Delaware Basin in western Texas became encircled by a reef complex in the Permian System, and near the end of the Permian this basin was filled by evaporite deposits. Beginning in the latest Permian time, the Sonoma orogeny resulted in continental accretion along the western margin of North America.

EXERCISES

1 What became of the Old Red Sandstone continent during late Paleozoic time?

2 If you could examine Late Carboniferous (Pennsylvanian) coal-bearing cycles in the field, how would you determine whether the coal deposits formed along a river or a shallow sea? (Hint: Refer to Figures 3-14 and 12-30.)

3 What groups of terrestrial plants that existed in late Paleozoic time survive today?

4 What justification is there for dividing the Carboniferous interval into the Mississippian and Pennsylvanian periods?

5 How did the history of glacial activity in Late Carboniferous time relate to the deposition of coal?

6 In eastern North America, mountain building progressed from New England to Texas during Paleozoic time. How does this pattern relate to continental movements? (Review the relevant parts of Chapters 7, 10, and 11, as well as the relevant part of this chapter.)

7 In what way may therapsids have been superior to the amphibians and reptiles that preceded them?

8 Why were evaporite deposits more widespread in Europe and North America than in most parts of Gondwanaland during Late Permian time?

ADDITIONAL READING

Bakker, R. T., "Dinosaur Renaissance," *Scientific American*, April 1975.

Benton, M. J., *Vertebrate Palaeontology*, Unwin Hyman, London, 1990.

Thomas, B. A., and R. A. Spicer, *The Evolution and Palaeobiology of Land Plants*, Croom Helm, London, 1987.

PART VI

THE MESOZOIC ERA

When the Mesozoic Era, or Age of Dinosaurs, began, all of the major continents of the world were joined in the supercontinent Pangaea. As the era progressed, Pangaea separated into many fragments, and tectonic activity created major mountain chains in many parts of the world, including the western margin of the Americas. Marine and terrestrial biotas, badly damaged by the mass extinction at the end of the Paleozoic, were impoverished at the start of the era. As they recovered, mollusks, swimming reptiles, and new kinds of fishes gained ascendancy in the oceans and dinosaurs soon dominated the land; mammals, which also evolved early in the era, remained small and unobtrusive. Huge flying reptiles and primitive birds appeared, and near the end of the era, flowering plants replaced conifers and their relatives as the dominant forms of terrestrial vegetation. Mass extinctions punctuated the Mesozoic history of life, and the one that marked the end of the era resulted in the disappearance of the dinosaurs.

Allosaurus, a huge carnivorous dinosaur of Jurassic age that roamed the American West, reached a length of about 12 meters (~40 feet) and its skull was nearly a meter (~3 feet) long. *(Chip Clark.)*

CHAPTER

13

The Early Mesozoic Era

The Mesozoic Era, or the "interval of middle life," began with the Triassic Period. The Triassic and the subsequent Jurassic Period together constitute slightly more than half of the era. Rocks representing these periods are especially well exposed and well studied in Europe.

Near the transition from the Paleozoic Era to the Mesozoic Era, the great supercontinent Pangaea took its final form, encompassing virtually all the major segments of Earth's continental crust. Pangaea was so large that much of its terrain lay far from any ocean and, as a result, became arid. During Jurassic time, however, sea level rose and marine waters spread rapidly over the land, leaving a more extensive record of shallow marine deposition than that of the Triassic System. Then, later in early Mesozoic time, Pangaea began to fragment, and before the end of the Jurassic Period, Gondwanaland was once again separate from the northern landmasses.

Life of early Mesozoic time differed substantially from that of the Paleozoic Era. For many groups of animals, recovery from the Late Permian biotic crisis was sluggish, but by the end of the Triassic Period, mollusks had reexpanded to become more diverse than they had been during the Paleozoic Era. Their success has continued to the present time. The marine ecosystem was also transformed during Triassic and Jurassic time by the addition of both modern reef-building corals and large reptiles, which joined fishes as swimming predators. On the land, gymnosperm floras that had conquered the land during the Permian Period continued to dominate landscapes, and flying reptiles and birds appeared as well. The most

Dinosaur tracks of Jurassic age in the Painted Desert in Arizona. *(Tom Bean.)*

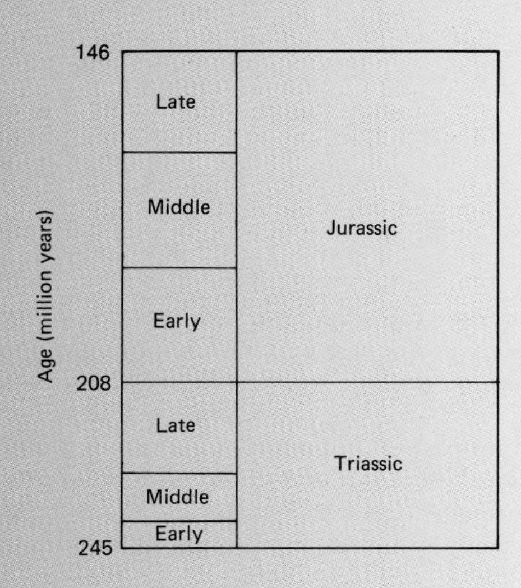

MAJOR EVENTS OF EARLY MESOZOIC TIME

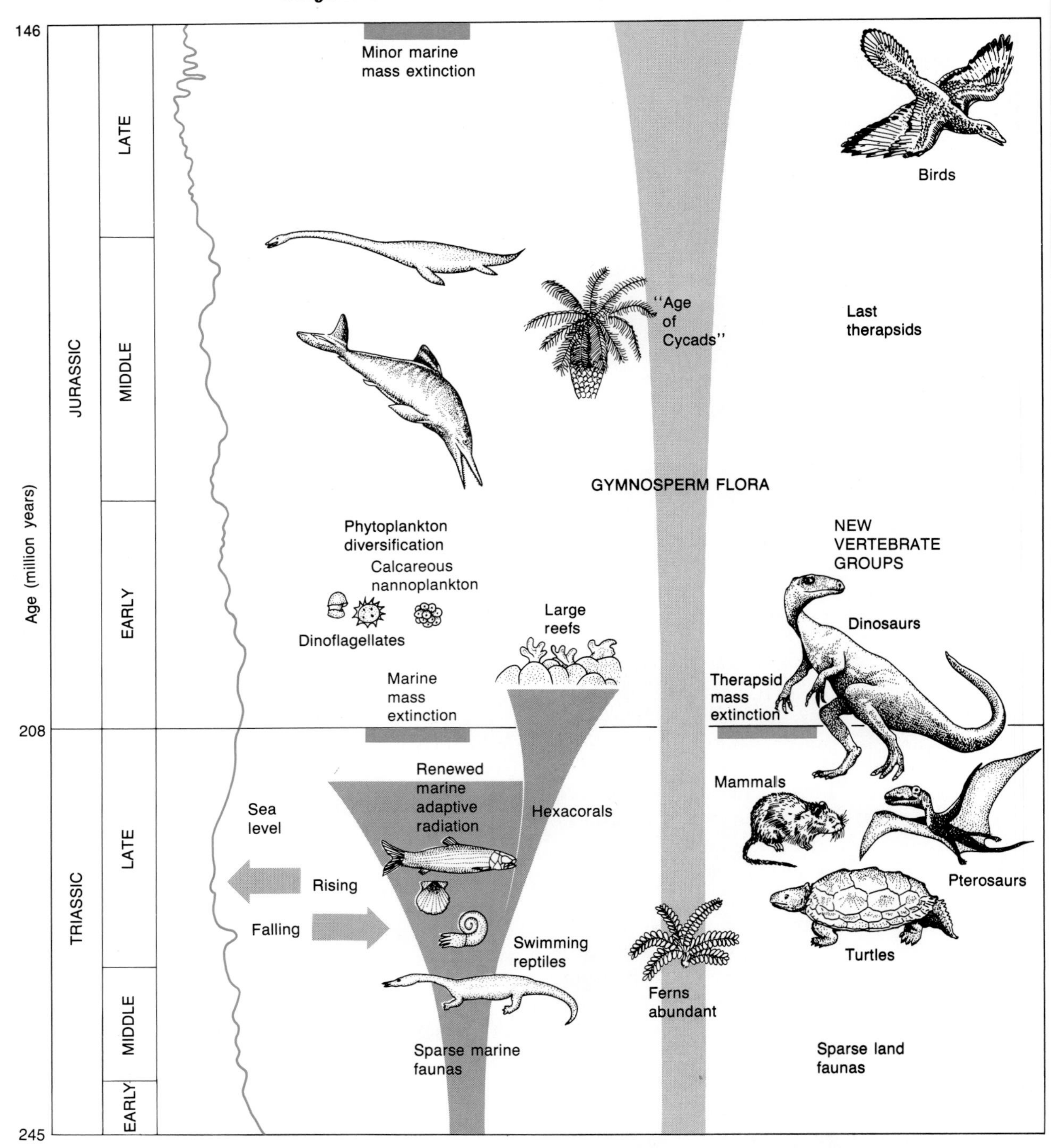

Sea level rose gradually during early Mesozoic time. Ferns flourished during Triassic time, but gymnosperms continued to dominate terrestrial environments. After the Late Permian crisis, marine and terrestrial faunas were impoverished, but adaptive radiations soon replenished animal life. Mass extinctions weaker than the terminal Permian crisis struck at the end of the Triassic and Jurassic periods. Several new groups of vertebrate animals, including dinosaurs, occupied Late Triassic terrestrial environments, and birds were present before the end of Jurassic time. In the oceans, adaptive radiation gave mollusks a major ecologic role. Hexacorals evolved in mid-Triassic time and by Early Jurassic time had produced reefs resembling those of the modern world. Large predatory reptiles evolved in the Triassic Period and flourished in the Jurassic Period as well.

dramatic event in the terrestrial ecosystem, however, was the emergence and diversification of the dinosaurs. Mammals also evolved during that time, but they remained small and relatively inconspicuous throughout the Mesozoic Era.

The Triassic System is bounded by the terminal Permian extinction below and by another extinction above. It was the unique fauna of this system that led Friedrich August von Alberti to distinguish the Triassic in 1834. Alberti originally named the system the Trias for its natural division in Germany into three distinctive stratigraphic units.

The Jurassic System also originated with a more abbreviated name, Jura, a label that was borrowed from a portion of the Alps in which the system is especially well exposed. The Jurassic was not formally established by a published proposal, however; instead, it gradually came to be accepted as a valid system during the first half of the nineteenth century, when its many distinctive marine fossils were widely investigated.

LIFE IN THE OCEANS: A NEW BIOTA

By the end of the great extinction that brought the Paleozoic Era to a close, several previously diverse groups of marine life had vanished and others had become rare. Gone were fusulinid foraminifera, lacy bryozoans, rugose corals, and trilobites. Most common in Lower Triassic rocks are mollusks. The ammonoids made a dramatic recovery after almost total annihilation; although only two ammonoid genera are known to have survived the Permian crisis, Lower Triassic rocks have yielded more than 100 genera of ammonoids. The adaptive radiation that produced these genera seems to have issued from the single genus *Ophiceras*, which appears to have been a descendant of *Xenodiscus* (see Figure 12-1). Other groups of marine life were slower to recover, but by Late Triassic time the seas were once again richly populated with animals.

Seafloor Life

Bivalve and gastropod mollusks were less severely affected by the Permian extinction than were many other groups. In fact, bivalves, like ammonoids, are frequently found in Lower Triassic rocks, although their diversity is somewhat limited. Both bivalves and gastropods expanded in number and in variety to become among the most important groups of early Mesozoic marine animals. As in the Paleozoic Era, some of the bivalves burrowed in the seafloor, while others rested on the sediment surface (Figure 13-1).

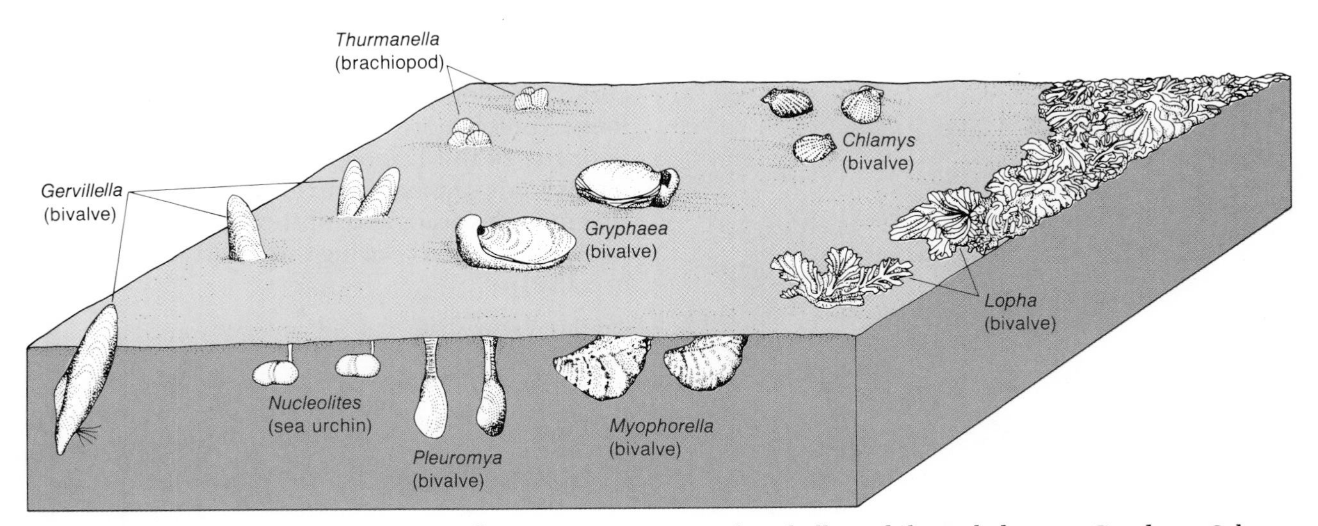

FIGURE 13-1 Life of a Late Jurassic seafloor. As in Paleozoic time, many animals lay exposed on the seafloor, but some of them were of new types, such as the irregularly shaped oyster *Lopha*, which cemented itself to other shells, and the coiled oyster *Gryphaea*. Other new animals, including sea urchins such as *Nucleolites*, lived within the sediment. *(Modified from F. T. Fürsich, Palaeontology 20:337–385, 1977.)*

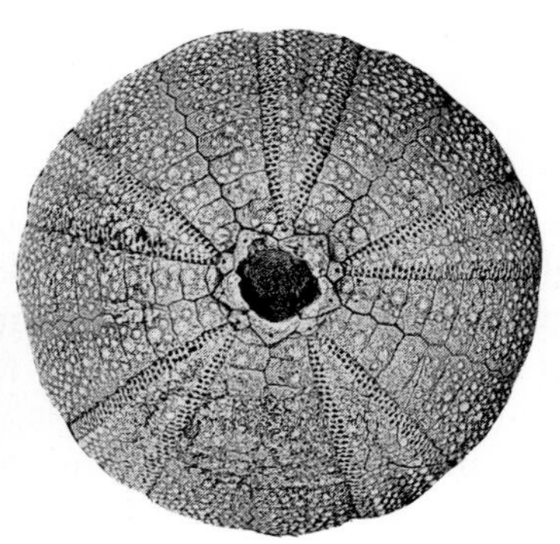

FIGURE 13-2 Side and bottom views of the Jurassic sea urchin *Leioechinus*. Like other nearly circular sea urchins, this animal lived on the surface of the sediment. In life, movable spines were attached to the numerous knobs by ball-and-socket joints. These spines served for defense and for locomotion. This animal was about the size of a doughnut. *(P. M. Kier, Smithsonian Institution.)*

Other than ammonoids and bivalves, only brachiopods make a modest showing in Lower Triassic rocks; all other marine invertebrates are rare. The brachiopods also diversified during the Triassic and Jurassic periods, but they subsequently declined until today they are very sparse. As we shall see in Chapter 14, their decline can probably be attributed to the appearance of new kinds of predators in late Mesozoic time.

Sea urchins, which had existed in limited variety during the Paleozoic Era, diversified greatly during the first half of the Mesozoic Era (Figure 13-2). Some of the new forms that emerged at this time were surface dwellers, like most of the Paleozoic sea urchins, but others lived within the sediment as actively burrowing deposit feeders (Figure 13-1).

Early in Mesozoic time, the place of the extinct Paleozoic corals was taken by the group that is still successful today—the **hexacorals** (Figure 13-3). This tentacle-feeding group of corals includes solitary species that resemble the solitary rugose corals of the Paleozoic Era (Figure 10-13) as well as colonial reef builders. Early reeflike structures of Middle Triassic age were low mounds that stood no more than 3 meters (~10 feet) above the seafloor. Most of these mounds were built by only a few kinds of organisms. By latest Triassic time, reefs were larger, some having been constructed by more than 20 species.

Some of the early coral mounds grew in relatively deep waters, so it appears that the earliest hexacorals, unlike the corals that form large tropical reefs today, did not live in association with symbiotic algae. Perhaps it was not until latest Triassic or Early Jurassic time, when hexacorals began to form large reefs, that this important symbiotic relationship was established.

Because of the success of bivalve and gastropod mollusks, sea urchins, and reef-building hexacorals, seafloor life of the Late Jurassic world looked much more like that of today than had seafloor life of the Paleozoic Era. Still missing were many kinds of modern arthropods, but the group that includes crabs and lobsters got off to a modest evolutionary start during the Jurassic Period (Figure 13-4).

FIGURE 13-3 Triassic hexacorals. The fragments of colonies shown here are 3 to 4 centimeters (~1.5 inches) across. *(George Stanley.)*

Pelagic Life

Presumably many kinds of planktonic organisms in Triassic and Jurassic seas left no fossil record, but a few types are abundant in early Mesozoic rocks. Acritarchs left a meager record, which indicates that they had not recovered from the great extinction near the end of the Devonian Period. In contrast, the **dinoflagellates**—which had existed

FIGURE 13-4 An early lobsterlike animal belonging to the Jurassic genus *Cycleryon*. Note how weak the claws of this animal were in comparison with those of a modern lobster or crab. *(Field Museum of Natural History, Chicago.)*

FIGURE 13-5 Calcareous nannoplankton from the Upper Jurassic Series of England. Many of the calcareous discs shown here are united just as they were when they formed armored shields around spherical algae cells. (See Figure 2-26*C*.) *(D. Noël.)*

even in the Paleozoic Era — underwent extensive diversification during mid-Jurassic time and remain an important group of producers in modern seas (Figure 2-26). The **calcareous nannoplankton,** another important group of living algae, made their first appearance in earliest Jurassic time (Figure 13-5). Today these floating algae are concentrated in tropical seas (Figures 2-26 and 2-28), and their armor plates rain down on the deep sea to become important constituents of deep-sea sediments.

Higher in the food chain, the ammonoids and the belemnoids played major roles as swimming predators. The ammonoids' evolutionary recovery after the Permian crisis led to great success throughout the Mesozoic Era. Individual ammonoid species, however, survived for relatively brief intervals, often a million years or less, and so they are extremely useful as index fossils for Mesozoic rocks (Figure 13-6). The belemnoids, which were squidlike relatives of the ammonoids, also pursued prey by jet propulsion (Figure 13-7). They evolved in late Paleozoic time but remained inconspicuous until the Mesozoic Era, at which time many types evolved. Conodonts, the toothlike structures that are now known to have belonged to eel-like animals (Figure 10-6), have also proved useful in the correlation of Triassic rocks, but by Jurassic time, conodont-bearing animals no longer existed.

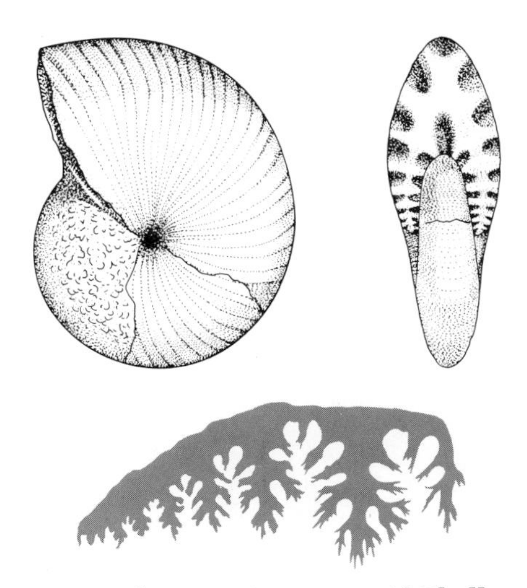

FIGURE 13-6 The Jurassic ammonoid *Phylloceras*. The suture pattern is shown below the shell. (The suture is the juncture between the convoluted internal partitions, or septa, and the coiled outer shell.)

A

B

FIGURE 13-7 Belemnoids, squidlike cephalopod mollusks that were related to ammonoids but lacked external shells. *A.* Like ammonoids, they were predators that swam by jet propulsion. *B.* The most commonly preserved part of a belemnoid is the cigar-shaped counterweight. This heavy structure acted to offset the buoyant effect of gas within the shell, thereby maintaining balance. *(A. Field Museum of Natural History, Chicago. B. Chip Clark.)*

Paleozoic ray-finned bony fishes gave rise to forms that were successful in early Mesozoic time but were still more primitive than most of their modern-day descendants. The scales that covered the bodies of these fishes, for example, were diamond-shaped structures that overlapped slightly or not at all (Figure 13-8*A*), in sharp contrast to the circular, strongly overlapping scales of nearly all modern bony fishes. Presumably the primitive diamond-shaped scales were less protective than the modern kind. Other features that distinguished early Mesozoic bony fishes from their modern counterparts were skeletons that consisted partly of cartilage rather than entirely of bone; relatively simple, primitive jaws; and tails that were highly asymmetrical, like those of all Paleozoic bony fishes. Some early Mesozoic bony fishes had teeth shaped like rounded pegs, which served to crush durable items of food—probably small shellfish. Bony fishes underwent many changes during the Mesozoic Era, and by the end of the era, few species with these primitive traits remained. One especially useful feature that developed during this time was the swim bladder, a sac of gas that allows advanced fishes to regulate their buoyancy. The swim bladder evolved from the lung, which was present in some primitive fishes.

Sharks were also well represented in early Mesozoic seas. One particularly prominent group, the hybodonts (Figure 13-9), had already diversified by the end of the Paleozoic Era. Some had teeth that were adapted for crushing shellfish, like those of the bony fish shown in Figure 13-8*C*. The living Port Jackson shark of Australia, a descend-

A

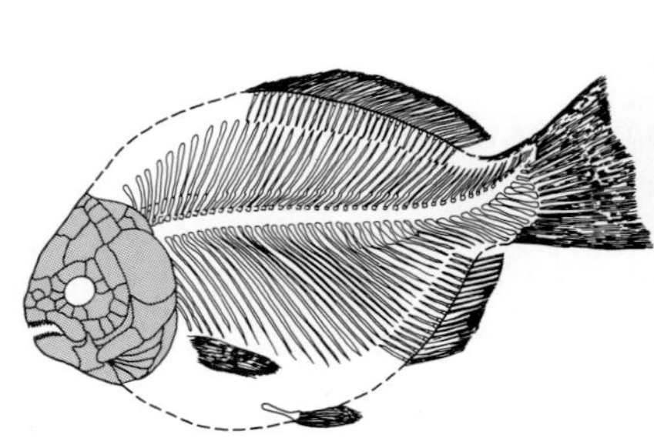

B

C

FIGURE 13-8 The Jurassic fish *Dapedius*. *A.* Unlike modern fishes, this early form had scales that barely overlapped one another. *B.* It also had asymmetrical skeleton supports within its tail (note the upturned spinal column). *C.* This particular genus had knobby teeth for crushing shellfish, like those in the roof of a similar fish's mouth. *(Museum Hauff, Holzmaden, Germany.)*

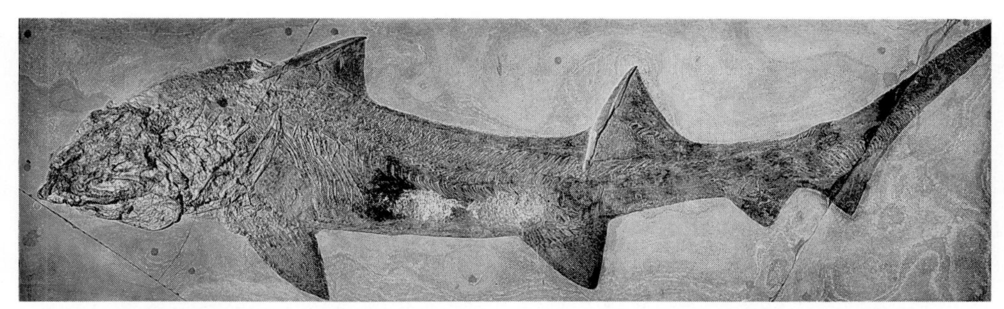

FIGURE 13-9 A hybodont shark *(Hybodus)* from the Lower Jurassic Series of Germany. This animal, which was about 2.2 meters (~7 feet) in length, resembled the bony fish of Figure 13-8 in that its teeth were adapted for crushing shellfish. *(Museum für Geologie und Paläontologie, Tübingen, Germany.)*

ant of the hybodonts, has similar teeth and feeds on mollusks. While the hybodonts declined in diversity late in Mesozoic time, more modern sharks appeared. Mackerel sharks, for example, evolved during the Jurassic Period, as did the family that includes the modern tiger shark.

Many reptiles that resembled the popular conception of sea monsters emerged in early Mesozoic seas. Among them were the **placodonts,** which, like many early Mesozoic fishes, were blunt-toothed shell crushers (Figure 13-10). The placodonts' broad, armored bodies gave them the appearance of enormous turtles. Cousins of the placodonts were the **nothosaurs** (Figure 13-11), which have been found in Early Triassic deposits and seem to have been the first reptiles to invade the marine realm. Nothosaurs had paddlelike limbs resembling those of modern seals. It seems likely that, like modern seals, they were not fully marine, but lived along the seashore and periodically plunged into the water to feed on fishes. Although the placodonts and nothosaurs did not survive the Triassic Period, a group of more fully aquatic reptiles evolved from the nothosaurs in mid-Triassic time, and these reptiles, known as **plesiosaurs,** played an important ecologic role for the remainder of the Mesozoic Era. Plesiosaurs apparently fed on fishes and, in Cretaceous time, attained the proportions of modern predatory whales, reaching some 12 meters (~40 feet) in length. The limbs of plesiosaurs were winglike paddles that propelled these animals through the water in much the same way that birds fly through the air (Figure 13-12).

By far the most fishlike reptiles of Mesozoic seas were the **ichthyosaurs,** or "fish-lizards," many of whose species must have been top predators in marine food webs. Superficially, ichthyosaurs bear a closer resemblance to modern dolphins, which are marine mammals, than to fishes;

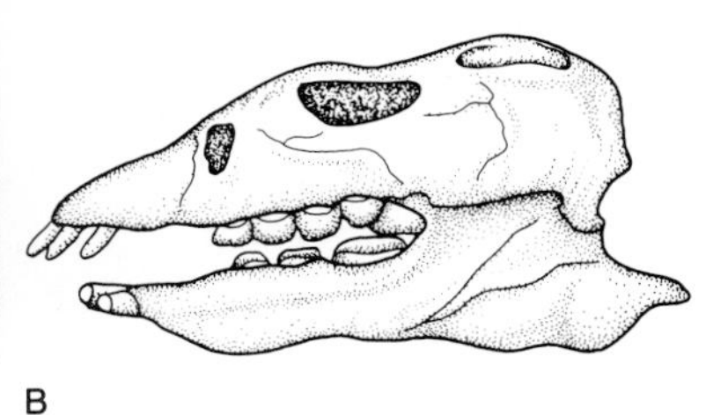

FIGURE 13-10 Reconstruction of a Triassic placodont. This aquatic reptile *(A)*, which crushed shelled marine invertebrates of the sea floor with large, rounded teeth *(B)*, was about 1.5 meters (~5 feet) long. *(A. Dan Varner.)*

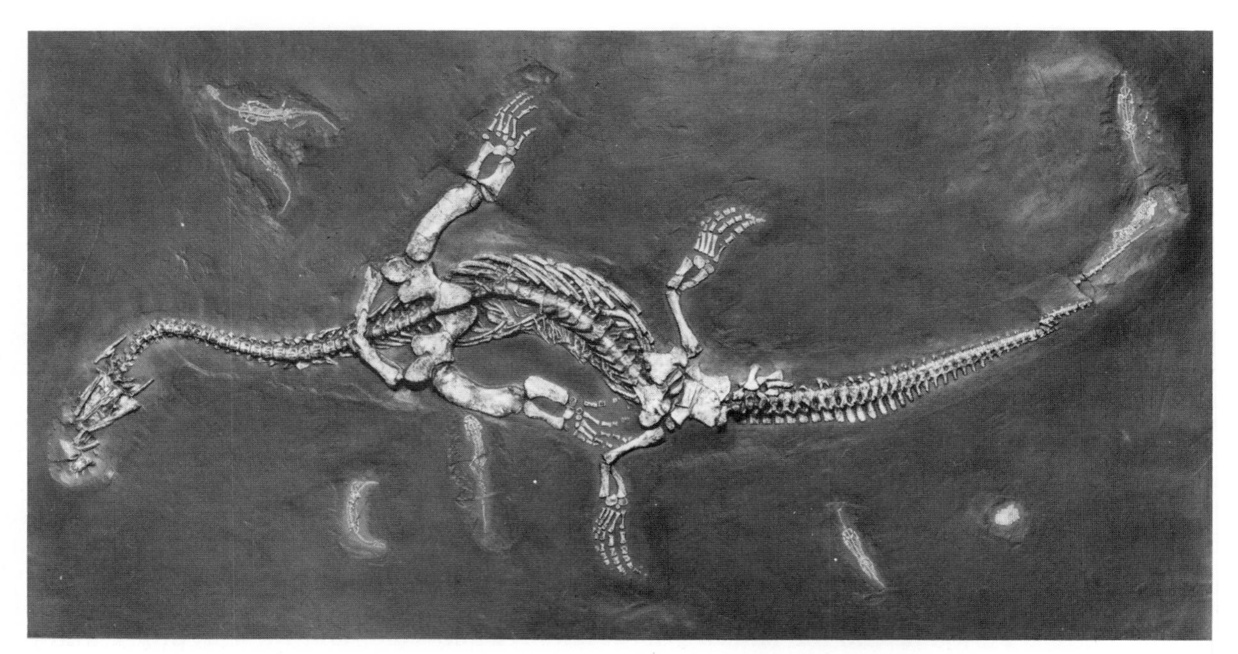

FIGURE 13-11 The nothosaur *Ceresiosaurus*, which was about 2.2 meters (~7 feet) long, preserved with small nothosaurs of a different family in the Middle Triassic Muschelkalk of Germany. Like modern seals, these animals probably fished along the shore. *(Museum für Geologie und Paläontologie, Tübingen, Germany.)*

outlines of skin preserved in black shales under low-oxygen conditions show the dolphinlike profiles of some ichthyosaurs (Figure 1-13). The ichthyosaurs, however, had upright tail fins rather than the horizontal pair of rear flukes that propel dolphins through the water. In addition, the extension of the backbone into the ichthyosaur tail bent downward, in contrast to the upward curve that characterized early Mesozoic bony fishes (Figure 13-8). Large eyes supplemented other adaptations of ichthyosaurs for rapid swimming in the pursuit of prey. Ichthyosaurs were fully marine and thus could not easily lay eggs; instead, they bore live young. In fact, skeletons of ichthyosaur embryos have been found within the skeletons of adult females.

FIGURE 13-12 Late Jurassic (Oxfordian) plesiosaurs from England mounted in swimming position. Note the paddle-like limbs. These two animals illustrate the two body types of plesiosaurs. *Cryptoclidus*, above, has a long neck and a short head, whereas *Peloneustes*, below, has a short neck and a long head. *Cryptoclidus* is about 3 meters (~10 feet) long. *(Museum für Geologie und Paläontologie, Tübingen, Germany.)*

FIGURE 13-13 A marine crocodile, *Steneosaurus*, in the Lower Jurassic Posidonia Shale of Holzmaden, Germany. The animal was more than 2.5 meters (~8 feet) in length. Adjacent to the skeleton on the left is a cluster of stones that were located in the gut of the animal, where they served to grind up food. *(Museum für Geologie und Paläontologie, Tübingen, Germany.)*

Surprising as it may seem, the last important group of early Mesozoic marine reptiles to evolve were the early **crocodiles,** which, as we will see, were related to the dinosaurs. Although crocodiles evolved in Triassic time as terrestrial animals, some were adapted to the marine environment by earliest Jurassic time (Figure 13-13). In fact, some crocodiles became formidable ocean-going predators whose finlike tails were well adapted for rapid swimming.

TERRESTRIAL LIFE

The presence of dinosaurs during the Mesozoic Era gave the biotas of large continents an entirely new character, but Mesozoic land plants were also distinctive. Because these plants were positioned at the bases of the food webs to which dinosaurs belonged, we will review them first.

Land Plants: The Mesozoic Gymnosperm Flora

Unlike terrestrial animals, land plants do not appear to have undergone a dramatic mass extinction at the close of the Paleozoic Era. As we learned in Chapter 12, the decline of the late Paleozoic floras began long before the end of the Permian Period. In effect, the transition from the late Paleozoic kind of flora to the Mesozoic kind of flora began before the start of the Mesozoic Era.

Among the groups that decreased in diversity long before the end of Permian time were the lycopod trees, which formed coal swamps, and the sphenopsid and cordaite trees, which inhabited higher ground. Persisting into the Mesozoic Era in greater numbers were ferns and seed ferns. Seed ferns, however, were reduced in abundance and apparently failed to survive into Jurassic time. Ferns, of course, survived, but diverse and abundant as they are today, they are nowhere near as prevalent as they were in Triassic time. Ferns, in fact, dominate Triassic fossil floras (Figure 13-14).

Most of the trees that stood above Triassic ferns belonged to three groups of gymnosperms that had already become established during the Permian Period. The most diverse of these three groups was the one comprising the **cycads** and **cycadeoids;** they were followed by the conifers,

FIGURE 13-14 A fossil fern from the Upper Triassic Chinle Formation of New Mexico (×0.7). *(U.S. Geological Survey.)*

A

B

FIGURE 13-15 Leaves of the living ginkgo, *Ginkgo biloba (A)*, and a similar leaf of early Cenozoic age *(B)*. *(A. Dave Watts/NHPA. B. Smithsonian Institution.)*

which we have already discussed (p. 38), and the **ginkgos.** All three of these groups survive to this day, but the cycads are rare, and there is only one living species of ginkgo remaining on Earth (Figure 13-15).

The trees that belonged to these three dominant groups are united as gymnosperms because they were all characterized by exposed seeds. The seeds of pines and other conifers, for example, rest on the projecting scales of their cones. There is a reason for this configuration. Whereas flowering plants, which did not evolve until Cretaceous time, can attract insect and bird pollinators, most gymnosperms rely primarily on wind to carry their pollen from tree to tree. With the possible exception of the pine family, all of the modern conifer families were present in early Mesozoic time. The few modern species of the less familiar cycads are tropical trees that superficially resemble palms (Figure 2-17). Cycadeoids, which were similar in form and closely related to the cycads, are extinct. The trunks of these plants are well known as early Mesozoic fossils. The single living species of ginkgo looks more like a hardwood tree (that is, like an oak or a maple) than a conifer, and, like hardwoods, it sheds its leaves seasonally. This surviving species of ginkgo is a true living fossil whose record extends back some 60 million years to the Paleocene Epoch (Figure 13-15), early in the Cenozoic Era.

Cycads, cycadeoids, conifers, and ginkgos formed the forests of the Jurassic Period, but the

FIGURE 13-16 Reconstruction of a Mesozoic landscape. Cycads, cycadeoids, and ferns appear in the foreground and conifers along the horizon. *(Drawing by Z. Burian under the supervision of Professor J. Augusta.)*

cycads were so dominant that the Jurassic interval has been called the Age of Cycads. Ferns of the Jurassic Period were less conspicuous as undergrowth than those of Triassic time. Both Triassic and Jurassic landscapes, however, would have looked more familiar to us than Paleozoic landscapes, largely because of the presence of conifers that closely resembled modern evergreens (Figure 13-16). Even so, the absence of flowering plants such as grasses and hardwood trees would have made Mesozoic floras appear strange to a modern observer.

Terrestrial Animals: The Age of Dinosaurs Begins

Unfortunately, local fossil records for land animals have been found to span the boundary between the Permian and Triassic systems in only two regions: the Karroo Basin of South Africa and an area of Russia near the Ural Mountains. Significantly, the fossil records of these two regions tell the same story. Just below the Permian-Triassic boundary in both regions, most of the dozens of genera of Late Permian mammal-like reptiles disappeared suddenly from the fossil record, marking a major mass extinction. What remained at the start of Triassic time were a few predatory genera and the large herbivore *Lystrosaurus,* famous for its fossil occurrence on many of the now widely dispersed fragments of Gondwanaland (Figure 6-12).

Four-Legged Animals Although the mammal-like reptiles rediversified during the Triassic Period to play an important ecologic role once more, they barely survived into the Jurassic Period. Nonetheless, they left an important legacy in the form of the true mammals, which evolved from them near the end of the Triassic Period. **Mammals,** which are endotherms with hair and which suckle their young, are the dominant large animals of modern terrestrial habitats, but they remained small and peripheral throughout the Mesozoic Era. Apparently no species grew larger than a house cat. Their problem seems to have been that the **dinosaurs** also evolved during the Late Triassic interval and quickly rose to dominance. (Mammals will be discussed at length in Chapter 14. The fossil record for mammals is better in the Cretaceous System than in the Triassic or Jurassic.)

Early dinosaurs' larger size (Figure 13-17) may have given them an advantage over primitive

FIGURE 13-17 Triassic animals of the genus *Lagosuchus* intimidating an early mammal of smaller body size. *Lagosuchus,* which was about 30 centimeters (~ 1 foot) tall, was either a primitive dinosaur or a thecodont that closely resembled the earliest dinosaurs. Thecodonts were the ancestors of dinosaurs. *(Drawing by Gregory S. Paul.)*

FIGURE 13-18 Terrestrial life of Late Triassic time in Argentina. The plants are of the widespread genus *Dicroidium.* The largest animals depicted here are thecodonts of the genus *Saurosuchus* that were about 7 meters (~25 feet) long. Confronting them are three small, primitive dinosaurs of the genus *Herrerasaurus.* The dead animal is the rynchosaurian reptile *Scaphonyx.* Two small thecodonts of the genus *Ornithosuchus* are scampering off in the foreground. In the left foreground is the long-legged primitive crocodile *Trialestes. (Drawing by Gregory S. Paul.)*

mammals. They inherited their locomotory capacity from their ancestors, the **thecodonts,** which evolved during the Triassic Period. Some thecodonts were adapted for speedy two-legged running in the fashion of ostriches and other flightless birds, but all thecodonts probably spent much time standing or walking on all fours. The upper portion of the legs of many thecodonts stood beneath their bodies rather than sprawling slightly out to the side as they did in mammal-like reptiles. This feature, which facilitated running, was passed on to the dinosaurs and seems to have been a key to the dinosaurs' success.

The first dinosaurs resembled bipedal thecodonts (that is, thecodonts that traveled on their hind legs), but the dinosaurs' skulls were differently formed and their teeth were more highly developed. Dinosaurs (formally termed Dinosauria) did not become gigantic before the end of the Triassic, but during Triassic time they did reach lengths of more than 6 meters (~20 feet). Figure 13-18 displays a small species along with thecodonts that stood on all fours, large mammal-like reptiles, and a primitive long-legged crocodile. The crocodiles, like the dinosaurs, evolved from thecodonts in Late Triassic time.

Two other important groups, both of which are familiar in the modern world, also appear to have become established in Triassic time. One was the **frogs,** which were and remain amphibians of small body size. The oldest known fossil displaying the form of a modern frog is of earliest Jurassic age (Figure 13-19), but froglike skeletons have also been found in Triassic rocks. The other modern group was the **turtles,** although the earliest turtles lacked the ability to pull their heads and tails fully into their protective shells.

By Late Triassic time, when mammal-like reptiles had declined in diversity, they lived alongside increasing numbers of dinosaurs, the still-diverse thecodonts, and other, mainly smaller amphibians and reptiles. A few kinds of large amphibians persisted as well. The stage was thus set for the ascendancy of the dinosaurs to a dominant position for the remainder of the Mesozoic Era.

The Rise of the Dinosaurs Unfortunately, the known fossil record of Early Jurassic time is too

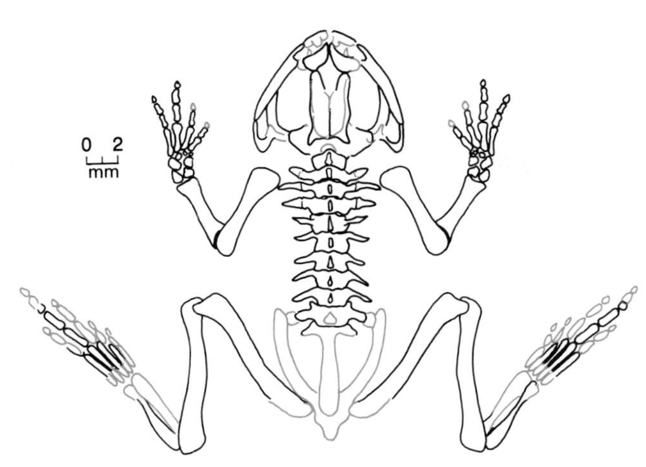

FIGURE 13-19 A partly restored skeleton of a frog belonging to the genus *Vieraella* from the Lower Jurassic Series of Argentina. This is one of the oldest true frogs known from the fossil record. *(After R. Estes and O. A. Reig, in J. L. Vial [ed.], Evolutionary Biology of the Anurans, Columbia University Press, New York, 1973.)*

FIGURE 13-20 Herbivorous dinosaurs from the Morrison Formation of Utah. *A. Diplodocus,* an animal that stretched to a length of more than 26 meters (~87 feet) from nose to tail. *B. Stegosaurus,* an armored herbivore with spikes on its tail. *(Smithsonian Institution.)*

poor to permit us to piece together the details of the dinosaur's great rise to dominance. Nonetheless, fossil remains of huge dinosaurs even in Lower Jurassic rocks indicate that the dinosaurs evolved rapidly; the oldest dinosaur giants are found in Australia.

Dinosaurs fall into two groups, which are characterized by different pelvic structures (Figure 1-7). All of the "bird-hipped" (**ornithischian**) dinosaurs were herbivores, whereas the "lizard-hipped" (**saurischian**) group included both herbivores and carnivores. In both groups there were species that traveled on two legs and others that moved about on all fours. The largest of all the dinosaurs were the **sauropods,** lizard-hipped herbivores that moved about on all fours. Some interesting aspects of dinosaur biology are presented in Box 13-1.

By Late Jurassic time, both bird-hipped and lizard-hipped dinosaurs were quite diverse. The most spectacular Jurassic assemblage of fossil dinosaurs in the world is found in the Upper Jurassic Morrison Formation, which extends from Montana to New Mexico. At Como Bluff, Wyoming, dinosaur bones are so common that a local sheep herder constructed a cabin of them because they were the most readily available building materials! The dinosaurs of the Morrison Formation, which include more than a dozen genera, are representative of the kinds of dinosaurs that lived throughout the world during Late Jurassic time. Several of the common Morrison species are shown in Figures 13-20 and 13-21.

Creatures That Took to the Air Late in the Triassic Period, vertebrate animals invaded the air for the first time as the **pterosaurs** came into being. These animals had long wings and hollow bones that served to facilitate flight. Some species had long tails as well (Figure 13-22). The structure of the pterosaur skeleton reveals that these reptiles were capable of flying, but the great length of the wings suggests that most species flapped their wings primarily when taking off and then, once

FIGURE 13-21 Reconstruction of the dinosaur fauna of the Morrison Formation. *Camarasaurus* walks toward the right in the background; *Diplodocus* (Figure 13-20*A*) is drinking water; at the right is *Stegosaurus* (Figure 13-20*B*). Reclining on the left is *Allosaurus*, a large carnivorous dinosaur, about 6 to 7 meters (~20 to 25 feet) long. In the foreground are dragonflies, pterosaurs, turtles, and a crocodile. The trees are conifers. *(Drawing by Gregory S. Paul.)*

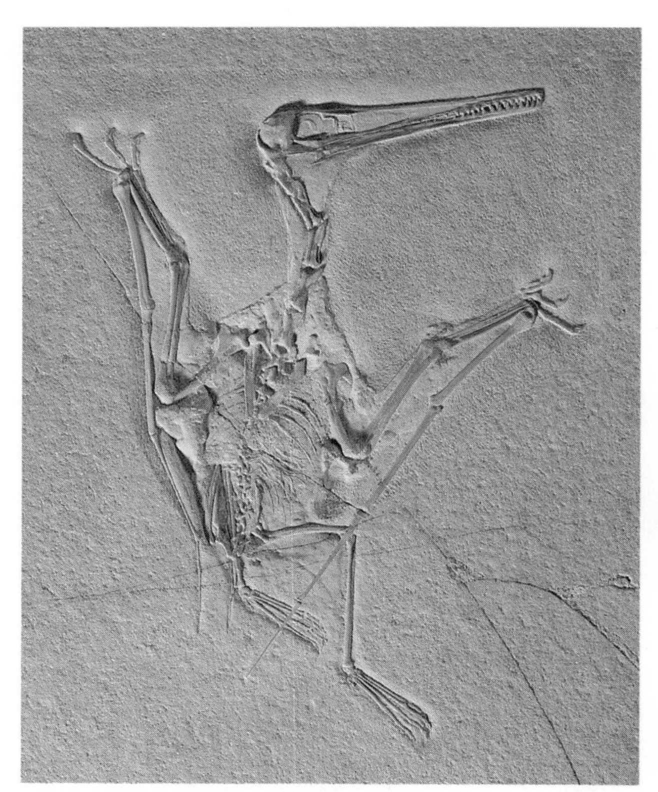

FIGURE 13-22 The pterosaur, or "flying lizard," *Pterodactylus.* This skeleton was preserved intact in the Solnhöfen Limestone of Germany; it measures about 60 centimeters (~2 feet) long. Many of the bones are hollow, like those of a bird. *(Franz Hoeck, Bayerische Staatssammlung für Paläontologie und Geologie München.)*

airborne, soared on air currents without much flapping. The behavior of pterosaurs when not in flight has been widely debated. Most species appear to have been able to walk and also to climb adeptly with the aid of their hooklike claws.

The oldest known fossil birds are of Late Jurassic age. The first clue to their existence was a feather discovered in 1861 in the fine-grained Solnhöfen Limestone of Germany, followed a few months later by the discovery of an entire skeleton of the species to which the feather belonged (Figure 13-23). This feathered animal was given the name *Archaeopteryx,* which means "ancient wing." *Archaeopteryx* had a skeleton so much like that of a dinosaur that it would be regarded as one were it not for its plumage. This animal is a classic missing link — in this case, the link between birds and their flightless ancestors. The teeth, large tail, and clawed forelimbs of *Archaeopteryx*, which are absent from advanced birds, reflect its dinosaur ancestry. *Archaeopteryx* lacks a breastbone and so is assumed to have possessed weak flying muscles. It was probably a clumsy flier by the standards of modern birds. Unfortunately, the hollow bones of birds are fragile, and no other bird bones have been found within the Jurassic System. There are unpublished reports of bird bones in Upper Triassic rocks. If verified, these finds would place the

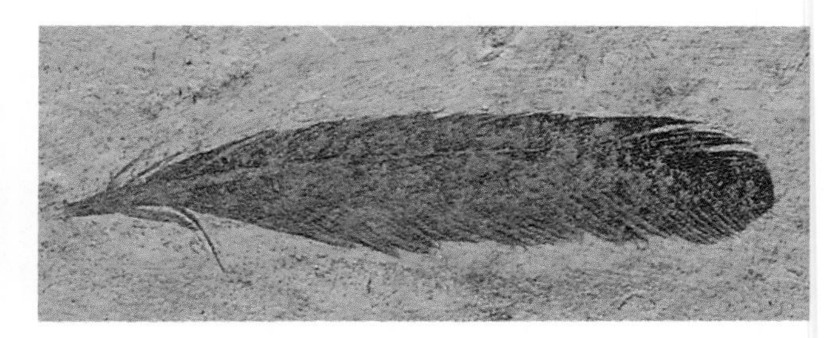

A

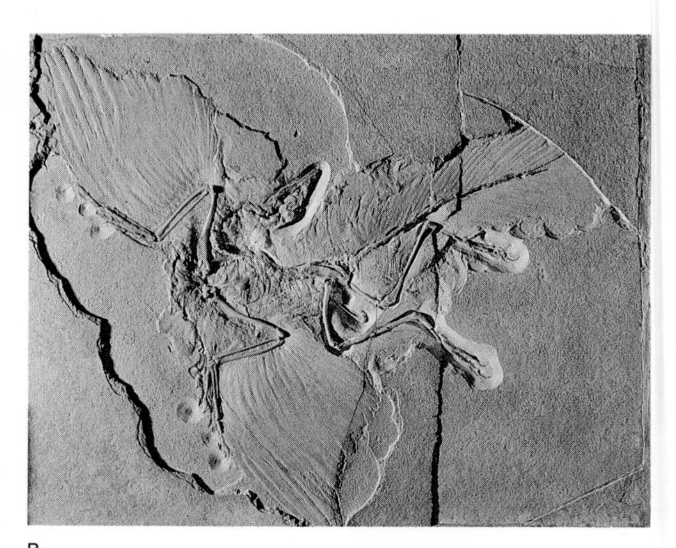

B

FIGURE 13-23 Fossil remains of *Archaeopteryx lithographica*, the oldest known bird, from the Upper Jurassic Solnhöfen Limestone of Germany. This animal was about the size of a crow. The existence of a bird during Solnhöfen deposition was first indicated by the discovery of a feather *(A)*. The asymmetry of the feather suggests that it aided in flight; flightless living birds have feathers that are symmetrical about the central shaft. The bird itself was soon found, and impressions of long feathers are clearly visible around it in the fine-grained limestone *(B)*. Despite the feathers, *Archaeopteryx* had a skeleton and teeth similar to those of dinosaurs. *(A. J. H. Ostrom, Peabody Museum of Natural History, Yale. B. Walther-Arndt-Fonds, Fordererkreis der naturwissenschaftlichen Museen Berlins e.V.)*

origin of birds much farther back in geologic time than is now believed. *Archaeopteryx* would then represent a primitive type of bird that happened to survive for a long interval of time.

PALEOGEOGRAPHY

About the time the Mesozoic Era began, all of the major landmasses of the world became united as the supercontinent Pangaea (Figure 13-24). Near the end of Triassic time, Pangaea began to break apart, but continental movement is so slow that even by the end of Jurassic time, the newly forming continental fragments were hardly separated. Thus, throughout the early Mesozoic Era, Earth's continental crust was concentrated on one side of the globe. Sea level rose slightly during Early Triassic time (see Major Events, p. 338). As in Late Permian time, however, the bulk of the continental crust during the Triassic Period stood above sea level, forming one vast continent. At the start of Triassic time, the Tethys, sometimes called the Tethyan Seaway, was an embayment of the deep sea projecting into the portion of equatorial Pangaea that today constitutes the Mediterranean. Later in Triassic and Jurassic time, rifting extended the Tethys between Eurasia and Africa and all the way westward between North and South America to the Pacific.

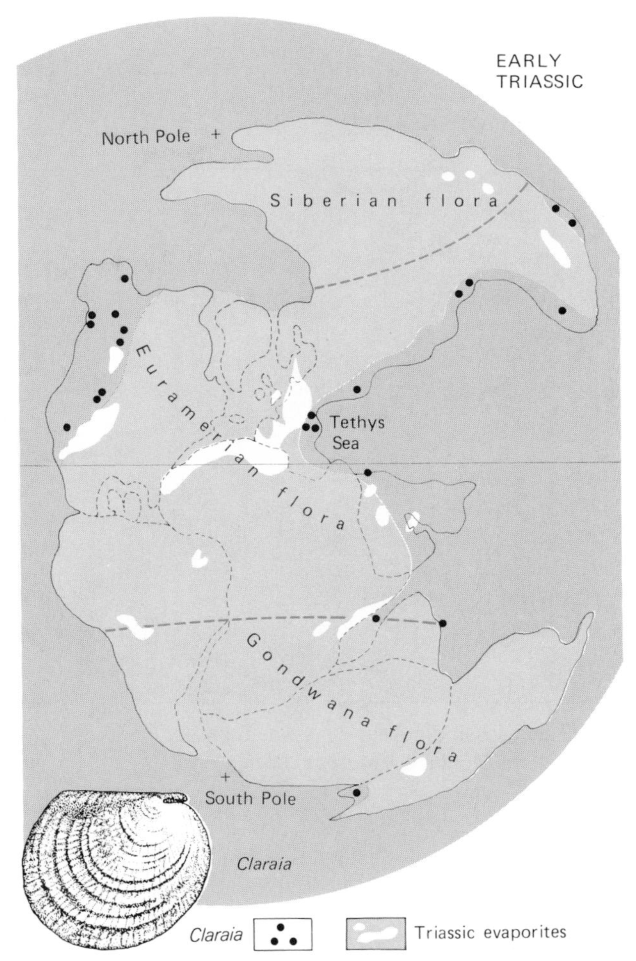

FIGURE 13-24 World geography of Early Triassic time. The Euramerian flora occupied a broad, warm belt across the middle of Pangaea, and the Siberian and Gondwana floras occupied regions to the north and south. The bivalve mollusk genus *Claraia (lower left)* was broadly distributed along both the eastern and western borders of Pangaea. The widespread evaporites illustrated here represent the entire Triassic Period, which was relatively warm and dry even at high latitudes. *(Modified from G. E. Drewry et al., Jour. Geol. 82:531–553, 1974.)*

Pangaea During the Triassic Period

Although the dominant land plants of the Triassic Period differed from those of the Permian, the distributional pattern of floras on Pangaea remained much the same; a Gondwana flora existed in the south and a Siberian flora in the north (Figure 13-24). The Euramerian flora grew under warmer, drier conditions at low latitudes; in fact, unusually extensive deposition of evaporites attests to the presence of arid climates far from the equator. This condition may have resulted in part from the sheer size of Pangaea, which was so large that many of its regions lay far from the low-standing oceans.

The survival of only a small percentage of Permian species into the Triassic Period resulted in some striking biogeographic distributions. In the oceans, the scalloplike bivalve genus *Claraia* ranged over an enormous area (Figure 13-24) and is assumed to have occupied the seafloor even in deep water. On land, the large herbivorous therapsid *Lystrosaurus* ranged over large areas of the globe; *Lystrosaurus* has been found in the fossil records of several continents that represent fragments of Gondwanaland (Figure 6-12). It appears that at the start of the Triassic Period, few therapsids preyed on *Lystrosaurus*, so its populations

Box 13-1
Who Were the Dinosaurs?

A nest of fossil dinosaur eggs was found in the Cretaceous of Montana. *(Museum of the Rockies.)*

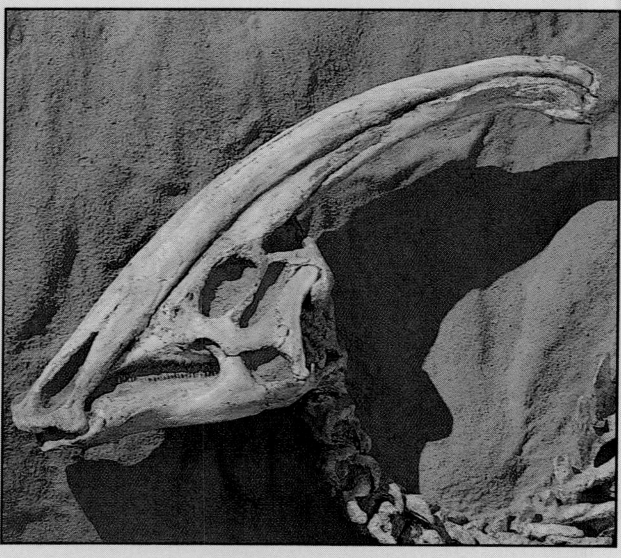

The skull crest of *Parasauralophus,* which was about one meter (~3 feet) long, probably served as a resonating chamber for trumpeting. *(Royal Tyrrell Museum of Paleontology/Alberta Culture and Multiculturalism.)*

As researchers bring increasing knowledge and experience to bear on their examination of dinosaur remains, many earlier notions of how these creatures looked and behaved are being debunked. It is now evident, for example, that not all dinosaurs were of massive proportions; many, in fact, were less than 1 meter (~3 feet) long. Moreover, while dinosaurs have often been portrayed as hulking, lumbering creatures, it is now known that many were as agile as ostriches, which are famous for their great speed. The orientations of dinosaur limbs in their sockets indicate that the legs were positioned almost vertically beneath the body, and fossil trackways of dinosaurs confirm that this posture was typical. The left and right tracks are nearly in line, signifying that both feet were positioned almost beneath the center of the body. Most fossil trackways also reveal long strides for the size of the individual tracks, suggesting that dinosaurs tended to move rapidly.

Perhaps even more surprising, dinosaurs appear to have been social animals. Each of the various duck-billed dinosaurs, for example, had a tall, crested skull, which may have functioned like the resonating chambers of a trumpet. Not only did many species of dinosaurs probably communicate by sound, but some of their well-preserved trackways show that large groups of animals sometimes traveled together.

It has long been known that dinosaurs laid eggs. Some of their eggs have remained in the circular pattern in which they were carefully laid; often, however, several circles of eggs were buried on top of one another. The eggs were tapered toward one end, which the mother thrust into the soil, and the babies hatched by breaking through the tops of the eggs. Nests of baby dinosaurs found in the Upper Cretacious of Montana—clusters of juvenile skeletons with broken eggshells in depressions—shows that the eggs were not abandoned. Like living birds and crocodiles, dinosaurs cared for their young, which grew rapidly. Having hatched at a length of just a few inches, they reached about 1.5 meters (~5 feet) in length before

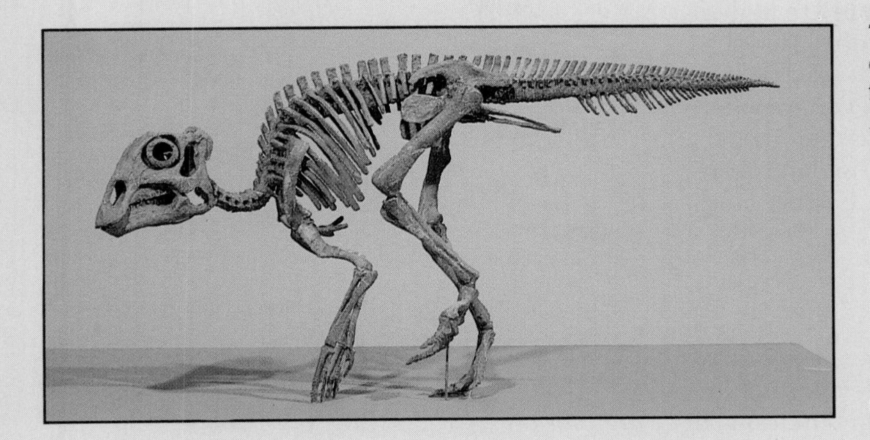

This *Maiosaura* hatchling, about 50 centimeters (20 inches) long, was found in a nest. *(Museum of the Rockies.)*

leaving the nest at the end of the warm season, perhaps only 3 or 4 months after hatching.

An important question has been posed with regard to the largest dinosaurs: How did the relatively small jaws of these enormous animals chew up enough food to live on? The answer is that these giant herbivores used their mouths and jaws only for gathering and swallowing plant food. In the animals' intestinal tracts were "gizzard stones," like those of birds but much larger, which helped the animals grind up coarse material after they had swallowed it.

The greatest controversy about dinosaurs, however, has related to their metabolism. Dinosaurs have traditionally been classified as reptiles, and it has thus been assumed that they were ectothermic (or cold-blooded). It has recently been argued, however, that dinosaurs were actually endothermic, or warm-blooded. The newer idea stems in part from evidence that dinosaurs were very active animals and in part from indications that they were more successful during the Mesozoic Era than were warm-blooded early mammals or the mammal-like reptiles, which had had a long evolutionary head start in the race for dominance on the land. The argument here is that dinosaurs could not have competed with mammals if they had not been capable of sustaining a high rate of activity for long intervals of time while hunting or fleeing from predators. Ectothermic animals, including modern reptiles, have little endurance (p. 314). Additional evidence derives from the fact that in communities of dinosaurs, predators usually made up less than 10 percent of the volume of living species, as is the case for living and fossil mammal communities. In communities of ectothermic animals, predators commonly represent 40 percent or so of the volume of living tissue. Having low metabolisms, they need little food, and many can sustain themselves on small populations of prey animals. The low percentage of predators in many dinosaur communities argues for a closer resemblance to mammal communities than to reptile communities. The microscopic structure of their bones has also been debated. Endothermic animals differ from ectothermic animals in bone structure, and it appears that the dinosaurs' bone structure was of the endothermic type. Finally, the rapid growth of juvenile dinosaurs points to a high metabolism. Modern reptiles grow more slowly than dinosaurs did.

Whatever the details of dinosaurs' metabolism may have been, we at least recognize that even large dinosaurs were active, highly adapted animals rather than simple, lumbering beasts. For all we know, many dinosaurs might have fared well in the modern world had they been given the chance. The sudden extinction of the dinosaurs at the end of the Cretaceous Period appears to have been more of an unfortunate accident than an indication of biological inferiority.

were huge. As we saw earlier, the therapsids underwent renewed radiation during the Triassic Period, and partway through the period thecodonts and dinosaurs began their expansion. Even when these vertebrate groups radiated to high diversities, most of their species were also wide-ranging. Many families of Triassic terrestrial vertebrates are found as fossils on several modern continents. In fact, it has been remarked that the Triassic is the only period whose fossil land vertebrates clearly indicate that all of Earth's continents were connected.

The Breakup of Pangaea

The most spectacular geographic development of the Mesozoic Era was the fragmentation of Pangaea, an event that began in the Tethyan region. As the Triassic Period progressed, the Tethyan Seaway spread farther and farther inland, and eventually the craton began to rift apart. The Tethys subsequently became a deep, narrow arm of the ocean separating what is now southern Europe from Africa. During the Jurassic Period, this rifting propagated westward, ultimately separating North and South America. South America and Africa, however, did not separate to form the South Atlantic until the Cretaceous Period; in fact, all of the Gondwanaland continents remained attached to one another until Cretaceous time. North America began to break away from Africa in mid-Jurassic time. Interestingly, this rifting generally followed the old Hercynian suture. Rifting occurred as some of the arms of a series of triple junctions joined, tearing Pangaea in two (Figure 6-32).

The rifting that formed the Atlantic had another important consequence. When continental fragmentation begins in an arid region near the ocean, evaporite deposits often form in that region (p. 152). Thus, as rifting began in Pangaea, extension produced normal faults between Africa and the northern continents (Figure 13-25); zones bounded by such faults sank, and water from the Tethys to the east periodically spilled into the trough and evaporated. Evaporites that were precipitated in this trough are now located on opposite sides of the Atlantic, both in Morocco and offshore from Nova Scotia and Newfoundland.

During Middle and Late Jurassic time, one arm of rifting passed westward between North

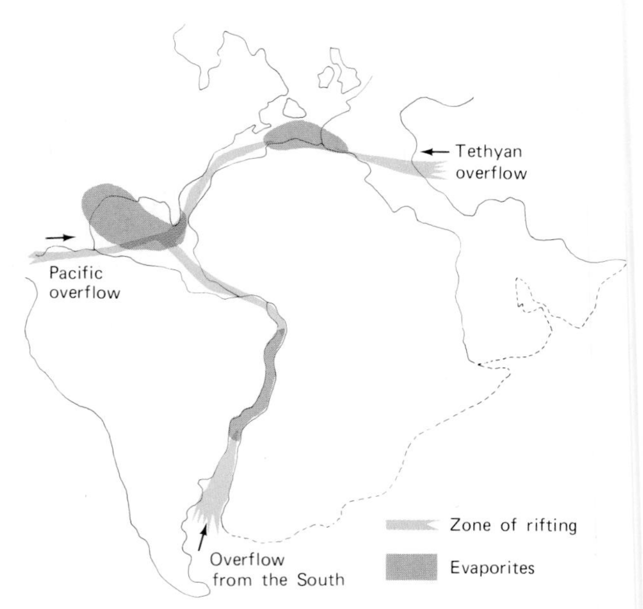

FIGURE 13-25 Early Mesozoic evaporites. Evaporites accumulated during the early stages of rifting that formed the Atlantic Ocean as oceans overflowed intermittently into newly forming fault basins. *(After K. Burke, Geology 3:613–616, 1975.)*

and South America, giving rise to the Gulf of Mexico. Early intermittent influxes of seawater into this rift, apparently from the Pacific Ocean, caused great thicknesses of evaporites to accumulate. Today these evaporites, which are known as the Louann Salt, lie beneath the Gulf of Mexico and in the subsurface of Texas. Because its density is low, the Louann Salt has in some places pushed up through younger sediments to form salt domes (Figure 13-26), many of which are associated with valuable reservoirs of petroleum and sulfur. The rifting that forms the South Atlantic did not begin until Early Cretaceous time, when the record shows that salt deposits formed after seawater spilled inland from the south (Figure 13-25).

The Jurassic World

Although sea level underwent only minor changes during Late Triassic and Early Jurassic time, it subsequently rose, with minor oscillations, until Late Jurassic time. Then, very late in the Jurassic Period, it underwent more rapid oscillations but remained at a high level, causing epicontinental seas to flood large areas of North America and Europe late in Jurassic time (Figure 13-27). Long

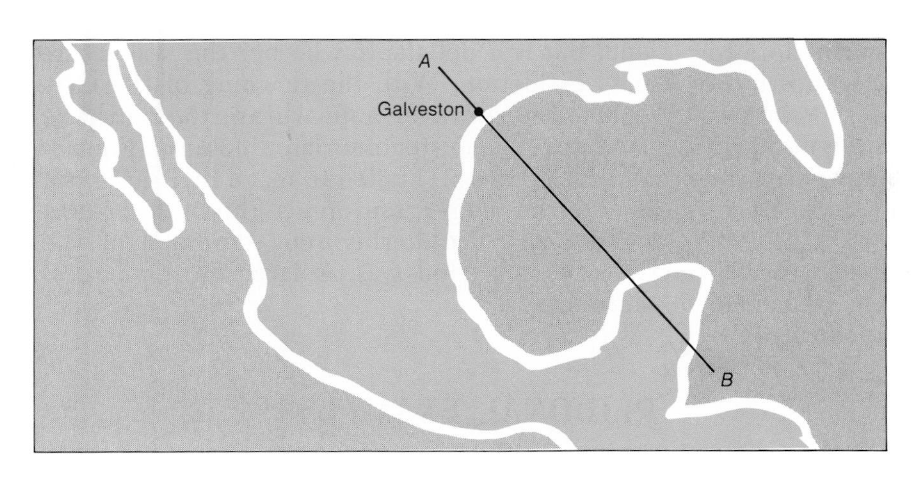

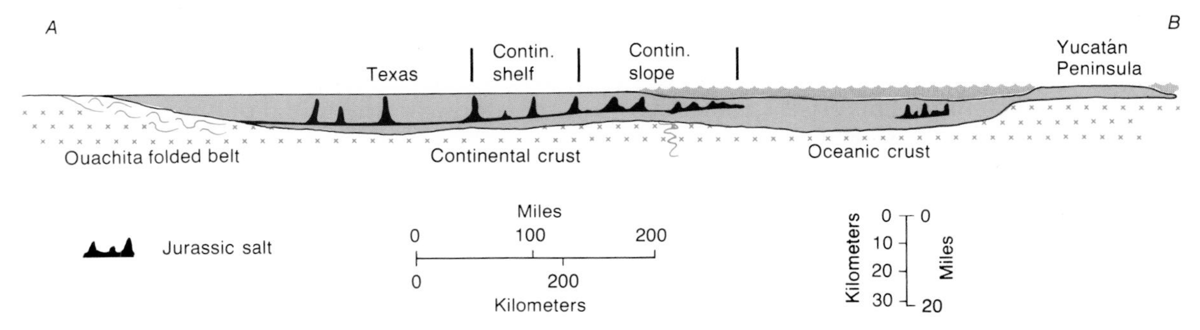

FIGURE 13-26 Position of Jurassic salt within the sediments beneath the Gulf of Mexico. The salt deposits, which rise into domes in some places because of their low density, accumulated when the Gulf of Mexico began to form by continental rifting. *(Modified from O. Wilhelm and M. Dwing, Geol. Soc. Amer. Mem. 83, 1972.)*

before the advent of plate-tectonic theory it was recognized that there were two biogeographic provinces of marine life in Europe during the Jurassic Period: the southern province, which was centered in the Tethys and is designated the Tethyan realm, and the northern province, which is labeled the Boreal realm. Figure 13-27 shows the positions of these provinces when sediments of the Oxfordian Stage were laid down at the time of maximum worldwide transgression for the Jurassic Period.

Coral reefs were largely restricted to the Tethyan realm, as were limestones and certain groups of mollusks. These observations suggest that the Tethyan was largely a tropical province. The transition from the Tethyan to the Boreal province resembled that which we see today from the tropical conditions of southern Florida, with its carbonates and coral reefs, to the subtropical conditions of northern Florida, where siliciclastic sediments prevail and reefs are lacking.

There is no doubt that temperature gradients from equator to pole were gentle throughout the Jurassic Period. Plants that appear to have required warmth (the Euramerian flora of Figure 13-27) occupied a broad belt extending to about

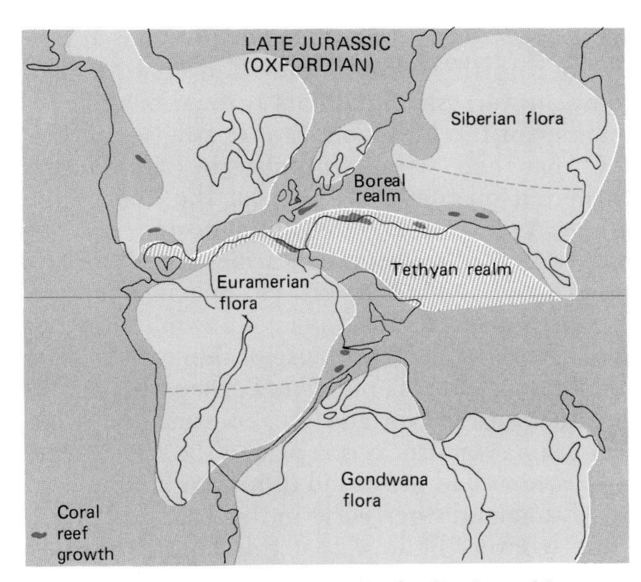

FIGURE 13-27 Late Jurassic (Oxfordian) world geography. The three floras persisted from Triassic time. The Tethyan marine realm, which was characterized by tropical life, including reef corals, extended all the way from the eastern Pacific across the newly forming Mediterranean and the newly forming Gulf of Mexico to the western Pacific. The cooler Boreal realm lay to the north.

60° north latitude. Even the Gondwana flora to the south and the Siberian flora to the north included groups of ferns whose modern relatives cannot tolerate frost. The high-latitude floras do not seem to have been tropical, however; they contained few cycads, and modern cycads are restricted to warm regions (Figure 2-17). Thus it appears that the large landmass that began to fragment during the Jurassic Period was bathed in tropical climates well to the north and south of the equator, even though temperatures were somewhat cooler toward the poles.

Mass Extinctions

The Triassic Period ended with one of the heaviest mass extinctions of all time. This crisis struck both on the land and in the sea. In the marine realm, about 20 percent of all families of animals suffered extinction. Conodonts and placodont reptiles (Figure 13-10) died out altogether. So did most species of bivalves, ammonoids, plesiosaurs (Figure 13-12), and ichthyosaurs (Figure 1-13), although all of these groups recovered in Jurassic time. The terrestrial victims included most genera of mammal-like reptiles and large amphibians.

The primary beneficiaries of the extinction on the land were the dinosaurs, which radiated rapidly during the Jurassic and then continued to dominate terrestrial habitats throughout the remainder of the Mesozoic Era. It now seems evident that there were two pulses of Late Triassic extinction on the land—one at the end of the Norian Age, the final stage of the period, and one at the end of the preceding Carnian Age. The timing of extinction in the seas is less clear, but many genera died out during the final few million years of the Norian Age. The cause of the Late Triassic extinctions remains unknown. Perhaps it is significant that late in Norian time, conifers and other groups of gymnosperms replaced the *Dicroidium* flora, which had prevailed in lowland habitats of Gondwanaland since early in the Triassic Period. Some workers believe that this floral transition signaled a climatic change, but this change may have entailed an increase in aridity rather than a lowering of temperatures.

In any event, the biosphere then remained relatively stable for millions of years. At the end of the Jurassic Period there was moderately heavy extinction of life both in the oceans and on the land, but it is debatable whether this was a true mass extinction. With the dawning of the Cretaceous Era, however, animal life on the land had a new aspect. The stegosaurian dinosaurs (Figures 13-20*B* and 13-21) failed to make the transition, as did the larger sauropods (Figures 13-20*A* and 13-21). No herbivorous dinosaur of the Cretaceous Period was as large as the largest sauropods.

REGIONAL EXAMPLES

In viewing regional events of early Mesozoic age, we will first focus on depositional basins in eastern North America—structures that illustrate events associated with the initial opening of the Atlantic Ocean. We will then examine the origin of the tectonic episodes that ultimately produced the mountains now standing in the American West.

Atlantic Fault Basins

During Early and Middle Triassic time, erosion subdued the Appalachian Mountains, which were centrally located in Pangaea. In Late Triassic time, long, narrow depositional basins bounded by faults developed on the gentle Appalachian terrain (Figure 13-28). These basins formed when Pangaea was splintered by normal faults on either side of the great rift that began to divide the continent, forming the Atlantic Ocean (Figure 6-32). One of the largest of these basins extended from New York City to northern Virginia and received sediments known as the Newark Supergroup. Here, during a Late Triassic and Early Jurassic interval of subsidence, the nonmarine sediments of the Newark Supergroup accumulated to a thickness of nearly 6 kilometers (~4 miles). Early Mesozoic basins resembling those of eastern North America are also found in Africa and South America, but these contain thick evaporite deposits. It was in the early phase of rifting that water from the Tethys periodically spilled into these southern basins to form vast salt deposits (Figure 13-25).

The best locations for investigating the basin sediments are in the eastern United States. One particularly well-studied basin passed through present-day Connecticut and Massachusetts,

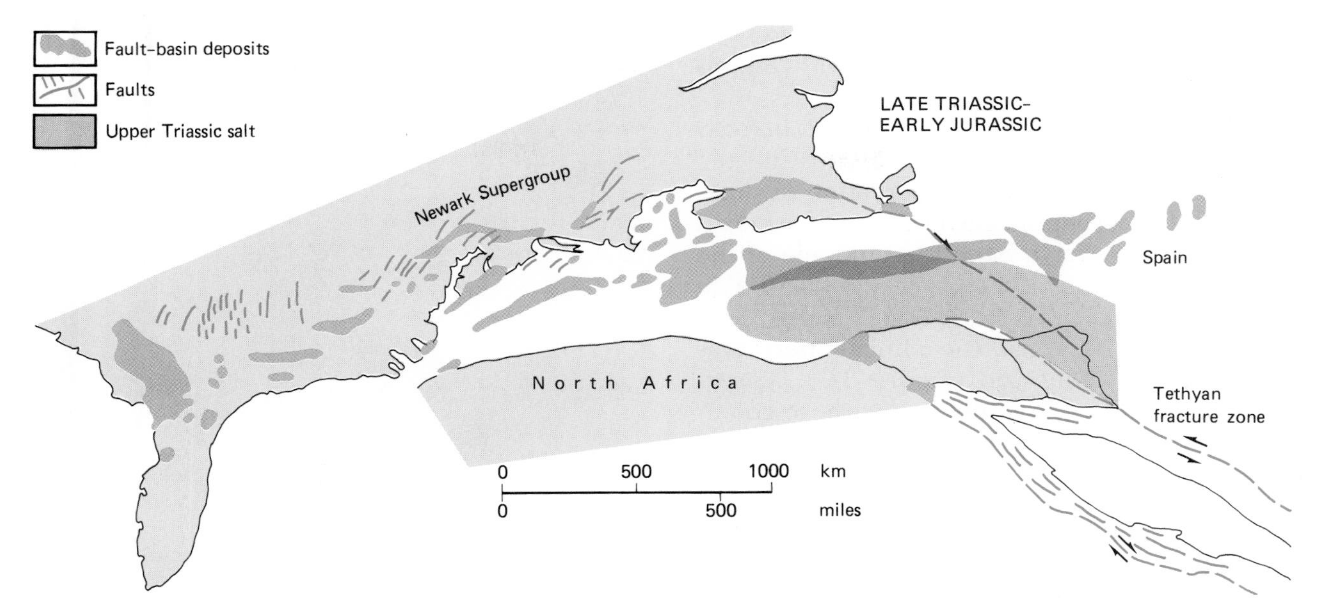

FIGURE 13-28 Geologic features in eastern North America and nearby regions during Late Triassic and Early Jurassic time. In eastern North America, block faulting produced elongate depositional basins, most of which paralleled the enormous rift that eventually formed the Atlantic Ocean. Salt deposits accumulated from the sporadic westward spilling of seawater from the Tethyan Seaway, where the Mediterranean was forming as a result of Africa's movement in relation to Europe. *(After W. Manspeizer et al., Geol. Soc. Amer. Bull. 89:901–920, 1978.)*

bounded on the east by a large normal fault along which the basin subsided continually while sediments accumulated from an eastern source area (Figure 13-29*A*). Several types of depositional environments existed within this basin. Coarse conglomerates that wedge out to the west accumulated on alluvial fans that spread from the eastern fault margin. Many sand-sized sediments of the basin are stream deposits. The fact that most of these deposits are composed of red arkose suggests that deposition in this area was rapid, since apparently there was little time for feldspars to disintegrate to clay. Well-laminated mudstones floored the lakes in the basin center. Cycles now visible in the sediments reflect expansion and contraction of the lakes, which for the most part must have been quite shallow. During some dry intervals, evaporite minerals were precipitated from the shrinking waters, but abundant fossil fish remains indicate that at other times the waters were hospitable to life. In fact, freshwater fishes underwent spectacular adaptive radiations in some of the larger lakes. These radiations resembled those that have occurred very recently in the great African lakes (Figure 5-13). It is interesting to note

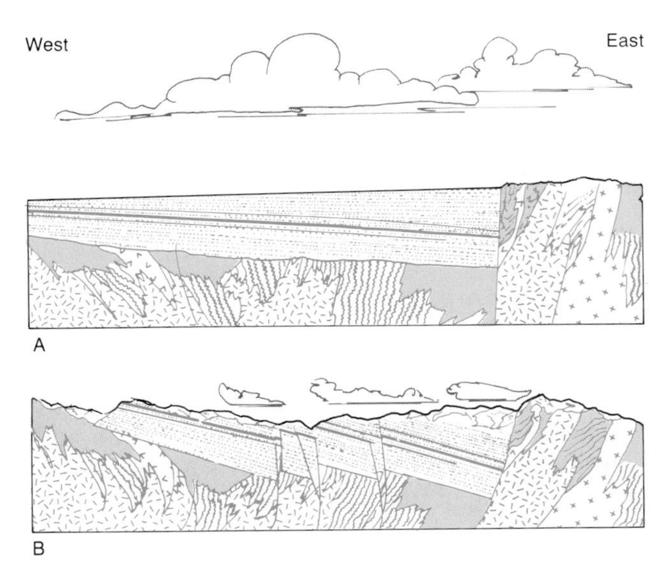

FIGURE 13-29 Cross sections of the large early Mesozoic fault basin that passes through central Connecticut, where the Newark Supergroup was deposited (Figure 13-28). *A.* The basin late in its depositional history, when great thicknesses of sediment had accumulated. As the basin subsided, lavas welled up periodically, forming dikes and sills, and gravels from the uplands to the east spread into the basin as alluvial fans. *B.* The eventual destruction of the basin by extensive faulting.

that these modern African lakes occupy rift valleys much like those in which the Newark Supergroup was deposited (Figure 6-31).

Although dinosaur tracks are common in rocks representing lake margins, conditions in the basins seldom favored the preservation of dinosaur skeletons, except in eastern Canada where a number of bones have been found. Some of the ancient soils in the basin that extends through Connecticut and Massachusetts contain caliche nodules, indicating that the climate here was warm and seasonally arid (Figure 3-1). Apparently bones decayed rapidly under these conditions, so that relatively few were preserved as fossils.

Periodically mafic magmas welled up through faults, forming dikes and widespread sills within the basin. One of the largest of these sills forms the Palisades along the Hudson River near New York City. At least some of the North American basins continued to subside until Early Jurassic time, when deposition ended with a final episode of faulting. After this time, the basins apparently moved so far westward along with the North American plate that they were no longer affected by the mid-Atlantic rifting. The fact that some of the basins are located several hundred kilometers from the present margin of North America (Figure 13-28) indicates how extensive the fracturing of a large continent can be; in most such instances, many small breaks and ruptures occur rather than a clean separation from sea to sea.

FIGURE 13-30 Silicified logs that have weathered out of the Triassic Chinle Formation in the Petrified Forest of Arizona. *(Tom Bean.)*

Western North America

Throughout the Triassic Period, much of the American West was the site of nonmarine deposition. Shallow seas expanded and contracted along the margin of the craton but for the most part remained west of Colorado. During Middle and Late Triassic time, western North America was especially free of marine influence.

Terrestrial and Marine Environments As in the Permian Period, the climate here remained largely arid. At times, however, there was sufficient moisture to permit the growth of large trees belonging to the Euramerian flora. The series of river and lake sediments in Utah and Arizona that are collectively known as the Chinle Formation, for example, erodes spectacularly in some places to reveal the well-known Petrified Forest of Arizona (Figure 13-30). In southwest Utah the Chinle is overlain by the Wingate Sandstone, a desert dune deposit. Over it lies a river deposit called the Kayenta Formation, on top of which rests the Navajo Sandstone. The Navajo, also a desert dune deposit, ranges upward in the stratigraphic sequence from approximately the position of the Triassic-Jurassic boundary. The Navajo is famous for its large-scale cross-bedding in the neighborhood of Zion National Park (Figure 3-10*C*).

During Middle and Late Jurassic time, as sea level rose throughout the globe (Major Events, p. 338), waters from the Pacific Ocean spread farther inland in a series of four transgressions, each more extensive than the last. The first such transgression went no farther than British Columbia and northern Montana, but the last, which is known as the Sundance Sea, spread eastward to the Dakotas and southward to New Mexico (Figure 13-31). Eventually, as mountain building progressed along the Pacific coast in Late Jurassic time, the Sundance Sea retreated.

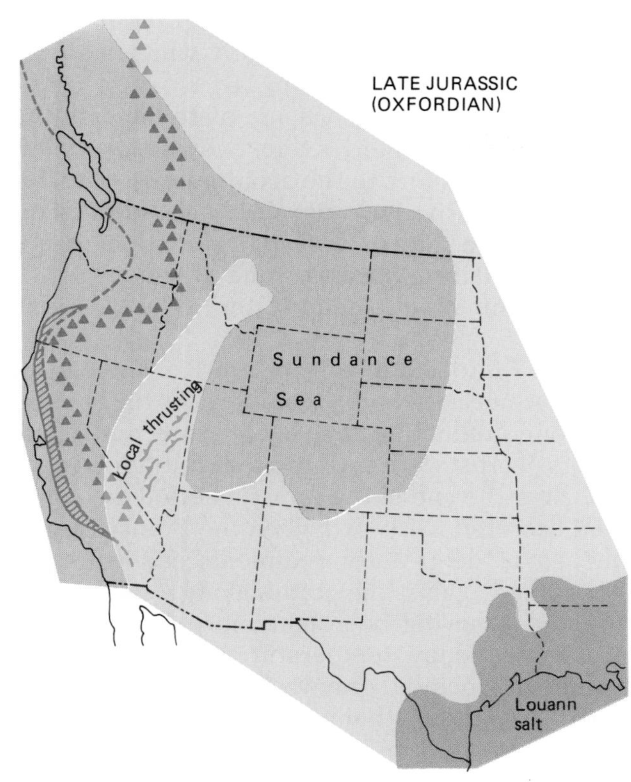

FIGURE 13-31 Geologic features of western North America during Late Jurassic (Oxfordian) time, when the Sundance Sea flooded a large interior region from southern Canada to northern Arizona and New Mexico. In mid-Triassic time, the western margin of the continent had ridden up against a subduction zone. As a result, during the Jurassic Period, a belt of igneous activity extended for hundreds of kilometers parallel to the Pacific coast. At this time, thrust faulting was largely limited to the state of Nevada.

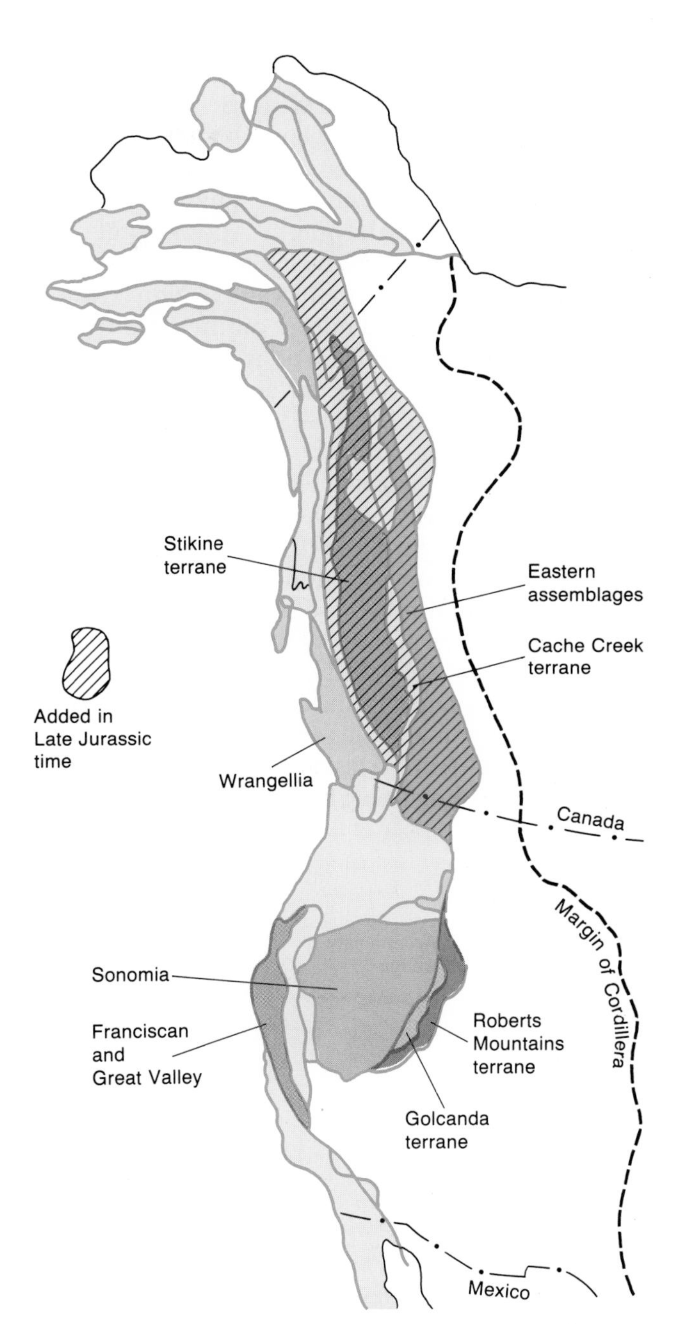

FIGURE 13-32 Exotic terranes in western North America (outlined in heavy colored lines). Early in the Triassic Period, the Sonomia and the Golconda terranes were sutured to the Roberts Mountains terrane. In Late Jurassic time a large composite terrane was sutured to Canada. It consisted of the Stikine terrane, the Cache Creek terrane, and the so-called Eastern assemblages. All these terranes had been united during the Triassic Period before colliding with North America. *(Adapted from J. B. Saleeby, Ann. Rev. Earth and Planet. Sci. 15:45–73, 1983.)*

Subduction and the Accretion of New Terranes

During the Mesozoic Era, the western margin of North America expanded by the addition of numerous island arc terranes and other microplates (Figure 13-32), in a manner analogous to the addition of exotic terranes to eastern North America in Paleozoic time, during episodes of mountain building in the Appalachian region (p. 175).

This mode of continental accretion actually began earlier. Recall that the Antler orogeny of Devonian and Early Carboniferous time entailed the collision of the Klamath island arc with the western margin of North America (Figure 11-33). This event added a sliver of exotic terrane, the Roberts Mountains terrane, to the western margin of North America (Figure 11-32).

In late Paleozoic time, after the Antler episode of accretion, the Golconda arc approached the Pacific margin of North America. Early in the Triassic Period this arc collided with the North American continent, just as the Klamath arc had done in the earlier Antler orogeny. The suturing of the Golconda arc, during what is known as the **Sonoma orogeny,** differed from the Antler orogeny in one important way, however: Rather than adding a narrow slice of island arc terrane to North America, it attached a broad microcontinent, known as Sonomia (Figure 13-33). Today Sonomia constitutes southeastern Oregon and northern California and Nevada (Figure 13-32). Squeezed in between Sonomia and the Roberts Mountains terrane was the Golconda terrane, formed largely of the accretionary wedge associated with the Golconda arc (Figure 13-33).

After the Sonoma orogeny ended, early in the Triassic Period, there was a brief interlude of tectonic quiescence along the west coast of North America. Then, in mid-Triassic time, the continental margin once again came to rest against a subduction zone and began an orogenic episode that extended from Alaska all the way to Chile. Mountain building along the Pacific coast of North America during the Mesozoic Era resembled the growth of the Andes to the south, which has continued to the present day (p. 161).

Subduction of the oceanic plate beneath the margin of North America thickened the continental crust by leading to the accumulation of intrusive and extrusive igneous rocks. The oldest intrusives of the Sierra Nevada Mountains were emplaced during Jurassic time (Figure 13-34); larger volumes were added later in the Mesozoic Era.

The Mesozoic history of the Pacific coast of North America is highly complex. At times more than one subduction zone lay offshore, and exotic slivers of crust were added to the continental margin. Near the end of the Jurassic Period the continent accreted westward when the Franciscan sequence of deep-water sediments and volcanics was forced against the craton along a subduction zone after having been metamorphosed at high pressures and low temperatures. The Franciscan sediments include graywackes and dark mudstones, together with smaller amounts of chert and limestone.

Before becoming attached to North America, the Franciscan sequence constituted an accretionary wedge, whose sediments were deformed and metamorphosed along the subduction zone at high pressures and relatively low temperatures; they represent a mélange (Figure 6-36). When

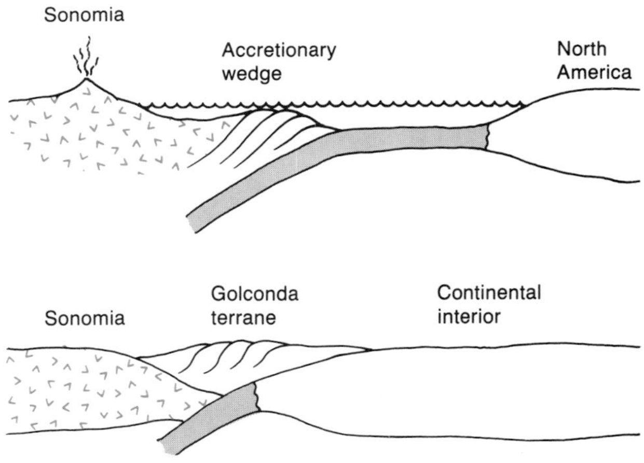

Sonomia
Accretionary wedge
North America
Sonomia
Golconda terrane
Continental interior

FIGURE 13-33 The accretion of Sonomia and the Golconda terrane to the western margin of North America. The eastern portion of the microcontinent of Sonomia was formed by the eruptions of an island arc. The Golconda terrane formed from an accretionary wedge that was squeezed between Sonomia and North America as the western margin of North America became wedged against the island arc that bordered Sonomia. (Figure 13-32 shows the location of Sonomia and the Golconda terrane in North America today.)

FIGURE 13-34 The Sierra Nevada Mountains at Yosemite National Park, California. Many of the granitic rocks that formed the Sierra Nevada were emplaced during the Jurassic Period. *(Peter Kresan.)*

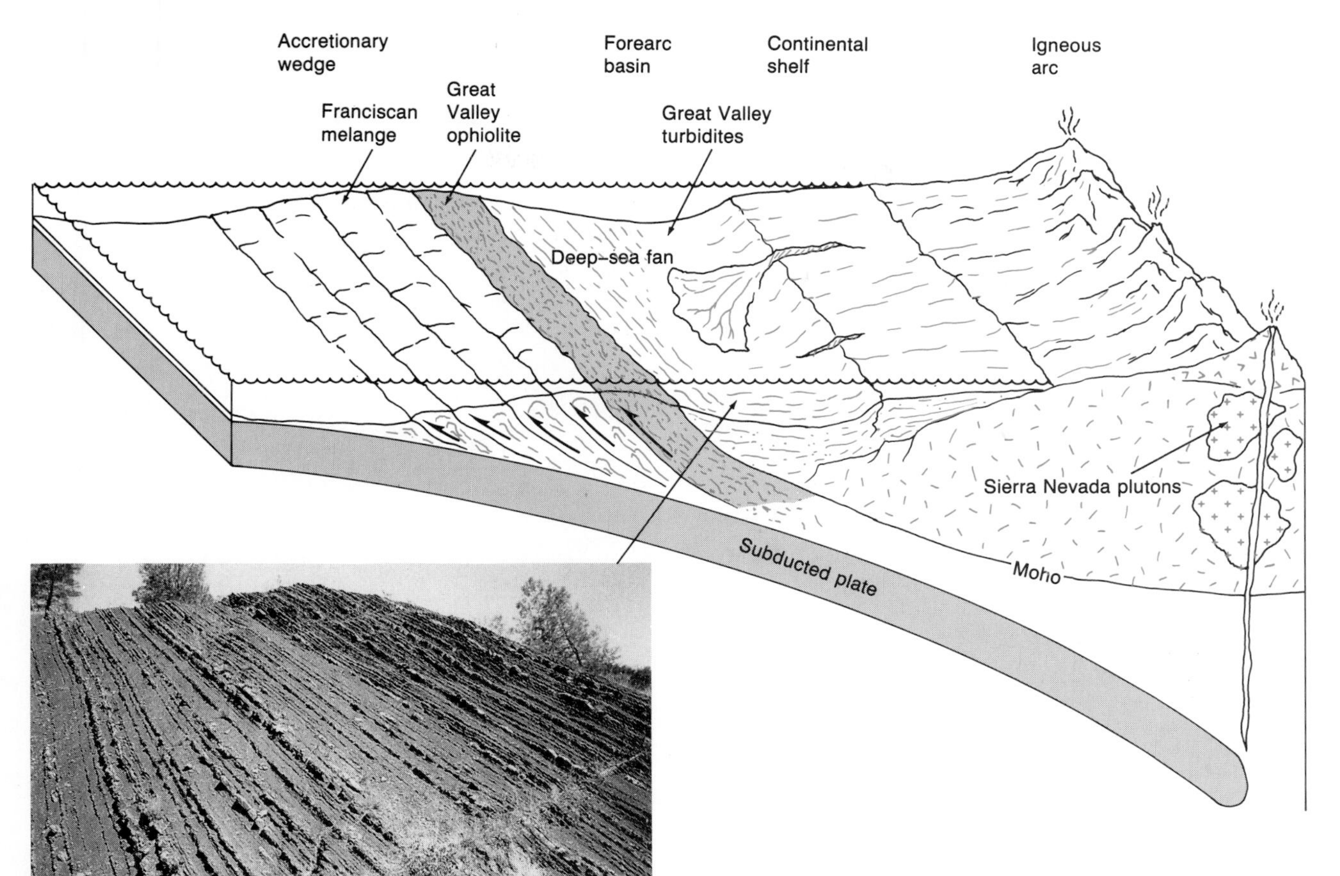

FIGURE 13-35 Block diagram depicting the Pacific margin of northern California in Jurassic time. The Franciscan mélange formed an accretionary wedge along the marginal subduction zone (see Figure 6-36). The Great Valley ophiolite was a zone of seafloor that was squeezed up along the eastern margin of the accretionary wedge, and the Great Valley sequence formed in Late Jurassic time as turbidites on deep-sea fans and in adjacent environments. The photograph shows turbidites that now lie along the western margin of the Sacramento Valley in California, where they have been tilted to a high angle by tectonic activity. The accretion of the Franciscan and Great Valley terranes to the continental margin during Late Jurassic and Cretaceous time extended North America westward (Figure 13-32). Today the Great Valley sequence still occupies a low region, the Central Valley of California, which consists of the Sacramento Valley in the north and the San Joachin Valley in the south. West of the Central Valley, portions of the Franciscan mélange have been elevated as part of the Coast Ranges. *(Adapted from R. K. Suchecki, Jour. Sedim. Petrol. 54:170–191, 1984.)*

the continental margin eventually collided with the accretionary wedge, the Franciscan rocks were piled up against the continent, along with the Great Valley sequence of deep-sea turbidites, which accumulated in the forearc basin (Figure 13-35). This Late Jurassic event approximately coincided with eastward folding and thrusting from the Sierra Nevada uplift. These tectonic events of Jurassic age are collectively known as the **Nevadan orogeny.** Orogenic activity that is related to the Nevadan orogeny, although it is not generally assigned this name, continued well into the Cretaceous Period.

Farther north, from northern Washington State to southern Alaska, a large exotic terrane collided with the margin of North America, resulting in substantial westward accretion. This exotic terrane was actually a composite block, formed of several smaller terranes (Figure 13-32). These smaller terranes include quite different suites of Paleozoic rocks and fossils, indicating that the terranes were once separate entities. On the other hand, the terranes share rock units of Triassic age, an indication that they were a single unit during the Triassic Period. The entire composite terrane was then accreted to North America late in Jurassic time, along the subduction zone that bordered the continent.

Deposition of a Foreland Basin To the south, in the western United States, the eastward thrusting and folding of Late Jurassic time greatly altered patterns of deposition as far east as Colorado and Wyoming. The Sundance Sea spread over a broad foreland basin east of the mountains. This represented the most extensive marine incursion since late Paleozoic time (Figure 13-31). In latest Jurassic time, however, the folding and thrust faulting that extended over Nevada, Utah, and Idaho produced a large mountain chain. The elevation of the land and the shedding of clastics eastward from the mountains drove back the waters of the Sundance Sea, leaving only a small inland sea to the north (Figure 13-36).

FIGURE 13-37 Excavation of dinosaur fossils in the Morrison Formation. These are partly intact skeletons that remain imbedded in the rock at Dinosaur National Monument, Utah. *(Dinosaur Nature Association.)*

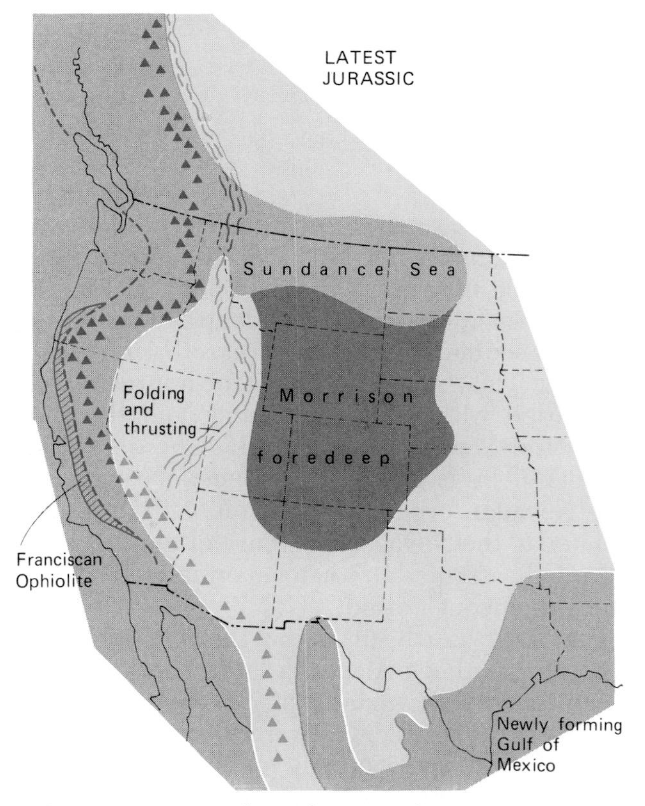

FIGURE 13-36 Geologic features of western North America during latest Jurassic time. A fold-and-thrust belt now extended for hundreds of kilometers roughly parallel to the coastline but far inland. Tectonic activity had driven the Sundance Sea from the western interior, leaving a foreland basin where the nonmarine Morrison Formation accumulated.

What remained in Colorado, Wyoming, and adjacent regions was a nonmarine foreland basin in which molasse deposits accumulated. Apparently, on the gentle profile of the foreland basin, even the lowest depositional environments were above sea level, because there was no initial deposition of marine flysch. The molasse of the foreland basin was deposited in rivers, lakes, and swamps, creating the famous Morrison Formation, which has yielded the world's most spectacular dinosaur faunas (Figure 13-37; see also Figures 13-20 and 13-21). The dinosaur skeletons in this formation are usually disarticulated, but often as many as 50 or 60 individuals are found together in one small area, an indication that these fossils may have accumulated during floods.

The Morrison Formation consists of sandstones and multicolored mudstones deposited over an area of about 1 million square kilometers. Caliche soil deposits indicate that the climate was seasonally dry during at least part of the Morrison depositional interval, while the scarcity of crocodiles, turtles, and fishes suggests that many of the lakes may have been saline at this time. The dinosaurs are found in deposits representing all of the Morrison environments — rivers, lakes, and swamps. This broad environmental distribution suggests that none of the species — not even the huge sauropods (Figure 13-20A) — was adapted specifically for a life of wading in large bodies of water. The Morrison Formation spans the last 10 million years or so of the Jurassic Period and is overlain by the nonmarine Cloverly Formation of Early Cretaceous age, which contains a completely different fauna of dinosaurs, apparently because of major extinctions at the end of the Jurassic Period.

CHAPTER SUMMARY

1 Triassic and Jurassic seas lacked important Paleozoic groups such as fusulinid foraminifera, rugose corals, and trilobites. Important groups of marine life that were present during this time included bivalve gastropods, ammonoid mollusks, brachiopods, sea urchins, hexacorals, bony fishes, sharks, and swimming reptiles.

2 Gymnosperms were the dominant group of plants in Triassic and Jurassic landscapes.

3 Although mammals originated in Triassic time, dinosaurs were much more successful than mammals during the Mesozoic Era. Flying reptiles evolved in the Triassic Period, and birds emerged in Jurassic time.

4 A mass extinction that was less severe than the terminal Permian event marked the end of the Triassic Period, and moderately heavy extinction occurred at the end of Jurassic time.

5 Very early in the Triassic Period, nearly all of Earth's continental crust was consolidated in the supercontinent Pangaea, and even at the end of Jurassic time, all of the continents remained close together. Evaporites mark zones where Pangaea began to rift apart early in the Mesozoic Era.

6 Fault-block basins that formed in eastern North America received thick deposits of sediment during the rifting episode that eventually formed the Atlantic Ocean between this continent and Africa.

7 The Sundance Sea invaded western North America during the Jurassic Period but was expelled by uplifting and sediment influx associated with the Nevadan orogeny. Dinosaur fossils are abundantly preserved in the molasse sediments that were then deposited in the vicinity of Utah.

EXERCISES

1 What important groups of Paleozoic marine animals were absent from Triassic seas?

2 What two kinds of flying vertebrates evolved during early Mesozoic time?

3 How did reefs formed by hexacorals during the Jurassic Period differ from those formed during Triassic time?

4 What suggests that dinosaurs had relatively advanced behavior of the sort that might explain their evolutionary success?

5 What was the geographic setting in which the most spectacular known assemblage of Jurassic dinosaurs was preserved?

6 In what areas is there evidence that oceans started to form in Triassic and Jurassic time? What is that evidence?

7 What plate-tectonic change might explain the eastward migration of the zone of igneous activity in western North America during the Jurassic Period?

8 By what mechanism did western North America expand westward during early Mesozoic time?

ADDITIONAL READING

Bakker, R. T., *The Dinosaur Heresies*, William Morrow and Company, New York, 1986.

Norman, D., *An Illustrated Encyclopedia of Dinosaurs*, Crescent Books, New York, 1985.

Paul, G. S., *Predatory Dinosaurs of the World*, Simon and Schuster, New York, 1988.

Thomas, B. A., and R. A. Spicer, *The Evolution and Palaeobiology of Land Plants*, Croom Helm, London, 1987.

CHAPTER 14

The Cretaceous World

Throughout most of Cretaceous time the seas stood much higher than they do today. Widespread shallow marine deposits on continental surfaces, together with nonmarine and deep-sea sediments, reveal that in many ways the Cretaceous Period was an interval of transition. Some Cretaceous sediments are lithified, like nearly all those of older systems; many others, however, consist of soft muds and sands, like most deposits of the younger Cenozoic Era. Fossil biotas of the Cretaceous Period also display a mixture of archaic and modern features. They include members of important extinct taxa, such as dinosaurs and ammonoids (groups that failed to survive the Cretaceous Period), as well as important modern taxa, such as flowering plants and the subclass of fishes that is the most diverse in the world today. It was during the Cretaceous Period that continents moved toward their modern configuration. At the start of the period the continents were tightly clustered, and Gondwanaland was prominent in the south. By the end of Cretaceous time, however, the Atlantic Ocean had widened and Gondwanaland had fragmented into most of its daughter continents.

The Cretaceous System was first described formally in 1822. For many years before that, however, Cretaceous rocks had been recognized as constituting a stratigraphic interval distinct from the Jurassic rocks below and the sediments above, now labeled Cenozoic. The name Cretaceous derives from *creta,* the Latin word for chalk, which is a soft, fine-grained kind of limestone that accumulated over broad areas of the Late Cretaceous seafloor.

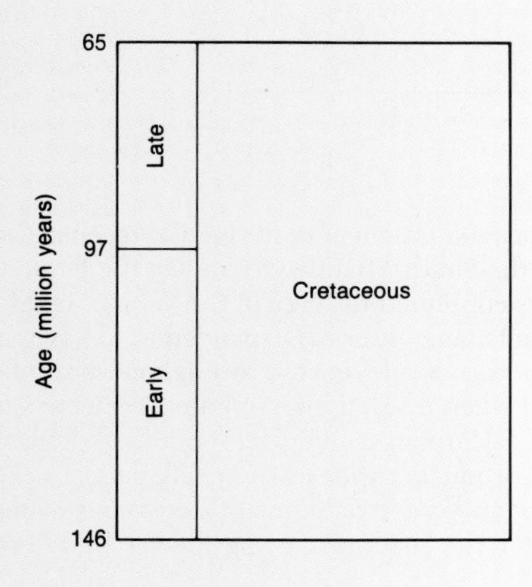

Chalk, the soft, powdery rock that is unusually abundant in the Upper Cretaceous Series in many areas. The chalk deposits shown here stand above the coastline of southeastern England, where they form the famous White Cliffs of Dover. *(David Woodfall/NHPA.)*

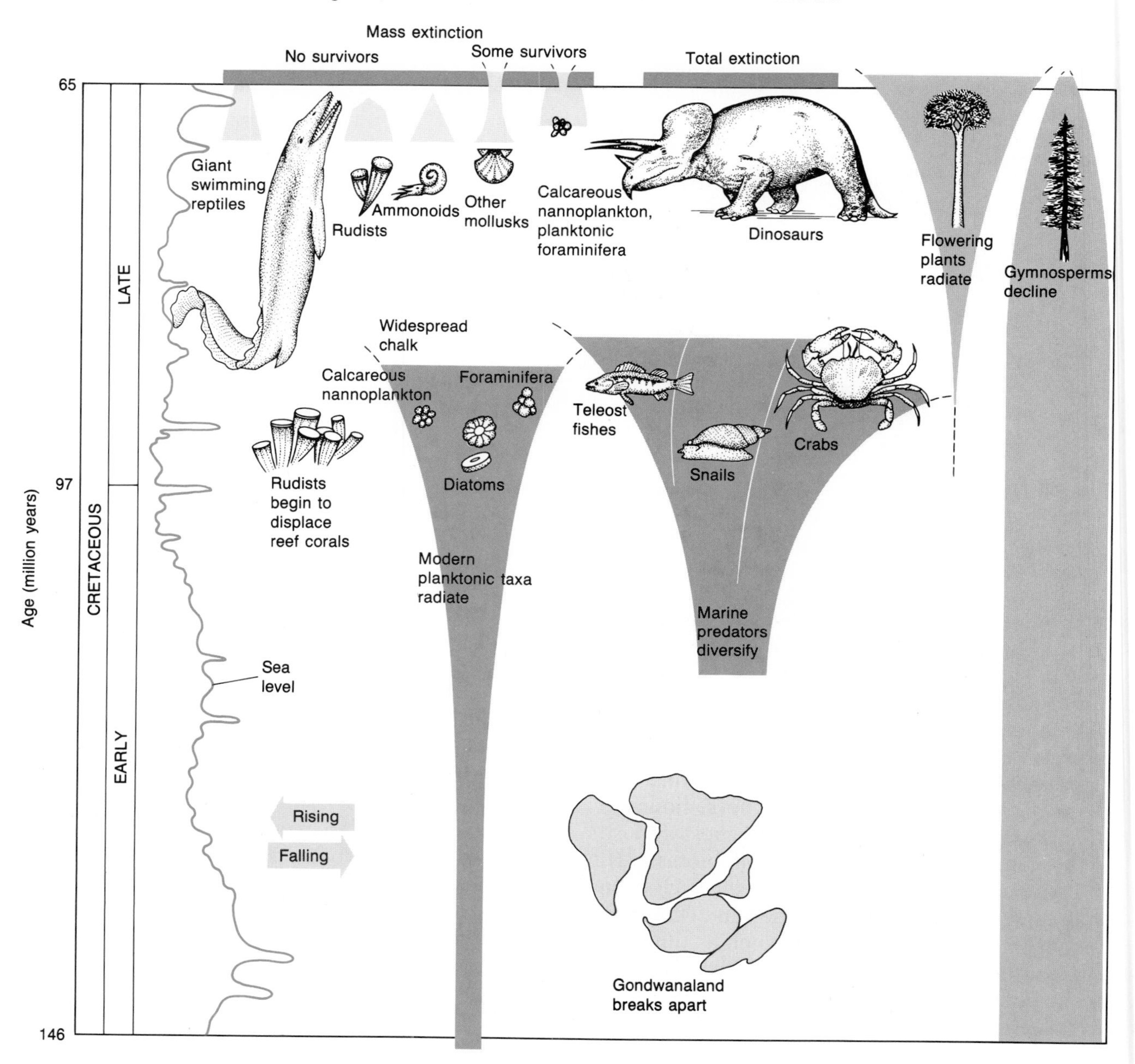

Gondwanaland fragmented during Cretaceous time, forming the South Atlantic Ocean. On the land, while dinosaurs continued to reign in the animal world, flowering plants (angiosperms) expanded at the expense of gymnosperms. Sea level rose throughout most of the period, and when it stood high in Late Cretaceous time, plates of calcareous nannoplankton rained down on the seafloor, producing widespread chalk deposits. The Late Cretaceous adaptive radiation of calcareous nannoplankton yielded the chalk-producing species, and two other modern planktonic groups, diatoms and foraminifera, diversified at the same time. Teleost fishes originated in mid-Cretaceous time and radiated along with two carnivorous groups that had originated earlier in the Mesozoic Era: the crabs and the predatory snails. In mid-Cretaceous time, rudist bivalves became the dominant builders of organic reefs, but they died out at the close of the Cretaceous Period along with numerous other groups, including the swimming reptiles and the dinosaurs.

LIFE

Life of the Cretaceous Period, both in the seas and on land, was a curious mixture of modern and archaic forms. In the marine realm, strikingly modern types of bivalve and gastropod mollusks populated Late Cretaceous seas along with enormous coiled oysters and other now-extinct sedentary bivalves. Diverse fishes of the modern kind occupied the same waters as a variety of ammonoids, belemnoids, and reptilian sea monsters — none of which have any close living relatives. On the land, floras changed from the Mesozoic type, which were dominated by gymnosperms, to the modern type, in which flowering plants predominate. Many groups of vertebrate animals that are still extant also evolved at this time: snakes and modern types of turtles, lizards, crocodiles, and salamanders. Dinosaurs, however, continued to rule the terrestrial ecosystem. Of all the modern groups of terrestrial vertebrates present during Cretaceous time, only the crocodiles approached the dinosaurs in bodily proportions. Mammals remained very small by modern standards.

Pelagic Life

The appearance of new groups of single-celled organisms gave marine plankton a thoroughly modern character by the end of Cretaceous time. The primary change among the phytoplankton was the evolutionary expansion of the diatoms. Diatoms may have existed during the Jurassic Period, but they did not radiate extensively until mid-Cretaceous time. Together with dinoflagellates and, in warm seas, calcareous nannoplankton (Figure 13-5), diatoms must have accounted for most of the photosynthesis that occurred in the Cretaceous seas, as they do today (p. 43). Today, as we have seen (p. 75), diatoms are the dominant contributors to the siliceous oozes of the deep sea, and their accumulation in deep-sea sediment was well under way before the end of the Cretaceous Period.

Higher in the pelagic food web, the modern planktonic foraminifera diversified greatly for the first time. This group, known as the globigerinaceans, has a meager fossil record in Jurassic rocks; not until the upper part of the Lower Cretaceous System is it well enough represented to be of great value in biostratigraphy (Figure 14-1).

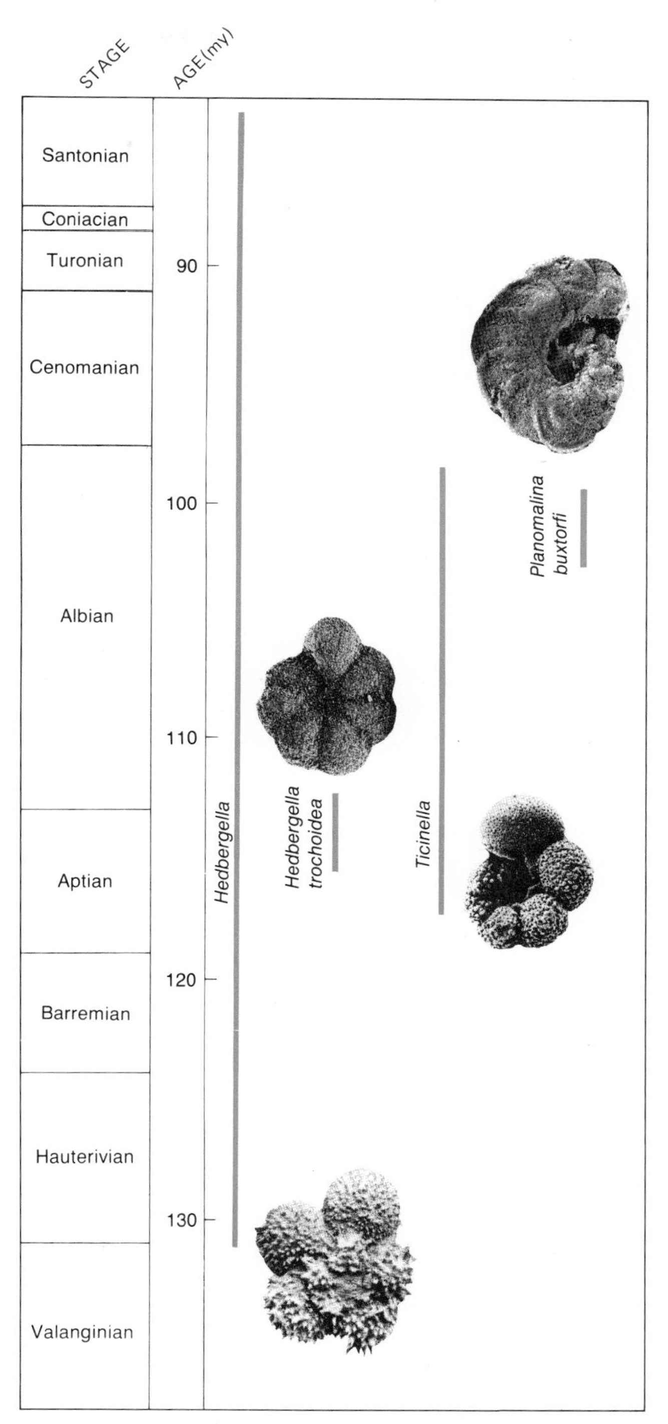

FIGURE 14-1 Early planktonic foraminifera (Globigerinacea). Cretaceous stages are shown at the left. These species average about $\frac{1}{2}$ millimeter ($\frac{1}{50}$ inch) in diameter. *(After A. Boersma.)*

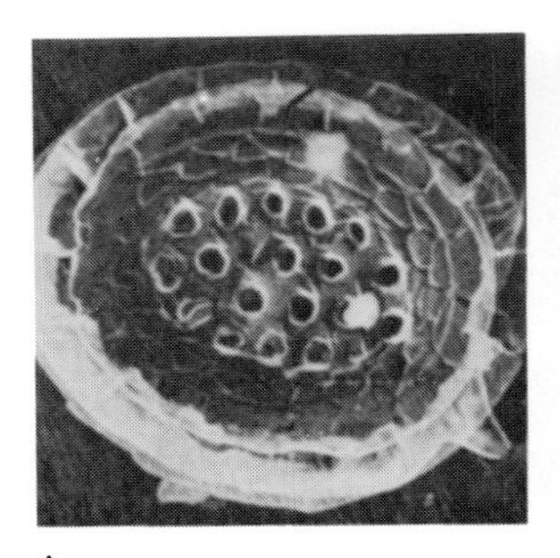

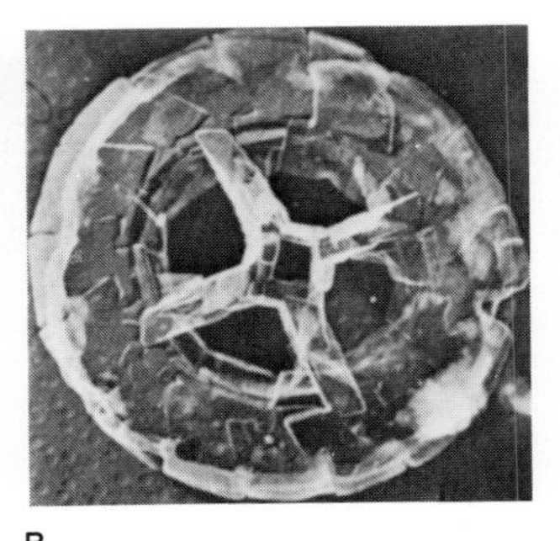

FIGURE 14-2 Plates of calcareous nannoplankton that represent the genera *Cribrosphaerella* (*A*, ×9750) and *Deflandrins* (*B*, ×7000). *(M. Black.)*

FIGURE 14-3 Reconstruction of the Late Cretaceous shallow sea represented by fossil faunas at Coon Creek, Tennessee. A large coiled ammonoid rests on the seafloor, surrounded by numerous swimming and feeding members of the straight-shelled ammonoid genus *Baculites* (see also Figure 14-4). *(Field Museum of Natural History, Chicago.)*

Late Cretaceous adaptive radiations of two of the single-celled planktonic groups altered depositional patterns in the pelagic realm: Since mid-Cretaceous time, both the globigerinacean foraminifera and the calcareous nannoplankton have contributed vast quantities of calcareous sediment to oceanic areas, whereas before about 100 million years ago, little or no calcareous ooze was present on the deep seafloor.

During Late Cretaceous time calcareous nannoplankton were so abundant in warm seas that the small plates that armored their cells accumulated in huge volumes as the fine-grained limestone commonly known as chalk (Figure 14-2). Cretaceous chalk is, in fact, widely used for writing on blackboards. The most famous chalk deposits in the world crop out along the southeastern coast of England, where they are known formally as the Chalk (p. 366). Similar but less massive chalks formed in shallow seas in Kansas and nearby regions and along the Gulf Coast of the United States. Why calcareous nannoplankton were so productive in the Cretaceous Period is uncertain, but never before or since have they yielded chalk deposits of such great extent or thickness.

Still higher in the pelagic food web of Late Cretaceous time, the ammonoids and belemnoids persisted as major swimming carnivores (Figure 14-3). The ammonoids serve as valuable index fossils for the Cretaceous System, just as they do for the Triassic and Jurassic. Among the Cretaceous ammonoids were many species with straight, cone-shaped shells and others with strangely coiled shells (Figure 14-4).

New on the scene in Cretaceous time were the **teleost fishes,** a subclass that is today the dominant group of marine and freshwater fishes. Teleosts are characterized by such features as symmetrical tails, round scales, specialized fins, and short jaws that are often adapted to take particular kinds of food. By Late Cretaceous time, a wide variety of teleosts already existed, including the largest species known from the fossil record (Figure 14-5). This group also included close relatives of the modern sunfish, carp, and eel, as well as members of the families that include salmon, pompanos, and the vicious South American piranha. Similarly, Cretaceous sharks resembled present-day forms; in fact, all of the living families of sharks had evolved by the end of the Cretaceous Period.

Most of the top carnivores of Cretaceous pelagic habitats, however, were not at all modern. Whereas whales of one kind or another have occupied the "top carnivore" adaptive zone during most of the Cenozoic Era, reptiles were the largest marine carnivores until the end of Cretaceous time. Ichthyosaurs and marine crocodiles were rare by this time, but plesiosaurs still thrived,

FIGURE 14-4 Cretaceous ammonoids. *A. Tragodesnoceras* (~×1) (rear and side views). *B. Nostoceras,* a strangely coiled form (~×1). *C. Baculites,* a straight-shelled form, shown in the reconstruction depicted in Figure 14-3. The specimen on the left is an internal filling that shows the complex folded septa, or partitions, within the shell (~×1). *(U.S. Geological Survey.)*

some exceeding 10 meters (~35 feet) in length. Plesiosaurs are depicted in Figure 14-6 along with other members of the Late Cretaceous pelagic community of the western interior of the United States. Huge marine lizards known as **mosasaurs** were probably the most formidable marauders of Cretaceous seas; some grew to be longer than 15 meters (~45 to 50 feet). Although there is direct evidence that mosasaurs attacked ammonoids (Figure 14-7), the reptiles' pointed teeth were not well adapted for crushing shells, so ammonoids probably did not form a major part of their diet.

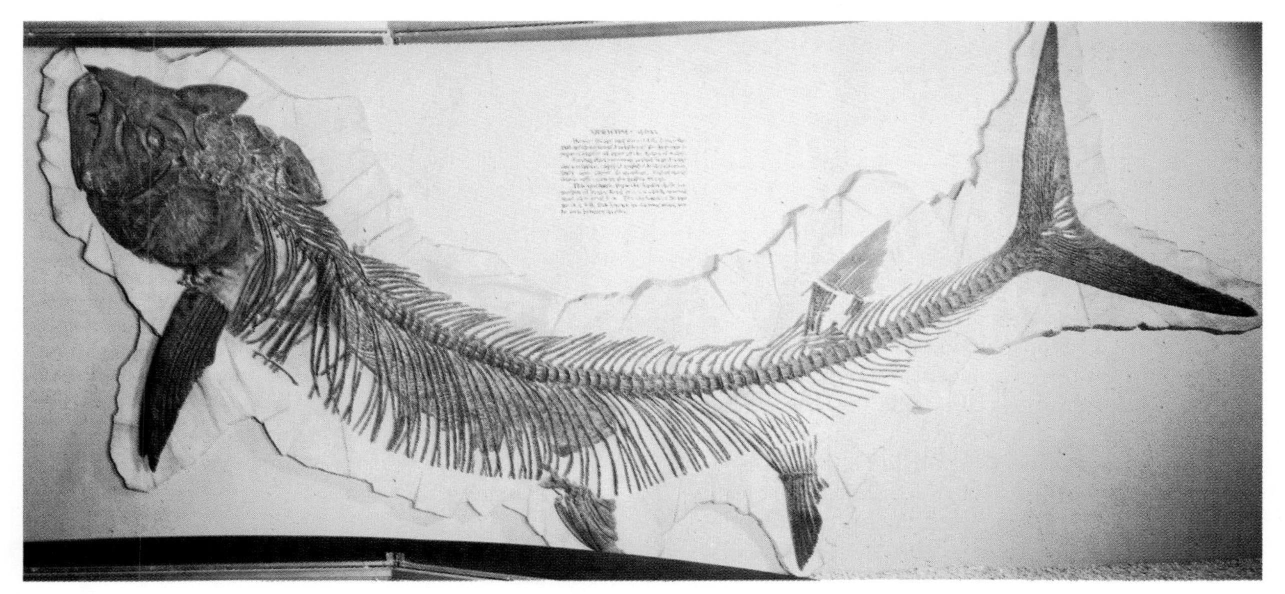

FIGURE 14-5 ***Xiphactinus,* a Cretaceous fish.** At about 5 meters (~16 feet) in length, this is the largest known teleost. A careful look will reveal that the animal shown here died with a good-sized fish in its belly. *(Smithsonian Institution.)*

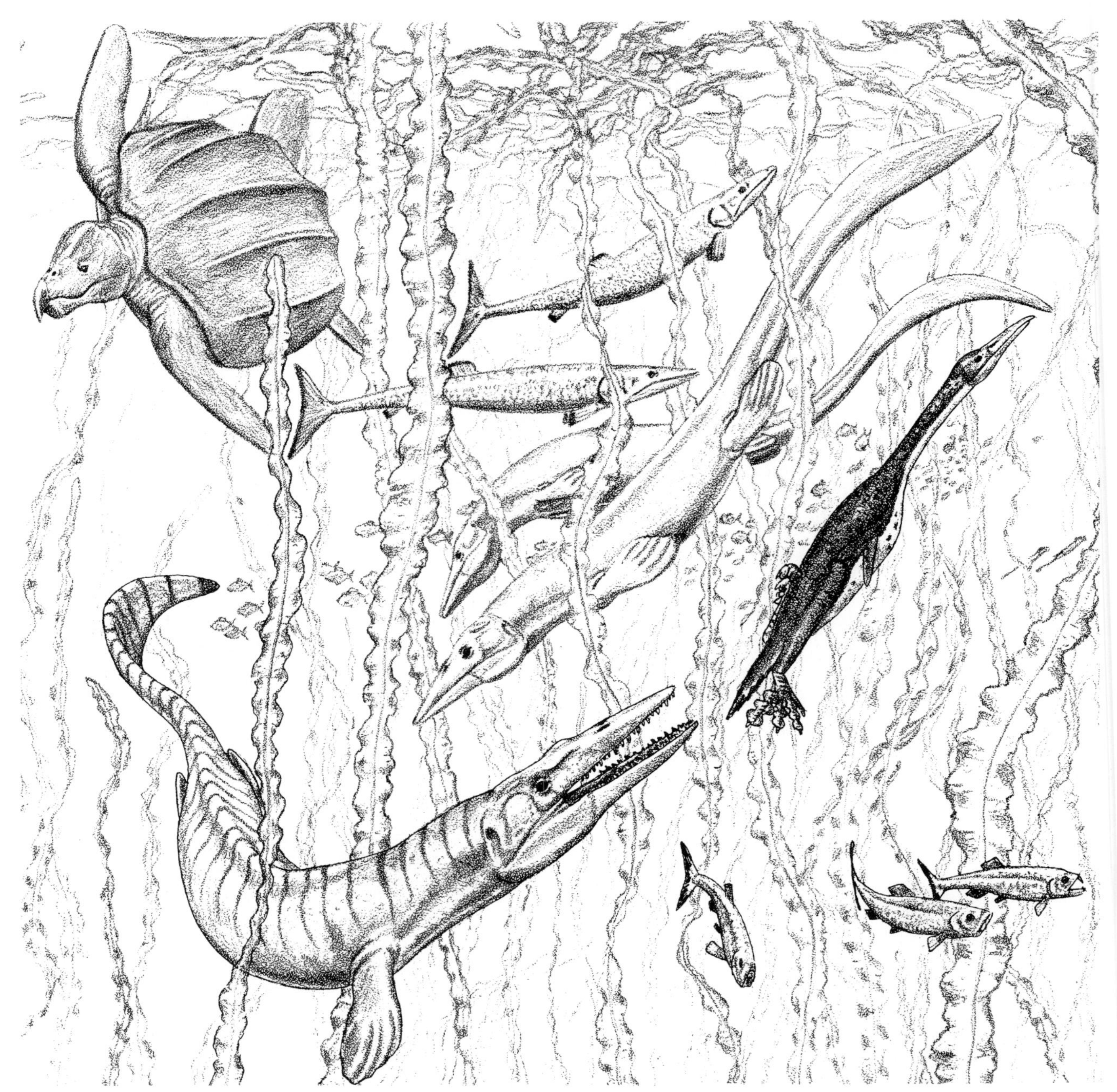

FIGURE 14-6 A reconstruction of marine life preserved in the Upper Cretaceous Pierre Shale of the western interior of the United States. The animals are shown swimming in a bed of kelp, which are algae of large proportions. The giant turtle on the left is *Archelon,* which reached a length of almost 4 meters (~12 feet). The striped animal at the lower left is the mosasaur *Clidastes,* and beyond it is a pair of mosasaurs of the genus *Platecarpus. Clidastes* is in pursuit of the diving bird *Hesperornis.* The teleost fishes are *Cimolichthyes* (the pikelike pair near the turtle) and *Enchodus* (the small fishes on the lower right). *(Drawings by Gregory S. Paul.)*

Less important animals in the marine ecosystem included the flightless diving bird *Hesperornis* and the marine turtles that evolved during the Cretaceous Period (Figure 14-6). The large feet and small wings of *Hesperornis* (Figure 14-8A) were adapted for swimming and their sharp,

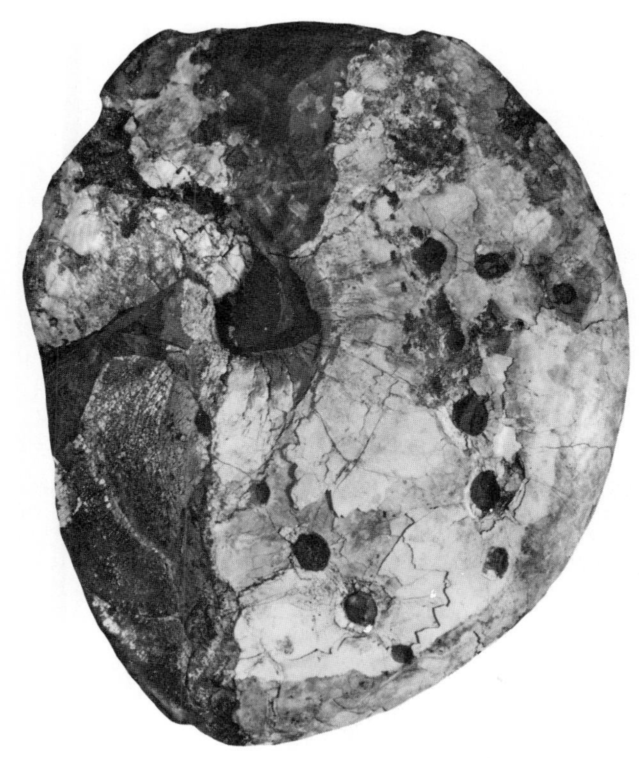

FIGURE 14-7 A Late Cretaceous ammonoid that was bitten by a mosasaur. This fossil, which was collected in South Dakota, displays 16 distinct bites. *(University of Michigan.)*

A

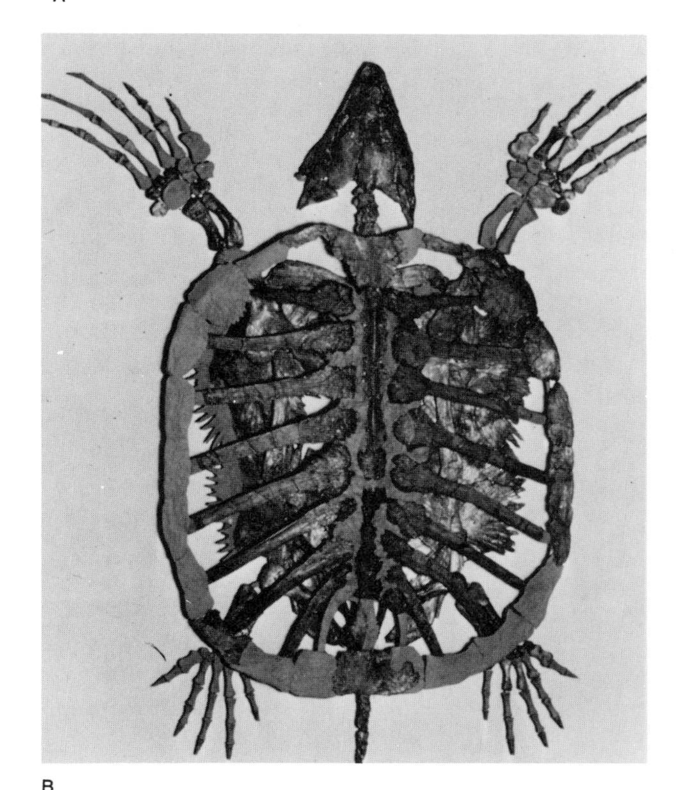

B

FIGURE 14-8 The Late Cretaceous diving bird *Hesperornis* and marine turtle *Archelon*. *(A)* The wings of this flightless animal had been almost entirely lost, but large, paddlelike feet were present, as were teeth for catching fish. *Hesperornis* was about 1.2 meters (~4 feet) long. *(B)* This animal reached a length of almost 4 meters (~12 feet). *(Smithsonian Institution.)*

backward-directed teeth for catching slippery fishes. The marine turtle *Archelon* (Figure 14-8*B*) grew to a length of nearly 4 meters (~12 feet).

Seafloor Life

Life on the seafloor began to take on a modern appearance during the Cretaceous Period. One noteworthy feature was the decline of the brachiopods, which had suffered greatly in the mass extinction at the end of the Paleozoic Era but had experienced a moderate expansion again early in Mesozoic time. Certain other groups of seafloor life that had been successful in the Jurassic Period continued to hold their own during Cretaceous time. Among them were the sea urchins and the hexacorals, which diversified but underwent no startling adaptive changes. Still other major groups retained many of their Jurassic families and genera but also produced new representatives that have survived to the present. Some of them are described below.

Foraminifera A large percentage of the families of bottom-dwelling foraminifera in existence today appeared during the Cretaceous Period, so that this group had a modern aspect (Figure 14-9).

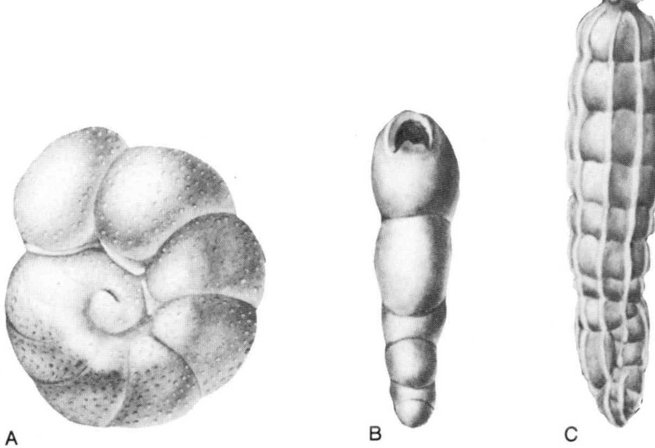

FIGURE 14-9 New genera of bottom-dwelling foraminifera of Cretaceous age. *A. Anomalinoides* (×68). *B. Pleurostomella* (×48). *C. Siphogenerinoides* (×46). *(H. Tappan.)*

Bryozoans The most common modern bryozoans are the cheilostomes, which typically encrust marine surfaces, including the hulls of boats, in the form of low-growing mats. Cheilostomes originated in Jurassic time but did not enjoy success until the Late Cretaceous, when they expanded to include more than 100 genera (Figure 14-10).

FIGURE 14-10 *Rhiniopora*, a genus of Cretaceous cheilostome bryozoans that grew over the substratum as a crust (×25). A colony consisted of many interconnected individuals, each of which emerged from a small opening to feed. *(E. Voigt.)*

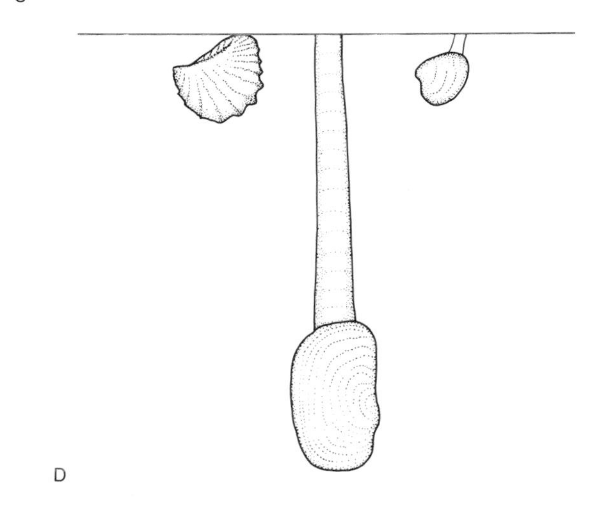

FIGURE 14-11 Burrowing bivalves of Cretaceous age. Like similar living animals, they obtained food and oxygen from water currents. *A. Scabrotrigonia* (×0.7) lacked siphons for channeling their water currents. *B* and *C. Panope* (×0.7) and *Aphrodina* (×1) possessed fleshy, tubelike siphons; the deflection of the line of muscle attachment within the shell allowed the siphons to be withdrawn. *D.* The fossils are all shown in living orientations in the line drawing. *(Smithsonian Institution.)*

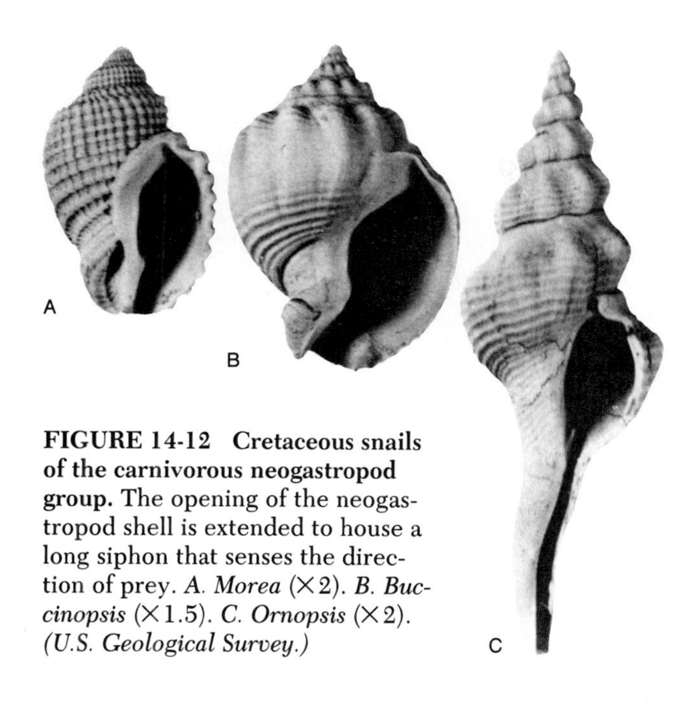

FIGURE 14-12 Cretaceous snails of the carnivorous neogastropod group. The opening of the neogastropod shell is extended to house a long siphon that senses the direction of prey. *A. Morea* (×2). *B. Buccinopsis* (×1.5). *C. Ornopsis* (×2). *(U.S. Geological Survey.)*

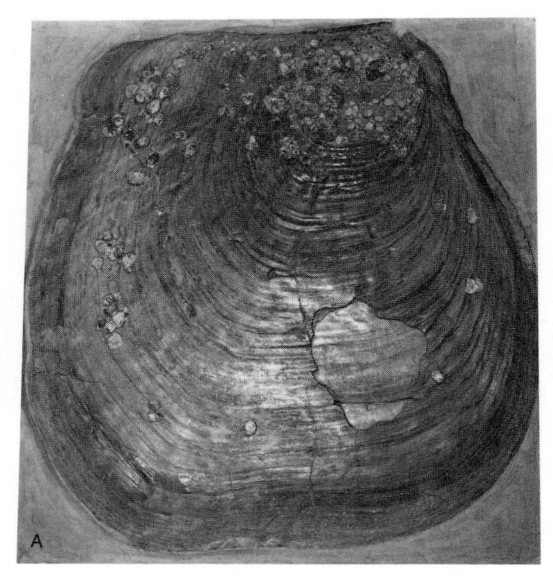

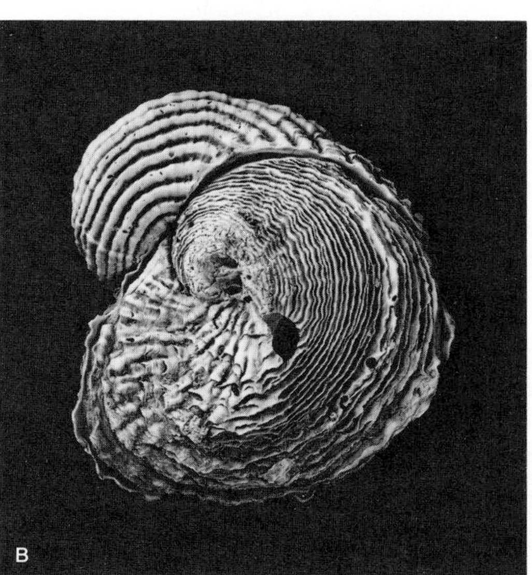

FIGURE 14-13 Cretaceous bivalve mollusks that lived on the surface of sediment. *A. Inoceramus*, a giant animal that was nearly 1 meter (~3 feet) long; some members of this species measuring more than half again this size are found in Kansas. *B. Exogyra*, a coiled oyster (×0.5). *(A. Sternberg Memorial Museum, Fort Hays State University. B. Smithsonian Institution.)*

Burrowing Bivalve Mollusks Early Cretaceous burrowing bivalves resemble those of the Jurassic, but by the end of the period new genera had appeared as well, including many that pumped water in and out of their shells through fleshy siphons (Figure 14-11).

Gastropod Mollusks, or Snails During the Cretaceous Period the aptly named Neogastropoda, or "new snails," produced many modern families and genera. Unlike most earlier snails, these animals are generally carnivorous, feeding on such prey as worms, bivalves, and other snails. Some live in the sediment, others on the sediment surface. Many modern seashells popular with collectors belong to neogastropod species (Figure 14-12).

Crabs A more or less modern type of crab had evolved during the Jurassic Period, but even more modern groups appeared during the Cretaceous Period.

Surface-dwelling Bivalve Mollusks Among bivalve mollusks living on the surface of the substratum, for example, coiled oysters and other groups that had existed during Jurassic time evolved species of enormous size (Figures 14-13 and 14-14). Of these large forms the rudists were of special significance because they lived like corals, forming large tropical reefs. Despite their success during the Cretaceous Period, the rudists did not survive into the Mesozoic Era.

Before their demise the rudists apparently flourished at the expense of reef-building corals. Shallow-water reefs built in Early Cretaceous time, like those built during the Jurassic Period

FIGURE 14-14 Reef building rudist bivalves. This is a polished slab of latest Cretaceous reef limestone from Jamaica that displays cross-sections of small rudists (average diameter ~5 centimeters [2 inches]). *(E. G. Kauffman and C. C. Johnson.)*

and in the modern world, were formed primarily by hexacorals and coralline algae. In mid-Cretaceous time, however, rudist bivalves assumed the dominant role in tropical reef growth. Rudists were mollusks with a cone-shaped lower shell and a lidlike upper shell (Figure 14-14). These curious animals attached themselves to hard objects (often other rudists) and grew upward, some reaching heights of more than 1 meter. Nearly all shallow-water reefs of Late Cretaceous age are dominated by rudists, which seem to have defeated the corals temporarily in competition for space. Most likely the rudists, like reef-building corals, grew rapidly by feeding on symbiotic algae that lived and multiplied in their tissues (Box 2-1). Only after the end of the Cretaceous Period, when rudists, like dinosaurs, became extinct, did corals prevail on reefs once more.

The Rise of Modern Marine Predators

Many of the general changes that occurred in bottom-dwelling marine life during Jurassic and Cretaceous time seem to have been related to the great expansion of modern types of marine predators. Among the new predators were the advanced teleost fishes (Figure 14-5), modern kinds of crabs, and carnivorous gastropods (Figure 14-12). Many of these new predators were efficient at penetrating shells: fishes by biting, crabs by crushing or peeling with their claws, and some of the gastropods by drilling holes.

The contrast between Paleozoic and Mesozoic predation on the seafloor is exemplified by the absence during Paleozoic time of large arthropods with crushing claws and by the virtual absence of holes drilled by predators in fossilized Paleozoic brachiopods and bivalve shells.

The decline of the brachiopods and the stalked crinoids, both of which were moderately well represented in early Mesozoic seas, can probably be attributed to the diversification of modern predators. The few species of stalked crinoids that survive today live in deep water; in shallow waters, predation by fishes is probably too severe to permit their existence. Similarly, more species of brachiopods today live in temperate seas than in tropical seas, where predation by crabs, fishes, and snails is severe. By the end of the Mesozoic Era, relatively few immobile species of animals lived on the surface of the seafloor in the mode typical of many groups of Paleozoic brachiopods (Figure 10-13). The ability to swim or to burrow actively appears to have been the best defense against predation, except for species that had defensive spines or unusually heavy protective shells.

Flowering Plants Conquer the Land

The greatest change in terrestrial ecosystems during the Cretaceous Period was the ascendancy of the **flowering plants** (angiosperms), although gymnosperm floras resembling those of Triassic and Jurassic age continued to dominate the land well into Cretaceous time. The most conspicuous change during Early Cretaceous time was in the types of gymnosperms that predominated: Conifers became the most numerous species, and the Age of Cycads came to a close. This development, too, was short-lived. About 100 million years ago, midway through the Cretaceous Period, the first angiosperms made their appearance on Earth, and during Late Cretaceous time they surpassed the conifers in diversity. Today there are some 200,000 species of flowering plants, including many types of grasses, weeds, wildflowers, and hardwood trees. In contrast, there are only about 550 modern conifer species, although some, including pines, firs, and spruces, are conspicuous in the modern landscape. The success of the flowering plants is one of the most fascinating chapters in the history of life. As we will see, this story includes several important episodes, some of which unfolded during the Cretaceous Period; others took place during Cenozoic time.

The term *flowering plant* can be misleading, because not all so-called flowering plants have showy flowers; all of them do, however, possess the kinds of reproductive structures that are found within showy flowers. The key reproductive feature that distinguishes angiosperms from gymnosperms (naked-seed plants) is the enclosure of the seed.

The Earliest Floras Of some 500 living families of flowering plants, only about 50 are now believed to be represented by Upper Cretaceous fossils. Some of these surviving Cretaceous groups are so common—among them are the sycamore trees *(Platanus)*, the hollies, the palm trees, the oak family, the walnut family, and the family that today includes birches and alders—that to a modern observer lacking a detailed knowledge of botany, forests of latest Cretaceous time would look relatively familiar (Figure 14-15). The open, unforested areas, however, would look quite strange; missing altogether would be grasses—the kind of vegetation that now characterizes meadows, prairies, and savannahs.

An early phase of the adaptive radiation of flowering plants is documented by fossils of the Atlantic Coastal Plain in Maryland. Here, within a sedimentary interval representing only about 10 million years of mid-Cretaceous time, both fossil leaves and fossil pollen increase in variety and in complexity of form (Figure 14-16). The early leaves have simple, smooth outlines, and their supporting veins branch in irregular patterns. Later leaves include varieties with many marginal lobes and veins that follow more regular geometric patterns. Probably the more regular patterns gave the leaves greater strength to withstand tearing.

FIGURE 14-15 A fossil of a Cretaceous member of the flowering plant group that is very similar to species of the modern world. The leaf represents *Platanus*, the genus of modern sycamores. *(Smithsonian Institution.)*

Secrets of Success The reasons why angiosperms diversified during the Late Cretaceous while gymnosperms declined are quite evident. One great advantage of the flowering plants is their ability to provide a food supply for their seeds. By a process known as double fertilization, one fertilization event produces a seed within the ovary and a second fertilization event, also within the ovary, produces a supply of stored food for the seed: The nutritional part of a kernel of corn or a grain of wheat is an example. The rapid manufacture of this food supply allows for the quick release of a well-fortified seed. Because gymnosperms lack this double-fertilization mechanism, it takes much longer for the parent plants to supply their seeds with enough food to enable the progeny to survive on their own. As a result, most gymnosperms have reproductive cycles of 18 months or longer. In contrast, thousands of species of flowering plants grow from a seed and then release seeds of their own in just a few weeks. Rapid colonization has been one of the secrets of the ecologic success of the flowering plants. This trait may have given flowering plants their initial foothold on Earth. It appears that most early species lived in unstable habitats, such as the margins of rivers, where the ability to colonize rapidly is a special advantage: These species were ecologic opportunists (p. 28).

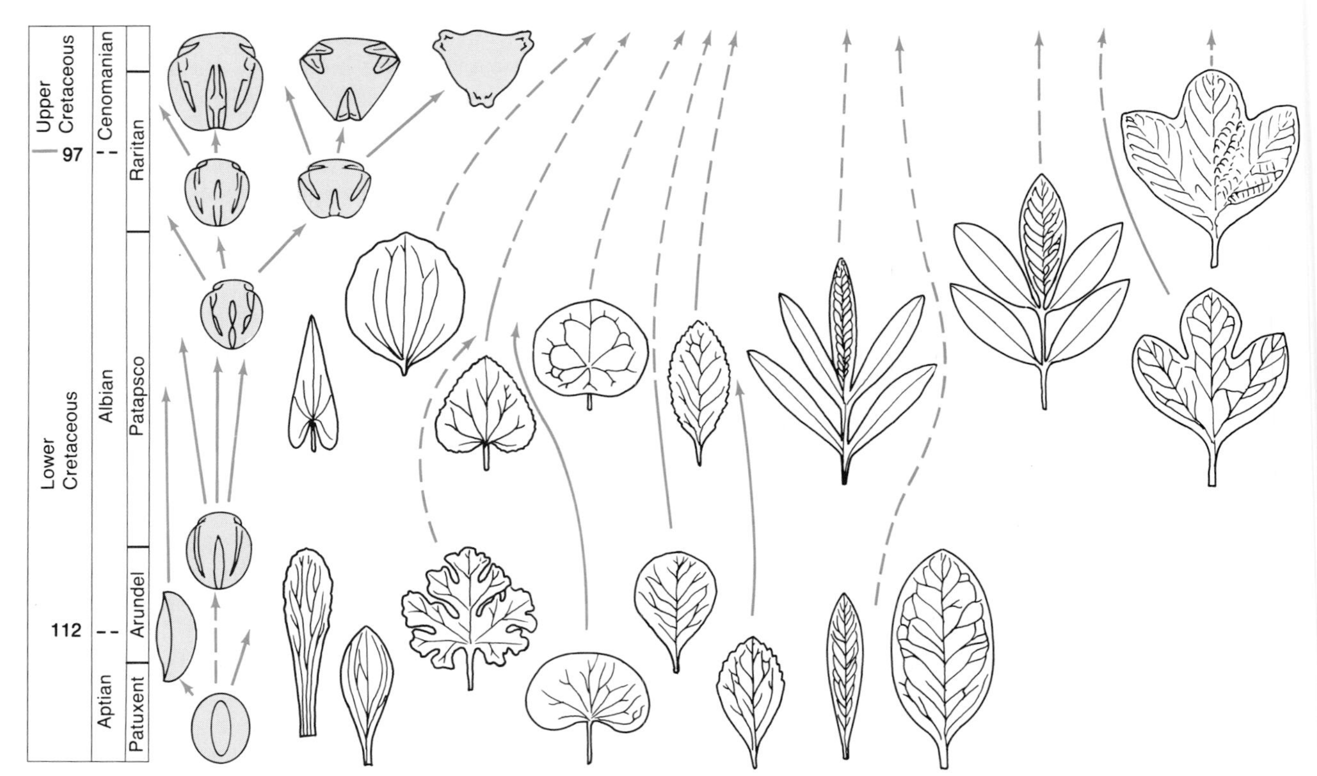

FIGURE 14-16 The pattern of initial adaptive radiation of flowering plants. These fossil leaves and pollen are found in formations (Patuxent through Raritan) of the Cretaceous Potomac Group of Maryland. Both pollen *(left)* and leaves exhibit an increase in complexity and variety of form through time. *(After J. A. Doyle and L. J. Hickey, in C. B. Beck [ed.], Origin and Early Evolution of the Angiosperms, Columbia University Press, New York, 1976.)*

A second reproductive mechanism of flowering plants that has contributed enormously to their success is the flowers' attraction of insects. Insects benefit from the nutritious nectar that the flowers provide, and the flowers benefit because the insects unknowingly carry pollen from one flower to another, fertilizing the plants on which they feed. This attraction is often specialized: A particular kind of insect feeds upon a particular kind of plant, providing a unique mechanism for speciation. If a flower of a new shape, color, or scent develops within a small population of plants, the flower may attract a different kind of insect than the one that visited its ancestors. The plants with the new kind of flower will thus be reproductively isolated from their ancestral species; in other words, the new forms will become a new species. In general, new kinds of insects create opportunities for the development of new species of plants (with new kinds of flowers); and similarly, new kinds of plants create feeding opportunities for new species of insects. This reciprocity has apparently accelerated rates of speciation in both flowering plants and insects. High rates of speciation have permitted the frequent development of new adaptations. Thus the symbiotic (mutually beneficial) relationship between flowering plants and insects has played a major role in the great success of both groups since mid-Cretaceous times.

Large Dinosaurs and Small Mammals

Owing to a patchy fossil record, Early Cretaceous vertebrate faunas are not well known, but various new kinds of dinosaurs are represented; these appear to be precursors of the Late Cretaceous dinosaurs that are well known from deposits of Wyoming, Montana, Alberta, and Asia.

FIGURE 14-17 The Late Cretaceous duck-billed dinosaur *Edmontosaurus*. Duck-billed dinosaurs of many species lived together in the American West. *(Smithsonian Institution.)*

FIGURE 14-19 *Albertosaurus*, a carnivorous dinosaur that lived during Late Cretaceous time. This animal, which stood about 4.2 meters (~ 14 feet) tall, was typical of large carnivorous dinosaurs in that its small forelimbs apparently were not used for grasping prey; instead, the enormous jaws must have served this function. *(Royal Tyrrell Museum/Alberta Culture and Multiculturalism.)*

In the American West, dinosaurs formed a community that has been compared with the modern mammal faunas of the African plains. Instead of antelopes, zebras, and wildebeests, there were many species of duck-billed dinosaurs (Figure 14-17). These fast-running herbivores probably traveled in herds and may have trumpeted signals to one another by passing air through complex chambers in their skulls (see Box 13-1). In place of rhinoceroses, there were horned dinosaurs with beaks and teeth for cutting harsh vegetation (Figure 14-18). Sharing the Late Cretaceous plains with these herbivores were fearsome predators, including the largest carnivorous land animals of all time, *Albertosaurus* (Figure 14-19) and *Tyrannosaurus*. Here, too, were terrestrial crocodiles (Figure 14-20) that grew to

FIGURE 14-18 The Late Cretaceous horned dinosaur *Triceratops*. This rhinolike genus appears to have been the last of the dinosaurs to disappear, having survived to the very end of the Cretaceous Period. *(Smithsonian Institution.)*

FIGURE 14-20 Head of a huge terrestrial crocodile, *Phobosuchus*, which probably fed on Late Cretaceous dinosaurs of small and medium size. The length of the head equaled the height of a large man. *(British Museum of Natural History.)*

FIGURE 14-21 Reconstruction of a Late Cretaceous fauna of Alberta. On the left is the armored herbivorous dinosaur *Edmontonia* in front of the duck-billed herbivore *Kritosaurus.* The duck-billed herbivores to their right belong to the genus *Corythosaurus.* The ferocious carnivore to the right of center is *Tyrannosaurus;* it confronts horned dinosaurs of the genus *Chasmosaurus,* with the large head shields, and *Monoclonius,* with the long horn. Passing overhead in the foreground are pterosaurs (flying reptiles) of the genus *Quetzalcoatlus.* The

water birds flying in the distance have feathered wings, in contrast to the naked wings of the pterosaurs. *(Drawing by Gregory S. Paul.)*

the remarkable length of 15 meters (~45 to 50 feet). Huge predatory dinosaurs of all types were typically represented by far fewer species and individuals than the herbivorous dinosaurs; presumably these predators played an ecologic role resembling that of modern African lions and hyenas, which are also much less numerous than their hoofed prey.

A general evolutionary trend toward large body size is evident in the Upper Cretaceous fossil record of the American West: Not only among the carnivorous group but also among the duckbills and horned dinosaurs, the species of largest body size were some of the last to evolve. Along with these giants, however, smaller species remained, so that there was a large variety of species toward the end of the period.

The skies above the plains where the dinosaurs roamed were populated—perhaps sparsely—by the two groups of flying vertebrates that had evolved earlier in the Mesozoic: the birds and the flying reptiles (Figure 14-21). Most of the Cretaceous birds were probably wading birds and shorebirds that lived like modern herons and cranes; there were no songbirds of the kind that surround us today. Flying reptiles were among the most spectacular of all Cretaceous animals. While they may have relied heavily on passive transport, soaring on the wind, it appears that many, if not all, species at times flapped their wings in flight. The largest known species, represented by fossils from the uppermost Cretaceous of Texas, is estimated to have had a wingspan of at least 11 meters (~35 feet). Members of this species, like modern vultures, may have soared through the sky in search of carrion in the form of dinosaur carcasses, to which their size would have been appropriately scaled.

Living a less conspicuous existence on the ground were several groups of verebrate animals with a future: amphibians (frogs and salamanders), reptiles (snakes, lizards, and turtles), and mammals. Frogs and salamanders, you will recall, had been present since the Jurassic Period, and by the end of the Cretaceous, some of their modern families had evolved. Lizards and turtles were present even earlier in the Mesozoic Era, and by the time the Cretaceous Period drew to a close, many of their modern families had come into being.

Snakes are a younger group. These limbless reptiles did not originate until the Cretaceous Period, and their great evolutionary expansion did not come about until the Cenozoic Era; all Cretaceous snakes were of the primitive group that includes the present-day boa constrictors and pythons.

Another Cretaceous vertebrate group with an especially bright future was the mammals, which differ from reptiles in maintaining their body temperature at a nearly constant level. Their fur is an adaptation associated with this condition: It serves as insulation. In addition, nearly all mammals also give birth to live young, which they nurse. Stages in the evolution of the skeleton during the evolution of early mammals are shown in Figure 14-22. Despite having evolved as early as mid-Triassic time, mammals remained quite small in body size throughout the Mesozoic Era. As Box 14-1 explains, only after the dinosaurs disappeared did the mammals diversify markedly to develop the wide variety of adaptations that are so familiar to us today.

PALEOGEOGRAPHY

Because the Cretaceous System has undergone less metamorphism and erosion than older geologic systems, it is represented on modern continents by an extensive record of shallow marine and nonmarine sediments and fossils. In addition, Cretaceous sediments and fossils are widespread in the deep sea, in contrast to the sparse deep-sea records for the Triassic and Jurassic periods; this difference reflects the fact that movements of plates across Earth's surface are rapid enough so that a large percentage of deep-sea sediments older than the Cretaceous System have by now been swallowed up along subduction zones. The relative abundance of Cretaceous sediments in the ocean basins and on the continents helps us to interpret paleogeographic patterns and global events of the period. Additional information is drawn from the Upper Cretaceous fossil record of flowering plants. As we saw in Chapter 2, these organisms are particularly sensitive to climatic conditions.

Sea Level, Climates, and Ocean Circulation

In the course of the Cretaceous Period there was a global rise of sea level, with only minor interruptions (Major Events, p. 368). As a result, sea level

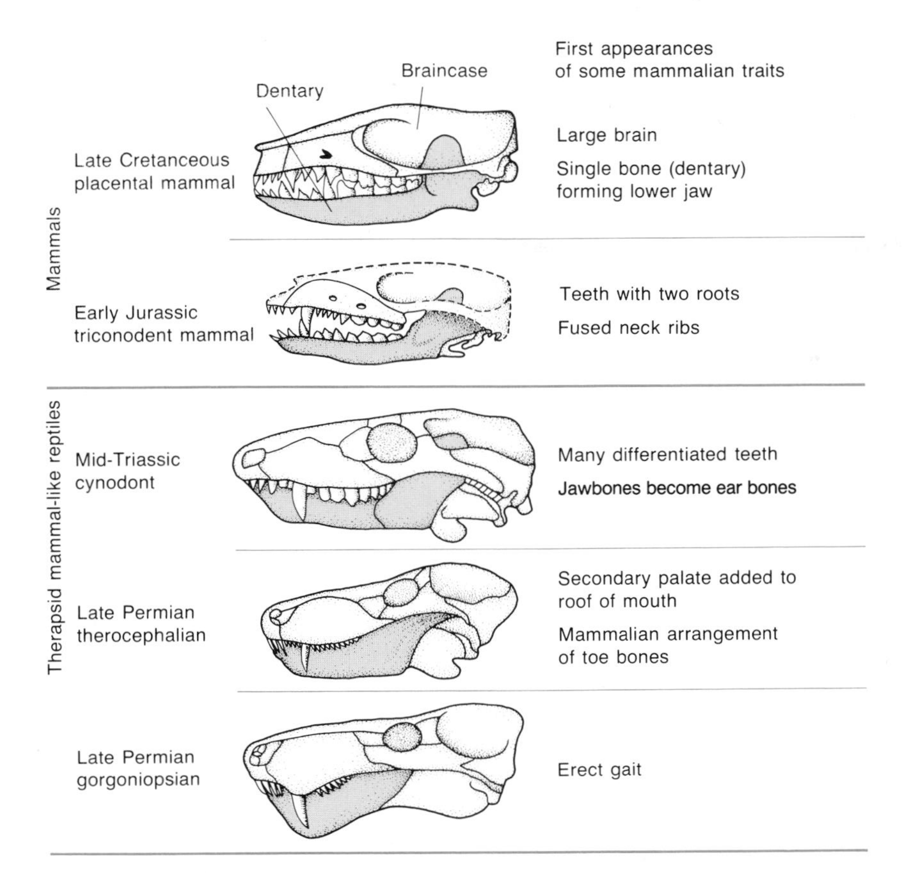

FIGURE 14-22 Stages in the evolution of mammals from mammal-like reptiles. Among the most important changes were the evolution of more highly differentiated, specialized teeth; enlargement of the brain; modification of jawbones into ear bones; and reduction of the number of bones forming the jaw to one, the dentary (shown in color). *(R. E. Sloan.)*

stood perhaps as high throughout most of the Late Cretaceous as at any other time in the Phanerozoic history of Earth, and continents were extensively blanketed by marine deposits.

During the Cretaceous Period, temperatures changed in different ways in different places, but both oxygen isotopes and fossil plant occurrences suggest that climates grew generally warmer during the first part of the period; at the end of Early Cretaceous time, about 100 million years ago, the average temperature on Earth reached a level perhaps higher than it has ever been since. Temperatures then generally declined during Late Cretaceous time.

During the middle part of the Cretaceous Period, there were intervals when black muds covered large areas of shallow seafloor (Figure 14-23). A connection seems to exist between global temperature, water circulation, and the presence or absence of black muds in the sedimentary record. These muds are known to form where bottom waters are depleted of oxygen. It appears that extensive black muds of this kind accumulated in shallow seas when unusually poor circulation within ocean basins led to the stagna-

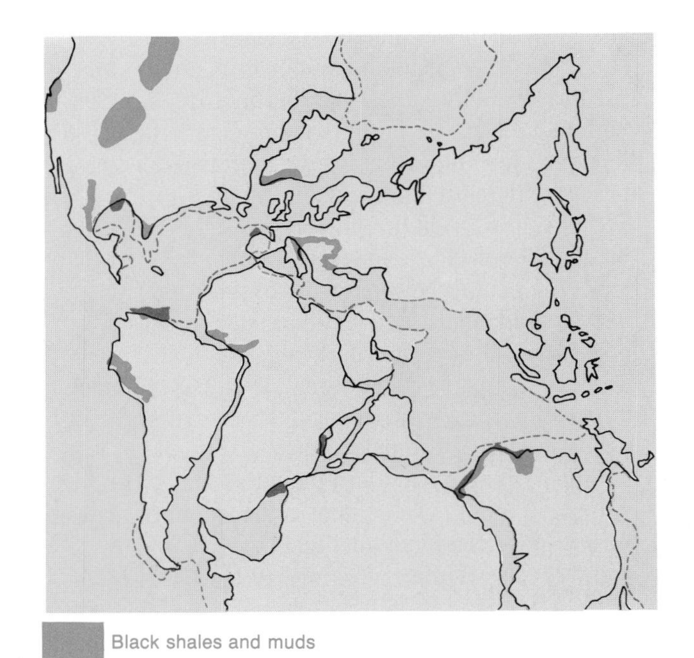

FIGURE 14-23 Widespread occurrence of black shales and muds of Aptian and Albian age on modern continents and continental shelves. *(After A. G. Fischer and M. A. Arthur, Soc. Econ. Paleont. and Mineral. Spec. Publ. No. 25:19–50, 1977.)*

Box 14-1
The Meek Did Inherit the Earth

Workers who opened a quarry at Stonesfield, England, in 1822 soon discovered a one-ton dinosaur and a tiny mammal in Jurassic strata. These early fossil discoveries told the most important story about Mesozoic terrestrial vertebrates that could not swim or fly: Huge dinosaurs ruled Earth, and no mammal grew larger than a modern domestic cat. The great success of the mammals would come only after the dinosaurs had vanished.

The oldest known mammals are of Late Triassic age. The bones that formed the joint between the jaw and the skull establish their identity. Different bones form this joint in mammal-like reptiles. With bodies only about 15 centimeters (6 inches) long and pointed snouts, the earliest mammals resembled modern shrews. Their fossil remains reveal a remarkable amount about their mode of life. Their pointed, cutting teeth show that they were carnivorous, and their small size would have restricted them to a diet made up largely of insects. Their mouth structure indicates that they were endothermic: They had a secondary palate, the bony structure that in all mammals separates the nasal air passages from the mouth so that they can breathe while they eat. Endothermy entails such a high metabolic rate that breathing cannot be interrupted for long. Reptiles, in contrast, can suspend breathing temporarily during their meals. Fossil skulls reveal that the brains of the earliest mammals were large for their overall size, and that large regions of the brain were associated with hearing and smell. The fact that these senses are particularly useful after dark suggests that these small creatures were nocturnal, avoiding the much larger dinosaurs, which presumably were active in daylight. Early mammals appear also to have suckled their young, as modern mammals do. Here the pattern of tooth development is the evidence. Lower vertebrates have functional teeth early in life, and many species replace worn or lost teeth more than once. Mammals, in contrast, do not have functional teeth for quite some time, because early in life their only food is their mother's milk. Because mammals do not need teeth until long after birth, they generally have only two sets of teeth, the baby teeth that appear in infancy and the adult teeth that replace them. Finally, the earliest mammals had rear feet that were adapted for grasping; this characteristic points to a life of tree climbing.

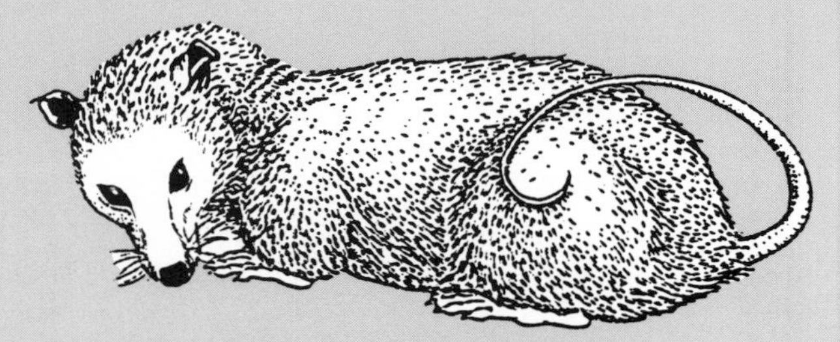

Purgatorius was about the size of a rat.

Although mammals remained small and relatively inconspicuous until the end of the Cretaceous, they did diversify to a degree. The first herbivorous forms, which evolved during the Jurassic Period, had gnawing teeth like those of modern rodents. By Late Cretaceous time, the two large modern groups of mammals were present. These were the placental forms, which include most modern species, and the marsupials, which carry their young in pouches and are the dominant group in Australia today (p. 118).

Among the small Cretaceous mammals was the genus *Purgatorius.* This animal would have had no special interest for a human observer during its lifetime, but from our modern perspective it had great significance. *Purgatorius* belonged to a group of animals that was ancestral to modern primates, including humans, and some workers even assign this group to the primate order. When a mass extinction at the end of the Cretaceous Period swept away the dinosaurs, many small mammals were fortunate enough to survive. Among them was *Purgatorius,* whose fossil remains are found in very early Cenozoic deposits. With the oppressive dinosaurs gone, the surviving mammals proliferated in a spectacular adaptive radiation, which produced the great diversity of mammals in the modern world. Had the group of ratlike animals to which *Purgatorius* belonged died out with the dinosaurs, we humans would never have evolved.

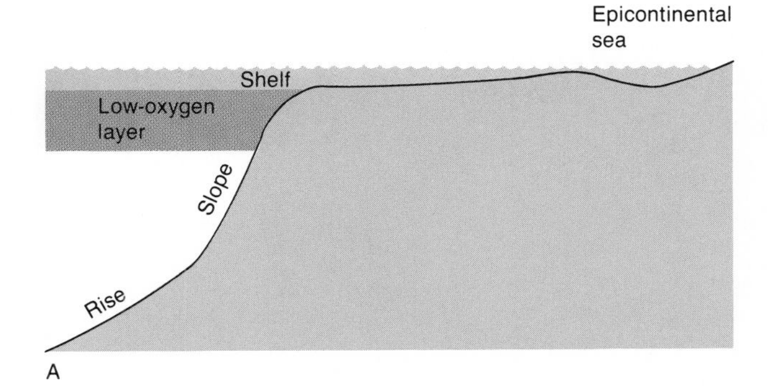

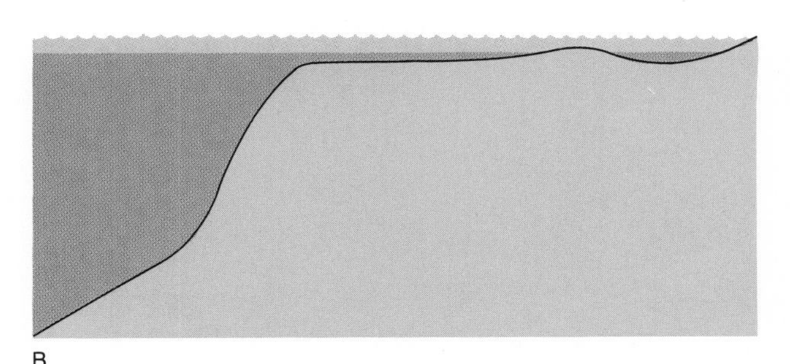

FIGURE 14-24 Expansions of the low-oxygen layer in the ocean at a time when the deep sea is warm. At times like the present, when cold water at the poles is sinking to the deep sea, it supplies deep waters with oxygen, so that the low-oxygen layer is relatively thin *(A)*. When polar waters are warmer and thus are of low density, they do not sink, so they fail to supply oxygen to the deep sea. Under these conditions, the warm waters below the depth of wave activity are relatively stagnant and the low-oxygen layer thickens, extending even into some epicontinental seas *(B)*. *(Modified from A. G. Fischer and M. A. Arthur, Soc. Econ. Paleont. and Mineral. Spec. Publ. No. 25:19–50, 1977.)*

tion of much of the water column. As Figure 14-24 indicates, these waters may at times have spilled over from oceanic areas into shallow seas, leading to the epicontinental deposition of black muds. At other times in Earth's history, including the present, cold waters in polar regions have sunk to the deep sea and spread along the seafloor toward the equator, carrying with them oxygen from the atmosphere (p. 40). The light color of the sediments that represent these intervals in cores that have been taken from the deep seafloor indicates the presence of oxygen during deposition. Extensive black muds were deposited when polar regions were too warm for oxygen-rich surface waters to descend and spread toward the equator. Thus the widespread accumulation of black muds provides still more evidence that the middle portion of the Cretaceous Period was a particularly warm interval; not even the waters of the deep sea were cold.

New Continents and Oceans

Although Pangaea had begun to break apart early in the Mesozoic Era, the smaller continents that had formed from the supercontinent remained tightly clustered at the beginning of the Cretaceous Period. The continued fragmentation of Pangaea and the dispersion of its daughter continents were among the most important developments in global geography during Cretaceous time. Especially dramatic was the breakup of Gondwanaland, two stages of which are shown in Figures 14-25 and 14-26. At the start of the Cretaceous Period, Gondwanaland, still intact, was barely attached to the northern continent. By the end of the period, South America, Africa, and peninsular India had all become discrete entities; of the present-day continents that represent fragments of Gondwanaland, only Antarctica and Australia remained attached to each other.

Widespread Evaporites The fragmentation and separation of continents during Cretaceous time caused new oceans to form and narrow oceans to widen. Most notable were the Early Cretaceous openings of the South Atlantic, the Gulf of Mexico, and the Caribbean Sea. As we have seen, evaporites had formed during the Jurassic Period, when marine waters spilled into the rifts that later widened to form the Gulf of Mexico and South Atlantic (Figure 13-25). Early in the Cretaceous Period, these basins were connected with the rest of the world's oceans, but the connections were narrow, and evaporites accumulated along the basin margins in the restricted bodies of water that resulted (Figure 14-25).

In fact, evaporites were deposited over a large portion of the Early Cretaceous globe (Figure 14-25), a condition that reflected the warm, equable climates of the interval. These climates also permitted the growth of coral reefs as far as

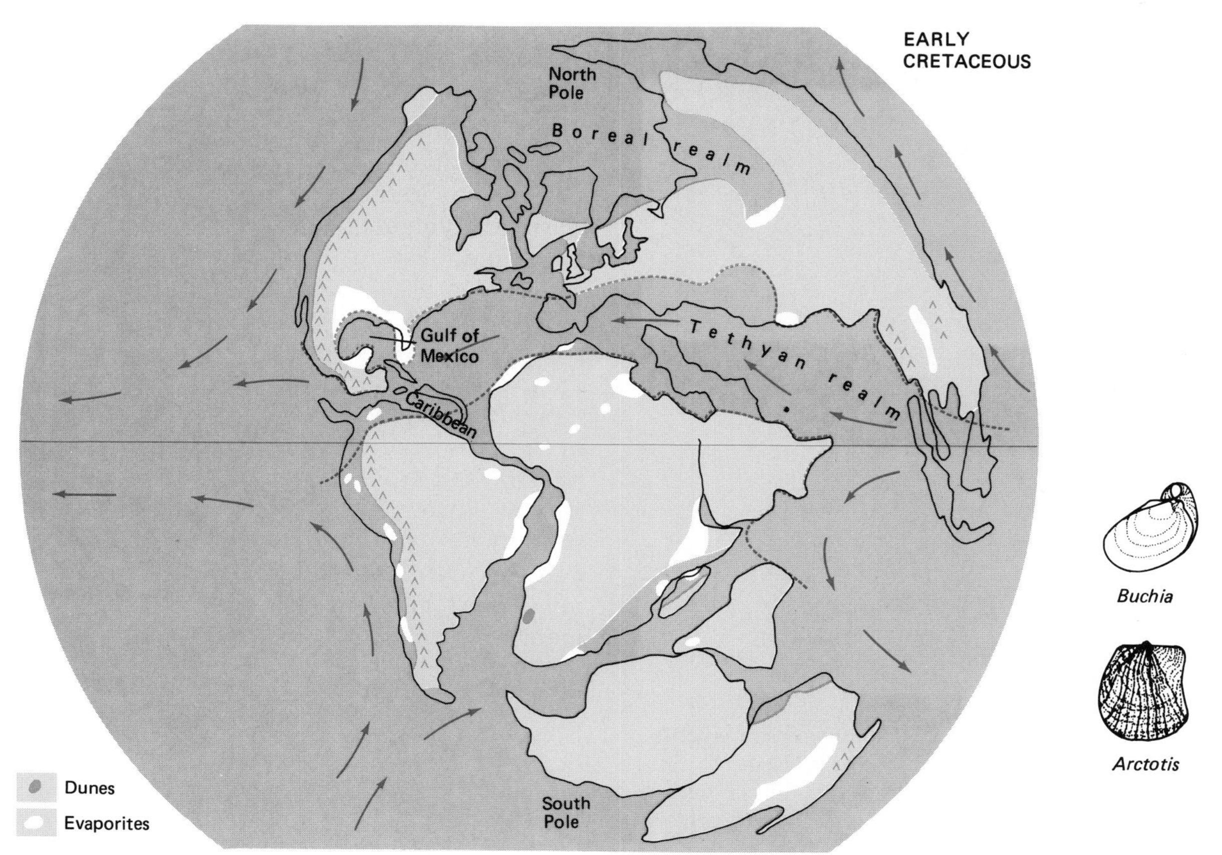

FIGURE 14-25 Global geography during Early Cretaceous time. Gondwanaland was beginning to break apart, but North America, Greenland, and Eurasia remained connected to one another. Climates were warm, and evaporites accumulated even far to the north and south of the tropical Tethyan realm. *Buchia* and *Arctotis* are two genera of marine bivalves that inhabited cool seas both north and south of the Tethyan realm. *(Partly after A. G. Smith and J. C. Briden, Mesozoic and Cenozoic Paleocontinental Maps, Cambridge University Press, Cambridge, England, 1977, and G. E. Drewry et al., Jour. Geol. 82:531–553, 1974.)*

30° from the equator. Further evidence of warm temperatures at high latitudes is the presence of fossil leaves of breadfruit trees in Cretaceous deposits of Greenland (Figure 14-27); similarly, warm-adapted plant species are found in the fossil record of northern Alaska. Latitudinal temperature gradients were so gentle that the tradewind belts probably extended farther north and south of the equator than they do today, resulting in the widespread accumulation of evaporites as their dry air swept over the land (p. 35).

Seaways and the Distribution of Life A dominant feature of the Cretaceous world was the great Tethys Seaway, where the trade winds drove surface waters westward without obstruction by large landmasses. Animals largely confined to the fully tropical Tethyan region included reef-forming corals and rudists. As in Jurassic time, the Tethys was an essentially tropical belt where carbonate deposition prevailed. During Late Jurassic time, however, only shallow Tethyan seas connected Caribbean waters with those of the Pacific (Figure 13-27); during the Cretaceous Period, the separation of North and South America provided a deep oceanic passage for these waters (Figures 14-25 and 14-26).

As in the Jurassic Period, the largely nontropical Boreal realm lay to the north of the tropical Tethys. Several genera of marine bivalves were largely restricted to the Boreal realm and to an equivalent high-latitude region in the Southern Hemisphere. Apparently they were intolerant of the warm temperatures of the Tethyan realm (Figure 14-26).

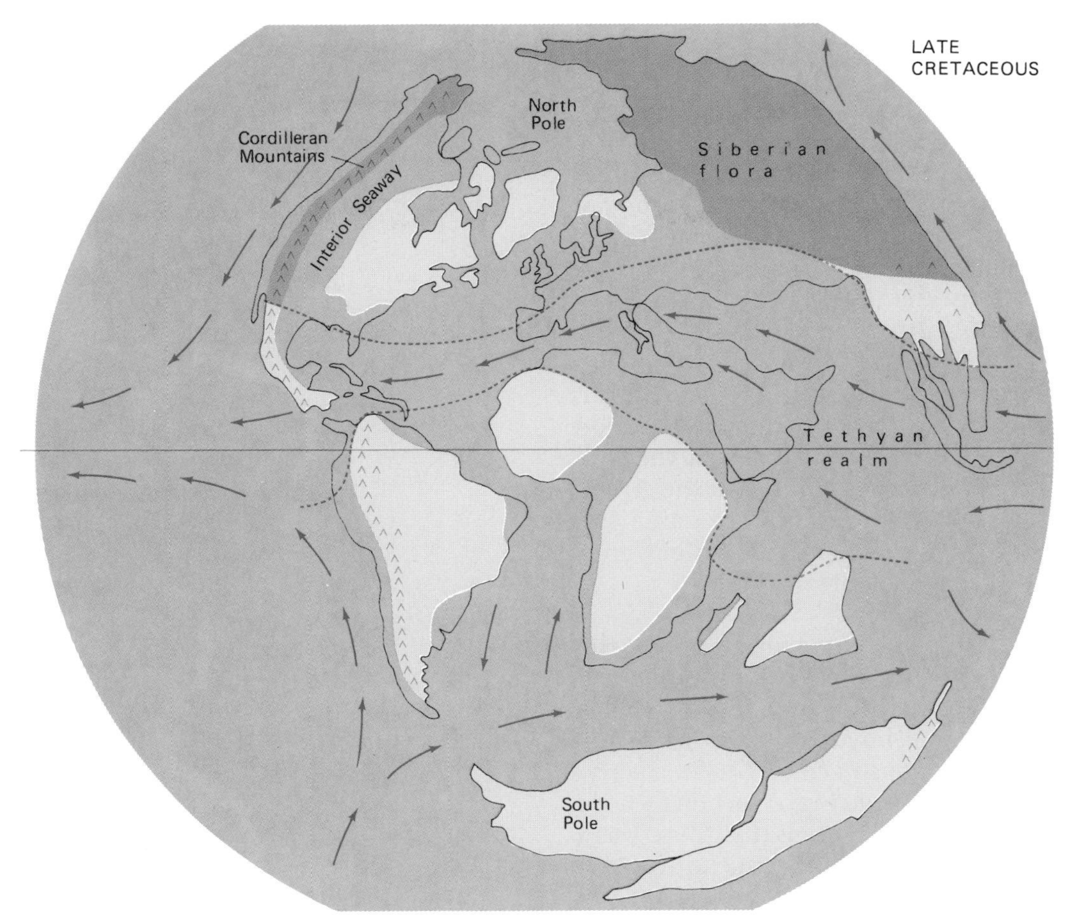

FIGURE 14-26 Global geography during Late Cretaceous time. The Siberian flora occupied not only Siberia but also the northern part of the narrow island formed by the Cordilleran mountain chain. Sea level stood higher in relation to most land areas than it had during Early Cretaceous time. The tropical Tethyan realm remained. *(Partly after A. G. Smith and J. C. Briden, Mesozoic and Cenozoic Paleocontinental Maps, Cambridge University Press, Cambridge, England, 1977.)*

A

B

FIGURE 14-27 A fossil breadfruit leaf *(A)* from the Cretaceous System of Greenland, a continent that was much warmer during Cretaceous time than it is today, and a modern breadfruit leaf *(B)*. *(A. The Swedish Natural History Museum. B. Gerry Ellis/Wildlife Collection.)*

Early in Cretaceous time the Arctic Ocean remained largely isolated from the Atlantic and supported a distinct marine fauna. Its isolation ended later in the period, when an important episode of continental rifting split the huge landmass of the Northern Hemisphere into North America, Greenland, and Eurasia (Figure 14-28).

It was not only rifting that connected the Atlantic and Arctic oceans during Late Cretaceous time. The progressive elevation of sea level that characterized most of the Late Cretaceous Period formed a narrow seaway that spread from the Gulf of Mexico all the way across Alaska to the Arctic Ocean (Figure 14-26). Later we will examine patterns of deposition within this great body of water, known as the Cretaceous Interior Seaway.

Greenhouse Warming? Intensification of the greenhouse effect is perhaps the simplest mechanism that can be invoked to explain the remarkable warming of global climates during the middle portion of the Cretaceous Period. Volcanic activity is one of the primary sources of carbon dioxide in the atmosphere, and it happens that there was a massive outpouring of lava beneath the western Pacific Ocean during the middle portion of the Cretaceous Period, between about 125 and 80 million years ago. Rates of seafloor spreading were very high during this interval, and a large volume of oceanic crust was formed (see Figure 6-27). It appears likely that this major episode of volcanic activity released a relatively large volume of carbon dioxide to the atmosphere, strengthening the greenhouse effect and warming climates all over the world. What reversed the trend toward the end of the Cretaceous remains uncertain.

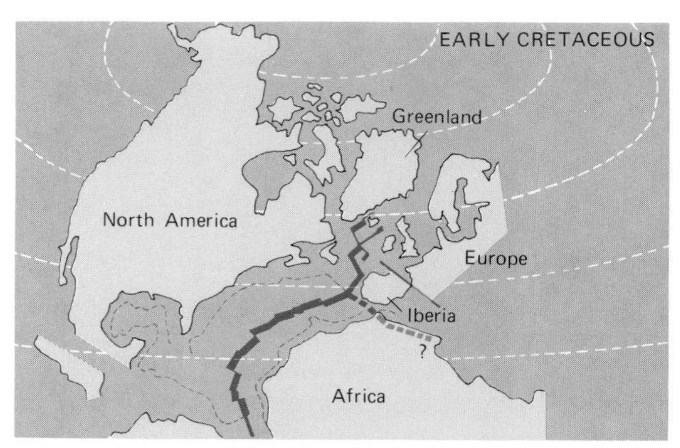

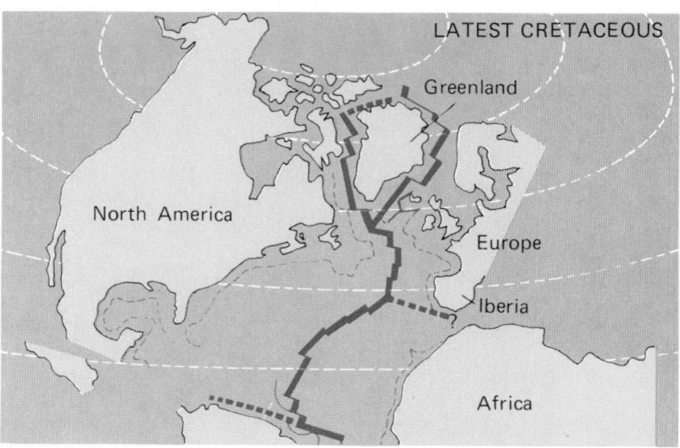

FIGURE 14-28 The opening of the North Atlantic during the Cretaceous Period. As North America and Africa moved apart, the Mid-Atlantic Rift was propagated northward along two branches, leaving Greenland as a separate landmass between North America and Eurasia. *(After J. G. Sclater et al., Jour. Geol. 85:509–552, 1977.)*

The Terminal Cretaceous Extinction

The Mesozoic Era came to a dramatic end with a mass extinction that was quite sudden on a geologic scale of time. Many forms of life that had played major ecologic roles for tens of millions of years disappeared. The most prominent in the minds of modern humans were the dinosaurs, but many other important groups of animals and plants died out as well. Flowering plants, which now dominated most terrestrial landscapes, suffered heavy losses. In the ocean, ammonoids disappeared, as did reptilian "sea monsters," including mosasaurs, plesiosaurs, and giant turtles (Figure 14-6). At the base of the food web, the calcareous nannoplankton declined to such an extent that they produced a much smaller total volume of chalk during the Cenozoic Era than they had during Late Cretaceous time. Most species of planktonic foraminifera also died out, and on the seafloor many groups of mollusks disappeared, including the reef-building rudists (Figure 14-14). The extinction of the rudists, like earlier extinctions of the Phanerozoic Eon, exemplified the fragility of reef ecosystems in general (see Box 2-1).

Discoveries during the past few years have intensified interest in the extinction that ended the Cretaceous Period. They have convinced many geologists that the extinction resulted from the cataclysmic collision of Earth with one or more large extraterrestrial objects.

An Extraterrestrial Cause? In 1981 a team of workers led by the physicist Luis Alvarez and his son Walter, a geologist, discovered an abnormally high concentration of the element iridium precisely at the level of the Cretaceous–Paleogene boundary in the stratigraphic section at Gubbio, Italy (Figure 14-29). Soon a comparable "iridium anomaly" was found at the same stratigraphic level in many other regions of the world. Since iridium is very rare on Earth but fairly abundant in meteorites, the Alvarez team advanced the hypothesis that a large meteorite struck Earth at the end of the Cretaceous, producing a great explosion that dispersed dust relatively rich in iridium high into the atmosphere. The dust from such an explosion would have spread around the globe and then settled to produce a layer of iridium-rich sediment in nearly all depositional environments. On a geologic scale of time, these events would have been instantaneous.

A meteorite of average composition would have had to be about 10 kilometers (6 miles) in diameter to produce the total amount of iridium that forms the worldwide anomaly. Unfortunately, the chemical composition of comets remains unknown, because these extraterrestrial bodies consist largely of frozen elements that vaporize when they strike Earth. Thus it is unknown whether comets, like meteorites, contain an unusually large concentration of iridium. To avoid specifying whether it was a meteorite or comet that may have brought the Mesozoic Era to a close, many geologists have employed the term **bolide** to designate the extraterrestrial object in question. A bolide may be either a meteorite or a comet.

Other geologists suggested a second hypothesis for the widespread iridium anomaly at the top of the Cretaceous System: a massive episode of volcanic activity. This was a reasonable suggestion because iridium is more abundant in Earth's asthenosphere than in its lithosphere, and many large volcanoes erupt lava that contains material from the asthenosphere. Furthermore, there was a massive eruption of lava in southern India very close to the end of the Cretaceous Period. Nonetheless, two additional kinds of evidence favor the idea that the source for the widespread iridium anomaly was extraterrestrial. Each is a type of grain that can form only under very intense pressure, such as the pressure that results when a large extraterrestrial body collides with Earth.

The first kind of grain displays groups of parallel welded factures that formed under enormous pressure (Figure 14-30). Geologists have recently found these "shocked" grains in many parts of the

FIGURE 14-29 The band of clay at the Cretaceous-Cenozoic boundary near Gubbio, Italy. The concentration of the rare element iridium is about 30 times higher in this clay than in sediments above and below. A rich assemblage of calcareous nannoplankton is present below the clay, but the assemblage above is sparse. *(W. Alvarez.)*

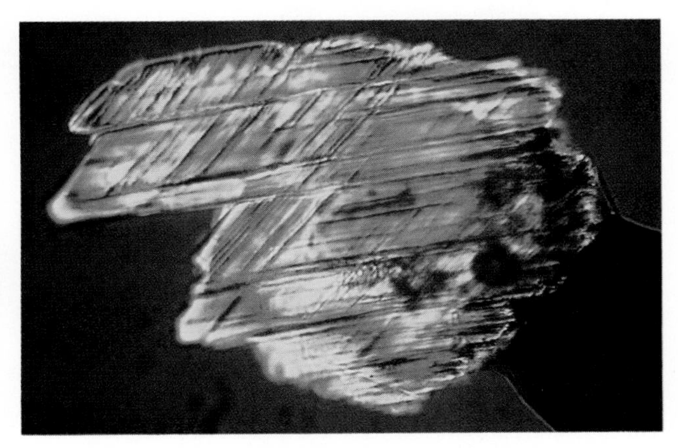

FIGURE 14-30 Shocked quartz grain from uppermost Cretaceous deposits in Montana, showing sets of planar lamellae. *(Glen A. Izett, U.S. Geological Survey.)*

world at the level of the iridium anomaly. They occur at sites where meteorites are known to have formed craters but have never been found at the sites of massive volcanic eruptions.

The second kind of grain is a "microspherule," a nearly spherical grain that resembles window glass in its molecular structure (Figure 14-31). This means that it is largely uncrystallized: It cooled so rapidly after having been liquefied that its chemical elements failed to assemble into a consistent geometric pattern. Microspherules, like shocked grains, occur where meteorites are known to have struck Earth. They formed when droplets of rock, liquefied by the enormous heat generated during the impact, were thrown into the atmosphere, where they cooled very quickly.

Intensely shocked mineral grains and glassy microspherules have now been found at the very top of the Cretaceous System in many parts of the world. Even if explosive volcanic eruptions could have produced these particles, no volcanic eruption would be powerful enough to eject such large particles so high into the atmosphere that they could spread many thousands of kilometers.

Opinions have varied as to how the impact of a large bolide would disturb environments on Earth. An impact in the oceanic realm would have different consequences than one on land or in a shallow sea. Even a bolide that made an oceanic landing would plunge through the water column and penetrate the lithosphere. Here are some consequences that many scientists have predicted for a bolide 10 kilometers in diameter:

FIGURE 14-31 Microspherules from a thin clay layer at the terminal Cretaceous boundary in Wyoming. These grains, which have since been chemically altered, were apparently once glassy structures that cooled rapidly from droplets of molten rock. *(Bruce F. Bohor, U.S. Geological Survey.)*

1 *Perpetual night* Dust particles would blow high into the atmosphere, spread around the world, and screen out nearly all sunlight. Much of the dust would remain aloft for several months, preventing plants from conducting photosynthesis.

2 *Months of global refrigeration* By darkening Earth, the atmospheric dust would plunge the entire planet into cold, wintery weather for several months.

3 *Delayed greenhouse warming* A bolide that landed in the ocean would send into the atmosphere not only small solid particles but also vast quantities of water vapor, which would remain long after the dust had settled. Not only would global cooling then come to an end, but the water vapor would trap solar radiation, thereby intensifying the greenhouse effect (p. 31). In other words, life might have been subjected first to a severe cold snap and then to abnormal warmth.

4 *Acid rain* The huge amount of energy released by the impact of a large bolide would cause oxygen and nitrogen in the atmosphere to combine chemically to produce oxides of nitrogen. When combined with water in the atmosphere, these oxides would produce acid rain. Acid rain might have been so severe and widespread as to cause the extinction of some species.

5 *Wildfires* Particles of soot have been found at the level of the iridium anomaly in some areas of the world. They may be the products of wildfires that raged across terrestrial environments, having been ignited by a great fireball at the site of the impact. Soot is present in many Cretaceous sediments, however, and it is possible that the soot at the top of the Cretaceous System is simply the product of normal wildfires caused by lightning.

In contemplating the consequences of a large bolide impact, scientists have gained new insights into global disturbances in general. For example, the idea that atmospheric dust would refrigerate the world after an impact led to the conclusion that a large-scale nuclear war would produce a nuclear winter. The dust and soot from explosions and fires caused by nuclear explosions would darken and cool the planet, causing massive death of humans and other forms of life far from areas of nuclear attack.

Fossils and the Timing of Extinction When it was first suggested that a bolide impact caused the terminal Cretaceous extinction, some aspects of the fossil record seemed to conflict with this idea. Several fossil groups appeared to have died out over several million years, not within a few hours, days, or months. Few species of dinosaurs, for example, were found in the uppermost few meters of Cretaceous sediment in Montana and nearby regions of Canada, where the fossil record of latest Cretaceous dinosaurs is the best in the world (p. 78). A decline in the number of preserved species upward in the record, toward the terminal Cretaceous boundary, seemed to indicate that dinosaurs died out gradually during the final few million years of the Cretaceous. The impact hypothesis, however, stimulated paleontologists to scour the fossil record here much more thoroughly than they had done before. The result was the discovery that many species of dinosaurs once thought to have died out long before the end of the Cretaceous may well have survived to the very end: Their fossils have now been found within a meter of the iridium anomaly. The lesson here is that the imperfection of the fossil record—or our imperfect knowledge of the record—can fool us. The failure of the record to yield fossils of certain species above a given level does not necessarily mean that these species had died out by the time the seemingly unfossiliferous sediments were deposited. Environments where the last members of the species lived may not be represented in the sediments that we have available for study. We may also have failed to find some fossils that are actually preserved in these sediments. The new discoveries of species close to the boundary indicate that this second problem caused us to underestimate the number of extinctions that occurred right at the end of the Cretaceous Period.

The excellent Cretaceous fossil record of flowering plants in the western United States also seemed until recently to indicate a gradual decline toward the end of the period, but more thorough collecting has documented a sudden extinction. Floras exhibit few losses below the iridium anomaly, but many species make their final fossil appearance very close to it. Furthermore, at the level of the anomaly there is a so-called fern spike: Pollen suddenly becomes very rare in the sediment, and in an interval just a few centimeters thick, spores of ferns heavily dominate the assemblage of microfossils representing terrestrial plants. This pattern suggests that communities of flowering plants died out suddenly and were replaced in the landscape by a heavy growth of ferns. When a fire or volcanic eruption wipes out a forest today, ferns readily invade the cleared land. Thus ferns are ecologic opportunists (p. 28). Above the fern spike and iridium anomaly are fossil pollen and leaves of flowering plants that represent a flora quite different from the kind that existed during Late Cretaceous time. Many Cretaceous groups of broad-leafed evergreen plants have disappeared. The postextinction flora is dominated by deciduous taxa—ones that lose their leaves every autumn. These taxa are better adapted to cold winters than are broad-leafed evergreens. This pattern of change is consistent with the idea that a bolide impact abruptly cooled Earth by loading the upper atmosphere with dust.

Despite all these new discoveries, there is evidence that some important taxa that failed to survive into the Cenozoic Era were on the decline or actually disappeared more than a million years before the iridium anomaly formed. Some inoceramid bivalves, a very diverse and abundant group until the final two million years or so of Cretaceous time, are conspicuous in the fossil record because they are very abundant or very large (Figure 14-13*A*). There is little chance that these forms survived in latest Cretaceous time and have been overlooked in the record. The rudist bivalves may not have disappeared altogether until the very end of the period, but they were definitely on the decline. Recall that these were the dominant reef builders in Late Cretaceous seas. Reef faunas of latest Cretaceous age were impoverished, consisting of relatively few species. Thus the histories of these two important groups of bivalves suggest that the marine ecosystem was in decline in some ways before the cataclysm that produced the iridium anomaly. The record of planktonic foraminifera suggests that this group may also have suffered a pulse of extinction slightly before the end of the Cretaceous and then another right at the end.

Although evidence is now mounting that many species of dinosaurs survived until the very end of the Cretaceous Period, the fossil record suggests that the dinosaur community, too, was under environmental stress somewhat earlier.

About three-quarters of all fossils in the strata just below the terminal Cretaceous boundary represent horned dinosaurs (Figures 14-18 and 14-21). Recall that biological communities that live in hostile environments are typically low in diversity (p. 28). This can mean either that they contain a small number of species or that they contain quite a few species but most are rare.

The evidence that heavy extinction was going on in the marine realm before the very end of the Cretaceous, especially in the environmentally sensitive reef community, and that the dinosaur community was also stressed raises an interesting possibility. Perhaps a bolide impact under these conditions was much more devastating to life than it would have been if it had occurred when ecosystems were healthy.

How Many Impacts Occurred and Where? Another possibility being considered is that more than one bolide struck Earth. At issue here is whether iridium anomalies and layers containing abundant shocked minerals occur at more than one level in uppermost Cretaceous strata in some areas of the world. Some workers claim to have found these features at more than one level, but their claims have yet to be confirmed.

Geologists are currently searching for one or more craters that might represent a Late Cretaceous impact. A crater that seismic studies have revealed beneath Cenozoic strata at Manson, Iowa, has been suggested as a likely candidate. Geologists have drilled into the rocks in which the structure occurs and brought to the surface metamorphic minerals that were probably formed by an impact at approximately the time the Cretaceous Period ended. The problem is that the Manson crater is not large enough to have been produced by a bolide that released the total volume of iridium estimated to have settled to Earth at that time. The second candidate is an apparent crater of approximately the right age in the continental shelf off the Yucatan Peninsula, in the Gulf of Mexico. This structure is larger, but it has not yet been shown certainly to be an impact crater. Very large microspherules at the terminal Cretaceous boundary in nearby areas suggest that an impact may indeed have occurred in this region. The reasoning here is that the largest microspherules produced by an impact would be too heavy to travel far; only small ones could be blown high into the atmosphere and spread thousands of miles before falling to Earth. Of course, it is possible that the small crater-like structure in Iowa and the larger one in the Gulf of Mexico formed from separate impacts, with the latter producing the large iridium anomaly and most of the extinction at the end of Cretaceous time.

The Geographic Pattern of the Crisis It appears that cooling of climates played a major role in the terminal Cretaceous extinction. Not only the reef-forming rudists but numerous other tropical species of marine life died out with the dinosaurs. Extinctions were less severe in seas far from the equator, where species were adapted to frigid conditions. The same pattern is evident for flowering plants on the land. As we have seen, broad-leafed evergreen species unable to tolerate cold winters were preferentially victimized in North America. Floras in northern Canada were dominated by cold-adapted, deciduous species, and they experienced lighter extinction.

There appears to have been another general geographic pattern of extinction that has yet to be studied in detail. As we have seen, land plants at moderately high latitudes in the Northern Hemisphere underwent sudden heavy extinction. In sharp contrast, no dramatic changes have been detected in floras of Australia or nearby regions of the Southern Hemisphere. Light losses in the Southern Hemisphere, if confirmed by further study, may reflect the location of one or more bolide impacts in the Northern Hemisphere. There was, however, total extinction of many groups of animals that were widespread in both hemispheres, including ammonoids in the seas and dinosaurs on the land. Furthermore, it is not clear that the destructive effects of a bolide impact could be restricted to the hemisphere where it occurred. This geographic puzzle awaits further research.

The Aftermath Most of the groups of animals and plants that survived the terminal Cretaceous crisis at reduced diversity expanded again during the Cenozoic Era. For a time, however, life on Earth was impoverished in many ways. Among the survivors, the calcareous nannoplankton exhibit an especially interesting ecologic pattern. Immediately after many species in this group died out or became quite rare at the very end of the Cretaceous, a few species blossomed to great abundance in the ocean. Perhaps these ecologic op-

FIGURE 14-32 The calcareous nannoplankton species *Braarudisphaera bigelowi* (×3135). *(From B. U. Haq, in B. U. Haq and A. Boersma [eds.], Introduction to Micropaleontology, Elsevier, New York, 1978.)*

portunists (p. 28) were especially tolerant of abnormal conditions. After new species evolved during the Cenozoic Era, the opportunists declined in abundance. One species, however, survived for more than 150 million years: *Braarudisphaera bigelowi* (Figure 14-32) exists even today. The fossil record shows that this form has occasionally spread to the open ocean and undergone population explosions, apparently because unusual conditions have briefly favored it over other species, but today it is restricted to marginal marine lagoons.

Plants on the land exhibited another notable pattern. In North America, deciduous flowering plants—the kind that survived the crisis with relatively little extinction—dominated the landscape for almost 10 million years. Possibly climates remained altered for much of this time. It appears, however, that broad-leafed evergreens took a long time to rediversify even after favorable conditions returned. Evolution does not always replace extinct forms of life quickly, even on a geologic scale of time. Thus deciduous forms remained dominant long after the crisis.

Of all taxa, it was the mammals that underwent the most spectacular diversification early in the Cenozoic Era. Their many strategies for exploiting the absence of the dinosaurs are explored in Chapter 15.

REGIONAL EXAMPLES

The great worldwide elevation of the seas that began near the end of Early Cretaceous time produced much of the Cretaceous record exposed on modern continents. This record tells many of the regional stories that follow. Mountain building continued in western North America and produced an enormous foreland basin, which became flooded by the seaway that extended from the Gulf Coast to the Arctic. The Gulf Coast itself was fringed by rudist reefs, and a rudist-rimmed carbonate bank also stretched along a large segment of the adjacent Atlantic coast until midway through the Cretaceous Period, when it gave way to the deposition of mud and sand that continues today. On the other side of the Atlantic, siliciclastic deposition early in the Cretaceous Period was followed by the widespread accumulation of chalk in Europe.

Cordilleran Mountain Building Continues

During Cretaceous time an important change took place in the pattern of igneous activity in western North America. Subduction of the Franciscan complex along the western margin of the continent continued, as did the associated igneous activity (Figures 13-35 and 14-33). By Late Cretaceous time, however, although volcanic and plutonic activity persisted in the Sierra Nevada region, the northern igneous activity had become concentrated to the east, in Nevada and Idaho (Figure 14-34). This pattern contrasted with that of the Late Jurassic Epoch, when igneous activity in the north had been centered near the coast, in northern California and Oregon (Figure 13-31). The likely explanation for the eastward migration of igneous activity in the northern United States is that the angle of subduction there had decreased. In mid-Cretaceous time, the subducted plate in this region probably began to pass downward beneath the continent at a low angle; the subducted crust therefore failed to sink deep enough to melt until it had extended far inland (Figure 14-34). The fold-and-thrust belt in front of the mountainous igneous regions also shifted inland in the northern United States. By Late Cretaceous time, folding and thrusting extended eastward as far as the Idaho-Wyoming border (Figure 14-34).

A major episode of igneous activity and eastward folding and thrusting coincided approximately with the Cretaceous Period; although this episode was not entirely divorced from earlier and later tectonic activity, it is separately identified as the Sevier orogeny. East of this belt lay a vast foreland basin, which in Late Cretaceous time was occupied by a narrow seaway stretching from the Gulf of Mexico to the Arctic Ocean.

The orogenic belt that occupied western North America during the latter half of Cretaceous time was unusually broad, apparently because of low-angle subduction. In its development

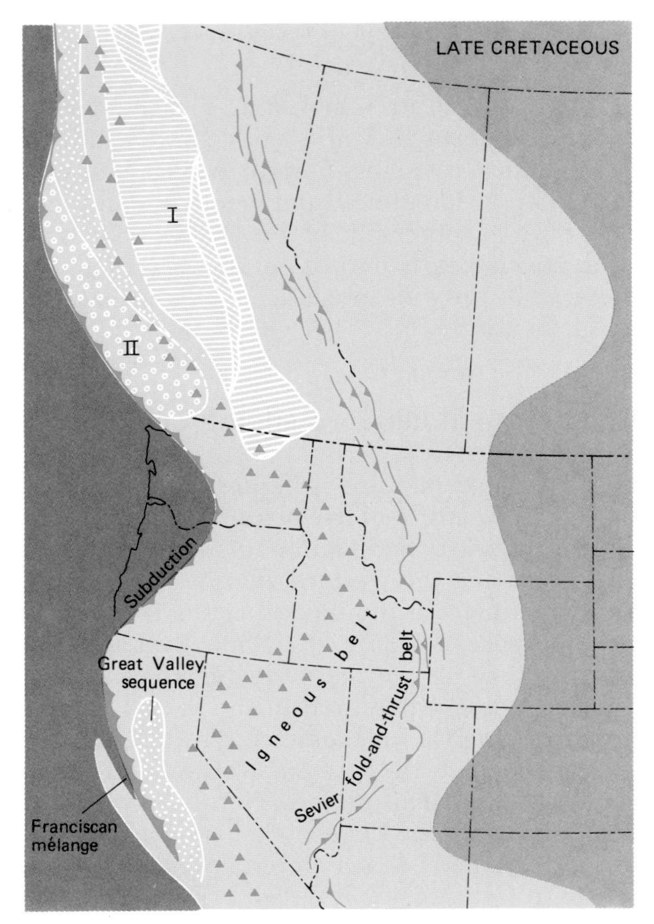

FIGURE 14-33 Late Cretaceous geologic features of western North America. Subduction produced the Franciscan mélange in California. North of California, igneous activity resulting from subduction was located far to the east of the continental margin; this activity, together with the folding and thrusting to the east, represented the latter part of the Sevier orogeny. In Canada, the margin of the continent consisted of two blocks of exotic terrane (I and II) that had been sutured to North America earlier in the Mesozoic Era; each of these blocks consisted of two or more slivers of crust that were welded together to form the block before it was attached to North America.

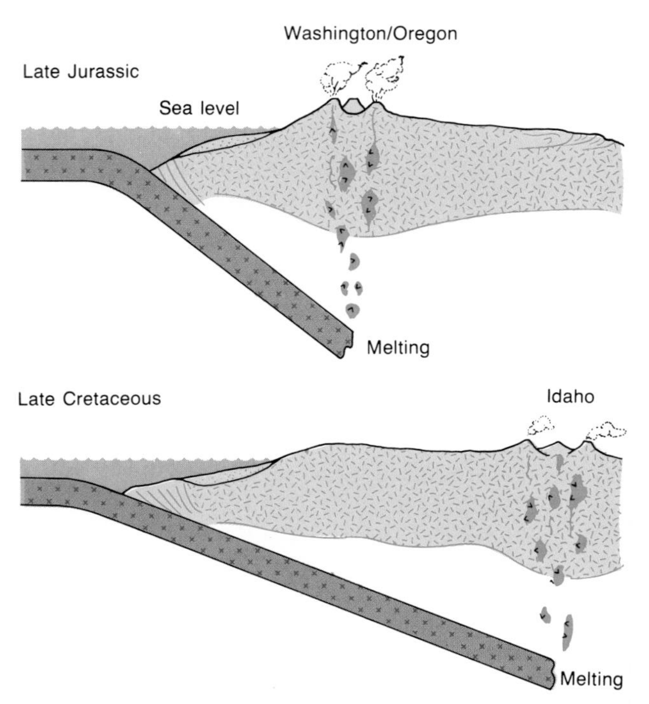

FIGURE 14-34 A likely explanation for the eastward migration of igneous activity in the Cordillera during Cretaceous time. As suggested here, the subducted plate began to pass downward at a reduced angle, so that it reached the depth of melting only after passing far to the east.

of a foreland basin and certain other features, however, the orogenic belt was typical. Like the modern Andes (Figure 7-5), for example, it was symmetrical: The Franciscan deformation at the continental margin (Figure 14-33) was mirrored on a larger scale by the Sevier folding and faulting east of the belt of igneous activity.

The Mesozoic history of western Canada is far more complicated. Recall that during the Jurassic Period, a sizable microcontinent was sutured to this region of North America (Figure 13-32). This theme of continental accretion continued into the Cretaceous Period, when a small microcontinent was attached along the western margin of the first. This new landmass, like the one accreted during the Jurassic, was a composite of two or more terranes. They had become amalgamated during the Jurassic Period and were attached to North America during Cretaceous time.

The Gulf Coast and North American Interior Seaway

During the latter part of Early Cretaceous time, a spectacular array of marine environments developed in the Cordilleran foreland basin. This foreland basin extended from the Gulf Coast all the way to the Arctic Ocean, which was probably much warmer than it is today, judging by nearby Alaskan floras.

Shortly before the end of the Early Cretaceous, during Albian time, Arctic waters spread southward, flooding a large area of western North America with the Mowry Sea (Figure 14-35). This body was named for the Mowry Formation that accumulated within it, which consists mostly of **oil shale,** a well-laminated shale in which dark layers rich in fish bones and scales alternate with thicker, lighter layers. The Mowry Sea formed as a part of the great mid-Cretaceous marine transgression that resulted in the deposition of black shales on many continents (Figure 14-24). To the south, the Gulf of Mexico was originally part of the tropical Tethyan realm (Figure 14-26). Rudist reefs flourished around its margin.

The Mowry Sea made brief and intermittent contact with the Gulf of Mexico before the end of Early Cretaceous time, but an enduring connection was established at the start of the Late Cretaceous. The result of this contact was an enormous inland sea, the Cretaceous Interior Seaway, that occupied the foreland basin to the east of the Sevier orogenic belt. Until just before the end of the Cretaceous Period, this seaway extended from the Gulf of Mexico to the Arctic Ocean (Figure 14-36). Most of the sediments deposited here were shed from the Cordilleran Mountains that formed to the west. The history of the seaway is especially well understood because of excellent stratigraphic correlations based on abundant fossil ammonoids; in addition, ash falls from volcanic eruptions to the west provided numerous marker horizons, many of which can be dated radiometrically (Figure 4-8).

Barrier islands bounded much of the seaway, with lagoons standing behind them. The lagoons were bordered by broad swamps; on the western margin of the seaway the swamps gave way to alluvial plains, which were succeeded near the mountains by alluvial fans.

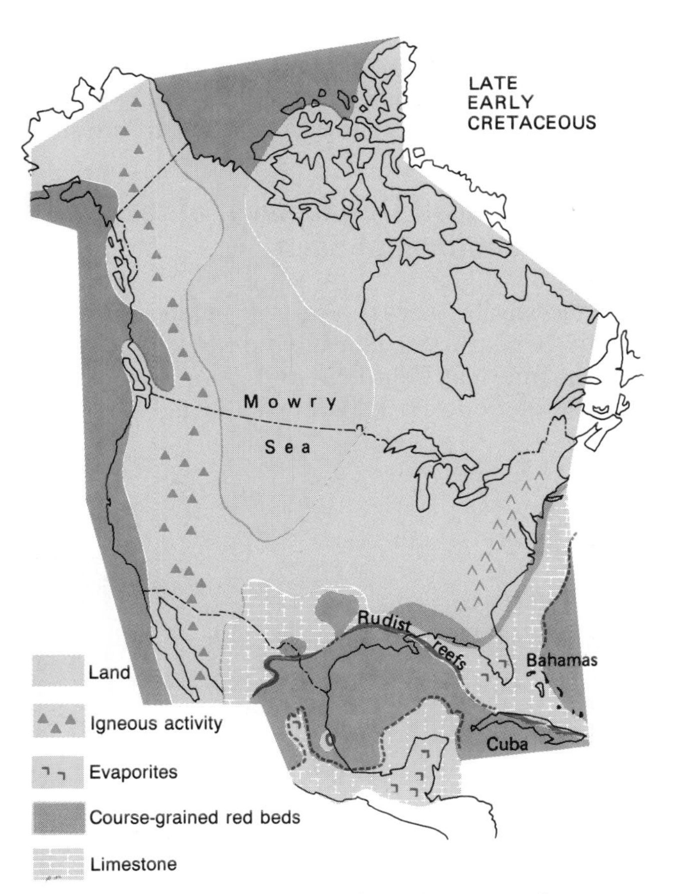

FIGURE 14-35 Geography of North America later in Early Cretaceous time. The Mowry Sea, where black muds were deposited, spread southward from the Arctic Ocean. A carbonate platform bordered by rudist reefs encircled the Gulf of Mexico, and carbonate deposition extended far to the north along the East Coast.

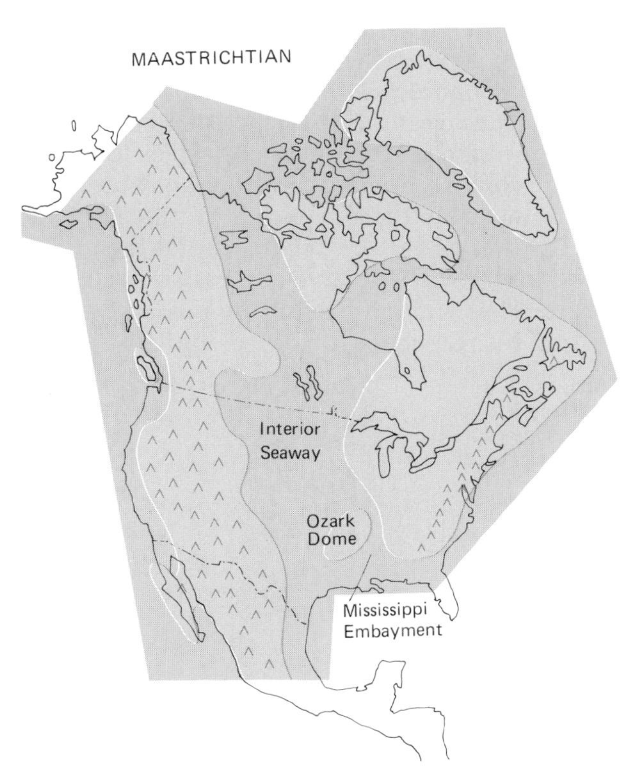

FIGURE 14-36 Flooding of North America during Maastrichtian time, just 2 or 3 million years before the great regression at the end of the Cretaceous Period. Until the regression, the Cretaceous Interior Seaway stretched from the Gulf of Mexico to the Arctic Ocean. The Mississippi Embayment occupied the area now occupied by the southern segment of the Mississippi River; at times this embayment was connected to the seaway by a passage along the east side of the Ozark Dome, a highland centered in Arkansas. *(After G. D. Williams and C. R. Stelck, Geol. Soc. Canada Spec. Paper No. 13, 1975.)*

The western shoreline of the seaway shifted back and forth, primarily in response to the rate of sediment supply. At all times conglomeratic sediments were shed eastward from the neighboring mountains as clastic wedges, but at times of particularly active thrusting or uplift these wedges prograded especially far to the east (Figure 14-37). Because of the great weight of sediments on the western side of the seaway, subsidence was more rapid there than farther east. Along the western margin, in nonmarine environments, Late Cretaceous dinosaurs left a rich fossil record.

The Upper Cretaceous strata of the interior seaway represent two large depositional cycles, one of which is illustrated in Figure 14-38. Each cycle consists of an interval of transgression followed by an interval of regression. In addition to changing rates of sediment supply, global changes in sea level and the changing rate of subsidence of the seaway floor must have influenced these patterns of transgression and regression. At times of low supply and maximum lateral expansion of the seaway, chalks were laid down in the center. The most famous of these deposits is the Niobrara Chalk (Figure 14-39), which occupies the middle of the upper transgressive-regressive cycle. The Niobrara has yielded beautifully preserved fossil vertebrates (Figures 14-5, 14-6, and 14-8).

To the south, Late Cretaceous seas spread from the Gulf of Mexico into the Mississippi Embayment, which was partially separated from the interior seaway by the Ozark Dome (Figure 14-36). Near the end of the period, siliciclastic sediments were shed into the eastern and western extremities of the Mississippi Embayment—into the region of Georgia from the ancestral eastern Appalachians and into the region of Tennessee from the ancestral Mississippi River. Between these two areas of siliciclastic deposition, large volumes of chalk accumulated in what is now central Alabama.

Just before the end of the Cretaceous Period, the seas retreated southward from the interior seaway and the Mississippi Embayment, and a new pulse of mountain building began along the western margin of the interior seaway. This Laramide orogeny, which continued well into the Cenozoic Era, is discussed in Chapter 15. Except for a brief and less extensive incursion just after the beginning of the Cenozoic Era, the seas have never returned to the western interior of North America.

The East Coast: Development of the Modern Continental Shelf

Seismic studies reveal a great thickness of sediments beneath the continental shelf in eastern North America (Figure 14-39). These sediments consist of deposits laid down during the early

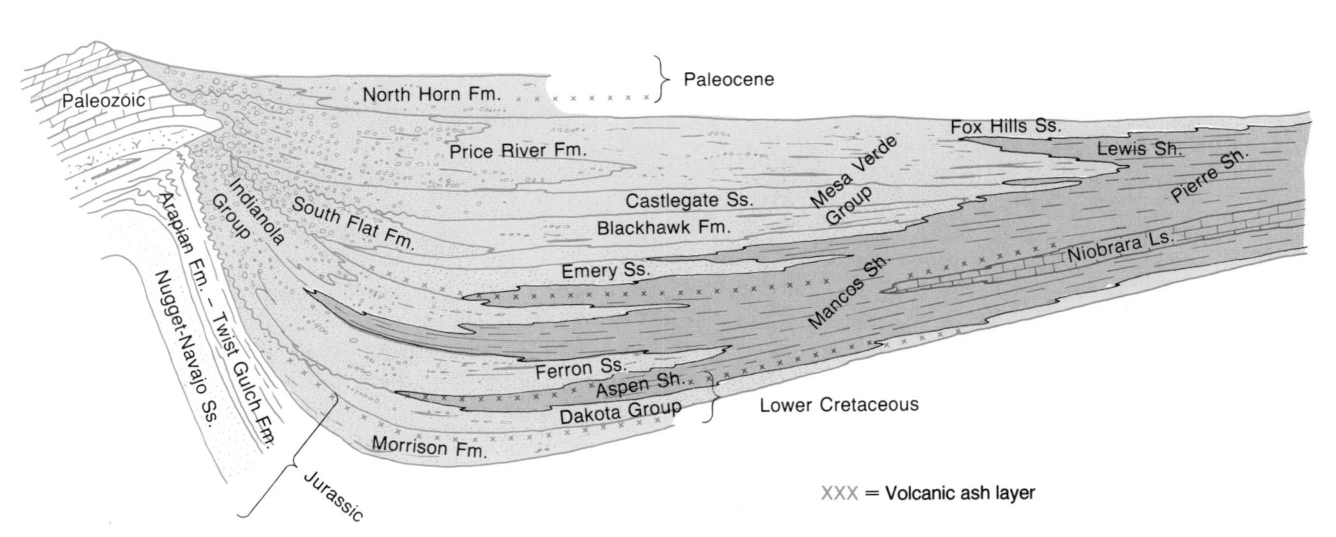

FIGURE 14-37 Cross section of Upper Cretaceous sediments in central Utah. These sediments were deposited in the foreland basin east of the Sevier orogenic belt. Clastic wedges in the west pass eastward into finer-grained marine sediments. *(After R. L. Armstrong, Geol. Soc. Amer. Bull. 79:429–458, 1968.)*

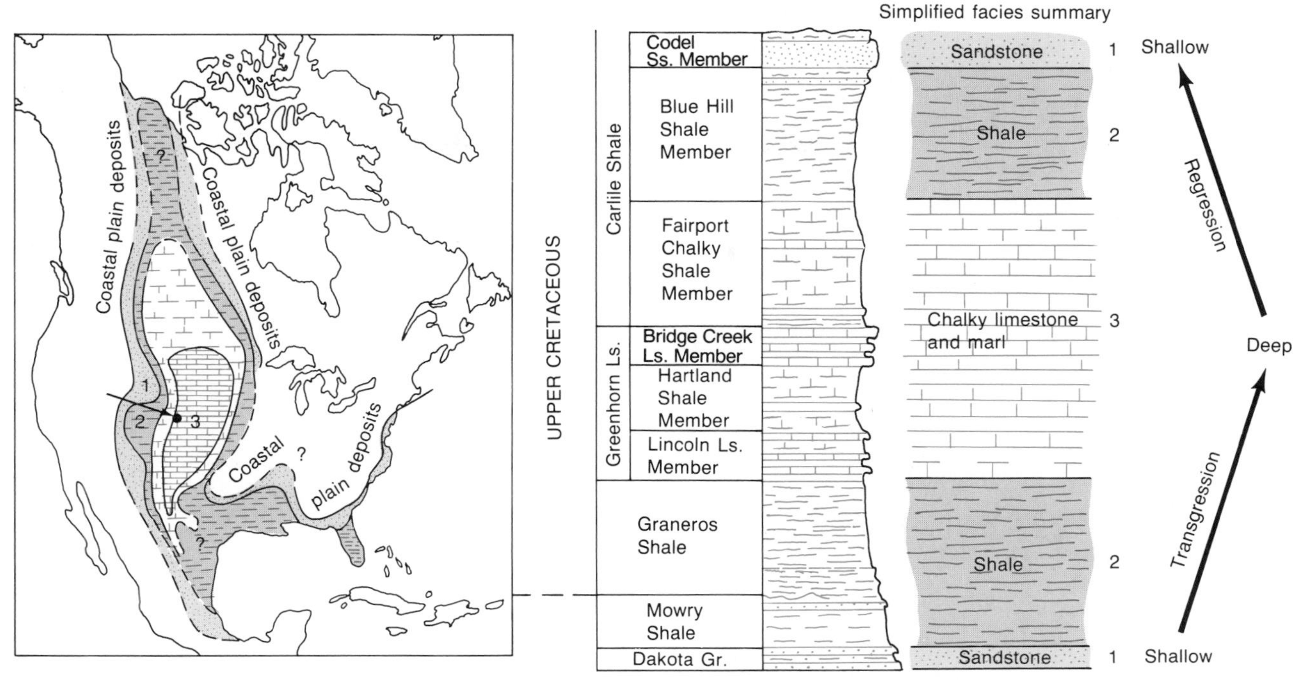

FIGURE 14-38 The early Late Cretaceous "Greenhorn" depositional cycle of the North American interior seaway. The three facies in the simplified facies summary are shown in map view on the left. The stratigraphic section represents the vicinity of eastern Colorado *(arrow on map)* where the cycle developed by an oscillation of the shoreline (transgression and regression). During the transgression, the area of chalky limestone and marl (limey clay) deposition in the center of the basin (facies 3) expanded; first facies 2 and then facies 3 spread into eastern Colorado. During the regression, the area of deposit of facies 3 contracted, and facies 2 and then facies 1 shifted into eastern Colorado. *(Modified from E. G. Kauffman, The Mountain Geologist 6:227–245, 1969.)*

Mesozoic episodes of rifting that formed the modern Atlantic Ocean. At the base are fault basin deposits like those of the Newark Supergroup that are exposed on the continent to the west (Figure 13-28). Next come large thicknesses of Jurassic carbonates that accumulated in the narrow, young Atlantic Ocean as passive margin deposition commenced (Figure 14-39). Above the Jurassic carbonates are more carbonates from the Early Cretaceous interval, when, under much warmer climatic conditions than exist today, reef-rimmed carbonate banks bordered the ocean from Florida to New Jersey (Figure 14-35).

Before the end of Early Cretaceous time, reef growth gave way to deposition of predominantly siliciclastic sediments. This change marked the beginning of the growth of the large clastic wedge that forms the modern continental shelf. The clastic wedge consists largely of sands and muds from the Appalachian region laid down in nonmarine and shallow marine settings. Off the coast of New Jersey (Figure 14-39) the total thickness of Cretaceous and Cenozoic sediments is approximately 3 kilometers (~2 miles). Only to the south, in southern Florida, did carbonate deposition persist to the present. Here about 3 kilometers of sediments, consisting mainly of carbonates, accumulated during the Cretaceous Period, and another 2 kilometers or so were added during the Cenozoic Era.

The clastic wedge apparently began to develop because of a renewed uplift of the Appalachian Mountain belt to the west, after it had been largely leveled during the Jurassic. Cretaceous sediments of the wedge lie mostly beneath the ocean or beneath younger deposits on the land; some, however, are exposed in two belts of the Coastal Plain Province. One stretches from New Jersey to Maryland and Virginia and the other is in the Carolinas. Like the younger Coastal Plain sediments, these Cretaceous deposits remain largely unconsolidated.

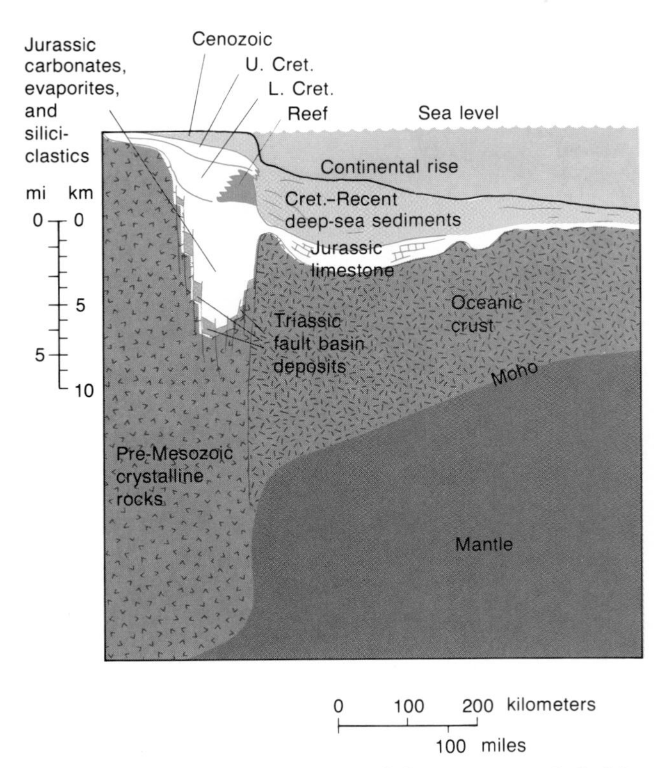

FIGURE 14-39 Cross section of the continental shelf and deep sea off the coast of New Jersey. Here the opening of the Atlantic is recorded by early Mesozoic sediments deposited within fault-bounded basins. During Early Cretaceous time, a reef-rimmed carbonate platform extended this far north under the influence of Tethyan ocean currents (see Figure 14-35). During Late Cretaceous time, carbonate deposition gave way to siliciclastic deposition, which has predominated to the present day. *(After R. E. Sheridan et al., The Geology of Continental Margins, Springer-Verlag, New York, 1974.)*

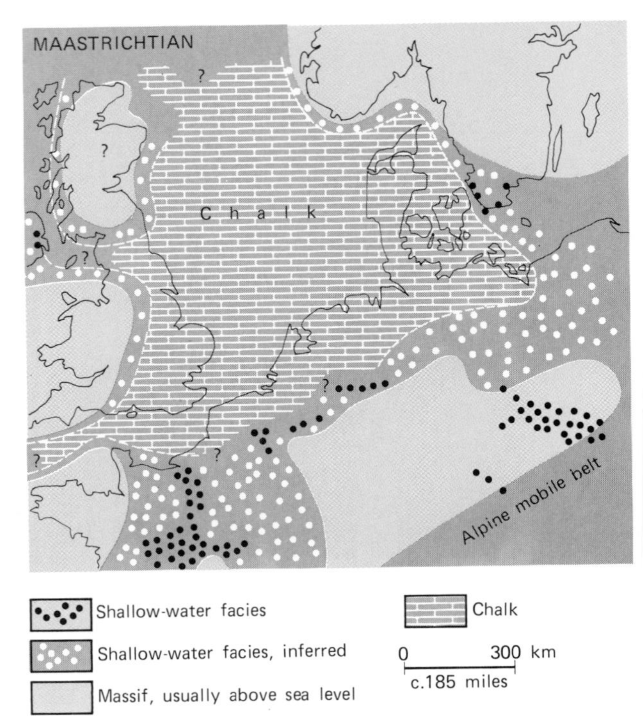

FIGURE 14-40 Paleogeography of northwestern Europe during Maastrichtian time. Chalk deposition was centered in the North Sea basin. Several stable blocks (massifs) formed islands around which marginal facies developed; these facies consisted mainly of coarse limestones but included siliciclastics as well. *(Modified from E. Hakansson et al., Spec. Publ. Internat. Assoc. Sedimentol. 1:211–233, 1979.)*

The Chalk Seas of Europe

To many geologists the term *chalk* refers specifically to the soft, fine-grained limestones of western Europe, although, as we have seen, chalky rocks are found elsewhere, including the Cretaceous System of North America. The Cretaceous chalks of Europe, which are spectacularly displayed as the White Cliffs of Dover and other coastline cliffs of southeastern England (p. 366), accumulated throughout almost all of Late Cretaceous time, when high sea levels flooded much of western Europe.

Except in the newly forming Alps to the south, tectonic activity was largely absent from western Europe during Late Cretaceous time, when the chalk accumulated. Here and there, massifs stood relatively high as islands or, during transgressions, as shallow seafloors (Figure 14-40). Chalk accumulated almost continuously in the basins that surrounded the massifs. The lower portions of the chalk often contain clay, but the remainder is relatively pure calcium carbonate, consisting of minute skeletal debris that is about 75 percent planktonic in origin. In fact, the rock remains soft because calcareous nannoplankton, which secrete the mineral calcite, are the dominant components. Calcite is less easily dissolved and reprecipitated as cement that hardens rock than is aragonite, which is the form of calcium carbonate secreted by many other organisms.

Although the chalk seas of Europe were unusually deep for epicontinental bodies of water, the general presence of oxidizing conditions along the seafloor is demonstrated by the widespread occurrence of a fauna of bottom-dwelling species, including bryozoans, ostracods, foraminifera, brachiopods, bivalves, echinoids, and soft-bodied burrowers. Knowing the typical thickness of the European chalk and the total time elapsed during its deposition, we can estimate that it accumulated at a rate perhaps as great as 15 centimeters (~6 inches) per 1000 years. The productivity of the calcareous nannoplankton here and in the interior seaway of North America was apparently greater than that of calcareous nannoplankton in any region of the oceans today.

CHAPTER SUMMARY

1 With the evolutionary expansion of the dinoflagellates, diatoms, and calcareous nannoplankton during the Cretaceous Period, the phytoplankton assumed a modern character. Similarly, the diversification of the planktonic foraminifera contributed to the modernization of the zooplankton.

2 Calcareous nannoplankton raining down on the seafloor produced thick deposits of chalk in western Europe and elsewhere.

3 On the seafloor, predators such as crabs, teleost fishes, and carnivorous snails diversified markedly during Cretaceous time, thereby altering the nature of the marine ecosystem.

4 In mid-Cretaceous time, rudist bivalves temporarily displaced corals as the primary builders of organic reefs.

5 On the land, angiosperms, or flowering plants, displaced gymnosperms as the most diverse group of plants.

6 At the end of the Cretaceous Period, mass extinction eliminated the ammonoids, rudists, marine reptiles, and dinosaurs and devastated many other groups of organisms. The impact of a bolide from outer space probably caused this biotic crisis.

7 Gondwanaland broke apart during the Cretaceous Period, forming the South Atlantic and other oceans.

8 Sea level rose throughout Early Cretaceous time, flooding much of Europe and spreading over western North America from the Gulf of Mexico to the Arctic Ocean. The average temperature of Earth was higher than it is today.

9 In western North America, orogenic activity shifted eastward from its Jurassic position. To the east of the orogenic belt, an interior seaway stretched from the Gulf of Mexico to the Arctic Ocean.

10 World climates cooled in Late Cretaceous time, but before this happened, they were so warm that rudist reefs flourished in southern Europe and extended from the Gulf of Mexico northward along the Atlantic coast to New Jersey.

EXERCISES

1 What accounts for the abundance of chalk in the Cretaceous System?

2 What were the most important groups of swimming predators in Cretaceous seas?

3 What evidence do land plants offer about the nature of Late Cretaceous climates?

4 What evidence suggests that a large extraterrestrial object struck the earth at the end of the Cretaceous Period?

5 What conditions may account for the formation of widespread black shales in seas that spread over continental surfaces at certain times during the Cretaceous Period?

6 What modern continents that were once part of Gondwanaland remained attached to each other at the end of the Cretaceous Period?

7 Why did thick siliciclastic deposits accumulate in the western interior of North America during Cretaceous time?

8 By what mechanism may increased volcanism have produced very warm climates during the Cretaceous Period?

ADDITIONAL READING

Alvarez, W., and F. Asaro, "What Caused the Mass Extinction? An Extraterrestrial Impact," *Scientific American,* March 1991.

Bakker, R. T., *The Dinosaur Heresies,* William Morrow and Company, New York, 1986.

Courtillot, V. E., "What Caused the Mass Extinction? A Volcanic Eruption," *Scientific American,* March 1991.

PART VII

THE CENOZOIC ERA

The Cenozoic Era came to be known as the Age of Mammals because of the rapid diversification of these animals after the dinosaurs had disappeared. Modern snakes, frogs, and songbirds made their appearance during Cenozoic time, as did grasses and weedy plants. Marine life underwent only modest changes, with the notable exception of the whales, which originated and diversified during the era. Youthful mountains such as the Alps, Himalayas, and Rockies that stand tall today had Cenozoic origins.

During mid-Cenozoic time climates cooled in many areas, the deep sea grew frigid, and a mass extinction struck life on land and in the sea. Climates and deep-sea waters have since remained cooler than they were early in the era, and about 3 million years ago continental glaciers formed in the Northern Hemisphere, initiating a glacial age that has continued to the present. This glacial age has witnessed the appearance of the human species and the evolution of its various modern cultures.

Two sifakas in a tree represent a primitive group of primates that is now restricted to the island of Madagascar and in danger of extinction. These animals in many ways resemble early Cenozoic primates from which monkeys, apes, and humans ultimately evolved. *(Frans Lanting/Minden Pictures.)*

CHAPTER

15

The Paleogene World

The close of the Cretaceous Period marked a major transition in Earth's history. Never again were calcareous nannoplankton so abundant that they formed massive chalk deposits; scarcely any belemnoids survived, and ammonoids, rudists, and marine reptiles were gone from the seas. What remained were marine taxa that persist as familiar inhabitants of modern oceans, among them bottom-dwelling mollusks and teleost fishes. On the land, the flowering plants of the Paleogene resembled those of latest Cretaceous time in many ways, but animal life changed dramatically. Taking the place of the dinosaurs were the mammals, which were universally small and inconspicuous at the start of the Paleogene interval but in many ways resembled modern mammals by the period's end.

The most profound geographic change during Paleogene time was a refrigeration of Earth's polar regions, which resulted in a chilling of the deep sea and ultimately in widespread glaciation (a phenomenon to be discussed in Chapter 16). Paleogene mountain-building events in western North America foreshadowed Neogene uplifts of such ranges as the Sierra Nevada and Rocky Mountains. In southern Europe the Paleogene elevation of the Alps was of comparable significance.

For the most part, the sediments that record these and other Cenozoic events are unconsolidated, or soft, although most Neogene carbonates and some siliciclastics are lithified.

Terrestrial Eocene sediments of pinkish color rest on top of marine Cretaceous sediments of tan color in Badlands National Park of South Dakota. *(Tom Bean.)*

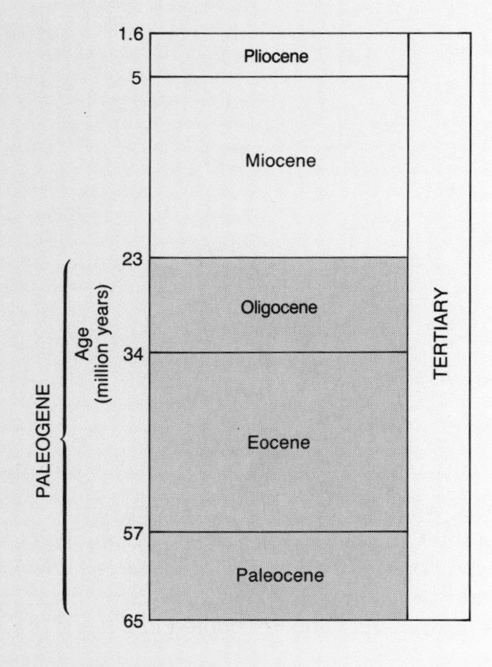

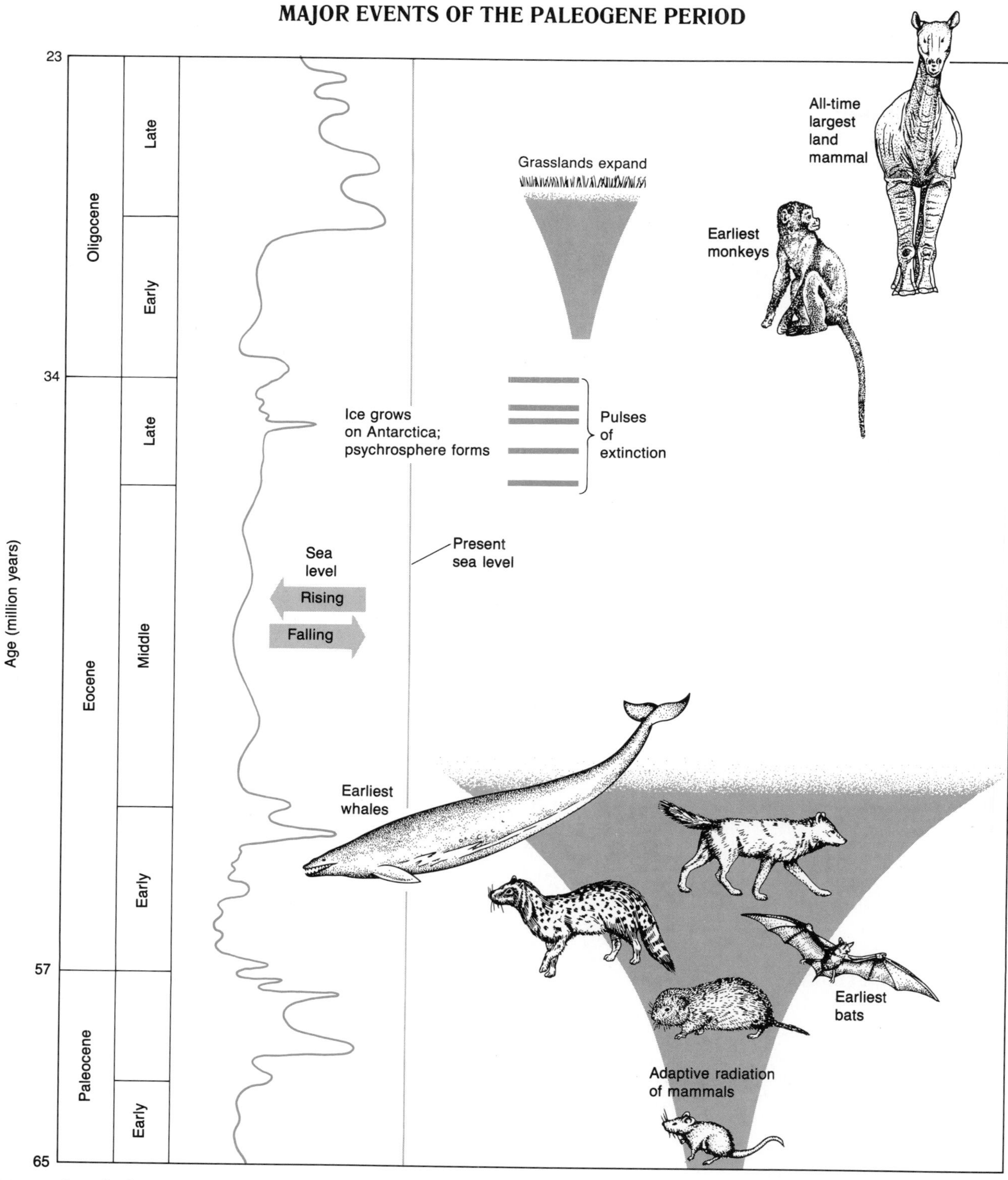

Mammals, which were small and of limited diversity early in the Paleocene Epoch, radiated rapidly. Near the end of the Eocene Epoch, Antarctica cooled down, and cold water began to settle to the deep sea, forming the psychrosphere. Climates cooled in many parts of the world at this time, and pulses of extinction removed a number of taxa. Sea level stood considerably higher than it stands today until late Oligocene time, when it dropped nearly to its present level.

FIGURE 15-1 Teleost fishes from the Eocene Series of Monte Bolca, Italy. These fishes had evolved a great variety of body forms by this time. *(Field Museum of Natural History, Chicago.)*

THE CENOZOIC TIME SCALE

Early in the history of modern geology the marked difference between Mesozoic and Cenozoic biotas was readily apparent. Subdivision of the Cenozoic Era itself is less clear-cut. Today many geologists divide the era into two periods: the Paleogene Period, which includes the Paleocene, Eocene, and Oligocene epochs, and the Neogene Period, which includes the Miocene, Pliocene, Pleistocene, and Recent (or Holocene) epochs. This Paleogene-Neogene classification has become increasingly popular during the past two decades. Traditionally, however, the Cenozoic Era has been divided into two periods of quite different lengths: the Tertiary, which encompasses the interval from the Paleocene through the Pliocene, and the Quaternary, which includes only the Pleistocene and Recent epochs (an interval of less than 2 million years). The advantage of the Paleogene-Neogene division is that it yields two periods of more similar duration.

The first formally recognized Paleogene epoch was the Eocene, which Charles Lyell established in 1833 on the basis of deposits found in the Paris and London basins. Lyell named and described the Eocene Series in his great book *Principles of Geology*, a work that also served to popularize the uniformitarian view of geology (p. 3). It was not until 1854 that Heinrich Ernst von Beyrich distinguished the Oligocene Series from the Eocene in Germany and Belgium on the basis of fossils, and then, in 1874, W. P. Schimper established the Paleocene Series on the basis of distinctive fossil assemblages of terrestrial plants in the Paris Basin.

WORLDWIDE EVENTS

Paleogene life is so familiar to us that it requires no special introduction; its most interesting features are the major expansions and contractions of certain taxonomic groups in all parts of the globe. Nor do we need a special section on Paleogene paleogeography, because the Paleogene lasted no more than 42 million years — a span in which paleogeographic changes occurred on such a small scale that they are best considered under the heading of Regional Events.

Evolution of Marine Life

The present marine ecosystem is, for the most part, populated by groups of animals, plants, and single-celled organisms that survived the extinction at the end of the Mesozoic Era to expand during the Cenozoic. Many benthic foraminifera, sea urchins, cheilostome bryozoans, crabs, snails, bivalves, and teleost fishes survived in sufficiently large numbers to assume prominent ecologic positions in Paleogene seas (Figure 15-1). Perhaps

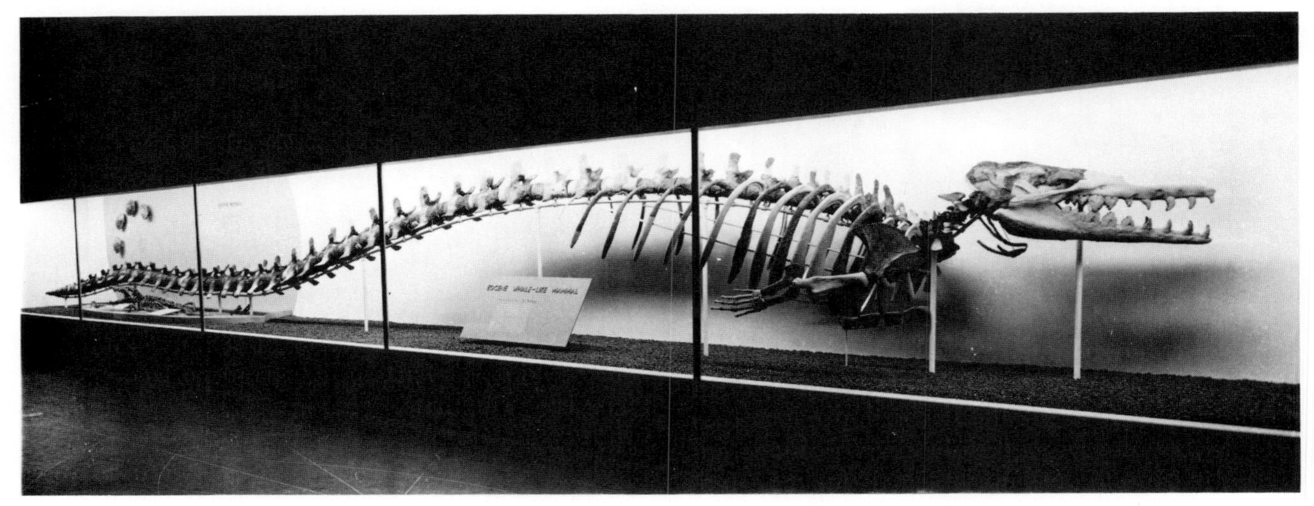

FIGURE 15-2 The Eocene whale *Basilosaurus*, a carnivore endowed with large teeth. This genus—not enormous by modern whale standards—attained a length of about 14 meters (~47 feet). *(Smithsonian Institution.)*

the biggest beneficiaries of the terminal Cretaceous extinction, however, were the reef-building corals, which had relinquished their dominant reef-building role to the rudists in mid-Cretaceous time but reclaimed it after the rudists' extinction. Unfortunately, few Paleocene coral reefs have been found. The rarity of Paleocene coral reefs also seems to indicate that reefs did not recover quickly from the mass extinction at the end of the Cretaceous. During the warm Eocene Epoch that followed the Paleocene, however, corals became widespread once again.

Calcareous nannoplankton, which had suffered severe losses at the end of the Cretaceous Period, rediversified somewhat during the Paleogene. These forms as well as the diatoms and dinoflagellates, which were not so adversely affected, have accounted for most of the ocean's productivity throughout the Cenozoic Era, just as they did in Cretaceous time.

Although many elements of Paleogene marine life closely resembled those of Late Cretaceous age, some forms were dramatically new. Perhaps the most distinctive marine organisms of this period were the whales, which evolved during the Eocene Epoch from carnivorous land mammals and quickly achieved success as large marine predators (Figure 15-2). Joining the whales as replacements for the reptilian "sea monsters," the top carnivores of the Mesozoic Era, were enormous sharks (Figure 15-3). Unlike

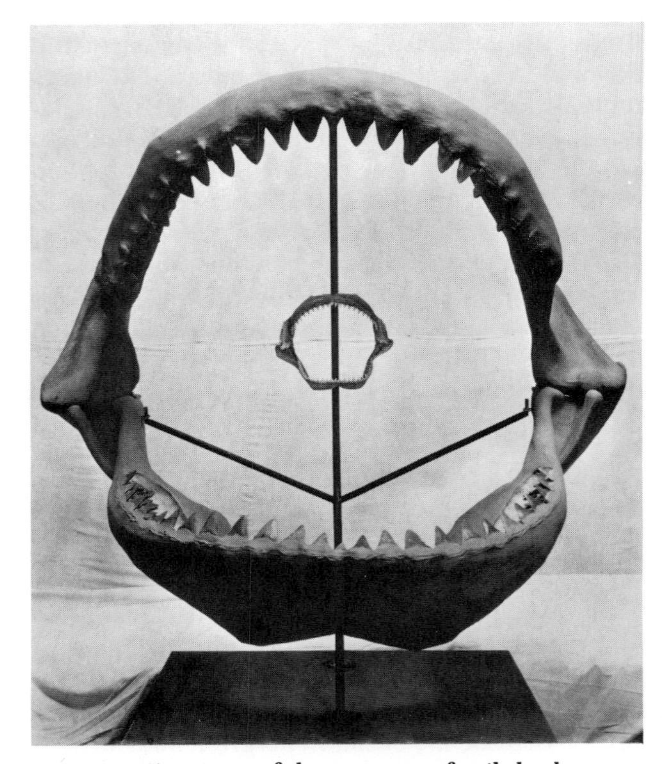

FIGURE 15-3 Jaws of the enormous fossil shark *Carcharodon* from the Eocene Epoch engulf the jaws of a modern shark. The jaws of the fossil shark were more than 2 meters (>6.5 feet) across. *(Field Museum of Natural History, Chicago.)*

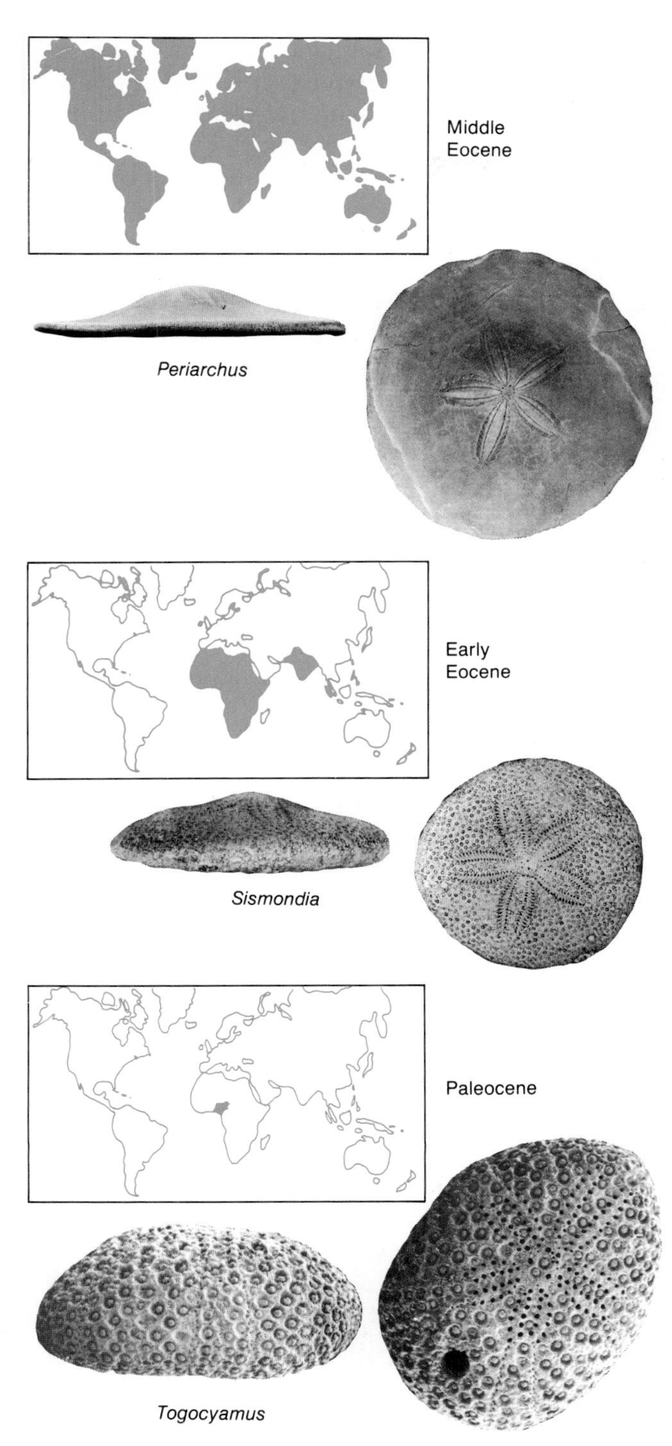

FIGURE 15-4 Specimens illustrating the evolution of the sand dollars from biscuit-shaped Paleocene ancestors *(Togocyamus)*, which are known only from Africa. True sand dollars *(Periarchus)* were present throughout the world by middle Eocene time. *Sismondia* is an intermediate genus that has been found in Africa and India. These animals were the size of small cookies. *(P. M. Kier, Smithsonian Institution.)*

the whales, however, sharks descended from similar creatures that lived during Cretaceous time.

It would appear that the marine ecosystem expanded during Paleogene time to include new niches along the fringes of the oceans. Sand dollars, for example, which are the only sea urchins able to live along sandy beaches, evolved at this time from biscuit-shaped ancestors (Figure 15-4), and new kinds of bivalve mollusks also invaded exposed sandy coasts. Both of these new groups have successfully inhabited shifting sands by virtue of their ability to burrow quickly into the sand again after they have been dislodged. Other newcomers to the ocean margins were the penguins, a group of swimming birds of Eocene origin, and possibly the pinnipeds, a group that includes walruses, seals, and sea lions. It is widely believed that the pinnipeds evolved before the beginning of the Neogene Period, although this group left no known Paleogene fossil record.

Evolution of Terrestrial Plants

The story of land plants is quite different. The transition to the Paleogene was apparently not marked by any drastic change in the character of terrestrial floras; instead, the great radiation of flowering plants simply continued. In the process, modern families of flowering plants evolved. By the beginning of Oligocene time, some 34 million years ago, about half of all genera of flowering plants were ones that are alive today, and although many modern plant genera had not yet evolved, forests had taken on a distinctly modern appearance. The oldest known rose species, for example, is of Late Eocene age (Figure 15-5).

One major evolutionary event that did take place during the Paleogene interval was the origin of the **grasses.** Although these usually low-growing flowering plants were present before the end of the Paleocene, they did not reach their full ecologic potential until late Oligocene and Miocene times. Early grasses were apparently confined to wooded or swampy areas. The mode of growth of early grasses, like that of the modern sedges that form marshlands along continental coastlines, did not allow their leaves to grow continuously and thus to recover from heavy grazing by animals of the sort that inhabit open country in large numbers. It was only an adaptive breakthrough—the origin of the continuous growth

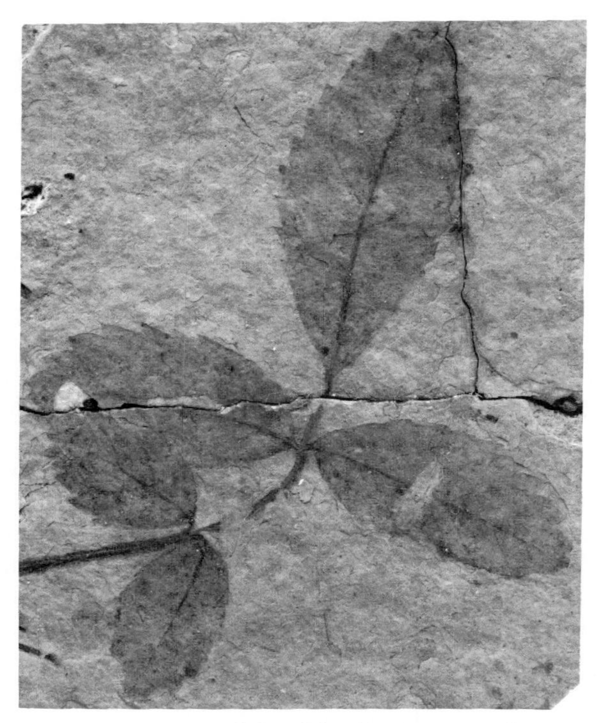

FIGURE 15-5 One of the oldest known rose plants. This specimen was preserved in fine-grained sediments of the Florissant Formation of Colorado. Fossils of the same species, *Rosa hilliae,* have also been found in the upper Eocene Green River Formation of Utah. It probably produced insect-attracting flowers, although these blossoms would not have been as beautiful, by our standards, as those that modern breeders have developed. *(Smithsonian Institution.)*

process, which forces us to cut our lawns every week or two — that ultimately enabled grasses to invade open country with great success. Once they were able to survive the effects of heavy grazing by animals, grasses quickly spread over vast expanses to form grasslands.[1]

Early Paleogene Terrestrial and Freshwater Animals

In Chapter 5 and Box 14-1 we saw that the mammals, having inherited the world from the dinosaurs, underwent a remarkably rapid adaptive radiation during the early part of the Cenozoic Era. It was probably through both competition and predation that the dinosaurs had prevented the mammals from undergoing any great evolutionary expansion during Mesozoic time.

At the start of the Paleocene Epoch, when they first had the world to themselves, most mammals were small creatures that resembled modern rodents; no mammal seems to have been substantially larger than a good-sized dog. Furthermore, most mammal species tended to remain generalized in both feeding and locomotory adaptations; those that dwelt on the ground generally retained a primitive limb structure that caused the heels of the hind feet and the "palms" of the front feet to touch the ground as the animals moved about. Perhaps 12 million years later, however, by the end of early Eocene time, mammals had diversified to the point at which most of their modern orders were in existence (Figure 15-6). Bats already fluttered through the night air (Figure 15-7), for example, and, as we have seen, large whales swam the oceans.

Also included among the Paleocene mammals were groups that had survived from Cretaceous time (Box 14-1), such as marsupials, multituberculates, and placental mammals called insectivores. Other Paleocene mammals have been assigned by some experts to the Primates, the order to which modern humans belong. Although these small animals were quite different from monkeys, apes, or humans in many respects, by early Eocene time they did climb with grasping hind limbs and forelimbs that foreshadowed our own hands and feet (Figure 15-8). By mid-Paleocene time, true mammalian carnivores — members of the living order Carnivora — had emerged (Figure 15-9). This is the order to which nearly all living carnivorous placental mammals belong. By the end of Paleocene time, the earliest members of the horse family had evolved as well; these animals were no larger than small dogs (Figure 15-10), but by the end of the epoch, larger herbivorous mammals, some the size of cows, had appeared.

The variety of mammals continued to increase in the Eocene Epoch. The number of mammalian families doubled to nearly 100, approximating that of the world today. In addition, more modern varieties of hoofed herbivores appeared. Most animals of this kind are known as **ungulates,** and they are divided into odd-toed ungulates (living horses, tapirs, and rhinos) and even-toed ungulates, or cloven-hoofed animals (cattle, antelopes, sheep,

[1] This idea has been proposed informally, but not yet published, by Leo J. Hickey of Yale University.

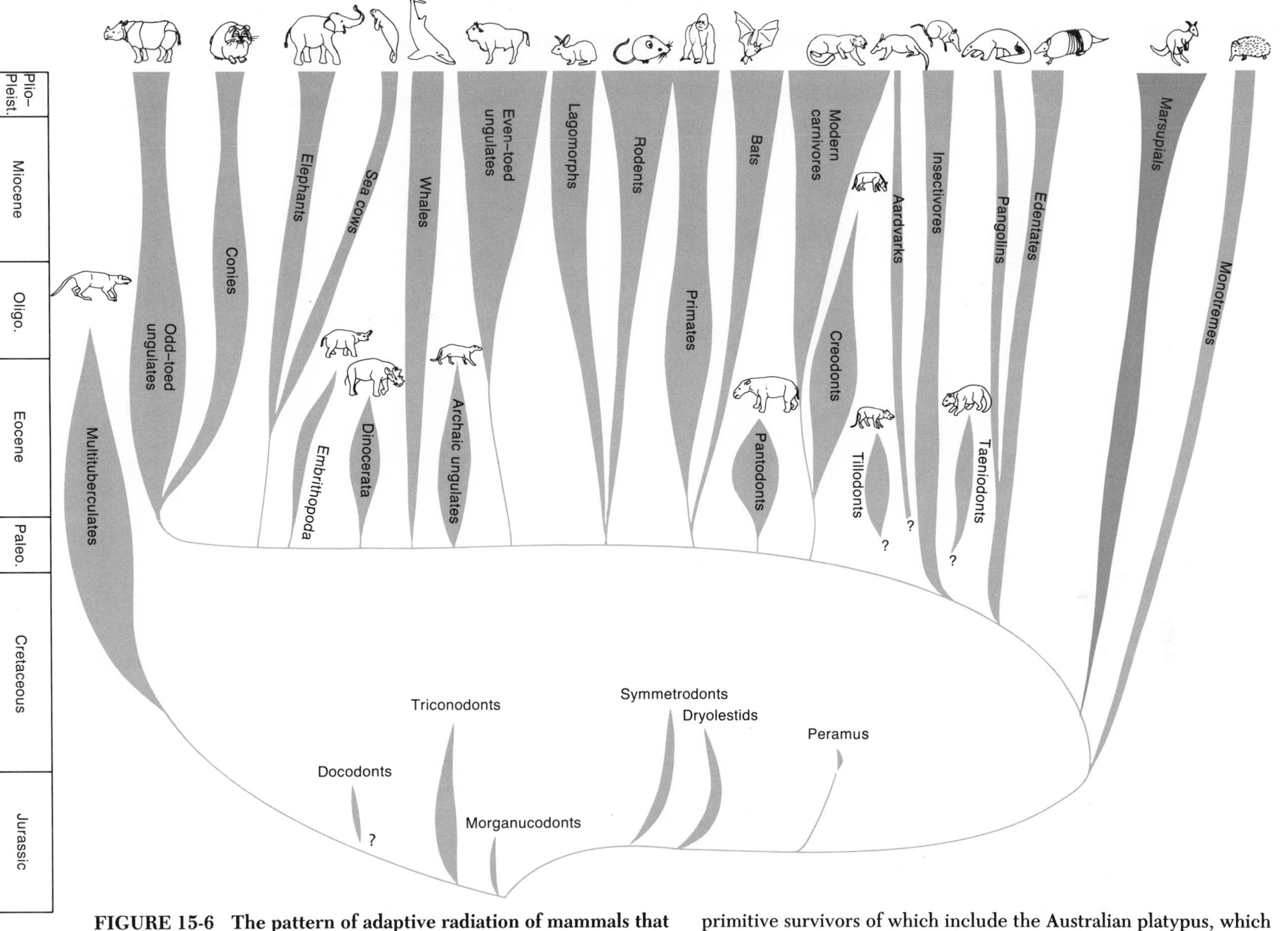

FIGURE 15-6 The pattern of adaptive radiation of mammals that gave the Cenozoic Era its informal name, the Age of Mammals. It is thought that the two large groups of modern mammals, the placentals and the marsupials, had a common ancestry in the Cretaceous Period. The multituberculates, which failed to survive into Neogene time, apparently evolved separately—as did the Monotremata, primitive survivors of which include the Australian platypus, which lays eggs instead of giving birth to live young. For the placental mammals, each vertical bar represents an order. Note that most of the new Cenozoic orders had already evolved by the beginning of the Eocene Epoch, about 10 million years after the dinosaurs disappeared. *(D. R. Prothero.)*

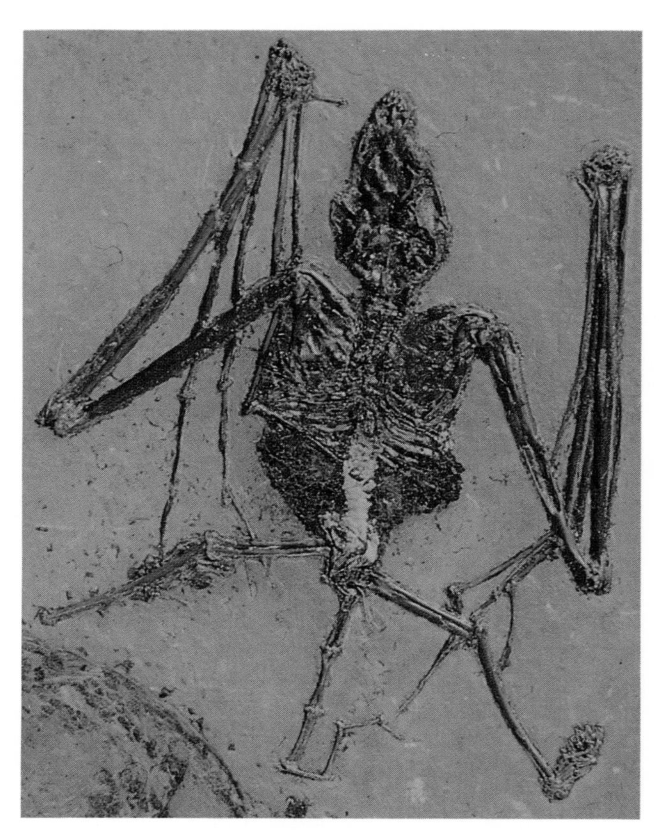

FIGURE 15-7 A complete skeleton of an Eocene bat from the Green River Formation of Wyoming. *(Senckenberg Museum.)*

FIGURE 15-8 Reconstruction of the early primate *Cantius*, a small genus of early Eocene time. This arboreal animal had large toes on its hind feet and nails much like our own, and it apparently jumped from limb to limb. *(R. T. Bakker.)*

FIGURE 15-9 Reconstruction of a mid-Paleocene biota of New Mexico. The large trees represent the sycamore genus *Platanus*, which has survived to the present time. A small insectivore *(Deltatherium)* rests on a small branch in the center of the drawing. On the ground, mesonychid carnivores of the genus *Ancalagon* feed on the small crocodile *Allognathosuchus. Ancalagon* is closely related to the mesonychids that gave rise to the whales. Ferns and sable palms (or fan palms) constitute the undergrowth in the background. *(Drawing by Gregory S. Paul.)*

FIGURE 15-10 ***Hyracotherium*** **("eohippus"), the earliest genus of the horse family.** This animal, which was present in Late Paleocene and Eocene time, was no larger than a small dog. It had four toes on each front foot and three on each hind foot. *(Field Museum of Natural History, Chicago.)*

FIGURE 15-12 ***Moeritherium*****, an early member of the elephant group.** This elongate animal stretched to a length of about 3 meters (~10 feet). During the Eocene Epoch it probably wallowed in shallow waters and grubbed for roots or other low-growing vegetation.

goats, pigs, bisons, camels, and their relatives). The odd-toed ungulates expanded before the even-toed group did, but primitive even-toed ungulates were also present early in the Eocene (Figure 15-11); of the modern even-toed types, camels and relatives of the present-day chevrotain (the Oriental "mouse deer") evolved before the end of the epoch. In addition, the earliest members of the **elephant** order appeared during early Eocene time; *Moeritherium*, the earliest genus well known from the fossil record, was a bulky animal, about 3 meters (~10 feet) long, with rudimentary tusks and a short snout rather than a fully developed trunk (Figure 15-12). The **rodents,** which had originated in the Paleocene, continued to diversify as well, but they may have attained their success at the expense of the archaic multituberculates, which were also specialized for gnawing seeds and nuts. As the rodents expanded during the Eocene, the multituberculates declined; they finally became extinct early in the Oligocene Epoch.

Among the animals that preyed on these herbivorous mammals were groups that originated in the Paleocene Epoch. They include the superficially doglike mesonychids and the diatrymas, which were huge flightless birds with powerful clawed feet and enormous slicing beaks (Figure 15-13). The diatrymas disappeared toward the end of the Eocene Epoch. At that time the mesonychids were joined by primitive members of three familiar modern carnivore groups: the dog, cat, and weasel families.

The monstrous diatrymas were not the only birds of the Eocene, but flying birds were much less diverse then than they are today. Most species were shore birds that waded in shallow water when they were not in flight (Figure 15-14). Not yet present were many other modern kinds of birds, including the songbirds that are so numerous today.

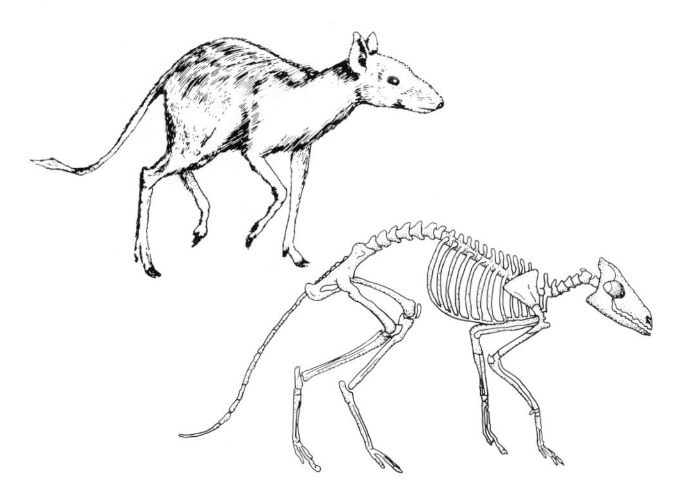

FIGURE 15-11 ***Diacodexis*****, an early even-toed ungulate, or cloven-hoofed herbivore.** *Hyracotherium* (Figure 15-10), in contrast, was an early odd-toed ungulate. The limb structure of *Diacodexis* shows that it was an unusually adept runner and leaper for early Eocene time. *(After K. D. Rose, Science 216:621–623, 1984.)*

FIGURE 15-13 Large terrestrial predators that had evolved by Early Eocene time. The animals that superficially resemble dogs are giant mesonychids of the genus *Pachyhyaena,* which were the size of small bears. The flightless birds guarding their chicks are members of the genus *Diatryma,* which stood about 2.4 meters (~8 feet) tall. *(Drawing by Gregory S. Paul.)*

FIGURE 15-14 The long-legged Eocene duck *Presbyornis.* The large numbers of fossils of this wading bird found in Green River sediments of Wyoming and Utah suggest that it lived in enormous colonies. *Presbyornis* left tracks revealing that webbed feet at the end of its long legs supported it when it walked in mud. Along with the tracks are lines of probing marks made by the beak while the bird searched for food. Its legs were too long to allow it to swim, but modern ducks have inherited its webbed feet and employ them as paddles to propel them through the water. *(After drawing by J. P. O'Neill. Photograph by the author.)*

FIGURE 15-15 Complete skeleton of an Eocene frog that belongs to a still-living family. This specimen comes from the Green River Formation of Wyoming. *(L. Grande.)*

FIGURE 15-16 An Eocene ant from the Green River Formation. This insect, which is about 2 centimeters ($\frac{4}{5}$ inch) in length, belongs to the family Formicidae, which has undergone such a great adaptive radiation that today it includes about 16,000 species. *(L. Grande.)*

As for other forms of vertebrate life, reptiles and amphibians were relatively inconspicuous during Paleogene time. The first record of the Ranidae, the largest family of living frogs, is in the Eocene Series (Figure 15-15), but the fossil record of this group of fragile animals is not good, so we do not know precisely when the Ranidae originated or attained high diversity.

There is no question that the insects took on a modern appearance with the origin of several modern families in Paleogene time (Figure 15-16). Several Oligocene forms, which have been preserved with remarkable precision in amber, closely resemble living species.

Mammals of the Oligocene Epoch

Mammals became increasingly modern during the Oligocene Epoch. As many Eocene families died out, many living (Recent) groups expanded. The horse family had disappeared from Eurasia during the Eocene, but a few horse species, including members of the three-toed genus *Mesohippus* (Figure 15-17), survived in North America. Other

FIGURE 15-17 Reconstruction of an Oligocene horse belonging to the genus *Mesohippus*. This three-toed animal stood about $\frac{1}{2}$ meter ($\sim$20 inches) tall at the shoulder. *(Field Museum of Natural History, Chicago.)*

FIGURE 15-18 ***Indrichotherium*, which, as far as we now know, was the largest mammal ever to exist on Earth.** This Oligocene giant from Asia belonged to the rhinoceros family and stood about 5.5 meters (~ 18 feet) at the shoulder. This is the height of the top of the head of a good-sized modern giraffe. *(Drawing by Gregory S. Paul.)*

FIGURE 15-19 Titanotheres, which constituted a family of odd-toed ungulates that were related to the rhinos but had blunt horns rather than sharp ones. Titanotheres were abundant in the American West during the Oligocene Epoch. *(Field Museum of Natural History, Chicago, neg. CK-12T.)*

odd-toed ungulates that enjoyed greater success during the Oligocene Epoch than in previous intervals were the rhinos, which included the largest land mammal of all time (Figure 15-18), and the rhinolike titanotheres (Figure 15-19). Many more large mammals populated the land during the Oligocene Epoch than during the Eocene.

As the Oligocene Epoch progressed, odd-toed ungulates were outnumbered for the first time by even-toed ungulates, including deerlike animals and pigs, which became especially diverse.

Among the carnivores, the dog, cat, and weasel families, which had their origins in Eocene time, radiated during the Oligocene Epoch and produced more advanced forms, including large saber-toothed cats (Figure 15-20), bearlike dogs, and animals that resembled modern wolves.

An especially important aspect of the modernization of mammals during the Oligocene Epoch was the appearance of monkeys and apelike primates. The genus *Aegyptopithecus*, an arboreal animal the size of a cat, had teeth resembling those of an ape but a head and a tail resembling those of a monkey (Figure 15-21). In Neogene time, before the arrival of humans, apes attained considerable diversity in Africa and Eurasia.

Climatic Change and Mass Extinction

During the second half of the Paleogene Period a mass extinction struck both on the land and in the sea. This catastrophe was not so severe as some

FIGURE 15-20 The Oligocene cat *Dinictis*, which was approximately the size of a modern lynx. This animal possessed advanced adaptations for running and springing upon prey. Its canine teeth were elongated for stabbing, but *Dinictis* was not a member of the true saber-toothed cat group. *(Drawing by Gregory S. Paul.)*

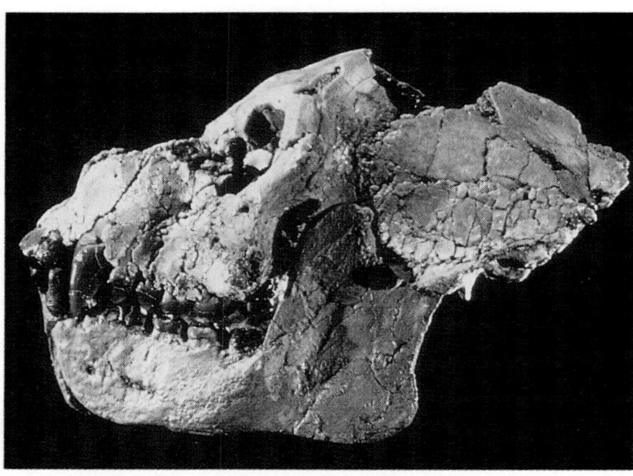

FIGURE 15-21 The Oligocene primate *Aegyptopithecus*, whose name reflects its discovery in Egypt. The skull resembles that of a monkey, but the teeth are apelike. The brain of *Aegyptopithecus* was unusually large for the size of the animal, perhaps reflecting a high level of intelligence for the Oligocene world. *(Drawing by S. F. Kimbrough. Photograph by E. Simons.)*

earlier ones. It did eliminate many genera and species, but relatively few higher taxa disappeared. Actually, it was not a single event but a series of pulses of extinction, apparently caused by changes in climatic conditions.

Early Eocene Warmth Fossil floras reveal that in Early Eocene time, before the pulses of extinction, Earth's climate was relatively warm. Southeastern England, for example, which was positioned very close to its present high latitude, was cloaked in a tropical jungle (Box 15-1). How did warm conditions develop so far from the equator? It is possible that a high concentration of carbon dioxide in the atmosphere strengthened the greenhouse effect at this time. Another possibility is that warm ocean currents carried a large amount of heat from low latitudes to high latitudes. The most likely source of the warm water was the region of the dry trade winds (p. 35). Here high rates of evaporation may have produced waters that were relatively dense because of their high salinity. Such waters would have descended below the surface and spread out, carrying some of their heat to high latitudes. There the upward mixing of these water masses would have warmed the surface waters. Winds passing over them would have warmed the land.

Global Climatic Change Fossil plants attest to climatic changes on the land after the Early Eocene interval of global warmth. Flowering plants (angiosperms) are commonly viewed as the thermometers of the past 100 million years. As we have seen, their value derives in part from the strong correlation between mean annual temperature and percentage of species within a flora that have leaves with entire margins (Figure 2-18). Data on the leaf margins of fossil floras of North America reveal that after Early Eocene time, major climatic changes took place throughout the world. There were three pulses of cooling, each more severe than the one before (Figure 15-22).

The events that took place in the deep sea provide an explanation for the episodes of cooling. Study of deep-sea cores reveals that at this time there was an increase in the ratio of oxygen 18 to oxygen 16 in the skeletons of the foraminifera in equatorial regions and near Antarctica. This increase apparently reflected the first growth of glacial ice on and adjacent to Antarctica, with the lighter oxygen isotope (^{16}O) accumulating preferentially in this ice. The growth of glacial ice on Antarctica was a prelude to the greater expansion of high-latitude ice sheets near the end of the Cenozoic Era, during what is informally known as the recent Ice Age.

It was during the Late Eocene cooling episode that the psychrosphere — the cool bottom layer of the ocean (p. 43) — came into being as cold, dense polar water began to sink to the deep sea. The short interval of isotopic change that characterizes deep-sea cores suggests that the psychrosphere formed in less than 100,000 years; the

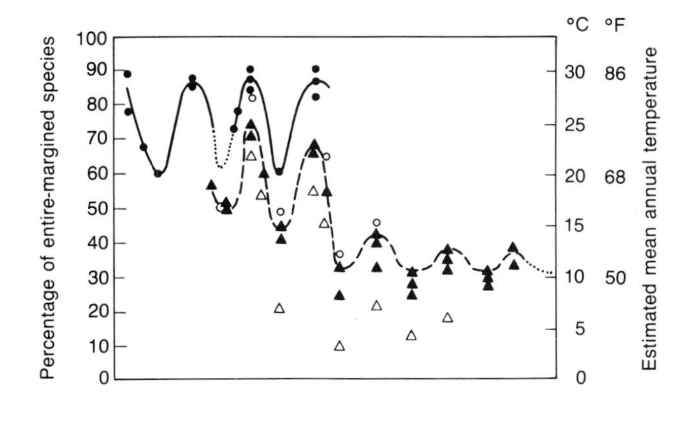

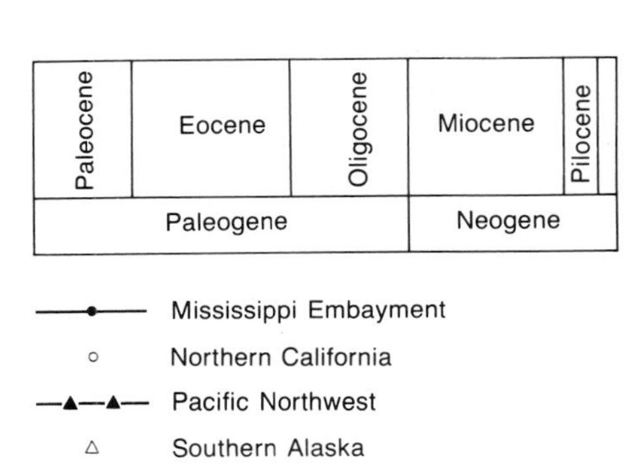

FIGURE 15-22 Estimated changes in temperature in four areas of North America in the course of Cenozoic time, based on percentages of entire-margined leaves in fossil floras. The fact that the curves follow parallel paths where they overlap in time suggests that the trends hold for large areas of Earth's surface. *(After J. A. Wolfe, American Scientist 66:694–703, 1978.)*

temperatures of bottom waters over large areas of the deep sea dropped by perhaps 4 to 5°C (~7 to 9°F). Later we will examine geologic changes that caused the psychrosphere to form.

Mass Extinction Deep-sea cores provide the most detailed record of the pattern of late Paleogene extinction—a record that consists of microfossils. The record of planktonic foraminifera reveals five successive pulses of change between about 40 and 31 million years ago (during Middle and Late Eocene time). The final pulse of extinction appears to have coincided with the final episode of cooling in North America, as indicated by changes in terrestrial floras (Figure 15-22). Each sudden change caused a few species of foraminifera to become less abundant or disappear altogether. The overall result was a reduction in diversity and a change in the composition of the fauna. The species that became extinct were mostly spiny forms that were adapted to warm conditions.

The pattern of extinction of calcareous nannoplankton is not yet documented in great detail, but during the final 7 million years of Eocene time the total number of species in the world fell from about 120 to just 40 or so. Calcareous nannoplankton are predominantly warm-water forms, so it is not surprising that this group should have been severely affected by global cooling.

Extinction was also heavy on the seafloor during the Eocene-Oligocene transition. Many areas lost species that were adapted primarily to warm conditions.

The fossil record of mammals in western North America reveals two episodes of heavy extinction during the general interval of biotic crisis: the first a few million years before the end of the Eocene Epoch and the second at the very end of the epoch. The first event eliminated many mammalian species and a modest number of genera. The second was less severe, but it did sweep away the last of the huge, rhinolike titanotheres (Figure 15-19). These crises seem to have coincided with the widespread climatic changes revealed by the changes in land plants.

It seems evident that climatically induced changes in the terrestrial flora played a major role in the mammalian extinctions. These floral changes resulted from increased aridity as well as from cooling of the climate. Many regions became more arid about 34 million years ago, because at the end of the epoch sea level declined dramatically from its relatively elevated position late in Eocene time (Figure 15-23 and Major Events, p. 404), leaving broad continental areas far from the ocean and poorly supplied with moisture. Presumably climatic cooling at this time was associated with a sudden expansion of glacial ice on Antarctica, which caused sea level to drop.

The effect of cooling and increased aridity was profound. The Eocene Epoch had been a time when moist tropical and subtropical forests cloaked much of North America and Eurasia. During the Oligocene Epoch, savannahs—grassy plains with scattered trees and shrubs (p. 36)—spread across large areas of major continents. At the same time, moist subtropical and tropical forests became confined to low latitudes, where they remain to this day in the form of jungles and rain

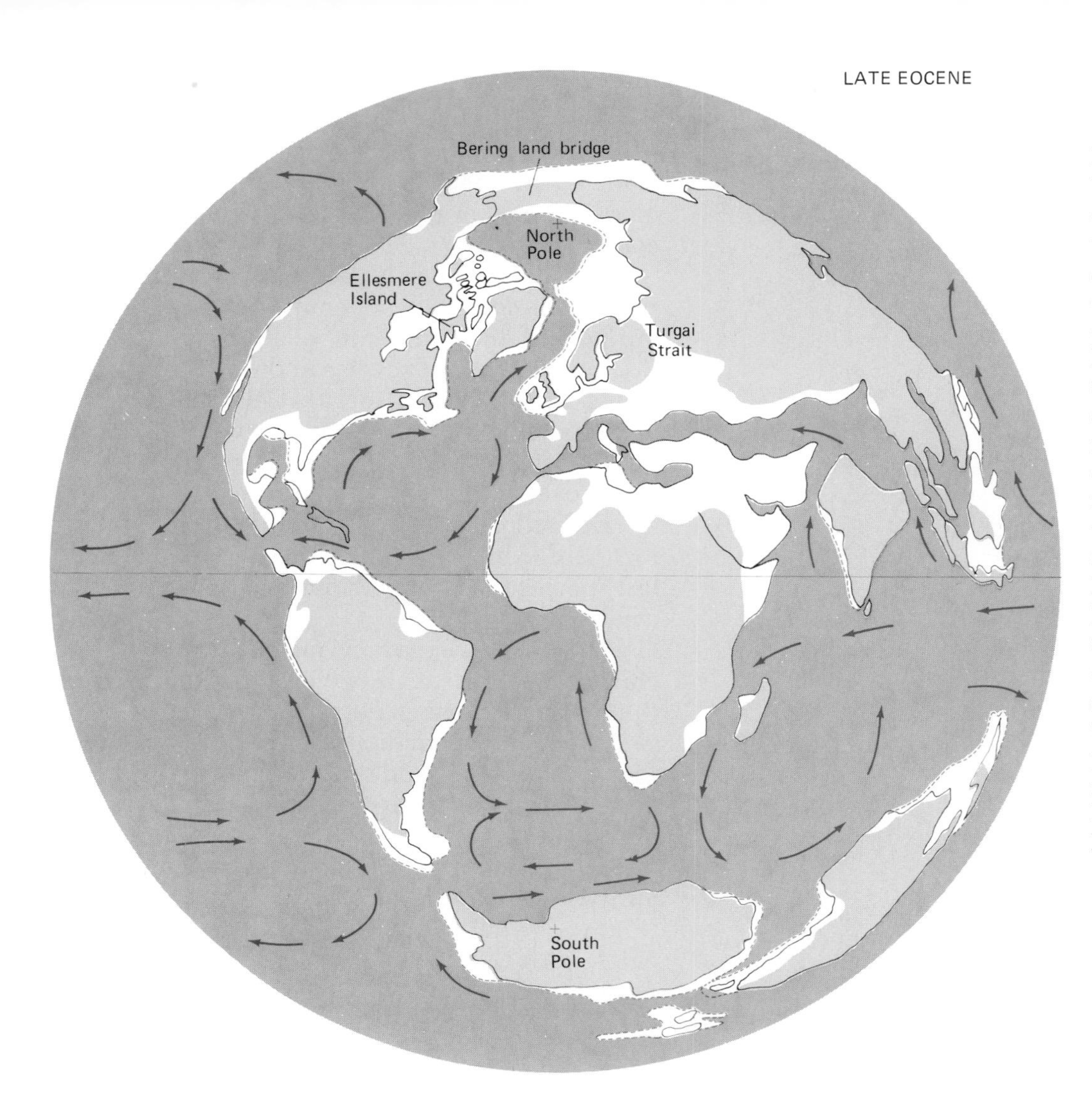

FIGURE 15-23 World geography during Late Eocene time. Substantial areas of continental crust were inundated during this interval. Relatively warm climates prevailed even as far north as Ellesmere Island, where a diverse fauna of terrestrial plants and vertebrate animals flourished in middle Eocene time. The Bering land bridge, a neck of continental crust, stood above sea level throughout the Eocene interval, permitting animals to migrate between North America and Eurasia.

forests. Because many of the evolutionary changes associated with this climatic transition took place in Miocene time, we will pick up this story in Chapter 16.

REGIONAL EVENTS

In examining important regional events of the Paleogene Period, we will travel to the ends of Earth—first to the South Pole, where Antarctica became separated from Australia and developed its icy cover, and then toward the North Pole, where land areas of North America and Eurasia were more closely connected than they are today. Then we will look at the history of the Cordilleran region of North America, where the Laramide orogeny yielded many structures that remain conspicuous in the Rocky Mountains today. We will also examine the nature of deposition along the Gulf Coast of North America, where large volumes of petroleum are trapped in soft Paleogene sediments.

Antarctica and the Origin of the Psychrosphere

The only major event of continental rifting that took place exclusively during the Cenozoic Era was the separation of Australia from Antarctica, which occurred close to the end of the Eocene Epoch. With the inception of this rifting early in Eocene time, Australia began to move northward, and the resulting alteration of wind and water circulation near the South Pole had consequences that extended over much of the planet.

Even before its separation from Australia, Antarctica had been centered over the South Pole, but it had remained warm because its shores were bathed in relatively warm waters from lower

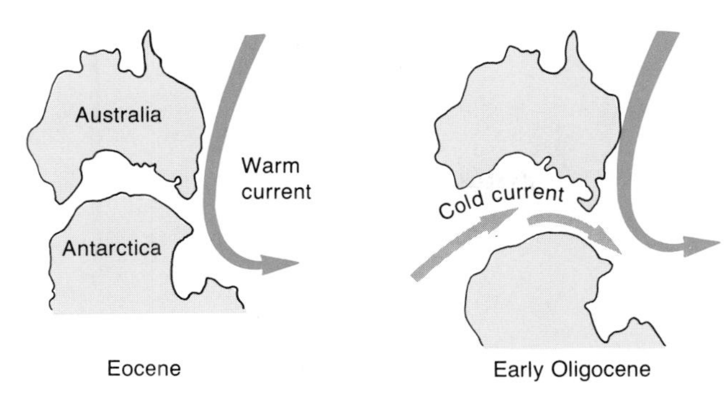

FIGURE 15-24 The separation of Australia from Antarctica at the end of Eocene time allowed a cold current to form along the margin of Antarctica, deflecting a warm current from the region. *(After J. P. Kennett and C. C. von der Borch, Initial Reports of the Deep Sea Drilling Project 90:1493–1517, 1986.)*

latitudes. When Australia broke away and drifted northward, a cold current formed between the two continents, deflecting the warm current that flowed toward Antarctica from Australia (Figure 15-24). This was the first change to cause cooling in Antarctica. Then, during the Oligocene Epoch, the circumpolar current (Figure 2-20) became fully established near the sea surface despite the fact that a continental connection remained between Antarctica and Australia. Thus Antarctica and the waters adjacent to it, being largely isolated in a polar position, cooled down even more.

In response to this cooling, the first sea ice began to form around Antarctica near the end of the Eocene. Waters of near-freezing temperature sank to the deep sea and spread northward, forming the psychrosphere. The direct result was an extinction of bottom-dwelling foraminifera and larger inhabitants of the deep seafloor adapted to temperatures well above freezing. At the same time, climates deteriorated throughout the world.

The global climatic change may have resulted almost entirely from the spread of cool polar waters toward the equator. It is also possible, however, that for some currently unknown reason there was a decrease in the greenhouse effect of Earth's atmosphere.

Changes in oxygen isotopes within seawater, as recorded in fossil foraminifera, reveal that ice built up on Antarctica during Oligocene time. Once the Antarctic "cooling system" was in place, the mean annual temperatures at high latitudes never again reached their Eocene levels; in fact, the climatic deterioration in Late Eocene and Oligocene time fostered further deterioration later in Cenozoic time. The recent Ice Age of the Northern Hemisphere (described in Chapter 16) would never have occurred if global temperatures had not already been lowered.

The Top of the World: Changing Positions of Land and Sea

Major geologic and geographic changes also took place during the Paleogene Period within 30° latitude of the North Pole (Figure 15-25). Many of these changes can be read from the ages of various segments of the deep seafloor. The Arctic deep-sea basin was in existence during the Cretaceous Period, but until nearly the end of Cretaceous time it remained separated from the Atlantic, except by way of shallow seas, because North America, Greenland, and Eurasia were still united as a single landmass. Then, as we saw in Chapter 14, the Mid-Atlantic Rift proceeded northward along two forks, splitting Greenland from North America on the west and from Eurasia on the east. Finally, in mid-Paleogene time, the pattern of plate movement simplified as the western arm of rifting ceased to be active, ending the relative movement of Greenland away from North America. Since that time, the rifting of the northern Atlantic has been limited to the area between Greenland and Scandinavia. The plate movements in this area, like those in the south polar region, contributed to the origin of the psychrosphere. Early in Paleogene time the Arctic Ocean was isolated from larger oceans to the south and thus retained its frigid waters. Then, late in Eocene time, rifting between Greenland and Scandinavia proceeded far enough to allow a channel to open up between the Arctic basin and the Atlantic, causing dense, frigid Arctic waters to spill into the larger ocean as part of the psychrosphere.

Meanwhile, continental crust has continued to separate the Arctic and Pacific basins; Alaska and Siberia have remained connected by a stretch of continental crust despite the fact that this connecting segment now lies submerged beneath the Bering Sea. During much of the Cenozoic Era, this segment, which is known as the Bering land bridge, stood above sea level and served as a land corridor between North America and Eurasia.

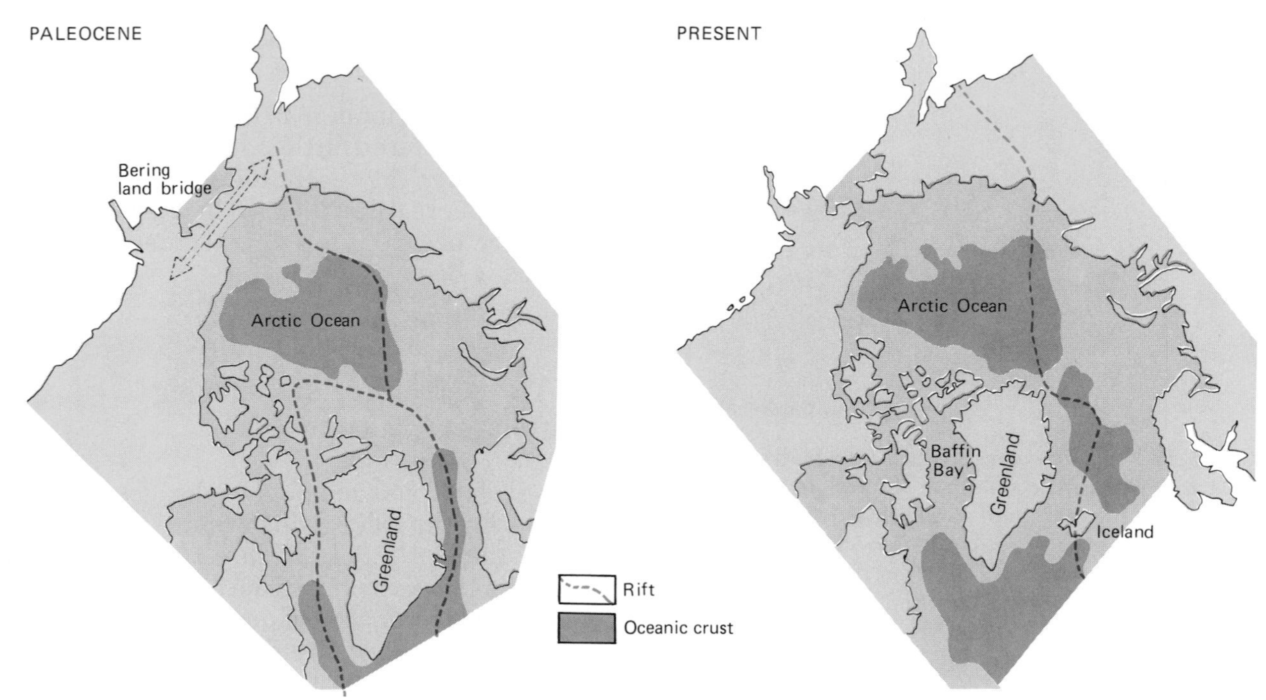

FIGURE 15-25 Changes in the pattern of rifting in the North Atlantic during the Cenozoic Era. In mid-Paleogene time, rifting ceased between Greenland and North America but continued between Greenland and Eurasia, where it persists today. *(After E. M. Herron et al., Geology 2:227–280, 1974.)*

This corridor remained open throughout Paleogene time, allowing mammals and land plants to migrate between Asia and North America. As we will see, however, inundations of the sea during the Neogene Period frequently prevented the exchange of species between the old and new worlds. Chapter 16 will show, however, that humans first reached the Americas by crossing the Bering land bridge during its most recent interval of existence, more than 10,000 years ago.

Tectonics of Western North America

Mountain-building activity in the Cordilleran region of North America continued into the Paleogene Period, but with a number of changes. Figure 15-26, which summarizes the orogenic history of the eastern Cordilleran region, shows that the Sevier episode spanned almost the entire Cretaceous Period. In latest Cretaceous time, however, a different style of tectonic activity was

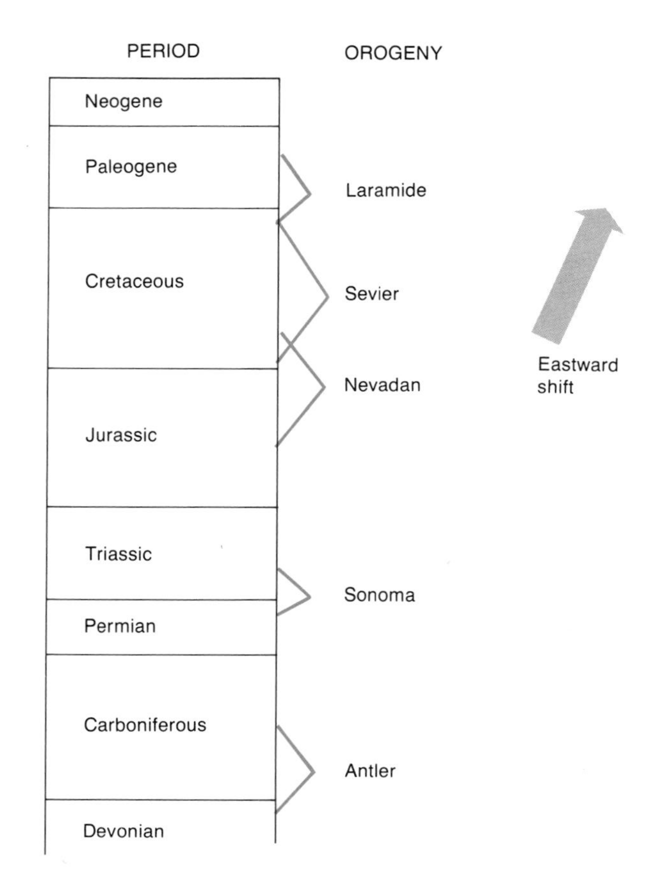

FIGURE 15-26 Summary of major orogenic events in the eastern Cordilleran region. Between Jurassic and Paleogene time, orogenic activity migrated eastward (see Figure 14-34).

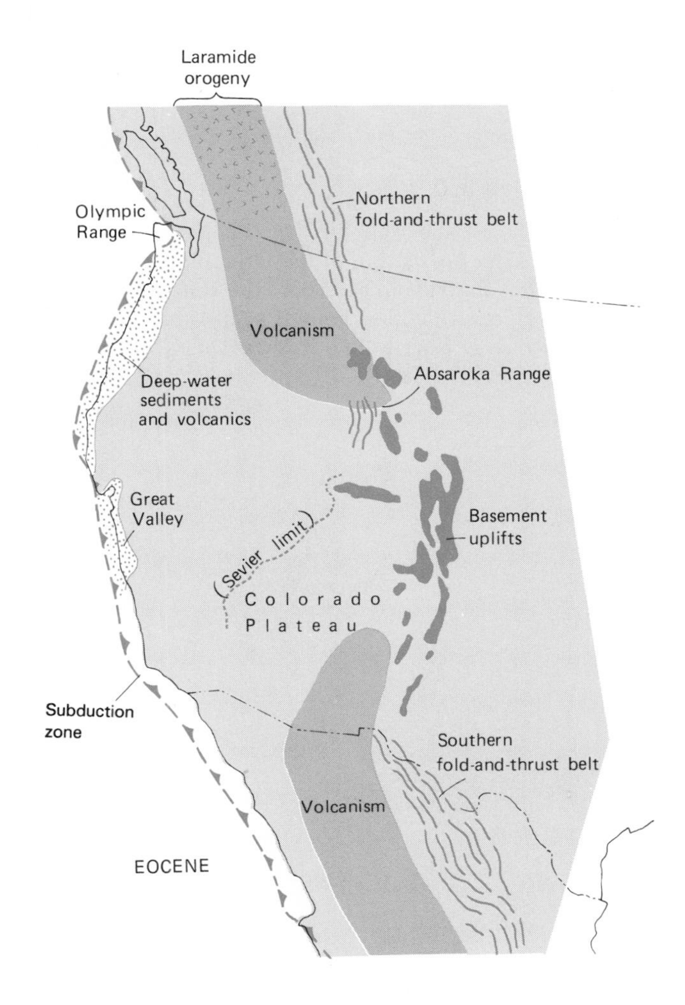

FIGURE 15-27 Geologic features of western North America during Eocene time. Subduction continued along the west coast. Marine sediments were deposited in the Great Valley of California, and deep-water sediments and volcanics accumulated in a forearc basin to the north in Washington and Oregon. Farther inland in the north and south, the Laramide orogeny produced a band of volcanism and, farther inland still, a belt of folding and thrusting. In Colorado and adjacent regions, however, the orogeny was expressed as a series of crystalline uplifts that extended far to the east of the Cretaceous Sevier orogenic belt. They may have formed by a slight clockwise rotation of the Colorado Plateau in relation to the continental interior.

initiated in the American West. This activity persisted through the Paleocene and well into Eocene time.

The Laramide Orogeny The episode characterized by the new tectonic style is known as the **Laramide orogeny.** The northern and southern segments of the Laramide were typical of orogenies in general. In the north, extending from the United States into Canada, there remained an active belt of igneous activity and, inland from it, an active fold-and-thrust belt (Figure 15-27). Thrust sheets of enormous proportions are spectacularly exposed in the Canadian Rockies (Figure 15-28).

FIGURE 15-28 The Lewis Thrust Fault in the Laramide fold-and-thrust belt of the northern Rockies. In this view the fault is exposed in the side of Summit Mountain, Glacier National Park, Montana. The upper half of the mountain is formed of Precambrian rocks that were thrust over the lighter-colored rocks below, which are of Cretaceous age. *(U.S. Geological Survey.)*

A similar pattern of tectonism persisted both in the southern United States and in Mexico.

The unusual features of the Laramide orogeny were in the central part of the western United States, where a broad area of tectonic quiescence extended from the Great Valley of California to the Colorado Plateau (Figure 15-28). East of this inactive region there was a strange pattern of tectonism in which large blocks of underlying crystalline rock were uplifted in a belt extending from Montana to Mexico. The largest of these blocks were centered in Colorado, where the Ancestral Rocky Mountains had been uplifted more than 200 million years earlier, late in the Paleozoic Era (p. 328).

The Pattern of Subduction What created the unusual tectonic pattern that characterized the central part of the Cordilleran region during the Laramide orogeny? Note that the Paleogene uplifts were, for the most part, positioned well to the east of Sevier orogenic activity (Figure 15-27), and recall in addition that an eastward migration of orogenic activity culminating in the Sevier orogeny had taken place during the Mesozoic Era. The widely favored explanation for the earlier eastward shift applies to the Laramide shift as well: A central segment of the subducted plate that passed beneath North America began to penetrate the mantle at a still lower angle, extending a great distance eastward before becoming deep enough to melt and to create igneous activity within the overlying crust (Figure 14-34).

Farther west, along the coast, the Great Valley of California continued to receive marine sediment, while northern California and the Sierra Nevada region remained as highlands. A separate basin in Washington and Oregon received deep-water sediments and layers of pillow lava. Here the Olympic Range began to form along a sharp inward bend of the subduction zone (Figures 15-27 and 15-29).

The Eastern Uplifts and Basins The eastern belt of uplifts stretches from Montana to New Mexico. Because this is the region in which the central and southern Rocky Mountains developed during Neogene time, Paleogene events that preceded the uplift of this segment warrant special attention. Deformation here began in latest Cretaceous time with the origin of north-south-trending ranges and basins (Figure 15-30); in Utah and Wy-

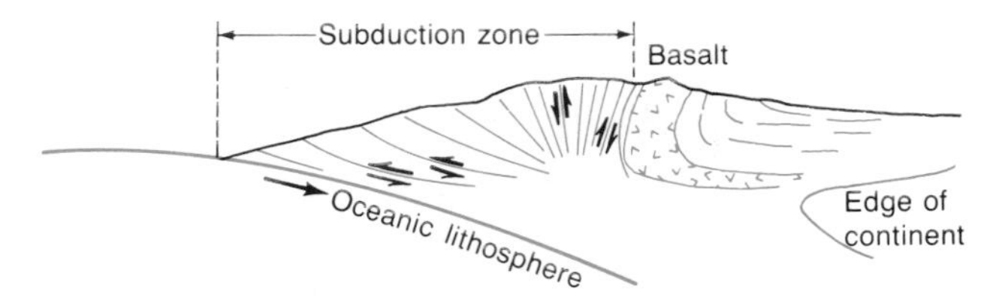

FIGURE 15-29 Formation of the Olympic Range in an embayment of the Pacific border of the state of Washington. Subduction piled sediments against basaltic rocks. *(After D. E. Kari and G. F. Sharman, Geol. Soc. Amer. Bull. 86:377–389, 1975.)*

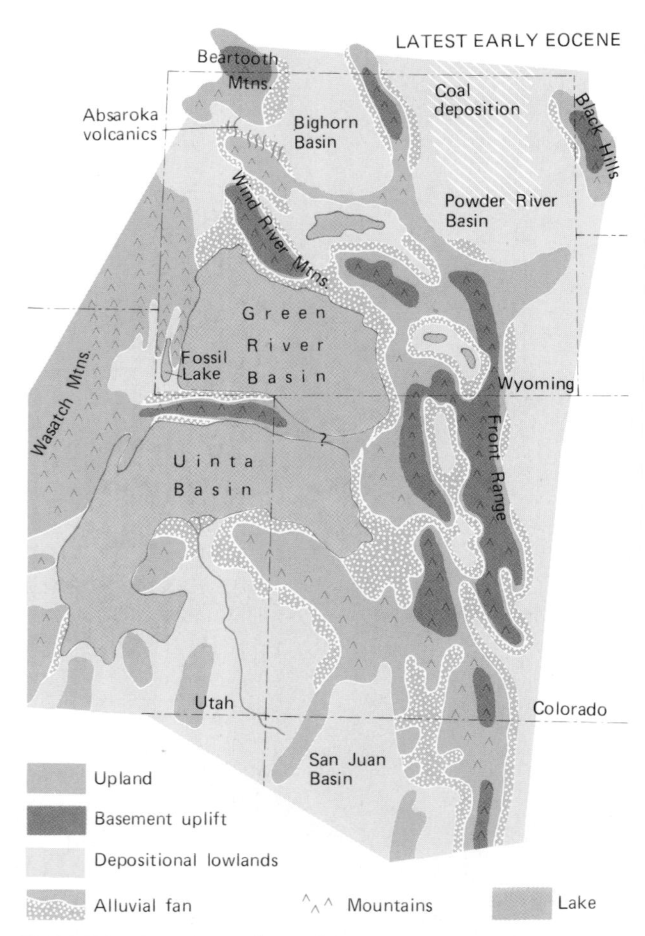

FIGURE 15-30 Geologic features associated with the Laramide uplifts of Colorado and adjacent regions at the end of Early Eocene time. The cores of major uplifts consist of Precambrian crystalline rock. The Green River Formation, which is well known for its oil shales and splendid fossils, was accumulating in the Green River and Uinta basins. To the north, volcanism formed the Absaroka Mountains, where Yellowstone Park is now located. The Black Hills of South Dakota represent the easternmost uplift.

oming these structures lay along the eastern margin of the northern fold-and-thrust belt. The uplift farthest to the east formed the Black Hills of South Dakota (p. 232). In Colorado, many ranges were formed by the elevation of large bodies of rock along thrust faults. It has been suggested that these uplifts were produced by a slight clockwise rotation of the Colorado Plateau, which behaved as a rigid crustal block, absorbing some of the convergence along the subduction zone to the west (Figure 15-27).

Today, of course, many peaks and ridges of the central and southern Rocky Mountains stand at very high elevations; the Front Range uplift, for example, rises far above the high plains of eastern Colorado (Figure 12-35). It is inappropriate, however, to evaluate the effects of the Laramide orogeny simply by viewing the Rocky Mountains today, because, as we will see, the high elevations of the modern Rockies reflect post-Laramide uplift. During the Laramide orogeny, erosion nearly kept pace with uplift in most areas of the United States, so that the regional topography remained less rugged than it is today. Basins in front of elevated areas were receiving large volumes of rapidly eroding material.

By the end of Early Eocene time the regional north-south pattern had weakened, and individual basins were experiencing independent histories. Most of these basins received alluvial and swamp deposits with abundant fossil mammal remains, and at times some were occupied by lakes. Near the beginning of Eocene time, sediments of the Wasatch Formation were laid down in and around rivers and swamps within the Bighorn, Green River, and Uinta basins.

Later in Early Eocene time, lakes came to occupy most of the areas within the basins, and these lakes survived, sometimes at reduced size, throughout much of Eocene Epoch (Figure 15-30). The famous Green River deposits accumulated in and around their margins. Plant remains in these sediments, including fossil palms, reveal that climates in this region were much warmer in Eocene time than they are today, being subtropical at times. These lake deposits, which are extremely well laminated, have commonly been termed *oil shales*, because algal material within them has broken down to yield vast quantities of petroleum. Unfortunately, this petroleum is disseminated throughout the rock and thus has proved difficult to extract. Nonetheless, the Green River deposits — the largest body of ancient lake sediments known — may eventually serve as a valuable source of fuel. The fine undisturbed lamination of the Green River lake deposits accounts for the remarkable preservation of a host of animal and plant fossils, including delicate creatures such as bats (Figure 15-7), frogs (Figure 15-15), and insects (Figure 15-16).

The Yellowstone Hot Spot Another interesting group of rocks found in the region of Paleogene basins and uplifts are the volcanics that form the Absaroka Range in western Wyoming and Montana. A large portion of Yellowstone National Park lies within the Absarokas; here the still-active geysers and hot springs serve as evidence

FIGURE 15-31 Petrified stumps standing upright in the Absaroka volcanics at Specimen Ridge, Yellowstone National Park. Some logs here have diameters of about 1 meter (~3 feet). *(National Park Service, U.S. Department of the Interior.)*

Box 15-1
Global Warming in the Eocene

Just how warm was the Eocene world? This question is especially pertinent as we contemplate the substantial warming that humans will soon inflict on Earth by adding carbon dioxide to the atmosphere.

The climatic history of England tells us that warm conditions spread to very high latitudes during the Eocene. Today England has a stable but relatively cool climate. It lies farther north than any American state but Alaska. It was similarly positioned during Early Eocene time, yet it was cloaked by a tropical jungle that shared many taxa with the modern equatorial region of Malaysia. Needless to say, England has never been this warm again.

Even more northerly regions were bathed in balmy climates during the Eocene. Fossil palm leaves are known from as far north as southern Alaska. These palms were part of a diverse subtropical flora that spread from Asia around the northern rim of the Pacific to North America as northern climates heated up early in the Cenozoic Era.

An especially interesting Eocene flora comes from Ellesmere Island, which lies well within the Arctic Circle, north of eastern Canada. Even this biota is subtropical, sharing many taxa with Eocene biotas of western North America and Eurasia. Late in the Eocene, more than 40 percent of the plant species of the Ellesmere flora had leaves with smooth margins—an indication that the average temperature resembled that of southern California today. Also present were large tortoises and alligators, which require warm winter temperatures. On Ellesmere Island today, temperatures hover around −18°C (about 0°F) throughout the year. In late Eocene time, however, the island remained above freezing even in winter.

Temperatures near the equator were not much warmer during Eocene time than they are today, if indeed they were warmer at all. If they had risen as much as temperatures at high latitudes, few plants or animals could have tolerated the heat. Instead, the temperature gradients from the equator to the poles were relatively gentle, because ocean currents transferred an enormous amount of heat toward the poles. It is even possible that so much warm water flowed away from the low latitudes that equatorial regions were slightly cooler than they are today. The tropical and subtropical zones of the Eocene world were so broad that Earth's average tempera-

Fossil palm frond from the Eocene Green River Formation of Wyoming, far north of where palms live today. *(Chip Clark.)*

that igneous activity has not ceased completely. Recall that Yellowstone seems to represent a hot spot in which igneous activity is localized (p. 149). During Eocene time this area stood at the eastern margin of the volcanic belt of the Pacific Northwest (Figure 15-27). Volcanism at this time was episodic, and all of the volcanic episodes were catastrophic, destroying entire forests and, we must assume, the animal life within them. Fossilized leaves, needles, cones, and seeds reveal the presence of lowlands with subtropical vegetation. Today the remnants of the Eocene forests can be seen at high elevations in Yellowstone National Park, where trees are preserved upright as stumps

ture must have been higher than it is today, but we do not know how much higher. Perhaps this condition resulted from greenhouse warming, but if so, we have not yet discovered what caused it.

The greenhouse warming of the near future will produce climatic patterns very different from those of the Eocene, because the warm currents that flow toward the poles today are very weak. Even if, for some reason, warming is once again more intense at high latitudes than nearer the equator, the poles will still remain relatively cool.

During Eocene time the temperature gradient was so gentle that the tropical zone stretched nearly halfway from the equator to the north pole and the subtropical zone spanned all, or nearly all, of the region north of the tropics. Not only is the equator-to-pole temperature gradient much steeper today, but most regions display much greater seasonal contrasts in temperature. In addition, larger areas at mid-latitudes are perennially or seasonally arid. The result is that a traveler to the Arctic from any point at the equator will encounter at least five distinctive floras (p. 34).

The complex climatic and floral patterns of the modern world will amplify the impact of future greenhouse warming on life. The problem is that today many regions lie near boundaries between floras. Warming, and the changes it produces in patterns of precipitation, will transform the floras, and hence the faunas, of these numerous marginal regions. Some species in these regions will suffer from exposure to new competitors or predators, or from confinement to smaller habitats. Many agricultural and recreational areas will be adversely affected as well. Unfortunately, we cannot yet reliably predict the regional patterns of climatic and biotic change that various degrees of greenhouse warming will produce.

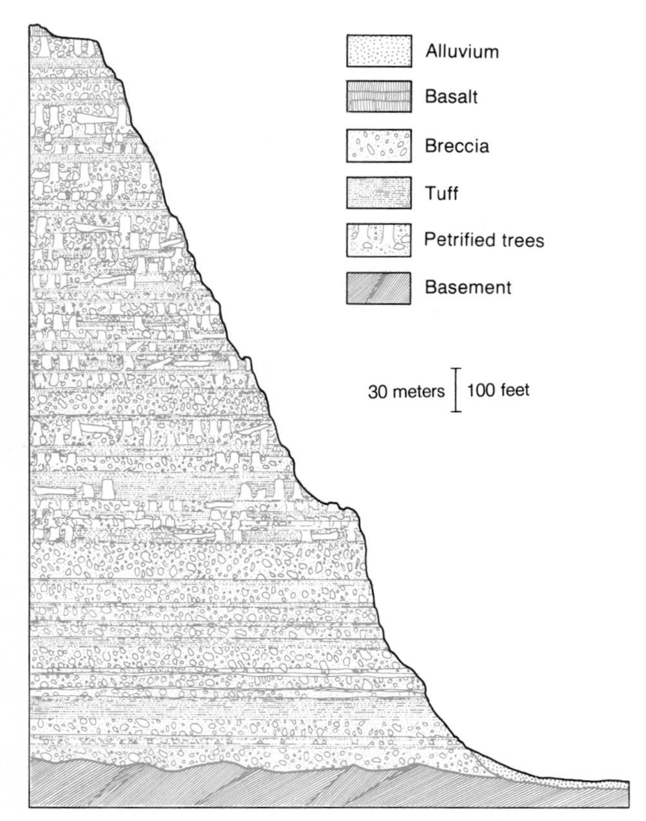

FIGURE 15-32 Succession of 27 petrified forests now exposed along the Lamar Valley in Yellowstone National Park. Each of these Eocene forests was destroyed by volcanic eruption. *(After E. Dorf, Scientific American, April 1964.)*

buried by lavas, mud flows, and flood-deposited volcanic debris (Figure 15-31). More than 20 successive forests, all of them killed in this way, can be identified (Figure 15-32).

As the Eocene Epoch drew to a close, the level of volcanism declined sharply in every part of the northwestern United States except Oregon and Washington. In addition, most of the depositional basins that lay between Montana and New Mexico were filled with sediment by the end of Eocene time. The Laramide orogeny was completed, and the uplands that it had produced were largely leveled; thus, as the Oligocene Epoch dawned, a monotonous erosion surface stretched across western and central North America, interrupted by only a few isolated hills.

The Present-Day Rockies Where, then, did the modern Rocky Mountains come from? As we will see in Chapter 16, the Rockies are the product of a renewed uplift during Neogene time. In many areas in the Rockies today, a person can look into the distance and view a flat surface formed by the tops of mountains (Figure 15-33). This is what remains of the broad erosional surface that existed

FIGURE 15-33 The so-called subsummit surface formed by the flat-topped Rocky Mountains as it appears northwest of Colorado Springs. Pikes Peak is a rare peak that rises high above this surface. The surface developed near the end of the Eocene Epoch, when nearly all of the highlands produced by the Laramide orogeny had been leveled. The subsummit surface has since been dissected by erosion after secondary uplift during Neogene time. *(U.S. Geological Survey.)*

at the end of the Eocene Epoch, but the surface now stands high above the Great Plains as a result of Neogene uplift.

During the Oligocene Epoch, although most of the Laramide uplifts had been leveled, a thin veneer of deposits spread as far east as South Dakota. The Badlands of South Dakota consist of rugged terrain carved mostly from Eocene and Oligocene deposits (p. 402). These deposits have yielded rich faunas of fossil mammals, and changes in the nature of their fossil soils reveal that aridity increased here, as it did in many parts of the world. Forests of Late Eocene age gave way to open woodlands and finally, in Late Oligocene time, to still drier savannahs.

The Gulf Coast

Unlike the Cordilleran region, the Gulf Coast of North America has remained an area of tectonic quiescence throughout the Cenozoic Era. As the Paleocene Epoch progressed, marine waters spread far inland along the Gulf for the last time during the Phanerozoic Eon (Figure 15-34). In the North Atlantic, waters spread all the way to the Dakotas in the form of the Cannonball Sea, named for a thin stratigraphic unit known as the Cannonball Formation. Later the seas retreated northward, but to the south, in the Mississippi Embayment, a thick sequence of Eocene marine sediments accumulated. In the Oligocene Epoch the seas withdrew to the approximate position of the present shoreline of the Gulf of Mexico, and then spread inland, but less far than they had during most of Eocene time. The total thickness of Paleogene sediments near the present coastline of

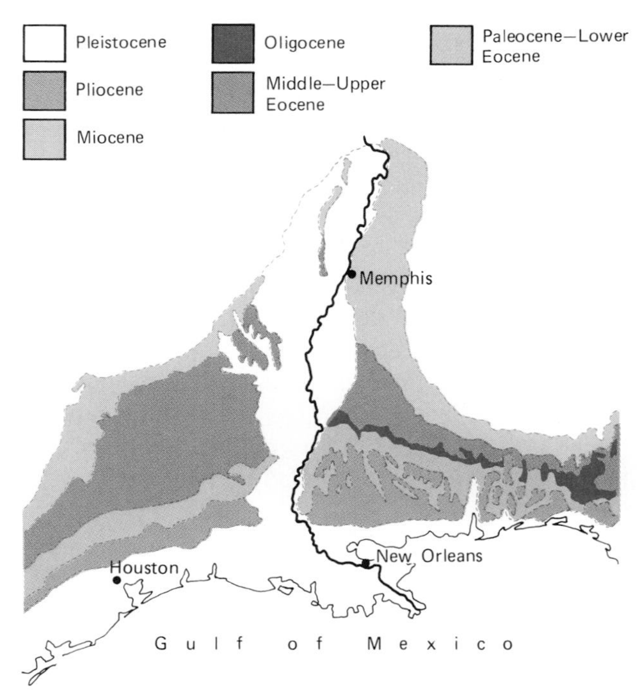

FIGURE 15-34 The segment of the Gulf coastal plain known as the Mississippi Embayment. Here marine sediments of Paleogene age were deposited farther inland than those of Neogene age.

the Gulf of Mexico exceeds 5 kilometers (~3 miles), largely because of the enormous quantities of sediment carried to the region by the Mississippi River system. The Cenozoic clastic wedge in this region, though largely buried, has been studied in great detail during the successful search for petroleum there.

CHAPTER SUMMARY

1 Marine life of the Paleogene Period resembled that of the modern world.

2 Plants on land also resembled those of the present time, except that grasses were rare until late in Paleogene time.

3 Terrestrial vertebrate animals were more primitive than those of the present. Although most of the mammalian orders alive today existed in Paleogene time, many Paleogene families and genera are now extinct. Most species of birds were wading animals that resembled storks or herons.

4 Fossil floras show that in Early Eocene time very warm climates extended to high latitudes.

5 Late in Eocene time, polar regions cooled, and cold, dense waters began to descend to the deep sea to form the psychrosphere—an ocean layer that has persisted to the present time.

6 Climates cooled in at least some regions during Late Eocene time, and cooling and drying persisted into Oligocene time. Partly as a result of these trends, grasslands replaced forests in many parts of the world. Many species became extinct both on the land and in the sea.

7 Both north and south polar regions underwent plate-tectonic changes in mid-Paleogene time. In the south, Australia broke away from Antarctica, which was left isolated over the South Pole. In the north, Europe and North America moved farther away from Greenland.

8 In western North America the Laramide orogeny produced northern and southern fold-and-thrust belts that were separated by a curious zone of uplifts. By the end of the Eocene Epoch, the Laramide orogeny had ended, and the mountains it had produced had been subdued by erosion.

EXERCISES

1 What groups of animals that played important roles in Late Cretaceous ecosystems were absent from the Paleocene world?

2 Which group changed more from the beginning of the Paleogene Epoch to the end—marine or land animals? Explain your answer.

3 Which animals took the place of Mesozoic reptiles in Paleogene oceans?

4 How does the fossil record of flowering plants reveal climatic change during Paleogene time?

5 What is the psychrosphere? How and when did it come into being?

6 How did the location of the Laramide orogeny differ from that of the Cretaceous Sevier orogeny? What might explain this change?

7 What did the region of the modern Rocky Mountains look like at the end of the Eocene Epoch?

8 What is the origin of the geologic features that intrigue tourists at Yellowstone National Park?

ADDITIONAL READING

Kennett, J., *Marine Geology,* Prentice-Hall, Inc., Englewood Cliffs, N.J., 1982.

Pomerol, C., *The Cenozoic Era,* John Wiley & Sons, Inc., New York, 1982.

Savage, R. J. G., and M. R. Long, *Mammal Evolution: An Illustrated Guide,* Facts on File, Inc., New York, 1986.

Wolfe, J., "A Paleobotanical Interpretation of Tertiary Climates in the Northern Hemisphere," *American Scientist* 66:694–903, 1978.

CHAPTER

16

The Neogene World

Because it includes the present, or Recent Epoch, the Neogene Period holds special interest for us. It was during the Neogene that the modern world took shape — that is, the ecosystem acquired its present configuration and important topographic features assumed the forms we are familiar with today.

No mass extinction marked the close of the Paleogene and the start of the Neogene. During the scant 24 million years of the Neogene, however, life and Earth's physical features have changed significantly. The most far-reaching biotic changes were the spread of grasses and weedy plants and the modernization of vertebrate life. Snakes, songbirds, frogs, rats, and mice expanded dramatically, too, and humans evolved from apes. The Rocky Mountains and the less rugged Appalachians took shape during Neogene time, as did the imposing Himalayas. The Mediterranean Sea dried up and rapidly formed again. The most widespread physical change on Earth, however, was climatic. Glaciers expanded across large areas of North America and Eurasia late in Neogene time. Although this glacial interval is commonly thought of as corresponding to the Pleistocene Epoch, it actually began late in the Pliocene Epoch and presumably will continue long into the future. The spreading of glaciers is normally episodic, and what is formally referred to as the Recent Epoch is almost certainly only the latest of numerous intervals between pulses of severe glaciation. Today ice caps in the far north remain poised to spread southward, as they have done repeatedly during the past 2 million years or so.

All four Neogene epochs — the Miocene, Pliocene, Pleistocene, and Recent — were named by Charles Lyell, who introduced the terms in 1833 in his *Principles of Geology*. Lyell distinguished the epochs of the Neogene Period on the basis of his observations of marine strata and fossils in France and Italy, noting that about 90 percent of the molluscan species found in Pleistocene strata are still alive in modern oceans, but that

These two perched boulders were transported to their present position, on a mountaintop in Yosemite National Park, by a Pleistocene glacier. *(Art Wolfe.)*

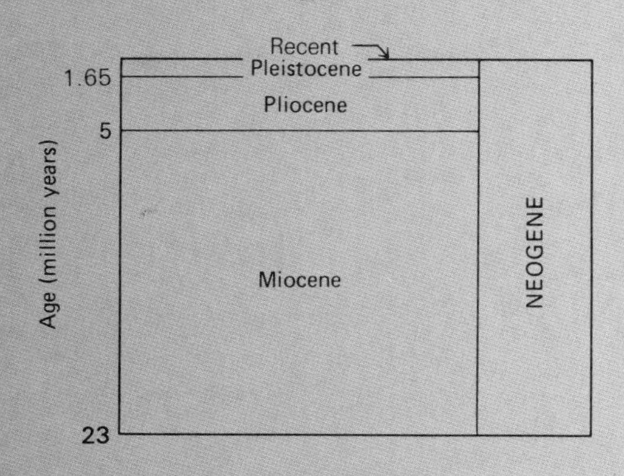

MAJOR EVENTS OF THE NEOGENE PERIOD

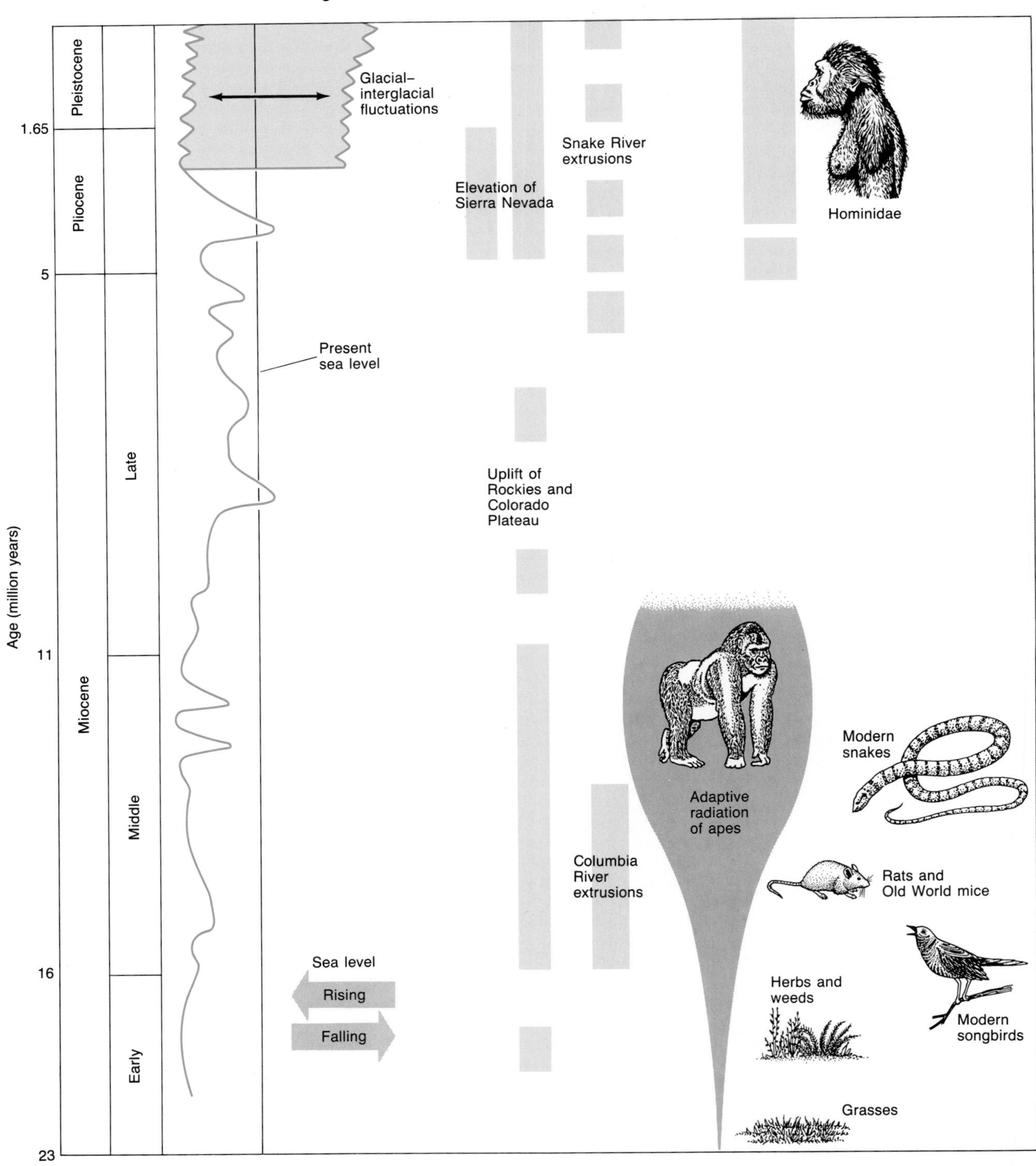

The general trend toward cooler and drier climates during the Miocene caused grasses, herbs, and weeds to diversify and replace forests. In response, seed-eating rats and mice proliferated, as did modern songbirds, which appeared in Middle Miocene time. The largest group of modern snakes, most of which prey on rodents or on eggs and chicks of birds, diversified a bit later. Apes radiated during the Miocene and then declined, but gave rise to the Hominidae (human family) near the start of the Pliocene. Uplift of the Rocky Mountains and Colorado Plateau during the Miocene intensified the trend toward drier conditions in the central United States. Similarly, uplift of the Sierra Nevada in the Pliocene produced arid conditions in the Basin and Range province, which lay in its rain shadow. North of the Basin and Range, massive igneous extrusions formed the Columbia River Plain during the Miocene and the Snake River Plain during the Pliocene. The modern Ice Age began during the Pliocene, about 2.5 million years ago, producing major fluctuations in sea level as glaciers waxed and waned in the northern hemisphere.

Pliocene strata contained fewer surviving species and that Miocene strata contained fewer still. It was not until later in the nineteenth century, however, that glacial deposits were recognized on land and found to correlate with the marine Pleistocene record.

WORLDWIDE EVENTS

In general, the animals and plants that inhabit Earth today are representative of more ancient Cenozoic life, so once again we are dealing with a period that requires no special introduction. As we examine Neogene forms of life in the context of the world ecosystem, we will see how this ecosystem changed as a result of profound climatic fluctuations. Paleogeography, however, has undergone little change during the brief Neogene interval.

Life in Aquatic Environments

Charles Darwin observed long ago that invertebrate life tends to evolve less rapidly than vertebrate life; thus the short Neogene Period has produced only modest evolutionary changes in invertebrate life.

Adaptive Radiations in the Sea Not surprisingly, the most dramatic evolutionary development in Neogene oceans has been the expansion of a group of vertebrate animals—the whales. During the Miocene Epoch a large number of whale species came into existence (Figure 16-1); among them were the earliest representatives of the groups that include modern sperm whales, which are carnivores with large teeth, and modern baleen whales, which feed by straining zooplankton from seawater. The specialized whales that we know as dolphins also made their first appearance early in Miocene time.

At the other end of the size spectrum for pelagic life, the globigerinacean foraminifera, which had suffered greatly in the mass extinction at the end of the Eocene, expanded again early in the Miocene Epoch. Neogene globigerinaceans serve as valuable index fossils for oceanic sediments. On the seafloor, evolutionary changes from Paleocene time were relatively minor.

Expansion of Freshwater Diatoms Unlike the marine diatoms, which began their evolutionary expansion in the Mesozoic Era, the dominant group of freshwater diatoms, the Pennales, did not evolve until early in Cenozoic time. By Miocene time, the Pennales comprised about 2000 known species (Figure 16-2) and had already assumed the role they play today as primary freshwater producers both in the planktonic realm and at the bottoms of lakes and rivers.

Life on the Land

Climatic changes exerted a profound influence over Neogene terrestrial biotas, and the geographic and evolutionary modifications of biotas preserved in the fossil record help us reconstruct these changes. As in Late Cretaceous and Paleogene times, flowering plant fossils represent our best gauge of climatic shifts.

Flowering Plants: Climatic Deterioration and an Explosion of Herbs In the world of plants, the Neogene Period might be described as the Age of Herbs. Herbs, or herbaceous plants, are small, nonwoody plants that die back to the ground after releasing their seeds. (Defined in this way, herbs

FIGURE 16-1 Reconstruction of the marine fauna represented by fossils of the Middle Miocene Calvert Formation of Maryland. Representing the whale family were early baleen whales *(Pelocetus)*, which strained minute zooplankton from the sea *(center)*; long-snouted dolphins *(Eurhinodelphis—lower left)*; and short-snouted dolphins *(Kentriodon—upper right)*. Sharks include the six-gilled shark *Hexanchus (lower right)*. *(Drawing by Gregory S. Paul.)*

include many more plants than the few that we use to season our food.) The recent success of herbs is primarily the result of the worldwide deterioration of climates during Oligocene and Miocene times, when cooler, drier conditions caused forests to shrink and opened up new environments to plants such as herbs and grasses, which prefer open habitats and can withstand low rainfall (Figure 16-3). Today there are some 10,000 species of grasses alone.

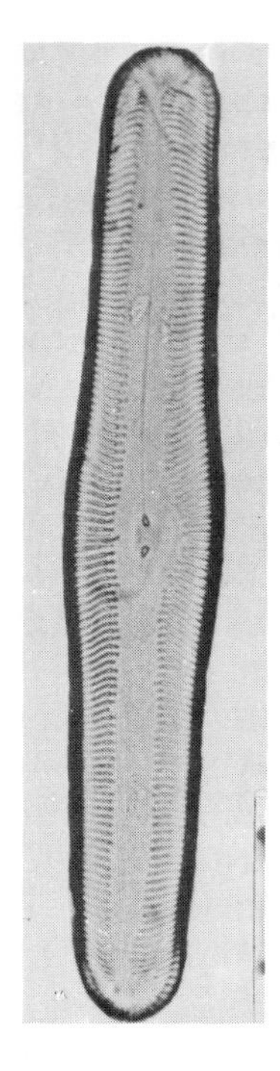

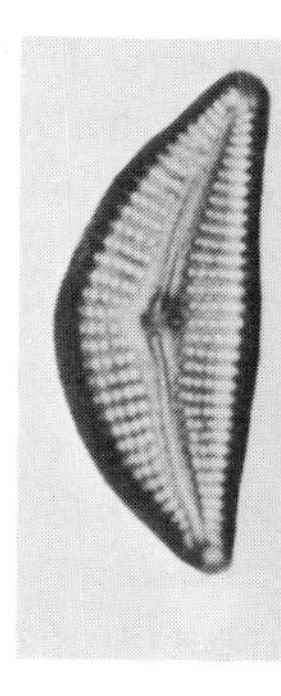

FIGURE 16-2 Miocene species of freshwater diatoms. *(G. W. Andrews, U.S. Geol. Surv. Prof. Paper No. 683A, 1970.)*

The Compositae, an important family of herbs that includes such seemingly distinct members as daisies, asters, sunflowers, and lettuces, appeared near the beginning of the Neogene Period, only 20 or 25 million years ago, yet today this family contains some 13,000 species, including the plants that ecologists refer to as weeds (p. 29). As any gardener knows, weeds are exceptionally good invaders of bare ground. They may not compete successfully against other plants to retain the space they invade, but they soon disperse their seeds to other bare areas that have been cleared by fires, floods, or droughts and spring up anew.

Tectonic events caused areas in eastern Africa, southern Asia, western North America, and South America to receive less rainfall than they had known during the Paleocene Epoch, and at the same time these areas felt the effects of global climatic trends. The presence in the Southeast Pacific of ice-rafted, coarse sediments of early Miocene age shows that by this time Antarctic glaciers had begun to flow to the sea. Cores of deep-sea sediment also reveal that the belt of siliceous diatomaceous ooze that encircled Antarctica (Figure 3-36) simultaneously expanded northward at the expense of carbonate ooze, which tends to accumulate where climates are warmer. This polar cooling shifted the distribution of oceanic plankton at the beginning of the Early Miocene Age. Ever since that time, assemblages of species at high latitudes have differed greatly from those at low latitudes.

Analyses of leaf margins from terrestrial floras of North America reveal only modest fluctuations during Neogene time, a trend that contrasts sharply with the dramatic drop in temperatures that marked the Eocene-Oligocene interval (Figure 15-22). A slight cooling trend began early in the Oligocene Epoch, however, and continued until the Ice Age began, late in the Pliocene Epoch. During this interval, climates not only became slightly cooler but also grew drier and more

FIGURE 16-3 A remnant of true prairie in Iowa. This is the kind of grassland that greeted early settlers of the American West. *(John Shaw/Tom Stack and Assoc.)*

seasonal. As a result, dry grasslands expanded into areas that had once been open woodlands and dense forests.

Modernization of Terrestrial Vertebrates Because we are naturally interested in the origins of large mammals, we often ignore the great success of smaller creatures. In fact, the Neogene Period might well be called the Age of Frogs, the Age of Rats and Mice, the Age of Snakes, or the Age of Songbirds, because all four of these groups have undergone tremendous adaptive radiations over the past few million years (see Major Events, p. 430).

Many late Neogene food webs include plants, herbivores, and carnivores that these adaptive radiations produced. This ecologic relationship indicates that the adaptive radiations of plants stimulated those of herbivores, and that these radiations in turn stimulated the diversification of snakes. Many species of rats and mice dig burrows in dry terrain and eat the seeds of grasses and herbs, for example. To a large extent, the success of these small rodents during Neogene time resulted from that of both the grasses and the Compositae and also, more fundamentally, from the drying and cooling of climates that favored these plants. Perhaps we can attribute the success of modern frogs and toads, whose species number about 2000, to their remarkable ability to catch insects through the quick protrusion of their long tongues. In any case, the snakes have obviously flourished largely because of the proliferation of frogs and rodents; few other predators can pursue small rodents down their burrows without digging. Before the start of the Neogene Period, there were few snakes except for members of the primitive boa constrictor group. Today, however, the more advanced snakes of the family Colubridae include about 1400 species, many of which are poisonous.

Also poorly represented before Neogene time were the passerine birds, or songbirds and their relatives, which are highly conspicuous today. These birds have also benefited from the diversification of seed-bearing species of herbs but, like frogs, they owe some of their success to the fact that they are well equipped to capture flying insects. It is probable that many types of flying insects were not heavily preyed upon until groups of passerine birds took to the air.

Of course, groups of large animals also developed their modern characteristics during the Neogene Period. Among the herbivores, for example, the horse and rhinoceros families dwindled after mid-Miocene time in a continuation of the general decline of the odd-toed ungulates. Meanwhile, the even-toed, or cloven-hoofed, ungulates expanded, especially through the adaptive radiation of both the deer family and the family called the Bovidae, which includes cattle, antelopes, sheep, and goats. The giraffe family and the pig family also radiated during the Miocene Epoch, but the number of species in these families has since declined. Similarly, many types of elephants, including those with long trunks, experienced great success during the Miocene and Pliocene intervals but later declined. Today only two elephant species survive: the large-eared African elephant and the smaller, more docile Indian elephant — the species that is commonly trained to perform in circuses.

Carnivorous mammals also assumed their modern character in the course of the Neogene Period; this group included the dog and cat families, both of which had appeared during Paleogene time. The bear and hyena families were other important Miocene additions to the carnivore group.

Many of the Neogene mammal groups expanded successfully because of the spread of savannahs and open woodlands (Figure 16-4). Several radiating herbivore groups, such as the antelopes and cattle, included many species that were well adapted for long-distance running over open terrain and that grazed on harsh grasses with the aid of high-crowned, continuously growing teeth. Many grasses contain tiny fragments of silica, and only continuously growing teeth can tolerate the resulting wear. Also on the increase were the groups of rodents that are adapted for burrowing in prairies and the elephants, which require open country simply to move around. As we might expect, the diversification of herbivores in savannahs and woodlands in turn fostered the success of groups of carnivores that were well adapted for attacking herbivores in open country — groups such as hyenas, lions, cheetahs, and long-legged dogs.

Ultimately, however, the greatest change in the terrestrial ecosystem was wrought by primates, simply because the ecologically disruptive

FIGURE 16-4 Reconstruction of the so-called *Hipparion* fauna of Asia. This diverse fauna occupied open country in Asia about 10 million years ago, in Late Miocene time. *Hipparion* is the galloping horse *(center)*. The elephant on the left, with downward-directed tusks, is *Dinotherium*. In the foreground, short-legged hyenas of the genus *Percrocuta* look on from their den. *(Drawing by Gregory S. Paul.)*

humans belong to this group. As a group, primates tend to favor forests over savannahs; in fact, most live in trees. As we have seen, monkeys were present by Oligocene time; the oldest group includes the so-called Old World monkeys, which now live in Africa and Eurasia. Before the end of the Oligocene interval, however, a distinctive group of monkeys reached South America. How they got there remains uncertain. These New World monkeys, which differed from their Old World counterparts in that most possessed prehensile (or grasping) tails, probably had a separate evolutionary origin. In any event, monkeys on both sides of the Atlantic underwent adaptive radiations during Neogene time.

Apes, which evolved in the Old World, flourished for a time but have since declined in number of species. We will discuss apes and apelike animals when we examine the origins of humans, which belong to the same superfamily, the Hominoidea (Figure 1-6). The most recent phases of human evolution have taken place within the climatic context of the recent Ice Age (Pleistocene Epoch); therefore, before we discuss the Hominoidea, it is appropriate to examine the major global events of this fascinating interval of geologic time and of the Pliocene Epoch that preceded it.

Late Neogene Climatic Change

The cooling and drying of climates in Oligocene and Miocene time caused grasses and weedy plants to expand their coverage of the land, and Early Pliocene time saw modest climatic changes as well. Late Pliocene and Pleistocene time, in contrast, was marked by strong, rapid climatic fluctuations in the Northern Hemisphere—changes that characterized the modern Ice Age.

Pliocene Equability Stratigraphic unconformities in various parts of the world indicate that at the very end of the Miocene Epoch, between 6 and 5 million years ago, global sea level fell by perhaps as much as 50 meters (~165 feet). This drop in sea level, as we will see, isolated the Mediterranean Sea from other oceans and caused it to dry up temporarily. This event probably resulted from removal of water from the ocean by the sudden expansion of glaciers in Antarctica.

As the Pliocene Epoch got under way, about 5 million years ago, sea level rose again, leaving marine deposits inland of coastlines in such areas as California, eastern North America, and countries bordering the North Sea and the Mediterranean. Fossil faunas and floras also reveal that global climates during this time were more equable than they are today; pollen analyses, for example, indicate that southeastern England was subtropical, or nearly so, and that northern Iceland enjoyed a temperate climate. Especially in the Northern Hemisphere, however, this warm interval came to a sudden close with the start of the modern Ice Age, between 3 and 2.5 million years ago. Almost certainly the glacial interval that ensued has not yet ended.

Pleistocene Continental Glaciation A wide variety of evidence documents the recent Ice Age, revealing details of the timing and geographic distribution of continental glaciation (Figure 16-5).

1 *Erratic boulders* Large rocks that sit on Earth's surface far from exposures of the bedrock from which they have broken, so-called **erratic boulders** are too large to have been transported by rivers, and it is difficult to imagine that any agent other than continental glaciers might have transported them (p. 428).

FIGURE 16-5 Reconstruction of glaciers as they existed during a typical glacial interval of the Pleistocene Epoch. Large continental glaciers were centered in North America, Greenland, and Scandinavia. *(After a drawing by A. Sotiropoulos in J. Imbrie and K. P. Imbrie, Ice Ages, Enslow Publishers, Short Hills, N.J., 1979.)*

2 *Glacial till* A mixture of boulders, pebbles, sand, and mud that has been plowed up, transported, and then deposited by glaciers, till is difficult to confuse with sediments deposited by other mechanisms, especially where it rests at Earth's surface and forms ridges known as moraines or is associated with outwash deposits (p. 55). Such moraines form much of Cape Cod, Massachusetts, where they extend into the marine realm. Retreating glaciers commonly left terminal moraines behind them, and when these glaciers melted back, they often left shallow basins in which water accumulated behind the moraines. The Great Lakes of North America occupy such basins; they did not exist before the modern Ice Age.

3 *Depression of the land* Earth's crust remains depressed in regions that lay beneath large glaciers a few thousand years ago. Hudson Bay, the only epicontinental sea that exists today in North America, occupies such a depressed region in eastern Canada (Figure 9-1). Scandinavia is similarly depressed because it too was burdened with a large ice cap until about 10,000 years ago, when the crust began to rebound.

4 *Glacial scouring* Glaciers smoothed the sides of mountains that they scraped past. Mount Monadnock, in New Hampshire, stood partially above surrounding ice sheets, as some mountains of Antarctica do today (Figure 16-6). The lower part of Mount Monadnock, which was smoothed by flowing glaciers, stands in sharp contrast to the upper part, which remains rugged. Small glaciers, known as alpine or mountain glaciers, left their mark along valleys within the Rockies and other mountain chains, where they flowed during the recent Ice Age. Here the most spectacular products of glaciation are U-shaped valleys that they sculpted from valleys that were once shaped like a V (Figure 9-23).

5 *Lowering of sea level* One important effect of each major expansion of ice sheets during the Pleistocene Epoch was a profound lowering of sea level as great quantities of water were locked up on the land. During major glacial expansions, most of the surfaces that now form continental shelves stood above sea level. Rivers cut rapidly downward through the soft sediments of many continental shelves to form valleys that exist today as submarine canyons, having been excavated further by submarine turbidity currents (Figure 3-34). During some glacial episodes, sea level dropped slightly more than 100 meters (~330 feet) below its present position.

FIGURE 16-6 Dark rock of the Prince Charles Mountains projecting above the surface of the modern Antarctic ice cap. Many North American mountains were partly buried in ice during the Pleistocene Epoch. *(D. Parer and E. Parer-Cook/AUSCAPE International.)*

Today there remain only two ice caps of the sort that expanded to cover broad areas many times during the Pleistocene Epoch. One of these modern ice caps covers much of Greenland (Figure 2-14) and the other covers nearly all of Antarctica (Figure 16-6). Today about three-quarters of the world's fresh water is locked up in glacial ice, and most of it belongs to the Antarctic ice cap. It may seem impressive that glaciers now contain about 25 million cubic kilometers of ice, but it has been estimated that the volume of ice was nearly three times as great during glacial advances of the Pleistocene Epoch, with larger ice sheets averaging about 2 kilometers (~1.2 miles) thick. The total volume of ice has been calculated from the volume of water that was removed from the ocean to lower sea level slightly more than 100 meters. Shelves of ice projected into the sea, and these shelves, together with the icebergs and pack ice that broke loose from them, spread over half the world's oceans.

6 *Migration of species* The alternations of Pleistocene glacial and interglacial intervals have caused climatic belts and the floras and faunas that occupy them to shift over distances measured in hundreds of kilometers. Thus fossils of mammals

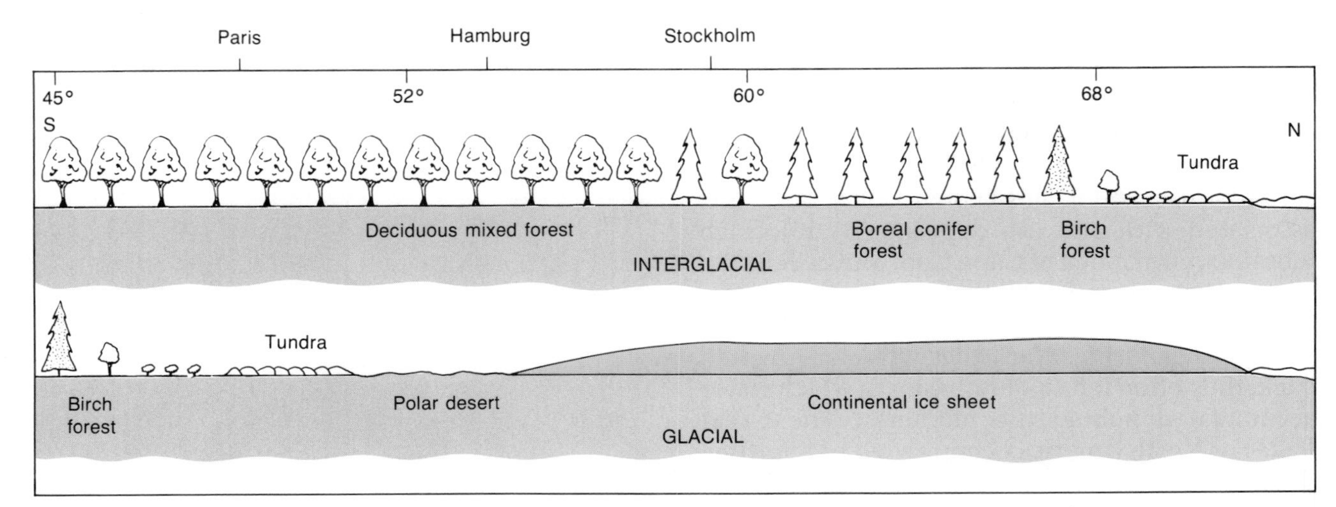

FIGURE 16-7 North-south migration of vegetation in Europe during the Pleistocene Epoch. During glacial intervals *(lower diagram)* continental glaciers spread southward to the vicinity of Hamburg, and tundra shifted to the latitude of Paris. *(After T. Van Der Hammen, in K. K. Turekian [ed.], The Late Cenozoic Glacial Ages, Yale University Press, New Haven, Conn., 1971.)*

such as the muskrat, which today does not range south of Georgia, reveal that climates in Florida were cool when glaciers pushed southward into the northern United States. Other fossil occurrences, such as those of hippopotamuses in Britain, show that during at least some interglacial intervals, climates were warmer than they are today.

One of the most useful fossil indicators of Pleistocene climates is the pollen of terrestrial plants. Pollen assemblages reveal climatic change by indicating the shifting of floras to the north or south. Figure 16-7 shows the southward movement of floras in Europe by about 20° latitude during the most recent glacial interval there.

During glacial intervals, some mountaintops that stood above ice sheets served as refuges for plant and animal species that could not live on the ice. Some species that were stranded during the most recent glacial advance could not migrate across warmer lowlands when the glaciers melted back and thus today remain isolated from other members of their species, which generally live farther north. In *On the Origin of Species* Charles Darwin pointed to the isolation of these still-marooned populations as evidence of continental glaciation.

Larger refuges between major centers of ice accumulation also existed during glacial intervals. The most famous of these refuges was an area known as Beringia, which was so designated because it included the region of the Bering Strait. During glacial episodes, regression of the seas turned the Bering Strait into a land corridor between Asia and North America, and it was by this landbridge that not only lower mammals but also the first humans entered the New World. Ironically, Beringia, which was hospitable to terrestrial

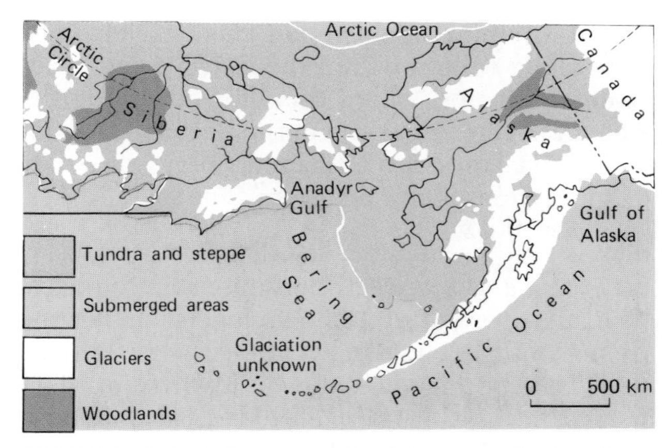

FIGURE 16-8 The geography of Beringia during the most recent glacial interval (the Wisconsin, or Riss-Würm, Stage). Though positioned at a high latitude, Beringia was dry and free of large glaciers. *(After D. M. Hopkins, in D. M. Hopkins [ed.], The Bering Land Bridge, Stanford University Press, Stanford, Calif., 1967.)*

FIGURE 16-9 Reconstruction of the mammalian fauna that occupied the steppe in the Alaska portion of Beringia, about 12,000 years ago, during the most recent glacial interval. Of 61 species depicted here, 11 are extinct; among them are the woolly mammoth, the American mastodon, the long-horned bison, a lion-like cat, and a saber-toothed cat. *(Mural by J. H. Matternes in the Smithsonian Institution.)*

mammals during the height of glaciation (Figures 16-8 and 16-9), included portions of Siberia and Alaska — areas that we now view as inhospitable to most species but that happened to remain unglaciated when other parts of the Northern Hemisphere were covered with ice.

The Chronology of Glaciation The Pleistocene Epoch is often thought of as the modern Ice Age, but the Ice Age actually began long before the end of the Pliocene Epoch. The most detailed chronology of the glaciation comes from oxygen-isotope ratios of foraminiferan skeletons preserved in deep-sea sediments. In Chapter 15 we saw that this ratio is greatly altered by the growth of continental glaciers; glaciers lock up water preferentially enriched in oxygen 16, the lighter isotope, which evaporates from the ocean at a higher rate than the heavier isotope, oxygen 18 (p. 416). When continental glaciers expand, the ratio of oxygen 18 to oxygen 16 increases in seawater and in foraminiferan skeletons. For foraminiferans that live in waters cooled by such glaciation, this isotopic shift is augmented by a second factor: the slight elevation of the ratio of isotopes incorporated in foraminiferan skeletons with decreasing temperatures (p. 48).

About 3.2 or 3.1 million years ago there was a slight increase in the ratio of oxygen 18 to oxygen 16 in the skeletons of planktonic foraminiferans in many geographic areas. This change resulted from a brief episode of widespread cooling, which terrestrial floras also document. In northwestern Europe, for example, several subtropical species of land plants, including palms, disappeared. Glaciers deposited tills in Iceland at this time. This climatic change signaled the beginning of the modern Ice Age. Nonetheless, it was reversed within about 100,000 years. Fossil biotas indicate that for the next half-million years or so, climates were at times quite warm again, even at high northern latitudes. For example, fossils reveal that sea otters occupied the Arctic Ocean. Thus it seems evident that large continental ice caps had not yet formed.

Continental glaciers did come into existence about 2.5 million years ago. Oxygen-isotope ratios in planktonic foraminiferans shifted markedly at this time. In addition, deep-sea deposits of this age in the North Atlantic record the first occurrence of sand grains released by melting icebergs. In other words, continental glaciers were now flowing to the North Atlantic, releasing sediment-laden blocks of ice. Terrestrial floras underwent dramatic changes as well. About 2.5 million years ago, northwestern Europe lost the last of many subtropical plant taxa that it had shared with Malaya earlier in Pliocene time. It was not only cooling of climates that affected floras, but also increased aridity. Many regions became drier because sea level fell as continental glaciers expanded, and broad continental areas came to lie far from oceans that had once supplied them with moisture.

Tills of Late Pliocene and Pleistocene age provide a poor chronology of the pattern of glaciation, for two reasons. First, the tills can seldom be dated directly; generally they lack both guide fossils and radioactive materials that might be used for absolute dating. Second, most tills have been obliterated by advances of glaciers after they were deposited. Far more revealing are the oxygen-isotope ratios of planktonic foraminiferans, which indicate that there were major expansions and contractions of glaciers throughout the Ice Age. Oscillations of ice sheets produced, at their extremes, what are termed **glacial maxima** and **glacial minima** (Figure 16-10). Today we live during the glacial minimum established when glaciers melted back between about 15,000 and 10,000 years ago. The most recent glacial maximum occurred between about 35,000 and 10,000 years ago, during the Wisconsin Stage. During this glacial maximum, sea level dropped to at least 100 meters (~330 feet) below its present level. This interval was preceded by the Sangamon glacial minimum, about 125,000 years ago, when sea level stood slightly higher than it does today.

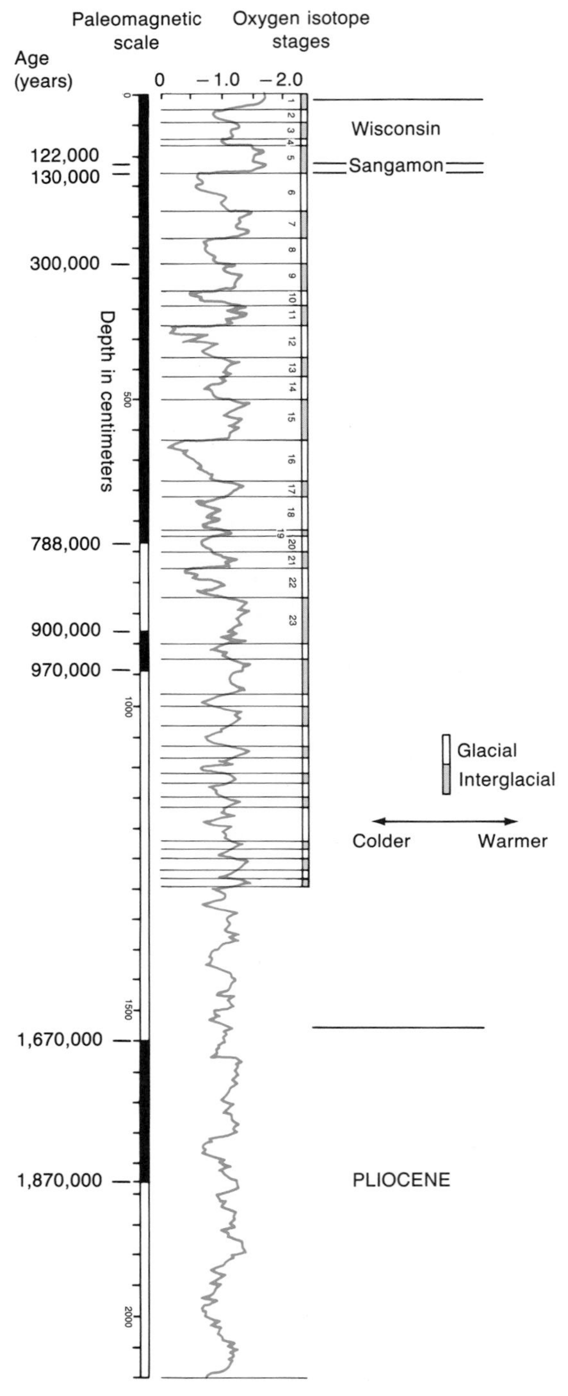

FIGURE 16-10 Oxygen-isotope fluctuations for planktonic foraminifera within a deep-sea core. Increases in the relative abundance of oxygen 18, indicated by shifts of the graph toward the left, indicate buildup of ice sheets and cooler temperatures. Oxygen-isotope stages (1–23) represent glacial and interglacial intervals. Two formal stages (Wisconsin and Sangamon) are shown on the right. The paleomagnetic time scale *(left)* provides dates for several levels in the core. *(After N. J. Shackelton and N. D. Opdyke, Geol. Soc. Amer. Mem. 145, 449–464, 1976.)*

The Nature of Glacial Expansions and Contractions Because of its relative recency, the last interval of glacial expansion (the late Wisconsin) has left the best record of glacial deposits. Furthermore, since this interval extended from about

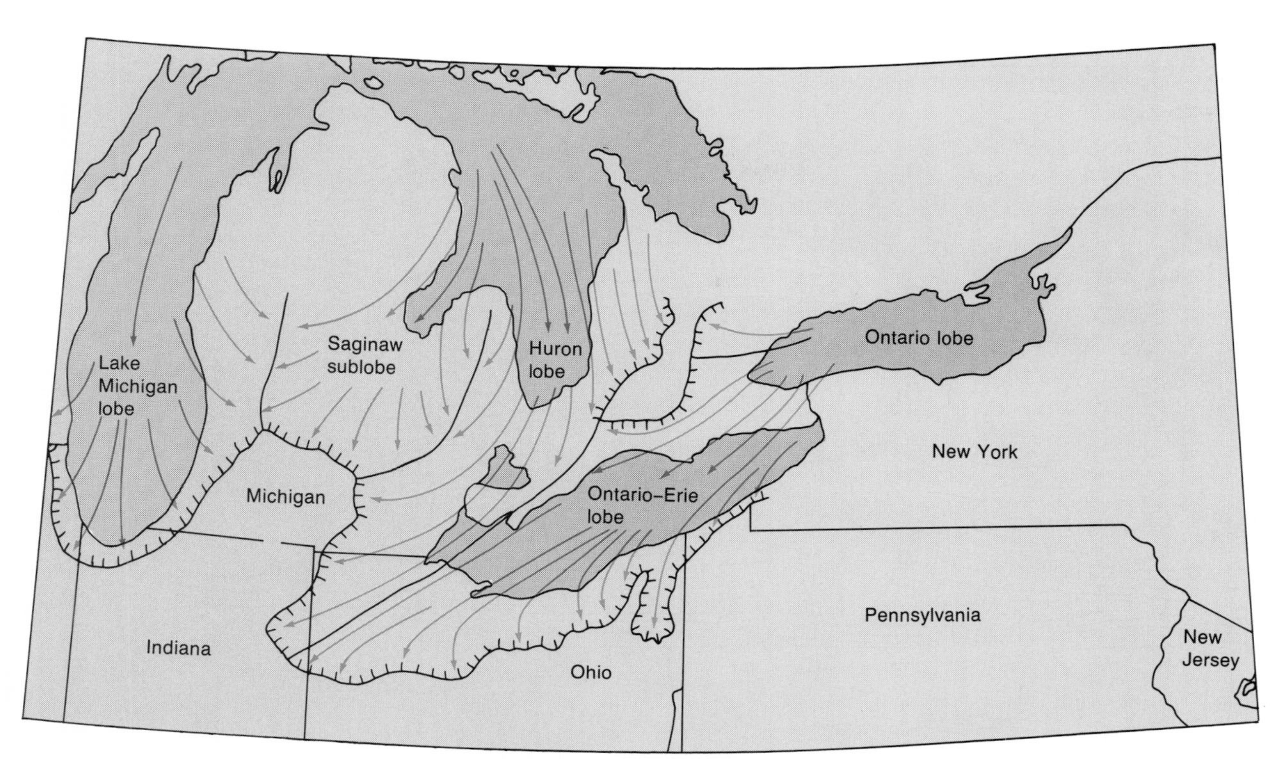

FIGURE 16-11 Glacial lobes in the Great Lakes region 14,800 years ago. Terminal moraines dated by the radiocarbon technique define lobe positions. *(After D. S. Fullerton, Quat. Sci. Rev. 5:23–36, 1986.)*

35,000 to 10,000 years ago, its organic deposits, including fossilized wood, are well within the range of radiocarbon dating. Detailed stratigraphic studies in the region of the Great Lakes have revealed that the late Wisconsin glacial advance was a complex event, consisting of pulses of glacial expansion separated by partial retreats. In addition, individual lobes of the North American ice sheets did not always expand and contract at the same rate. Ultimately, however, all the glacial lobes retreated into large basins that became the Great Lakes when the ice sheets melted back into Canada (Figure 16-11).

A major international project called **CLIMAP** was formed in 1971 to reconstruct the oceanographic conditions and climates that characterized the late Wisconsin glacial interval. Many of CLIMAP's conclusions have been based on studies of the geographic distributions of fossil planktonic foraminifera and calcareous nannoplankton that have been obtained from deep-sea cores but represent species that remain alive today. Knowledge of the environmental requirements of these planktonic species has permitted paleontologists to reconstruct the distributions of water masses associated with particular temperatures, and these temperatures, combined with information on the distribution of terrestrial floras and faunas, have made it possible to map the distribution of Pleistocene vegetation and environments on land (Figure 16-12).

During the Wisconsin glacial interval, north-south temperature gradients steepened in the Northern Hemisphere both in shallow seas and on land. Winter temperatures fell only slightly in most tropical areas but plummeted at latitudes north of 30° in the Northern Hemisphere. Other patterns of climatic change, however, were more complex. Some areas that lay only a few degrees south of continental glaciers, for example, became much wetter than they are today. Among them was the Great Basin in the American West, which drew rainfall from glaciated terrains to the north and accumulated numerous lakes in areas that are now arid (Figure 16-13). Subtropical deserts, such as portions of the Sahara near the Mediterranean, also became wetter in their northern regions. In contrast, environments of low rainfall—steppes, semideserts, savannahs, and dry grasslands—stretched in a broad belt across Eurasia

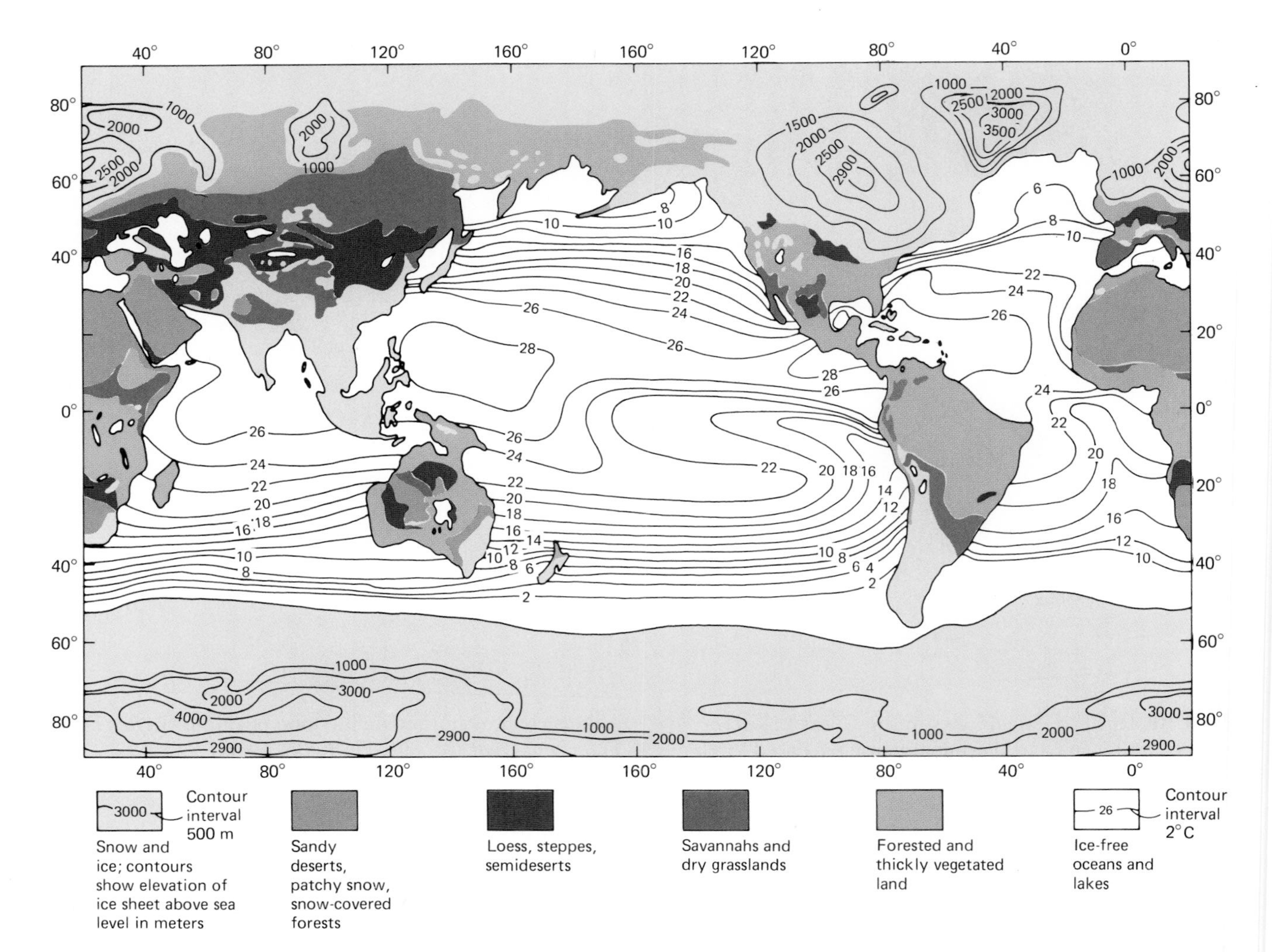

FIGURE 16-12 Geographic features reconstructed from the height of the most recent glacial interval, about 18,000 years ago, when sea level was about 85 meters (~280 feet) below its present level. Temperatures (in degrees centigrade) are estimates for the month of August. *(Modified from CLIMAP, Science 191:1131–1144, 1976.)*

south of glaciers. At the same time, the tropical rain forests of South America and Africa, which lay relatively far from the lowered oceans, shrank back in areas where drying conditions came to prevail, eventually giving way to vegetation that required less moisture.

In Europe the Alps and the Pyrenees developed glaciers of their own under the cold climatic conditions, and the east-west trend of these cold barriers blocked the southward migration of many species that were moving ahead of continental glaciers advancing from the north. Many forms of life, such as magnolia trees, were caught in this glacial trap and disappeared from Europe, but they still managed to survive in North America, where southward migration to Florida and Central America remained unimpeded.

Antarctica, which was positioned over the South Pole, accumulated a continental glacier long before the beginning of the Pleistocene Epoch, but the isolation of this island continent after the breakup of Gondwanaland prevented its glaciers from spreading over a vast continental area in the manner of the late Paleozoic glaciers of Gondwanaland. Nonetheless, snow and ice accumulated over the southern tip of Africa, an eastern segment of Australia, and a large part of southern South America (Figure 16-12). A large lake occu-

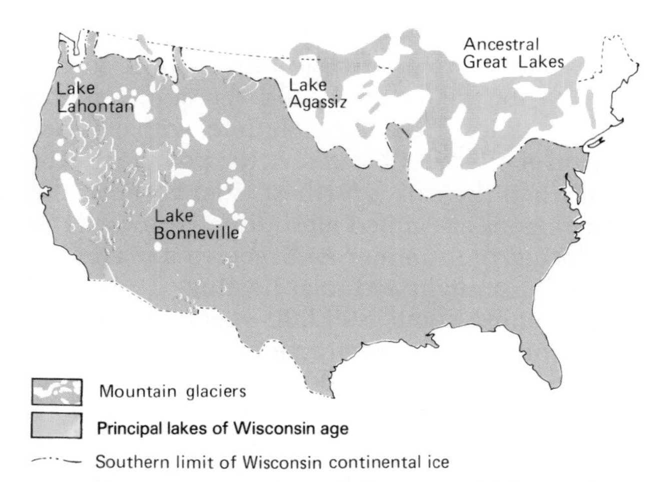

FIGURE 16-13 Locations of glaciers and lakes in the United States during the most recent glacial interval. Lake Agassiz and the ancestral Great Lakes formed to the south of the continental glacier as it retreated northward near the end of the glacial interval. *(C. B. Hunt, Natural Regions of the United States and Canada, W. H. Freeman and Company, New York, 1974.)*

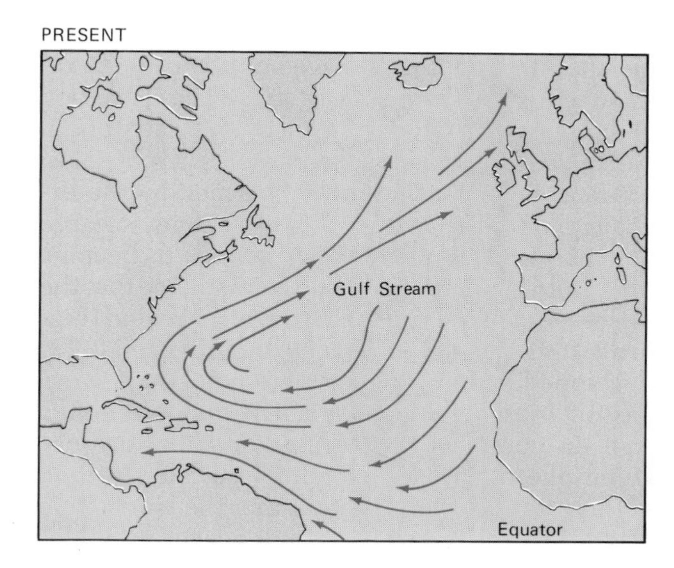

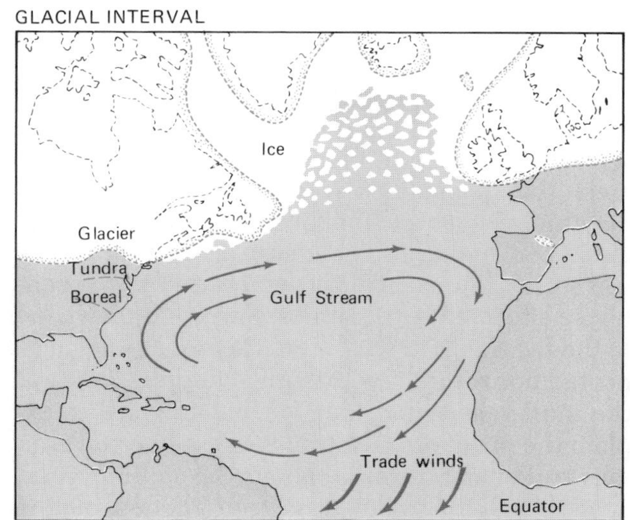

FIGURE 16-14 The Atlantic Ocean today and during the most recent Pleistocene glacial interval.

pied central Australia, and much of this continent was wetter and more heavily vegetated than it is today.

Three large glacial centers developed in the Northern Hemisphere—one in North America, one in Greenland, and one in Scandinavia (Figure 16-5). Because the northern Atlantic Ocean was adjacent to these three glacial centers, it was more profoundly affected by the Pleistocene glaciers than any of the world's other major oceans except the Arctic, and land areas adjacent to the Atlantic also suffered marked climatic change. During glacial intervals, pack ice must have choked large areas of the North Atlantic (Figure 16-14), just as it now occupies bays adjacent to northern Canada. Farther south, along the east coast of North America, glaciers flowed southward to New Jersey, and tundra occupied what is now Washington, D.C.

The Causes of Glaciation What factors have brought about Pleistocene glaciation? When we consider this question, it is important to understand that both the expansion and the contraction of glaciers are unstable processes; once either process has begun, it usually accelerates automatically. Ice reflects a much greater percentage of warming sunlight than land or water does: Its albedo is higher. So if a small amount of climatic cooling generates glacial ice, this ice reflects more sunlight than was reflected before the glacier developed, and this leads to further cooling. The glacier then expands, so even less sunlight is absorbed; and so on. In the same manner, climatic warming melts ice, resulting in exposure of light-absorbing land, which produces even more warming. When we ask why glaciers expand and contract, then, to a large extent we are asking why a new trend (expansion or contraction of ice) gets started. A small change in climatic conditions can ultimately have an enormous effect.

What might have initiated the Ice Age is a question that geologists have not been able to resolve with any degree of certainty. The formation of ice earlier in geologic history can be attributed partly to the movement of continents across poles; whereas polar seas are often warmed by the exchange of water with warmer regions, a large landmass centered over a pole tends to become quite cold. Apparently it is no coincidence that the major ice caps of both the Ordovician Period (Figure 10-30) and the late Paleozoic (Figure 12-22) developed at times when large continents were passing over the South Pole. The Pleistocene Ice Age, in contrast, was focused in the Northern Hemisphere, with the North Pole itself centered in the Arctic Ocean. It is true that the North Pole was partly isolated from warmer oceans by neighboring landmasses, but this condition existed for millions of years before ice sheets formed and does not in itself provide the answer.

It has been suggested that the sudden elevation of mountain ranges might have triggered the Ice Age by creating high-altitude mountain glaciers that expanded by reflective cooling and eventually lowered global temperatures. It has also been hypothesized that a reduction in the sun's energy output or a decrease in the greenhouse effect of Earth's atmosphere might have led to the Ice Age. These conjectures are neither supported nor refuted by existing evidence. Perhaps the most reasonable idea yet put forward to explain the onset of glaciation is that when the Isthmus of Panama formed, about 3.5 million years ago, the Gulf Stream was strengthened by the northward deflection of the equatorial current. Moisture that the warm Gulf Stream waters thus supplied to northern regions led to an increase in snowfall here and to a buildup of ice caps. We will return to this idea when we review Neogene events in the Atlantic Ocean.

Glacial Cycles Just as interesting as the cause of the modern Ice Age is the source of the glacial oscillations that have characterized the Ice Age since its inception. It is now generally agreed that changes in Earth's relationship with the sun have caused these oscillations. Figure 16-10 indicates that these oscillations were more frequent during the early part of the Ice Age than later on.

Early in the Ice Age, glacial oscillations seem to have corresponded to the so-called **tilt cycle** of Earth's axis of rotation. The axis is always tilted slightly away from vertical with respect to the plane of Earth's orbit around the sun, but the angle of tilt oscillates through time, with a periodicity of about 41,000 years. At the point in the tilt cycle when the axis is farthest from vertical, the polar regions are aimed most directly toward the sun during the summer and receive a maximum amount of sunlight and solar heating.

Beginning about 800,000 years ago, glacial oscillations became less frequent, shifting to a periodicity of 90,000 to 100,000 years. This new periodicity appears to have corresponded to changes in the shape of Earth's orbit. These changes result from oscillations in the gravitational pull of other planets on Earth, which have a periodicity of about 92,500 years. When the orbit changes so as to bring Earth closest to the sun, it receives more solar heat than it does when it is farther away. It is not known why, about 800,000 years ago, Earth's orbital oscillations came to govern the expansion and contraction of glaciers, overshadowing the tilt cycle.

Astronomical arguments have been carried even further in attempts to explain smaller-scale climatic cycles, such as changes in sea level. Thorium dating of coral reefs that now stand about 6 meters (~20 feet) above sea level has revealed that the world's oceans stood much higher about 125,000 years ago than they do today. Sea level subsequently dropped but rose quickly about 105,000 years ago, and then it dropped and rose once more about 82,000 years ago. The second and third high stands were lower than the first, but it is not known precisely how far sea level fell in between the high stands. Terraces formed by reef growth are found at many localities, including New Guinea (Figure 16-15), Barbados (an island in the West Indies), and the Florida Keys. They appear to record minor oscillations that occurred during a single larger oscillation—the major interglacial interval known as the Sangamon.

It has been suggested that the minor oscillations recorded in the fossil-reef terraces represent an astronomical cycle known as the **precession of the equinoxes.** Precession is a slight wobble in the axis of Earth's rotation caused by the gravitational pull exerted by the sun and the moon on this planet. The wobble causes each of Earth's hemispheres to experience its longest and shortest days at a different position in its elliptical orbit each

FIGURE 16-15 Reef-built terraces fringing the Huon Peninsula of New Guinea. These high terraces represent interglacial epochs, when sea level stood relatively high throughout the world. *(A. Bloom.)*

year. Winter in the Northern Hemisphere is likely to be coldest when, on the longest day of the year (June 21), Earth happens to be positioned at the point in its elliptical orbit that lies farthest from the sun. Earth is in this position once every 22,000 years—the time required for completion of one precession cycle. It is probably no accident that Earth's 125,000-, 105,000-, and 82,000-year-old high stands of sea level were also separated by intervals of approximately 22,000 years. Thus it appears that precession of the equinoxes imposes small-scale cycles on larger (92,500-year) cycles that have exerted such profound control over glaciation during the past 800,000 years.

Climatic cycles, which are reflected in glacial movements, have also occurred on smaller scales. Between about A.D. 1500 and 1850, for example, the world experienced a minor episode of refrigeration known as the Little Ice Age, which reached its peak about 1700. During this period such areas as Scandinavia and New England suffered bitter winters and short summers that caused numerous major crop failures. George Washington and his troops at Valley Forge were actually fortunate, in 1777 and 1778, to have experienced a winter that was relatively mild for the times. Studies of glacial moraines show that over the last 10,000 years (since the last major glacial interval) the world has experienced three other cold intervals comparable to the Little Ice Age. The warm interval preceding the first of these cold periods peaked about 7000 years ago; global temperatures were warmer then than they have been at any time since.

On a still smaller scale, numerous cycles are evident in climatic records of the last hundred years or so (Figure 16-16). Beginning around the turn of the century, the average annual temperature at Earth's surface showed a rising trend, reaching a maximum rate of increase in 1938. Since that time the rate of change has followed a zigzag course downward, and the average annual

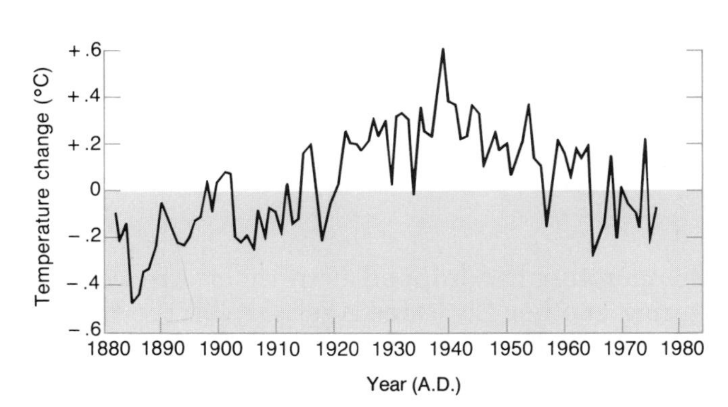

FIGURE 16-16 Changes in average annual temperature over the past 100 years. *(After J. M. Mitchell, in Energy and Climate, National Academy of Sciences, Washington, D.C.)*

Box 16-1
Environmental Warnings from the Geologic Record

The segment of the geologic record that chronicles the latter part of Pleistocene time has recently surprised scientists with evidence about rates and patterns of geographic change that had not been predicted earlier. As it turns out, climates have changed very rapidly again and again, and biological communities have been shuffled and reshuffled, rather than persisting as stable assemblages of species.

Fossil pollen collected from lake deposits is an especially useful clue to the past. Burrowing animals are relatively rare in lakes (p. 54), so such deposits provide a nearly continuous pollen record of changes in nearby plant communities. Because particular communities of plant species grow in particular terrestrial habitats today, biologists once tended to assume that these communities existed during the Pleistocene and migrated as units when climates shifted. The detailed record of vegetational changes in North America since the most recent glacial maximum, about 18,000 years ago, paints quite a different picture. Species have not shifted in concert, but have changed their geographic distributions independently. The communities of today are not long-standing associations that have evolved a stable ecologic structure. These associations are momentary; they have existed only a short time and will change markedly in the near future. Consider two familiar kinds of trees, pines and oaks. Today in eastern North America, pines are concentrated in the northern Great Lakes region and central Canada and also far to the south, in Florida and nearby areas. Oaks are generally uncommon in these two regions, but they are abundant in the zone between them. Fossil pollen reveals that, in sharp contrast, pines and oaks thrived in association with each other 18,000 years ago, during the most recent glacial maximum. While glaciers spread across the northeastern United States, trees of both kinds flourished only in a small region that extended northward from Florida. For them and for plants generally, patterns of distribution have changed rapidly and continuously over the past 18,000 years.

Annual bands of ice are visible in this glacier in the Andes of South America. *(Lonnie G. Thompson, Byrd Polar Research Institute, Ohio State University.)*

Other kinds of data provide a precise chronology of environmental change. Where climatic conditions limit growth rates of plants, annual rings in the trunks of very old trees provide a year-by-year history of climatic change. Cold conditions produce narrow rings because the growing season is short.

temperature has dropped many times. Are we entering another Little Ice Age? Or will the trend reverse itself so that we will climb further out of the cold period that reached a climax about 1700? Unfortunately, attempts to answer these questions represent little more than conjectures.

One factor that concerns many scientists is the possibility that the burning of wood and fossil fuels for energy, a process that adds carbon dioxide to the atmosphere, is causing Earth's climate to warm up considerably as a result of the greenhouse effect (p. 31). If Earth's temperature were to rise by only a few degrees, the partial melting of glaciers would raise the oceans so high that the major port cities of the world would be flooded. It is possible that human activities are already producing significant greenhouse warming. The sudden nature of climatic changes during the Pleisto-

Radiocarbon dating gives approximate ages even for stumps of long-dead trees, whose rings then document rates of climatic change. Distinct annual layers also extend downward in a glacier to a depth where the pressure of overlying ice obliterates them. The layers of many glaciers are visible to the naked eye because a relatively large amount of dark dust from the atmosphere accumulates with snow during the dry season every year. Where there are no fresh breaks in a glacier to reveal a stratigraphic record of layering, drilling can provide cores for study. The isotopic composition of oxygen in ice layers, like that of oxygen in marine plankton, records the expansion and contraction of a glacier. Annual layers in an ice core from Greenland reveal that the last glacial maximum gave way to warm conditions during the remarkably brief interval of 20 years. Similarly, an ice core from a glacier in the Andes of Peru indicates that the Little Ice Age ended in this region slightly before 1880, during an interval of only about 3 years.

These revelations alert us to the possibility that human-induced global warming may quickly push the world's climates across a critical threshold. If ice shelves in Antarctica suddenly melted, for instance, sea level would rise rapidly everywhere. At the same time, sudden changes in patterns of precipitation will affect agriculture and water supplies. Today we can predict neither the precise environmental changes that will take place when the world's temperature rises nor the complex effects that various possible patterns of environmental change would have on life; but more detailed study of the environments and life of the recent past should improve our predictions.

cene and the complex impact of these changes on biotas gives us cause for alarm as we contemplate our potential to alter Earth's climate (Box 16-1).

We have been discussing how humans may be influencing the climate without first having discussed where our species came from to begin with. After reviewing regional events of Neogene time, we will step back and review what is known about human origins.

REGIONAL EVENTS

We will begin our regional tour of the Neogene Period by reviewing the history of the western United States, which is highlighted not only by the elevation of imposing mountains that form part of our scenery today—the Cascade Range, the Sierra Nevada, and the Rocky Mountains—but also by climatic changes that resulted from the uplifting of these mountains. Next we will examine events in and around the Atlantic Ocean; among our topics here will be the origin of the modern mountainous topography of the Appalachians, the uplift of the Isthmus of Panama, the origin of the Caribbean Sea, and the cooling of the Atlantic with the onset of the Ice Age. We will then review the Neogene history of the African continent and the famous African rift valleys, which have contributed valuable fossil remains of human ancestors.

Development of the American West

The pre-Neogene history of mountain building in the Cordilleran region, described in earlier chapters, is summarized in Figure 15-26. By late Paleogene time, uplifts resulting from the final mountain-building episode of the western interior, the Laramide orogeny, had been largely subdued by erosion, which set the stage for the Neogene events that produced the Rocky Mountains. In the broad region west of the Rockies, the Neogene Period was a time of widespread tectonic and igneous activity, which built most of the mountains standing there today.

Provinces of the American West Lying between the Great Plains and the Pacific Ocean are several distinctive physiographic provinces that have taken shape largely in Neogene time, primarily as a result of uplift and igneous activity. Let us briefly review the present characteristics of these provinces before considering how they have come into being (Figure 16-17).

The lofty, rugged peaks of the Rocky Mountains, some of which stand more than 4.5 kilometers (~14,000 feet) above sea level, could only be of geologically recent origin. We have seen that the widespread subsummit surface of the Rockies was all that remained of the Laramide uplifts by the end of the Eocene Epoch about 40 million

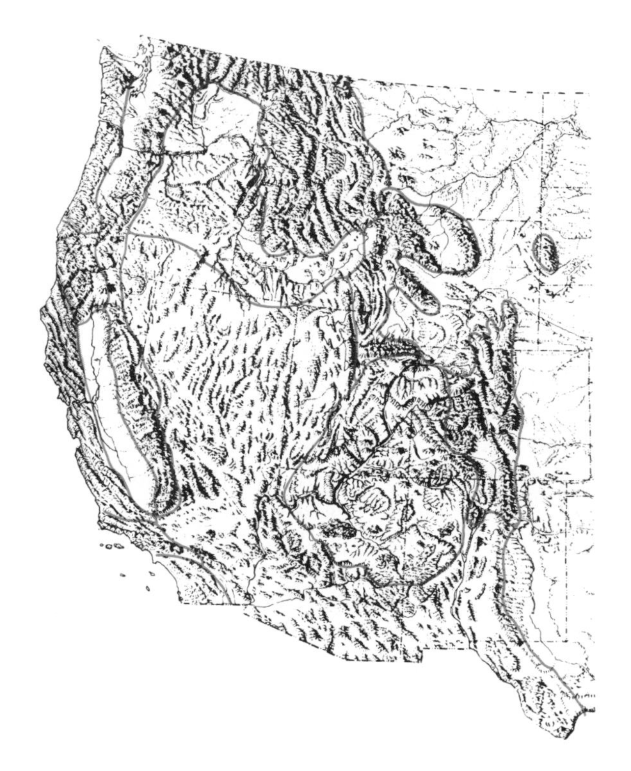

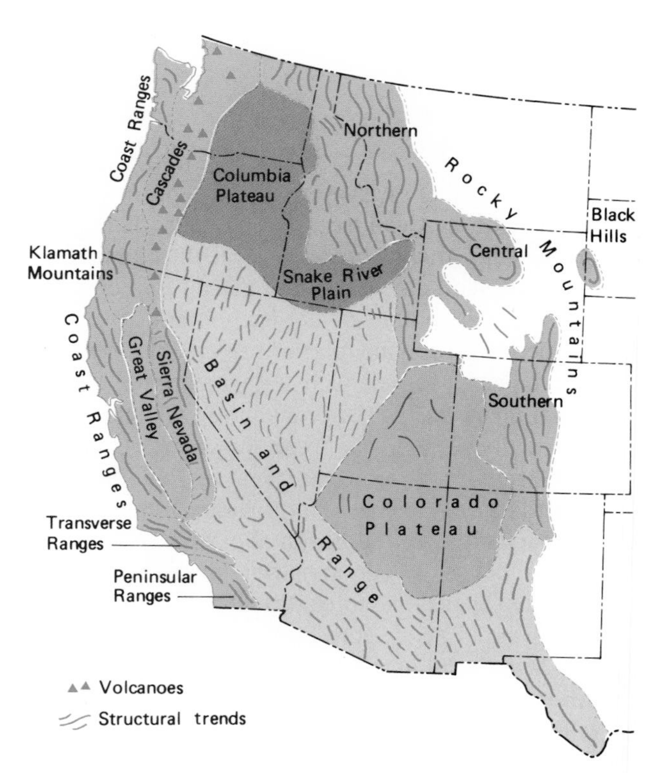

FIGURE 16-17 Major geologic provinces of western North America. The map at the left shows the relations of the provinces to topographic features. *(Topographic map based on U.S. Geological Survey, National Atlas of the United States of America.)*

years ago (Figure 15-33). One question we must answer, then, is how the Rocky Mountain region became mountainous once again.

Centered adjacent to the Rockies in the "four corners" area where Colorado, Utah, New Mexico, and Arizona meet is the oval-shaped Colorado Plateau, much of which stands about 1.5 kilometers (~1 mile) above sea level. The Phanerozoic sedimentary units here are not intensively deformed. Some, however, are gently folded in a steplike pattern, and others, especially to the west, are offset by block faults (Figure 16-18). Cutting through the plateau is the spectacular Grand Canyon of the Colorado River (see Figure AII-14), but about 10 million years ago neither the Colorado Plateau nor the Grand Canyon existed.

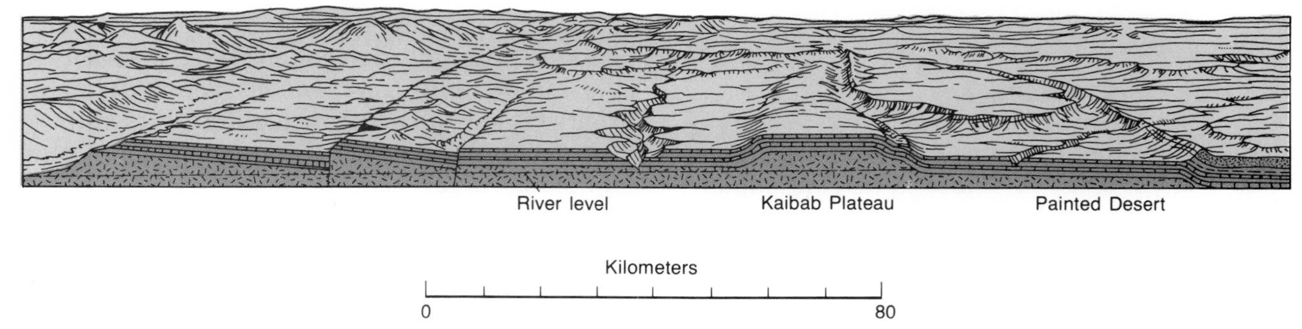

FIGURE 16-18 Block diagram of the western part of the Colorado Plateau north of the Grand Canyon. This high-standing region is characterized by block faulting *(left)* and gentle, steplike folds *(right)*. *(After P. B. King, The Evolution of North America, Princeton University Press, Princeton, N.J., 1977.)*

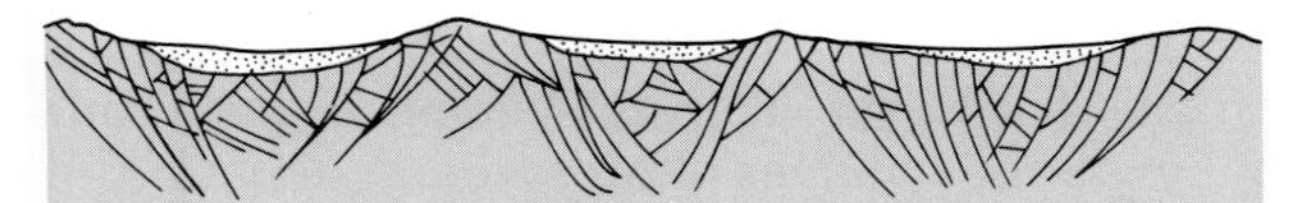

FIGURE 16-19 The possible pattern of the block faulting in the Basin and Range Province that might have been responsible for lateral extension of the crust.

The origin of these features forms another part of our story.

West of the Rockies and the Colorado Plateau, within the belt of Mesozoic orogeny, lies the Basin and Range Province (Figure 16-17). This is an area of north-south-trending block-fault valleys and intervening ridges (Figure 16-19)—features of Neogene origin. A large area of this province forms the Great Basin, an arid region of interior drainage (p. 56). Volcanism has been associated with some faulting episodes, and sediment eroded from the ranges blankets the valleys to depths ranging from a few hundred meters to about 3 kilometers (~2 miles). The thickness of Earth's crust in the Basin and Range Province ranges from about 20 to 30 kilometers, in contrast to thicknesses of 35 to 50 kilometers in the Colorado Plateau. The thinning and block faulting in the Basin and Range Province point to extension of the crust by at least 65 percent and perhaps by as much as 100 percent.

Farther north, centered in Oregon, is a broad area covered by volcanic rocks of the Columbia River and Snake River plateaus (see Figure AI-8). Today the climate here is cool and semiarid; only about one-quarter of the plateau area is cloaked in forest and woodland, and sagebrush and drier conditions characterize about half of the terrain. In Oligocene time, however, lavas had not yet blanketed the region, and, as remains of fossil plants reveal, a large forest of redwood trees grew there.

Along the western margin of the Columbia Plateau stand the lofty peaks of the Cascade Range (Figure 16-20). These are cone-shaped volcanoes that represent the volcanic arc associated with subduction of the Pacific plate along the western margin of the continent. Volcanism began here in Oligocene time and continues to the present, as the recent eruptions of Mount St. Helens attest (Figure 1-1).

The Cascade volcanic belt passes southward into the Sierra Nevada Range, a mountain-sized fault block of granitic rocks. The plutons forming the Sierra Nevada were emplaced in east-central California during Mesozoic time, before igneous activity at this latitude shifted inland. As we will see, however, the present topography of the Sierra Nevada is of Neogene origin. This mountain range is unusual in that, throughout its length of some 600 kilometers (~350 miles), it is not

FIGURE 16-20 Mount Rainier above Mount Stewart, two of the high peaks of the Cascade Range in Oregon. *(Art Wolfe.)*

breached by a single river. This is why it represented such a formidable obstacle to early pioneers attempting to reach the Pacific.

The Sierra Nevada Range stands between the Basin and Range Province to the east and the Great Valley of California to the west (Figure 16-21). The Great Valley is an elongate basin containing large volumes of Mesozoic sediment (the Great Valley Sequence; see Figure 13-35) eroded from the plutons of the Sierra Nevada region long before the modern Sierra Nevada formed by block faulting (Figure 13-34). Resting on top of these sediments are Cenozoic deposits, some of which accumulated during marine invasions of the Great Valley and others during times of nonmarine sedimentation.

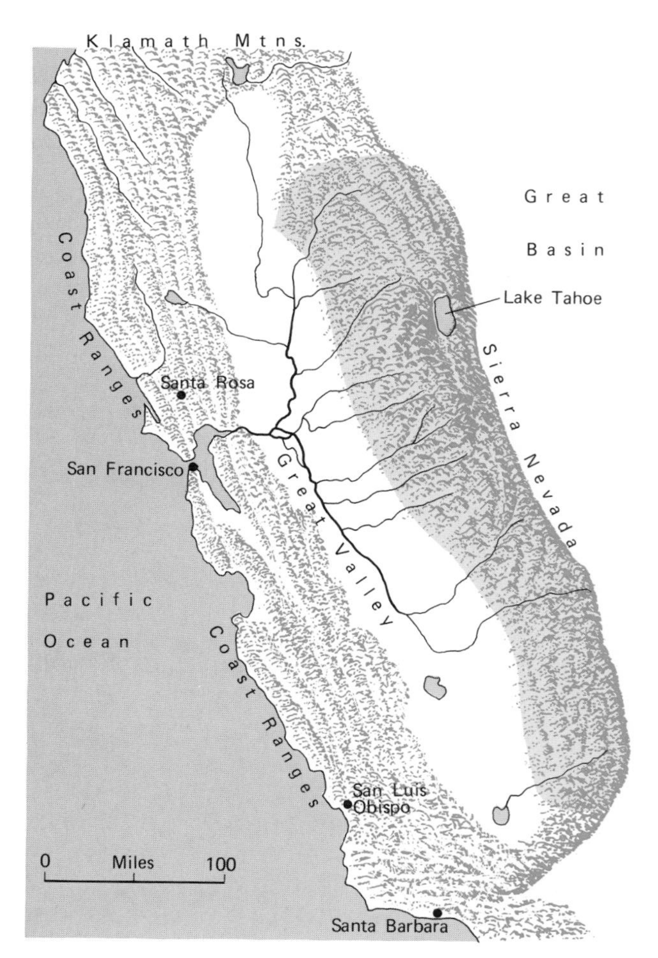

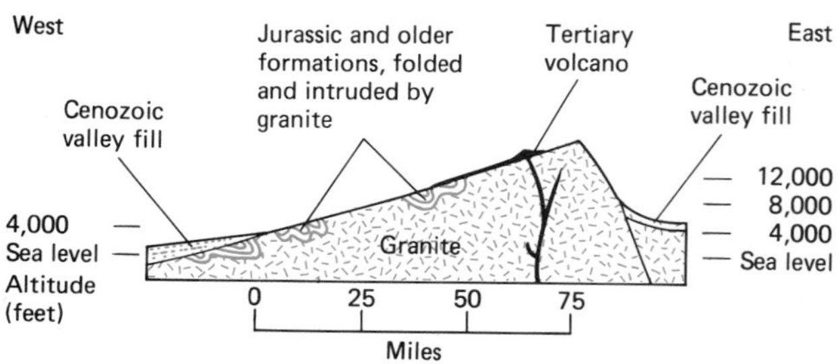

FIGURE 16-21 The Sierra Nevada fault block of California and Nevada. Sediments of the Great Valley lap onto the gentle western slope of the Sierra Nevada. *(After C. B. Hunt, Natural Regions of the United States and Canada, W. H. Freeman and Company, New York, 1974.)*

West of the Great Valley are the California Coast Ranges, which consist of slices of crust that include crystalline rocks representing Mesozoic orogenic activity, Franciscan rocks of deep-water origin (Figure 16-22), and Tertiary rocks. To the south, the Transverse and Peninsular ranges are formed of similarly faulted and deformed rocks, but these ranges lie inland of the main belt of Franciscan rocks in the region of intensive Mesozoic igneous activity. Striking features of all of these mountainous terrains are the great faults that divide the crust into sliver-shaped blocks. The longest and most famous of these faults is the San Andreas, which extends for about 1600 kilometers (1000 miles). Until the great San Francisco earthquake of 1906, it was not widely recognized that the San Andreas was still active. The earthquake of 1906 was produced by a sudden horizontal movement of up to 5 meters (~ 16 feet) along the fault. Geologic features cut by the San Andreas Fault show that its total movement during the past 15 million years has amounted to about 315 kilometers (190 miles). Continued movement at this rate for the next 30 million years or so would bring Los Angeles northward to the latitude of San Francisco—through which the fault passes. As we will see, the faulting and uplifting of the Coastal Ranges of California are probably related not only to the Neogene uplift of the Sierra Nevada but also to the origins of the Basin and Range topography to the east.

The Olympic Mountains of Washington have quite a different history. These relatively low mountains, which lie to the west of the Cascade Volcanics, consist of oceanic sediments and volcanics that were deformed primarily during Eocene time in association with subduction along the continental margin (Figure 15-29).

Development of the American West: The Miocene Epoch Geologic features of the far west in

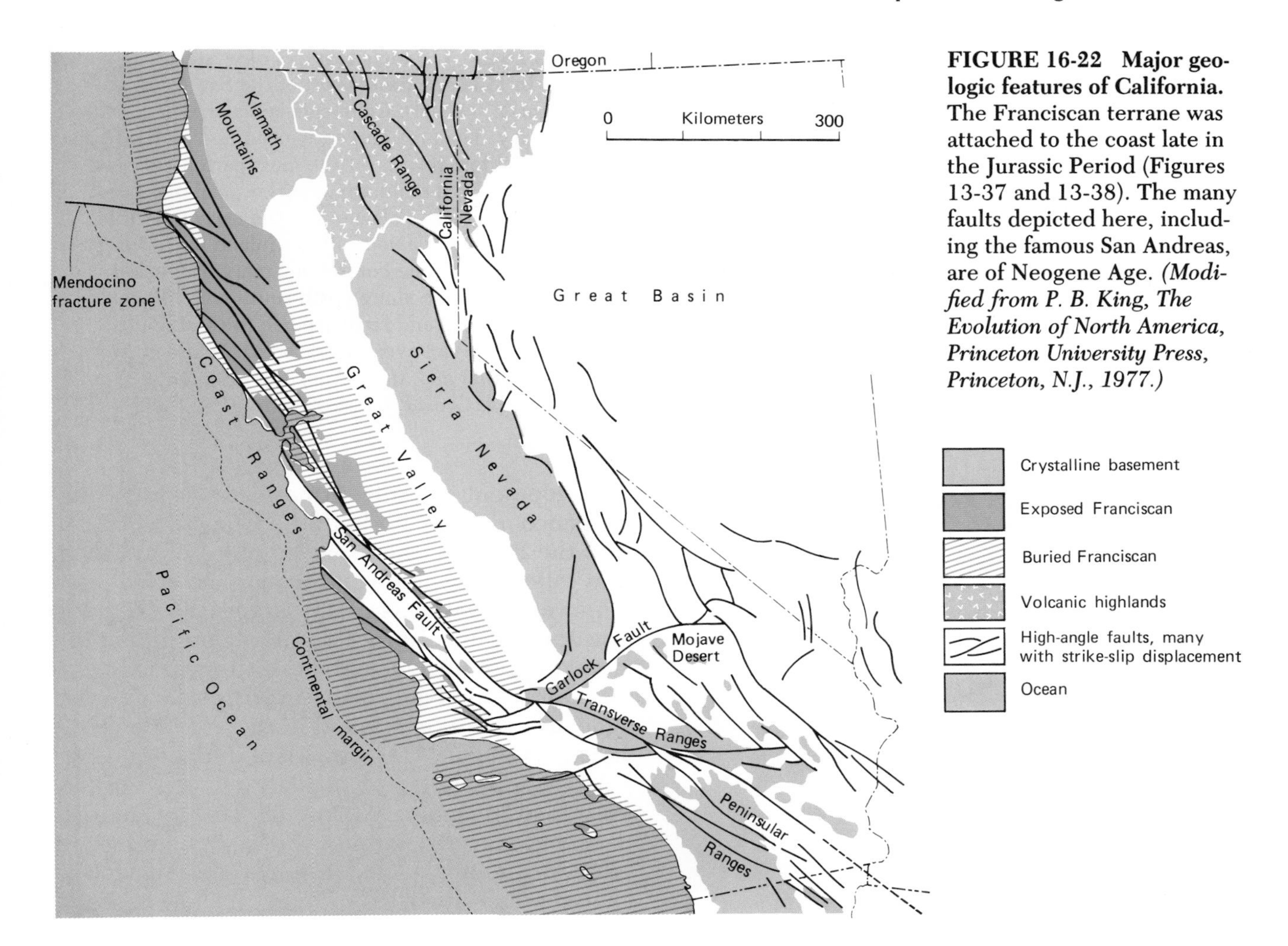

FIGURE 16-22 Major geologic features of California. The Franciscan terrane was attached to the coast late in the Jurassic Period (Figures 13-37 and 13-38). The many faults depicted here, including the famous San Andreas, are of Neogene Age. *(Modified from P. B. King, The Evolution of North America, Princeton University Press, Princeton, N.J., 1977.)*

Miocene time are shown in Figure 16-23. Subduction continued beneath the continental margin in the northwestern United States, and the resulting volcanic arc produced peaks in the Cascade Range, where volcanism continues today. To the south, in California, the mid-Miocene interval was a time of faulting and mountain building; elements of the modern Coast Ranges and other nearby mountains were raised, and the seas were driven westward. Meanwhile, as in Paleogene time, the Great Valley remained a large embayment, and during Miocene time it received great thicknesses of siliciclastic sediments, most of which were shed from the region of the modern Sierra Nevada. Although the block faulting that eventually produced the Sierra Nevada did not begin until Pliocene time, volcanoes of the southern end of the Cascade Island Arc were shedding sediments westward, from the position where the Sierra Nevada later stood.

Before the Sierra Nevada rose appreciably, the Basin and Range Province began forming to the east. Volcanism began during Paleogene time, and the Basin and Range topography began to form near the beginning of the Miocene. To the north of the Basin and Range Province, great volumes of basalt spread from fissures. Most of the great Columbia Plateau formed by outpourings of lava between about 16 and 13 million years ago (see Figure AI-8); individual basalt flows of this plateau range in thickness from 30 to 150 meters (~100 to 500 feet), and in places the total accumulation reaches about 5 kilometers (~3 miles).

One of the most important Miocene events in the Cordilleran region was a broad regional uplift that affected the Basin and Range Province, the Colorado Plateau, and the Rocky Mountains. Today even the basins of the Basin and Range Province nearly all stand at least 1.3 kilometers (~$\frac{3}{4}$ mile) above sea level. Thus it is remarkable

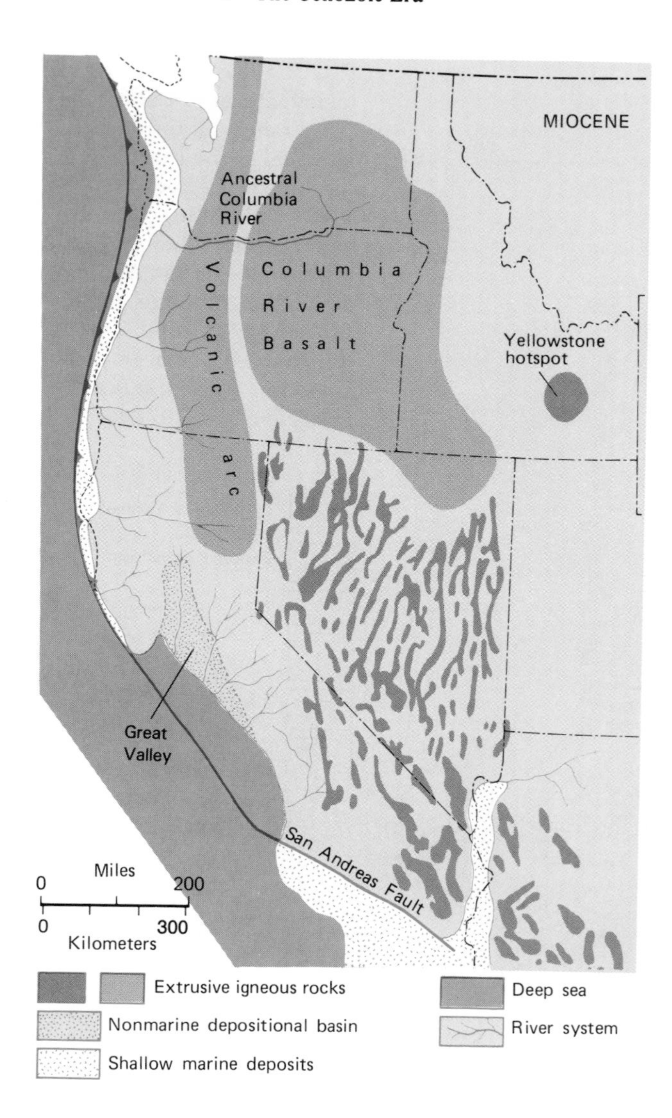

FIGURE 16-23 Geologic features of western North America in Miocene time. West of the San Andreas Fault, coastal southern California lay farther south than it does today. The area currently occupied by the Great Valley of California was, for the most part, a deep-water basin from which a nonmarine depositional basin extended to the north. Volcanoes of the early Cascade Range formed along a volcanic arc inland from the subduction zone along the continental margin. Igneous rocks were extruded along north-south-trending faults in the Great Basin, and farther north the Columbia River Basalts spread over a large area. *(Modified from J. M. Armentrout and M. R. Cole, Soc. Econ. Paleont. and Mineral. Pacific Coast Paleogeog. Symp. 3:297–323, 1979.)*

that mid-Cenozoic fossil floras of this region are characterized by species that could have lived only at low altitudes. Geologists have reconstructed even more precise histories of uplift for the Colorado Plateau and Rocky Mountains by studying the time at which rivers have cut through well-dated volcanic rocks. Many of the rivers of these regions existed before uplift began in the Miocene Epoch, and they cut rapidly downward as the land rose, producing deep gorges. A large part of the Grand Canyon, for example, was incised during the rapid elevation of the Colorado Plateau between about 10 and 8 million years ago.

Uplift in the Rockies began slightly earlier, in Early Miocene time, and terrain that now forms the southern Rockies has since risen between 1.5 and 3.0 kilometers (~1 to 2 miles).

Development of the American West: Plio-Pleistocene Time In both the Colorado Plateau and the Rockies, the Miocene pulse of uplifting was followed by a lull and then by an episode of renewed elevation; in fact, large-scale uplifting was the dominant process in the Cordilleran region during Pliocene and Pleistocene time.

During the Pliocene and Pleistocene epochs, igneous activity continued in the volcanic provinces of Oregon, Washington, and Idaho (Figure 16-24). Many of the scenic volcanic peaks of the Cascades, including Mount St. Helens (Figure 1-1), have formed within the past 2 million years or so. Beginning in late Miocene time and continuing sporadically to the present, the flow of basalt from fissures has produced the Snake River Plain, which amounts to an eastward extension of the Columbia Plateau (Figure 16-24).

Faulting and deformation continued in California during the Pliocene and Pleistocene epochs. Since the beginning of the Pliocene Epoch, about 5 million years ago, the sliver of coastal California that includes Los Angeles has moved northward on the order of 100 kilometers (~60 miles). The Great Valley has, of course, remained a lowland to the present day, but during Pliocene and Pleistocene time it became transformed from a marine basin into a terrestrial one. Early in the Pliocene Epoch, seas flooded the basin from both the north and the south, but as the epoch progressed, uplift associated with movement along the San Andreas Fault eliminated the southern connection. Eventually nonmarine deposition prevailed throughout the Great Valley, which is now one of the world's richest agricultural areas as well as a site of large reservoirs of petroleum.

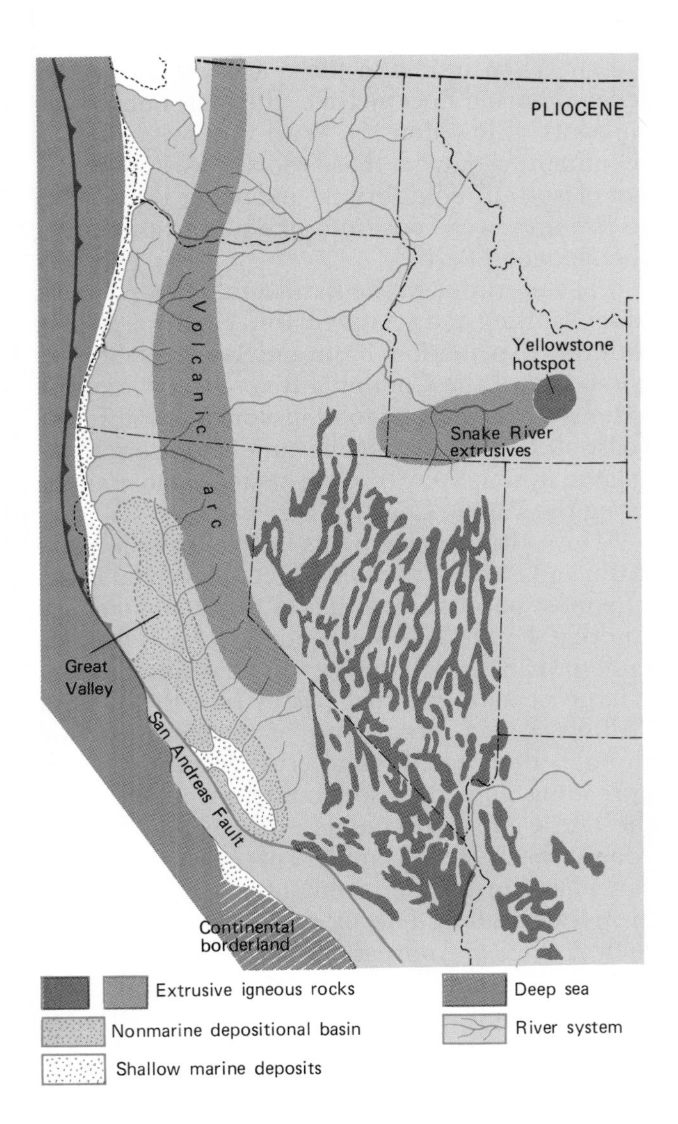

FIGURE 16-24 Geologic features of western North America in Pliocene time. The Great Valley of California was a shallow basin that received nonmarine sediments except where a shallow sea flooded its southern portion. The volcanic arc continued to form volcanoes of the Cascade Range, and igneous rocks continued to be extruded along faults in the Great Basin. The Snake River extrusives spread over a large area of southern Idaho west of Yellowstone. *(Modified from J. M. Armentrout and M. R. Cole, Soc. Econ. Paleont. and Mineral. Pacific Coast Paleogeog. Symp. 3:297–323, 1979.)*

Meanwhile, areas to the east of the Great Valley underwent major uplift. The Sierra Nevada had experienced considerable tilting during Miocene time, but fossil plants of Miocene age, preserved on the crest of the Sierra Nevada, were still types that could not have lived as much as a kilometer above sea level. Thus it was not until Pliocene time that the Sierra Nevada (Figure 16-25) became elevated to its present height, which exceeds 4.3 kilometers (~14,000 feet). The consequences of this uplift were enormous for the Basin and Range Province to the east, where uplift continued from Miocene time on a smaller scale. Sitting in the rain shadow of the Sierra Nevada, this area, which in Miocene time had been covered by evergreen forests, came to be carpeted by savannah vegetation. This trend toward increasing aridity, which was compounded by the global trend toward drier climates, continued into very late Neogene time, when the Great Basin became a desert.

FIGURE 16-25 The eastern face of the Sierra Nevada. This is a fault scarp that is partly dissected by youthful valleys. The view is from the Owens Valley, Inyo County, California. *(W. C. Mendenhall, U.S. Geological Survey.)*

The Colorado Plateau and Rocky Mountains, where uplift had slowed in Late Miocene time, experienced rapid elevation once again. Streams that had been established millions of years earlier cut rapidly downward as the uplift proceeded. The Rockies attained most of their modern elevation during Pliocene time, and the result was the origin of deep canyons that now carry streams through tall mountain ranges (Figure 12-34). The Colorado Plateau also rose again during Pliocene and Pleistocene time, and the Colorado River responded by cutting swiftly downward. In fact, it appears that much of the Grand Canyon of the Colorado (see Figure AII-14) formed during just the past 2 or 3 million years.

The renewed elevation of the Rocky Mountains left the Great Plains to the east in a partial rain shadow. Sediments derived from the rejuvenated Rockies spread eastward, forming the Pliocene Ogallala Formation. Caliche nodules are abundant in many parts of the Ogallala, indicating the presence of seasonally arid climates (p. 52). The Ogallala is a thin, largely sandy unit that lies buried under the Great Plains from Wyoming to Texas, and it serves as a major source of groundwater. Unfortunately, this is ancient water that is not being renewed as rapidly as it is drawn from Earth. As a result, severe water shortages may one day strike many areas of the central United States.

It is important to recognize that the rain-shadow effects of both the Sierra Nevada and the Rockies were superimposed on the larger global trend toward drier, cooler climates that has characterized the post-Eocene world. Then, with the onset of the Pleistocene Epoch, frigid conditions brought glaciation to mountainous regions of the western United States just as they foster glaciation in Alaskan mountains today (Figure 2-15). The Sierra Nevada, for example, was heavily glaciated, as were portions of the Rocky Mountains (Figure 16-13). Today broad U-shaped valleys in both mountain systems testify to the scouring activity of Pleistocene glaciers (Figure 9-23).

Possible Mechanisms of Uplift and Igneous Activity in the American West What has led to the many tectonic and igneous events of Neogene time in the American West? It seems likely that the secondary uplift of the Colorado Plateau and the Rocky Mountains, which took place long after the Laramide orogeny, may to a large extent represent simple isostatic adjustment (Figure 1-17). When uplifts in these areas were largely leveled during Eocene time, they left behind felsic roots of low density. With the weight of the mountains removed, these roots were apparently out of isostatic equilibrium, and hence they began to rise up toward positions of equilibrium during the Neogene Period.

The elevation of the Basin and Range, with its block faulting and relatively thin crust, requires a different explanation. Basin and Range events, the spreading of the Columbia River Basalt, and the extensive faulting and folding along the California coast all began in Miocene time and seem to be related in some way to plate-tectonic movements along the Pacific Coast.

How these movements have produced the Basin and Range Province is a controversial issue. The most popular idea relates to the famous San Andreas Fault. The San Andreas is a transform fault (p. 128) associated with the East Pacific Rise, a large oceanic rift that passes into the Gulf of California and breaks up as it passes inland through thick continental crust (Figure 16-26). Spreading along the rise should have ceased when the rise came into contact with the subduction zone at the western boundary of the North American plate; instead, movement must have been propagated along one or more transform faults such as the San Andreas, passing along the continental margin. Crustal shearing adjacent to a strike-slip fault such as the San Andreas will automatically cause extensional faulting similar to that of the Great Basin. This hypothesis is deficient in one regard: It fails to account for the broad elevation of the Basin and Range Province during Neogene time.

We are more certain about the general pattern of tectonism along the Pacific Coast. North America encountered the Pacific plate near the beginning of the Miocene Epoch. Movements along the San Andreas and other faults that have formed since that time account for the complex slivering and deformation in the Coast Ranges and neighboring areas (Figure 16-22).

The Western Atlantic Ocean and Its Environs

Although the margins of the Atlantic Ocean were relatively quiescent during the Neogene Period, they did experience mild vertical tectonic move-

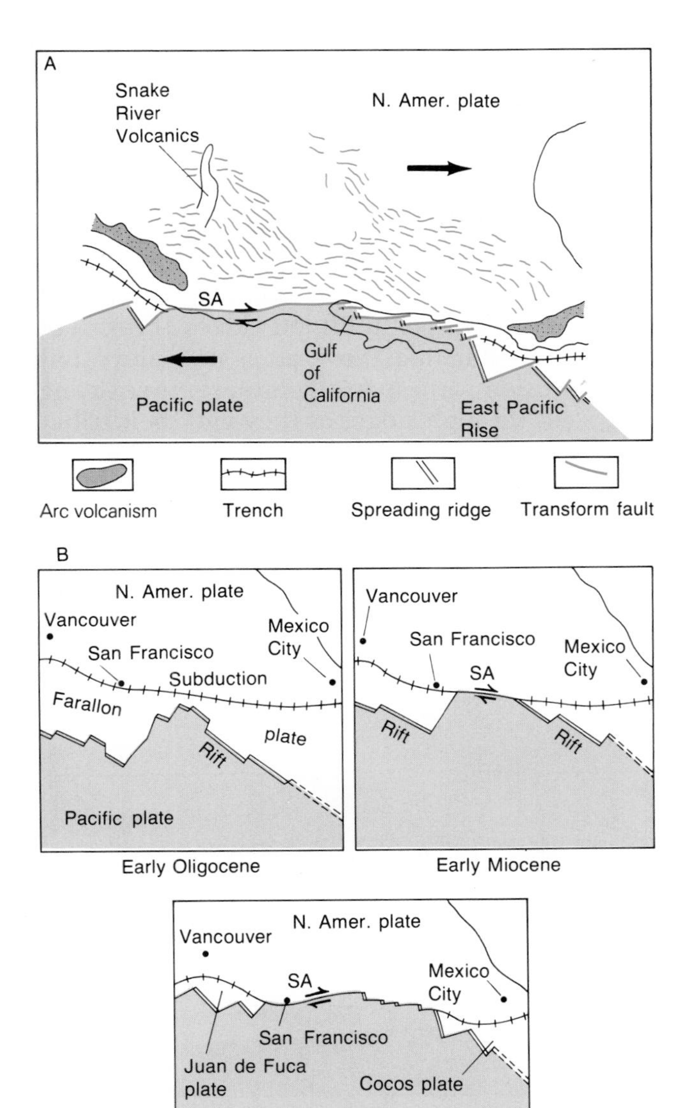

FIGURE 16-26 Plate-tectonic features that may account for the Basin and Range structure in western North America. *A.* The present situation, with the spreading ridge known as the East Pacific Rise passing into the Gulf of California. The East Pacific Rise abuts against the North American continent and is offset westward along the San Andreas Fault (SA); it appears that shearing forces resulting from relative movement of terrain on either side of the San Andreas Fault *(heavy arrows)* pull the crust apart, producing the north-south-trending faults of the Basin and Range Province. *B.* The second alternative. The rift zone between the Pacific and Farallon plates encountered the thick crust of North America along the subduction zone that bordered the continent. Unable to pass inland, the rift was divided along a strike-slip fault (the San Andreas). *(Based on J. H. Stewart, Geol. Soc. Amer. Mem. 152:1–31, 1975. After Atwater.)*

ments—and these movements, together with more profound changes in sea level, had major effects on water depths and shoreline positions.

Global sea level has never stood as high during the Neogene Period as it did during much of Cretaceous or Paleogene time. For this reason, Neogene marine sediments along the Atlantic Ocean stand above sea level in only a few low-lying areas. Among the most impressive of the Miocene deposits found here are those of the Chesapeake Group, which form cliffs along the Chesapeake Bay in Maryland. The Chesapeake Group accumulated in the Salisbury Embayment during a worldwide high stand of sea level between about 16 and 14 million years ago. The Salisbury Embayment is one of several downwarps of the American continental margin (Figure 16-27). Inhabiting the waters of the Salisbury Embayment was a rich fauna that included many large vertebrates, especially whales, dolphins, and sharks (Figure 16-1). Most of the fossils of baleen whales represent juvenile animals, which suggests that the embayment was a calving ground. Perhaps

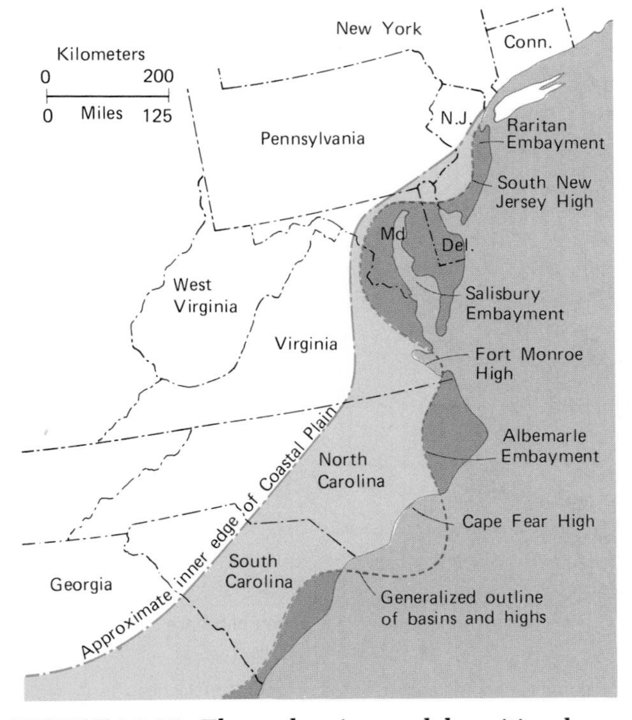

FIGURE 16-27 Elevated regions and depositional embayments along the mid-Atlantic coast during the Miocene Epoch. *(After J. P. Menard et al., Geol. Soc. Amer. Northeast-Southeast Sections, Field Trip Guidebook 7a, 1976.)*

sharks were numerous because the young whales were especially vulnerable prey. Land-mammal bones are also found here and there in the Chesapeake Group, indicating that the waters of the embayment were shallow. Pollen from nearby land plants settled in the Salisbury Embayment, leaving a fossil record that shows a warm temperate flora near the base of the Chesapeake Group slowly giving way upward in the sedimentary sequence to a flora adapted to slightly cooler conditions.

The Chesapeake Group and other buried deposits of earlier age to the south consist primarily of siliciclastic sediments shed from the Appalachians to the west. Erosional features associated with the Appalachians reveal that these ancient mountains have a complex history. Like the modern Rockies, the existing topographic mountains that we call the Appalachians are the product of secondary uplift. The Appalachian orogenic belt was largely leveled by erosion long before the end of the Mesozoic Era. In many regions there were three or more additional intervals of uplift and erosion during the Cenozoic Era. The erosion that followed intervals of uplift left ridges of resistant folded rock standing above elongate valleys, but when erosion was especially intense, preexisting rivers cut through ridges as they and their tributaries carved out the valleys (Figure 16-28).

The timing of Mesozoic uplift in the Appalachian region appears to have varied from place to place but remains poorly dated. It does appear, however, that a widespread interval of intense uplift in Middle Miocene time was responsible for produced the modern topography, characterized by ridges of resistant rocks, such as sandstone, and valleys of easily eroded rock, including shale.

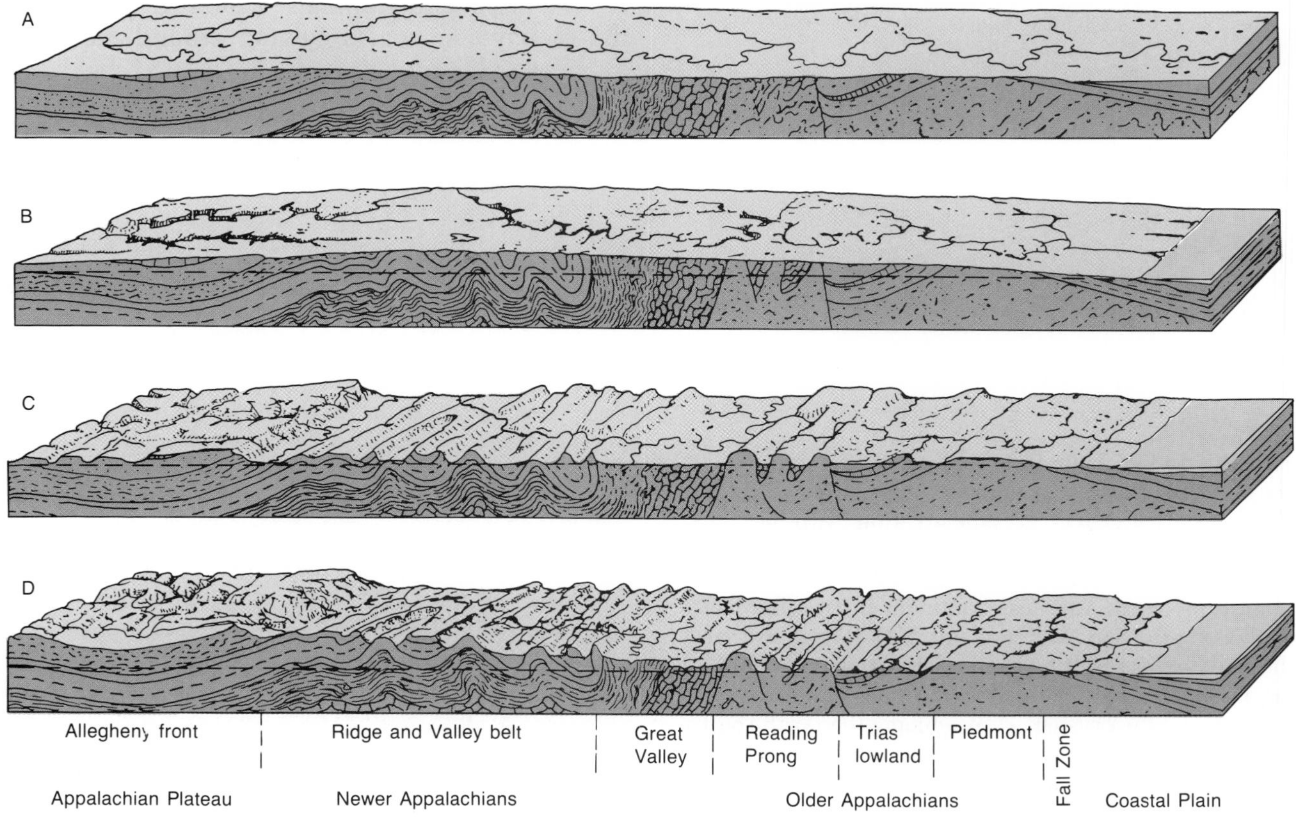

FIGURE 16-28 The way episodic uplift has rejuvenated topography in the Valley and Ridge Province of the Appalachians. *A* depicts the modest amount of relief that characterized some regions early in the Cretaceous Period. Later intervals of uplift and erosion *(B–D)* have produced the modern topography, characterized by ridges of resistant rocks, such as sandstone, and valleys of easily eroded rock, including shale. *(After D. W. Johnson, Stream Sculpture on the Atlantic Slope, Columbia University Press, New York, 1931.)*

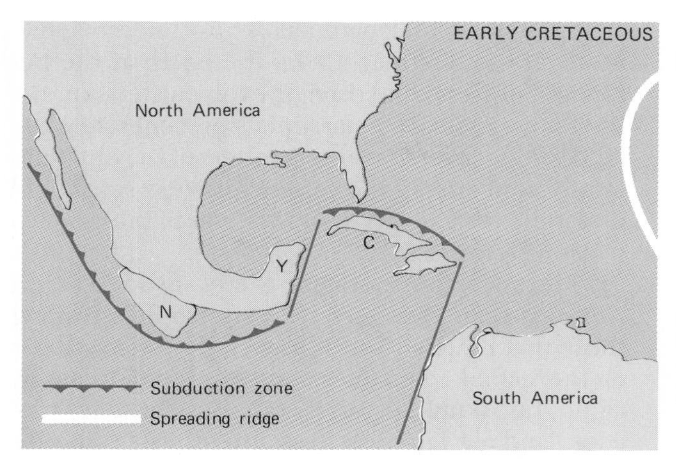

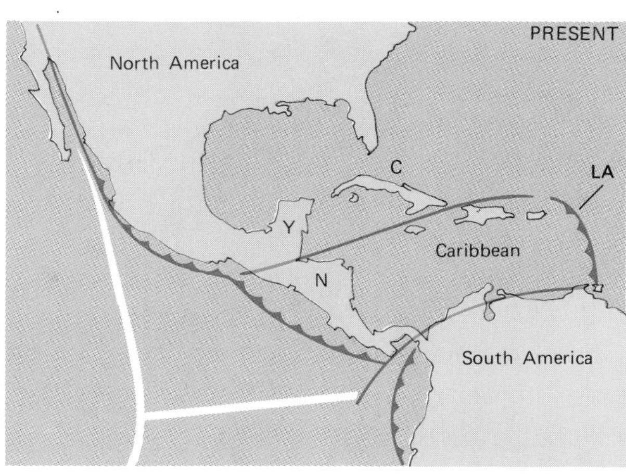

FIGURE 16-29 Tectonic development of the Caribbean Sea as a segment of Pacific oceanic crust that has overridden Atlantic crust along a subduction zone that now extends northward from South America east of the Lesser Antilles. N = Nicaragua; Y = Yucatan; C = Cuba; LA = Lesser Antilles. *(After G. W. Moore and L. Del Castillo, Geol. Soc. Amer. Bull. 85:607–618, 1974.)*

the erosion that produced the unusually thick deposits of the Chesapeake Group in the Salisbury Embayment (Figure 16-27).

Birth of the Caribbean Sea

Although the Caribbean Sea is now an embayment of the Atlantic Ocean, it was once connected to the Pacific (Figure 16-29). During the Cretaceous Period, the floor of the Caribbean, which consists of oceanic rocks, was a small segment of the Pacific plate that was pushing toward the Atlantic, but during the Cenozoic Era, the Caribbean seafloor has lain along the north coast of South America while the Atlantic plate has been subducted beneath it. The Caribbean plate became a discrete entity late in Cenozoic time, when a new subduction zone came to connect the subduction zone bordering North America with the one bordering South America. The Greater Antilles—Cuba, Puerto Rico, Jamaica, and Hispaniola—represent an ancient mountain belt that is actually the southern end of the North American Cordillera. The Lesser Antilles represent an island arc west of the subduction zone, together with islands formed by deformation associated with subduction. The Yucatan Peninsula is a broad carbonate platform that lies to the west of the Caribbean, and, as we have seen, the Bahamas are an ancient carbonate platform lying to the north (Figure 3-29).

The Great American Interchange of Mammals

During the Pliocene Epoch the waters of the Caribbean became separated from those of the Pacific when tectonic activity formed the narrow Isthmus of Panama. Deep-sea cores reveal that different species of planktonic foraminifera lived on opposite sides of the isthmus about 3.5 million years ago, suggesting that the isthmus was forming by this time.

The most profound effect of the newly formed isthmus lay in its role as a land bridge that allowed mammals to migrate between North and South America. Before this time, although a few species had passed from one continent to the other early in the Neogene Period—perhaps by swimming or floating on logs—the terrestrial faunas of North and South America had remained largely separate from each other. South America had been a great island continent and, like Australia, was populated by many marsupial mammals. The marsupials of Australia and South America share common ancestors that populated these continents as well as Antarctica when all three were part of a single Mesozoic landmass; in fact, Eocene mammal faunas resembling those of South America occur in Antarctica. By Pliocene time, however, when the isthmus developed, South American marsupials differed greatly from Australian marsupials, and the South American fauna included several groups of placental mammals whose ancestors had reached the continent from the north at the beginning of Cenozoic time or even earlier. Among the South American marsupials present when the land bridge formed were members of the opossum family, and among the placentals were sloths and armadillos that dwarf their relatives in the modern world (Figure 5-5).

More North American species invaded South America than vice versa (Figure 16-30). Among those that reached South America were members of the camel, pig, deer, horse, elephant, tapir, rhino, rat, skunk, squirrel, rabbit, bear, dog, raccoon, and cat families. Migrating in the opposite direction were monkeys, anteaters, armadillos, porcupines, opossums, and other, less familiar animals.

The formation of the Isthmus of Panama was associated with a Plio-Pleistocene pulse of orogeny that elevated the Andean mountain ranges to

FIGURE 16-30 Animals that took part in the great faunal interchange between North and South America when the Isthmus of Panama was elevated, connecting the continents. The animals shown in North and Central America are immigrants from the south; they include armadillos, sloths, porcupines, and opossums. The animals shown in South America are immigrants from the north; among them are rabbits, elephants, deer, camels, and members of the bear, dog, and cat families. More animals migrated southward than northward. *(Drawing by M. Hill Werner, courtesy of L. G. Marshall.)*

the south by 2 to 4 kilometers (~ 1.5 to 2.5 miles). Large areas of South America thus came to lie in the rain shadow of the Andes. Dry conditions intensified during Pleistocene glacial episodes, at which time parts of the great Amazonian rain forest were fragmented into small pockets separated by grassland barriers or by poorly wooded areas. Even today some groups of rain-forest-dwelling birds and lizards remain clustered in the areas where they were concentrated during the Pleistocene Epoch.

Tectonic, Climatic, and Biotic Changes in Africa

On the opposite side of the Atlantic, Africa underwent profound geographic and biotic changes related to Neogene tectonic activity and global climatic change. The African landmass remained unusually stable for hundreds of millions of years, in part because most of its area lay well within Gondwanaland for a long time. Currently, however, the strength of the African craton is being tested by continental rifting, and it is not certain whether this great continent will survive or be torn asunder.

Near the beginning of Miocene time, two large domes, the Ethiopian Dome and the Kenyan Dome, began to rise up in eastern Africa—a process that was followed by volcanism and crustal rifting (Figure 16-31). The Ethiopian Dome in the north was the site of the large three-pronged rift (triple junction) that split the Arabian Peninsula from the rest of Africa, allowing the Red Sea and the Gulf of Aden to open in Early Miocene time (p. 150). The southern prong was crossed by the western rift, which eventually propagated all the way to the southwestern margin of Africa. Since early in the Miocene Epoch, the ever-deepening valleys of this rift system have received large quantities of sediment and have also harbored some large lakes.

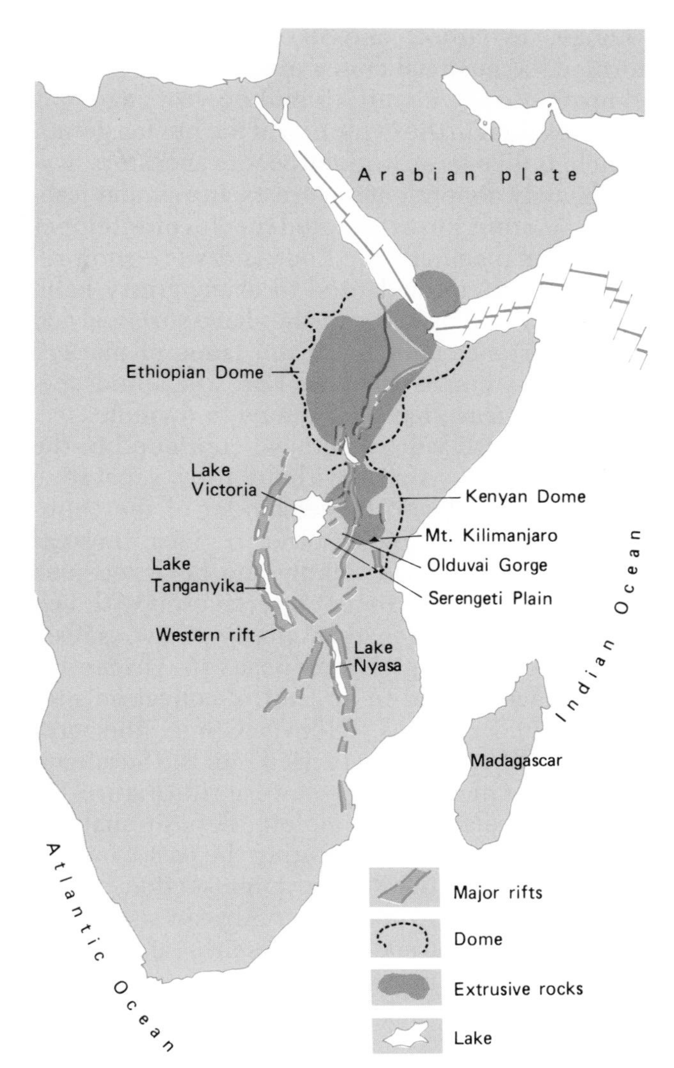

FIGURE 16-31 Features associated with rifting in eastern Africa. The Arabian plate is splitting off from Africa along two prongs of a three-pronged rift centered in an area of crustal doming, part of which forms the Ethiopian Dome, where extrusive rocks are widespread. The third prong extends into continental Africa, forming the rift valleys, some of which harbor large lakes. *(After E. R. Oxburgh, in W. W. Bishop [ed.], Geological Background to Fossil Man, Scottish Academic Press, Edinburgh, 1970.)*

Neogene tectonism in Africa has had important biotic consequences that were enhanced by trends toward more arid climatic conditions—as were similar changes in western North America and in South America. The emergence of wind barriers, together with the global trend toward cooler, more arid conditions, caused grassy terrain to replace most forests in eastern Africa—a change analogous to that in western North America. With respect to the moist winds from the South Atlantic, much of eastern Africa now lies in the rain shadow of the Kenyan Dome, east of the western rift. East of the eastern rift, tall volcanic mountains such as scenic Mount Kilimanjaro trap moisture from winds moving inland from the Indian Ocean. In response to this vegetational change, evolution and extinction have transformed the mammal faunas of Africa. The pulse of climatic change about 2.5 million years ago may have resulted in the evolution of the human genus, which, unlike its australopithecine ancestors, was not heavily dependent on forests. In a similar fashion, numerous forest-adapted species of antelopes and other mammals died out and were soon replaced by species adapted to drier, grassy habitats. Many of these new species have survived and are important elements of the fauna of modern African savannahs (p. 24). Forest-adapted species, in contrast, have continued to dwindle.

Rifting and volcanism have continued to the present day in Africa without fully separating eastern Africa from the remainder of the continent. Like rift valleys of other regions, those of Africa have subsided rapidly and have received large volumes of sediment interlayered with volcanic rocks. Gorges that streams have subsequently cut through the deposits in African rift basins have proved to be fruitful collection sites for hominid remains. Olduvai Gorge, the most famous of these sites, is incised into the Serengeti Plain at the margin of the eastern rift (Figure 16-32). The relatively complete depositional sequence at Olduvai representing the past 2 million years has made the stratigraphic section here a standard against which other African sequences are often compared. The presence of volcanic rocks that can be dated by radiometric and fission-track techniques (p. 86) and that can be correlated paleomagnetically further enhances the value of the Olduvai section as a standard for worldwide comparison.

FIGURE 16-32 Olduvai Gorge, a locality in East Africa famous for fossil remains and artifacts of members of the human family. (For its location in present-day Tanzania, see Figure 16-31.) *(R. L. Hay.)*

Destruction of the Tethyan Seaway

The collision of the African plate and southern India with Eurasia during the Cenozoic Era destroyed what remained of the Tethyan Seaway. Today only vestiges of the seaway remain in the form of the isolated Mediterranean, Black, Caspian, and Aral seas.

A Salinity Crisis in the Mediterranean At the end of the Miocene Epoch the Mediterranean Sea underwent spectacular changes. The first strong hint that geologists had of these changes was the discovery in 1961 of pillar-shaped structures in seismic profiles of the Mediterranean seafloor. These structures looked very much like the salt domes of Jurassic age in the Gulf of Mexico (Figure 13-26), but if the strange features were indeed salt domes, the salt could have formed only by evaporation of the Mediterranean. In 1970 the presence of evaporites here was confirmed by drilling that brought up anhydrite in cores of latest Miocene age. The idea that the Mediterranean had somehow turned into a shallow hypersaline basin was confirmed by the discovery of halite (rock salt) near the center of the eastern Mediterranean Basin.

Further evidence that the Mediterranean shrank by evaporation at the end of Miocene time was the discovery of deep valleys filled with Pliocene sediments lying beneath the present beds of such rivers as the Rhone in France, the Po in Italy, and the Nile in Egypt. Rivers such as the Rhone and the Nile were already flowing into the Mediterranean earlier in the Miocene Epoch, and when the waters of the sea fell, the rivers cut deep canyons. In attempting to find solid footing for the Aswan Dam, Soviet geologists discovered a canyon buried beneath the present Nile delta and judged it to rival the modern Grand Canyon of Arizona in size!

Clearly, what happened at the end of the Miocene Epoch was that the single narrow connection between the Mediterranean Sea and the Atlantic Ocean nearly closed—probably as a result of a lowering of sea level in the Atlantic. Rates of evaporation similar to those of the Mediterranean region today would dry up an isolated sea as deep as the Mediterranean in a mere thousand years. During the crisis enough water must have flowed into the Mediterranean from the Atlantic to keep it from drying up altogether.

All of this happened between about 6 million years ago, when the eastern passage to the Atlantic closed, and 5 million years ago, when the Mediterranean basin refilled with deep water. Five-million-year-old deep-water microfossils in sediments on top of evaporites attest to the refilling. Apparently the connection with the Atlantic was enlarged again when the natural barrier at Gibraltar was suddenly breached. It has therefore been suggested that the first Atlantic waters must have been carried into the deep basin by a waterfall that would have dwarfed Niagara Falls.

Events to the East Farther east, the Tethyan Seaway was interrupted during Miocene time by the attachment of the Indian Peninsula to the Eurasian plate (p. 163), where molasse that shed southward from the newly forming Himalayas produced a lengthy and relatively complete fossil record for mammals. The famous Siwalik beds of Pakistan and India, for example, provide a nearly continuous record for the interval from 11 to 1 million years ago (Figure 7-9). The Siwalik beds document the composition of the rich faunas that occupied the spreading savannahs of late Miocene and Pleistocene age (Figure 16-4). The great rivers of eastern Asia, which flow from the Himalayas to the sea, also formed in Miocene time during the uplift of the Himalayan region. Because of the high relief and abundant rainfall of the region, the Indus and Ganges of India and the several large rivers of Indochina contribute huge volumes of sediment to the ocean each year.

HUMAN EVOLUTION

In addition to the single species that now constitutes the human family, the superfamily Hominoidea currently consists of just four species of the ape family — the common chimpanzee, the pigmy chimpanzee, the gorilla, and the orangutan — together with six species of the gibbon family (Figure 1-6). The human family, Hominidae, did not evolve from the modern ape family, Pongidae; instead, the two families have followed independent lines of evolution. Although it is possible that Hominidae and Pongidae evolved independently from a single family of primitive apes, the case is not clear.

Early Apes in Africa and Asia

Although an extensive Plio-Pleistocene fossil record has been uncovered for the Hominidae, very few known fossil remains of any age represent the Pongidae. Furthermore, the fossil record of the superfamily Hominoidea in latest Miocene time (8 to 5 million years ago) is almost entirely barren. Sediments representing this interval in Africa, where apes and human ancestors evolved, are rare and poorly studied.

Farther back in the Miocene series are fossils of two extinct hominoid families of species that we can loosely call apes. Among these early forms must be ancestors of both modern apes and modern humans, but the evolutionary connections are not yet understood. The oldest fossils of these early apes come from African sediments about 20 million years old. Both of these extinct families first spread from Africa to Eurasia about 15 to 16 million years ago. This was just a few million years after Africa, having moved thousands of kilometers northward after the breakup of Gondwanaland, finally collided with Eurasia and allowed the exchange of mammals between the two landmasses. Not only hominoids made their way northward to Eurasia during mid-Miocene time, but also many other previously isolated groups of African mammals, including elephants and giraffes (both of which have since become extinct in Eurasia).

The early apes underwent large adaptive radiations in both Africa and Eurasia. Some species were so large that they must have spent most of their time on the ground, but others were probably arboreal (tree-climbing) animals.

Throughout most of Miocene time the Old World was much more heavily populated with apes than Africa is today. By the end of the Miocene, however, only a single genus of primitive apes seems to have survived. This was the aptly named *Gigantopithecus*, a gorilla-sized creature that lived on into the Pleistocene. After the decline of apes, very close to the boundary between the Miocene and Pliocene epochs, there was an evolutionary event of great significance: the emergence of the earliest hominids from some unknown group of apes. These hominids constitute a distinct subfamily within the Hominidae and are informally termed the australopithecines. They are of special interest to modern humans because we are their only living descendants.

The Australopithecines

The australopithecines include *Australopithecus* and the closely related genus *Paranthropus*. Members of the two genera weighed roughly the same as chimpanzees (the males were much larger than the females) and their brain was only slightly larger than that of a chimp. It appears that, like chimpanzees, they were active tree climbers. When they were on the ground, however, they walked upright like humans rather than on all fours in the manner of apes (Figure 16-33). We will first discuss *Australopithecus* because it was the immediate ancestor of our genus, *Homo*. A bit later we will take a closer look at *Paranthropus*. This genus also evolved from *Australopithecus* and outlived it, surviving alongside *Homo* until about a million years ago, when it finally disappeared.

FIGURE 16-33 Reconstruction of the oldest known species of the human family, *Australopithecus afarensis*. The low forehead, heavy brow, and projecting lower face of this species are all evident here. *(After a reconstruction by J. H. Matternes.)*

Experts recognize two similar species of *Australopithecus*. *Australopithecus africanus* occupied southern Africa and *Australopithecus afarensis*, northeastern Africa. Although these animals did not evolve from modern apes, they were in many respects intermediate in form between apes and humans, resembling one group more in some ways and the other group more in others. The skull was more apelike. In addition to having a relatively small brain, *Australopithecus* resembled apes in having front teeth that were large in relation to the size of their molars (Figure 16-34). The face was also apelike, with heavy bony ridges above the eye sockets and a large, projecting jaw (Figure 16-35). In fact, the bony ridges strengthened the skull for the attachment of heavy muscles that operated the large jaw. These features and the large molars indicate that *Australopithecus* fed largely on coarse food, probably fruits and seedpods. Apparently it did not fashion stone tools, and presumably lacked the intelligence to do so.

Australopithecus was much shorter than an average modern human. Females appear to have averaged slightly more than a meter (~ 3.5 feet) in height, and males were about 30 percent taller (4.5 feet). The famous skeleton called Lucy, found in Ethiopia in 1973, was in life a female of the

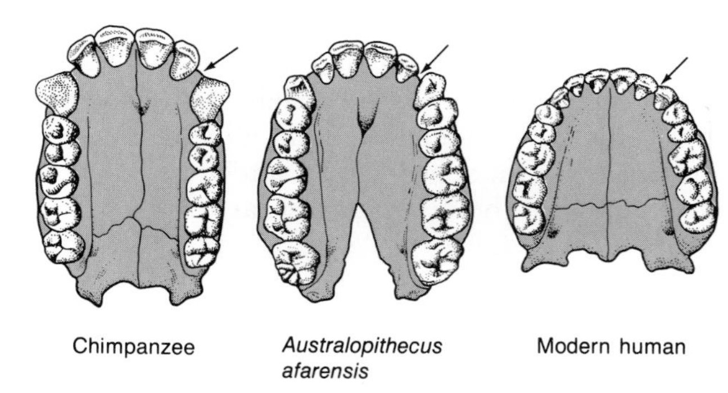

FIGURE 16-34 The upper teeth of *Australopithecus afarensis*, of a chimpanzee, and of a modern human. The chimpanzee has larger canines (eyeteeth) than humans, and a gap separates each canine from the adjacent incisor *(arrow)*. The incisors of the ape are also large in relation to the molars. In these important features, *Australopithecus afarensis* is intermediate between apes and humans. *(After D. Johanson and M. Edey, Lucy: The Beginning of Humankind, Simon and Schuster, New York, 1981.)*

FIGURE 16-35 The skull of *Australopithecus africanus*, revealing a much lower forehead and a more projecting lower face than those found in modern humans. *(Chip Clark.)*

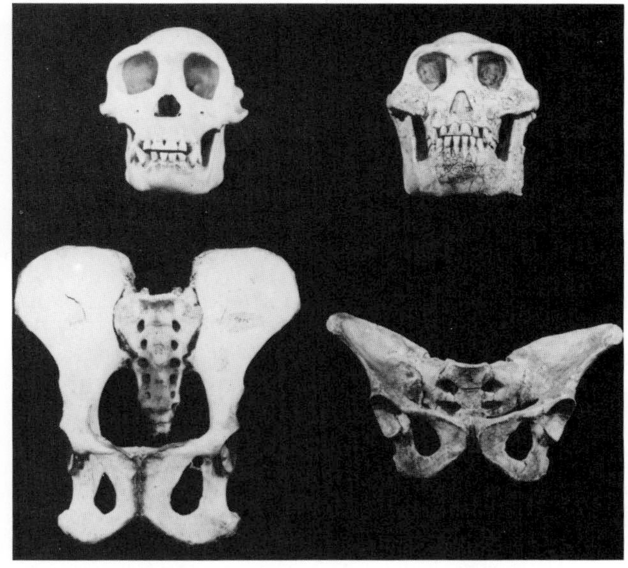

FIGURE 16-36 Skeletal features of *Australopithecus afarensis (right)* and of a modern chimpanzee *(left)*. The skulls of the two animals are more similar than the pelvises. The wide pelvis of *Australopithecus* resembles that of a modern human, in whom it serves as an adaptation for upright posture. *(D. C. Johanson.)*

species *Australopithecus afarensis*. Lucy's bones and those of less fully preserved individuals reveal that the australopithecine body, unlike the skull, more closely resembled a human's than an ape's. The pelvis was broad, for support of the body in a vertical posture, unlike the narrow, elongate pelvis of an ape (Figure 16-36). Tracks beautifully preserved in volcanic ash in the geographic region occupied by *Australopithecus afarensis* provide direct evidence of upright, two-legged walking. In fact, they are remarkably similar to the footprints of modern humans (Figure 16-37).

Although *Australopithecus* walked much the way we do, it appears to have spent a great deal of time in trees. Its strong wrists and long, curved fingers and toes resembling those of an ape would have been useful for climbing. It also resembled apes in having an upward-directed shoulder joint that would have helped it to climb well. Its legs were shorter than ours in relation to its body weight. These legs and the long toes would have been useful for climbing but detrimental to endurance in running on the ground. In short, the locomotory adaptations of *Australopithecus* represented an adaptive compromise between the need to move efficiently on the ground and the

FIGURE 16-37 Tracks made in volcanic ash at Laetolil, Tanzania, more than 3 million years ago. *(Photograph by J. Reader, courtesy of the National Geographic Society.)*

need to climb trees adeptly. It was a less capable climber than a chimpanzee and a less capable walker and runner than a modern human.

It is not difficult to understand why *Australopithecus* would have climbed trees frequently. First of all, many of the fruits and seedpods that it ate must have grown on trees. Second, in order to avoid predators, it must have needed to sleep in trees and occasionally to flee into them during the day. A band of *Australopithecus* individuals probably slept in a grove of trees and fed in and around it during the day, staying close enough to the grove to flee into the trees if predators approached. When the local food supply dwindled, they would have been forced to migrate across dangerous open country to a new grove. Today this behavior characterizes both baboons, which are very large monkeys, and chimpanzees. *Australopithecus* would have been as defenseless against lions and hyenas as are these other large primates. Like them, it lacked advanced weapons and was slower on the ground than large four-legged herbivores such as antelopes and zebras.

Australopithecus survived for the better part of 2 million years, from at least 4 million to about 2.3 million years ago (Figure 16-38). From this genus, sometime between 2.5 and 2 million years ago, the modern human genus, *Homo*, evolved.

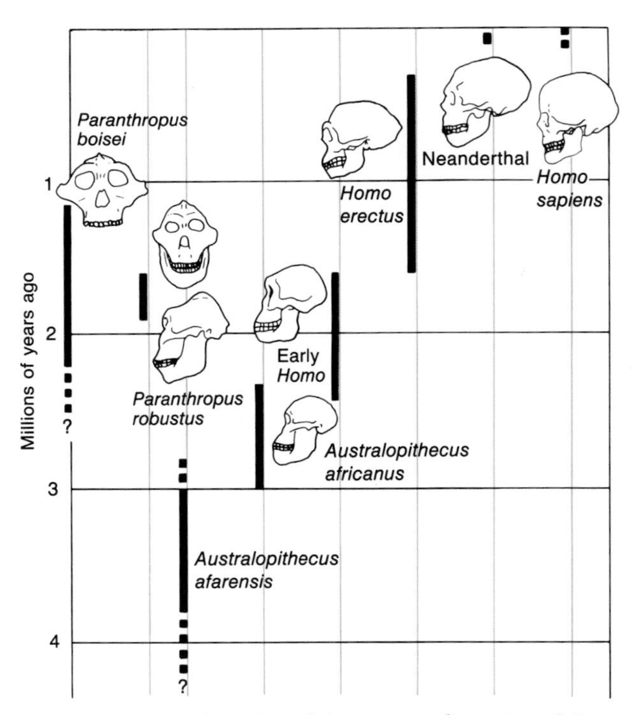

FIGURE 16-38 Stratigraphic ranges of species of the Hominidae (human family) as currently recognized from fossil data. The two *Paranthropus* species represented on the left were not ancestors of modern humans.

The Human Genus Makes Its Appearance The oldest bones thus far assigned to *Homo* are about 2.4 million years old. We do not know which of the two species of *Australopithecus* was its immediate ancestor. It appears that by about 2 million years ago, two or more species of *Homo* were in existence. Because the taxonomy of these early forms is not yet well established, it is convenient to group them all under the informal label "early *Homo*."

Some fossil skulls of early *Homo* reveal a large brain capacity, one of *Homo*'s trademarks. The average volume of an *Australopithecus* skull is about 450 cubic centimeters; the volume of the early *Homo* skull shown in Figure 16-39, in contrast, is about 760 cubic centimeters, and fragments of less well-preserved skulls indicate brain capacities well above 800. The skull of early *Homo* exhibits additional features that make it more human in form than the skull of *Australopithecus*. The teeth, for example, are smaller.

Fossil remains of the pelvis and thigh bone of early *Homo* do not differ greatly from those of modern humans. These bones appear to have belonged to large individuals that spent almost all of their time on the ground.

Early *Homo* appears to have put its large brain to good use in the manufacture of stone tools. In fact, some types of fossils included within this group have been assigned to a species named *Homo habilis*, or "handy man," because stone tools have been found in some of the deposits from which such fossils have been collected. These tools include sharp flakes of stone and many-sided "core" stones that are what was left after the flakes were broken away (Figure 16-40). Such tools are referred to as Oldowan, because they were first found at Olduvai Gorge, the site of many important fossil hominid discoveries (Figure 16-32). The oldest known skull of *Homo* and also the oldest known Oldowan tools, which come from farther north, in Ethiopia, are about 2.4 million years old. Thus it appears likely that early *Homo* was in existence by this time.

Perhaps australopithecines, like modern chimpanzees, supplemented their largely vegetarian diet by devouring small animals, but meat apparently formed a much larger part of early *Homo*'s diet. Scratched bones of other mammals

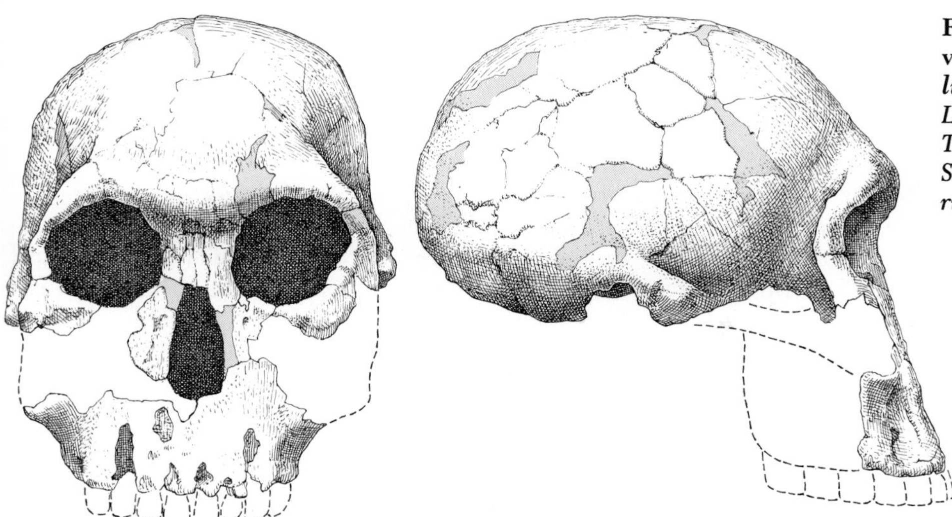

FIGURE 16-39 Front and side views of the skull of *Homo habilis*. *(From A. Walker and R. E. F. Leakey, "The Hominids of East Turkana." Copyright 1978 by Scientific American, Inc. All rights reserved.)*

found in association with Oldowan stones indicate that early *Homo* used its stone tools to sever meat from bones. It is uncertain whether most of this meat was the product of active hunting or of scavenging on carcasses killed by other animals. In any event, the rather abrupt appearance of *Homo* requires an explanation. *Australopithecus* had existed with little evolutionary change for at least 1.5 million years before it gave rise to the human genus. One theory holds that the sudden global climatic change that occurred about 2.5 million years ago led to the origin of *Homo*. This explanation relates to the way *Homo*'s large brain evolved.

Brain Size, Climbing, and the Shrinkage of Forests The size of early *Homo*'s brain can be attributed mainly to a change in the pattern of infant development. The brains of all newborn primates (including monkeys, apes, and humans) account for about 10 percent of their total body weight — a very large proportion. The brains of all these species grow rapidly before birth. In monkeys and apes, however, this high rate of brain growth slows dramatically shortly after birth, so that the brains of adults are only moderately larger than those of newborns. In *Homo*, however, the brain continues to grow rapidly for about a year after birth. This is the main reason why adult humans have such large brains. The size of the human brain increases only moderately after the age of 1, but an average year-old human infant already has a very large brain capacity — more than twice that of an adult chimpanzee.

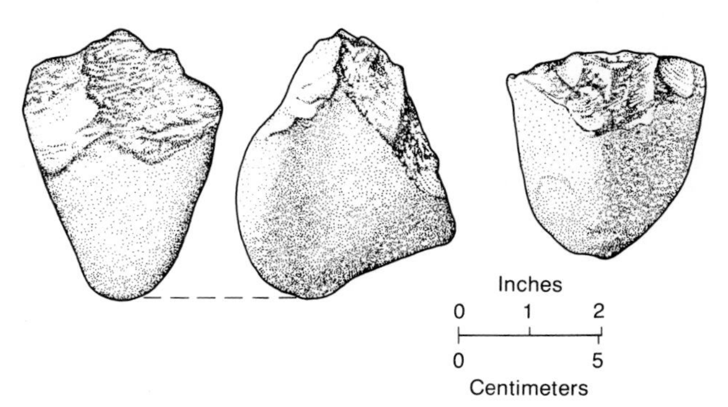

FIGURE 16-40 Simple stone implements from Olduvai Gorge, Tanzania, representing the Oldowan culture that existed about 2.5 million years ago. These are "core" implements from which flakes were struck. *(After K. P. Oakley, Man the Toolmaker, British Museum of Natural History, London, 1950.)*

The extension of the high rate of brain growth through the first year of life was achieved through a general evolutionary delay of maturation. Our permanent teeth do not replace our baby teeth, for example, until we are much older than apes are when they undergo these changes. A key feature of the human pattern of development is that it produces great intelligence at a very early age, so that small children can engage in relatively advanced learning.

We humans develop so slowly that our infants are helpless much longer than the infants of any other species of mammals. The disadvantage of the need to provide years of care to our offspring is more than offset, however, by the enormous advantages that result from the expansion of the brain: Humans are much more intelligent than any other species of mammals, and for this reason we are able to cope with a host of environmental problems. We are neither physically powerful nor fleet of foot, yet we have come to dominate Earth.

Nonetheless, it is quite evident why australopithecines did not evolve a large brain for more than 1.5 million years of existence. As we have seen, they were required to spend a significant portion of their lives in trees. A mother who needed both arms free to climb trees could not have carried a helpless infant about with her. In other words, the large human brain could evolve only after human ancestors stopped climbing trees habitually. Why did they finally stop doing so? Recall that climates became cooler and drier over broad regions of the world about 2.5 million years ago, when continental glaciers expanded in the Northern Hemisphere. Fossil pollen reveals that at this time terrestrial floras in Africa underwent a dramatic change. Forests shrank and grasslands expanded. This is exactly the kind of change that would be expected to force animals such as australopithecines to abandon dependence on tree climbing. Presumably the australopithecines faced an ecological crisis: When they no longer had trees to climb, they lost important sources of food and were suddenly more exposed to such predators as lions and hyenas. Probably many populations of *Australopithecus* died out during this crisis, but apparently at least one population evolved into *Homo.* A large brain was so valuable for avoiding predators and developing advanced hunting techniques that its value overshadowed the problems that resulted from the prolonged helplessness of their infants.

The fossil record of antelopes indicates the impact that the shrinkage of forests at the start of the modern Ice Age had on mammals in general. About 2.5 million years ago, many species that were adapted to forests died out, and species adapted to grassy environments spread over broad areas. It appears that human ancestors experienced comparable changes. The youngest fossils assigned with certainty to the genus *Australopithecus* are about 2.5 millions years old. Although at least one small-brained species with an *Australopithecus*-like body survived to the end of Pliocene time, *Australopithecus*-like forms were gone by early in the Pleistocene Epoch. By this time, too, early *Homo* had given rise to a species that was remarkably similar to modern humans. It has been named *Homo erectus.*

Homo Erectus, Our Recent Ancestor

About 1.6 million years ago or slightly earlier, *Homo erectus* evolved from early *Homo.* This new species seems not to have differed greatly from early *Homo* but it was slightly more similar to modern humans. In addition, *Homo erectus* was the first hominid species to have migrated beyond Africa. Known as *Pithecanthropus* before its similarities to modern humans were fully acknowledged, *Homo erectus* lived not only in Africa and Europe but also in China, where it has been referred to as "Peking man," and in Java, where it has been called "Java man." Dates for fossil skulls

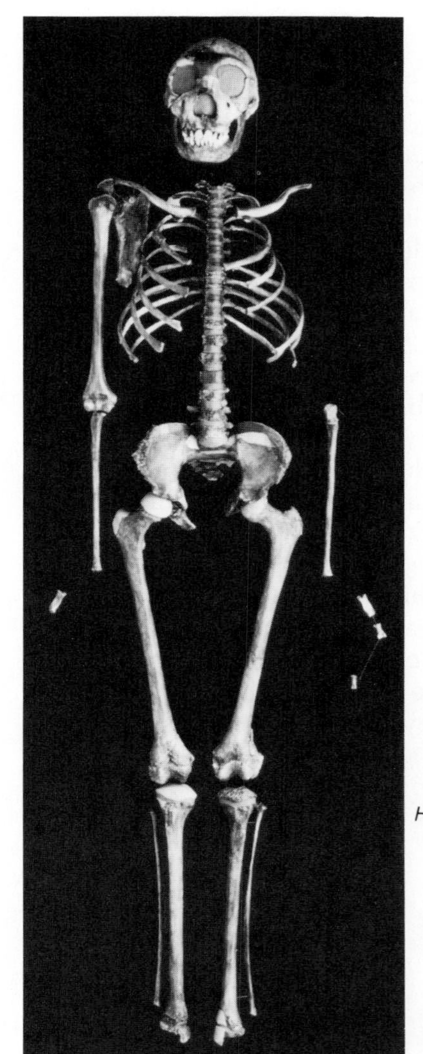

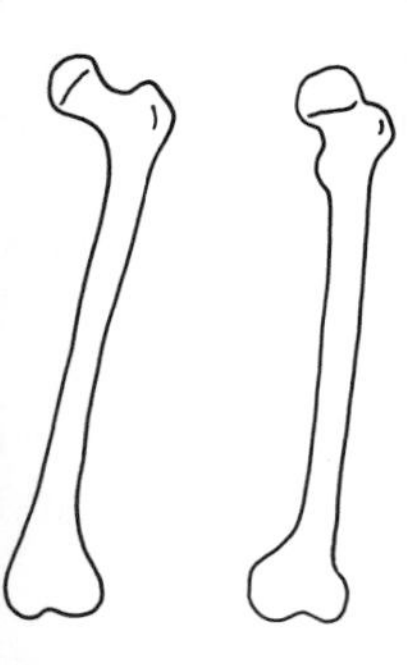

FIGURE 16-41 Nearly complete skeleton of an 11- or 12-year-old *Homo erectus* boy from strata 1.6 million years old. His height was about 5 feet 3 inches (1.6 meters), and had he lived to adulthood, he would have grown to a height of about 6 feet (1.8 meters). On the right, the neck of this boy's femur is seen to be much longer than that of a femur belonging to a typical modern human, who has a wider pelvis.

of *Homo erectus* encompass a long interval of time, from about 1.6 million years ago to perhaps 300,000 years ago.

The recent discovery in Africa of the 1.6-million-year-old skeleton of an 11- or 12-year-old boy has revealed the remarkable resemblance between *Homo erectus* and modern humans (Figure 16-41). This species had a slightly smaller brain than ours but it resembled us in body size. The pelvis of *Homo erectus* was narrower than ours, and here we see why this extinct species had a smaller brain. The size of the pelvis limits the size of the brain in hominids, because a baby's head must pass through its mother's pelvis during birth. The narrow pelvis of *Homo erectus* would not have permitted the birth of a modern baby of average size. Evolution has provided us with a mechanism to develop a brain larger than *Homo erectus*'s: Our larger pelvis allows our brain to grow larger before birth, so we have a head start in the growth of the adult brain.

There is some evidence, however, that the brain size of *Homo erectus* increased somewhat over the course of the species' existence. In some individuals it approached the average for modern humans. The braincase of *Homo erectus* was also relatively longer and lower than ours; furthermore, its forehead was low and interrupted by brow ridges that were even larger than those of the two slender species of *Australopithecus*. These features, though modified, were inherited from *Australopithecus*, as were the projecting mouth and heavy lower jaw—but the mouth and jaw as well as the cheekbones were less highly developed than those of *Australopithecus*. The teeth of *Homo erectus* were smaller than those of *Australopithecus*, and though they are larger than the teeth of modern humans, they bear a close resemblance to them. *Homo erectus* was also a toolmaker. Its stone culture, known as **Acheulian culture,** included the production of magnificent hand axes (Figure 16-42).

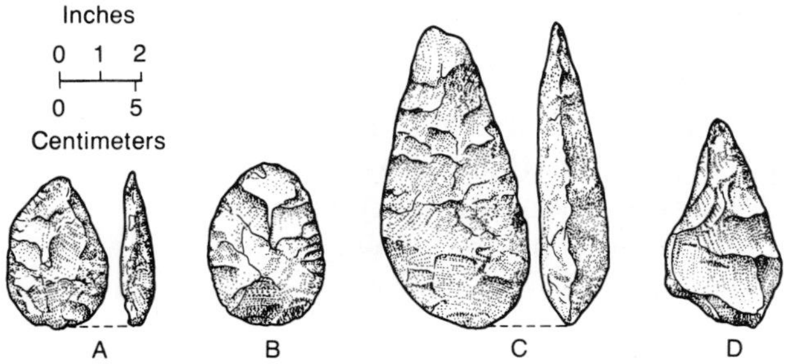

Inches
0 1 2
0 5
Centimeters
A B C D

FIGURE 16-42 Tools of the widespread Acheulian culture attributed to *Homo erectus*. *A.* Twisted oval tool from Saint-Acheul near Amiens, France. *B.* Oval hand ax from south of Wadi Sidr, Israel. *C.* Large hand ax from Orgesailie, Kenya. *D.* Hand ax from Hoxne, Suffolk, England. *(After K. P. Oakley, Man the Toolmaker, British Museum of Natural History, London, 1950.)*

The Robust Australopithecines

Recall that some australopithecines were assigned to the genus *Paranthropus*. These forms are informally called robust australopithecines, in recognition of the enormous bony ridges over their eyes and their massive jaws and teeth (Figure 16-43). They evolved from *Australopithecus*, apparently through natural selection associated with a diet that included very coarse food, such as roots or thick leaves. It is not certain from which species they evolved. The oldest robust skeleton is dated at about 2.5 million years, but shortly after 2 million years ago, two species existed, *Paranthropus robustus* in southern Africa and *Paranthropus boisei* in east-central Africa. Despite the massive structure of their skulls, these animals did not differ greatly from *Australopithecus* in stature or

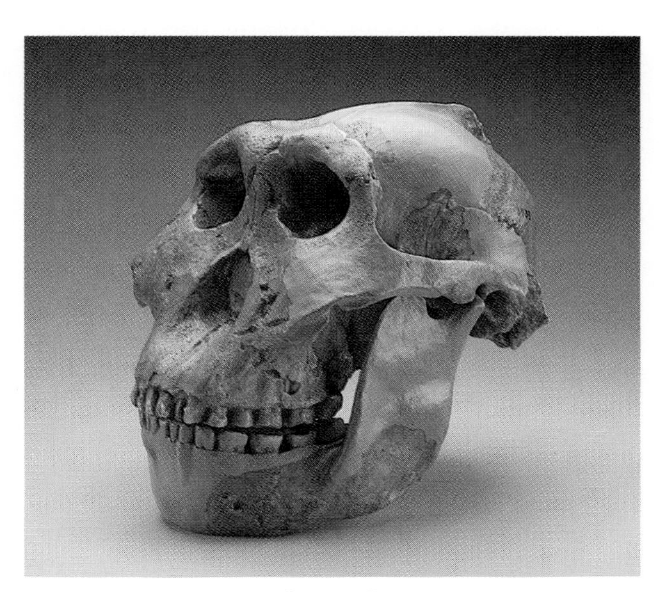

FIGURE 16-43 *Paranthropus boisei*, a heavy-browed species with massive jaws and teeth that served for grinding up vegetable matter. *Paranthropus robustus* was somewhat less robust than this species. *(Chip Clark.)*

physique. *Paranthropus boisei* lived until about 1 million years ago, well into the Pleistocene Epoch. The survival of both species for some time after the sudden environmental change perhaps resulted from their ability to feed on a wide range of foods, which may have allowed them to avoid dangerous journeys between sparse groves of trees in search of a meal.

Was Neanderthal One of Us?

Unfortunately, sediments in Africa representing the interval above the youngest specimens of *Homo erectus* have yielded a poor suite of fossil remains. As a result, we are not certain exactly what evolutionary pathway led from *Homo erectus* to our species, *Homo sapiens.* The interval of poor fossil remains extends from about 400,000 to 100,000 years ago, and most of the partial skulls found within this interval have robust brow ridges and low foreheads reminiscent of *Homo erectus.*

In sediments about 100,000 years old, the hominid fossil record improves with the appearance of well-preserved fossils of the creature known as Neanderthal (Figure 16-38). Traditionally, Neanderthal has been regarded as a variety or subspecies of our species *(Homo sapiens neanderthalensis),* but there is some justification for recognizing it as a separate species *(Homo neanderthalensis).* The record of this humanoid creature extends from Spain to central Asia, ranging in time up to about 35,000 years ago. Enough of its bones and artifacts have been found in caves to suggest that Neanderthal frequently took shelter in caves.

Neanderthal is so designated because its bones were first found in the Neander Valley of Germany in 1856, three years before Charles Darwin wrote *On the Origin of Species.* This creature differed from modern humans in a number of skeletal features. Neanderthal resembled *Homo erectus* in its long, low skull with prominent brow ridges, projecting mouth, and receding chin (Figure 16-38). On the other hand, its brain was quite large—slightly larger, on average, than that of modern humans. The slightly larger brain is by no means indicative of superior intellect, however; Neanderthal's body was somewhat more massive than ours, though slightly shorter, and may have required some extra brain cells simply for the purpose of motor control. Its bones were also more dense than ours, and it probably moved more slowly than we do.

Neanderthal also differed from modern humans in the presence of a gap between its cheek teeth and the back of its jaw and in certain distinctive features of the shoulder blade, pelvis, and hand. The finger bones of Neanderthal have stronger attachment areas for tendons than those found in modern humans, suggesting that this creature's grip was stronger than ours.

The distribution of the two forms in time and space suggests that Neanderthal and *Homo sapiens* may in fact have been reproductively isolated. They simultaneously occupied a region of the Middle East from at least 110,000 to 90,000 years ago. Radiocarbon dating tells us that Neanderthals died out during the most recent glacial interval. They vanished from eastern Europe about 40,000 years ago and from western Europe and the Middle East perhaps 5000 years later. At this time the Cro-Magnon people, who were anatomically almost identical to modern humans, spread throughout Europe. The fact that Cro-Magnon suddenly replaced Neanderthal in Europe suggests that the two did not interbreed successfully and thus may well have been distinct species.

Where did the biologically modern populations of *Homo sapiens* known as Cro-Magnon originate? Africa now appears to be the most likely site. Skulls recently discovered in African sediments more than 100,000 years old have heavy brow ridges and a flattened braincase, but they also display minor features that more closely resemble traits of modern humans.

Perhaps one population of *Homo erectus* was ancestral to our species while another evolved into Neanderthal. In any event, Neanderthals developed a distinctive stone culture known as **Mousterian,** characterized by flakes of stone fashioned into knives, scrapers, and projectile heads—far more sophisticated tools than the Acheulian implements of *Homo erectus.* The Cro-Magnon culture that followed at first incorporated Mousterian elements but then emerged as a more sophisticated culture known as **Late Neolithic,** in which a variety of specialized tools were invented.

Even more innovative was the artwork of the Cro-Magnon people, which seems to reflect a new use of the powers of imagination. Their magnificent cave paintings, primarily of animals, can still be admired in France and Spain (Figure 16-44), while other artifacts show that their artistic efforts included modeling in clay, carving friezes, decorating bones, and fabricating jewelry from teeth and shells.

FIGURE 16-44 A cave painting of the Cro-Magnon people. *(Ferrero/Labat/AUSCAPE International.)*

One intriguing fact is that, whatever their mental powers may have been, Neanderthals did have religion. Burial sites reveal that Neanderthals sometimes prepared their dead for a future life by interring them with flint tools and cooked meat (Figure 16-45). In the Zagros Mountains of Iraq, a Neanderthal man who died after a skull injury was buried in a bed of boughs and flowers that can be identified from the pollen they left beneath the skeleton.

FIGURE 16-45 Reconstruction of the burial of a young Neanderthal, whose skeleton shows that the head was cradled on one arm. Also present are the charred bones of animals that apparently were left as food for the dead individual in an afterlife. *(Painting by Z. Burian under the supervision of Professor J. Augusta, Professor J. Filipa, and Dr. J. Moleho.)*

Human Expansion and the Extinction of Large Mammals

Anthropologists continue to debate not only how the various races of *Homo sapiens* spread throughout the world but also whether all of these races evolved from Cro-Magnon or simply share with Cro-Magnon features that characterize all older populations of *Homo sapiens.* One particularly noteworthy question regarding the geographic spread of *Homo sapiens* is: When did our species first reach the Americas? This question is intertwined with another: What caused the extinction of many large terrestrial mammals near the end of the last glacial advance, about 11,000 years ago? As a result of this extinction event, the world today is impoverished with regard to large mammals; only the faunas of the African savannahs and open woodlands give us a hint of what the rich Pliocene terrestrial ecosystems must have been like, not only in Africa but also in Eurasia and North America. In fact, the modern African fauna, which is rich by today's standards, is impoverished in comparison with the diverse faunas that characterized the Pliocene and Pleistocene epochs. Deterioration of the climate reduced the diversity of animal species in Eurasia during the Pleistocene Epoch, but even the fossil record of that epoch has yielded a fauna of more than 50 genera of Bovidae (cattle, antelope, sheep, goats, and related hoofed animals).

Within the rich Plio-Pleistocene faunas were many species of immense proportions. Among the huge Pleistocene species were an elephant-sized bison of the American plains whose horns spread more than 2 meters (~7 feet), or three times the

FIGURE 16-46 ***A.*** **Reconstruction of "Folsom man" hunting giant bison.** ***B.*** **The original Folsom point, found between the fossil ribs of a giant bison.** *(Denver Museum of Natural History.)*

span for the living species; a North American beaver *(Castoroides)* that was nearly as large as a black bear; horses of the modern genus *Equus,* which evolved to the size of our artificially bred Clydesdale draft horses; and mammoths (members of the elephant genus *Mammuthus*) that stood about 4.5 meters (~ 15 feet) at the shoulder, making them about 30 percent taller than an average African elephant.

Radiocarbon dating of archaeological sites now indicates that modern humans reached the Americas earlier than 30,000 years ago, but their weapons and hunting skills may have been too primitive to have caused the extinctions of large mammals. Some anthropologists have argued that these extinctions occurred at the hands of later humans armed with newly developed weapons for which large, conspicuous mammals were easy marks. Because they were represented by few individuals to begin with, such species were eventually hunted out of existence. This theory suggests that a wave of humans swarmed into North America about 11,000 years ago, by way of the Bering land bridge, and were such successful big-game hunters that their populations expanded and spread rapidly — perhaps by about 16 kilometers (~ 10 miles) per year — in their efforts to find new game. Radiocarbon analyses of wood associated with stone weapons reveals that humans at this time employed sophisticated projectile systems in the New World; they are known to have launched flint tips of many types at the ends of lances and may also have propelled darts with throwing sticks. The famous Clovis and Folsom projectile points represent this stage in the development of weaponry (Figure 16-46).

Whether humans armed with advanced stone weapons destroyed numerous species of large mammals remains controversial. Nonetheless, as modern humans cause the extinction of numerous kinds of animals and plants every decade, it is sobering to reflect that *Homo sapiens* may have begun to destroy other species of mammals more than 10,000 years ago, long before the origins of cities, guns, and chain saws.

CHAPTER SUMMARY

1 Invertebrate life in the oceans underwent only minor changes during Neogene time, but whales radiated rapidly.

2 On land, grasses and herbs, benefiting from a cooling of climates, diversified and occupied more territory during Miocene time.

3 Terrestrial mammals assumed their modern character during Neogene time, and frogs, rats and mice, snakes, and songbirds underwent major adaptive radiations.

4 Apes radiated during the Miocene Epoch but have since dwindled in number of species.

5 The modern Ice Age began about 2.5 million years ago with the origin of continental glaciers in the Northern Hemisphere. Climates in many regions became cooler and drier, and grassy habitats expanded while forests shrank.

6 During the past 1.65 million years, continental glaciers have expanded and contracted more than 18 times, with expansions lowering sea level by as much as 100 meters. Glaciers are now in a contracted state, but they will probably expand again.

7 During Neogene time the American West has been affected by extensive volcanism, and major uplifts have produced the Sierra Nevada, the Rocky Mountains, and the Colorado Plateau.

8 In the western Atlantic region, the Caribbean Sea, which is bounded on the east by an island arc, developed its modern configuration during Neogene time, and the Isthmus of Panama was uplifted, permitting extensive biotic interchange between North and South America.

9 East of the Atlantic, the Mediterranean Sea became weakly connected to this larger ocean about 6 million years ago and briefly became hypersaline. Soon thereafter reenlargement of the connection allowed the Mediterranean to fill with normal marine waters again.

10 During Neogene time, Africa has been fragmenting along spreading zones known as rift valleys, and in early Miocene time it moved northward into contact with Eurasia, allowing biotic interchange between the two continents.

11 *Australopithecus,* the oldest known genus of the human family, evolved at least 4 million years ago. It possessed a small body, a relatively small brain, and apelike teeth. *Homo,* the modern human genus, evolved in Africa at least 2 million years ago, perhaps as a consequence of the shrinkage of forests. More than 100,000 years ago, humans of the modern type were present.

EXERCISES

1 In what ways did mammals become modernized during the Neogene Period?

2 What factors influence the isotopic composition of oxygen in skeletons of marine organisms? Which of these factors has been dominant during the Pleistocene Epoch?

3 What changes in the geographic distribution of animals did the uplift of the Isthmus of Panama produce?

4 How did the Rocky Mountains develop their present configuration in the course of the Neogene Period?

5 What was the climatic effect of the uplift of the Sierra Nevada mountain range?

6 What kinds of volcanic activity occurred in the American West during Neogene time?

7 How did climatic change about 2.5 million years ago alter the general distribution of terrestrial vegetation?

8 How did the Appalachian Mountains develop their present configuration in the course of Neogene time?

9 How did members of the genus *Australopithecus* differ from modern humans?

ADDITIONAL READING

Imbrie, J., and K. P. Imbrie, *Ice Ages,* Enslow Publishers, Short Hills, N.J., 1979.

Kennett, J., *Marine Geology,* Prentice-Hall, Inc., Englewood Cliffs, N.J., 1982.

Lewin, R., *Human Evolution: An Illustrated Introduction,* W. H. Freeman and Company, New York, 1984.

Pomeral, C., *The Cenozoic Era,* John Wiley & Sons, Inc., New York, 1982.

Savage, R. J. G., and M. R. Long, *Mammal Evolution: An Illustrated Guide,* Facts on File, Inc., New York, 1986.

Sutcliffe, A. J., *On the Track of Ice Age Mammals,* Harvard University Press, Cambridge, Mass., 1985.

Epilogue

On our voyage through geologic time, we have seen continents grow as they have crept over the asthenosphere, and we have seen them break apart along great rifts where new oceans have formed. We have seen how mountain systems have risen up and worn down, how seas have flooded continents and receded, and how, from humble Archean beginnings, life expanded to form a complex marine ecosystem and then to conquer the land. Through the operation of these and other processes, Earth has taken its present shape and its inhabitants have assumed their modern character.

PLATES CONTINUE TO MOVE

Turning our gaze far into the future, we can predict major geologic events from what has gone before. About 30 million years from now, the movement of the narrow sliver of crust that joins the rest of North America along the San Andreas Fault will have brought Los Angeles to a position alongside San Francisco. In 10 million years, barring renewed uplifts, the rugged Rocky Mountains and Alps will have been subdued by erosion, but the Himalayas will probably still stand tall, owing to continued wedging of the Indian Peninsula beneath them. Though Earth's crust is unusually thick in Africa, that continent may fragment along its present rift valleys in another several million years; alternatively, convective cells in the mantle may shift, leaving the continent intact.

WHAT IS OUR EVOLUTIONARY FUTURE?

Whether our species will observe major plate-tectonic changes of the future is uncertain. It is sobering to contemplate that few species of our class (the Mammalia) have survived more than 3 to 4 million years. While it would be nice to believe that our intelligence and cultural flexibility will earn us an exceptionally long stay on Earth, the enormous destructive potential of our nuclear weaponry must dampen our optimism.

A question that warrants less concern but nonetheless attracts much interest is what the evolutionary future of our species might be, however long we may ward off extinction. Our species has experienced very little anatomical change, at least in Europe, during the past 40,000 years. Will this stability continue? In Chapter 5 we noted that some paleontologists believe that most species, in fact, change relatively little once they have originated by rapid evolution from populations of preexisting species; other paleontologists believe that most species evolve at substantial rates throughout their existence. This controversy, however, has little bearing on the future of biological evolution of the human species. Our species, unlike all others, evolves culturally as well as biologically, and our cultural evolution has interfered with our biological evolution, making our biological evolution different from that of all other species. For example, many physically impaired individuals who would not survive in nature are sustained by society—especially by modern medicine—and given the opportunity to produce children. On the other hand, our society discourages reproduction of certain individuals who carry genetic features that impair their health and might impair the health of their offspring. Selection pressures in modern societies differ not only from those of other species but also from those that affected more primitive human societies long ago. Physical strength and speed are not now at such a high premium as they were when members of our species wielded simple weapons against wild animals and human foes. *Homo sapiens* actually has

FIGURE E-1 Logging produced this clear-cut region in a forest of the Nimkish Valley of British Columbia. *(Tom Bean.)*

the potential to direct its own evolution as it sees fit by practicing artificial breeding of the sort that has produced such divergent breeds of domestic dogs as the Saint Bernard, Chihuahua, and greyhound. And if the prospect of selective breeding in our species were not disturbing enough in itself, we must also contemplate the ethical questions raised by modern advances in genetic engineering. By these techniques, scientists can introduce entirely new genetic features to the human species.

WHAT WILL HAPPEN TO THE GLOBAL ECOSYSTEM?

The Northern Hemisphere emerged from the last Pleistocene glacial episode about 10,000 years ago. An average interval between major glacial expansions of the Pleistocene lasted many times 10,000 years. It seems, then, that our species, which endured the most recent advance of continental glaciers, need not fear another major glacial advance for many millennia. Nonetheless, climatic changes on a smaller scale could quickly disrupt our civilization. During the present interglacial interval, humans have built cities far to the north, which are vulnerable to even a minor glacial advance. Moreover, a cool interval comparable to the Little Ice Age of the past few centuries would impair agriculture in large areas of Russia, Scandinavia, and Canada. On the other hand, as we have already noted, a warming climatic trend could melt enough polar ice to flood coastal cities; such warming is likely to result from fuel-burning practices that add carbon dioxide to the atmosphere, accentuating the greenhouse effect.

During the next half century we may damage life on Earth even more drastically by altering and destroying natural habitats than by intensifying the greenhouse effect. As we have seen, our destruction of such habitats as coral reefs, wetlands, and rain forests is eliminating populations and entire species. Not only are the disappearing forms of life aesthetically pleasing, but many are medically, agriculturally, or industrially valuable — or have the potential to become so. An understanding of ancient ecosystems places present and likely future events in perspective. Such understanding, for example, enhances our appreciation of the unfortunate state of grassland faunas today. During Pliocene and Pleistocene time, diverse faunas of mammals populated grasslands in North America, Asia, and Africa. Large elements of these faunas are now extinct in North America and Asia, owing in part to climatic fluctuations during the glacial age and in part to human hunting and habitat destruction. Humans alone are responsible for the near extinction of the American bison late in the last century. African savannahs still host a rich mammalian fauna, but it is seriously imperiled.

As we face the future, it is important for us to understand not only how our activities may alter or destroy habitats but also what will happen to life as a result. Lessons from the past will serve us well here — lessons about the kinds of species that have always been vulnerable to extinction, for example, and lessons about the particular ways in which environments and life have changed during the past few thousand years to produce our present ecosystems.

CONTINUING EDUCATION

Most of the lessons of this book illuminate less perplexing issues. We have traced the geologic histories of many regions and have reviewed the major developments in the history of life. In the perspective thus gained, all the features of our planet's landscape, from small-scale local outcrops to jagged mountain chains and sandy coastal plains, take on new meaning. Fossils, which are abundant in nearly all regions of the world except crystalline Precambrian shields, also come to life. You can supplement your new understanding of these phenomena by making use of the many regional libraries and state and federal agencies from which information about local rocks and fossils is available. Museums and parklands also offer special glimpses of geology and paleontology. In summary, the world around you will be a richer store of information now that you comprehend, in broad outline, the evolution of Earth and life through time.

APPENDIX I

Minerals and Rocks

The rocks and soil on which we stand are aggregates of mineral grains. As defined in Chapter 1, a mineral is a naturally occurring solid element or compound whose atoms are organized in a particular configuration. (A compound is a substance that consists of chemically combined elements.) This appendix introduces some characteristics of important minerals and rocks, beginning with a review of the basic properties of minerals.

PROPERTIES OF MINERALS

A chemical **element** is a fundamental substance made up of a particular kind of atom that, in its most stable state, has equal numbers of electrons and protons. The electrons have a negative charge, while the protons have a positive charge. Electrons surround the atom's nucleus, where protons reside along with neutrally charged neutrons. Some kinds of atoms tend to lose one or more of their outermost electrons to other kinds of atoms. Atoms that lose their electrons are said to become positively charged ions, whereas those that gain electrons are called negatively charged ions. Positively and negatively charged ions are attracted to each other in the way that hair is sometimes attracted to a synthetic fabric that is drawn over it. The attachment that results, which is called an ionic bond, is essential to the formation of mineral compounds. Uncharged atoms can also combine in other ways; for instance, they can share electrons in a process called covalent bonding. When different kinds of atoms bond together, they form compounds whose characteristics sometimes differ dramatically from those of their constituent elements in pure form.

Stable chemical configurations such as those that characterize most minerals cannot exist unless two conditions have been met: First, positively and negatively charged ions must be combined in the proper proportions so that there is no charge imbalance; and second, these positive and negative ions must be of relative sizes that enable them to fit snugly together to form a solid structure. The geometrically consistent internal structure of a mineral is known as a **crystal lattice** (Figure AI-1).

The physical properties of minerals, including their hardness, density, external form, and pattern of breakage, are determined by their chemical composition and crystal lattice structure. The strength of a mineral's chemical bonds, for example, is the primary determinant of that mineral's hardness. Thus diamond, which is composed solely of the element carbon, is the hardest mineral on Earth because all of its carbon atoms are strongly bonded to one another. The mineral

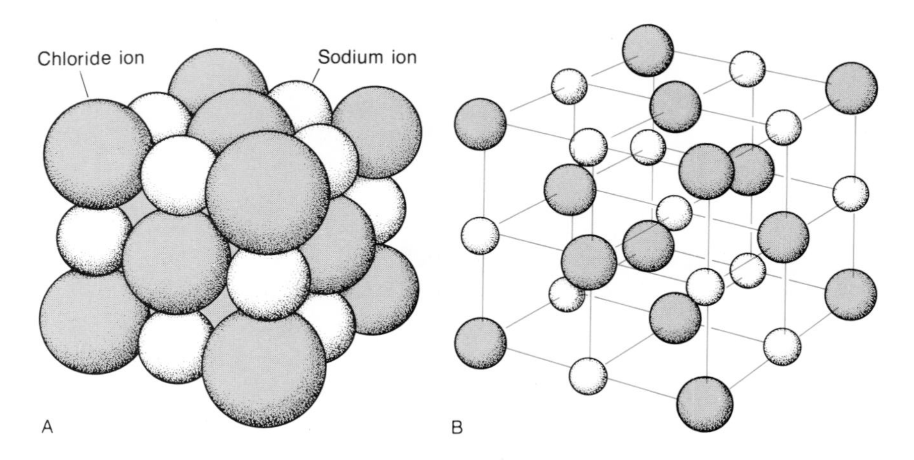

FIGURE AI-1 The structure of the mineral halite, also known as rock salt. Halite is composed of equal numbers of positive sodium ions and negative chloride ions, which fit together to form a cubic crystal lattice. *A.* The actual position of the ions. *B.* The cubic pattern is evident in the expanded lattice.

graphite also consists of pure carbon, but because its carbon atoms are arranged in layers that are weakly attached to one another, it is very soft—soft enough to be used as the "lead" in pencils.

The density of a mineral—or its mass per unit of volume—is partially determined by the types of atoms of which it is constituted. An iron atom, for example, is heavier than an aluminum atom because it contains more protons and neutrons, and compounds of iron are therefore more dense than compounds of aluminum. The density of a mineral is also determined to some extent by the degree to which its atoms are packed together. Figure AI-2, for example, shows two crystal structures of the mineral **olivine,** which is thought to be the primary component of Earth's mantle. At a depth of about 400 kilometers (~250 miles) beneath Earth's surface, seismic waves abruptly accelerate, indicating that the mantle abruptly increases in density at this point. Laboratory experiments reveal that as temperature and pressure are increased to the values that are thought to exist at this depth, olivine alters from its less dense form to its more dense form as a result of changes in the mineral's crystal lattice. This phenomenon probably explains the seismic acceleration that takes place within the mantle at a depth of 400 kilometers.

Minerals may vary slightly in their chemical composition, but only within specified limits. Such variation usually results when one element in a mineral's crystal lattice is substituted for an-

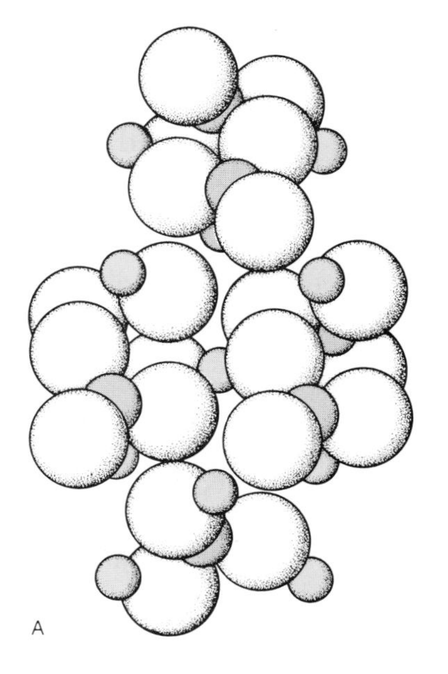

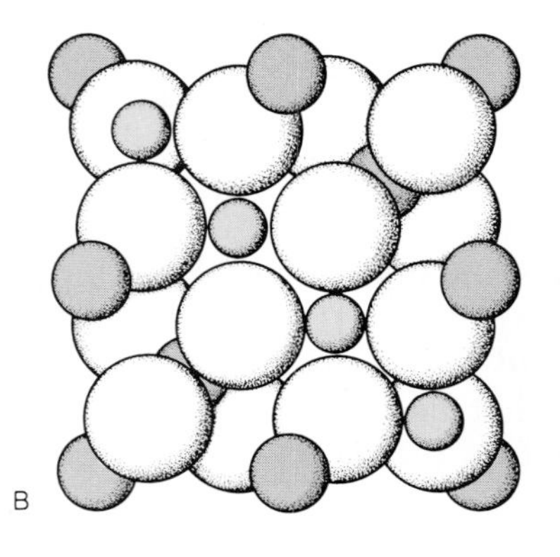

FIGURE AI-2 The structure of olivine (a mixture of Mg_2, SiO_4, and Fe_2SiO_4), which is an important mineral of Earth's mantle. The less dense crystal form of the mineral *(A)* exists under lower pressures (or at shallower depths) than the more dense crystal form *(B)*. The large ions are oxygen, the small ions are silicon, and the medium-sized ions are magnesium (or iron).

other whose atoms are roughly the same size. Iron, for example, frequently substitutes for magnesium in some crystal lattices, and strontium often substitutes for calcium in others.

A crystal's external form and the way it breaks also reflect its internal structure, as illustrated by the mineral **halite** (NaCl), which forms from the evaporation of natural waters (Figure AI-1). In halite, which is informally known as rock salt, sodium and chlorine ions form natural cubic units within the crystal, maintaining the cubic shape as the crystal grows. Halite also breaks along planes of weakness known as **cleavage planes,** which lie at right angles to one another following zones of weak attachment between planes of atoms. In other minerals, characteristic external forms and breakage patterns also reflect internal crystal structure. Unlike halite, however, many minerals do not ordinarily break along cleavage planes; moreover, cleavage planes in many minerals do not parallel crystal surfaces.

A few groups of minerals are especially abundant on Earth because they are important rock-forming minerals. We will now review several of these groups, whose properties and roles in the formation of rock are summarized in Table AI-1.

TABLE AI-1 Major mineral groups

	Chemical properties	*Physical properties*	*Rock-forming contribution*	*Comments*
Silicates	SiO_4 tetrahedra are the basic units	Mostly hard, except for mica and clay minerals; most have a glassy or pearly luster	Dominant mineral group in igneous, sedimentary, and metamorphic rocks	Most crystallize at high temperatures and occur in sediments only as detritus
Carbonates	Positive ions attached to CO_3	Soft, light-colored	Mostly sedimentary, but also form marble, a metamorphic rock	Include calcite, aragonite, and dolomite
Sulfates	Positive ions attached to SO_4	Soft, light-colored, water-soluble	Most rock-forming varieties are sedimentary	Form large sedimentary evaporite deposits, including gypsum and anhydrite
Halides	Positive ions attached to negative ions of elements such as chlorine (Cl) or bromine (Br)	Soft, light-colored, water-soluble	Most rock-forming varieties are sedimentary	Form large sedimentary deposits, including halite (rock salt)
Oxides	Metallic ions combined with oxygen	Soft to hard	Mostly sedimentary, but many varieties are present in igneous and metamorphic rocks	Some, including the iron materials magnetite and hematite, are major ore minerals
Sulfides	Metallic ions combined with sulfur	Soft to medium hard, often with a metallic luster	Have a minor role in rock forming	Many are important ore minerals that form at high temperatures

Tetrahedron of four oxygens with a silicon hidden in the center

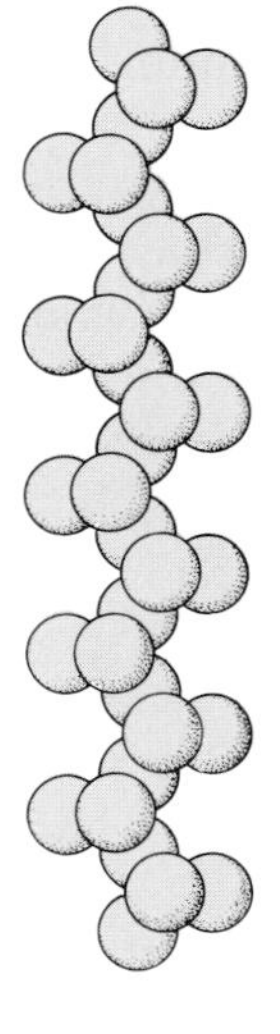

Pyroxene (single chain)

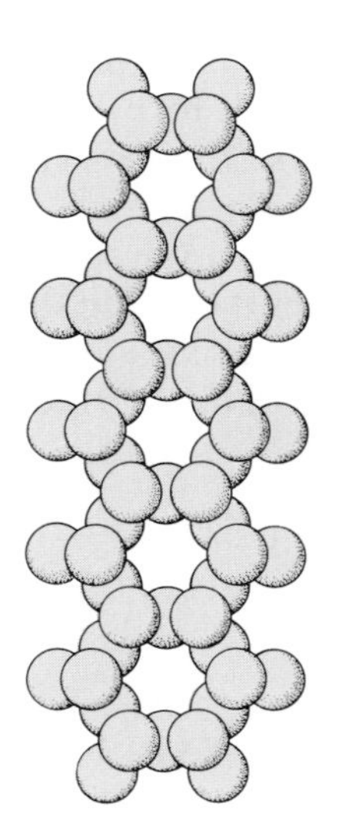

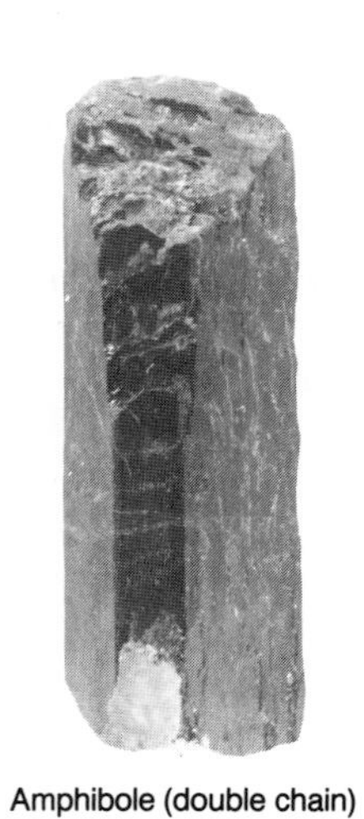

Amphibole (double chain)

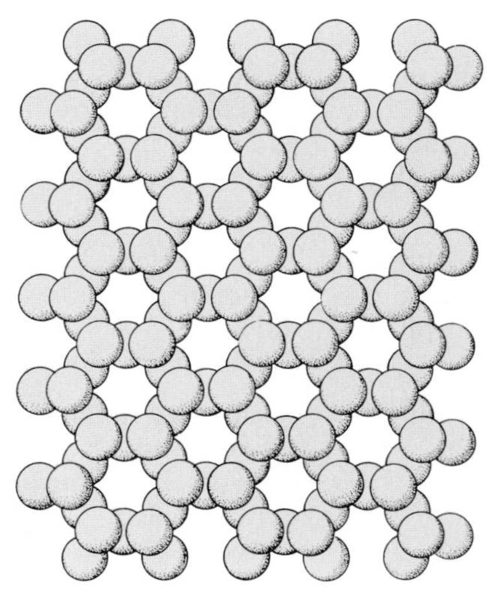

Mica, *left*, and clay, *right* (sheet)

Quartz, *left,* and pink feldspar, *right*

FIGURE AI-3 Rock-forming silicate minerals. Diagrams show the arrangements of tetrahedra in five important rock-forming mineral groups. (Other ions are omitted for simplicity.)

In amphiboles and pyroxenes, the silicate tetrahedra are assembled into long chains that are bonded together by ions of iron, calcium, or magnesium positioned between them. The iron and magnesium make these minerals dense and often dark in color.

In mica and clay minerals, the silicate tetrahedra are more fully connected to form two-dimensional sheets that are bonded together by sheets of aluminum, iron, magnesium, or potassium. Because the bonds between these sheet silicates are weak, micas and clays cleave into thin flakes. Clay minerals are especially weak and almost always occur naturally as small flakes.

The feldspars are the most common group of minerals in Earth's crust, and quartz is second. Quartz is the simplest silicate mineral in chemical composition, consisting of nothing but interlocking silicate tetrahedra. Because each oxygen is shared by two adjacent tetrahedra, for the mineral as a whole there are only two times rather than four times as many oxygens as silicons (the ratio in a single tetrahedron). Hence the chemical formula of quartz is SiO_2. Quartz is very hard because its silicons and oxygens are tightly bonded.

Feldspars differ from quartz in that their structure includes both silicate tetrahedra and tetrahedra in which aluminum takes the place of silicon. Ions of one or more additional types (potassium, sodium, and calcium) also fit into the framework in differing proportions. Unlike quartz, feldspars display good cleavage. Feldspars are slightly softer than quartz, and they also weather chemically much more readily in nature. *(Clay [kaolite], W. D. Keller; quartz and pink feldspar, Chip Clark; other photographs by the author.)*

FAMILIES OF ROCK-FORMING MINERALS

The mineral groups known as **silicates** are the most abundant minerals in Earth's crust and mantle. In silicates, four negatively charged oxygen atoms form a tetrahedral structure around a smaller, positively charged silicon atom. Figure AI-3 illustrates how silicate tetrahedra unite in various ways, usually with other atoms, to form the most important silicate minerals of Earth's crust. Not only are silicates the primary constituents of the igneous rocks of the crust, but they also represent the most important minerals of crustal sedimentary and metamorphic rocks. Silicates tend to form at the high temperatures that exist deep within Earth's crust and mantle and that also characterize molten rock that reaches the surface from great depths.

Carbonate and **sulfate minerals** are also important rock formers, but, unlike silicates, most of these minerals form at low temperatures near Earth's surface. Carbonates are constructed of one or more positive ions, such as calcium, magnesium, and iron, bonded to the negative carbonate ion. The carbonate ion is a composite ion formed of one carbon linked to three oxygens (CO_3). **Calcite** ($CaCO_3$) is the most abundant carbonate mineral (Figure AI-4). **Aragonite,** another important carbonate mineral, has the same chemical composition as calcite, but it differs in its crystal structure; thus aragonite's physical and chemical properties also differ from those of calcite. The shells of corals, mollusks, and many other marine organisms consist of calcite, aragonite, or both. **Dolomite** resembles calcite, but half of the calcium ions are replaced by magnesium. Sulfates are formed of positive ions (such as calcium, iron, or

FIGURE AI-4 The mineral calcite ($CaCO_3$). Excellent cleavage causes calcite crystals to break into rhombic fragments. *(Chip Clark.)*

FIGURE AI-5 The composition of igneous rocks. In contrast to felsic rocks, mafic rocks contain no quartz and not much more than 50 percent silica. Note that plutonic rocks, which crystallize slowly within Earth, are more coarse-grained than volcanic rocks, which crystallize more rapidly at Earth's surface. All of the specimens shown here would fit in the palm of your hand. *(Photographs by the author.)*

strontium) that are attached to negative sulfate ions, each of which consists of sulfur bonded by four oxygens (SO_4). Many sulfates also form at low temperatures near Earth's surface.

Although oxides make up only a small percentage of the large bodies of rock on Earth, these minerals form many important ore deposits. Rocks whose primary components are the oxides magnetite (Fe_3O_4) and hematite (Fe_2O_3), for example, yield most of the iron that is put to human use. Many other ore minerals are **sulfides,** which are compounds of metals such as copper or iron in combination with sulfur.

TYPES OF ROCKS

Rocks are classified on the basis of their size, their composition, and the arrangement of their constituent grains. We will now examine the main types of rocks that belong to each of the fundamental groups of rocks—igneous, sedimentary, and metamorphic.

Igneous Rocks

When igneous rocks of Earth's crust are classified on the basis of their chemical composition, most fall into two major groups—**felsic** and **mafic.** The term *felsic,* which is derived from *feldspar,* the silicon-rich mineral seen in Figure AI-3, is used as a general designation for all silicon-rich igneous rocks. Such rocks are generally light-colored and of low density. Granite is the most abundant felsic igneous rock; two kinds of feldspar constitute about 60 percent of its volume (Figure AI-5). Not surprisingly, granite and other felsic rocks also contain high percentages of quartz. Continental crust (Figure 1-16) is predominantly felsic, containing large volumes of granite-like rocks. Mafic rocks, in contrast, are relatively low in silicon and contain no quartz. Such rocks are, however, rich in magnesium and iron; hence their name, which derives from the *ma* in *magnesium* and the *f* in *ferrous.* This abundance of magnesium and iron makes mafic rocks, such as **gabbro** (Figure AI-5), darker and heavier than felsic rocks. Mafic rocks form most of the oceanic crust, while **ultramafic rocks,** which are even lower in silicon, form the mantle below the crust. Olivine (Figure AI-2) is one of the primary constituents of ultramafic rocks.

Igneous rocks can also be classified according to grain size. This classification system is especially useful in that grain size reflects the rate at which igneous rocks cooled from a molten state. If a rock cools slowly, its crystals can grow large, thus producing a coarse-grained rock, whereas rapid cooling "freezes" molten rock into small crystals that yield a fine-grained rock. The fine-grained equivalent of granite is called **rhyolite,** and the fine-grained equivalent of gabbro is called **basalt.** The dense rock that forms the oceanic crust consists primarily of basalt.

Most molten rock that cools within the crust or at Earth's surface comes from the mantle; as this molten rock rises, in the form of a blob or plume, it melts the crustal rock with which it comes into contact. Molten rock found within Earth is known as **magma,** but when it appears at Earth's surface through an opening called a **vent,** it is called **lava.** Some magma cools within Earth and thus never reaches the surface. Because cooling here usually takes place slowly, the result is nearly always an igneous rock of coarse grain size, such as granite or gabbro. Lava, in contrast, usually cools rapidly at Earth's surface, and the result is usually a fine-grained igneous rock such as rhyolite or basalt.

Magma that cools within Earth forms bodies of rock that are sometimes referred to as **intrusions**—so called because they often displace or melt their way into preexisting rocks—or **plutons. Sills,** on the other hand, are sheetlike or tabular plutons that have been injected between sedimentary layers, and **dikes** are similarly shaped plutons that cut upward through sedimentary layers or crystalline rocks (Figures AI-6 and AI-7).

In contrast to intrusive, or plutonic, rocks, which form from magma at depths within Earth, extrusive, or volcanic, rocks form when lava cools

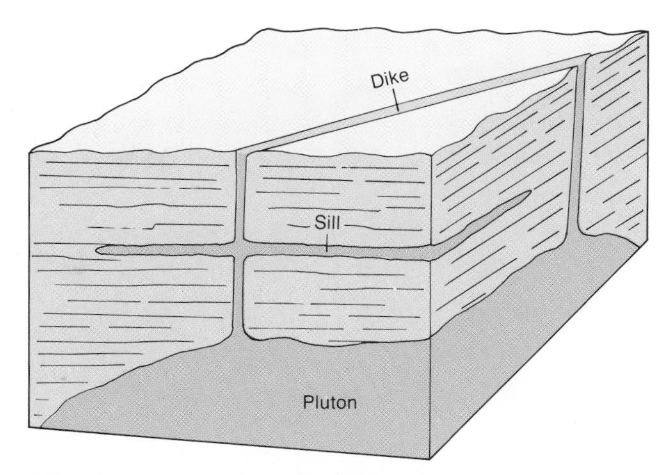

FIGURE AI-6 The configurations of intrusive bodies of igneous rock. A pluton is a large body of irregular shape; a dike is a tabular body that cuts across layered rocks; and a sill is a tabular body that has been intruded between layers.

FIGURE AI-7 A dike of basalt cutting through the Old Red Sandstone Formation south of Wemyss Point, Scotland. *(British Institute of Geological Sciences.)*

FIGURE AI-8 Flows of Miocene basalt along the Columbia River near Vantage, Washington. The cliff is about 330 meters (~1100 feet) high. Figure 16-23 shows the widespread distribution of these so-called Columbia River Basalts. *(Washington Department of Conservation and Development.)*

at Earth's surface. Some volcanic rocks form simply by flowing out of cracks called **fissures,** from which they spread over large areas. The volcanic rocks that spread widely from fissures are almost always mafic, because felsic lavas are more viscous than mafic lavas. Mafic extrusive rocks that have flowed widely are often referred to as **flood basalts** (Figure AI-8).

Other lavas erupt from tubes or fissures to build the cone-shaped structures that we call volcanoes. Volcanoes that form oceanic crust seldom produce tall, narrow cones but instead form broad cones called **shield volcanoes.** This shape results from the mafic composition of oceanic lavas, which gives the lavas a low viscosity that allows them to spread rapidly in all directions (Figure AI-9). Lava that emerges from the crust beneath the sea cools rapidly in a way that gives its surface a hummocky configuration, creating rock known as **pillow basalt** (Figure AI-10). On continents, hot magma that rises toward the Earth's surface often melts surrounding felsic rock, thereby mixing with the felsic magma. Because felsic lava is more viscous than mafic lava, volcanoes on the surfaces of continents tend to take the form of tall cones. An example is Mount St. Helens, which erupted in the state of Washington in 1980 (Figure 1-1). At the summit of most volcanoes is a hollow crater that forms after an eruption, when the unerupted lava sinks back down into the vent and hardens.

Volcanic rocks can also form in ways that do not involve the cooling of flowing lava. Some volcanic eruptions, for example—including that of Mount St. Helens in 1980—are explosive, hurling solid fragments of previously formed volcanic rock great distances. These fragments range in size from dust to blocks several meters across. Loose debris of various sizes settles to form rock known as **tuff.** Although tuff is deposited in the same manner as sedimentary rock, it is usually classified as volcanic simply because it consists of volcanic particles. In fact, some tuffs form from hot grains that melt together as they settle, but others harden after water percolating through them precipitates cement. Rocky material, however, is not all that issues from volcanic vents. Gases of many kinds are also emitted from volcanoes along with rock fragments, and in volcanic areas such as Yellowstone Park, steamy geysers shoot up from sites where underground water is heated against magma and vaporized.

FIGURE AI-9 Recently cooled lava in Hawaii, showing ropy structure. *(Peter Kresan.)*

FIGURE AI-10 Pillow basalts that formed at Española, one of the Galápagos islands, more than 3 million years ago, when lava cooled beneath the sea. Most of the "pillows" are less than 1 meter (~3 feet) across. *(Photograph by the author.)*

Sediments and Sedimentary Rocks

Most sediments that form rocks belong to one of three categories—siliciclastic, chemical, or biogenic.

To understand what **siliciclastic sediments** are, one must first understand what a clast is. A **clast** is a particle of rock that has been transported; **clastic rocks** are aggregates of clasts. Clasts are the solid products of erosion (see below) and are also referred to as **detritus**, and the adjective *detrital* is often applied to clastic rocks that have been transported and deposited by wind, water, or ice. Siliciclastic sedimentary rocks, then, are detrital rocks that are composed of silicate clasts. Clasts composed of quartz and clay are particularly common at Earth's surface because they are supplied in vast quantities by the breakdown of preexisting rocks under conditions in which many other solid materials are destroyed. Often, however, grains that are mineral aggregates and grains consisting of feldspars and other individual minerals also survive this breakdown process to become components of siliciclastic rocks.

Chemical sediments are created by precipitation from natural waters; most accumulate near the site where they originally formed. Many chemical sediments form by the evaporation of bodies of water.

Biogenic sediments consist of mineral grains that were once parts of organisms. Some of these grains are pieces of organic skeletons, such as snail shells or "heads" of coral, and others are the tiny, complete skeletons of single-celled creatures. Most biogenic sediments, however, consist of the skeletal remains of a variety of organisms rather than just one or two.

We will soon examine these three types of sedimentary rocks in greater detail, beginning with the siliciclastic group. First, however, it is necessary to consider where the clasts of siliciclastic rocks come from.

Erosion and Sediment Production As we have seen, rocks are not indestructible; those that are located at or near Earth's surface can be transformed by chemical reactions or broken by external forces. The products of this decay ultimately move away from the site of origin under the influence of gravity, wind, water, or ice. As we noted in Chapter 1, the term **erosion** is used to describe all of the processes that cause rock to loosen and move downhill or downwind. *Erosion* refers both to the chemical processes that loosen particles of rock and to the physical processes that break rock apart and transport the resulting particles.

The term **weathering** is applied to those aspects of erosion that take place before transport. There are physical as well as chemical weathering processes. Ice, snow, water, and earth movements are the primary agents of physical weathering. Water expands when it freezes, and when it freezes in cracks and crevices within rocks, it exerts such tremendous pressure that it can often split rocks apart. Earth movements, which are discussed in Appendix II, can also fracture rock.

Destructive chemical processes constitute the most pervasive kind of weathering, and water and watery solutions act as its primary agents. Water at Earth's surface, for example, readily converts feldspar to clay, carrying away some ions in the process. Because feldspars are the most abundant minerals in the igneous rocks of continents (Figure AI-3), clay is also very abundant at Earth's surface. Like micas, clay minerals are sheet silicates (Figure AI-3). Clay differs from mica, however, in that the molecular structure of its sheets is weak. Thus clay minerals do not form large flakes, such as those that typify mica; it is the minuteness of their flakes that gives clay sediments their fine-grained texture.

FIGURE AI-11 Weathering of a large boulder of gabbro in Mesa Grande, San Diego County, California. As minerals in the gabbro undergo chemical weathering, layers of rock "spall off" and crumble to clay and other minerals. *(W. T. Schaller, U.S. Geological Survey.)*

Quartz, in contrast to feldspar, is quite resistant to weathering, a characteristic that accounts for the abundance of sand on Earth's surface. Most sand consists of quartz grains that are similar in size to, or slightly smaller than, the quartz grains of granites and similar crystalline rocks.

When the feldspar grains of a crystalline rock weather to clay, the rock crumbles, releasing both the flakes of clay and the grains of quartz as sedimentary particles (Figure AI-11). Rainfall or the meltwaters of snow or ice wash many of these particles into streams, from which they are carried to larger bodies of water. Eventually many of the particles settle from the waters of rivers, lakes, or oceans as sediment, which in time may become hard sedimentary rock.

Both water and oxygen take part in the weathering of mafic rocks, converting these iron-rich minerals to iron oxide minerals that resemble rust. In the process, silica is carried away in solution. Because mafic rocks form by solidification of magmas at higher temperatures than felsic rocks, mafic minerals at Earth's surface are generally less stable than felsic minerals and therefore undergo more rapid chemical weathering. Mafic minerals are rarely seen in the beach sands that fringe oceans, because most of these minerals weather to crumbly iron oxide either within the rock where they formed or close to the area in which that rock was exposed to air and water. In the same manner, feldspars are rarely found on sandy beaches, because most turn to clay either within or close to their parent rocks.

Siliciclastic Sedimentary Rocks Siliciclastic sedimentary particles and rocks are classified according to grain size, as Figure AI-12 illustrates. In this definitional system, the term **clay** denotes particles that are smaller than $\frac{1}{256}$ millimeter. Although this quantitative definition may seem redundant and contradictory, because clays also constitute a family of sheet silicate minerals, the use of the term *clay* in reference to both mineralogy and particle size seldom creates confusion. Nearly all clay mineral particles are of clay size and nearly all particles of clay size are clay minerals. **Mud** is a term that embraces aggregates of clay, silt, or some combination of clay and silt. Material of **sand** size ranges from $\frac{1}{16}$ millimeter to 2 millimeters, and **gravel** includes all particles larger than sand, including **granules, pebbles, cobbles,** and **boulders.**

When sediment settles from water, coarse-grained particles settle faster than fine-grained particles, as can be seen when sediments of various grain sizes are mixed with water in a tall glass container and allowed to settle (Figure AI-13). Clay settles so slowly in water that in nature very little of it falls from rapidly moving water such as that of a river channel or a wave-ridden sandy shore of the ocean. Instead, most of it is deposited in calm waters such as those of lakes, quiet lagoons, and the deep sea.

Not only do coarse sediments settle more quickly from water than do fine sediments, but they are also less easily picked up or rolled along surfaces by moving water. Gravel, for example, tends to be deposited near its source area (the area in which it was originally produced by erosion); thus gravel usually accumulates along the flanks of the mountains from which it is eroded, seldom reaching the center of large oceans.

Siliciclastic rocks are classified according to composition as well as grain size. Rocks formed largely of clay-sized particles, for example, consist primarily of flakes of clay minerals, because few other minerals are abundant in nature at such a small size. When rocks such as these exhibit a tendency to break along bedding surfaces, they

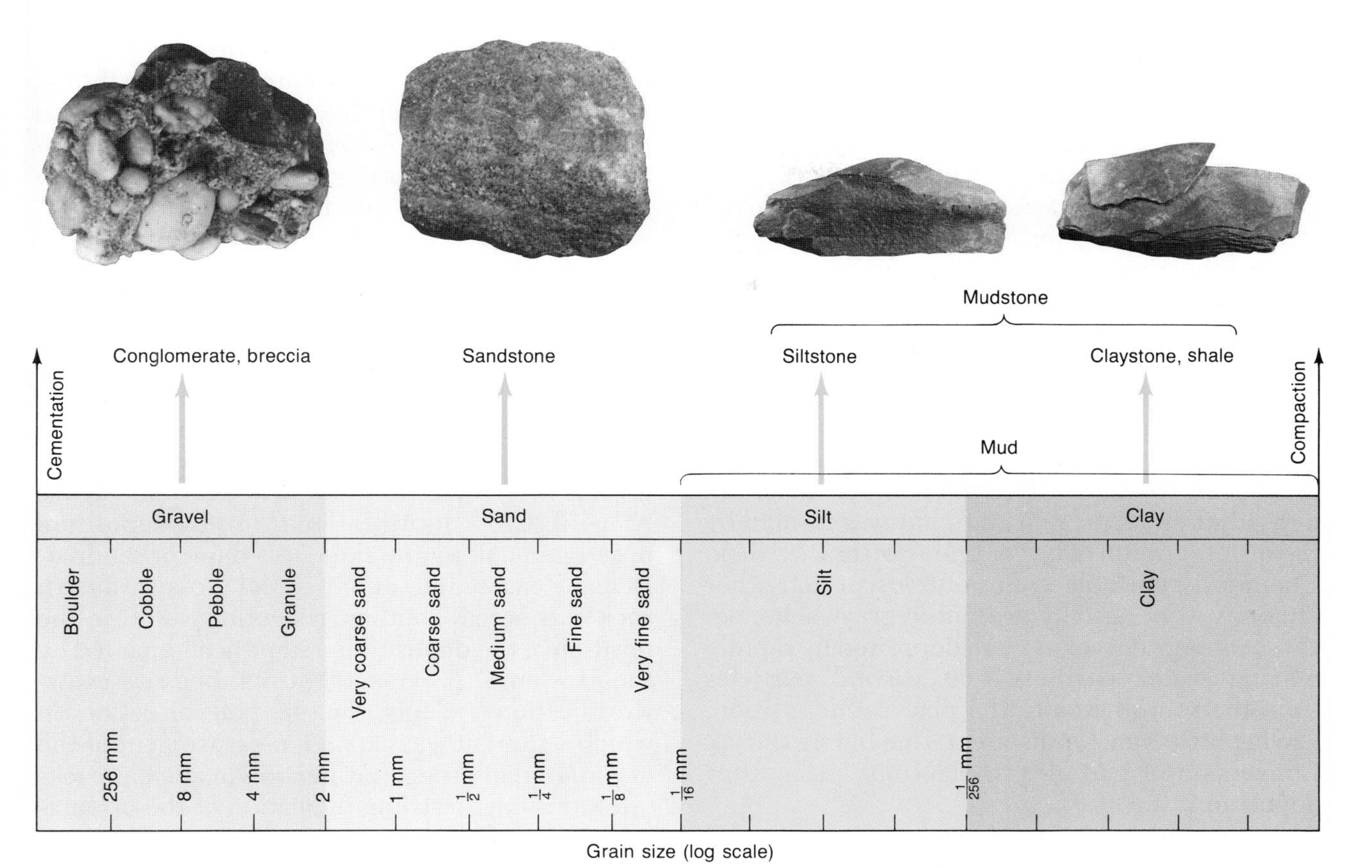

FIGURE AI-12 Classification of sedimentary rocks according to grain size. Sediments range in size from clay to silt, sand, and gravel; gravel is divided into granules, pebbles, cobbles, and boulders. Gravelly rocks are called conglomerates when their pebbles and cobbles are rounded and breccias when they are angular; these rocks normally contain sand as well. Rocks in which sand dominates are called sandstones. Rocks in which silt dominates are called siltstones. Rocks formed of clay are called claystone if they are massive and shale if they are fissile (or platy). Siltstones, claystones, and shales are all varieties of mudstone. All of the specimens shown here would fit in the palm of your hand. *(Photographs by the author.)*

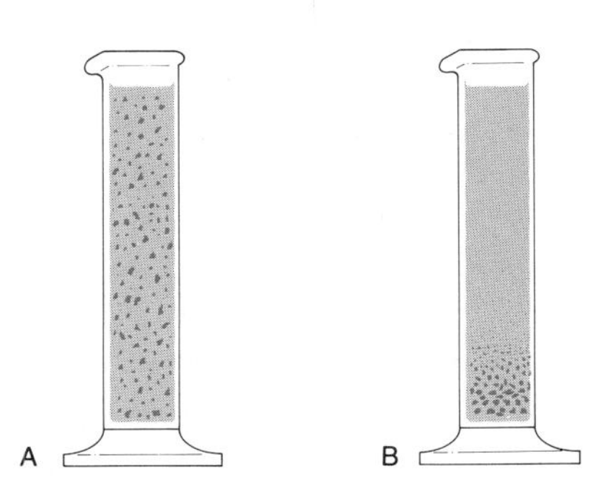

FIGURE AI-13 The settling pattern of sediment after it is suspended in water *(A)*. The coarsest sediment settles most quickly and therefore ends up at the bottom of the deposit; the finest sediment settles last *(B)*.

are said to be **fissile** and are called **shale.** Not all rocks composed primarily of clay are fissile, however, because fissility results from the horizontal alignment of flakes of clay during deposition. Thus **claystone,** a rock that is identical in composition to shale, exhibits little or no fissility because of the irregular orientation of its clay particles.

Most sedimentary grains of silt and sand size are composed of quartz. Even though quartz is very hard, quartz grains suffer abrasion as they bounce and slide downstream along the floors of rivers. In the process, they tend to become smaller and more rounded. Because so many sand-sized grains are quartz, the word **sandstone** is sometimes automatically interpreted to mean quartz sandstone. There are, however, several other kinds of siliciclastic rocks that consist largely of sand-sized grains. One of these is **arkose,** whose primary constituents, grains of feldspar, often give the rock a pinkish color. Arkose usually accumulates only in proximity to its parent rock, soon after the feldspar grains are released by partial weathering and erosion.

Another important rock in which sand-sized particles normally predominate is **graywacke,** which is so designated because it is usually dark gray. Graywacke consists of a variety of sedimentary particles, including sand-sized and silt-sized grains of feldspar and dark rock fragments and also substantial amounts of clay. Most of the clay in graywacke was not carried to the environment of deposition in its present state but was formed by the disintegration of larger grains within the rock. Chemically unstable grains of feldspar and other minerals were initially present in graywacke, because most graywackes were deposited by rapidly moving water currents that carried particles from the source area to the place of deposition, leaving little time for disintegration before burial. Conversion of particles to clay took place after deposition.

A rock that contains large amounts of gravel is termed a **conglomerate** if the gravel is rounded and a **breccia** if it is angular. In both cases, however, sand or granules nearly always fill the spaces between the pieces of gravel. Grains are said to be *poorly sorted* when they are of mixed sizes, and the implication is that moving water did not separate the grains well according to size before they were deposited. Sand along a beach, in contrast, has usually been washed and transported by water currents and waves, and thus it tends to be *well sorted*—that is to say, it tends to consist of particles whose size range is very narrow. Most of the particles in a handful of beach sand are likely to be either medium- or fine-grained.

Siliciclastic grains vary not only in size and chemical composition but also in the manner in which they are arranged within rocks. When the cross section of a sedimentary rock is examined, distinct divisions are revealed. These divisions, which are called **beds** if they are thicker than 1 centimeter and **laminations** if they are thinner, usually represent discrete depositional events. Often the grains at the base of one bed are either coarser or darker and heavier than those of the bed below—or else these grains differ conspicuously in some other way that reflects a later and slightly different origin. As described in Chapter 3, bedding patterns, which are termed **sedimentary structures,** reflect modes of deposition and provide useful tools for interpreting the environments in which ancient sediments were deposited.

A **graded bed** is one in which grain size decreases from the bottom to the top (Figure AI-14*A*). This pattern usually results from the normal settling process that characterizes sediments of mixed grain size, with the coarser sediment settling more rapidly than finer sediment (Figure AI-13).

When beds or laminations are deposited at an angle from the horizontal, the pattern that they form is referred to as **cross-stratification** (Figure AI-14*B*). Cross-stratification forms by deposition of sediment along the slope of a dune or sandbar. In many cases, one set of parallel cross-strata in a rock cuts across another, reflecting a shift in the position of the depositional slope accompanied by the erosion of previous deposits. Because cross-stratification reflects the general direction in which sediment was carried, measurement of the orientation of cross-stratification in ancient rocks often provides a strong indication of the orientation of an ancient river or shoreline.

Ripples, which are small elongate dunes that form at right angles to the direction of wind or water movement, can also reveal the direction of water movement in ancient environments. Ripples often produce cross-stratification as well (Figure AI-14*C*).

A variety of processes transform soft siliciclastic sediments into hard rock. These processes, which affect sediments after deposition but before metamorphism, are collectively referred to as **diagenesis.** Both physical and chemical diagenetic processes harden, or lithify, sediment. The primary physical process of lithification is **compaction,** a process in which grains of sediment are squeezed together beneath the weight of overlying sediment. The extent of compaction varies with grain size. Sand grains, for example, are deposited closely packed, so sand-sized sediments experience only a small amount of compaction. When clay particles are deposited, on the other hand, they are often separated by films of water that represent as much as 60 percent of the total sediment volume. This is why muddy sediments usually experience a great deal of compaction after burial, as water is squeezed from them. In the process, the sediments become hardened.

A B C

FIGURE AI-14 Sedimentary structures. *A.* A graded bed, with gravel at the base (5 centimeters [~2 inches]). *B.* Cross-stratification in a sandstone. *C.* Ripples on a bed of sandstone. *(Photographs by the author.)*

Chemical diagenesis takes many forms, one of the most important of which is the lithifying process of **cementation.** In this process, minerals crystallize from watery solutions that percolate through the pores between grains of sediment. The cement that is thus produced may or may not have the same chemical composition as the sediment. Sandstone, for example, is often cemented by quartz but more commonly by calcite. Cement is most easily studied by examination of thin sections, which are slices of rock ground so thin that they transmit light and thus can be examined by microscope (Figure AI-15). Cementation is less extensive in clayey sediments than in clean sands because after clays undergo compaction, they are relatively impermeable to mineral-bearing solutions. Clean sands are initially much more permeable than clays, but their porosity is reduced—and sometimes virtually eliminated—by pore-filling cement.

FIGURE AI-15 Cement bordering quartz grains in sandstone as seen microscopically in a thin section. Here the rounded sand grains appear gray or white in polarized light, depending on their orientation, and granular calcite cement between the grains is iridescent. *(Peter Kresan.)*

Cement sometimes gives sedimentary rocks their color. This is often the case for red siliciclastic sedimentary rocks called **red beds.** Some red beds are mudstones, some are sandstones, and some are conglomerates, but almost all derive their color from iron oxide, which acts as a cement.

Chemical and Biogenic Sedimentary Rocks The grains that form chemical and biogenic sedimentary rocks were not produced by erosion but instead were precipitated within the bodies of water in which they accumulated. Some chemical and biogenic sediments are difficult to distinguish from one another, so we will consider both types.

The most common chemical sedimentary rocks are **evaporites,** which form from the evaporation of seawater or other natural water. Many evaporites are massive, well-bedded deposits that consist of vast numbers of crystals (Figure AI-16). Among the most important evaporites are **anhydrite** (calcium sulfate, $CaSO_4$) and **gypsum** (calcium sulfate with water molecules attached, $CaSO_4 \cdot H_2O$). It is important to note, however, that the terms *anhydrite* and *gypsum* refer both to the minerals with these names and to rocks that are composed largely of these minerals. *Halite* is

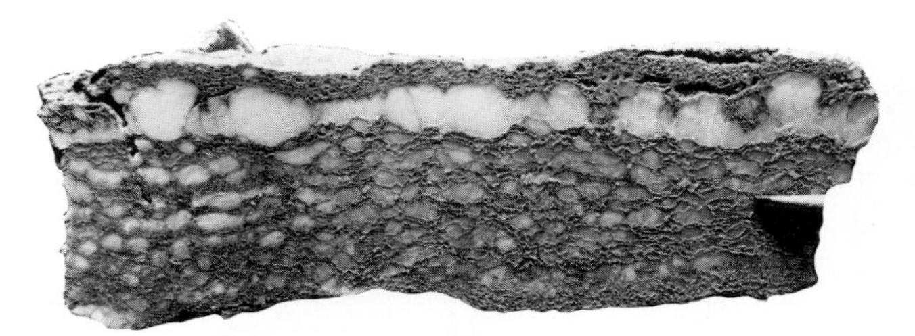

FIGURE AI-16 Nodular gypsum rock of Miocene age from Sicily. *(L. A. Hardie.)*

another term that refers both to a mineral and to an evaporite rock. It is the presence of sodium chloride in large amounts that makes seawater salty, so it is no surprise that halite (Figure AI-1) should accumulate in large quantities when seawater evaporates. Halite deposits have great economic value, providing us both with table salt and with the rock salt that is used to melt ice on highways.

Evaporites are readily precipitated from water but are also readily dissolved, so they do not survive long at Earth's surface except in arid climates. When evaporites are buried far beneath younger deposits, however, they are protected from potentially destructive groundwater and thus can survive for long geologic intervals.

Other types of chemical sediments are less abundant than evaporites. Among the most important of these are chert, phosphate rocks, and iron formations. **Chert,** which is also called **flint,** is composed of extremely small quartz crystals that have been precipitated from watery solutions. Some cherts occur as bedded rocks, while others appear as irregular, rounded masses called **nodules** (Figure AI-17). Typically, impurities give chert a gray, brown, or black color. Chert also breaks along curved, shell-like surfaces; American Indians took advantage of this feature when they fashioned chert into arrowheads. Some cherts form within bedded rock such as limestone when nodules or layers grow from silica-rich solutions that have moved through the rock. On the other hand, some bedded cherts are thought to have formed by direct precipitation of silica (SiO_2) from seawater, and others are biogenic deposits that result from the accumulation on the seafloor of the microscopic skeletons of single-celled organisms. These skeletons consist of a type of silica that differs from quartz in that it is amorphous (or noncrystalline). During diagenesis, water percolates through deposits of these skeletons, converting them to very hard chert that consists of minute interlocking quartz crystals (Figures AI-17 and AI-18). Cherts older than 100 million years or so have suffered extensive chemical diagenesis; thus many are difficult to identify as biogenic or chemical.

Phosphate rocks consist primarily of calcium phosphate. Although they are of marine origin, their precise mode of formation is not known. It would appear that these rocks form in areas where marine life is abundant, because this is thought to be the source of their phosphate.

Iron formations are complex rocks that usually consist of oxides, sulfides, or carbonates of

FIGURE AI-17 A chert nodule. *(Photograph by the author.)*

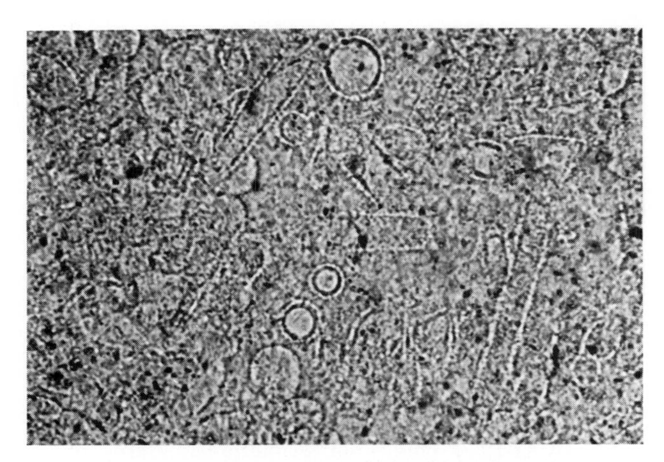

FIGURE AI-18 Photomicrograph of lithified siliceous ooze that forms the Caballos Formation of mid-Paleozoic age. "Ghosts" of sponge needles and radiolarians are still visible. *(E. F. McBride.)*

iron, often in association with chert. Iron formations are widespread only in very old Precambrian rocks, and many form important iron ore deposits. They are examined in greater detail in Chapter 9.

Limestones include both chemical and biogenic bodies of rock. Because they are not as soluble in water as evaporites, limestones are much more common at Earth's surface, where they are quarried extensively for the production of building stone, gravel, and concrete. Although ancient limestones consist primarily of the mineral calcite (Figure AI-4), many of their grains were initially composed of the mineral aragonite. (You will recall that aragonite has the same chemical composition as calcite but differs in its crystal structure.) Aragonite can form at the temperatures and pressures that exist at Earth's surface, but it is relatively unstable under these conditions and in time becomes transformed into calcite. Little or no aragonite remains in most limestones older than a few million years.

Dolomite is a carbonate mineral that is relatively uncommon in modern marine environments but common in many ancient rocks. As mentioned earlier, it differs from calcite in that half of the calcium ions of the crystal lattice are replaced by magnesium ions. In fact, much dolomite has formed by the chemical alteration of calcite. When dolomite is the dominant mineral of an ancient rock, the rock is also called dolomite. Because limestones and dolomites are often intimately associated and frequently intergrade with one another in mineral composition, they are sometimes referred to collectively as **carbonate rocks.** Similarly, unconsolidated sediments consisting of aragonite, calcite, or both minerals are often called **carbonate sediments.**

Carbonate sediments form in two ways: by the direct precipitation of aragonite needles from seawater and through the activities of organisms. Many types of marine life, including corals and most mollusks, grow shells or other kinds of skeletons that consist of aragonite or calcite. At death or even earlier, these organisms contribute skeletal material to the seafloor as sedimentary particles. Some of these particles retain their original sizes, while others diminish in size through breakage or wear. The product of this biological contribution is an array of carbonate particles that are similar to siliciclastic grains and, like siliciclastic grains, can be classified according to size. Thus we speak of carbonate sands and carbonate muds, and we find that carbonate sands display many of the sedimentary structures that we also see in siliciclastic sands, including cross-stratification. Most carbonate particles that are sand-sized or larger can be seen to be skeletal particles (Figure AI-19), but it is often difficult to determine the origin of mud-sized material. Even **aragonite needles,** which are the primary components of carbonate muds, are produced both by direct precipitation and by the collapse of carbonate skeletons, especially of algae. In ancient fine-grained limestones, aragonite needles have been transformed to tiny calcite grains. The resulting granular texture reveals little about the configuration or mode of origin of the original carbonate particles.

Calcium carbonate is precipitated only from seawater that contains relatively little carbon dioxide, and carbon dioxide is less soluble in warm water than in cold water. Thus carbonate sediments accumulate primarily in tropical seas, where winter water temperatures seldom drop below 18°C (~64°F). This condition not only explains the direct precipitation of carbonate minerals in tropical or near-tropical seas but also accounts for the fact that few organisms that live in cold water secrete massive skeletons of calcium

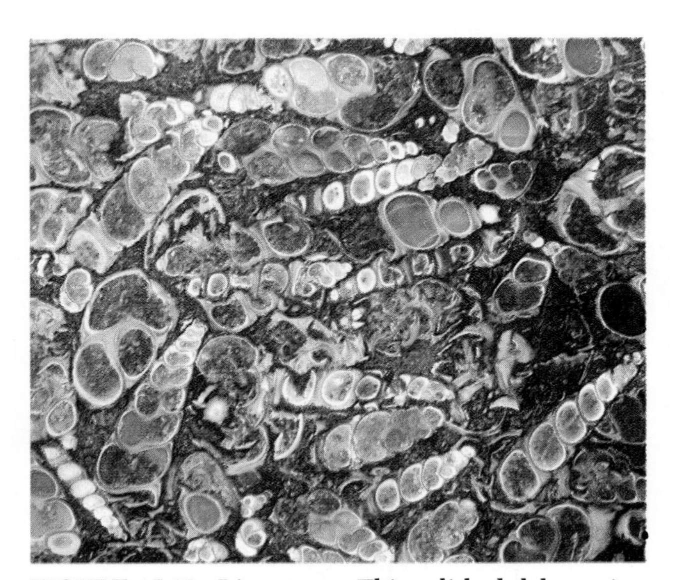

FIGURE AI-19 Limestones. This polished slab consists primarily of Eocene snails, which average 2–3 centimeters (~1 inch) in length. *(Peter Kresan.)*

carbonate. Carbonate sediments can also form in freshwater habitats, usually as a result of the carbonate-secreting activities of certain algae.

Unlike siliciclastic clays, carbonate rocks do not compact greatly after burial. Instead, they harden primarily by cementation, in the same manner as siliciclastic sands. Carbonate sediments are nearly always cemented by carbonate minerals simply because rich sources of such cements are close at hand.

Metamorphic Rocks

Metamorphic rocks form by the alteration of other rocks at temperatures and pressures that exceed those normally found at the Earth's surface. Metamorphism alters both the composition and the texture of all kinds of other rocks—igneous rocks, sedimentary rocks, and rocks that are already metamorphic.

There are three fundamental types of metamorphism: dynamic, regional, and contact. **Dynamic metamorphism,** which results from movements of earth, causes rock to fracture and even pulverize along zones of movement and then welds the resulting fragments together in new textures. In this type of metamorphism, pressure plays a greater role than temperature. **Regional metamorphism** transforms deeply buried rocks at high temperatures and pressures; as its name implies, this process operates over areas whose dimensions are measured in hundreds of kilometers. **Contact metamorphism** is caused by igneous intrusion, which "bakes" surrounding rock. This is usually a local phenomenon that may occur deep within Earth or near the surface. High temperature usually plays a larger role in contact metamorphism than high pressure.

We will take a closer look at the three classes of metamorphism and the rocks that they produce.

FIGURE AI-20 A cataclastic breccia formed by the fracturing of rock during an igneous intrusion. The large fragments consist of the metamorphic rock hornfels, which was produced by local heating of rocks by magma. Later, pressure from the igneous intrusion fragmented the hornfels to form the breccia. *(H. C. Granger, U.S. Geological Survey.)*

Dynamic Metamorphism When movements of the solid Earth or of magma cause rocks to bend intensively or to break apart, the rocks may be shattered or their grains squeezed into new shapes and orientations. The resulting rocks are called **cataclastics** (Figure AI-20). Most cataclastic rocks become lithified at the time of deformation, when heat and especially pressure are so intense that they cause the particles of these rocks to grow together and interpenetrate. Nonetheless, dynamic metamorphism is primarily a physical rather than a chemical process.

Regional Metamorphism For reasons that are discussed in Chapter 7, igneous activity usually extends along the length of an actively forming mountain chain. Along each side of such an igneous belt is a zone of regional metamorphism produced by high temperatures and pressures extending outward from the igneous belt. Most rocks in zones of regional metamorphism display a texture known as **foliation,** which is an alignment

FIGURE AI-21 Foliated metamorphic rocks. Slate *(A)* is a low-grade metamorphic rock, schist *(B)* is a medium-grade rock, and gneiss *(C)* is a high-grade rock. *(Photographs by the author.)*

of platy minerals caused by the pressures applied during metamorphism (Figure AI-21). **Gneiss** is a coarse-grained metamorphic rock whose intergrown crystals resemble those of igneous rock but whose minerals tend to be segregated into wavy layers. **Schist** consists largely of platy minerals such as micas, which lie roughly parallel to one another in such a way that the rock tends to break along parallel surfaces. **Slate** is a finer-grained rock in which aligned platy minerals produce fissility much like that of shale. The mineral alignment of slate, however, results from deformational pressures rather than from depositional orientation of the sort that produces the bedding of shale.

Metamorphic rocks of a particular texture can form over a wide range of temperature and pressure conditions. Mineral assemblages of metamorphic rocks, however, serve as critical "thermometers" and "barometers," varying with the temperature and pressure of metamorphism. White micas, for example, occur only in low- and medium-grade metamorphic rocks. (**Grade** is the word used to indicate level of temperature and pressure.) The green mica-like mineral **chlorite** is more restricted in its occurrence, because it is exclusively a low-grade mineral. In contrast, **pyroxenes** (Figure AI-3) are restricted to rocks of intermediate and high grade. The grade of metamorphism in a regional metamorphic zone typically declines away from the neighboring belt of igneous activity that supplied the heat for metamorphism.

Not all rocks in regional metamorphic zones are foliated. Some have homogeneous, granular textures, which indicate not only that their interlocking mosaics of crystals lack preferred orientations but also that certain minerals are not segregated into bands. **Marble** and **quartzite,** for example, are usually homogeneous, granular metamorphic rocks. Marble consists of calcite, dolomite, or a mixture of the two, and it forms from the metamorphism of sedimentary carbonates. Quartzite consists of nearly pure quartz, and it forms from the metamorphism of quartz sandstone. The simple mineralogical composition of marble and quartzite, together with their lack of platy minerals, prevents both of these rocks from exhibiting foliation even when they form under great pressure.

Contact Metamorphism Contact metamorphism is a more localized phenomenon, but it is like regional metamorphism in displaying a gradient: The grade of metamorphism declines away from the heat source. **Hornfels** usually forms in areas adjacent to local igneous intrusions. This is a fine-grained granular metamorphic rock of varying composition that is formed at high temperatures (Figure AI-20). Farther away are lower-grade metamorphic rocks that may be either granular or foliated (Figure AI-22).

The Upper Limit of Metamorphism When conditions become so hot that a rock melts, metamorphism ceases. Rocks that form when the resulting

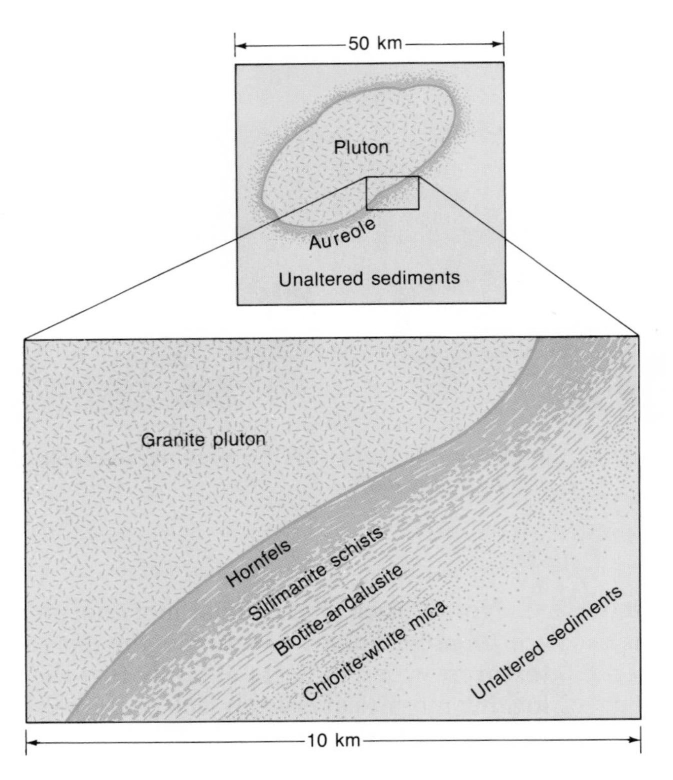

FIGURE AI-22 Contact metamorphism adjacent to a granite pluton. The heat from the magmatic intrusion that formed the pluton metamorphosed the sediments that were intruded. The sediments closest to the pluton were metamorphosed to hornfels, a fine-grained metamorphic rock that forms at high temperatures. A series of metamorphic zones with other mineral assemblages extends away from the pluton, with assemblages formed at progressively lower temperatures away from the pluton. *(After F. Press and R. Siever, Earth, W. H. Freeman and Company, New York, 1982.)*

molten material cools are then classified as igneous. Very high-grade metamorphic rocks that nearly pass through a molten state before cooling resemble coarse-grained igneous rocks in that they are granular rocks that display little or no foliation. Those that consist largely of quartz and feldspar are known as **granulites.** High-grade metamorphic rocks are difficult to distinguish from igneous rocks, and the origins of many high-temperature rocks are the subject of debate.

APPENDIX

Deformation Structures in Rocks

It has long been known that the rocks of Earth's crust and the fossils within them can be fractured and deformed on a very large scale. Especially where mountains have been uplifted, pieces of crust extending many square kilometers have been transported great distances, and large bodies of rock have been contorted in complicated ways. In Chapter 1 we noted that the study of these movements is called tectonics. Many of the forces that have caused such movements result from the motions of large plates of the lithosphere—motions that are considered in Chapters 6 and 7 under the label of plate tectonics. This appendix summarizes the kinds of geologic structures that result from the application of large forces to rock. It also illustrates how these visible structures allow us to decipher earth movements of the past.

BENDING AND FLOWING OF ROCKS

We can see from geologic outcrops that rocks have been warped, twisted, and folded, and that they have even flowed. Appendix I makes the statement that most rocks consist of discrete mineral grains. When forces are applied to a rock, the component grains may be affected in various ways; they may slide past one another, change shape, or break along parallel planes. (In the last case, a grain deforms in the way a deck of cards on a table changes shape when we push on one end of the deck near the top.) Internal deformation of a large body of rock by any of these mechanisms takes place very slowly, but when many of the grains are affected, the entire body of rock can undergo radical changes in shape in the course of millions of years. Such changes usually take place at great depths within Earth's crust.

One common type of large-scale rock deformation is referred to as **folding.** Compressive forces can shorten Earth's crust, creating folds (Figure 1-14) that come in many sizes. When folded sedimentary rocks are viewed with their oldest beds at the bottom and their youngest beds on top, the folds that are concave in an upward direction, with their vertexes at the bottom, are termed **synclines,** while those that are concave in a downward direction, with their vertexes at the top, are termed **anticlines** (Figure AII-1). Many rocks do not simply bend when they are folded; instead, material is displaced from one part of a bed toward another. When a large, complex body of rock is subjected to an external force, certain weak beds are often more intensely folded than other, more durable beds. Shales, for example, tend to be weak and to deform quite severely in comparison with massive sandstones and limestones that are subjected to the same forces.

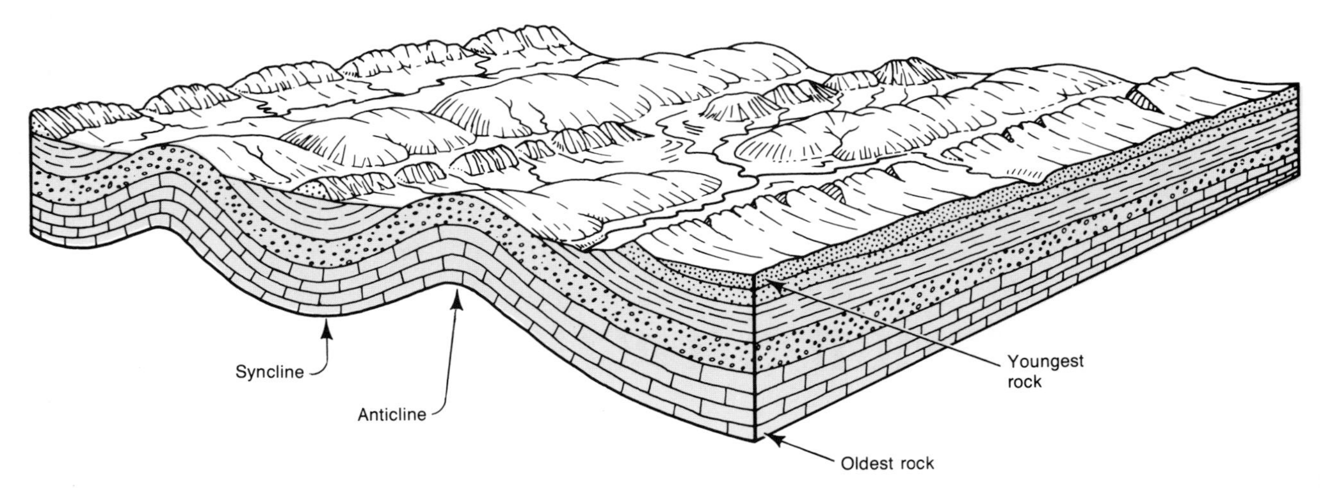

FIGURE AII-1 Block diagram of humplike folds, or anticlines, and troughlike folds, or synclines. *(After F. Press and R. Siever, Earth, W. H. Freeman and Company, New York, 1986.)*

Igneous and metamorphic rocks are relatively durable, but they, too, can be folded.

To describe shapes of folds in greater detail, geologists have developed special terminology. A tilted bed, for example, is said to have a **dip**— a term that describes the angle that the bed forms with the horizontal plane. In other words, the dip is the direction in which water would run down the surface of the bed. The **strike** of a bed, in contrast, is the compass direction that lies at right angles to the dip (Figure AII-2); strikes are always horizontal. It is sometimes said that the **regional strike** for a given area is in a particular geographic orientation — north–south, say. This does not mean that every strike in this area has the same orientation, only that most of the fold axes trend north–south, so that the strikes of most beds do too.

A fold is said to have an **axial plane,** which is an imaginary plane that cuts through the fold and divides it as symmetrically as possible. In actuality, many folds are asymmetrical, with one limb (or flank) dipping more steeply than the other. If either limb is rotated more than 90° from its original position, the fold is said to be **overturned** (Figure AII-3); a fold lying on one limb, with the axial plane nearly horizontal, is said to be **recumbent.**

The **axis** of a fold is the line of intersection between the axial plane and the beds of folded rock. Often the axis plunges, which means that it lies at an angle to the horizontal (Figure AII-4*A*).

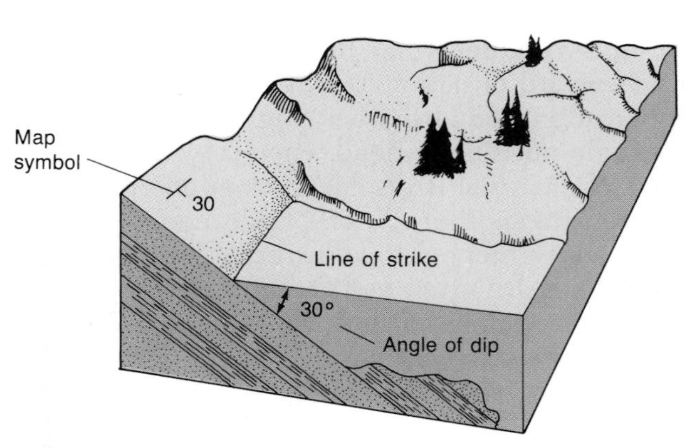

FIGURE AII-2 Block diagram illustrating the strike and dip of inclined beds.

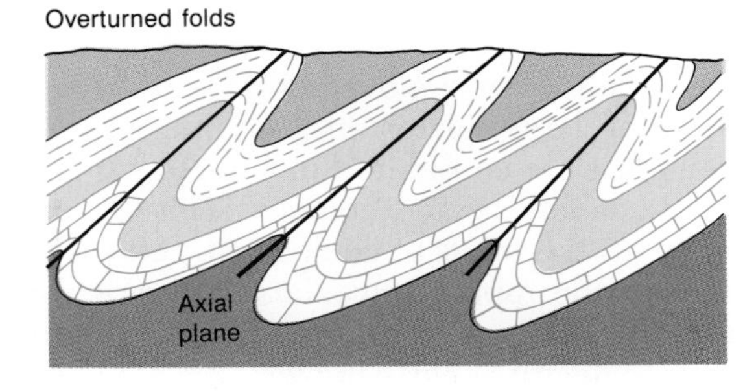

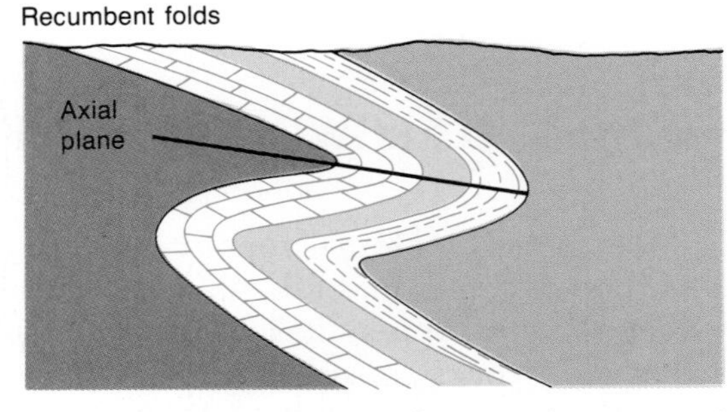

FIGURE AII-3 Cross-sectional views of overturned and recumbent folds. Both limbs of an overturned fold dip in the same direction. One limb of a recumbent fold is upside down. *(After F. Press and R. Siever, Earth, W. H. Freeman and Company, New York, 1986.)*

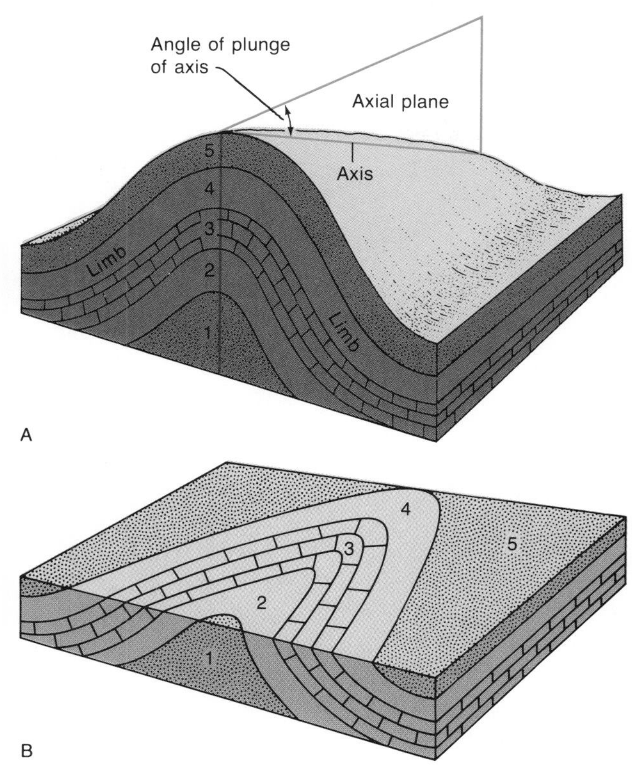

FIGURE AII-4 A plunging fold before *(A)* and after *(B)* truncation resulting from erosion. Note how erosion produces a curved outcrop pattern.

When a **plunging fold** is truncated by erosion, its beds form a curved outcrop pattern (Figures **AII-**4*B* and **AII-5**). A series of plunging folds then produces a scalloped surface pattern (Figure **AII-6**). Because most folds plunge (we could not expect many to be perfectly horizontal), this scalloped pattern is characteristic of regions where sedimentary rocks have been extensively folded.

Stratified rocks do not always bend into folds such as those just described. Local uplift of a sedimentary sequence can form a **dome,** and local depression can form a **basin.** Both of these structures, when eroded, yield concentric, circular bands of outcrop for stratified rocks. There is a simple way to distinguish the outcrop pattern of a dome from that of a basin. In a dome, the oldest beds lie in the center, whereas in a basin, it is the youngest beds that are centrally positioned (Figure AII-7). Some basins are enormous. The entire state of Michigan, for example, forms the central part of a structural basin (Figure AII-8). It is important to understand, however, that some structural domes form topographic basins when the oldest (central) beds happen to be weak so that they erode easily. This is true of the basin shown in Figure AII-9, in which many of the beds forming the flanks stand well above the heavily eroded

FIGURE AII-5 Aerial view of a plunging fold. The view is along the axis of the Virgin Anticline of southwestern Utah. Compare with Figure AII-4. *(J. S. Shelton, Geology Illustrated, W. H. Freeman and Company, New York, 1966.)*

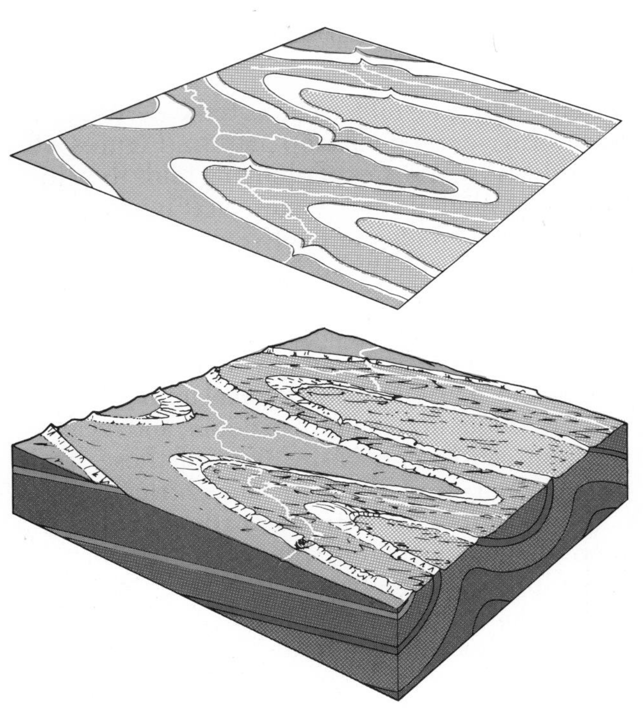

FIGURE AII-6 Block diagram *(below)* and map *(above)* showing the scalloped outcrop pattern of a series of plunging folds. Note that the resistant rock units form ridges on the eroded surface. *(After W. K. Hamblin and J. D. Howard, Exercises in Physical Geology, Burgess Publishing Company, Minneapolis, 1975.)*

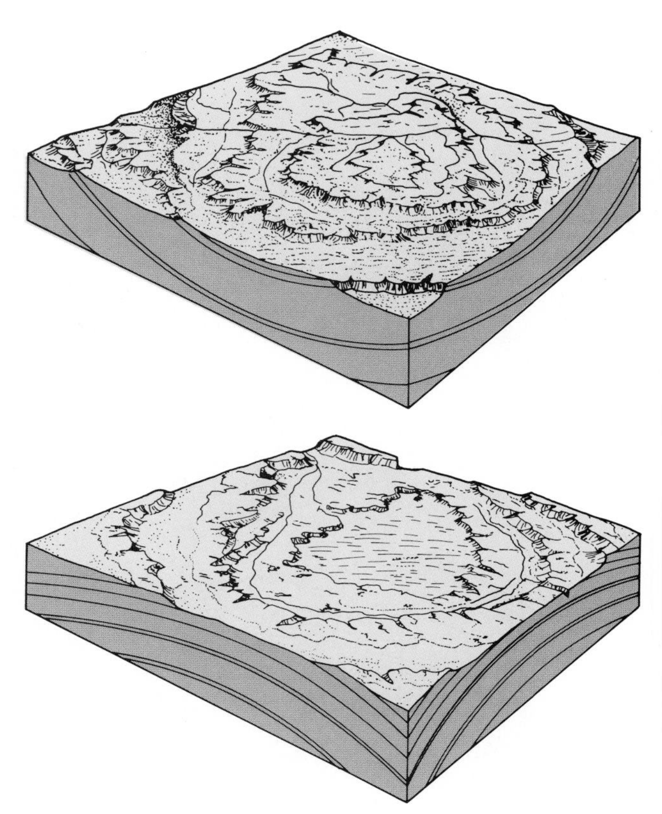

FIGURE AII-7 The concentric outcrop pattern of a structural basin *(below)* and of a structural dome *(above)*. The rocks in the center of the structural basin are relatively durable and thus have remained at a high elevation; those in the center of the structural dome are relatively weak and have been deeply eroded to form a topographic basin. *(After W. K. Hamblin and J. D. Howard, Exercises in Physical Geology, Burgess Publishing Company, Minneapolis, Minnesota, 1975.)*

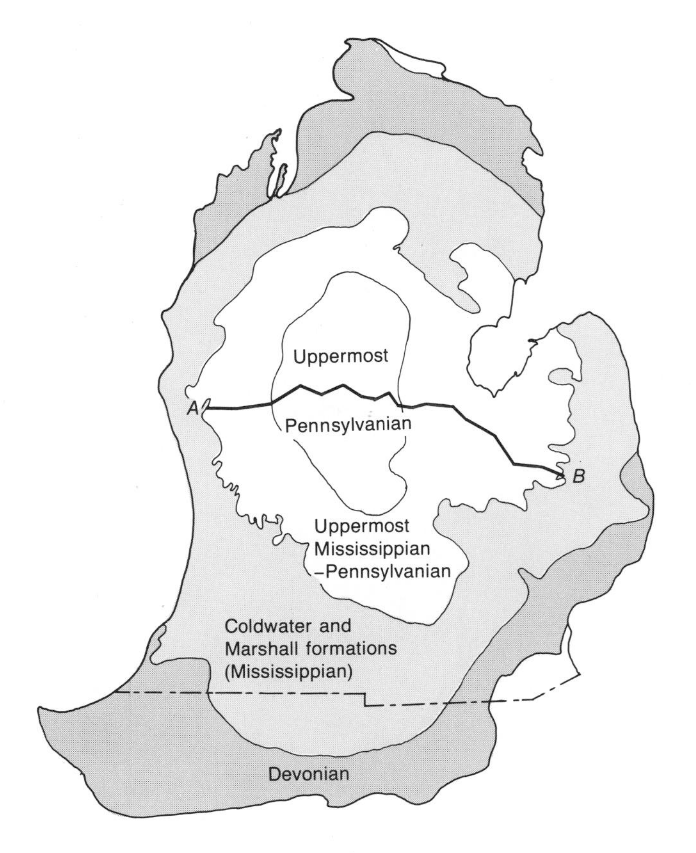

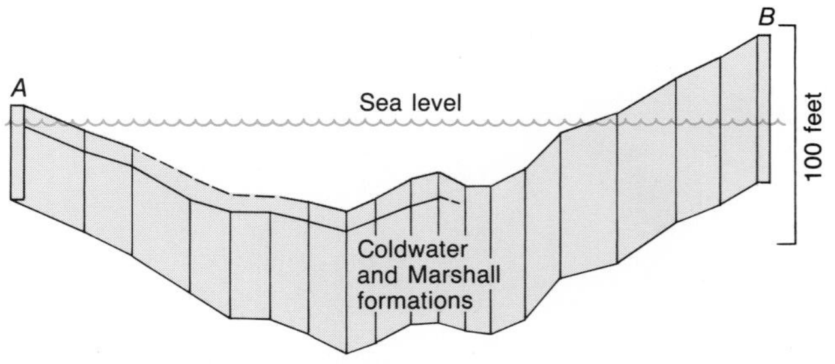

FIGURE AII-8 A simplified geologic map of the state of Michigan. The fact that the youngest rock units are in the center indicates that this is a structural basin. The cross section from *A* to *B* shown above illustrates the configuration of the Coldwater and Marshall formations, which occur throughout most of the state but are buried beneath younger deposits in the center. This cross section has been constructed from information obtained by drilling. *(After V. Brown Monnett, Amer. Assoc. Petrol. Geol. Bull. 32:629–688, 1948.)*

FIGURE AII-9 Structural dome near Rawlings, Wyoming. Here weak rocks in the center have been eroded deeply to form a topographic basin. *(J. S. Shelton, Geology Illustrated, W. H. Freeman and Company, New York, 1966.)*

"core." Similarly, the centers of structural basins sometimes stand at high elevations.

Figure AII-9 also shows that a dome can be oblong rather than circular. An extremely elongate dome amounts to an anticline that plunges in two directions. Similarly, a very long basin amounts to a syncline that has upturned ends.

BREAKING OF ROCKS

Rocks do not always bend or flow when they are heavily stressed. Sometimes they behave in a more brittle fashion and simply break. **Joints** are fractures in rock that can result from various kinds of stress. In some bodies of rock, joints are oriented randomly, but a particular source of stress often forms a set of nearly parallel joints. Two or more intersecting sets give many outcrops a blocky or columnar appearance. Joints are simple fractures, along which there is no appreciable movement of the opposing rock surfaces. **Faults,** on the other hand, are fractures along which measurable movement has occurred. A fault, like a bed of rock, has an orientation that can be described by a strike and a dip. Various kinds of faults are seen in Figure AII-10. **A normal fault** has a steep

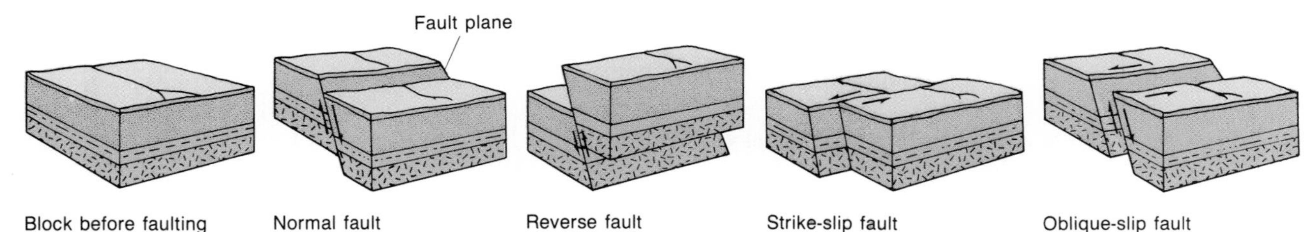

FIGURE AII-10 Several types of faults. *(After F. Press and R. Siever, Earth, W. H. Freeman and Company, New York, 1986.)*

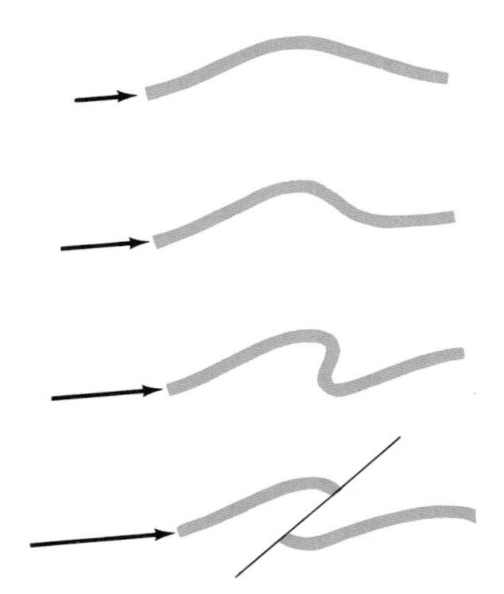

FIGURE AII-11 An overturned fold can give rise to a thrust fault as force continues to be applied. *(After F. Press and R. Siever, Earth, W. H. Freeman and Company, New York, 1986.)*

dip—one that is closer to vertical than horizontal—and the rocks above the fault plane move downward in relation to those below. This type of fault is frequently formed when extensional stresses spread rocks apart; the rocks break and then the upper block slides downward under the influence of gravity. A **reverse fault,** in contrast, usually results from compression, which causes the rock to fracture and the block above the fracture to slide up over the lower block. **Strike-slip faults** result from horizontal shearing forces—forces that break the rock at a high angle and move the blocks on each side of the break horizontally past each other. Rocks may move only a few centimeters along a strike-slip fault, or they may move hundreds of kilometers. An **oblique-slip fault** is intermediate in character between a strike-slip fault and a normal or reverse fault.

Low-angle reverse faults, termed **thrust faults,** can account for enormous earth movements when mountains form. Large, tabular segments of crust known as **thrust sheets** may be transported for tens or even hundreds of kilometers along fault planes that lie just a few degrees from horizontal.

Thrust faults are causally related to overturned folds. A force that compresses bedded rocks to form an overturned fold may eventually break the bedded rocks and move the upper limb of the fold along a thrust fault (Figure AII-11). Thus it is not surprising that belts of deformation within many mountain chains have been formed by a combination of folding and thrusting. Elongate zones of such structures are known as **fold-and-thrust belts** (Figure AII-12).

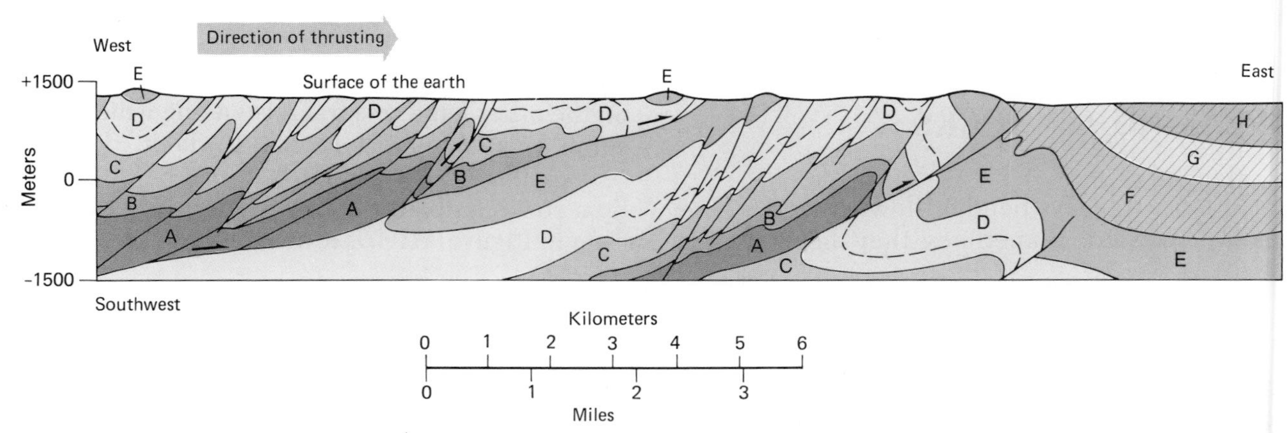

FIGURE AII-12 A fold-and-thrust belt of the Rocky Mountains southwest of Calgary, Alberta, Canada. Thrust faults, shown as dark lines, are intimately associated with overturned folds. Thrusting has been toward the east, and folds are overturned in the same direction. The oldest beds *(A)* are Lower Carboniferous, and the youngest *(H)* are Paleogene. *(After P. B. King, The Evolution of North America, Princeton University Press, Princeton, New Jersey, 1977.)*

Movement along most faults is sporadic. Between movements, stress within Earth builds up against friction along the fault plane. Finally the stress overcomes the friction, and a sudden movement takes place. After stress builds up again, the process is repeated, and movement thus occurs in small steps. Sudden movement along a fault causes Earth to shudder: an earthquake occurs. It is difficult to know just when stress will overcome friction along a fault, and this is why we cannot accurately predict an earthquake.

New faults have originated during recorded history, and measurable movement has also taken place recently along old faults. Active faults often produce visible linear features at Earth's surface (facing p. 128). Rocks are often fractured along fault planes, and fracturing tends to invite erosion where faults intersect Earth's surface. Thus faults may be expressed topographically as valleys. The Great Glen Fault, for example, runs the full width of Scotland and is generally marked by narrow valleys, one of which cradles Loch Ness, the lake famous for its apparently mythical monster.

STRUCTURAL CROSS SECTIONS

Sequences of folding and faulting leave records that can be deciphered to reveal Earth movements of the past. An example is seen in Figure AII-13, an idealized cross section of rocks within the Basin and Range Province of western North America. Here the sequence of events began with deposition of horizontal sediments on top of crystalline basement *(A)*, which was followed by folding and thrusting from the west *(B)*. Next, an erosion surface developed, truncating the folds *(C)*. Then a sheet of lava was spread over the erosion surface *(D)*. Finally, normal faults broke the terrain into a series of blocks *(E)*. We can reconstruct the sequence of these events by working backward. The normal faulting had to be the last event, because the normal faults cut through all other features—through the sheet of lava, the folds, and the thrust faults. The next-to-last event must have been the extrusion of the lava sheet, because it was laid down on an erosion surface that truncated all other features except the normal faults. This implies that the folding and thrusting were the first mode of deformation. By using this kind of logic, we can unravel local geologic history.

Paleozoic and Triassic sedimentary rocks that form the upper walls of the mile-deep Grand Canyon of Arizona provide us with a more complex exercise in the analysis of regional geologic history. Rocks of every Paleozoic period but the Ordovician and Silurian are present along this great valley (an unconformity separates Devonian rocks from Cambrian rocks). Figure AII-14 shows how the regional history can be read from the Grand Canyon walls.

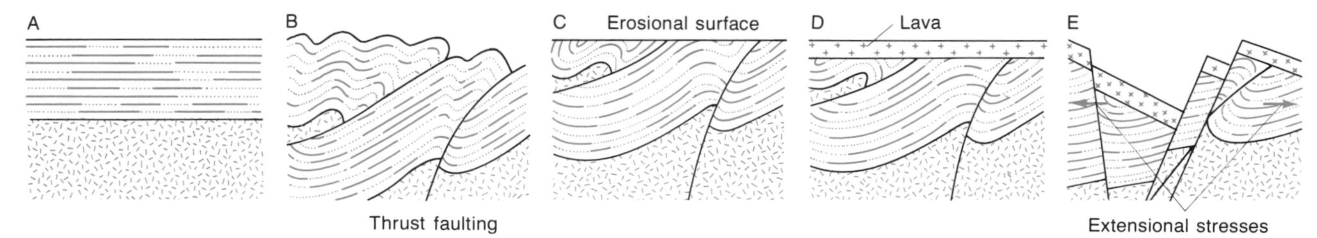

FIGURE AII-13 Stages in the development of the Basin and Range Province of western North America, according to W. M. Davis. *A.* Sedimentary units are deposited. *B.* The sedimentary deposits are folded and thrust-faulted by compressive forces. *C.* An erosion surface develops. *D.* Sheets of lava are spread over the erosion surface. *E.* Block faults break up the terrain. The final configuration *(E)* is all that geologists see, and from this they reconstruct the earlier events. *(P. B. King, The Evolution of North America, Princeton University Press, Princeton, New Jersey, 1977.)*

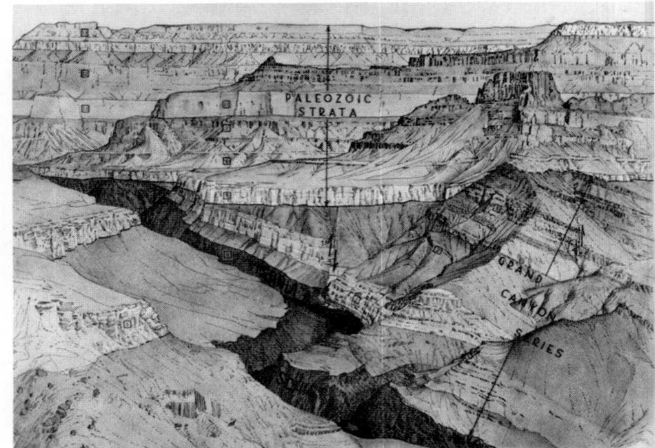
PALEOZOIC STRATA
GRAND CANYON SERIES

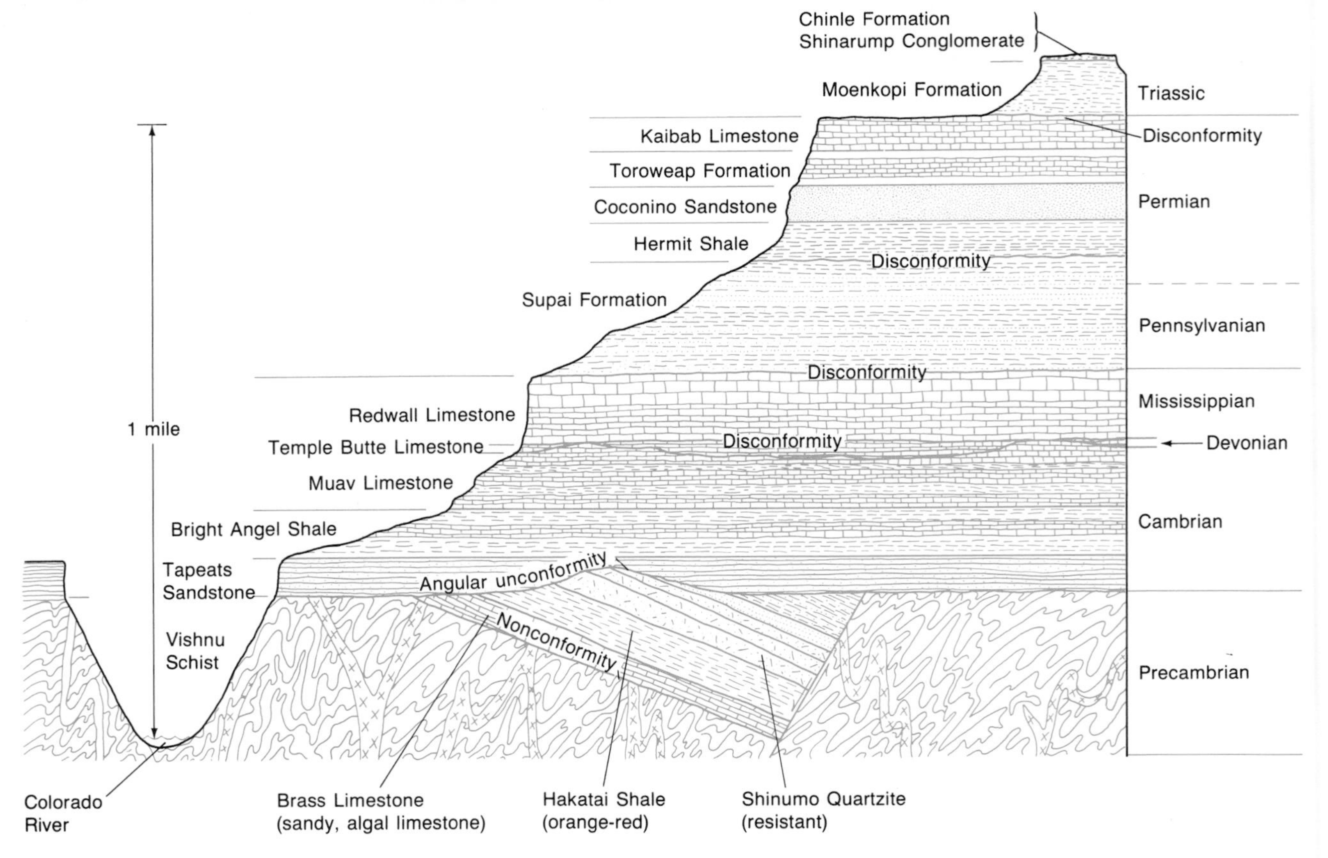
Chinle Formation
Shinarump Conglomerate
Moenkopi Formation
Triassic
Disconformity
Kaibab Limestone
Toroweap Formation
Coconino Sandstone
Permian
Hermit Shale
Disconformity
Supai Formation
Pennsylvanian
Disconformity
Mississippian
Redwall Limestone
1 mile
Temple Butte Limestone
Disconformity
Devonian
Muav Limestone
Bright Angel Shale
Cambrian
Tapeats Sandstone
Angular unconformity
Nonconformity
Vishnu Schist
Precambrian
Colorado River
Brass Limestone (sandy, algal limestone)
Hakatai Shale (orange-red)
Shinumo Quartzite (resistant)

FIGURE AII-14 The history of the Grand Canyon region. Here the sequence of Paleozoic and Triassic rocks rests right side up on Precambrian rocks and is separated from them by an angular nonconformity. In some places the lowest Cambrian unit, the Tapeats Sandstone, overlies the Precambrian sedimentary rocks, and in others it overlies a metamorphic unit, the Vishnu Schist. The Vishnu itself is occasionally intruded by granite rocks. The intimate relationship between the intrusions and the metamorphic structures of the Vishnu Schist indicates that intrusion and metamorphism took place at the same time. A summary of the events recorded in the Grand Canyon region is as follows:

Step 1. The earliest event that we can recognize is the deposition of the Vishnu sediments. We cannot reconstruct the properties of these sediments in detail.

Step 2. Next came the metamorphism of the Vishnu sediments to schist and the accompanying intrusion of the sediments by granitic rocks.

Step 3. The Precambrian sedimentary unit numbered 1 lies unconformably above the Vishnu. Step 3, then, is the development of an erosion surface on the Vishnu.

Step 4. The deposition of units 1 through 4.

Step 5. In places, faults have elevated units 1 through 4 so that they are in contact with the older Vishnu Schist. The normal, or block, faulting that had this effect must therefore have followed the deposition of units 1 through 4.

Step 6. After the block faulting, erosion planed off many irregularities in the topography.

Step 7. On the erosional surface thus formed, the Paleozoic and Triassic sediments were laid down, with a major unconformity separating Cambrian and Devonian deposition (the Muav and Temple Butte limestones) and with minor unconformities separating some younger units.

Later Steps. The section of rocks exposed in the Grand Canyon does not in itself reveal the history of the region after deposition of the Triassic sediments that cap the sedimentary sequence. Regional studies, however, show that Cenozoic uplift of the plateau over the past 2 million years or so led to the formation of the present topography.

APPENDIX

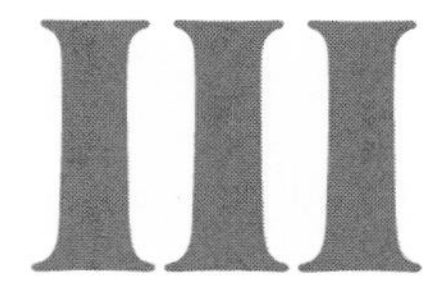

Classification of Major Fossil Groups

This is a compilation of important fossil taxa. Listed with most groups are figure numbers directing the reader's attention to illustrations of one or more fossils or living representatives. An examination of these figures and their captions in the context of this appendix will provide a comprehensive overview of fossil life. Bear in mind that because the fossil record is incomplete, many taxa undoubtedly lived longer than the time ranges indicated for them here. Also, classifications are subjective, and some workers employ classifications other than those adopted here.

KINGDOM MONERA

Includes all **prokaryotes,** which are single-celled organisms whose DNA is not organized into chromosomes or housed in a nucleus. Mitochondria and chloroplasts are lacking. Some experts divide the Monera into two kingdoms: one primitive and the other more advanced.

Phylum **"Fermenting bacteria"** Obtain energy by breaking down organic compounds. Cannot tolerate oxygen. *(Archean–Recent)*

Phylum **Thiopneutes (sulfate-reducing bacteria)** Obtain energy by converting sulfate to hydrogen sulfide (H_2S), which smells like rotten eggs. Live in mud. Cannot tolerate oxygen. *(Archean–Recent)*

Phylum **"Anaerobic photosynthetic bacteria"** The most primitive photosynthetic organisms. Some types tolerate very high or low temperatures, but few tolerate oxygen. *(Archean–Recent)*

Phylum **Cyanobacteria** (**cyanophytes,** also called blue-green bacteria) Photosynthetic cells that are threadlike or nearly spherical. Form algal mats and stromatolites. *(Archean–Recent)* Figures 3-30, 3-31, 8-21, 8-22, 8-23, 8-24, 9-5, 9-8, and 9-9; also opposite page 1, pages 210, 212, and 242

KINGDOM PROTOCTISTA

Single-celled and simple multicellular eukaryotes, including algae.

Phylum **Dinoflagellata** Floating algae, mostly single-celled. Characterized by two whiplike flagella. Important phytoplankton in lakes and in the ocean. Some species live as symbiotic individuals in the tissues of corals or other animals. *(Triassic–Recent)* Figure 2-26; also pages 338 and 368

Informal Group **Acritarcha** An artificial group of hollow, organic-walled fossils that apparently represent algae. *(Proterozoic–Recent)* Figures 9-10 and 10-9

Phylum **Haptophyta (calcareous nannoplankton)** Photosynthetic, single-celled floating algae whose nearly spherical cells are armored by shieldlike plates of calcium carbonate. Important phytoplankton in the tropics, where their plates accumulate on quiet seafloors. *(Paleozoic? Triassic–Recent)* Figures 2-26, 13-5, 14-2, and 14-32; also pages 338 and 368

Phylum **Bacillariophyta (diatoms)** Photosynthetic, single-celled algae, protected by pillbox-like skeletons composed of silica. Some species are planktonic and some are benthic; some are marine and others are lake dwellers. Important phytoplankton in lakes and cool seas. *(Cretaceous–Recent)* Figures 2-26 and 16-2; also page 368

Phylum **Rhodophyta (red algae)** Photosynthetic multicellular algae in which the chlorophyll is contained in reddish intracellular structures. Predominantly marine. Coralline red algae produce the massive algal ridge of modern coral reefs. *(Proterozoic? Paleozoic–Recent)*

Phylum **Chlorophyta (green algae)** Photosynthetic single-celled and multicellular algae whose chlorophyll gives them a green color. Some are calcareous and contribute vast quantities of calcium carbonate to the sedimentary record. *(Proterozoic? Paleozoic–Recent)*

Phylum **Foraminiferida (foraminifera)** Single-celled animal-like marine forms with skeletons (tests) through which projections of protoplasm extend. Some (superfamily **Globigerinacea**) are planktonic but most are benthic. Some experts unite them with the radiolarians and with *Amoeba* and its relatives in the phylum **Sarcodina.** *(Cambrian–Recent)* Figures 12-6, 12-7, 14-1, and 14-9; also page 80

Phylum **Actinopoda (radiolarians** and their relatives) Single-celled animal-like marine forms with radially symmetrical skeletons (tests) through which projections of protoplasm extend. The skeleton consists of silica or strontium sulfate. *(Cambrian–Recent)* Figures 3-37 and AI-18

KINGDOM PLANTAE

Plants. Multicellular, sexually reproducing eukaryotes with chloroplasts. More complex than algae. All plants except some mosses and liverworts have vascular tissue for fluid conduction. (Phyla marked with asterisks are **gymnosperms,** or vascular plants with naked seeds.)

Phylum **Bryophyta** (**mosses** and **liverworts**) Small plants that lack well-developed fluid-conducting tissues and live in moist environments. *(Late Paleozoic–Recent)*

Phylum **Psilophyta** Small vascular plants with simple stems. Mostly extinct. *(Middle Paleozoic)* Figures 11-16 and 11-19

Phylum **Lycopodophyta (lycopods — club mosses** and their relatives) Spore-bearing plants in which the leaves are arranged in a helical pattern around the stem. All living species are small and inconspicuous, but many extinct species, including the dominant plants of Carboniferous coal swamps, were large trees. *(Silurian–Recent)* Figures 11-17, 11-18, and 12-9; also page 298

Phylum **Sphenophyta** (**sphenopsids — horsetails** and their relatives) Spore-bearing plants in which the stem is divided into nodes that bear whorls of branches. Living members are small plants, but some Carboniferous species were trees. *(Devonian–Recent)* Figure 12-12

Phylum **Filicinophyta (ferns)** Spore-bearing plants with branching leaves that have spore-bearing organs on their lower surface. *(Devonian–Recent)* Figure 13-14; also pages 298 and 338

*Phylum **Cycadophyta** Primitive seed plants, most of which have the form of shrubs or trees.

Class **Pteridospermales (seed ferns)** Seed-bearing plants with fernlike foliage, some the size of bushes or trees. *(Middle Paleozoic–Jurassic)* Figures 6-3, 12-10, and 12-11; also page 298

Class **Cycadeoidales** (extinct relatives of **cycads**) Cycad-like plants of a primitive type. *(Triassic–Cretaceous)* Figure 13-16

Class **Cycadales** (modern **cycads**) Plants with column-like trunks and feather-like leaves. *(Triassic–Recent)* Figures 2-17 and 13-16; also page 338

Class **Cordaitophyta (cordaites)** Trees with tall trunks and elongate leaves. *(Late Paleozoic)* Figure 12-13

*Phylum **Ginkgophyta (ginkgos)** Deciduous trees with small, fan-shaped leaves that have regularly branching veins. Only a single species survives today. *(Early Mesozoic–Recent)* Figure 13-15

*Phylum **Coniferophyta (conifers**—cone-bearing gymnosperms, including **pines, spruces,** and **firs)** Shrubs and trees with cones and needle-shaped leaves. *(Late Paleozoic–Recent)* Figures 12-14, 13-16, and 13-21

Phylum **Angiospermophyta (flowering plants,** including **grasses** and **hardwood trees)** Seed plants in which the gamete-bearing generation is hidden within the flower. *(Cretaceous–Recent)* Figures 2-18, 5-2, 14-15, 14-16, 14-27, 15-5, 15-31, and 16-3; also Box 15-1 and pages 338 and 430

KINGDOM ANIMALIA

Animals. Multicellular organisms that obtain their nutrition by consuming other organisms.

Phylum **Porifera (sponges)** Very simple sedentary aquatic animals that strain food from water that they pump through pores in their walls. Supporting the tissues of most are spinelike spicules of calcite or silica that can be preserved as fossils. Include the **Stromatoporoidea,** which were important reef builders during the Paleozoic. *(Cambrian–Recent)* Figures 10-1*C*, 10-16, 12-27, and AI-18

Phylum **Archaeocyatha** Vase-shaped, double-walled organisms that attached to hard substrata to form reeflike structures. They probably sieved food from water that passed through holes in their skeletons. *(Cambrian)* Figure 10-11

Phylum **Coelenterata** or **Cnidaria** Saclike animals having inner and outer body layers with jelly-like material in between. Most are carnivores that capture food with stinging cells.

Class **Scyphozoa (jellyfishes)** Floating marine forms with limited swimming capacity. Most capture food with dangling tentacles. *(Precambrian–Recent)*

Class **Anthozoa (corals, sea anemones,** and **sea whips** and **fans)** Benthic forms whose tentacles extend upward.

Order **Rugosa (rugose corals,** or **tetracorals)** Corals with fourfold symmetry. Most species were solitary "cup corals." *(Cambrian–Permian)* Figures 10-13, 10-15A, 11-3, and 11-4

Order **Tabulata (tabulates)** Colonial, often reef-building corals with horizontal platforms (tabulae) in their skeletons. Some forms assigned to this group may actually have been sponges. *(Ordovician–Recent)* Figures 10-13, 10-16, 11-4, and 11-5

Order **Scleractinia (hexacorals)** Solitary or colonial corals with septa (vertical partitions) of the skeleton present in multiples of six. Most colonial species have symbiotic dinoflagellates in their tissues that play a role in the building of reefs. *(Triassic–Recent)* Figures 3-3 and 5-10; also Box 2-1 and page 338

Phylum **Bryozoa** or **Ectoprocta (moss animals)** Colonial aquatic animals in which the connected individuals are tiny structures with tentacles used to capture food suspended in the water. Many Paleozoic types grew upright as fanlike or branching stony colonies. Most post-Paleozoic types have less heavily calcified skeletons and encrust hard substrata. *(Ordovician–Recent)* Figures 10-13, 10-15*C*, 12-5, and 14-10

Phylum **Brachiopoda (lamp shells)** Double-valved marine animals that capture suspended food by means of a loop-shaped structure bearing tentacles.

Class **Inarticulata** Brachiopods that lack teeth. Since the Cambrian, they have included fewer species than the Articulata. *(Cambrian–Recent)* Figure 10-4

Class **Articulata** Brachiopods with hinge teeth; that is, teeth that lock their valves together. Most attach to the substratum by means of a fleshy stalk called a pedicle. *(Cambrian–Recent)* Figures 10-4, 10-13, 10-14, 10-15*C*, 10-19, 11-1, and 12-2

Phylum **Mollusca** A diverse group of invertebrates, most of which have a foot and a flaplike mantle that secretes a shell.

Class **Monoplacophora** Primitive crawling forms with cap-shaped shells. *(Cambrian–Recent)* Figure 10-1*A*

Class **Gastropoda (snails)** Mostly crawling forms with coiled shells and bodies that are twisted into a U shape. Some are marine, some freshwater, and some terrestrial. *(Cambrian–Recent)* Figures 10-13, 10-17, 11-2, 14-12, and **AI**-19; also page 368

Class **Bivalvia** (**bivalves—clams, mussels, oysters, scallops,** and their relatives) Mollusks with two shell halves (valves). Some are infaunal and others epifaunal. Some feed on suspended matter and others consume sediment, from which they digest organic matter. *(Cambrian–Recent)* Figures 5-19, 10-13, 11-2, 13-1, 13-24, 14-11, 14-13, 14-14, and 14-25; also page 368

Class **Cephalopoda** (**octopuses, squids, nautiloids,** and **ammonoids**) Carnivores that swim by jet propulsion. They capture prey with tentacles and tear it apart with a beak.

Subclass **Nautiloidea** Possess a chambered shell in which the partitions (septa) are attached to the wall along slightly curved lines. *Nautilus* is the only living genus. *(Cambrian–Recent)* Figures 10-13 and 11-6; also page 242

Subclass **Ammonoidea** Descendants of nautiloids in which the septa are attached to the shell wall along wavy lines. *(Devonian–Cretaceous)* Figures 11-6, 12-1, 13-6, 14-3, and 14-4; also page 270

Subclass **Coleoidea** (**squids, octopuses,** and their relatives, including the extinct **belemnoids**) Groups in which the shell is reduced or lacking. *(Late Paleozoic–Recent)* Figure 13-7

Phylum **Annelida** Segmented worms, some of which are marine, others freshwater, and still others (including earthworms) nonmarine. They are responsible for many of the burrows visible in marine sediments younger than about 700 million years. *(Late Precambrian–Recent)* Figures 9-19 and 10-9*B*

Phylum **Echinodermata** Marine animals having fivefold radial symmetry (most starfishes have five arms); internal skeletons that consist of plates of calcite; and small, elongate suction cups (tube feet) for feeding or locomotion. Figure 10-5

Class **Echinoidea (sea urchins)** Globe-shaped and disc-shaped species with movable spines. Most are grazers or deposit feeders. *(Ordovician–Recent)* Figures 10-18, 13-1, 13-2, and 15-4

Class **Stelleroidea (starfishes)** Mostly carnivorous forms that can pry open bivalves with their tube feet. *(Ordovician–Recent)* Figures 1-9 and 10-13

Class **Crinoidea (sea lilies)** Animals that use branching arms to capture food suspended in the water. Some attach to the seafloor by a stalk, but others are free-living and can swim awkwardly. *(Cambrian–Recent)* Figures 1-8, 10-13, 10-15*B*, 10-20, 12-3, and 12-4

Phylum **Onychophora** Creatures that resemble annelids in having segmented, flexible bodies and arthropods in having jaws derived from legs. They have unjointed legs and are probably evolutionary intermediates between annelids and arthropods. Living species are terrestrial, but Cambrian fossils occur in marine deposits. *(Cambrian–Recent)* Figure 10-8*C*

Phylum **Arthropoda** Forms with an external skeleton and jointed appendages.

Subphylum **Trilobitomorpha (trilobites)** Primitive marine arthropods with a three-lobed body (a central lobe and two lateral lobes) and hard skeletons. *(Cambrian–Permian)* Figures 7-17, 10-2, 10-3, and 10-12; also pages 240 and 242

Subphylum **Chelicerata** (**spiders, scorpions, horseshoe crabs, eurypterids,** and their relatives) Arthropods that possess six front segments that are united into a head as well as a pair of jointed pincers. *(Cambrian–Recent)* Figure 11-7; also page 238

Subphylum **Crustacea (crabs, lobsters, ostracods,** and numerous other groups, including important members of the zooplankton) A varied group of arthropods, mostly aquatic. *(Cambrian–Recent)* Figures 10-7 and 13-4; also page 368

Subphylum **Insecta (insects)** Includes more living species than all other animal taxa combined. *(Devonian–Recent)* Figures 1-10, 2-3, 12-16, 12-20, 13-21, and 15-16; also pages 270, 298, and 300

Phylum **Hemichordata** An inconspicuous group in the modern world, but it includes the Paleozoic graptolites (class **Graptolithina**). These were colonial animals, some of which floated and some of which were attached to the seafloor. *(Cambrian–Carboniferous)* Figure 4-2

Phylum **Chordata** Animals with a dorsal nerve chord and embryonic features that include gill slits in the throat region and a rod of cartilage (notochord) that, in some groups, becomes the vertebral column. Box 10-1

Class **Conodonta (conodonts)** These animals, long known only from their teeth, are now known from rare fossils of their bodies. They were apparently eel-shaped swimmers. *(Cambrian–Triassic)* Figure 10-6

Class **Agnatha (jawless fishes)** Include the living lamprey and extinct Paleozoic **ostracoderms.** *(Cambrian–Recent)* Figures 10-10, 11-8, 11-9, and 11-10

Class **Acanthodii** Primitive jawed fishes of small size with numerous spiny fins. *(Middle and late Paleozoic)* Figures 11-8 and 11-11

Class **Placodermi** Early jawed fishes. Most were heavily armored and a few were very large. *(Middle Paleozoic)* Figures 11-8, 11-12, and 11-13

Class **Chondrichthyes (sharks** and their relatives) Fishes with skeletons composed of cartilage. *(Middle Paleozoic–Recent)* Figures 11-8, 11-13, 12-3, 13-9, 15-3, and 16-1; also Box 11-1

Class **Osteichthyes (bony fishes)**

Subclass **Actinopterygii (ray-finned fishes)** Fishes with fins supported by numerous slender, raylike bones.

Infraclass **Chondrostei** Primitive ray-finned fishes with lungs, a primitive jaw, diamond-shaped scales, an asymmetrical tail, and a partly cartilaginous skeleton. Only a few species, including sturgeons, survive today. *(Devonian–Recent)* Figures 11-8 and 11-14

Infraclass **Holostei** Ray-finned fishes resembling the Chondrostei, but with the lungs transformed into a swim bladder and a slightly more advanced jaw. Only a few species, including the garpike and bowfins, survive today. *(Triassic–Recent)* Figure 13-8

Infraclass **Teleostei** The advanced group that includes most living species of ray-finned fishes. The jaws are highly developed, the scales are rounded and overlapping, the tail is symmetrical, and the skeleton is entirely bony. *(Cretaceous–Recent)* Figures 2-3, 5-13, 5-20, 14-5, 14-6, 15-1, and 16-1

Subclass **Sarcopterygii** Characterized by fleshy fins and the ability to breathe air.

Order **Crossopterygii (lobe-finned fishes)** Fishes with teeth and fin bones resembling those of primitive amphibians, to which they were ancestral. There is only one living genus. *(Devonian–Recent)* Figures 11-8, 11-15, 11-22, and 11-23

Order **Dipnoi (lungfishes)** Freshwater fishes with cylindrical bodies. Some can survive buried in mud for months. *(Devonian–Recent)* Figure 11-8

Class **Amphibia** The oldest terrestrial vertebrates, although they require water for reproduction.

Subclass **Labyrinthodontia** Early amphibians with solid skulls and complex internal tooth structure. Some were quite large. *(Devonian–Triassic)* Figures 11-22, 11-23, 12-17, and 12-20; also pages 270 and 300

Subclass **Salientia (frogs, toads,** and their relatives) Swimming and hopping groups. *(Early Mesozoic–Recent)* Figures 13-19 and 15-15

Subclass **Caudata (salamanders** and their relatives) Small amphibians that possess a tail.

Some remain in aquatic habitats as adults. *(Early Mesozoic–Recent)* Figure 5-18

Class **Reptilia (reptiles)** Vertebrate animals that have scales or armor and reproduce by amniote eggs.

Subclass **Anapsida** Reptiles whose skull has a solid roof.

Order **Cotylosauria** (stem reptiles) The earliest reptiles. *(Late Paleozoic–Triassic)* Figures 12-18 and 12-20; also page 300

Order **Chelonia (turtles)** Reptiles with few or no teeth and a protective shell. *(Early Mesozoic–Recent)* Figures 5-6, 13-21, 14-6, and 14-8*B*; also page 338

Subclass **Euryapsida** Reptiles with a large opening in the skull behind the eye. Mostly marine.

Order **Protorosauria** Ancestral euryapsids that lived on land. *(Permian–Mesozoic)*

Order **Ichthyosauria (ichthyosaurs)** Swimming reptiles, shaped like dolphins, that bore live young. *(Triassic–Cretaceous)* Figure 1-13

Order **Placodontia (placodonts)** Large, turtle-shaped animals that crushed mollusks with rounded teeth. *(Triassic)* Figure 13-10

Order **Sauropterygia** Swimming reptiles with paddle-like limbs and small heads, often on long necks. The primitive **nothosaurs** probably lived along the shore like seals, but the advanced **plesiosaurs** were fully aquatic. Figures 13-11 and 13-12

Subclass **Diapsida** Reptiles with two openings in the skull behind the eye.

Infraclass **Lepidosauria** Primitive diapsids and their descendants, including **lizards, snakes,** and the huge Cretaceous marine lizards called **mosasaurs.** Figures 14-6 and 14-7; also page 430

Infraclass **Archosauria** Advanced diapsids.

Order **Thecodontia** A varied group that was ancestral to the dinosaurs. Some types were able to run on two legs. Others superficially resembled crocodiles. *(Triassic)* Figures 13-17 and 13-18

Order **Crocodilia (crocodiles** and **alligators)**. Aquatic and terrestrial carnivores with flattened skulls. *(Triassic–Recent)* Figures 13-13, 13-18, 13-21, and 14-20

Order **Pterosauria (flying reptiles)** Winged reptiles with long skulls and short bodies. *(Mesozoic)* Figures 13-22 and 14-21; also page 338

(Group uncertain) **Dinosauria** Although traditionally classified as reptiles, dinosaurs may have been distinct enough to form a separate class, especially if they were endothermic.

Order **Saurischia (lizard-hipped dinosaurs)** This group included carnivores that traveled on two legs and often attained giant proportions, and even larger herbivorous sauropods that walked on all fours. *(Triassic–Cretaceous)* Figures 1-7, 13-18, 13-20*A*, 13-21, 14-19, and 14-21; also page 334

Order **Ornithischia (bird-hipped dinosaurs)** Herbivorous forms, some of which traveled on two legs (for example, duck-billed dinosaurs) and others that traveled on all fours. *(Triassic–Cretaceous)* Figures 1-7, 13-20*B*, 13-21, 14-17, 14-18, and 14-21; also Box 13-1

Subclass **Synapsida (mammal-like reptiles)** Terrestrial animals with an opening in the skull behind the eye and with highly differentiated teeth. Like the dinosaurs, they may deserve a taxonomic position apart from the Reptilia, with which they have traditionally been classified.

Order **Pelycosauria** Early mammal-like reptiles, including forms in which the vertebral spines were elongated to support a huge fin or sail of uncertain function. *(Late Paleozoic)* Figures 12-19 and 12-20; also page 300

Order **Therapsida** Advanced mammal-like reptiles, with legs positioned more fully beneath the body than the legs of pelycosaurs. The lower jaw was also formed largely of a single bone. May have been endothermic or partly so. *(Permian–early Mesozoic)* Figures 6-12, 12-21, 14-22, and 15-21; also page 300

Class **Aves (birds)** Endothermic flying vertebrates with feathers. *(Jurassic–Recent)*

Subclass **Archaeornithes** Primitive toothed birds, of which only *Archaeopteryx* is known. Closely resembled dinosaurs in skeletal form. *(Jurassic)* Figure 13-23; also page 338

Subclass **Neornithes** Modern birds. *(Late Mesozoic–Recent)* Figures 5-4, 5-7, 14-18A, 14-21, 15-13, and 15-14; also page 430

Class **Mammalia** Vertebrates that have hair and suckle their young.

Subclass **Eotheria** Small, primitive mammals that evolved from mammal-like reptiles. *(Early Mesozoic)* Figure 13-17; also page 338

Subclass **Prototheria** Egg-laying mammals, of which only the living echidna and platypus are known. *(Recent)* Figure 14-22

Subclass **Allotheria (multituberculates)** Small mammals with complex teeth for grinding food. *(Jurassic–Paleogene)*

Subclass **Theria** The group that includes nearly all modern mammals.

Infraclass **Pantotheria** Forms whose dentition suggests that they were ancestral to marsupial and placental mammals. *(Mesozoic)*

Infraclass **Metatheria (marsupials)** Mammals that rear their offspring in a pouch. *(Cretaceous–Recent)* Figures 5-15, 15-1, and 16-30

Infraclass **Eutheria (placentals)** The dominant group of modern mammals except in Australia, where marsupials prevail. *(Cretaceous–Recent)* See Figure 15-6 for a summary of the stratigraphic distribution of placental mammal orders; other placentals are represented in Figures 1-6, 1-12, 2-3, 2-5, 2-16, 3-2, 5-1, 5-5, 5-9, 5-10, 5-15, 5-16, 5-17, 5-21, 14-22, 15-2, 15-7, 15-8 through 15-13, 15-17 through 15-21, 16-4, 16-9, 16-30, and 16-33 through 16-36; also Box 5-1 and pages 400, 404, and 430

APPENDIX

Stratigraphic Stages

In many parts of the world the geologic record has been divided into stages. As discussed in Chapter 4, stages are time-stratigraphic units. For the most part, the stages recognized in Europe have become the standard stages with which stages defined elsewhere are correlated. Correlations remain imperfect, however, as do estimates of the absolute ages of stage boundaries. This appendix is a reference for students who encounter unfamiliar stage names in their studies. Figure AIV-1 lists major Paleozoic and Mesozoic stages that were first defined in Europe and shows how a number of North American stages are currently believed to correlate with them. Figure AIV-2 presents the same kind of information for Cenozoic stages, showing how European stages are thought to relate to American stages that are based on biostratigraphic zones for fossil land mammals.

	AGE (million years)	SYSTEM		SERIES	STAGE (European)	STAGE (North American)
CENOZOIC			QUATERNARY	PLEISTOCENE		
		NEOGENE	TERTIARY	PLIOCENE	*(See Figure AIV-2)*	
				MIOCENE		
	23	PALEOGENE		OLIGOCENE		
				EOCENE		
				PALEOCENE		
MESOZOIC	65	CRETACEOUS		UPPER	Maastrichtian	
					Campanian	
					Santonian	
					Coniacian	
					Turonian	
					Cenomanian	
				LOWER	Albian	
					Aptian	
					Barremian	
					Hauterivian	
					Valanginian	
					Berriasian	
	146	JURASSIC		UPPER	Tithonian	
					Kimmeridgian	
					Oxfordian	
				MIDDLE	Callovian	
					Bathonian	
					Bajocian	
					Aalenian	
				LOWER (LIAS)	Toarcian	
					Pliensbachian	
					Sinemurian	
					Hettangian	
	208	TRIASSIC		UPPER	Rhaetian	
					Norian	
					Carnian	
				MIDDLE	Ladinian	
					Anisian	
				LOWER	Scythian	
	245					

ERA	AGE (million years)	SYSTEM		SERIES	STAGE (European)	STAGE (North American)	(SERIES)
PALEOZOIC		PERMIAN		UPPER	Tatarian	Ochoan	
					Ufimian/Kazanian	Guadalupian	
				LOWER	Kungurian	Leonardian	
					Artinskian	Leonardian	
					Sakmarian	Wolfcampian	
					Asselian	Wolfcampian	
	290	CARBON-IFEROUS	PENNSYLVANIAN	UPPER	Stephanian	Virgilian	
					Westphalian	Missourian	
					Westphalian	Desmoinesian	
					Westphalian	Atokan	
					Westphalian	Morrowan	
	323		MISSISSIPPIAN	LOWER	Namurian	Springerian	
					Namurian	Chesterian	
					Visean	Meramecian	
					Visean	Osagean	
					Tournaisian	Kinderhookian	
	363	DEVONIAN		UPPER	Famennian	Chautauquan	
					Frasnian	Senecan	
				MIDDLE	Givetian	Erian	
					Eifelian	Erian	
				LOWER	Emsian	Ulsterian	
					Siegenian	Ulsterian	
					Gedinnian	Ulsterian	
	409	SILURIAN		UPPER	Ludlovian	Cayugan	
				LOWER	Wenlockian	Niagaran	
					Llandoverian	Medinan	
	439	ORDOVICIAN		UPPER	Ashgillian	(shaded area)	
					Caradocian	(shaded area)	
				LOWER	Llandeilian	Chazyan	Champlanian
					Llanvirnian	Whiterockian	Champlanian
					Arenigian		Canadian
					Tremadocian		Canadian
	510	CAMBRIAN		UPPER	Dolgellian	Trempealeauan	Croixan
					Maentwrogian	Franconian	Croixan
					Maentwrogian	Dresbachian	Croixan
				MIDDLE	Menevian		Albertan
					Solvan		Albertan
				LOWER	Lenian		Waucoban
					Atdabanian		Waucoban
					Tommotian		Waucoban
	570						

Detail of shaded area

STAGE (North American)	(SERIES)
Richmondian	Cincinnatian
Maysvillian	Cincinnatian
Edenian	Cincinnatian
Trentonian	Mohawkian
Black River	Mohawkian
Ashbyan	Mohawkian

FIGURE AIV-1 Major Paleozoic and Mesozoic stages of Europe and North America. Through correlation and absolute dating, efforts are being made to extend the European stages to all parts of the world.

AGE (million years)	EPOCH		STAGE (European)	STAGE (North American land mammal)
	PLEISTOCENE		*(See Table 18-1)*	RANCHOLABREAN
				IRVINGTONIAN
1.6	PLIOCENE	UPPER	PIACENZIAN	BLANCAN
		LOWER	ZANCLEAN	
5	MIOCENE	UPPER	MESSINIAN	HEMPHILLIAN
			TORTONIAN	
				CLARENDONIAN
		MIDDLE	SERRAVALLIAN	BARSTOVIAN
			LANGHIAN	
		LOWER	BURDIGALIAN	HEMINGFORDIAN
			AQUITANIAN	
23	OLIGOCENE	UPPER	CHATTIAN	ARIKAREEAN
				WHITNEYAN
		LOWER	RUPELIAN	ORELLAN
				CHADRONIAN
34	EOCENE	UPPER	PRIABONIAN	
				DUCHESNEAN
		MIDDLE	BARTONIAN	
			LUTETIAN	UINTAN
				BRIDGERIAN
		LOWER	YPRESIAN	WASATCHIAN
57	PALEOCENE	UPPER	THANETIAN	CLARKFORKIAN
			(UNNAMED)	TIFFANIAN
		LOWER	DANIAN	TORREJONIAN
65				PUERCAN

FIGURE AIV-2 Major Cenozoic stages of Europe and North America. The North American stages, which are currently only crudely correlated with the European stages, are based largely on fossil occurrences of land mammals in the Midwest and West. (Figure 4-17 illustrates the Cenozoic pattern of reversals of Earth's magnetic polarity—a pattern that is also widely employed for global correlation.) The most recent epoch, not listed in this diagram, is known as the Recent or Holocene. This brief epoch began at the end of the Pleistocene, roughly 10,000 years ago.

Glossary

Terms that are used in these definitions and that are also defined in this glossary are in many instances *italicized* for the reader's convenience. For descriptions of important minerals and more extensive discussions of rock types, the reader is referred to Appendix I. Similarly, a more detailed review of rock deformation and its terminology can be found in Appendix II. Taxonomic names of organisms are not included in this glossary; Appendix III outlines the major biological taxa, with references to illustrations in the text.

A

Abyssal plain The broad expanse of seafloor lying between about 3 and 6 kilometers (~ 2 to 4 miles) below sea level.

Accretionary wedge A body of *rocks* that have accumulated above an oceanic *plate* undergoing *subduction*. Slices of *mélange* pile up along *thrust faults* to form the wedge.

Acheulian culture The tool culture of *Homo erectus*, a species of the human family that lived during the Pleistocene Epoch.

Acritarchs An extinct group of apparently *eukaryotic phytoplankton* whose earliest representatives are in Proterozoic rocks.

Active lobe (of a delta) The site on a *delta* where functioning *distributary channels* cause the delta to grow seaward.

Active margin The border of a continent along which *subduction* occurs, producing igneous activity and deformation.

Actualism The interpretation of ancient *rocks* by applying the results of analyses of modern-day geologic processes in accordance with the principle of *uniformitarianism*.

Adaptation A feature of an organism that serves one or more functions useful to the organism.

Adaptive breakthrough An evolutionary innovation that affords a group of organisms a special ecologic opportunity and often leads to the *adaptive radiation* of that group.

Adaptive radiation The rapid origin of many new *species* or higher *taxa* from a single ancestral group.

Age, geologic The division of geologic time smaller than an *epoch*.

Albedo The percentage of solar radiation reflected from Earth's surface. This percentage is higher for ice than for land or water, and usually higher for land than for water.

Alluvial fan A low, cone-shaped structure that forms where an abrupt reduction in slope—for example, the transition from a highland area to a broad valley—causes a stream to slow down.

Amino acid One of the chemical building blocks of a *protein*. There are 20 amino acids, each a unique combination of carbon, hydrogen, oxygen, and nitrogen.

Amniote egg The type of egg laid by reptiles and birds, having a nutritious yolk and a hard outer shell to protect the embryo from the dry environment. The amniote egg is named for the amnion, a sac that contains the embryo.

Angular unconformity An *unconformity* separating horizontal *strata*, above, from older strata that had been tilted and eroded.

Anhydrite The mineral that consists of calcium sulfate ($CaSO_4$), or the *rock* composed of this *mineral*.

Anomaly, magnetic A local increase or decrease in the strength of Earth's *magnetic field* caused by the magnetism of nearby *sediments* or *rocks*.

Anticline A fold that is concave in a downward direction—that is, the vertex is the highest point.

Apparent polar wander A hypothetical migration of Earth's magnetic pole that would account for the changing orientation, with age, of *paleomagnetism* in *rocks* at a particular fixed location. Most geologists believe that it is the continents that have moved, not the magnetic pole—that polar wander is indeed only apparent, not real.

Aragonite A form of calcium carbonate that some organisms secrete to form skeletons and that is an important mineral in many *limestones*.

Aragonite needles Slender crystals of the mineral aragonite that constitute most *carbonate muds* in the modern ocean. Some of the needles form by direct precipitation from seawater and some by the collapse of the skeletons of organisms.

Arkose A *rock* consisting primarily of *sand*-sized particles of feldspar. Most arkose accumulates close to the source area of the feldspar, because feldspar weathers quickly to *clay* and seldom travels far.

Asthenosphere The *ultramafic* layer of Earth lying below the *lithosphere*. The asthenosphere is marked by low seismic velocities, suggesting that it is partly molten.

Atmosphere The envelope of gases that surrounds Earth.

Atoll A circular or horseshoe-shaped organic *reef* growing on a submerged volcano.

Autotroph (See *Producer*)

Axial plane An imaginery plane that cuts through a fold, dividing it as symmetrically as possible.

Axis of a fold The line of intersection between the *axial plane* of a fold and the *beds* of folded rock.

B

Backswamp A broad vegetated area that lies adjacent to a *meandering river* and becomes covered with water when the river overflows its banks.

Banded iron formation Alternating layers of iron-rich and iron-poor *rocks.* Most rocks of this type are older than about 2 billion years.
Barrier island An elongate island composed of *sand* heaped up by ocean waves that lies approximately parallel to the shoreline.
Barrier island–lagoon complex The set of marginal marine environments that consists of a *barrier island,* the *lagoon* behind it, and (usually) *tidal flats, marshes,* and sandy beaches.
Barrier reef An elongate organic *reef* that parallels a coastline and is large enough to dissipate ocean waves, leaving a quiet-water *lagoon* on its landward side.
Basalt A fine-grained, *extrusive, mafic igneous rock*; the dominant rock of oceanic *crust.*
Basin, structural A roughly circular depression of stratified *rocks.*
Bed A distinct sedimentary layer *(stratum)* thicker than 1 centimeter.
Bedding The arrangement of a *sedimentary rock* into discrete layers *(strata)* thicker than 1 centimeter *(beds).*
Bedding surface The surface between two sedimentary *beds.*
Benthic (or **benthonic**) **life** (See *Benthos)*
Benthos The bottom-dwelling life of an ocean or freshwater environment.
Big bang The enormous explosion that created the expanding universe.
Biogenic sediment *Sediment* consisting of *mineral* grains that were once parts of organisms.
Biogeography The study of the distribution of organisms on a geographic scale.
Biostratigraphic unit A body of *rock,* such as a *zone,* defined on the basis of its *fossil* content and having approximately time-parallel upper and lower boundaries.
Biota A collective term for all the animals and plants of an *ecosystem.*
Bolide A *meteorite* or comet that explodes upon striking Earth.
Boulder A piece of *gravel* larger than 256 millimeters (~ 10 inches).
Boundary stratotype A *stratigraphic section* where the boundary between two *chronostratigraphic units* is formally recognized.
Brackish water Water whose *salinity* is lower than that of normal seawater and higher than that of fresh water, ranging from 30 to 0.5 parts salt per 1000 parts water.
Braided stream A stream that has many intertwining channels separated by bars of coarse *sediment.* Braided streams develop where sediment is supplied to the stream system at a very high rate—on an *alluvial fan,* for example, or in front of a melting *glacier.*
Breccia A *rock* that resembles *conglomerate* in consisting of *clasts* of *gravel* surrounded by *sand*; in breccia the clasts are angular, whereas in conglomerate they are rounded.

C

Calcareous ooze Fine sediment on the seafloor at low latitudes, consisting of skeletons of single-celled *planktonic* organisms.
Calcite A form of calcium carbonate that some organisms secrete to form skeletons and that is an important mineral in many *limestones.*
Calcrete (See *Caliche)*
Caliche (calcrete) Nodular calcium carbonate that accumulates in the layer of *soil* below the topsoil in warm climates that are dry part of the year.
Carbonaceous chondrite A *stony meteorite* that contains carbon compounds.
Carbonate mineral A *mineral* in which the basic building block is a carbon atom linked to three oxygen atoms. Calcite, aragonite, and *dolomite* are the most abundant carbonate minerals found in *sediments* and *sedimentary rocks.*
Carbonate platform A marine structure that is composed largely of calcium carbonate and that stands above the neighboring seafloor on at least one of its sides.
Carbonate rock A *sedimentary rock* that consists primarily of *carbonate minerals.* The dominant mineral is nearly always either calcite, in which case the rock is *limestone,* or dolomite, in which case the rock is *dolomite.*
Carbonate sediment Unconsolidated *sediment* that consists primarily of *carbonate minerals,* usually aragonite or calcite.
Carbonization The mode of *fossilization* in which liquids and gases escape, leaving a residue of carbon on the surface of an *impression* of the organism.
Carnivore An animal that feeds on other animals or animal-like organisms.
Cataclastic rock *Metamorphic rocks* that resemble *breccias* or poorly sorted *sandstones* and that form by dynamic *metamorphism,* which breaks and reorients grains.
Catastrophism The outmoded doctrine that sudden, violent, and widespread events caused by supernatural forces formed most of the rocks that are visible at Earth's surface.
Cementation The *lithification* of *sediment* by the precipitation of *minerals* from watery solutions percolating through the sediment.
Chemical sediment A *sediment* created by precipitation of one or more *minerals* from natural waters.
Chemosynthesis The breakdown of simple chemical compounds within a cell for the production of energy. *Sulfate-reducing bacteria* exemplify this process.
Chert (flint) An impure *rock,* often gray, that consists primarily of extremely small quartz crystals precipitated from water solutions.
Chlorite A green silicate mineral that is an important constituent of *metamorphic rocks,* including those of Archean greenstone belts.
Chloroplast A body within a plant cell or plantlike cell that serves as the site of *photosynthesis* within the cell. Chloroplasts are apparently evolutionary descendants of cyanophytes that became trapped in other single-celled organisms.
Chromosome One of several elongate bodies in which *DNA* is concentrated within the nucleus of a cell.
Chronostratigraphic unit A formally named body of *rock,* also called a *time-stratigraphic unit,* representing a particular interval of time—an *erathem, system, series,* or *stage.*
Circumpolar current The circular flow of water around Antarctica, resulting from the juncture of the *westwind drifts* of the Atlantic, Pacific, and Indian oceans.
Clast A solid product of *erosion.* Clasts are sometimes referred to as *detritus* or detrital material.
Clastic rock A *rock* that is an aggregate of detrital material, or *clasts.*
Clastic wedge A wedge-shaped body of *molasse.*
Clay A member of the clay *mineral* family, which includes *silicates* that resemble micas.
Claystone A *sedimentary rock* that consists primarily of *clay* but that is not *fissile* like *shale.*
Coal Altered organic matter formed from stratified plant remains. It con-

tains more than 50 percent carbon and burns readily.
Coal measure The British term for a *cyclothem*.
Cobble A piece of *gravel* measuring between 8 and 256 millimeters (~ 10 inches).
Community, ecologic Populations of several *species* living in the same *habitat*.
Compaction (of sediment) The process in which grains of *sediment* are squeezed together beneath the weight of overlying sediment.
Competition, ecologic The condition in which two *species* vie for an environmental resource, such as food or space, that is in limited supply.
Components, principle of The principle that a body of *rock* is younger than any other body of rock from which any of its components are derived.
Concretion A hard, nodular structure formed in *sediment* or in a *sedimentary rock* by *diagenesis*.
Conglomerate A *rock* consisting of rounded *clasts* of *gravel* surrounded by *sand*.
Consumer, ecologic (heterotroph) Animals or animal-like organisms, which feed on other organisms.
Contact metamorphism Local *metamorphism* caused by *igneous intrusion* that "bakes" nearby *rocks*.
Continental accretion The marginal growth of a continent along a *subduction zone* by mountain building or by addition of a *microplate*.
Continental drift The movement of continents with respect to one another over Earth's surface.
Continental rise A more gently sloping region along the base of the *continental slope*. The continental rise is formed of *sediment* transported down the slope, often by *turbidity currents*.
Continental shelf An extension of a continental landmass beneath the sea, also called a continental margin.
Continental slope The sloping submarine portion of a continent, extending from the *continental margin* to the *continental rise* or the *abyssal plain*.
Convection Rotational flow of a fluid resulting from imbalances in density. Convection often occurs because the fluid below is heated and becomes less dense than the fluid above or because the fluid above is cooled and becomes more dense than the fluid below.
Convective cell One of a number of rotational units believed to operate within Earth's *mantle* as a result of *convection*.
Convergence, evolutionary The evolution of similar features in two or more different biological groups, or *taxa*.
Cope's rule The tendency for body size to increase during the evolution of a group of animals.
Core (of Earth) The central part of Earth below a depth of 2900 kilometers. It is thought to be composed largely of iron and to be molten on the outside with a solid central region.
Coring The process of inserting a tube into *sediments* or *rocks* and then extracting the tube along with a core, or plug, of material for study.
Coriolis effect The tendency of a current of air or water flowing over Earth's surface to bend to the right in the Northern Hemisphere and to the left in the Southern Hemisphere.
Correlation The use of *fossils* to establish that spatially separated *stratigraphic sections* are the same geologic age.
Craton The portion of a continent that has not experienced *tectonic* deformation since Precambrian or early Paleozoic time.
Cross-bedding *Cross-stratification* in which the individual strata exceed 1 centimeter in thickness.
Crosscutting relationships, principle of The principle that when one fault is offset by another, the second is younger than the first.
Cross-lamination *Cross-stratification* in which the individual strata are thinner than 1 centimeter.
Cross-stratification A sedimentary structure in which groups of *strata* lie at angles to the horizontal.
Crust The outermost layer of the *lithosphere*, consisting of *felsic* and *mafic rocks* less dense than the rocks of the *mantle* below.
Crystalline rocks *Igneous* or *metamorphic rocks*.
Cyanobacteria Photosynthetic *prokaryotes* that originated in Archean time and that form *stromatolites*.
Cycle, sedimentary A composite sedimentary unit that is repeated many times in succession within a given region. The unit includes two or more characteristic *beds* or groups of beds arranged in a characteristic vertical sequence that often reflects *Walther's law*.
Cyclothem Sedimentary *cycles* that include coal *beds*. Most cyclothems are of Late Carboniferous (Pennsylvanian) age.

D

Declination, magnetic The angle that a compass needle makes with the line running to the geographic North Pole, reflecting the fact that the magnetic pole (to which the compass needle points) does not coincide with the geographic pole.
Deep-focus earthquake An earthquake produced by movements along or within a subducted slab of *lithosphere* more than 300 kilometers (~ 190 miles) below Earth's surface.
Deep seafloor The *continental slope* and *abyssal plain*.
Degassing The loss of gases by Earth early in its history, when it became liquefied.
Delta A depositional body of *sand*, *silt*, and *clay* formed when a river discharges into a body of standing water so that its current dissipates and drops its load of *sediment*. This structure takes its name from the Greek letter Δ, which it resembles in shape.
Delta, river-dominated A *delta* that projects far out into the ocean because its construction from river-borne *sediment* prevails over the destructive forces of the sea.
Delta front The submarine slope, or forest, of a *delta* extending downward from the *delta plain*. The delta front is usually the site of accumulation of *silt* and *clay*.
Delta plain The upper surface of a *delta*, characterized by *distributary channels* and their *natural levees* and intervening swamps.
Desert A terrestrial environment that receives less than about 25 centimeters (10 inches) of rain per year and consequently supports only a few kinds of plants.
Detritus, sedimentary Loose material produced by *erosion*.
Diagenesis The set of processes, including solution, that alter *sediments* at low temperatures after burial.
Dike A sheetlike or tabular body of *igneous rock* that cuts through sedimentary layers or crystalline rocks.
Dip The angle that a tilted *bed* or *fault* forms with the horizontal.
Disconformity An *unconformity* above *rocks* that underwent *erosion* before the *beds* above the unconformity were

deposited. The *strata* above and below a disconformity are horizontal.
Distributary channels Channels on a *delta plain* that radiate out from the mainland, carrying river water to the ocean in several directions.
Diversity, biotic The variety of organisms living in an *ecosystem*.
DNA Deoxyribonucleic acid, the "double helix" molecule that carries chemically coded genetic information and is passed from generation to generation.
Dollo's law The rule that any substantial evolutionary change is virtually irreversible because genetic changes are not likely to be reversed in an order exactly opposite to the order in which they originally developed.
Dolomite A *mineral* that consists of calcium magnesium carbonate, with calcium and magnesium present in nearly equal proportions, or the *sedimentary rock* that consists largely of this mineral.
Dome, structural A blister-like uplift of stratified *rocks*.
Dropstone A stone dropped to the bottom of a lake or ocean from a melting body of ice afloat on the surface.
Dynamic metamorphism *Metamorphism* that entails the shattering of *rocks* and the deformation or reorientation of their grains.

E

Easterlies Winds that form near Earth's poles where cold, dense air descends and flows toward the west under the influence of the Coriolis effect.
Ecology the branch of biology concerned with the factors that govern the distribution and abundance of organisms in natural environments.
Ecosystem The organisms of an ecologic *community* together with the physical environment that they occupy.
Ectothermic "Cold-blooded": characterized by the physiological condition in which an animal's body temperature is controlled by the external environment.
Endothermic "Warm-blooded": characterized by the physiological condition in which an animal's body temperature is controlled internally and maintained at a more or less constant level.
Eon The largest formal unit of geologic time. There are three eons: the Archean, Proterozoic, and Phanerozoic.
Epicontinental sea A shallow sea formed when ocean waters flood an area of a continent far from the *continental margin*.
Epoch, geologic A division of geologic time shorter than a *period*.
Equatorial countercurrent The eastward-flowing global ocean current that carries the water that has been piled up by the *equatorial current*.
Equatorial current The global ocean current pushed westward along the equator by the *trade winds*.
Era, geologic A division of geologic time shorter than an *eon* but including two or more *periods*.
Erathem A time-stratigraphic unit consisting of all the rocks that represent a geologic *era*.
Erosion The group of processes that loosen *rock* and move pieces of loosened rock downhill.
Erratic boulder A boulder that a *glacier* has transported from its place of origin to a distant location.
Eukaryote (sometimes spelled **eucaryote**) An organism that consists of one or more *eukaryotic cells*. All organisms except bacteria and cyanophytes are of this type.
Eukaryotic cell (sometimes spelled **eucaryotic**) A cell characterized by a nucleus with *chromosomes*, *mitochondria*, and other complex internal structures. This is the kind of cell that forms higher organisms (all organisms but bacteria and cyanophytes).
Eustatic sea-level change A change in the level of all oceans throughout the world.
Evaporite A *mineral* or *rock* formed by precipitation of crystals from evaporating water.
Evergreen coniferous forest A high-latitude forest, often adjacent to *tundra* and always dominated by coniferous trees such as spruces, pines, or firs.
Exotic terrane A block of *lithosphere* that has been sutured to a much larger continent.
Exposure (See *Outcrop)*
Exterior drainage A drainage pattern in which lakes and rivers carry runoff from a region beyond the borders of that region.
Extinction The total disappearance of a *species* or higher *taxon*.
Extrusive igneous rock An *igneous rock* that has been erupted onto Earth's surface.

F

Facies The set of characteristics of a *rock* that represents a particular local environment.
Fault A surface along which *rocks* have broken and moved past one another.
Fauna A collective term for all the animals of an *ecosystem*.
Felsic rock A silicon-rich igneous *rock* that contains only a small percentage of iron and magnesium. *Granite* is the most abundant example. Felsic rocks dominante the crust of continents.
Fissile Having the property of fissility, or tending to break along *bedding* surfaces. This is a property of some *sedimentary rocks*, especially *shales*.
Fission-track dating The dating of a *rock* according to the number of fission tracks produced by the decay of uranium 238. In the process of decaying, uranium 238 atoms eject subatomic particles that leave microscopic tracks in the surrounding rock.
Fissure A crack in a body of *rock*, often filled with *minerals* or *intrusive igneous rock*.
Flint (See *chert*)
Flood basalt *Extrusive rocks* or *mafic* composition that have flowed widely over Earth's surface.
Flora A collective term for all the plants of an *ecosystem*.
Flysch Shales and *turbidites* that accumulate in deep water within a *foreland basin* bordering an active mountain system.
Focus, earthquake The point within Earth at which a rupture occurs, causing an earthquake.
Fold-and-thrust belt The tectonic zone of a mountain chain characterized by folds and *thrust faults* and positioned adjacent to the *metamorphic belt* and farther away than the metamorphic belt from the igneous core of the mountain chain.
Folding Tectonic bending of rocks into *anticlines* and *synclines* or other contorted configurations.
Foliation The alignment of platy *minerals* in *metamorphic rocks*, caused by the high pressure applied during *metamorphism*.

Food chain The sequence of nutritional steps in an ecosystem, with *producers* at the bottom and *consumers* at the top.
Food web The nutritional structure of an *ecosystem* in which more than one *species* occupies each level. Thus there are usually several *producer* species and several *consumer* species in a food web.
Forearc basin An elongate depositional basin that lies between an igneous arc and the associated *accretionary wedge*.
Foreland basin An elongate basin lying in front of an active mountain system and receiving *sediment* (primarily *flysch* and *molasse*) from the mountains.
Formation The fundamental *lithostratigraphic unit*. A body of *rock* characterized by a particular set of *lithologic* features and given a formal name.
Fossil The remains or tangible traces of an ancient organism preserved in *sediment* or *rock*.
Fossil fuel Condensed and altered organic matter than can be burned to supply energy for human use. Examples are coal, petroleum, and natural gas.
Fossil succession The vertical ordering of fossil *taxa* in the geologic record, reflecting the operation of evolution and *extinction*.
Fossilization The formation of a *fossil*.
Fringing reef An elongate organic *reef* that fringes a coastline and has no *lagoon* on its landward side.

G

Gabbro A *mafic igneous rock;* the coarse-grained, *intrusive* equivalent of *basalt*.
Gamete A sex cell (egg or sperm) that carries half the normal complement of *chromosomes* and combines with another sex cell to produce a new individual possessing the normal complement.
Gene A unit of inheritance consisting of a segment of *DNA* that performs a particular function. Many genes provide coded information for the synthesis of particular *polypeptides*.
Gene pool The sum total of the genetic components of a population.
Glacial maximum The point of an ice cap's maximum advance.
Glacial minimum The point of an ice cap's farthest retreat.
Glacier A large mass of ice that creeps over Earth's surface.
Gneiss A coarse-grained *metamorphic rock* resembling an *igneous rock* but with its component *minerals* segregated into wavy layers.
Graben A *fault*-block basin produced by the depression of a block of *crust* bounded by *normal faults*.
Grade, metamorphic A classification system based on the level of temperature and pressure responsible for *metamorphism*. *Metamorphic rocks* are divided into high-grade, medium-grade, and low-grade categories.
Graded bed A *bed* in which grain size decreases from bottom to top.
Gradualistic model The theory that most evolutionary change takes place in small steps within well-established *species*.
Granite A coarse-grained *felsic igneous rock* consisting of feldspar and quartz with minor *mafic* components. The most common *intrusive* rock of continental *crust*.
Granule A piece of *gravel* of small size (between 2 and 4 millimeters).
Granulite A high-*grade metamorphic rock* that nearly became molten during *metamorphism* and, as a result, resembles an *igneous rock* in being granular and lacking *foliation*.
Grassland (See *Savannah)*
Gravel *Sediment* larger than *sand* (larger than 2 millimeters).
Graywacke A *siliciclastic* rock that is poorly sorted but consists primarily of *sand*-sized grains. Some of the grains are dark, giving the rock its characteristic dark-gray color.
Greenhouse effect The warming of Earth's surface and of its lower atmosphere caused by the accumulation of carbon dioxide in the *atmosphere,* which acts in the same manner as the glass in a greenhouse, allowing solar radiation to pass to Earth's surface and then preventing much of the resulting heat from escaping from the lower atmosphere.
Greenstone belts Podlike bodies of *rock* characteristic of Archean terranes. They consist of volcanic rocks and associated *sediments* that have commonly been metamorphosed so that they have a greenish color.
Group A *lithostratigraphic unit* of a rank higher than *formation*.
Guide fossil (See *Index fossil)*
Guyot A flat-topped volcanic seamount in the deep sea. It appears that guyots form in shallow water when wave action truncates the upper part of a volcano and that they are transported to deeper water by lateral *plate* movement.
Gypsum A *mineral* that consists of calcium sulfate with water molecules attached ($CaSO_4 \cdot H_2O$), or the *rock* that consists primarily of this mineral.
Gyre A large-scale circular flow of winds or ocean currents.

H

Habitat An environment that supports life.
Half-life The time-required for a particular radioactive *isotope* to decay to half its original amount. This time is consistent for any isotope, regardless of the amount of the isotope present at the outset.
Halite The *mineral* that consists of sodium chloride (NaCl), popularly known as rock salt, or the *rock* that consists primarily of this mineral.
Herbivore An animal that feeds on plants or plantlike organisms.
Heterotroph (see *Consumer)*
Homogeneous accretion A possible mode of origin of Earth in which materials of many different densities would have aggregated haphazardly, without becoming layered according to density.
Homology The presence, in two different animal or plant groups, of organs that have the same ancestral origin but serve different functions.
Hornfels A fine-grained, granular *metamorphic rock* of varying composition formed by *contact metamorphism* under conditions of high temperature and low pressure.
Hot spot A small area of heating and igneous activity in Earth's crust where a *thermal plume* rises from the mantle. The Hawaiian Islands represent a hot spot.
Humus Organic matter in *soils*, formed largely by the decay of leaves, woody tissues, and other plant materials.
Hydrological cycle, global The continuous passage of water from the atmosphere to the land and to bodies of water and then back again to the atmosphere.

Hypersaline water Water that is higher in *salinity* than normal seawater (contains more than 40 parts salt per 1000 parts water).
Hypsometric curve A graph that displays the proportions of Earth's surface that lie at various altitudes above and various depths below sea level.

I

Igneous rock A *rock* formed by the cooling of molten material.
Impression The mode of *fossilization* in which the flattened imprint of an organism forms on a *bedding surface.*
Index fossil (guide fossil) A *species* or genus of *fossils* that provides for especially precise *correlation.* An ideal index fossil is easily distinguished from other *taxa,* is geographically widespread, is common in many kinds of *sedimentary rocks,* and is restricted to a narrow stratigraphic interval.
Inhomogeneous accretion A possible mode of origin of Earth in which materials would have condensed in a sequence determined by their density, causing concentric layers to form with the most dense materials at the center.
Interior drainage A drainage pattern in which lakes and rivers fail to carry runoff from a region beyond the borders of the drainage area. The rainfall is so light that streams and rivers are temporary, drying up at intervals.
Intertidal zone The belt that is alternately exposed and flooded as the *tide* ebbs and flows along a coast.
Intrusion A body of *igneous rock* that formed within Earth rather than at its surface.
Intrusive igneous rock A rock formed by the cooling of *magma* within Earth.
Intrusive relationships, principle of The principle that an *intrusive igneous rock* is always younger than the rock that it invades.
Iron formation Complex *sedimentary* or weakly *metamorphosed rock* that usually consists of oxides, sulfides, or *carbonates* of iron, often in association with *chert.*
Iron meteorite A *meteorite* in which iron is the primary component.
Island arc A curved chain of islands produced by volcanism at a site where *magma* rises through the *lithosphere* from a *subducted plate.*
Isostatic adjustment The mechanism whereby areas of Earth's *crust* rise or subside to keep the crust in gravitational equilibrium as it floats on the *mantle.* Thus, a mountain is balanced by a root of crustal material.
Isotope One of two or more varieties of an element that differ in number of neutrons within the atomic nucleus.

J

Joint A fracture in rock that results from various kinds of stress.

K

Key bed (marker bed) A sedimentary bed, such as a bed of volcanic ash, that is of nearly the same age everywhere and thus is useful for *correlation.*

L

Lagoon A ponded body of water along a marine coastline, usually landward of a *barrier island* or organic *reef.*
Lamination A distinct sedimentary layer *(stratum)* thinner than 1 centimeter.
Late Neolithic The tool culture of early modern humans, the Cro-Magnon people of Europe.
Laterite A soil rich in oxides of aluminum and of iron, which give it a rusty red color.
Lava Molten rock (*magma*) that has reached the surface of Earth.
Life habit The mode of life of an organism, or the way it functions within its *niche*—how it obtains nutrients or food, reproduces, and stations itself or moves about within its environment.
Limestone A *sedimentary rock* that consists primarily of calcium carbonate. Limestones may be *biogenic* or chemical in origin.
Limiting factor, ecologic An environmental condition, such as temperature, that restricts the distribution of a *species* in nature.
Lithification The consolidation of loose *sediment* by *compaction,* precipitation of *mineral* cement, or a combination of these processes to form a *sedimentary rock.*
Lithology The physical and chemical characteristics of a *rock.*
Lithosphere Earth's outer rigid shell, situated above the *asthenosphere* and consisting of the *crust* and upper *mantle.* The lithosphere is divided into *plates.*
Lithostratigraphic unit A body of characteristic *lithology* that is formally recognized as a *formation, member, group,* or *supergroup.*
Loess The wind-blown silt that accumulates in the frigid desert in front of a *glacier.* It is deposited by the *meltwaters* of the glacier.
Longshore current An ocean current that flows along a coast, often sweeping *sand* in a direction parallel to the coastline.

M

Mafic rock A dense, dark *igneous rock* that is relatively poor in silicon and rich in iron and magnesium. *Basalt,* the characteristic igneous rock of oceanic *crust,* is an example.
Magma Naturally occurring molten *rock.*
Magnetic field, Earth's The field of magnetism that results from motions of Earth's iron-rich outer core; these motions cause Earth to behave like a giant bar magnet, with a north and south pole.
Magnetic reversal A reversal of the polarity of Earth's *magnetic field,* as recorded in rocks magnetized by the field.
Mantle The zone of Earth's interior between the *core* and the *crust,* ranging from depths of approximately 40 to 2900 kilometers. It is composed of dense *ultramafic* silicates and divided into concentric layers.
Marble A homogeneous, granular *metamorphic rock* that consists of calcite, *dolomite,* or a mixture of the two and forms by the *metamorphism* of *limestone* or *dolomite.*
Maria (singular, **mare**) The large craters on the surface of the moon.
Marker bed (see *Key bed)*
Marsh, intertidal A *habitat* along the seashore that is dominated by low-growing plants and is alternately flooded by the *tide* and exposed to the air. The remains of the plants usually accumulate to form *peat.*
Mass extinction An episode of large-

scale *extinction* in which large numbers of *species* disappear in a few million years or less.

Meandering river A river that winds back and forth like a ribbon, depositing *sediment* on the inside of each curve and eroding sediment on the outside.

Mediterranean climate A climate characterized by dry summers and wet winters, often found along coasts lying about 40° from the equator. Much of California and much of the Mediterranean region of Europe have this kind of climate.

Mélange A chaotic, deformed mixture of *rocks*, such as often forms where subduction occurs along a *deep-sea trench*.

Meltwaters The waters that issue from the front of a melting *glacier*.

Member A *lithostratigraphic unit* of a rank lower than *formation*.

Metamorphic belt The metamorphic zone parallel to the long axis of a mountain chain and near the igneous core of the mountain chain. This is a zone of *regional metamorphism*.

Metamorphic rock A *rock* formed by *metamorphism*.

Metamorphism The alteration of *rocks* within Earth under conditions of high temperature and pressure.

Meteorite An extraterrestrial object that crashed into Earth after being captured by Earth's gravitational field.

Microplate A small lithospheric plate, usually of predominantly *felsic* composition.

Mid-ocean ridge The ridge on the ocean floor where oceanic *crust* forms and from which it moves laterally in each direction.

Mineral A naturally occurring inorganic solid element or compound with a particular chemical composition or range of compositions and a characteristic internal structure.

Mitochondrion (plural, **mitochondria**) A body within a *eukaryotic cell* in which complex compounds are broken down by oxidation to yield energy and, as a by-product, carbon dioxide. The mitochondrion is apparently an evolutionary descendant of a small bacterium that became trapped within a larger one.

Moho (See *Mohorovičić discontinuity*)

Mohorovičić discontinuity (Moho) The boundary between the *crust* and *mantle*, marked by a rapid increase in the velocity of *seismic waves*.

Molasse Nonmarine and shallow marine sediments — representing such environments as *alluvial fans*, river systems, and *barrier island–lagoon complexes* — that accumulate in front of a mountain system after heavy sedimentation from the mountains has driven deep marine waters from the *foreland basin* there.

Mold A *fossil* that consists of a three-dimensional imprint of an organism or part of an organism.

Moraine, glacial A ridge of *till* plowed up in front of a *glacier*.

Mousterian culture The tool culture of Neanderthal, a late Pleistocene human or humanoid creature.

Mud An aggregate consisting of *silt*- or *clay*-sized sediment or a combination of the two.

Mutation A chemical change in a genetic feature. Such changes provide much of the variability on which *natural selection* operates.

N

Natural levee A gentle ridge bordering a *meandering river* or a *distributary channel* of a *delta* and composed of *sand* and *silt* deposited by the river or distributary channel when it overflows its banks.

Natural selection The process recognized by Charles Darwin as the primary mechanism of evolution. The selection process, which operates on heritable variability, results from differences among individuals in longevity and in rate of production of offspring.

Nebula The dense cloud of matter that remains after a *supernova* explodes.

Nekton Fishes and other marine animals that move through the water primarily by swimming.

Niche, ecologic The ecologic position of a species in its environment, including its requirements for certain kinds of food and physical and chemical conditions and its interactions with other species.

Nodule An irregular rounded mass of *chert* or some other *mineral* embedded in a *sedimentary rock*. Most nodules are formed by diagenetic processes where they are found.

Nonconformity An *unconformity* separating bedded *rocks*, above, from *crystalline rocks*, below.

Normal fault A fault whose *dip* is steeper than 45° and along which the *rocks* above have moved downward in relation to the rocks below.

O

Oceanic realm The portion of the ocean that lies above the *deep seafloor*.

Oil shale A well-laminated *shale* in which dark, organic-rich layers alternate with thicker, lighter-colored layers. Petroleum can be removed from such rock, but not at a cost that is economically profitable at present.

Ooze, deep-sea Fine-grained sediment in the deep sea that consists of calcareous or siliceous skeletons of dead *planktonic* organisms.

Ophiolite A segment of seafloor that is elevated so as to rest on continental *crust*. An ophiolite usually includes *turbidites*, black *shales*, *cherts*, and *pillow basalts* along with *ultramafic rocks* from the *mantle*.

Opportunistic species A *species* that specializes in the rapid invasion of newly vacated *habitats*, where there is little competition from other species.

Original horizontally, principle of The principle enunciated in the seventeenth century that all *strata* are horizontal when they form. (A more accurate statement would be that almost all strata are initially more nearly horizontal than vertical.)

Original lateral continuity, principle of The principle that similar *strata* found on opposite sides of a valley or some other erosional feature were originally connected.

Orogenesis The process of mountain building

Orogeny An episode of mountain building.

Outcrop A portion of a body of *rock* that is visible at Earth's surface. (Some geologists restrict this term to rocks laid bare by natural processes and apply the term *exposure* to artificially exposed areas of rock.)

Outwash, glacial Well-stratified glacial *sediment* deposited by a stream of *meltwater* issuing from a melting *glacier*.

Overturned fold A fold in which at least one limb has been rotated more than 90° from its original position.

Oxygen cycle, global The continuous flow of oxygen through the *atmosphere*.

It is supplied to the atmosphere by the breakdown of H_2O in the upper atmosphere and by *photosynthesis* and is removed from the atmosphere by organic respiration and by the oxidation of various materials.

Oxygen sink A reservoir consisting of a chemical element or compound that combines readily with oxygen and thus removes it from the *atmosphere*. During the early part of Precambrian time, sulfur, iron, and other elements and compounds served as important oxygen sinks, preventing oxygen from accumulating in the atmosphere.

P

Paleogeography The geography of the geologic past.

Paleomagnetism The magnetism of a *rock*, developed from Earth's *magnetic field* when the rock formed.

Parasite An organism that derives its nutrition from other organisms without killing them.

Partial melting A process in which heating causes a mass of *rock* to become partially molten. Partial melting occurs because the various minerals that form a rock melt at different temperatures.

Particulate inheritance The presence of hereditary factors called *genes* that retain their identity while being passed on from parent to offspring.

Passive margin A *continental margin* that is not affected by *rifting, subduction, transform faulting,* or other large-scale tectonic processes, but that instead forms a shelf that accumulates *sediments.*

Patch reef (pinnacle reef) A small mound or reef growing in the *lagoon* behind a *barrier reef.*

Peat Sediment composed primarily of plant debris that has not metamorphosed to form *coal.*

Pebble A piece of *gravel* measuring between 4 and 8 millimeters.

Pelagic life Oceanic life that exists above the seafloor.

Pelagic sediment Fine-grained sediment that settles through the oceanic water column to the deep sea. Some of this sediment is *biogenic.*

Period, geologic The most commonly used unit of geologic time, representing a subdivision of an *era.*

Permineralization The mode of *fossilization* in which porous spaces within part of an organism (such as bony or woody tissue) become filled with *mineral* material.

Photic zone The upper layer of the ocean, where enough light penetrates the water to permit *photosynthesis.*

Photosynthesis The process in which plants and single-celled plantlike organisms employ the compound chlorophyll to convert carbon dioxide and water from their environment into energy-rich sugar, which fuels essential chemical reactions.

Photosynthetic bacteria Bacteria that employ *photosynthesis*, but with hydrogen sulfide taking the place of water in the production of sugar. Such bacteria are poisoned by oxygen and probably flourished in Archean time when the *atmosphere* lacked free oxygen.

Phylogeny A segment of the "tree of life" that includes two or more evolutionary branches.

Phytoplankton *Plankton,* or floating aquatic life, that is photosynthetic. Most phytoplankton *species* are single-celled algae.

Pillow basalt Basalt with a hummocky surface formed by rapid cooling of *lava* beneath water.

Pinnacle reef (See *Patch reef)*

Plankton Organisms that float in the ocean or in lake waters.

Plate A segment of the *lithosphere* that moves independently over Earth's interior.

Plate tectonics The study of the movements and interactions of lithospheric *plates.*

Playa lake A temporary lake in a region of *interior drainage.* When such a lake dries out, *evaporite* deposits form.

Plunging fold A fold whose axis plunges (lies at an angle to the horizontal) so that the *beds* of the fold have a curved *outcrop* pattern if they are truncated by erosion.

Pluton An *intrusion*, or massive body of *intrusive igneous rock.*

Point bar An accretionary body of *sand* on the inside of a bend of a *meandering river.*

Polypeptide An organic molecule containing a large number of *amino acids.* A *protein* is a large polypeptide.

Population A group of individuals that live in the same area and interbreed.

Precambrian shield The Precambrian portion of a *craton* exposed at Earth's surface.

Precession of the equinoxes A slight wobble in Earth's axis of rotation caused by the gravitational pull of the sun and moon.

Predation The eating of an animal by one of another species.

Prodelta The gently seaward-sloping bottom-set area of a *delta front* where *clay* accumulates in deep water.

Producer, ecologic A plant or plant-like organism that manufactures its own food.

Prograde To grow seaward by the accumulation of *sediment* or *sedimentary rocks. Deltas* often prograde, as do organic *reefs*. Progradation produces *regression,* or seaward migration, of the shoreline.

Prokaryote (sometimes spelled **procaryote)** An organism that consists of a *prokaryotic cell:* a member of the kingdom Monera, which includes bacteria and cyanophytes.

Prokaryotic (sometimes spelled **procaryotic) cell** A primitive cell in which there are no *chromosomes* and no nuclei, *mitochondria,* or certain other internal structures characteristic of the cells of higher organisms.

Protein (See *Polypeptide)*

Protozoan A single-celled animal-like *eukaryote.*

Pseudoextinction The disappearance of a species, not by dying out but by evolving to the point at which it is recognized as a different species.

Psychrosphere The deepest zone of the modern ocean, where waters are nearly freezing. Convection causes the cold waters in polar regions to descend to the deep sea and form the psychrosphere.

Punctuation model The theory that most evolutionary change occurs rapidly, through *speciation.*

Q

Quartzite A *metamorphic rock* formed by the *metamorphism* of quartz *sandstone* and consisting of almost pure quartz.

R

Radioactive decay The spontaneous breakdown of certain kinds of atomic nuclei into one or more nuclei of different elements, with a release of energy and subatomic particles.

Radiocarbon dating Radiometric dating by means of carbon 14, a radioactive *isotope* with a *half-life* so short that its decay can be used to date materials younger than about 70,000 years.
Radiometric dating Measurement of the amount of naturally occurring radioactive *isotopes* in *rocks* in relation to their daughter isotopes (products of *radioactive decay*) to ascertain the ages of the rocks.
Rain forest, tropical A jungle that develops in an equatorial region where heavy, regular rainfall results from the cooling of air that has ascended after being warmed and picking up moisture near Earth's surface.
Rain shadow A region that is on the downwind side of a mountain and receives little rain because the winds rise as they pass over the mountain, cooling and dropping most of their moisture before they reach the other side.
Recumbent fold A fold that rests on one of its limbs, with the axial plane nearly horizontal.
Red beds Sediments of any grain size that are reddish, usually because of the presence of iron oxide cement.
Redshift An increase in the wavelengths of light waves traveling through space.
Reef, organic A solid but porous *limestone* structure standing above the surrounding seafloor and constructed by living organisms, some of which contribute skeletal material to the reef framework.
Reef flat The flat upper surface of a *reef*, usually standing close to sea level (often in the *intertidal zone*).
Regional metamorphism The creation of *metamorphic rocks* over areas whose dimensions are measured in hundreds of kilometers. Usually this happens in association with mountain building.
Regional strike In deformed terrain, the prevailing orientation of fold axes or of the lines of outcrop of tilted *beds*.
Regression A seaward migration of a marine shoreline and of nearby environments.
Relict distribution The localized occurrence of a taxonomic group after it has died out throughout most of the geographic area that it previously occupied.
Remobilization *Regional metamorphism* and deformation that affect a segment of *crust* previously altered by similar processes.
Reverse fault A *fault* above which *rocks* have moved uphill in relation to rocks below.
Rhyolite A fine-grained, *felsic extrusive igneous rock* equivalent in mineralogical composition to *granite*.
Rift A juncture between two *plates* where *lithosphere* forms and the plates diverge.
Ripples Small dunelike structures formed on the surface of *sediment* by moving water or wind.
Rock An aggregate of interlocking or attached grains, each of which is typically composed of a single *mineral*.
Rock-stratigraphic unit (See *Lithostratigraphic unit)*

S

Salinity The saltiness of natural water. The salinity of normal seawater is 35 parts salt per 1000 parts water.
Sand Sediment ranging in size from 1/16 to 2 millimeters
Sand dune A hill of *sand* that has been piled up by the wind. The sand within a dune is characterized by *trough-bedding*.
Sandstone A *siliciclastic sedimentary rock* consisting primarily of *sand*—usually sand that is predominantly quartz.
Savannah A broad grassland, which typically forms where there is enough rainfall to sustain grass but not enough to sustain the trees that form woodlands or forests.
Scavenger An organism that feeds on other organisms after they are dead.
Schist A *metamorphic rock* that consists mainly of platy *minerals* such as micas that lie nearly parallel to one another so that the rock tends to break along parallel surfaces.
Sediment Material deposited on Earth's surface by water, ice, or air.
Sedimentary rock A *rock* formed by the consolidation of loose *sediment* or by precipitation from a watery solution.
Sedimentary structure A distinctive *bedding* pattern in a *sedimentary rock*.
Seed A reproductive structure of a plant—produced by the union of *gametes* and then released from the plant—that has the potential to grow into a new plant. The seed is actually a juvenile stage of the *spore*-bearing generation of the plant.
Seismic wave An oscillatory movement of Earth, caused naturally or by artificial means such as the setting off of an explosion.
Series A time-stratigraphic unit consisting of all the *rocks* that represent a geologic *epoch*.
Sexual recombination The mixing of *chromosomes* from generation to generation, which continually creates new genetic combinations and hence new kinds of individuals upon which *natural selection* can operate.
Shale A *fissile sedimentary rock* consisting primarily of *clay*.
Shelf break The edge of a *continental shelf*, where it meets the *continental slope*.
Shield volcano A low cone-shaped volcano of the type that often forms in oceanic areas, where *lavas* tend to be *mafic* and hence of low viscosity, so that they spread rapidly when they erupt.
Silicates The *mineral* group that includes the most abundant minerals in Earth's *crust* and *mantle*. The basic building block of silicates is a tetrahedral structure consisting of four oxygen atoms surrounding a silicon atom.
Siliciclastic sediment *Detrital sediment* consisting of *silicate minerals*. This is the most abundant kind of sediment of Earth.
Sill A sheetlike or tabular body of *intrusive igneous rock* injected between layers of sediment.
Silt *Sediment* in which articles range in size from $\frac{1}{256}$ to $\frac{1}{16}$ millimeter.
Siltstone A *siliciclastic sedimentary rock* consisting primarily of *silt*.
Slate A fine-grained *metamorphic rock* that is *fissile*, like the *sedimentary rock shale*, but whose fissility results from alignment of platy materials by deformational pressures rather than by depositional orientation of particles.
Soil Loose *sediment* that accumulates in contact with the *atmosphere*.
Solar nebula theory The theory that the solar system formed from a cloud of cosmic dust.
Speciation The origin of a new species from two or more individuals of a preexisting species.
Species A group of individuals that interbreed or have the potential to interbreed in nature and that do not breed with other interbreeding groups.
Spore A reproductive structure, not produced from *gametes*, that is released from a plant and has the potential to grow into a new plant.
Stabilization, orogenic Consolidation of the *sedimentary rocks* of an *orogenic belt* along the margin of a continent by *metamorphism*, folding, and *thrust faulting*.

Stage The time-stratigraphic unit ranking below a *series* and consisting of all the *rocks* that represent a geologic *age*.
Stony meteorite A *meteorite* of rocky composition.
Stony-iron meteorite A *meteorite* that consists of a mixture of metallic material (chiefly iron) and stony material.
Stratification The arrangement of *sedimentary rocks* in discrete layers (or *strata*).
Stratigraphic section A local *outcrop* or series of adjacent outcrops that displays the *rocks'* vertical sequence.
Stratum (plural, **strata**) A distinct sedimentary layer.
Strike The compass direction that lies at right angles to the *dip* of a tilted *bed* or *fault* (that is, the compass direction of a horizontal line lying in the plane of a tilted bed or fault).
Strike-slip fault A high-angle *fault* along which the *rocks* on one side move horizontally in relation to rocks on the other side with a shearing motion.
Stromatolite An organically produced sedimentary structure that consists of alternating layers of organic-rich and organic-poor *sediment*. The organic-rich layers have usually been formed by sticky threadlike algae, which have trapped the sediment of the organic-poor layers.
Subduction Descent of a slab of *lithosphere* into the *asthenosphere* along a *deep-sea trench*.
Subduction zone The region where *subduction* of the *lithosphere* occurs.
Substratum The surface—*sediment, rock,* or another organism—on which or within which a *benthic* aquatic organism lives.
Subtidal The belt positioned seaward of the *intertidal zone*.
Sulfate mineral A *mineral* in which the basic building block is a sulfur atom linked to four oxygen atoms. Most sulfate minerals are highly soluble in water and form by the evaporation of natural waters.
Sulfate-reducing bacteria Fermenting bacteria that obtain energy by converting sulfate compounds into sulfide compounds. These bacteria, which cannot tolerate oxygen, are common in the muds of swamps, ponds, and *lagoons*. Bacteria of this type seem to have flourished in Archean time, when the *atmosphere* held little oxygen.
Sulfides *Minerals* that are combinations of metals and sulfur. Many are valuable ores.
Supergroup A *litho-stratigraphic unit* of a rank higher than *group*.
Supernova An exploding star that casts off matter of low density.
Superposition, principle of The principle that in an undisturbed sequence of *strata,* the oldest strata lie at the bottom and progressively younger strata are successively higher.
Supratidal zone The belt along a coast just landward of the *intertidal zone* and flooded only occasionally, during storms or unusually high *tides*.
Surf zone The zone along a sandy beach where *waves* break.
Suture The juncture between two continents that have been united along a *subduction zone*.
Suturing The unification of two continents along a *subduction zone*.
Syncline A fold that is concave in an upward direction—that is, the vertex is the lowest point.
System A time-stratigraphic unit consisting of all the *rocks* representing a geologic *period*.

T

Tabulate-strome reef A reef formed by tabulate corals and stromatoporoid sponges during Ordovician, Silurian, or Devonian time.
Talus, reef The pile of rubble sloping seaward from the living surface of a *reef*.
Taxon (plural, **taxa**) **(taxonomic group)** A formally named group of related organisms of any rank, such as phylum, class, family, or species.
Taxonomic groups (See *Taxon*)
Taxonomy The study of the composition and relationships of *taxa* of organisms.
Tectonic cycle The characteristic cycle of deposition associated with an episode of mountain building. Shallow-water deposition along a *continental shelf* gives way to deep-water *flysch* deposition when a *foreland basin* forms, and this, in turn, gives way to nonmarine *molasse* deposition when heavy sedimentation drives deep marine waters from the *foreland basin*.
Tectonics The segment of *structural geology* concerned with largescale features such as mountains.
Temperate forest A forest dominated by deciduous trees (trees that lose their leaves in winter). This kind of forest typically grows under slightly warmer climatic conditions than *evergreen coniferous forests*.
Terrane A geologically distinctive region of Earth's *crust* that has behaved as a coherent crustal block.
Thermal plume A column of *magma* rising from the *mantle* through the *lithosphere*.
Thrust fault A low-angle *reverse fault*. During many mountain-building episodes, large slices of *rock (thrust sheets)* move hundreds of kilometers over rigid unrelated rocks.
Thrust sheet A large, tabular segment of the *crust* that moves along a *thrust fault* during mountain building.
Tidal flat A surface where mud or sand accumulates in the *intertidal zone*.
Tide A major movement of the ocean that results primarily from the gravitational attraction of the moon. Tides ebb and flow in particular regions as Earth rotates beneath a bulge of water created by the pull of the moon.
Till, glacial Heterogeneous *sediment* plowed up and deposited by a *glacier*.
Tillite Lithified *till*.
Tilt cycle The interval of 41,000 years during which the angle of Earth's axis of rotation with respect to its orbit around the sun oscillates and returns to its original position.
Time-stratigraphic unit (See *Chronostratigraphic unit)*
Top carnivore A carnivorous animal positioned at the top of a *food chain* or *food web*.
Topsoil The upper zone of many *soils*, consisting primarily of *sand* and *clay* mixed with *humus*.
Trace fossil A track, trail, or burrow left in the geologic record by a moving animal.
Trade winds Winds that in each hemisphere blow diagonally westward toward the equator between about 20 and 30° latitude of the equator. They result from a zone of high pressure that forms where air that has risen near the equator builds up.
Transform fault A *strike-slip fault* along which two segments of *lithosphere* move in relation to each other. Many transform faults offset *midocean ridges*.
Transgression A landward migration of a marine shoreline and of nearby environments.
Transpiration The emission of water by plants into the *atmosphere*, primarily through the leaves.
Trench, deep-sea A deep, elongate

trough along which a *plate* in the *lithosphere* is *subducted.*

Triple junction A point where three lithospheric *plates* meet.

Tropical climate A climate in which the average annual temperature is in the range of 18 to 20°C (64 to 68°F) or higher. Most tropical climates lie within about 30° of the equator.

Trough cross-stratification *Cross-stratification* of *sediments* in which one set of *beds* is truncated by erosion in such a way that the next set of beds laid down accumulates on a curved surface.

Tuff A *rock* deposited as a *sediment* but consisting of volcanic particles.

Tundra A terrestrial environment where air temperatures rise above freezing during the summer but a layer of soil beneath the surface remains frozen. Tundras are characterized by low-growing plants that require little moisture.

Turbidite A graded *bed*, often with poorly sorted *sand* at the base and *mud* at the top, formed when a *turbidity current* slows down and spreads out.

Turbidity current A flow of dense, *sediment*-charged water that moves down a slope under the influence of gravity.

Type section A formally designated local *stratigraphic section* where a *lithostratigraphic unit* such as a *formation* was originally defined.

U

Ultramafic rock A very dense *rock* that is even poorer in silicon and richer in iron and magnesium than a *mafic rock.* Ultramafic rocks characterize Earth's mantle.

Unconformity A surface between a group of sedimentary *strata* and the *rocks* beneath them, representing an interval of time during which erosion occurred rather than deposition.

Ungulate A hoofed herbivorous mammal. Odd-toed ungulates include horses and rhinoceroses. Even-toed ungulates include cattle, antelopes, sheep, deer, and pigs.

Uniformitarianism The principle that there are inviolable laws of nature that have not changed in the course of time.

Upwelling Ascent of cold water from the deep sea to the *photic zone,* usually providing nutrients for rich growth of *phytoplankton.* Upwelling is most common where ocean currents drag surface waters away from *continental margins.*

V

Varves Alternating layers of coarse and fine *sediments* that accumulated in a lake in front of a *glacier.* A coarse layer forms each summer, when streams of *meltwater* carry *sand* and *silt* to the lake. A fine layer forms each winter, when the surface of the lake is covered by ice, so that only *clay* and organic matter settle to the bottom.

Vascular plant A plant that has vessels in its stem for transporting water, nutrients, and food.

Vent An opening in Earth's *crust* through which *lava* emerges at the surface.

Vestigial organ An organ that serves no apparent function but was functional in ancestors of the organism that now possesses it.

Volcanic igneous rock (See *Extrusive igneous rock)*

W

Walther's law The principle that when depositional environments migrate laterally, *sediments* of one environment come to lie on top of sediments of an adjacent environment.

Wave, surface A wave on the surface of the ocean produced by the circular movement of water particles under the influence of the wind.

Weathering The various aspects of *erosion* that take place before transport. Some weathering is physical and some is chemical.

West-wind dirft The eastward-flowing ocean current near the North or South Pole that is created by the major ocean gyres and the reinforcing *westerly* winds.

Westerlies Winds that flow toward the northeast in each hemisphere between about 30 and 60° from the equator. These winds result from the same high-pressure zone that produces the *trade winds,* which flow in the opposite direction.

Z

Zone, biostratigraphic A body of *rocks* characterized by the presence of one or more *index,* or *guide, fossils.* The upper and lower boundaries of a zone are approximately the same ages everywhere.

Zooplankton *Plankton,* or floating aquatic life, that is animal-like in mode of nutrition (feeds on other organisms).

Index

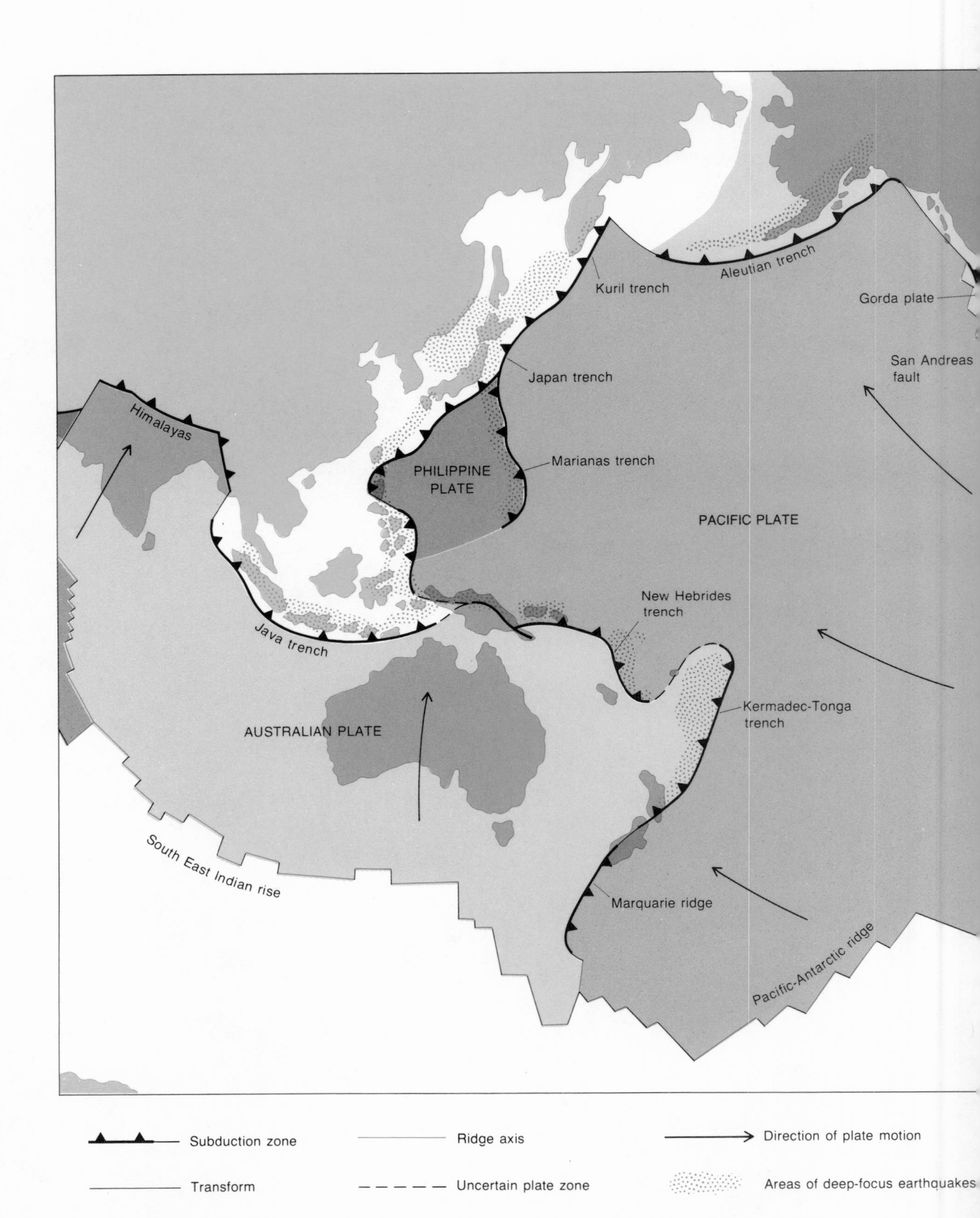

Aleutian trench
Kuril trench
Gorda plate
San Andreas fault
Japan trench
Himalayas
Marianas trench
PHILIPPINE PLATE
PACIFIC PLATE
New Hebrides trench
Java trench
Kermadec-Tonga trench
AUSTRALIAN PLATE
South East Indian rise
Marquarie ridge
Pacific-Antarctic ridge
Subduction zone
Ridge axis
Direction of plate motion
Transform
Uncertain plate zone
Areas of deep-focus earthquakes

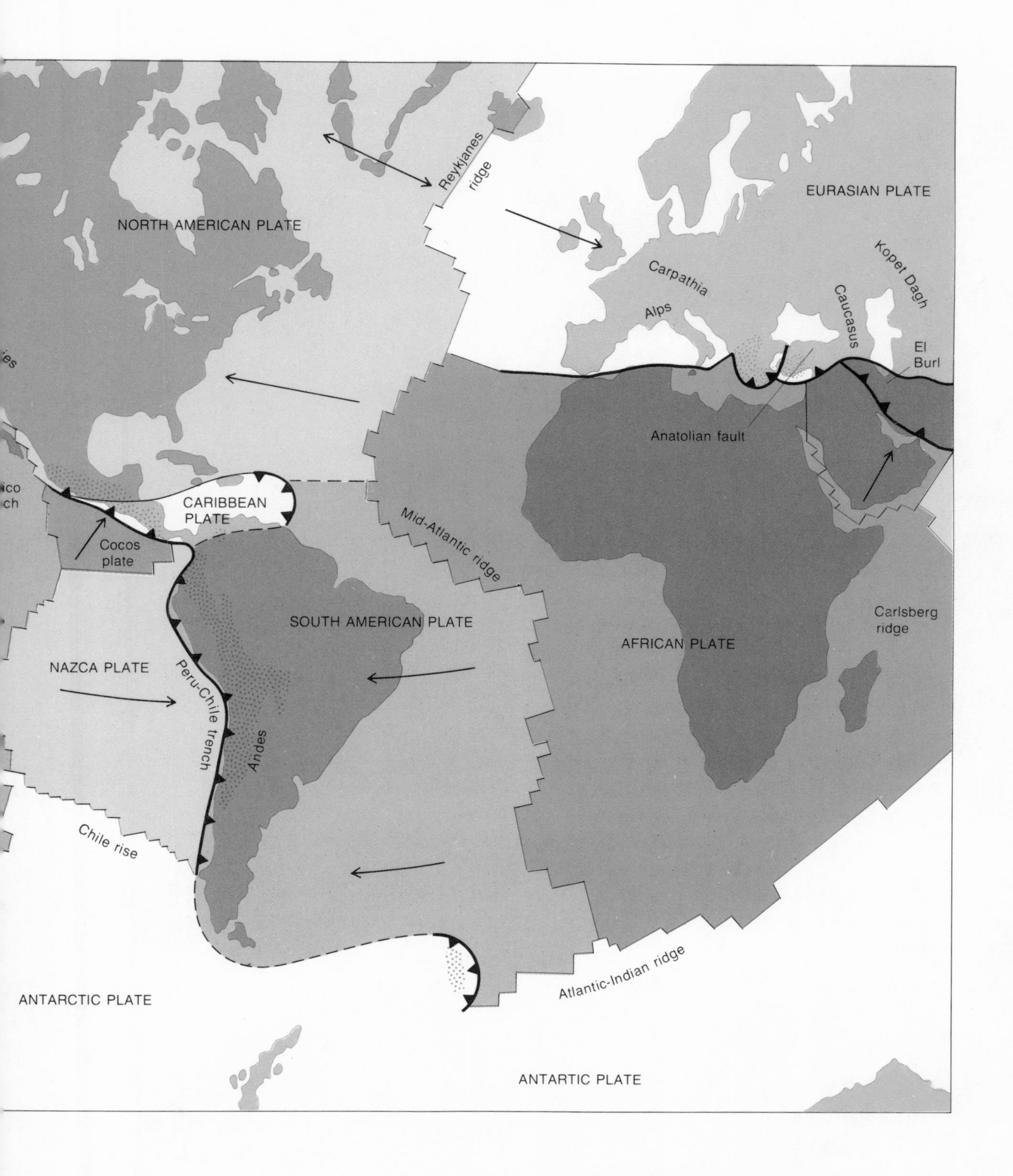

After "Plate Tectonics" by J. F. Dewey.